EARTH MANUAL

A WATER RESOURCES TECHNICAL PUBLICATION

A guide to the use of soils as
foundations and as construction materials
for hydraulic structures

SECOND EDITION

CBS Publishers & Distributors Pvt. Ltd.

New Delhi • Bengaluru • Chennai • Kochi • Kolkata • Mumbai
Hyderabad • Uttarakhand • Nagpur • Patna • Pune • Jharkhand

ISBN: 81-239-1328-1

First Indian Edition: 1990
Reprint: 2001

Published by **Satish Kumar Jain** and produced by **Varun Jain** for
CBS Publishers & Distributors Pvt. Ltd.,
4819/XI Prahlad Street, 24 Ansari Road, Daryaganj, New Delhi - 110002
delhi@cbspd.com, cbspubs@airtelmail.in • www.cbspd.com
Ph.: 23289259, 23266861, 23266867 • Fax: 011-23243014

Corporate Office: 204 FIE, Industrial Area, Patparganj, Delhi - 110 092
Ph: 49344934 • Fax: 011-49344935
E-mail: publishing@cbspd.com • publicity@cbspd.com

Branches:
• *Bengaluru:* 2975, 17th Cross, K.R. Road, Bansankari 2nd Stage,
  Bengaluru - 70 • Ph: +91-80-26771678/79 • Fax: +91-80-26771680
  E-mail: cbsbng@gmail.com, bangalore@cbspd.com
• *Chennai:* No. 7, Subbaraya Street, Shenoy Nagar, Chennai - 600030
  Ph: +91-44-26681266, 26680620 • Fax: +91-44-42032115
  E-mail: chennai@cbspd.com
• *Kochi:* Ashana House, 39/1904, A.M. Thomas Road, Valanjambalam,
  Ernakulum, Kochi • Ph: +91-484-4059061-65
  Fax: +91-484-4059065 • E-mail: cochin@cbspd.com
• *Kolkata:* 6-B, Ground Floor, Rameshwar Shaw Road, Kolkata - 700014
  Ph: +91-33-22891126/7/8 • E-mail: kolkata@cbspd.com
• *Mumbai:* 83-C, Dr. E. Moses Road, Worli, Mumbai - 400018
  Ph: +91-9833017933, 022-24902340/41 • E-mail: mumbai@cbspd.com

Representatives:
• Hyderabad: 0-9885175004   • Nagpur: 0-9021734563
• Patna: 0-9334159340   • Pune: 0-9623451994
• Jharkhand: 0-9811541605   • Uttarakhand: 0-9716462459

Printed at:
J.S. Offset Printers, Delhi (India)

PREFACE TO THE SECOND EDITION

The purposes of the Second Edition remain essentially the same as those which prompted the First Edition, as described in the latter's Preface. Constantly-changing concepts of soil mechanics—as evidenced by new research techniques and ideas, innovations in construction methods and equipment, and computer-generated solutions to previously insurmountable soils-analyses problems—make mandatory this Second Edition. To improve its readability and provide for the new material, the Manual has increased in size as those familiar with the First Edition will recognize.

The contributors to the Manual have held important the need for uniformity in terminology, so that all personnel—field and office alike—speak the same language. Much effort has been expended to achieve consistency of terms in the text and the 39 designations or procedures that comprise the appendix. This may be noted especially in the material on soil classification, and methods of logging and reporting; and types and methods of field explorations and investigations, and the tools and equipment required to obtain the desired information.

Although the Manual is primarily geared to the Reclamation organization, engineers and technicians of other governmental agencies, foreign governments, and private firms can, with modifications, utilize the information as a guide to their individual investigations, control of earth construction, and laboratory testing since emphasis is upon practical applications rather than upon complex theory. Users of the Manual should recognize that certain recommendations and values are the result of experience and cannot always be mathematically proved, nor should one attempt to. The Manual has been written as a guide and aid for the construction of a safe and stable structure with utmost concern for the safety of lives.

New material, not covered in the First Edition, includes material on: stabilized soils (soil-cement and asphaltic concrete), more complete information on field investigations and testing equipment in both chapter 2 and designation E–2, an expanded discussion on pipelines, and a newly developed designation, E–39, titled, "Investigations for Rock Sources for Riprap", which describes investigative and reporting procedures. In addition to the conversion factors in the First Edition, conversion curves are included to facilitate the increased utilization of metric units.

Major revisions center on designation E–16, which has been rewritten and retitled, "Measurement of Capillary Pressures in Soils", and designation E–17, "Triaxial Shear of Soils", which has been rewritten to conform to advanced developments in the procedure. Introduced in E–17 is the "Triaxial Shear Test with Zero Lateral Strain" referred to in modern soil mechanics texts as the K_0-test, which now can easily be performed through the use of the electronic computer.

Since the "Rapid Compaction Control" method, designation E–25, is being used extensively in 35 foreign countries as well as the United States, reorientation of the text material has been made for presenting the material in a manner more readily adaptable to both field and office use. More recently (1970), Australia has been granted permission by the Commissioner, Bureau of Reclamation, to incorporate the "Rapid Compaction Control" method in the Australian standards, "Testing Soils for Engineering Purposes". Designation E–12 has similarly been reoriented for ease in performing the relative density test in cohesionless soils.

Designations E–27 through E–35 covering "Instrument Installations" have been revised and updated to reflect changes in equipment and materials, techniques in installation procedures, and to clarify some of the methods of reading and reporting of data. To be commended are those dedicated field personnel who recognize inconsistencies or problems in the field related to "instruments" and who so often resolve the problems on-the-job. Reflected in these designations are many of their recommendations which have been offered unselfishly.

While environmental and ecological problems are major concerns of the Bureau of Reclamation, space and time limitations cause exclusion of discussion of views and policies regarding these highly important design considerations. It still remains the responsibility of each planner, investigator, designer, and constructor to consider these problems in his work.

There are occasional references to proprietary materials or products in this publication. These must not be construed as an endorsement since the Bureau cannot endorse proprietary products or processes of manufacturers or the services of commercial firms for advertising, publicity, sales, or other purposes.

Indicative of the monumental task involved in the preparation of this Second Edition is that some 90 persons—engineers, technicians, and those of other disciplines—from the Bureau of Reclamation in its Engineering and Research Center, Denver, Colo., constructively contributed to the content in some measure. The efforts of these people, some of whom are internationally acknowledged, are greatly appreciated.

Special recognition is given to H. J. Gibbs, Chief, Earth Sciences Branch, Division of General Research, and F. J. Davis, Supervisory Civil Engineer, Hydraulic Structures Branch, Division of Design and

Construction, for authoring much of the technical material, for their technical advice, and for their overall guidance. In addition, recognition is made of engineers C. W. Jones, W. Ellis, R. R. Ledzian, G. DeGroot, and P. C. Knodel, and technician R. C. Hatcher, all of the Division of General Research, and engineer W. W. Daehn of the Division of Design and Construction for their major contributions. Because illustrations are invaluable to a publication, recognition must be made to R. E. Glasco and Mrs. H. Fowler for their patience, guidance, and help in obtaining illustrations of the highest quality.

This Second Edition was edited and coordinated, and supplemental technical information and illustrations provided by H. E. Kisselman, general engineer, Technical Services and Publications Branch.

PREFACE TO THE FIRST EDITION

The need for an up-to-date formal edition of the Bureau of Reclamation's Earth Manual became evidenced by the many printings of the June 1951 tentative edition. More than 6,000 copies of that edition were distributed, and a worldwide demand for the publication continues.*

The tentative edition was a combination and revision of three early manuals; namely, the Earth Materials Laboratory Test Procedures, the Field Manual for Rolled Earth Dams, and the Earth Materials Investigation Manual. In the present formal edition, the Earth Manual has been completely rewritten and contains much material not covered in the earlier edition.

The manual provides current technical information on the field and laboratory investigations and construction control of soils used as foundations and materials for dams, canals, and many other types of structures built for Reclamation projects in the United States of America. It contains both standardized procedures that have been found desirable for securing uniform results throughout the Bureau, and general guidelines intended to assist but not to substitute for engineering judgment.

Chapter I describes the Unified Soil Classification System developed jointly by the Bureau of Reclamation and the Corps of Engineers, Department of the Army, from the system proposed by Professor A. Casagrande of Harvard University, and discusses the various properties of soils relating to engineering uses. Investigations of soils are covered in chapter II which describes the various stages of investigation corresponding to the stages of development of the Bureau projects, and gives technical information necessary for planning and executing explorations and for presenting the results.

Chapter III presents information on the control of construction from the soils standpoint, for both foundation treatment and compaction control of fills. In addition to a general treatment of the subject applicable to all types of earthwork, separate sections are devoted to problems of rolled earth dams, canals, and miscellaneous construction features. For each of these, information on design features and usual specifications provisions are given to provide control personnel with a background to assist in implementing the recommended control techniques.

*Since the tentative edition, 28,000 copies of the first formal edition and its revised reprint (first printing) have been printed and distributed.

vii

The appendix contains detailed procedures for sampling, classification, and field and laboratory testing of soils. Instructions for installing and obtaining information from instruments that measure pore-water pressures and displacements within and adjacent to earth embankments are also included. A tabulation of conversion factors commonly used in earth construction is included at the end of the appendix.

The appendix contains laboratory test procedures which are used in both the project laboratories and the central Bureau laboratory, and some of the more complex test procedures which are performed only in the central laboratory. The first group is presented for use by all Bureau laboratories, so that uniformity of test results will be obtained. The second group is presented for the purpose of recording the test procedures used in the central Bureau laboratory, so that the results of these tests, which are often contained in published reports, can be properly understood and interpreted.

Although the basic test procedures are standardized, it is emphasized that soil testing does not lend itself to strict standardization as normally applied to such construction materials as concrete, steel, and asphalt. Tests on soils must be carried out in such a way that natural conditions and operational conditions are fully accounted for during the test. In a simple test as for gradation determinations, the standard procedure may not take into account particle breakdown that may occur during excavation and compaction. Therefore, the test procedure may require changes to provide "as constructed" gradation data. The compaction and other properties of some soils may be changed by drying and rewetting. In such cases, the field construction conditions must be taken into account during the test.

When variations from standard procedures are required, the variations must be described when reporting results, so that proper interpretations can be made. The exact procedures to be used for consolidation and shear tests normally require considerable judgment to assure that natural and operational prototype conditions are being duplicated as closely as possible. For this reason, the basic procedures presented for these two tests are merely guides and the details of actual test procedures used are covered in individual reports.

There may be instances in which the procedures and instructions in the Earth Manual are at variance with contract specifications requirements. In such cases, it should be understood that the specifications take precedence. It is also pointed out that each employee of the Bureau is accountable to his immediate supervisor and should request advice concerning any doubtful procedure.

Unless otherwise noted, the terms and symbols used in the Earth Manual conform to those given in ASTM Designation D–653–58T,

"Tentative Definitions of Terms and Symbols Relating to Soil Mechanics," ASTM Standards 1958, part 4, page 1211. These terms are also given in Paper 1826, "Glossary of Terms and Definitions in Soil Mechanics," Journal of the Soil Mechanics and Foundations Divisions, ASCE, October 1958.

Some of the exploratory and test procedures in chapter II and in certain of the designations which are applicable to the investigations, design, and construction of small earth structures have also been included in the publication "Design of Small Dams," recently published (1960).

The Earth Manual represents the joint efforts of many engineers and technicians over a period of many years covering the preparation of the tentative edition, its three predecessor publications, and this first formal edition. The present edition (including the revised reprint) was prepared by engineers in the Earth Dams Section, Dams Branch, Division of Design, and the Soils Engineering Branch, Division of Research, in Denver, Colo., under the direction of Chief Engineer B. P. Bellport and his predecessors.

Major writing was performed by F. C. Walker, J. W. Hilf, and W. W. Daehn of the Earth Dams Section; and W. G. Holtz and A. A. Wagner of the Soils Engineering Branch. Contributing writers were F. J. Davis, E. E. Esmiol, H. J. Gibbs, and C. W. Jones. The draft was reviewed and helpful suggestions submitted by geologists and engineers in various other branches of the organization. Principal among these were W. H. Irwin, A. F. Johnson, H. G. Curtis (deceased), and C. E. McClaren. The contributions of these individuals, as well as those of the many other engineers and technicians who contributed in various ways to the publication, are gratefully acknowledged. The draft was edited and coordinated by J. W. Hilf, and final review and preparation for publication was performed by E. H. Larson, Head, Manuals and Technical Records Section, Technical and Foreign Services Branch.

There are many references throughout the text to the Assistant Commissioner and Chief Engineer and the Office of Assistant Commissioner and Chief Engineer. For expediency these references were not changed in the reprint to agree with recent organization changes (February 1963), and they should be interpreted as meaning Chief Engineer and Office of Chief Engineer, respectively. References to the Commissioner's Office, Denver, should also be interpreted as meaning the Office of Chief Engineer.

CONTENTS

Section   *Page*

Preface to the Second Edition _____ iii
Preface to the First Edition _____ vii

CHAPTER I—PROPERTIES OF SOILS

A. IDENTIFICATION AND CLASSIFICATION

1. General _____ 1
2. Soil components _____ 2
   (a) Size _____ 2
   (b) Gradation (Grain-size distribution) _____ 3
   (c) Shape _____ 4
3. Soil moisture _____ 4
4. Characteristics of soil components _____ 10
   (a) Gravel and sand _____ 10
   (b) Silt and clay (Fines) _____ 11
   (c) Organic matter _____ 13
5. Classification of soils _____ 14
   (a) General _____ 14
   (b) Field classification _____ 14
   (c) Laboratory classification _____ 16
6. Description of soils _____ 17
   (a) General _____ 17
   (b) Borrow materials _____ 17
   (c) Foundations for structures _____ 18
7. Properties of soil groups _____ 18

B. INDEX PROPERTIES

8. Definition _____ 22
9. Gradation _____ 23
10. Soil consistency _____ 25
11. Porosity and void ratio _____ 28
12. Specific gravity _____ 29

Section   *Page*

13. Water content or moisture content _____ 32
14. Density or unit weight _____ 33
15. Penetration resistance _____ 40
16. Unconfined compressive strength _____ 42
17. Soluble salts _____ 43

C. ENGINEERING PROPERTIES

18. General _____ 43
19. Shear strength _____ 44
    (a) General _____ 44
    (b) Direct shear _____ 45
    (c) Triaxial shear _____ 46
    (d) Pore-water pressure _____ 46
    (e) Capillary stresses _____ 48
    (f) Sliding resistance _____ 48
20. Compressibility _____ 49
    (a) General _____ 49
    (b) Control of compressibility _____ 51
    (c) Load-compression characteristics _____ 53
    (d) Load-expansion characteristics _____ 53
21. Permeability _____ 55
    (a) Definition _____ 55
    (b) Ranges of permeability _____ 59
    (c) Control of permeability _____ 59
    (d) Determination of permeability values _____ 60
22. Changes in soil properties _____ 61
23. Workability _____ 62
24. Frost action _____ 63

CHAPTER II—INVESTIGATION

A. STAGES OF INVESTIGATION

25. General _____ 65
26. Reconnaissance _____ 67
    (a) Objectives _____ 67
    (b) Sizes and depths of investigated areas _____ 67
27. Feasibility _____ 68
    (a) Objective _____ 68
    (b) Organization of the investigation _____ 69

Section *Page*

28. Specifications _____ 71
    (a) Scope _____ 71
    (b) Programing the exploration _____ 72

B. PRINCIPLES OF INVESTIGATIONS

29. General _____ 73
    (a) Objectives _____ 73
    (b) Classification of structure foundations _____ 74
30. Sources of map and photo information _____ 75
    (a) Topographic maps _____ 75
    (b) Geologic maps _____ 77
    (c) Agricultural soil maps _____ 80
    (d) Air photos _____ 84
31. Surface exploration _____ 87
    (a) General _____ 87
    (b) Fluvial soils _____ 88
    (c) Glacial deposits _____ 92
    (d) Eolian deposits _____ 94
    (e) Residual soils _____ 96
32. Subsurface exploration _____ 100
    (a) General _____ 100
    (b) Point structures _____ 101
    (c) Line structures _____ 101
    (d) Damsites _____ 106
    (e) Tunnels _____ 106
    (f) Borrow areas _____ 106
    (g) Selection of samples _____ 108
    (h) Field tests _____ 111
33. Exploration for materials having specific properties _____ 111
    (a) General _____ 111
    (b) Impervious materials _____ 112
    (c) Pervious materials _____ 113
    (d) Riprap and rockfill _____ 115
34. Materials for stabilized and modified soils _____ 119
    (a) General _____ 119
    (b) Compacted soil-cement _____ 120
    (c) Plastic soil-cement _____ 121
    (d) Asphaltic concrete _____ 121
    (e) Modified soil _____ 121

Section   *Page*

C. EXPLORATORY METHODS

35. General _____ 122
36. Test pits, trenches, and tunnels _____ 123
    (a) General _____ 123
    (b) Test pits _____ 124
    (c) Trenches _____ 125
    (d) Tunnels _____ 127
37. Auger borings _____ 129
38. Rotary drilling _____ 136
39. Drive-tube boring _____ 140
40. Miscellaneous methods _____ 142
    (a) Nonsampling borings _____ 142
    (b) Geophysical methods _____ 142
41. Field tests _____ 143
    (a) General _____ 143
    (b) Field permeability tests _____ 143
    (c) Field vane test _____ 143
    (d) Inplace density tests _____ 143

D. RECORDING AND REPORTING OF DATA

42. Maps _____ 144
43. Logging of exploratory holes _____ 146
    (a) Location of holes _____ 146
    (b) Identification of holes _____ 146
    (c) Log forms _____ 147
    (d) Description of soils _____ 155
    (e) Description of rock cores _____ 155
44. Subsurface sections _____ 157
45. Sampling _____ 158
46. Reports _____ 159

CHAPTER III—CONTROL OF EARTH CONSTRUCTION

A. PRINCIPLES OF CONSTRUCTION CONTROL

47. Importance of control _____ 165
48. Organization _____ 166
49. Specifications _____ 167
50. Inspection _____ 169

Section   *Page*

51. Field laboratory facilities _____ 171
52. Reports _____ 174

B. EARTHWORK

53. General _____ 176
54. Embankment _____ 176
    (a) Types of embankment _____ 176
    (b) Dumped fill _____ 176
    (c) Selected fill _____ 177
    (d) Equipment-compacted embankment _____ 178
    (e) Rolled earthfill _____ 178
    (f) Tractor-compacted embankment _____ 179
    (g) Blended earthfill _____ 181
    (h) Modified soil fill _____ 183
    (i) Hydraulic fill _____ 183
55. Linings and blankets _____ 184
    (a) General _____ 184
    (b) Rock blankets _____ 186
    (c) Sand and gravel or crushed rock blankets _____ 187
    (d) Impervious blankets and linings _____ 189
    (e) Topsoil blankets or zones _____ 189
56. Backfill _____ 190
    (a) General _____ 190
    (b) Compacted backfill _____ 191
57. Excavation _____ 191

C. FOUNDATIONS

58. General _____ 195
59. Bearing capacity _____ 195
60. Stability _____ 196
61. Settlement and uplift _____ 201
62. Deterioration _____ 202
63. Permeability _____ 202
64. Inadequate foundation conditions _____ 205
    (a) General _____ 205
    (b) Topsoil _____ 206
    (c) Swamp muck _____ 206
    (d) Silt and sand _____ 206
    (e) Talus and spoil piles _____ 211

Section   *Page*

    (f) Clays _____ 211
    (g) Soft or saturated materials _____ 211

D. ROLLED EARTH DAMS

65. Foundation treatment _____ 212
    (a) Design features _____ 212
    (b) Specifications provisions _____ 218
    (c) Control techniques _____ 220
66. Compacted earthfill _____ 221
    (a) Design considerations _____ 221
    (b) Specifications provisions _____ 223
    (c) Control techniques _____ 225
67. Compacted pervious fill _____ 234
    (a) Design considerations _____ 234
    (b) Specifications provisions _____ 236
    (c) Control techniques _____ 237
68. Rockfill and riprap _____ 238
69. Miscellaneous fills _____ 241
70. Instrument installations _____ 243
    (a) Instruments _____ 243
    (b) Installation of earth dam instrumentation _____ 243
    (c) Inspection _____ 244
    (d) Observations _____ 244
    (e) Record tests _____ 245
71. Records and reports _____ 247
    (a) Daily reports _____ 247
    (b) Periodic progress reports _____ 249
    (c) Final embankment construction reports _____ 250
    (d) Earth dam instrumentation reports _____ 253
72. Control criteria _____ 253

E. CANALS

73. Design features _____ 260
74. Specifications provisions _____ 266
    (a) General _____ 266
    (b) Subgrades and foundations for embankments and
         compacted earth lining _____ 266
    (c) Earth embankments and linings _____ 267

Section *Page*

    (d) Riprap, protective blankets, gravel fills, and gravel subbase _____ 269

75. Control techniques _____ 274

    (a) General _____ 274

    (b) Subgrades and embankment foundations _____ 274

    (c) Earth embankments and linings _____ 278

F. PIPELINES

76. Design features _____ 282

77. Specifications provisions _____ 283

    (a) General _____ 283

    (b) Pipeline excavation _____ 283

    (c) Backfill in pipe trenches _____ 283

    (d) Compacting backfill in pipe trenches _____ 286

    (e) Compacted backfill for bedding _____ 286

G. MISCELLANEOUS CONSTRUCTION FEATURES

78. Highways and railroads _____ 288

    (a) General _____ 288

    (b) Design features _____ 288

    (c) Earthwork specifications provisions _____ 289

    (d) Control techniques _____ 293

79. Miscellaneous structures _____ 294

    (a) General _____ 294

    (b) Structure foundations on soil or rock _____ 294

    (c) Pile and caisson foundations _____ 300

    (d) Transmission tower footings _____ 303

    (e) Backfill _____ 304

    (f) Filters _____ 305

H. STABILIZED SOILS

80. General _____ 309

81. Compacted soil-cement _____ 310

    (a) Design considerations _____ 310

    (b) Construction provisions _____ 310

    (c) Control techniques _____ 315

    (d) Control testing _____ 316

Section   *Page*

82. Plastic soil-cement ------------------------------------- 318
83. Asphaltic concrete ------------------------------------- 322
    (a) General ------------------------------------- 322
    (b) Design considerations ------------------------ 322
    (c) Construction provisions ---------------------- 324
    (d) Control testing ------------------------------ 325
84. Modified soil -- 326

APPENDIX

Procedures for Sampling, Classification, and Testing of Soils and Installation of Instruments

SAMPLING

Designation		Page
E-1	Disturbed sampling of soils	327
E-2	Undisturbed sampling of soils	341

CLASSIFICATION

E-3	Visual and laboratory methods for identification and classification of soils	387
E-4	Lists of laboratory equipment	408

LABORATORY TESTS

E-5	Preparation of soil samples for testing	419
E-6	Gradation analysis of soils	424
E-7	Soil consistency tests	435
E-8	Soluble salts determination of soils	448
E-9	Moisture determination of soils	450
E-10	Specific gravity of soils, aggregate, and density of irregular blocks of soil	453
E-11	Proctor compaction test (moisture-density relations of soil)	466
E-12	Relative density of cohesionless soils	479
E-13	Permeability and settlement of soils	491
E-14	Permeability and settlement of soil containing gravel	505
E-15	One-dimensional consolidation of soils	509
E-16	Measurement of capillary pressures in soils	521
E-17	Triaxial shear of soils	545

APPENDIX—Continued

Designation   *Page*

FIELD TESTS

E-18   Field permeability tests in boreholes _____   573
E-19   Field permeability test (well permeameter method) __   578
E-20   Inplace vane shear test _____   593
E-21   Field penetration test with split-tube sampler _____   603
E-22   Needle-moisture determination of soils _____   610
E-23   Field density of dry, gravel-free soils _____   610
E-24   Field density test procedure _____   613
E-25   Rapid compaction control _____   621
E-26   Vertical load-settlement relationship for individual piles   642

INSTRUMENT INSTALLATIONS

E-27   Instructions for installing and reading hydraulic-type
       twin-tube piezometers in earth dams _____   650
E-28   Instructions for installing and reading porous-tube pie-
       zometers _____.   686
E-29   Instructions for installing and reading internal vertical
       movement devices _____   699
E-30   Instructions for installing and reading internal horizontal
       movement devices _____   719
E-31   Instructions for installing and reading foundation settle-
       ment apparatus _____   730
E-32   Instructions for installing and reading measurement
       points—embankment _____   733
E-33   Instructions for installing and reading measurement
       points—concrete structures—outlet works conduits
       —conduit-type spillways _____   738
E-34   Instructions for installing and reading measurement
       points—concrete structures—chute and stilling basin
       of outlet works—chute-type spillways _____   741
E-35   Recording earthquake vibrations _____   745

ADDITIONAL FIELD AND LABORATORY TESTS

E-36   Field permeability test (shallow-well permeameter
       method) _____   747

APPENDIX—Continued

Designation   *Page*

E–37 Method for calibrating mechanical laboratory soil compactors -- 755

E–38 Compaction test for soil containing gravel (moisture-density relations) -------------------------- 760

E–39 Investigations for rock sources for riprap ----------- 775

Conversion Factors

Some conversion factors commonly used in earth construction - 782

Conversion Curves

Conversion curves to convert inches to centimeters and feet to meters -- 786

Conversion curves to convert gallons to liters and acre-feet to cubic meters --------------------------------------- 787

Conversion curves to convert square feet to square meters and acres to hectares ------------------------------------ 788

Conversion curves to convert second-feet to cubic meters per second and miles to kilometers ----------------------- 789

LIST OF FIGURES

Figure   *Page*

1   Well-graded gravel (GW), Vermejo project, New Mexico   5

2   Uniform, poorly-graded sand (SP), Cherry Creek Reservoir, Colorado _____ 6

3   Poorly-graded gravel (GP), Falcon Dam, Texas _____ 7

4   Typical shapes of bulky grains _____ 8

5   Flaky grains _____ 9

6   Elongated grains _____ 10

7   Unified soil classification chart _____ Facing 12

8   Engineering use chart _____ 20

9   Consistency limits _____ 26

10   Typical relationships between the liquid limit (LL), and plasticity index (PI), for various soils _____ 28

11   Soil properties nomograph _____ 31

12   Compaction and penetration resistance curves for various soils _____ 34

13   Effect of compactive effort on the compaction and penetration-resistance curves _____ 35

14   Relation of relative density and dry density (scaled to plot as a straight line) _____ 38

15   Maximum and minimum densities of typical sand and gravel soils _____ 39

16   Maximum dry density versus gravel content for compacted soils _____ 41

17   Effect of relative density on the friction factor for coarse-grained soils _____ 47

18   Results of an undrained triaxial shear test on lean clay with and without correction for pore-air pressure ____ 49

19   Compression characteristics of various compacted embankment soils based on field measurements _____ 52

20   Void ratio-load curve, compression index, and preconsolidation load _____ 54

21   Compressibility characteristics of a fine sand in relation to placement relative density _____.'__ 55

22   Load-expansion curves for two typical expansive clays __ 56

LIST OF FIGURES—Continued

Figure   *Page*

23   Typical example of the effect of placement moisture and density on an expansive clay _____   57

24   Examples of canal lining failures caused by expansion of clay soils _____   58

25   Relationship of permeability to gravel content for specimens (sand-gravel mixtures) of various relative densities _____   60

26   Ice lens formation between chalk layers _____   64

27   Status of 7½- and 15-minute topographic quadrangle mapping in the United States—1969 _____Facing   76

28   Status of large scale geologic mapping in the United States —1969 _____Facing   80

29   Status of published pedological soil maps—1971__Facing   80

30   Soil triangle of the basic soil textural classes _____   82

31   Rock strata illustrating folding in sedimentary rocks ____   88

32   Sinkhole plain indicating deep plastic soils over cavernous limestone, developed in humid climate _____   89

33   Aerial view and topography of an alluvial fan, a potential source of sand and gravel _____   90

34   Aerial view and topography of stream deposit showing river alluvium and three levels of gravel terraces ____   93

35   Aerial view and topography of terminal moraine of continental glaciation _____   95

36   Aerial view and topography of loess identified by smooth silt ridges; usually parallel, right-angle drainage patterns; and steep-sided, flat-bottomed gullies and streams _____   97

37   Photomicrograph showing typical open structure of silty loess _____   98

38   A 34-foot cut on ½:1 slope in loess formation—Franklin Canal, Nebraska _____   99

39   Depth of preliminary exploratory holes for point structures _____   102

40   Soil profile at a pumping plant _____   103

41   Depth of preliminary exploratory holes for canal, road, and pipeline alinement _____   104

42   Example of geologic profile along the centerline of a pipeline _____   105

43   Geologic map and section of a damsite _____   107

Figure   *Page*

44   Exploration for embankment materials—Location map
       and section for a typical damsite _____   109

45   Results of a major blast in a riprap quarry _____   117

46   Blast test in igneous rock investigated as a source for
       riprap _____   117

47   Talus slope of igneous rock proposed for riprap _____   118

48   Test pit cribbing _____   125

49   Excavation of hand-dug test pit in borrow area _____   126

50   Equipment-excavated test pit showing location of samples_   127

51   Trenching, a low-cost method of obtaining soil samples__   128

52   Shallow test trench excavated by bulldozer _____   128

53   Trench in steep abutment area excavated by bulldozer
       and backhoe _____   129

54   Exploring a borrow area with a hand auger _____   130

55   Types of hand augers (2-inch helical, 2- and 6-inch Iwan,
       and 6-inch Fenn (adjustable)) _____   131

56   Illustration of the helical, disc, and barrel types of
       machine-driven augers showing basic differences ____   132

57   Undisturbed sampling with a double-tube helical auger
       and description of equipment _____   133

58   Disc auger used to explore borrow areas of fine-grained
       soils _____   134

59   Bucket auger used in exploration of a borrow area con-
       taining gravel particles _____   135

60   Enclosed auger _____   135

61   Diamond drill rig used in foundation exploration _____   137

62   Core barrels used for obtaining samples of rock _____   138

63   Arrangement of cores in a core box to insure proper
       identification of samples _____   139

64   Geologic log of a drill hole _____   148

65   Log of a hand-dug test pit—For foundation investigation   149

66   Log of a test pit excavated by backhoe—For both borrow
       and foundation materials investigations _____   150

67   Log of an auger hole—For borrow materials investigation   151

68   Penetration resistance and drill hole data for subsurface
       exploration _____   152

69   Gradation and plasticity data for loessial soils in the
       Kansas-Nebraska area _____   156

70   Summary of field and laboratory tests for embankment
       materials report—Impermeable-type materials _____   163

LIST OF FIGURES—Continued

Figure   *Page*

71   Summary of field and laboratory tests for embankment materials report—Permeable-type materials _____   164

72   Examples of floor plans for field control laboratories ___   172

73   Typical field laboratories _____   173

74   Large-scale permeability apparatus in a field laboratory__   182

75   Sluiced backfill in cut-and-cover section of a power tunnel_   185

76   Placing a 3-foot blanket of riprap over an 18-inch cobble blanket in the tailrace of a powerplant _____   187

77   Vibratory consolidation of sand and gravel beneath a spillway slab _____   192

78   Consolidating backfill about a conduit by means of internal vibrators _____   193

79   A deposit of impervious material which overlies a deposit of pervious material _____   194

80   An example of an arc failure of a natural slope (landslide) resulting from excessive moisture entering the material _____   198

81   View of cutoff trench at Davis Dam, Arizona-Nevada __   199

82   View of cutoff trench at Twin Buttes Dam, Texas _____   200

83   View of a typical sinkhole (pothole) _____   203

84   Portable grout machine used to force a mixture of cement and water into holes drilled in the foundation _____   204

85   Cracking and settling of canal bank in dry, low-density silt _____   208

86   Ponding dry foundation of Trenton Dam, Nebraska, to facilitate consolidation _____   209

87   Criterion for treatment of relatively dry fine-grained foundations _____   210

88   Curtain grouting in right abutment of Granby Dam, Colorado _____   216

89   Temporary diversion channel through Bonny Dam, Colorado _____   218

90   Foundation preparation by washing to remove detrimental air-slaked and loose material at Twin Buttes Dam, Texas _____   220

91   Power tamping of earthfill at contact with irregular rock abutment _____   222

92   Separation plant at Meeks Cabin Dam, Wyoming, where impervious material containing oversize was screened into fractions above and below 5 inches _____   224

LIST OF FIGURES—Continued

Figure		Page
93	Placing, spreading, and compacting earthfill in a relatively restricted area on Blue Mesa Dam, Colorado	226
94	General view of placing, spreading, and compacting earthfill in a relatively unrestricted area on Twin Buttes Dam, Texas	227
95	Test section of residual soil in borrow area at Trinity Dam, California	228
96	Average field and laboratory compaction curves for three dam embankment soils	229
97	Elevating grader equipped with vertical cutting blade for mixing upper layer of silty clay with lower sand and gravel deposit	231
98	Summary of field and laboratory tests of compacted fill controlled by Proctor test	235
99	Test pit in compacted embankment, Bonny Dam, Colorado, being backfilled	236
100	Summary of field and laboratory tests of compacted fill controlled by relative density test	239
101	Examples of riprap being placed on dam embankment for upstream slope protection	242
102	Tabulation for recording embankment roller data	247
103	Frequency of readings—Earth dam instrumentation	248
104	Interoffice transmittal form	254
105	Typical laboratory tests for the development of shear planes	256
106	Statistical analysis of test results for density control of fill	258
107	Statistical analysis of test results for moisture control of fill	259
108	Plasticity criterion for impervious, erosion-resistant compacted earth linings	263
109	Typical canal section and compacted earth lining	264
110	Placing and compacting thick earth lining in Wellton-Mohawk Canal, Gila project, Arizona	264
111	Completed thick earth lining in Wellton-Mohawk Canal, Gila project, Arizona	265
112	Size limits for reasonably well-graded, pervious sand-gravel material	273
113	Compacted earthwork acceptance tests for construction control of miscellaneous structures—Relative density method	276

LIST OF FIGURES—Continued

Figure   *Page*

114   Compacted earthwork acceptance tests for construction control of miscellaneous structures—Proctor compaction method 277

115   Typical trench sections and compacted backfill for bedding requirements for pressure pipe 285

116   Typical trench sections and soil-cement bedding requirements for pressure pipe 287

117   Heavy equipment constructing access road to Flaming Gorge Dam site through Dutch John Gap 289

118   Completed access road to Crystal Dam site in Colorado 290

119   Earthwork operations for bridge abutments of Trenton Dam railroad relocation 290

120   Backfilling operations over culvert on Flaming Gorge Dam site access road 295

121   Riprap placing operations at bridge abutment at Trenton Dam railroad relocation 296

122   Criterion for predicting relative density of sand from the penetration resistance test 297

123   Examples of use of filter criteria for selecting materials 308

124   General view of Lubbock Regulating Reservoir showing soil-cement facing 1 year after construction 311

125   General view of soil-cement facing on Merritt Dam 6 years after construction 311

126   General view of soil stockpile and pugmill for mixing soil-cement—Lubbock Regulating Reservoir 312

127   Spreading soil-cement on placement area—Merritt Dam 313

128   General view of placing and rolling operations—Lubbock Regulating Reservoir 313

129   Power brooming to clean surface prior to placement of next lift—Lubbock Regulating Reservoir 314

130   Cement vane feeder and soil feed belt to pugmill—Cheney Dam 314

131   Summary of field and laboratory tests of compacted soil-cement 317

132   A comparison of bedding methods for pipe bedded to 0.37 outside diameter—Plastic soil-cement compared to compacted backfill 320

133   Special machine to place plastic soil-cement under pipe 321

134   Slip-form paver placing asphaltic concrete canal lining 322

135   Placing asphaltic concrete on upstream face of a rockfill dam ....

LIST OF FIGURES IN APPENDIX

Figure   *Page*

1–1   Sampling trench _____   329
1–2   Auger sampling _____   339
2–1   Initial steps to obtain a hand-cut, undisturbed block
        sample _____   347
2–2   Final steps in obtaining a hand-cut, undisturbed block
        sample _____   348
2–3   Method for obtaining a hand-cut, undisturbed cylin-
        drical sample _____   349
2–4   Typical drill rig for mechanical sampling _____   351
2–5   Samplers used for undisturbed soil sampling _____   352
2–6   Denison sampler (double-tube core barrel) _____   356
2–7   Denver sampler (double-tube core barrel) _____   357
2–8   Pitcher sampler (double-tube core barrel) _____   358
2–9   Double-tube auger sampler _____   359
2–10 Sampling with double-tube soil or core sampler _____   360
2–11 Thin-wall open drive sampler _____   364
2–12 Sampling with thin-wall open drive sampler _____   365
2–13 Dimensions of thin-wall sampling tubes _____   366
2–14 Thin-wall fixed-piston sampler (Hvorslev type) _____   370
2–15 Butter's thin-wall fixed-piston sampler (Hvorslev type) _   371
2–16 Thin-wall fixed-piston sampler (Osterberg type) ____   372
2–17 Sampling with fixed-piston sampler (Hvorslev type) __   373
2–18 Diamond Core Drill Manufacturers Association no-
        menclature _____   378
2–19 Core drill casing _____   379
2–20 Nominal dimensions for casing and accessories _____   380
2–21 Commercial standards for core bit sizes _____   381
2–22 Standard drill rod sizes _____   382
2–23 Undisturbed soil sampling data form _____   384
2–24 Thin-wall tube box _____   385
2–25 Shipping container for block samples _____   386
3–1   Unified soil classification chart _____Facing   390
3–2   Data form for visual classification of borrow area soil
        samples and foundation samples _____   392
3–3   Data form for visual classification to be used in the
        field _____   394
3–4   Reactions of a silty soil to shaking and squeezing
        (dilatancy test) _____   398
3–5   Graphic symbols for soils _____   407

LIST OF FIGURES IN APPENDIX—Continued

Figure		Page
5–1	Hopper and hand sieves for separating coarse aggregate and soil materials	420
5–2	Sample splitter	421
5–3	Sample preparation and gradation analysis data form	422
5–4	Composite sample preparation data form	423
6–1	Stirring apparatus, dispersion cup, hydrometer, and hydrometer cylinder	425
6–2	Sieve analysis equipment	426
6–3	Air dispersion tube	426
6–4	Gradation analysis curves	432
6–5	Synthetic gradation analysis computations	433
7–1	Equipment for the liquid limit test	436
7–2	Test for liquid limit	438
7–3	Data form for soil consistency tests	440
7–4	Soil consistency test (one-point liquid limit method)	442
7–5	Test for plastic limit	443
7–6	Apparatus for shrinkage test for soils	444
9–1	Drying oven	451
9–2	Moisture determination data	453
10–1	Apparatus for determining apparent specific gravity by the flask method	454
10–2	Recording data and computing the apparent specific gravity of soils	455
10–3	Portions of calibration chart for specific gravity flasks for 100-gram sample	457
10–4	Recording and computing specific gravity by the calibration method	460
10–5	Pycnometer (pint fruit jar with edges ground) and siphon can	461
10–6	Recording and computing density of irregular blocks or cores of soil by the suspended weight in air and water method	465
11–1	Performing compaction test—Manual method	467
11–2	Performing compaction test—Machine method	468
11–3	Performing compaction test and view of compactor mechanism	469
11–4	Recording data for compaction and penetration-resistance tests	471
11–5	Determining penetration resistance	472
11–6	Extruder for compaction specimens	473

LIST OF FIGURES IN APPENDIX—Continued

Figure		Page
11–7	Proctor compaction test and summary of physical properties	475
12–1	Vibratory table	480
12–2	Relative density test equipment—Dial indicator reading and check	481
12–3	Relative density test equipment	482
12–4	Relative density test method using a vibratory table—Calibration sheet	483
12–5	Relative density test method using a vibratory table—Computation sheet	485
12–6	Plotting results of relative density test	490
13–1	Standard 8-inch permeability apparatus	493
13–2	Standard 8-inch permeability test being performed in the laboratory	494
13–3	Recording preparation of sample data for permeability test	495
13–4	Recording compaction and settlement test data	496
13–5	Recording permeability test data	500
13–6	Permeability cylinders for nonloaded specimens	504
14–1	Standard 20-inch permeability apparatus	506
14–2	Standard 20-inch permeability test being performed in the laboratory	507
15–1	Pneumatically loaded one-dimensional consolidometer	510
15–2	Specimen container for one-dimensional consolidometer	511
15–3	Drawing of one-dimensional consolidometer specimen container	511
15–4	Cutting undisturbed specimen from an undisturbed block sample	512
15–5	Specimen placement data for one-dimensional consolidation test	513
15–6	Recording time-consolidation data for one-dimensional consolidation test	515
15–7	Summary of load data for one-dimensional consolidation test	518
15–8	Chart for simplifying computation of permeability by falling-head method	519
15–9	Load-consolidation curves	520
15–10	Time-consolidation curves	521
16–1	Test chamber for measuring negative pore pressure by the exposed end-plate method	524

LIST OF FIGURES IN APPENDIX—Continued

Figure		Page
16–2	Soil specimen on exposed ceramic end plate _____	526
16–3	Data tabulated from a pore-pressure test with an initial capillary pressure less than 5 atmospheres negative _____	528
16–4	Pore-pressure development for a soil with an initial capillary pressure less than 5 atmospheres negative_	530
16–5	Data tabulated from a pore-pressure test with an initial capillary pressure greater than 5 atmospheres negative	532
16–6	Pore-pressure development for a soil with an initial capillary pressure greater than 5 atmospheres negative	534
16–7	Pore-pressure test for μ_c during undrained compression	535
16–8	End plate with small air inlet at the center _____	537
16–9	Sample data sheet for triaxial shear and three-dimensional specimen placement _____	541
16–10	Sample data and computation sheets _____	542
17–1	Triaxial shear stress test data _____	547
17–2	Pore-air pressure measuring system with positive separation between soil and ceramic end plate _____	550
17–3	Triaxial shear test chamber _____	551
17–4	Dynamic triaxial testing equipment for testing soils finer than the No. 4 sieve size under pulsating vertical loads _____	553
17–5	Small triaxial shear machine for testing 1⅜- by 3-inch specimens _____	554
17–6	Triaxial shear machine for testing 2- by 5-inch specimens _____	555
17–7	High-pressure triaxial shear machine for testing 2- by 5-inch specimens _____	556
17–8	Equipment for testing 6- by 12-inch specimens using chamber pressures up to 1,000 pounds per square inch _____	557
17–9	Equipment for triaxial testing of 6- by 12-inch and 9- by 22½-inch specimens up to 250 pounds per square inch chamber pressure _____	558
17–10	The effect of capillary (suction) pressure on the cohesion and ϕ values _____	562
17–11	Least squares method of computations for coefficient of internal friction and cohesion _____	564
17–12	An element of soil which is laterally confined (no lateral strain) and compressed by a vertical pressure_	565

LIST OF FIGURES IN APPENDIX—Continued

Figure   *Page*

17–13   Specimen end plates _____   567
17–14   Soil specimen with end plates in place and enclosed
        with a rubber membrane _____   568
17–15   Soil specimen with girth gage _____   569
17–16   Triaxial apparatus with cylinder, cap, and piston in
        place _____   570
17–17   Triaxial machine with girth gage _____   571
17–18   Results of a typical K_o-test _____   572
18–1   An open-end pipe test for soil permeability which can
        be made in the field _____   574
18–2   The packer test for soil permeability _____   574
19–1   Drawing of well permeameter test apparatus _____   579
19–2   Typical well permeameter test installation _____   580
19–3   Hand and power augers for excavating test wells ____   581
19–4   Maximum permeability coefficients measurable with
        bob-float valves of the sizes shown _____   582
19–5   Data on soils and well for well permeameter test ____   584
19–6   Nomograph for determining coefficient of permeability
        from well test with a low water-table condition ____   586
19–7   Nomograph for determining coefficient of permeability
        from well test with a high water-table condition ____   587
19–8   Time and volume measurements for well permeameter
        test _____   588
19–9   Example of discharge-time curve for well permeameter
        test _____   589
19–10   Nomograph for computing minimum volume of water
        to be discharged during well permeameter test to
        determine test duration _____   590
19–11   Relationship between depth of water in test well and
        distance to water table in well permeameter test __   592
20–1   Inplace vane shear test apparatus _____   594
20–2   Modified vane for friction determination _____   595
20–3   Method of installation of the vane testing apparatus __   596
20–4   Rotary cutting bit for use with the vane shear test
        apparatus _____   598
20–5   Vane test data sheet _____   600
20–6   Plot of vane test results _____   601
20–7   Typical calibration curve for a vane shear test apparatus   602
21–1   Standard drilling equipment used to perform penetra-
        tion tests _____   605

LIST OF FIGURES IN APPENDIX—Continued

Figure		Page
21–2	A portable power-operated rig used to perform penetration tests	606
21–3	A portable hand-operated rig used to perform penetration tests in shallow holes	606
21–4	Automatic-trip drive hammer	607
21–5	Standard split-tube and split-liner penetration resistance samplers	608
21–6	Split-liner sampler	608
22–1	Form for reporting results of Proctor needle test	611
24–1	Sand pouring devices (A) and (B), and template (C), for inplace soil density tests	614
24–2	Cone-type sand pouring device for inplace soil density tests	615
24–3	Field density test record (including rapid compaction control)	617
24–4	Procedure for inplace density test	619
25–1	Laboratory equipment required for rapid compaction control	623
25–2	Rapid compaction control method for 7.50 pounds of moist soil (normal density range)	626
25–3	Rapid compaction control method for 7.50 pounds of moist soil (low density range)	628
25–4	Procedure for rapid compaction control method	630
25–5	Graphical parabola method	631
25–6	Rapid method of compaction control—Tabulation of coordinates of maximum density point	632
25–7	Rapid method of compaction control—Alternative method to eliminate drying requirement for soils close to optimum water content	634
25–8	Recording data for rapid compaction control test	636
26–1	Pile test with load applied by means of a hydraulic jack acting against a reaction member held by anchor piles	643
26–2	Pile test with load from weighted box or platform applied to pile by means of a hydraulic jack	644
26–3	Pile test with load from weighted box or platform supported directly by the pile	645
26–4	Recording information relative to pile driving	647
26–5	Recording pile driving test data	648
26–6	Recording pile-loading test data	649

Figure		Page
27–1	Hydraulic-type twin-tube piezometer installation	651
27–2	Hydrostatic pressure readings	653
27–3	Foundation-type piezometer tip (with ceramic disc)	654
27–4	Embankment-type piezometer tip (with ceramic disc)	655
27–5	Foundation-type piezometer tip (with pipe extension)	656
27–6	Compression-type coupling	659
27–7	Gage and valve connections	660
27–8	Foundation-type piezometer tip and installation details	661
27–9	Offset trench	663
27–10	Use of rakes to aline piezometer tubes during installation—Trinity Dam	665
27–11	Bentonite cutoff	666
27–12	Piezometer terminal well layout	667
27–13	Piezometer terminal well—Elevation of upstream wall and typical gage and valve installation	669
27–14	Piezometer terminal well—Elevations of left wall and downstream wall and layout of gages	670
27–15	Air traps for piezometer terminal wells	674
27–16	Example—Separate gage readings	680
27–17	Example—Master gage readings	681
28–1	Porous-tube piezometer	687
28–2	Sounder device and reel for porous-tube piezometer	690
28–3	Components of the water level sounder	691
28–4	Ohmmeter deflection when probe contacts water surface	691
28–5	Installing porous-tube piezometer	692
28–6	Time lag for 95 percent equalization	696
28–7	Time lag tests	697
28–8	Example—Porous-tube piezometer readings	698
29–1	Vertical movement and foundation settlement installation	700
29–2	Vertical movement device in rock-free soils—10-foot spacing	701
29–3	Measuring torpedo	703
29–4	Reading scale	704
29–5	Water level indicator	705
29–6	Reel for water level sounder—For use on internal movement device	706

LIST OF FIGURES IN APPENDIX—Continued

Figure   *Page*

29–7   Example—Internal vertical movement readings _ _ _ _ _   708
29–8   Vertical movement device—Installation in rock-free
         soils _   711
29–9   Vertical movement device—Installation in rocky soils_   712
29–10   Using power auger (air) for drilling 8-inch-diameter
         holes to top of 2-inch riser pipe of vertical movement
         device _   713
29–11   Using air to remove sand from around previously
         placed pipe cover of vertical movement device _ _ _ _   714
29–12   Hand tools used in setting units of vertical movement
         device _   715
29–13   Tamping of backfill around 2-inch riser pipe on vertical
         movement device _   716
29–14   1½-inch crossarm pipe being placed inside 2-inch riser
         pipe of vertical movement device _ _ _ _ _ _ _ _ _ _   717
29–15   Special wrench fabricated by contractor for installing
         2-inch-diameter pipe extensions in vertical move-
         ment device and in foundation settlement installation   717
29–16   Installation of vertical movement device in rocky soils
         —Trinity Dam _   718
30–1   Horizontal movement device _ _ _ _ _ _ _ _ _ _ _ _ _ _   721
30–2   Horizontal movement device installation in rock-free
         soils _   724
30–3   Horizontal movement device installation in rocky soils_   726
30–4   Installation of horizontal movement device in rock-free
         soils—Trinity Dam _   728
30–5   Example—Vertical movement readings _ _ _ _ _ _ _ _ _ _   729
31–1   Baseplate installation _   731
31–2   Example—Foundation settlement readings _ _ _ _ _ _ _ _   732
32–1   Embankment measurement point installation _ _ _ _ _ _ _   734
32–2   Embankment measurement point readings _ _ _ _ _ _ _ _   737
33–1   Conduit measurement point installation _ _ _ _ _ _ _ _ _   739
33–2   Example—Conduit measurement point readings _ _ _ _   740
34–1   Example—Measurement point readings—Chute-type
         spillways _   743
34–2   Example—Measurement point readings—Chute-type
         spillways _   744
35–1   Special housing located on an embankment in which
         strong-motion accelerograph is installed _ _ _ _ _ _ _ _   745

LIST OF FIGURES IN APPENDIX—Continued

Figure | *Page*

35–2 An example of an accelerograph used on structures located in seismic areas _____ 746
36–1 Shallow-well permeameter test equipment _____ 748
36–2 Field permeability test (shallow-well method) _____ 750
36–3 Shallow-well permeameter installed in the compacted earth lining of a canal side slope _____ 751
36–4 Curve showing relationship between kinematic viscosity ratio of water and temperature in degrees centigrade 752
36–5 Permeability determination of canal lining by permeameter well _____ 753
36–6 Computation of seepage through soil linings _____ 754
37–1 Apparatus for calibrating mechanical laboratory soil compactors _____ 756
38–1 Compaction machine for soil containing gravel _____ 761
38–2 Compaction apparatus for soil containing gravel ____ 762
38–3 Sample preparation equipment for compaction test for soil containing gravel _____ 763
38–4 Percent of Proctor maximum dry density in relation to percent of gravel _____ 764
38–5 Compaction test for soil containing gravel—Data sheet 1 _____ 765
38–6 Compaction test for soil containing gravel—Data sheet 2 _____ 768
38–7 Compaction test for soil containing gravel—Data sheet 3 _____ 770
38–8 Compaction test for soil containing gravel—Data sheet 4 _____ 771
38–9 Form for plotting compaction test curves _____ 774

LIST OF TABLES

Table | *Page*

1 Description of soils _____ 19
2 General suitability of soils for compacted backfill by consolidating processes _____ 192
3 Relation of soil index properties and probable volume changes for highly plastic soils _____ 212
4 Criteria for control of compacted dam embankments _____ 257

LIST OF TABLES—Continued

Table		Page
5	Materials and placing requirements for various fills used in canals	270
6	Permissible rock sizes for various thicknesses of riprap and coarse gravel	272

LIST OF TABLES IN APPENDIX

1–1	Samples of construction materials	328
1–2	Samples of foundation materials	330
1–3	Accessible methods of exploration	332
1–4	Nonaccessible methods of exploration	334
2–1	Minimum size of exploratory hole	342
2–2	Representative undisturbed sample sizes for foundation exploration	344
2–3	Approximate proportions of mud mixtures	354
2–4	General recommendations for thin-wall, open drive sampling	368
3–1	Check list for description of coarse-grained soils	402
3–2	Check list for description of fine-grained and partly-organic soils	403
3–3	Identification of consistency of fine-grained soils	404
10–1	Absolute density of water in grams per cubic centimeter	458
11–1	Points for curve of complete saturation	476

INDEX

Index		791

LIST OF TABLES—Continued

Table
5 Munsell and pitting equivalents for various fills used in
   embankments 270
6 Permissible rock sizes for various thicknesses of riprap and
   coarse gravel 277

LIST OF TABLES IN APPENDIX

1-1 Samples of exploration patterns ....
1-2 Samples of foundation materials ....
1-3 Accessible methods of exploration ....
1-4 Nonaccessible methods of exploration ....
2-1 Minimum size of exploratory hole ....
2-2 Representative, undisturbed sample sizes for foundation
     exploration ....
2-3 Approximate proportions of mud mixtures ....
2-4 General recommendations for thin wall, open drive
     sampling ....
3-1 Check list for description of coarse-grained soils ..
3-2 Check list for description of fine-grained and organic
     soils ....
3-3 Identification of consistency of fine-grained soils .
10-1 Absolute density of water in gram per cubic centimeter
11-1 Points for curve of complete saturation ....

INDEX

Index ....

Chapter I

PROPERTIES OF SOILS

A. Identification and Classification

1. General.—Most soils are a heterogeneous accumulation of mineral grains that are not cemented together. However, the term "soil" or "earth" as used by engineers includes virtually every type of uncemented or partially cemented inorganic and organic material found in the ground. Only hard rock which remains firm after exposure is wholly excluded. To the engineer engaged in design and construction of foundations and earthwork, the physical properties of soils such as their unit weight, permeability, shear strength, compressibility, and interaction with water are of primary importance. The suitability of soils for raising crops interests the agronomist, the details of their origin and distribution concern the geologist, and the mineralogical properties are studied by the petrographer. In the solution of special soil problems, the engineer consults with these scientists who are specialists in their respective fields.

It is advantageous to have a standard method of identifying soils and classifying them into categories or groups which have distinct engineering characteristics. This enables engineers in the design office and those engaged in field work to speak the same language, thus facilitating exchange of information and experiences. Logs of explorations containing adequate soil classifications and descriptions can be used in making preliminary estimates, in determining the extent of additional field investigations needed for final design, in planning an economical testing program, and in extending test results to additional explorations. For final design of important structures, however, visual soil classification must be supplemented by laboratory tests to determine performance characteristics of the soil, such as permeability, shear strength, and compressibility under expected field conditions. Knowledge of soil classification, including typical engineering properties of soils of the various groups, is especially valuable to the engineer engaged in prospecting for earth materials or investigating foundations for structures.

In 1952, the Bureau of Reclamation and the Corps of Engineers, with Professor A. Casagrande of Harvard University as consultant, reached

1

agreement on a modification of Professor Casagrande's Airfield Classification System, which they named the "Unified Soil Classification System." This system takes into account the engineering properties of soils; it is descriptive and easy to associate with actual soils; and it has the flexibility of being adaptable both to the field and to the laboratory. Probably its greatest advantage is that a soil can be classified readily by visual and manual examination without the necessity for laboratory testing. The Unified Soil Classification System is based on the size of the particles, the amounts of the various sizes, and the characteristics of the very fine grains.

A soil mass consists of solid particles and pore fluids. The solid particles generally are mineral grains of various sizes and shapes, occurring in every conceivable arrangement. These solid particles can be divided into various components, each of which contributes its share to the physical properties of the whole. Soil classification can best be understood by first considering the properties of these soil components and their interaction with water. Accordingly, sections 2, 3, and 4 describe the constituents of soil and introduce concepts used in the classification system. Section 5 gives the essentials of the classification scheme for soils found in nature as shown on the unified soil classification chart. In addition to proper classification, it is important to include an adequate description of the soil in reports or logs of explorations. This subject is discussed in section 6. Section 7 contains a comparison of the engineering properties of typical soils of each classification group.

2. Soil Components.—(a) *Size*.—Particles larger than 3 inches are excluded from the Unified Soil Classification System. The amount of oversized material, however, is of great importance in the selection of sources for embankment material; hence, logs of explorations must always contain information on quantity and size of particles larger than 3 inches. Rounded particles are called cobbles if they are between 3 and 12 inches in size, and boulders if they are greater than 12 inches in size. Angular particles above 3 inches in size are classified as rock fragments.

Within the size range of the system there are two major divisions; namely, the coarse grains and the fine grains. Coarse grains are those larger than the U.S. Standard Series No. 200 sieve size (0.074 mm.), and are further divided as follows:

Gravel (G), from 3 inches to No. 4 sieve (3/16 inch):
Coarse gravel—3 inches to ¾ inch
Fine gravel—¾ inch to No. 4 sieve
Sand (S), from No. 4 sieve to No. 200 sieve:
Coarse sand—No. 4 to No. 10 sieve
Medium sand—No. 10 to No. 40 sieve
Fine sand—No. 40 to No. 200 sieve.

For visual classification, ¼ inch is considered equivalent to the No. 4 sieve size, and the No. 200 sieve is about the smallest size particle that can be distinguished individually by the unaided eye.

Fine grains, or fines, are smaller than the No. 200 sieve size and are of two types: *silt* (M) and *clay* (C). Older classification systems defined clay variously as those particles either smaller than 5 microns (0.005 mm.) or 2 microns (0.002 mm.), and they defined silt as fines larger than the clay sizes. It is a mistaken idea, however, that typical engineering characteristics of silt and clay correspond to particular grain sizes. Natural deposits of rock flour that exhibit all the properties of silt and none of clay may consist entirely of grains smaller than 5 microns. On the other hand, a soil mass may consist mainly of particles larger than 5 microns, but containing small quantities of extremely fine, colloidal-sized particles and behave as a typical clay. Size distinction, therefore, is not made between silt and clay in the Unified Soil Classification System; rather, the two materials are differentiated by their behavior.

Organic material (O) is often a component of soil but it has no specific grain size. It is distinguished by the composition of its particles rather than by their sizes, which range from colloidal-sized particles of molecular dimensions to fibrous pieces of partly decomposed vegetable matter several inches in length.

(b) *Gradation (Grain-size Distribution).*—The amounts of the various sizes of grains present in a soil can be determined in the laboratory by means of sieving for the coarse grains and by sedimentation (hydrometer analysis) for the fines. The laboratory results are usually presented in the form of a cumulative grain-size curve. For soils consisting mainly of coarse grains the grain-size distribution reveals something of the physical properties of the material. On the other hand, the grain size is much less significant for soils containing a preponderance of fine grains. Gradations of coarse-grained soils are:

Well-graded (W)—Good representation of all particle sizes from largest to smallest

Poorly-graded (P)—Uniform, most particles about the same size; or skip (or gap) gradation, absence of one or more intermediate sizes.

In the field, the soil is first classified as either coarse-grained or fine-grained and, if classified as a coarse-grained soil, the soil is further classified as well-graded or poorly-graded by visual examination. In the laboratory, the gradation can be determined by the use of criteria based on the range of sizes and on the shape of the grain-size curve. The measure of size range is called the coefficient of uniformity, C_u, which is the ratio of the 60-percent-finer-than size (D_{60}) to the 10-percent-finer-than size (D_{10}). The shape of the grain-size curve is given by the coefficient of curvature, C_c, which is the ratio of the square of the 30-percent-finer-than size (D_{30})^2 to the product of (D_{60}) by (D_{10}). The limiting ranges for these values with respect to soil classification are given on the chart in figure 7. Photographs of typical gradations and corresponding grain-size curves are shown in figures 1, 2, and 3.

(c) *Shape.*—The shape of the particles has an important influence on the physical properties of a soil. The following shapes are most common:

Bulky or equidimensional grains.—These may be further described as rounded, subrounded, subangular, and angular as shown in figure 4. The coarse-grained components of a soil are usually of the bulky type, consisting chiefly of the minerals quartz and feldspar.

Flaky grains, also called platelike particles.—These are present in appreciable quantities in many fine-grained soils. Mica and some clay minerals have this shape which is mainly responsible for their high compressibility. See figure 5.

Elongated grains and fibers.—The most commonly encountered materials in this class are the clay mineral halloysite (fig. 6), asbestos, some types of volcanic ash, and organic soils such as peat.

3. Soil Moisture.—A typical soil mass has three constituents—soil grains, air, and water. In soils consisting largely of fine grains, the amount of water present in the voids has a pronounced effect on the soil properties. Three main states of soil consistency are recognizable:

(1) *Liquid state*—in which the soil is either in suspension or behaves like a viscous fluid;

(2) *Plastic state*—in which the soil can be rapidly deformed or molded without rebounding elastically, changing volume, cracking, or crumbling; and

(3) *Solid state*—in which the soil will crack when deformed or will exhibit elastic rebound.

The solid state is sometimes divided into a semisolid state and a solid state as shown on figure 9.

In describing these soil states, it is customary to consider only the fraction of soil smaller than the No. 40 sieve size (the upper limit of the fine sand component). For this soil fraction the water content, in percentage of dry weight, at which the soil passes from the liquid state into the plastic state is called the liquid limit, *LL*. Similarly, the water

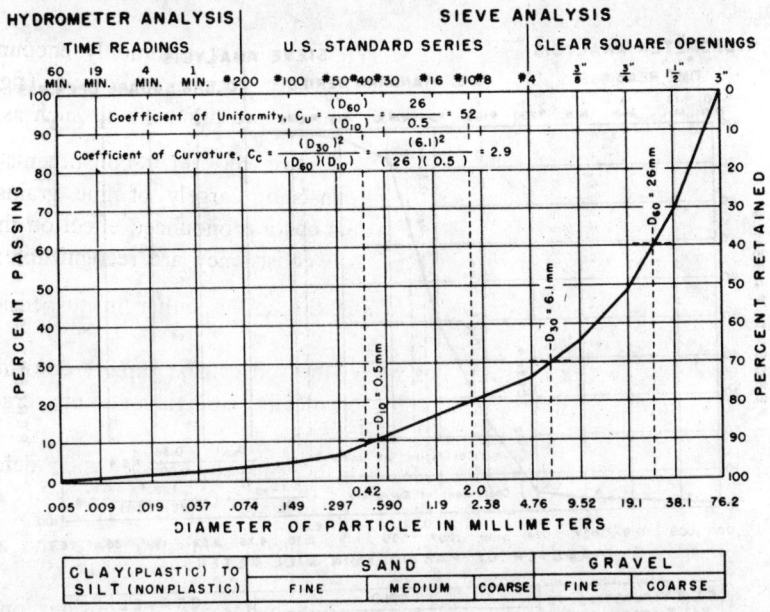

Figure 1.—Well-graded gravel (GW), Vermejo Project, New Mexico. P455-D-15105 and 101-D-569.

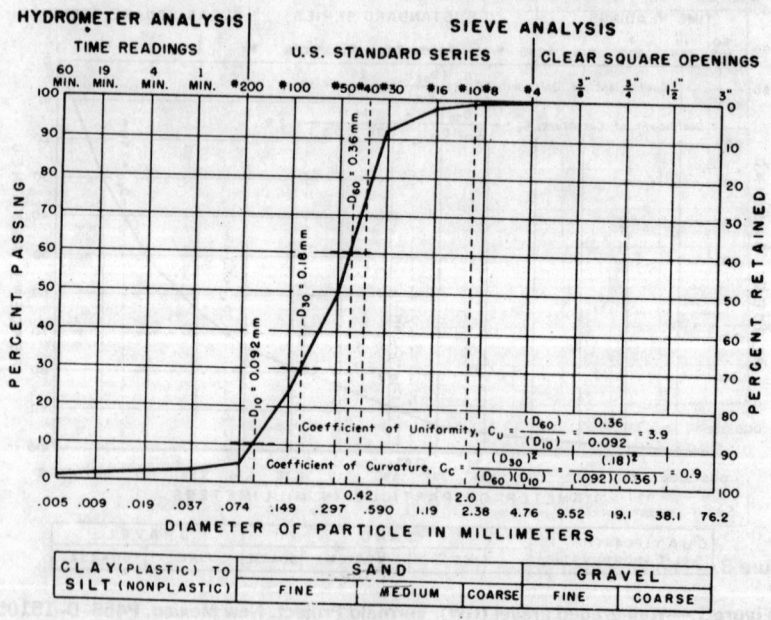

Figure 2.—Uniform, poorly-graded sand (SP), Cherry Creek Reservoir, Colorado.
PX–D–15105 and 101–D–570.

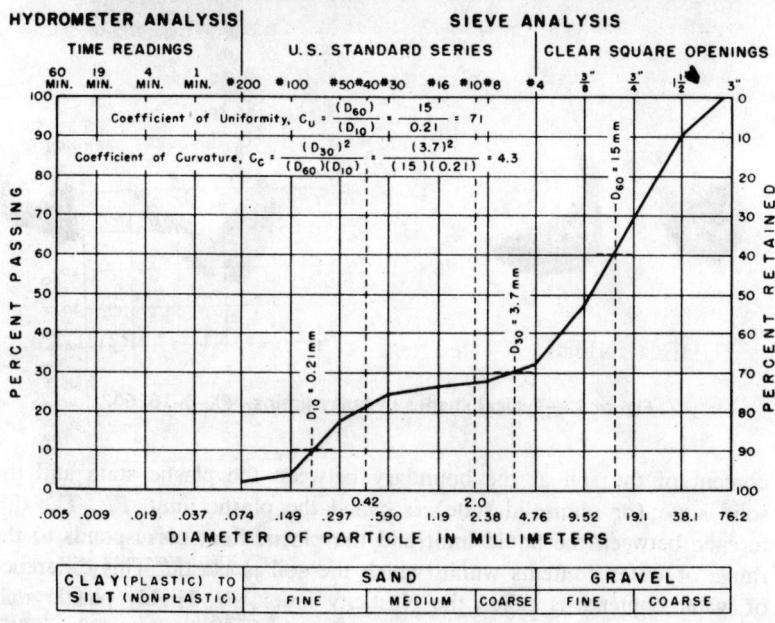

Figure 3.—Poorly-graded gravel (GP), Falcon Dam, Texas. POA–4–D–15106 and 101–D–571.

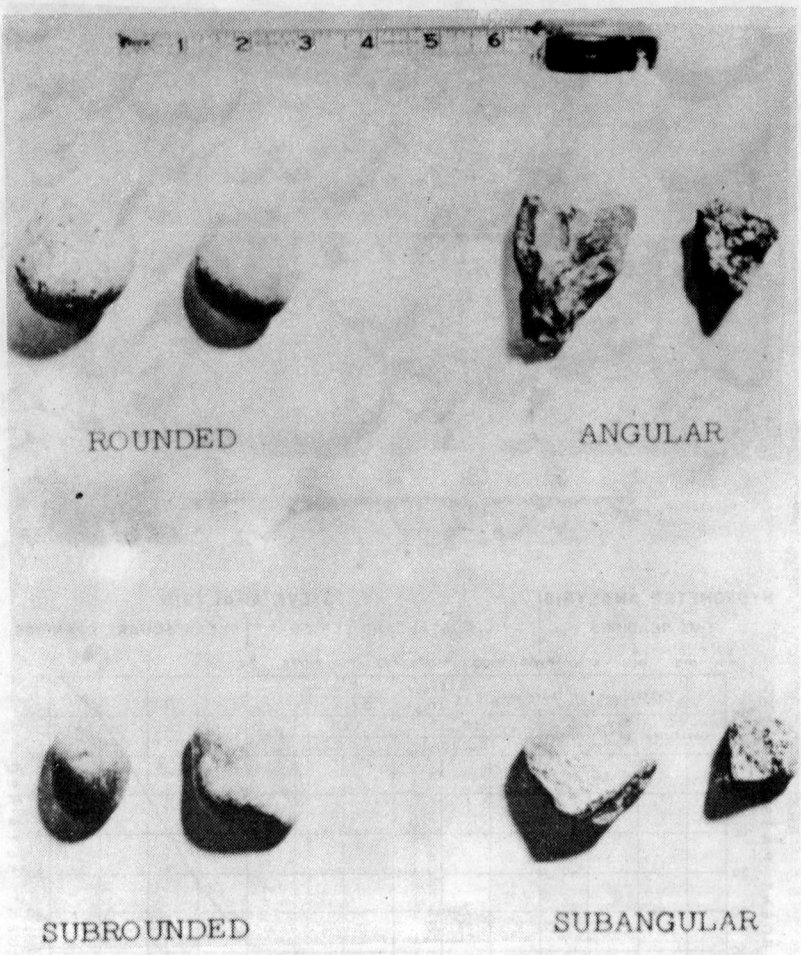

Figure 4.—Typical shapes of bulky grains. PX–D–16266.

content of the soil at the boundary between the plastic state and the solid state (or semisolid state) is called the plastic limit, *PL*. The difference between the liquid limit and the plastic limit corresponds to the range of water contents within which the soil is plastic. This difference of water contents is called the plasticity index, *PI*. Highly plastic soils have high *PI* values. In a nonplastic soil, the plastic limit and the liquid limit are the same, and the *PI* equals 0.

These limits of consistency, which are called "Atterberg limits" after a Swedish soils scientist, are used in the Unified Soil Classification

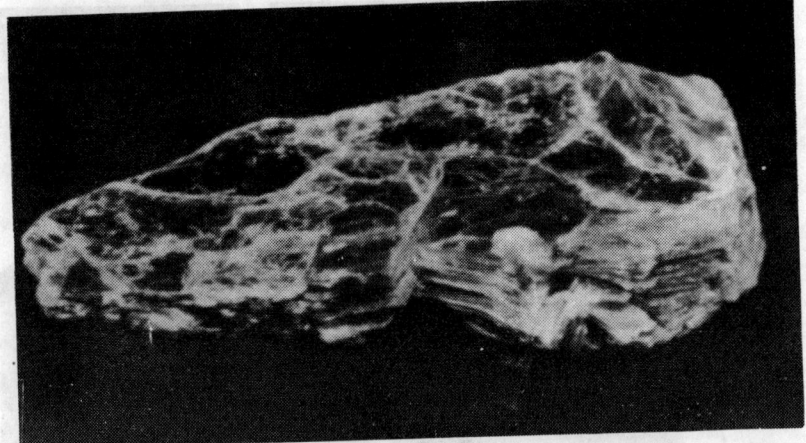

(A) MICA

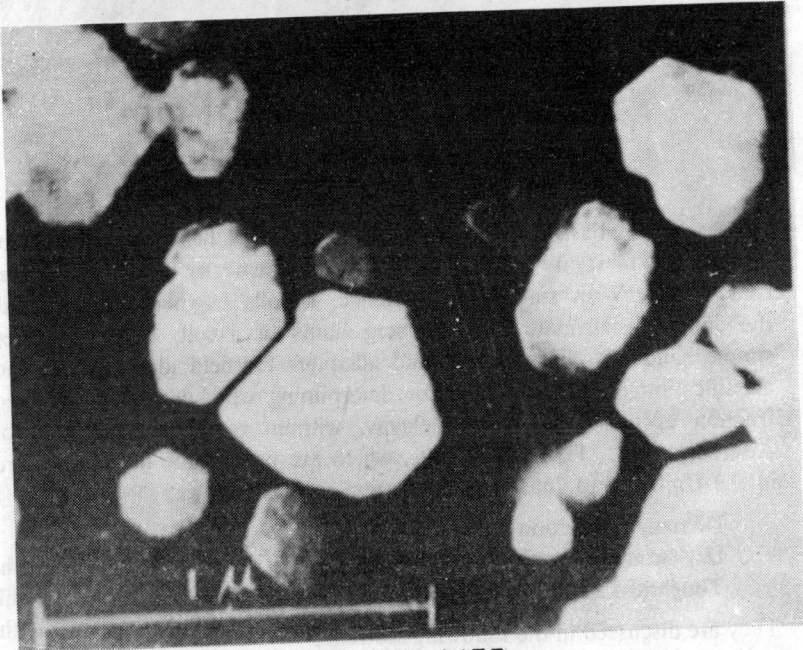

(B) KAOLINITE

1u = 1 micron = 0.001 mm.

Figure 5.—Flaky grains. E-2254-2NA and PX-D-16261.

HALLOYSITE
lu = l micron = 0.001 mm.

Figure 6.—Elongated grains. PX–D–16261.

System as the basis for laboratory differentiation between materials of appreciable plasticity (clays) and slightly plastic or nonplastic materials (silts). With sufficient experience, a soils engineer may acquire the ability to estimate the Atterberg limits of a soil. However, three simple hand tests have been found adequate for field identification and classification of fine soils and for determining whether the fine-grained fraction of a soil is silty or clayey, without requiring estimation of Atterberg limits. These hand tests, which are part of the field procedure in the Unified Soil Classification System, are as follows:

Dilatancy (reaction to shaking)
Dry strength (crushing characteristics)
Toughness (consistency near plastic limit).

They are discussed in the following section.

4. **Characteristics of Soil Components.**—(a) *Gravel and Sand.*—Both of the coarse-grained components of soil (gravel and sand) have essentially the same engineering properties, differing mainly in degree. The division of gravel and sand sizes by the No. 4 sieve is arbitrary and does not correspond to a sharp change in properties. Well-graded, compacted gravels or sands are stable materials. When devoid of fines the coarse-grained soils are pervious, easy to compact, and little affected

by moisture or frost action. Although grain shape and gradation as well as size affect the engineering properties of the coarse-grained soils, gravels are generally more pervious, more stable, and less affected by water or frost than are sands containing the same amount of fines.

As a sand becomes finer and more uniform it approaches the characteristics of silt, with corresponding decreased permeability and reduction in stability in the presence of water. Very fine sands are difficult to distinguish visually from silts. Dry sand, however, exhibits no cohesion (does not hold together) and feels gritty in contrast to the very slight cohesion and smooth feel of dried silt.

(b) *Silt and Clay (Fines).*—Even small amounts of fines may have important effects on engineering properties of the soils in which they are found. As little as 10 percent of particles smaller than the No. 200 sieve size in sand and gravel may make the soil virtually impervious, especially when the coarse grains are well graded. Also, serious frost heaving in well-graded sands and gravels may be caused by less than 10 percent fines. The utility of coarse-grained materials for surfacing roads can be improved by the addition of a small amount of clay to act as a binder for the sand and gravel particles.

Soils containing large quantities of silt and clay are the most troublesome to the engineer. These materials exhibit marked changes in physical properties with change of water content. A hard, dry clay, for example, may be suitable as a foundation for heavy loads so long as it remains dry but may turn into a quagmire when wet. Many of the fine soils shrink on drying and expand on wetting, which may adversely affect structures founded upon them or constructed of them. Even when the water content does not change, the properties of fine soils may vary considerably between their natural condition in the ground and their state after being disturbed. Deposits of fine particles which have been subjected to loading in geologic time frequently have a soil structure which gives the material unique properties in the undisturbed state. When the soil is excavated for use as a construction material or when the natural deposit is disturbed, for example, by driving piles, the soil structure is destroyed and the properties of the soil are changed radically.

Silts differ from clays in many important respects, but because of similarity in appearance they often have been mistaken one for the other, sometimes with unfortunate results. Dry, powdered silt and clay are indistinguishable from each other, but they are easily identified by their behavior in the presence of water. Recognition of fines as being either silt or clay is an essential part of the Unified Soil Classification System.

Silts are the slightly plastic or the nonplastic fines. They are inherently unstable in the presence of water and have a tendency to become "quick" when saturated and assume the character of a viscous fluid and commonly

FIELD IDENTIFICATION PROCEDURES FOR FINE GRAINED SOILS OR FRACTIONS

These procedures are to be performed on the minus No. 40 sieve size particles, approximately $\frac{1}{64}$ in. For field classification purposes, screening is not intended; simply remove by hand the coarse particles that interfere with the tests.

DILATANCY (Reaction to shaking)

After removing particles larger than No. 40 sieve size, prepare a pat of moist soil with a volume of about one-half cubic inch. Add enough water if necessary to make the soil soft but not sticky.

Place the pat in the open palm of one hand and shake horizontally, striking vigorously against the other hand several times. A positive reaction consists of the appearance of water on the surface of the pat which changes to a livery consistency and becomes glossy. When the sample is squeezed between the fingers, the water and gloss disappear from the surface, the pat stiffens, and finally it cracks or crumbles. The rapidity of appearance of water during shaking and of its disappearance during squeezing assist in identifying the character of the fines in a soil.

Very fine clean sands give the quickest and most distinct reaction whereas a plastic clay has no reaction. Inorganic silts, such as a typical rock flour, show a moderately quick reaction.

DRY STRENGTH (Crushing characteristics)

After removing particles larger than No. 40 sieve size, mold a pat of soil to the consistency of putty, adding water if necessary. Allow the pat to dry completely by oven, sun, or air drying, and then test its strength by breaking and crumbling between the fingers. This strength is a measure of the character and quantity of the colloidal fraction contained in the soil. The dry strength increases with increasing plasticity.

High dry strength is characteristic for clays of the CH group. A typical inorganic silt possesses only very slight dry strength. Silty fine sands and silts have about the same slight dry strength, but can be distinguished by the feel when powdering the dried specimen. Fine sand feels gritty whereas a typical silt has the smooth feel of flour.

TOUGHNESS (Consistency near plastic limit)

After removing particles larger than the No. 40 sieve size, a specimen of soil about one-half inch cube in size is molded to the consistency of putty. If too dry, water must be added and if sticky, the specimen should be spread out in a thin layer and allowed to lose some moisture by evaporation. Then the specimen is rolled out by hand on a smooth surface or between the palms into a thread about one-eighth inch in diameter. The thread is then folded and rerolled repeatedly. During this manipulation the moisture content is gradually reduced and the specimen stiffens, finally loses its plasticity, and crumbles when the plastic limit is reached.

After the thread crumbles, the pieces should be lumped together and a slight kneading action continued until the lump crumbles.

The tougher the thread near the plastic limit and the stiffer the lump when it finally crumbles, the more potent is the colloidal clay fraction in the soil. Weakness of the thread at the plastic limit and quick loss of coherence of the lump below the plastic limit indicate either inorganic clay of low plasticity, or materials such as Kaolin-type clays and organic clays which occur below the A-line.

Highly organic clays have a very weak and spongy feel at the plastic limit.

flow. Quick silts are often called "bulls liver" by construction men. Silts are fairly impervious, difficult to compact, and are highly susceptible to frost heaving.

Silt masses undergo change of volume with change of shape (the property of dilatancy), in contrast to clays which retain their volume with change of shape (the property of plasticity). The dilatancy property, together with the "quick" reaction to vibration, affords a means of identifying typical silt in the loose wet state. The dilatancy test is described on the classification chart (fig. 7) and is photographically illustrated in figure 3-4, designation E-3. When dry, silt can be pulverized easily under finger pressure (indicative of very slight dry strength) and will have a smooth feel between the fingers in contrast to the grittiness of fine sand.

Silts differ among themselves in size and shape of grains which are reflected mainly in the property of compressibility. For similar conditions of previous geologic loading, the higher the liquid limit of a silt, the more compressible it is. The liquid limit of a typical bulky-grained, inorganic silt is about 30 percent, while highly micaceous or diatomaceous silts (so-called elastic silts), consisting mainly of flaky grains, may have liquid limits as high as 100 percent. The differences in quicking and dilatancy properties afford a means of distinguishing in the field between silts of low liquid limits (L) and those of high liquid limits (H).

Clays are the plastic fines. They have low resistance to deformation when wet, but they dry to hard cohesive masses. Clays are virtually impervious, difficult to compact when wet, and impossible to drain by ordinary means. Large expansion and contraction with changes in water content are characteristic of clays. The small size, flat shape, and type of mineral composition of clay particles combine to produce a material that is both compressible and plastic. The higher the liquid limit of a clay, the more compressible it will be when compared at equal conditions of previous geological loading. Hence, in the Unified Soil Classification System, the liquid limit is used to distinguish between clays of high compressibility (H) and those of low compressibility (L). Differences in plasticity of clays are reflected by their plasticity indexes. At the same liquid limit, the higher the plasticity index, the more cohesive is the clay.

Field differentiation among clays is accomplished by the toughness test in which the moist soil is molded and rolled into threads until crumbling occurs, and by the dry strength test which measures the resistance of the clay to breaking and pulverizing. The dry strength and the toughness tests are shown on the classification chart (fig. 7). With a little experience in performing these tests, the low compressibility and low plasticity "lean" clays (L), can be readily distinguished from the highly plastic, highly compressible "fat" clays (H).

(c) *Organic Matter.*—Organic matter in the form of partly decomposed vegetation is the primary constituent of peaty soils. Varying amounts of finely divided vegetable matter are found in plastic and nonplastic sediments and often affect their properties sufficiently to influence their classification. Thus, we have organic silts and silt-clays of low plasticity and organic clays of medium to high plasticity. Even small amounts of organic matter in colloidal form in a clay will result in an appreciable increase in liquid limit (and plastic limit) of the material without increasing its plasticity index. Organic soils are dark gray or black in color and usually have a characteristic odor of decay. Organic clays feel spongy in the plastic range as compared to inorganic clays. The tendency for soils high in organic content to create voids by decay or to change the physical characteristics of a soil mass through chemical alteration makes them undesirable for engineering use. Soils containing even moderate amounts of organic matter are significantly more compressible and less stable than inorganic soils; hence, they are less desirable for engineering use.

5. Classification of Soils.—(a) *General.*—Soils in nature seldom exist separately as gravel, sand, silt, clay, or organic matter, but are usually found as mixtures with varying proportions of these components. The Unified Soil Classification System is based on recognition of the type and predominance of the constituents, considering grain size, gradation, plasticity, and compressibility. It divides soils into three major divisions: coarse-grained soils, fine-grained soils, and highly organic (peaty) soils. In the field, identification is accomplished by visual examination for the coarse grains and by a few simple hand tests for the fine-grained soils or fraction, as described in the classification chart (fig. 7) and designation E–3. In the laboratory, in addition to visual examination, the grain-size curve and the Atterberg limits can be used. The peaty soils (Pt) are readily identified by color, odor, spongy feel, and fibrous texture, and are not further subdivided in the classification system.

(b) *Field Classification.*—A representative sample of soil (excluding particles larger than 3 inches) is first classified as coarse-grained or fine-grained by estimating whether 50 percent, by weight, of the particles can be seen individually by the naked eye. Soils containing more than 50 percent of individual particles that can be seen are coarse-grained soils; soils containing more than 50 percent of particles smaller than the eye can individually distinguish are fine-grained soils. (The No. 200 sieve size particles are about the smallest that can be seen individually by the unaided eye.) If the soil is predominantly coarse-grained, it is then identified as being a gravel or a sand by estimating whether 50 percent or more, by weight, of the coarse grains are larger or smaller than the No. 4 sieve size (about ¼ inch).

If the soil is a gravel, it is next identified as being "clean" (containing little or no fines) or "dirty" (containing an appreciable amount of fines). For clean gravels final classification is made by estimating the gradation: the well-graded gravels belong to the GW group, and uniform and skip-graded gravels, the poorly-graded gravels, belong to the GP group. Dirty gravels are of two types: those with nonplastic (silty) fines (GM) and those with plastic (clayey) fines (GC). The determination of whether the fines are silty or clayey is made by the three manual tests for fine-grained soils.

If a soil is a sand, the same steps and criteria are used as for the gravels in order to determine whether the soil is a well-graded clean sand (SW), poorly-graded clean sand (SP), sand with silty fines (SM), or sand with clayey fines (SC).

If a material is predominantly (more than 50 percent by weight) fine-grained, it is classified into one of six groups (ML, CL, OL, MH, CH, OH) by estimating its dilatancy (reaction to shaking), dry strength (crushing characteristics), and toughness (consistency near the plastic limit), and by identifying it as being organic or inorganic. The test procedures and the behavior of the various groups of fine-grained soils for each of the hand tests are shown on the classification chart (fig. 7).

Soils typical of the various groups are readily classified by the foregoing procedures. Many natural soils, however, will have property characteristics of two groups, because they are close to the borderline between the groups either in percentages of the various sizes, or in plasticity characteristics. For this substantial number of soils, borderline classifications are used; that is, the two group symbols most nearly describing the soil are connected by a hyphen, such as GW–GC.

If the percentages of gravel and sand sizes in a coarse-grained soil are nearly equal, the classification procedure is to assume the soil is a gravel and then continue the classification procedure using the chart until the final soil group, say GC, is reached. Since it could have been assumed that the soil is a sand, the correct field classification is GC–SC, because the criteria for the gravel and sand subgroups are identical. Similarly, within the gravel or sand groupings, borderline classifications such as GW–GP, GM–GC, GW–GM, SW–SP, SM–SC, and SW–SM can occur.

Proper borderline classification of a soil near the borderline between coarse-grained and fine-grained soils is accomplished by classifying it first as a coarse-grained soil and then as a fine-grained soil. Such classifications as SM→ML and SC–CL are common.

Within the fine-grained division, borderline classifications can occur between low-liquid-limit soils and high-liquid-limit soils as well as between silty and clayey materials in the same range of liquid limits. For example, one may find ML–MH, CL–CH, and OL–OH soils; ML–CL, ML–OL, and CL–OL soils; and MH–CH, MH–OH, and CH–OH soils.

(c) *Laboratory Classification.*—Although most classifications of soil will be done visually and by simple hand tests, the Unified Soil Classification System has provided for precise delineation of the soil groups by gradation analyses and Atterberg limits tests in the laboratory. Laboratory classifications are often performed on representative samples of soils which are to be subjected to extensive testing for shear strength, compressibility, and permeability. They can also be used to advantage in training the field classifier of soils to improve his ability to estimate percentages and degrees of plasticity.

The grain-size curve is used to classify the soil as being coarse-grained or fine-grained, and if coarse-grained, into gravel or sand using the 50 percent criterion. Within the gravel or sand groupings, soils containing less than 5 percent finer than the No. 200 sieve size are considered "clean" and are then classified as well-graded or poorly-graded by their coefficients of uniformity and of curvature. In order for a clean gravel to be well-graded (GW), it must have *both* a coefficient of uniformity, C_u, greater than 4 and a coefficient of curvature, C_c, between 1 and 3; otherwise, it is classified as a poorly-graded gravel (GP). A clean sand having *both* C_u greater than 6 and C_c between 1 and 3 is in the SW group; otherwise, it is a poorly-graded sand (SP).

"Dirty" gravels or sands are those containing more than 12 percent fines, and they are classified as either silty or clayey by results of their Atterberg limits tests as plotted on the plasticity chart shown in figure 7. Silty fines are those which have a plasticity index, *PI*, less than 4 *or* which plot below the "A" line. Clayey fines are those which have a *PI* greater than 7 *and* which plot above the "A" line.

Coarse-grained soils containing between 5 and 12 percent fines are borderline cases between the clean and the dirty gravels or sands (GW, GP, SW, SP, and GM, GC, SM, SC). Similarly, borderline cases may occur in dirty gravels and dirty sands where the *PI* is between 4 and 7 (GM–GC, SM–SC). It is theoretically possible, therefore, to have a borderline case of a borderline case; but this refinement is not used, and the rule for correct classification is to favor the nonplastic one. For example, a gravel with 10 percent fines, a C_u of 20, a C_c of 2.0, and a *PI* of 6 would be classified GW–GM rather than GW–GC.

If a soil is determined to be fine-grained by using the grain-size curve, it is further classified into one of the six groups by plotting the results of Atterberg limits tests on the plasticity chart, with attention being given to the organic content. Inorganic, fine-grained soils with *PI* greater than 7 and above the "A" line are CL or CH, depending on whether their liquid limits are less than 50 percent or more than 50 percent, respectively. Similarly, inorganic, fine-grained soils with *PI* less than 4 or below the "A" line are ML or MH, depending on whether their liquid limits are less than or more than 50 percent, respectively. Fine-grained soils which fall above the "A" line but which have a plasticity index between 4 and 7 are classified ML–CL.

Soils below the "A" line that are definitely organic are classified as OL if they have liquid limits less than 50 percent and as OH if the liquid limits are above 50 percent. Organic silts and clays are usually distinguished from inorganic silts, which have the same position on the plasticity chart, by odor and by color. However, when the organic content is doubtful, the material can be ovendried, remixed with water, and retested for liquid limits. The plasticity of fine-grained organic soils is greatly reduced on ovendrying due to irreversible changes in the organic colloids. Ovendrying also affects the liquid limit of inorganic soils but to a much smaller degree. A reduction in liquid limit after ovendrying to a value less than three-fourths of the liquid limit before ovendrying is considered positive identification of organic soils.

6. Description of Soils.—(a) *General.*—Although the group symbols of the Unified Soil Classification System will indicate typical soils, there are important characteristics of soils which are not fully designated by symbols, yet are easily ascertained by the investigator. It is necessary, therefore, for the classifier of soils to use a word picture to describe the soil in addition to placing it in the proper group. This is of particular importance for soils that are being investigated for use inplace as foundations for structures. Here, the natural condition of the soil, such as its looseness or compactness, its structure, and its drainage characteristics, is equally as important as the classification of its constituents.

The purposes for which soils are investigated can be divided into two categories: (1) borrow materials for embankment or for backfill, and (2) foundations for structures. The emphasis on various features to be described depends on which of the categories is involved. For many structures large quantities of soil must be excavated to reach a desired foundation. In the interest of economy, maximum use should always be made of this excavated material in the construction of embankments and for backfill. A foundation area, therefore, often becomes a source of material, and the investigation of the area must take into account its potential dual usage. Descriptions of soils encountered in such explorations must, therefore, contain the essential information required both for borrow materials and for foundation soils.

(b) *Borrow Materials.*—Soils that are potential sources of borrow material for embankments must be described adequately in the log of the exploratory test pit or auger hole. Since these materials are destined to be disturbed by excavation, transportation, and compaction in the fill, their structure is less important than the amount and characteristics of the soil constituents. However, the recording of their natural water content is important. Very dry borrow materials require the addition of large amounts of moisture for compaction control; and very wet soils containing appreciable fines may require extensive processing in order to be usable. For simplicity, the natural water content of borrow soils should be reported as either *dry, moist, wet,* or *saturated.* However, caution should be exercised when dealing with borrow materials which have been described as saturated since complete saturation (all voids full of water) is not discernible visually, especially in clayey soils. A soil that is reported dry should be one that will definitely require the addition of moisture in order to compact it properly in a fill. A soil should be reported as being moist if it is reasonably close to the Proctor optimum water content as defined in section 13. Soils that are reported as wet should be obviously well beyond the optimum water content. Borrow-pit holes are logged so as to indicate divisions between soils of different classification groups. Within the same soil group, significant changes in moisture should also be logged.

(c) *Foundations for Structures.*—When soils are being explored as foundations for hydraulic structures, their natural structure, compactness, and water content are of outstanding importance. Logs of foundation explorations, therefore, must emphasize the inplace condition of the soil in addition to describing its constituents. The natural state of foundation soils is significant because bearing capacity and settlement under load vary tremendously with the consistency or compactness of the soil. Therefore, information that a clay soil is hard and dry or soft and moist is important. Changes in consistency of foundation soils due to moisture changes under operating conditions must be considered in the design. Correct field classification is needed so that the effect of these moisture changes on the properties of the foundation soils can be predicted.

Table 1 lists data that are needed to describe soils for borrow materials and for foundations. Under each of these categories, the information desired for coarse-grained soils and for fine-grained soils is indicated by an "X". All of these descriptive data are not always needed. Judgment should be used to include pertinent information, to avoid negative information, and to eliminate repetition. The items indicated by "XX" must always be reported. Examples of soil descriptions are given in the soil classification chart, figure 7.

7. Properties of Soil Groups.—As a consequence of the emphasis on soil behavior, it is possible to indicate the engineering properties typical of the various soil groups in the Unified Soil Classification System. Such a direct relation of properties to soil classification is of value

Table 1.—Description of soils

Items of descriptive data	Borrow		Foundation	
	Coarse-grained soils	Fine-grained soils	Coarse-grained soils	Fine-grained soils
	XX: Required information on all logs			
	X: Desired information			
Typical name (examples are shown in classification chart)	XX	XX	XX	XX
Approximate percentages of gravel and sand	X	X	X	X
Maximum size of particles (including cobbles and boulders)	XX	X	X	X
Shape of the coarse grains—angularity	X		X	
Surface condition of the coarse grains—coatings	X			
Hardness of the coarse grains—possible breakdown into smaller sizes	X		X	
Color (in moist condition for fine-grained soils and fraction of fines in coarse-grained soils)	X	X	X	X
Moisture (dry, moist, wet, saturated)	XX	XX	XX	XX
Organic content	X	X	X	X
Plasticity—degree (nonplastic, low, medium, high) and dilatancy, dry strength, and toughness for fine-grained soils and of the fine-grained fraction in coarse-grained soils	X	XX	XX	XX
Structure (stratification, lenses and seams, laminations, giving dip and strike and thickness of layer; honeycomb, flocculent, root holes, etc.)			XX	XX
Cementation—type	XX	X	XX	XX
Degree of compactness—loose or dense (excepting clays)	X	X	XX	XX
Consistency in undisturbed and remolded states (clays only)				XX
Local or geologic name	X	X	XX	XX
Group symbol	XX	XX	XX	XX

TYPICAL NAMES OF SOIL GROUPS	GROUP SYMBOLS	IMPORTANT ENGINEERING PROPERTIES			
		PERMEABILITY WHEN COMPACTED	SHEAR STRENGTH WHEN COMPACTED AND SATURATED	COMPRESSIBILITY WHEN COMPACTED AND SATURATED	WORKABILITY AS A CONSTRUCTION MATERIAL
WELL-GRADED GRAVELS, GRAVEL-SAND MIXTURES, LITTLE OR NO FINES.	GW	PERVIOUS	EXCELLENT	NEGLIGIBLE	EXCELLENT
POORLY-GRADED GRAVELS, GRAVEL-SAND MIXTURES, LITTLE OR NO FINES.	GP	VERY PERVIOUS	GOOD	NEGLIGIBLE	GOOD
SILTY GRAVELS, POORLY-GRADED GRAVEL-SAND-SILT MIXTURES	GM	SEMIPERVIOUS TO IMPERVIOUS	GOOD	NEGLIGIBLE	GOOD
CLAYEY GRAVELS, POORLY-GRADED GRAVEL-SAND-CLAY MIXTURES.	GC	IMPERVIOUS	GOOD TO FAIR	VERY LOW	GOOD
WELL-GRADED SANDS, GRAVELLY SANDS, LITTLE OR NO FINES.	SW	PERVIOUS	EXCELLENT	NEGLIGIBLE	EXCELLENT
POORLY-GRADED SANDS, GRAVELLY SANDS, LITTLE OR NO FINES.	SP	PERVIOUS	GOOD	VERY LOW	FAIR
SILTY SANDS, POORLY-GRADED SAND-SILT MIXTURES.	SM	SEMIPERVIOUS TO IMPERVIOUS	GOOD	LOW	FAIR
CLAYEY SANDS, POORLY-GRADED SAND-CLAY MIXTURES.	SC	IMPERVIOUS	GOOD TO FAIR	LOW	GOOD
INORGANIC SILTS AND VERY FINE SANDS, ROCK FLOUR, SILTY OR CLAYEY FINE SANDS WITH SLIGHT PLASTICITY.	ML	SEMIPERVIOUS TO IMPERVIOUS	FAIR	MEDIUM	FAIR
INORGANIC CLAYS OF LOW TO MEDIUM PLASTICITY, GRAVELLY CLAYS, SANDY CLAYS, SILTY CLAYS, LEAN CLAYS.	CL	IMPERVIOUS	FAIR	MEDIUM	GOOD TO FAIR
ORGANIC SILTS AND ORGANIC SILT-CLAYS OF LOW PLASTICITY.	OL	SEMIPERVIOUS TO IMPERVIOUS	POOR	MEDIUM	FAIR
INORGANIC SILTS, MICACEOUS OR DIATOMACEOUS FINE SANDY OR SILTY SOILS, ELASTIC SILTS.	MH	SEMIPERVIOUS TO IMPERVIOUS	FAIR TO POOR	HIGH	POOR
INORGANIC CLAYS OF HIGH PLASTICITY, FAT CLAYS.	CH	IMPERVIOUS	POOR	HIGH	POOR
ORGANIC CLAYS OF MEDIUM TO HIGH PLASTICITY	OH	IMPERVIOUS	POOR	HIGH	POOR
PEAT AND OTHER HIGHLY ORGANIC SOILS.	PT	—	—	—	—

Figure 8.—Engineering use chart. (Sheet 1 of 2.) 101-D-235(1).

RELATIVE DESIRABILITY FOR VARIOUS USES (NO. I IS CONSIDERED THE BEST)									
ROLLED EARTHFILL DAMS			CANAL SECTIONS		FOUNDATIONS		ROADWAYS		
							FILLS		
HOMOGENEOUS EMBANKMENT	CORE	SHELL	EROSION RESISTANCE	COMPACTED EARTH LINING	SEEPAGE IMPORTANT	SEEPAGE NOT IMPORTANT	FROST HEAVE NOT POSSIBLE	FROST HEAVE POSSIBLE	SURFACING
1	1	—	1	1	—	1	1	1	3
—	—	2	2	—	—	3	3	3	—
2	4	—	4	4	1	4	4	9	5
1	1	—	3	1	2	6	5	5	1
—	—	3 IF GRAVELLY	6	—	—	2	2	2	4
—	—	4 IF GRAVELLY	7 IF GRAVELLY	—	—	5	6	4	—
4	5	—	8 IF GRAVELLY	5 EROSION CRITICAL	3	7	8	10	6
3	2	—	5	2	4	8	7	6	2
6	6	—	—	6 EROSION CRITICAL	6	9	10	11	—
5	3	—	9	3	5	10	9	7	7
8	8	—	—	7 EROSION CRITICAL	7	11	11	12	—
9	9	—	—	8	8	12	12	13	—
7	7	—	10	8 VOLUME CHANGE CRITICAL	9	13	13	8	—
10	10	—	—	—	10	14	14	14	—
—	—	—	—	—	—	—	—	—	—

Figure 8.—Engineering use chart. (Sheet 2 of 2.) 101–D–235(2).

to the engineer who is faced with a variety of soil problems. Also, the investigator of earth foundations or borrow materials often desires to know how the soil he has classified compares with other kinds of foundation or construction materials. Therefore, the engineering use chart for soils, figure 8, has been prepared to provide this information.

Three important engineering properties of soils typical of each classification group are listed on the chart adjacent to the group symbol. These are: (1) permeability when compacted, (2) shear strength when compacted and saturated, and (3) compressibility when compacted and saturated. In addition, the workability as a construction material has been rated. Based on these properties, the workability, and on experience, the use chart also compares the soil groups in desirability for use in rolled earthfill dams, canal sections, foundations, and roadways. It should be recognized that the numerical ratings given in the chart are relative and are intended only as a guide to aid the investigator in comparing soils for various purposes.

B. Index Properties

8. Definition.—Engineers are continually searching for simplified tests that will increase their knowledge of soils beyond that which can be gained from visual examination without having to resort to the expense, detail, and precision required with the engineering properties tests. These simplified tests provide indirect information about the engineering properties of soils and are, therefore, called index tests. The most widely used tests of this nature are gradation for coarse-grained soils and soil consistency tests (designation E-7) for fine-grained soils. These have already been mentioned in defining the limits of the various groups of soils in the Unified Soil Classification System.

Moisture and density relationships are commonly used as index properties in evaluating foundations and for construction control. Other index tests include penetration resistance needle tests, field penetration tests with either the split-tube or split-liner samplers, and unconfined compression tests on undisturbed samples from fine-grained soil foundations to provide information concerning their strength. The data secured from the index tests, together with the visual descriptions, are often sufficient for design purposes for minor structures, particularly those which are unlikely to utilize the available soil properties to a large extent. This information is also used in making preliminary designs for the purposes of determining probable cost of a major structure and limiting the amount of detailed testing required. It is presumed that materials in a limited area with similar index properties will have similar engineering properties. It should be remembered, however, that the correlation between index properties and engineering properties is not perfect, and if the former are used in making designs, a liberal factor of safety should be included.

It is customary to express certain terms pertaining to the properties of soil as a percentage, whereas others are usually expressed as decimals. For example, degree of saturation, moisture content (water content)[1], and porosity are commonly written:

$$S = 85.6\%$$
$$w = 16.2\%$$
$$n = 34.3\%$$

On the other hand, void ratio, air void ratio, water void ratio, and specific gravity are expressed as:

$$e = 0.522$$
$$e_a = 0.075$$
$$e_w = 0.447$$
$$G_s = 2.760$$

In order to avoid confusion in computations involving these quantities, a simple rule should be followed: Always express quantities as a decimal in all computations. The answers can be given in percentages, provided the percent sign is used.

Example:

Given: Dry density (dry unit weight) $\gamma_d = 113.2$ p.c.f.

$$w = 16.2\%$$
$$G_s = 2.760$$
$$n = 34.3\%$$

To find degree of saturation, S:

$$S = \frac{e_w}{e} = \frac{wG_s}{\dfrac{n}{1-n}} = \frac{(0.162)(2.760)}{\dfrac{0.343}{1-0.343}} = 0.856 \text{ or } 85.6\% \qquad (1)$$

9. Gradation.—Gradation is a descriptive term which refers to the distribution and size of grains in a soil. It is determined by the gradation analysis of soils, designation E–6, and is presented in the form of a

---

[1] The terms "moisture content" and "water content" are used interchangeably in the Earth Manual.

cumulative grain-size curve in which particle sizes are plotted to a logarithmic scale with respect to percentage, by weight, of the total sample plotted to a linear scale (figs. 1, 2, and 3).

Certain terms and expressions have come into common usage when using the gradation curve. A particular diameter of particle is indicated by "D" with a numeral subscript which corresponds to a point on the curve equivalent to the percentage passing. In figure 1, $D_{10} = 0.5$ mm. This means that 10 percent, by weight, of the soil is composed of particles smaller than 0.5 mm. The 10 percent size (D_{10} size) is also called the "effective" size. This term was introduced by Hazen in connection with his work on sanitary filters. He found that the sizes smaller than the effective size affected the functioning of filters as much or more than did the remaining 90 percent of the sizes. Other sizes, such as D_{15}, D_{50}, and D_{85}, are also used in filter design. The sizes D_{10}, D_{30}, and D_{60} are used in defining the gradation characteristics of a soil. The gradation curve is used to designate the various components of the soil by grain size. Typical names and size ranges of the various soil components used by the Bureau of Reclamation are defined in the discussion of soil components, section 2, and are shown in figures 1, 2, and 3.

A soil is said to be well-graded when there is a good representation of all the particle sizes from the largest to the smallest. (See fig. 1.) A soil is considered to be poorly-graded if there is an excess or a deficiency of certain particle sizes within the limits of the minimum and maximum sizes or if the range of predominant sizes falls within three or less consecutive sieve size intervals on the gradation curve. (See fig. 2.) When there is an absence of one or more intermediate sizes between the maximum and minimum (fig. 3), the material is said to have a gap or skip gradation.

To determine whether a material is well graded or poorly graded, coefficients describing the slope and shape of the gradation curve have been defined as follows:

$$\text{Coefficient of uniformity, } C_u = \frac{D_{60}}{D_{10}} \tag{2}$$

$$\text{Coefficient of curvature, } C_c = \frac{(D_{30})^2}{D_{10} \times D_{60}} \tag{3}$$

To be well graded, the coefficient of curvature must be between 1 and 3, and in addition, the coefficient of uniformity must be greater than 4 for gravels and greater than 6 for sands.

A number of different upper size limits are frequently used in appraising and utilizing soils. It is generally assumed that the excluded portion of the total material will not have a significant influence on the engineering properties of the total material unless the portion excluded is more than 25 percent. However, the workability of a soil is often affected appreciably by the presence of so-called "oversize," and it is important, therefore, always to record the percentage that the sampled material represents of the total material. The following table lists the more important upper size limits for various applications:

Size	*Application*
12 inches	Sand, gravel, and cobble fill for construction
5 inches	Impervious rolled fill for construction
3 inches	Soil classification, large-scale laboratory testing, and some field construction
No. 4 sieve	Standard laboratory testing
No 40 sieve	Consistency limits
No. 200 sieve	Fine-grained soil fraction

Utilizing gradation in determining engineering properties is limited to coarse-grained materials. In poorly-graded uniform materials, the permeability increases approximately as the square of the effective size, (D_{10}). For such materials compressibility is normally small except for very fine sands. Shear strength consists almost wholly of internal friction and is more or less independent of size. Uniform materials are generally workable; that is, they are easily excavated and compacted. As the range in sizes of coarse-grained soils increases, there is a reduction in permeability, compressibility decreases, and shear strength increases. The workability of well-graded soil is good.

The amount and character of the fines influence the properties of coarse-grained materials. Permeability is reduced with increasing quantity of fines, rapidly with small quantities and at a slower rate with large quantities. Compressibility and shear strength are affected only slightly by small percentages of fines in the coarse-grained soils, but the effects increase with increase in fines. If the fines are clayey the cohesive strength of the soil increases with increasing clay content.

10. Soil Consistency.—The physical properties of most fine-grained soils, and particularly clayey soils, are greatly affected by the water content. In consistency a clay may be very soft, that is, a viscous liquid, or it may be very hard, having the properties of a solid, depending on its water content. In between those extremes, the clay may be molded and formed without cracking or rupturing the soil mass. In this condition, it is referred to as being plastic. Plasticity is an outstanding characteristic of clays. It is used to identify and distinguish clayey soils.

In 1911, a Swedish soils scientist, A. Atterberg, developed a series of hand-performed tests for determining the clay activity or plasticity of soil for agricultural uses. These tests are now known as the Atterberg limits tests. The complete series of tests has been grouped under designation E-7, "Soil Consistency Tests."

Four stages, or states, can be recognized for describing the consistency of a soil. These are: (1) the liquid state, (2) the plastic state, (3) the semisolid state, and (4) the solid state. These states of consistency are related to the water content. Although the transition between each of the states is gradual, test conditions have been established arbitrarily to delineate the water content as a precise point in the transition among the four stages of consistency. These water contents, which are determined by ovendrying procedures, are called the liquid limit, plastic limit, and shrinkage limit. As the tests are performed only on the soil fraction which passes a No. 40 sieve, the relation of this fraction to the total material must be considered when determining the consistency of the total material from these tests.

The significance of the limits and their relation to the stages of consistency can be explained by discussion of figure 9. As a very wet, fine-grained soil dries, it passes progressively through different stages of consistency. In a very wet condition, the mass will act like a viscous liquid which is referred to as the liquid state. As the soil dries, there is a reduction in volume of the mass which is very nearly proportional to the loss of water. When the moisture in the soil reaches a value equivalent to the liquid limit, the mass becomes plastic.

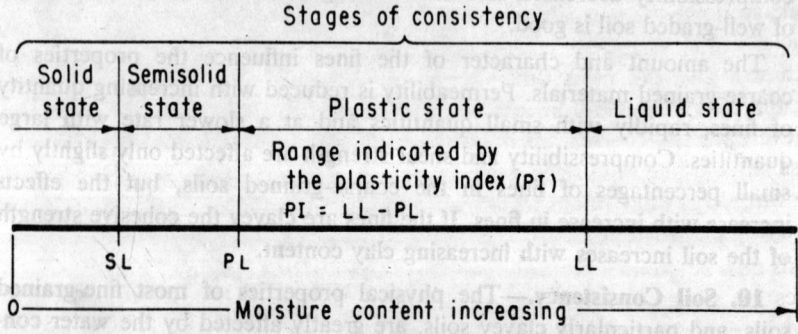

Figure 9.—Consistency limits. 101–D–161.

The liquid limit, LL, is defined as that water content expressed as a percentage of the dry weight of soil at which the soil first shows a small but definite shear strength as the water content is reduced. Conversely, with increasing moisture, it is that water content at which the soil mass just starts to become fluid under the influence of a series of standard shocks, as discussed in designation E–7.

As the water content is reduced below the liquid limit, the soil mass becomes stiffer and will no longer flow as a liquid. However, it will continue to be deformable, or plastic, without cracking until the plastic limit is reached.

The *plastic limit, PL,* is defined as that water content expressed as a percentage of the dry weight of soil at which the soil mass ceases to be plastic and becomes brittle, as determined by a procedure for rolling the soil mass into threads one-eighth inch in diameter. The plastic limit is always determined by reducing the water content of the soil mass.

The *plasticity index, PI,* is the difference between the liquid and plastic limits and represents the range of moisture within which the soil is plastic. (See figure 9 and plasticity chart on figure 7.) Silts have slight or no plasticity indices, while clays have higher indices. Plasticity index in combination with the liquid limit indicates how sensitive the soil is to changes in moisture.

As the water content of the soil is reduced below the plastic limit, the soil becomes a semisolid—that is, it can be deformed but considerable force is required and the soil cracks. This condition is referred to as a semisolid state. As further drying takes place, the soil mass will eventually reach a "solid" state at which no further shrinkage will occur. The water content filling the soil voids under this condition is called the *shrinkage limit, SL.* This is a water content below which a reduction in moisture will not cause a decrease in the volume of the soil mass. Below the shrinkage limit, the soil is considered to be a solid—that is, most of the particles are in very close contact and are very nearly in an arrangement which will give the most dense condition. In most fine-grained plastic soils, the plastic limit will be appreciably greater than the shrinkage limit. However, for the coarser fine-grained soils—that is, the soils containing coarse silt and fine sand sizes—the shrinkage limit will be near the plastic limit. Shrinkage limit values together with other index values are useful in identifying expansive soils.

Soils may be grouped according to their liquid limits and their plasticity indices on a plasticity chart as shown in figure 10. Such plots can be useful in predicting properties of soils by comparison with similar plots for tested soils.

As an index of engineering properties, soil consistency applies only to fine-grained soils. The variation of engineering properties generally is related to the four zones on the plasticity chart which determines the soil classification group. For soils plotting above the "A" line, permeability is very low, and its variation is unimportant. Compressibility increases with increasing liquid limit. For the same liquid limit, the greater the plasticity index the greater will be the strength at the plastic limit.

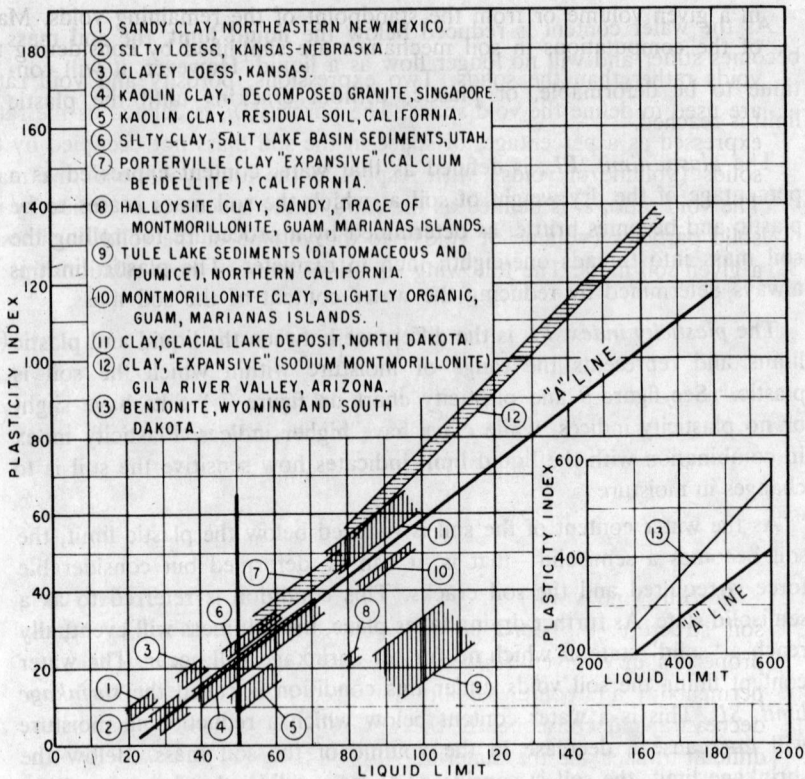

Figure 10.—Typical relationships between the liquid limit (LL) and the plasticity index (PI) for various soils. 101–D–170.

In addition to the coarse-grained groups, the engineering use chart, figure 8, indicates the engineering properties of fine-grained soil groups.

The degree of consistency of a fine-grained cohesive soil can be determined by its relative consistency, C_r, which defines the water content of the soil in relation to the liquid limit and the plastic limit of the same material. The equation for relative consistency is:

$$C_r = \frac{LL - w}{LL - PL} = \frac{LL - w}{PI} \qquad (4)$$

It is usually expressed as a percentage. A soil with relative consistency of 0 is at its liquid limit, and a soil at 100 percent relative consistency is at its plastic limit.

11. Porosity and Void Ratio.—In the evaluation of a soil, one may examine it either from the standpoint of the amount of solids contained in a given volume or from the standpoint of the remaining voids. Many of the computations in soil mechanics are simplified by considering the voids rather than the solids. Two expressions, porosity and void ratio, are used to define the void space. The porosity, n, is defined as the ratio, expressed as a percentage, of space in the soil mass not occupied by the solids (volume of voids) with respect to the total volume of the mass. The void ratio, e, is defined as the ratio of the space not occupied by the solid particles (volume of voids) to the volume of the solid particles in a given soil mass. The following equations express these relationships:

$$n = \frac{V_v}{V_t} = \frac{e}{1+e} \tag{5}$$

$$e = \frac{V_v}{V_s} = \frac{n}{1-n} \tag{6}$$

where: $V_t =$ total volume,
$V_v =$ volume of voids, and
$V_s =$ volume of solids.

Porosity and void ratio are measures of the state or condition of a soil structure. As porosity and void ratio decrease, the engineering properties of a given soil become more dependable with decreases in permeability and compressibility and an increase in strength. As porosity decreases, and consequently the void ratio decreases, it becomes more difficult to excavate the material. At a given water content it is necessary to increase compactive effort to obtain a decrease in porosity. However, similar properties may be obtained in different soils at widely different conditions of porosity. Engineering properties of a soil do not vary directly with its porosity; the relationship is generally complex.

12. Specific Gravity.—In the investigation of a soil, the most easily visualized condition involves the volume occupied by soil solids, V_s, the volume occupied by soil moisture, V_w, and the volume occupied by air in the soil mass, V_a. However, most measurements are more readily obtained by weight. To correlate weight and volume, a factor called specific gravity is required. Specific gravity is defined as the ratio between the unit weight of a substance and the unit weight of water at 4° C. There are several different types of specific gravity in common use. Those used by the Bureau of Reclamation are: absolute specific gravity, apparent specific gravity, and several types of bulk specific gravities. These values are obtained by the methods outlined in designation E–10.

The *absolute specific gravity* is determined by analyzing the amount and kind of mineral constituents present in the soil. For this test, all the coarse grains are pulverized to at least finer than the No. 200 sieve size so that all the impermeable pores or voids in the coarser grains are exposed. The specific gravity determined in this manner is the highest value which can be obtained for a given soil.

The *apparent specific gravity* is determined on the soil particles as they occur naturally. Voids which exist within the grains and which cannot be filled with water are referred to as "impermeable voids." The apparent specific gravity may equal the absolute specific gravity, but is usually smaller depending on the percentage of impermeable voids present. Apparent specific gravity ranges from 2.50 to 2.80 for most soils, with a majority of soils having an apparent specific gravity very near 2.65. Unless specifically stated to the contrary, the term "specific gravity" in this manual is assumed to mean the "apparent specific gravity." It is used to compute many of the important soil properties involving volume determinations, such as porosity, void ratio, and degree of saturation as shown in the following equations:

$$n = 1 - \frac{\gamma_d}{62.4G_s} \tag{7}$$

$$e = \frac{62.4G_s}{\gamma_d} - 1 \tag{8}$$

$$S = \frac{V_w}{V_v} = \frac{wG_s}{e} \tag{9}$$

where: n = porosity,
 e = void ratio,
 S = degree of saturation,
 G_s = apparent specific gravity of the soil,
 γ_d = dry density in pounds per cubic foot,
 V_w = volume of water,
 V_v = volume of voids, and
 w = water content in terms of ratio of weight of water to dry weight of soil.

A quick check of soil computations can be made using the nomograph in figure 11 (8- by 10½-inch prints of this nomograph are available from the Engineering and Research Center, Denver, Colo. 80225).

The *bulk specific gravity* (*saturated surface dry*) is the specific gravity with the permeable or surface voids of the particles filled with water. This value is smaller than the value determined for apparent specific gravity unless there are no permeable voids in the particles. The bulk specific gravity, saturated surface dry, of aggregates is used for concrete mix design and also for quality tèsts for riprap and rock fill materials. In the field density test procedure, the specific gravity of the gravel

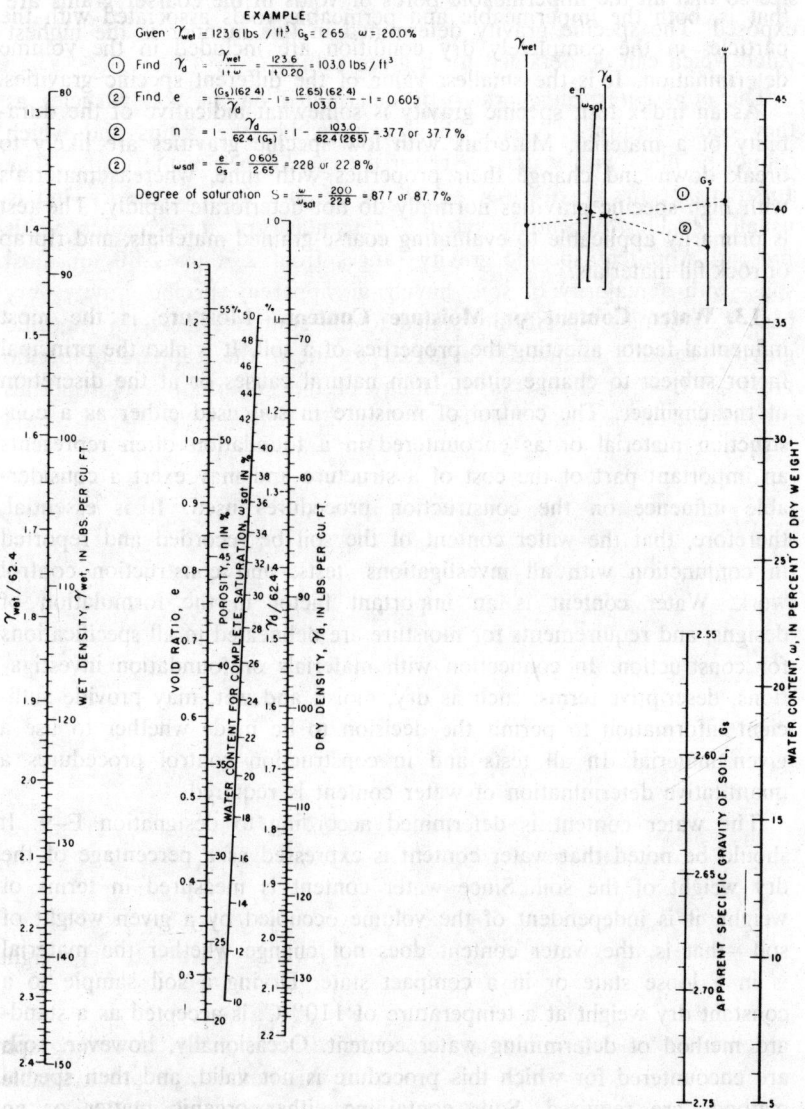

EXAMPLE

Given γ_{wet} = 123.6 lbs./ft.³ G_s = 2.65 ω = 20.0%

① Find γ_d = $\dfrac{\gamma_{wet}}{1+\omega}$ = $\dfrac{123.6}{1+0.20}$ = 103.0 lbs./ft.³

② Find e = $\dfrac{(G_s)(62.4)}{\gamma_d}$ - 1 = $\dfrac{(2.65)(62.4)}{103.0}$ - 1 = 0.605

② n = 1 - $\dfrac{\gamma_d}{62.4\,(G_s)}$ = 1 - $\dfrac{103.0}{62.4\,(2.65)}$ = .377 or 37.7 %

② ω_{sat} = $\dfrac{e}{G_s}$ = $\dfrac{0.605}{2.65}$ = .228 or 22.8%

Degree of saturation S = $\dfrac{\omega}{\omega_{sat}}$ = $\dfrac{.200}{.228}$ = .877 or 87.7%

Figure 11.—Soil properties nomograph. 101–D–177.

particles is determined on a bulk surface-dry basis, but the permeable voids may not be entirely filled with water; hence, the expression *bulk specific gravity (wet surface dry)* is used. The *bulk specific gravity (ovendry)* is, in effect, the minimum dry unit weight of the particles distributed throughout the entire or effective volume of the particles;

that is, both the impermeable and permeable voids associated with the particles in the completely dry condition are included in the volume determination. It is the smallest value of the different specific gravities.

As an index test, specific gravity is somewhat indicative of the durability of a material. Materials with low specific gravities are likely to break down and change their properties with time, whereas materials with high specific gravities normally do not deteriorate rapidly. The test is primarily applicable to evaluating coarse-grained materials, and riprap or rock fill materials.

13. Water Content or Moisture Content.—Moisture is the most influential factor affecting the properties of a soil. It is also the principal factor subject to change either from natural causes or at the discretion of the engineer. The control of moisture in soil used either as a construction material or as encountered in a foundation often represents an important part of the cost of a structure and may exert a considerable influence on the construction procedures used. It is essential, therefore, that the water content of the soil be recorded and reported in conjunction with all investigations, tests, and construction control work. Water content is an important factor in the formulation of designs, and requirements for moisture are delineated in all specifications for construction. In connection with materials or foundation investigations, descriptive terms, such as dry, moist, and wet, may provide sufficient information to permit the decision to be made whether to use a given material. In all tests and in construction control procedures a quantitative determination of water content is required.

The water content is determined according to designation E-9. It should be noted that water content is expressed as a percentage of the dry weight of the soil. Since water content is measured in terms of weight, it is independent of the volume occupied by a given weight of soil—that is, the water content does not change whether the material is in a loose state or in a compact state. Drying a soil sample to a constant dry weight at a temperature of 110° C. is accepted as a standard method of determining water content. Occasionally, however, soils are encountered for which this procedure is not valid, and then special methods are required. Soils containing either organic matter or an appreciable amount of salts require special treatment. Ovendrying sometimes causes changes in soil properties, and therefore ovendried materials should not be used in laboratory tests unless specifically required.

Optimum water content, w_0.—In 1933, Mr. R. R. Proctor showed that the dry density of a soil obtained by a given compactive effort depends on the amount of water the soil contains during compaction. He pointed out that for a given soil and a given compactive effort there is one water content, called optimum water content or optimum moisture content, that will result in a maximum dry density of the soil, and that water contents both greater and smaller than this optimum value will result in dry densities less than the maximum (see fig. 12(C)). Optimum water content, as defined for Bureau work, is based on the compactive effort described in the standard Proctor compaction test, designation E–11. Figure 13 shows the variation of optimum water content when the compactive effort is varied from the standard.

Absorbed moisture, w_a.—The moisture in saturated surface-dry gravel and cobble-sized particles is called absorbed moisture. Depending on the moisture conditions in a soil containing gravel, sufficient moisture may not always be available for the gravel to take up the maximum amount. The actual amount of absorbed moisture is determined experimentally by separating the gravel from the soil and determining the wet and dry weight of the gravel. (See designation E–24.)

The water content of a coarse-grained soil containing gravel can be determined by the following relation:

$$w_t = w_s(1 - P) + w_g P \qquad (10)$$

where: w_t = water content of the total material in percent of the dry weight of the total material,

w_s = water content of the soil less than No. 4 sieve size in percent of the dry weight of this soil,

w_g = water content of the gravel in percent of the dry weight of the gravel, and

P = ratio of gravel to total material by dry weight.

Engineering properties change so much with water content and to different degrees with different soils, that this factor as an index is used only to assist in the interpretation of other index factors. It is used primarily in evaluating soils in their natural state, both as foundations and as construction materials sources.

14. Density or Unit Weight.—The weight of a unit volume of soil is its most easily determined property. Consequently, it has become the basic parameter to which all other performance characteristics are related. The relationships between density and other soil properties are as a rule complex, but in engineering practice it is generally assumed that simple relationships exist. A large number of qualified expressions for density are in common use. To avoid confusion, it is important that the kind of density reported is clearly delineated.

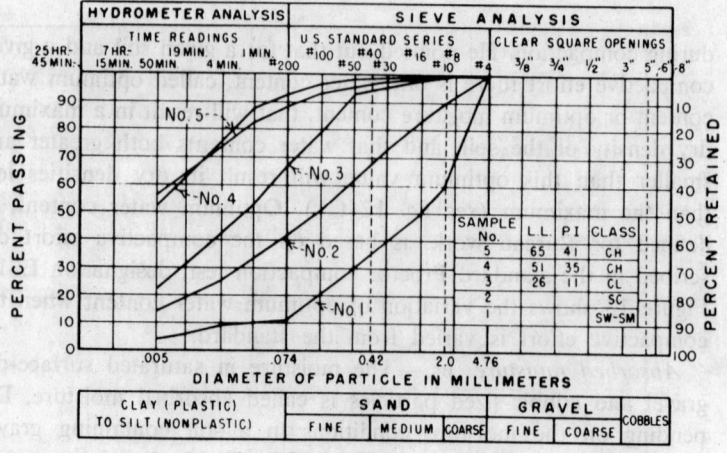

A. GRADATION CURVES

SAMPLE No.	L.L.	PI	CLASS
5	65	41	CH
4	51	35	CH
3	26	11	CL
2			SC
1			SW-SM

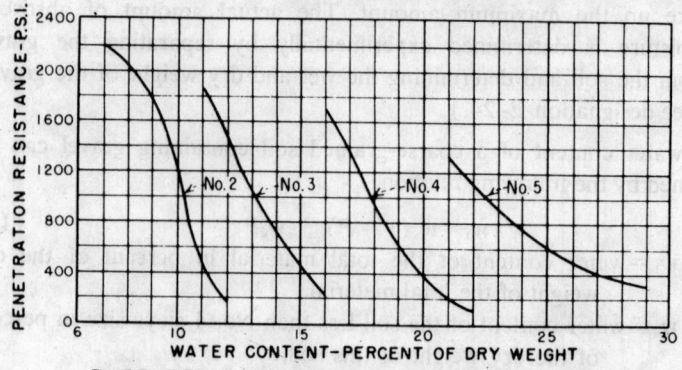

B. PENETRATION RESISTANCE—MOISTURE CURVES

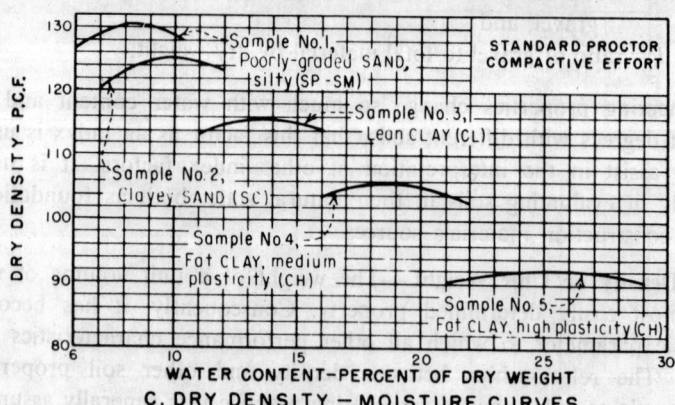

C. DRY DENSITY — MOISTURE CURVES

Figure 12.—Compaction and penetration resistance curves for various soils.
101-D-175.

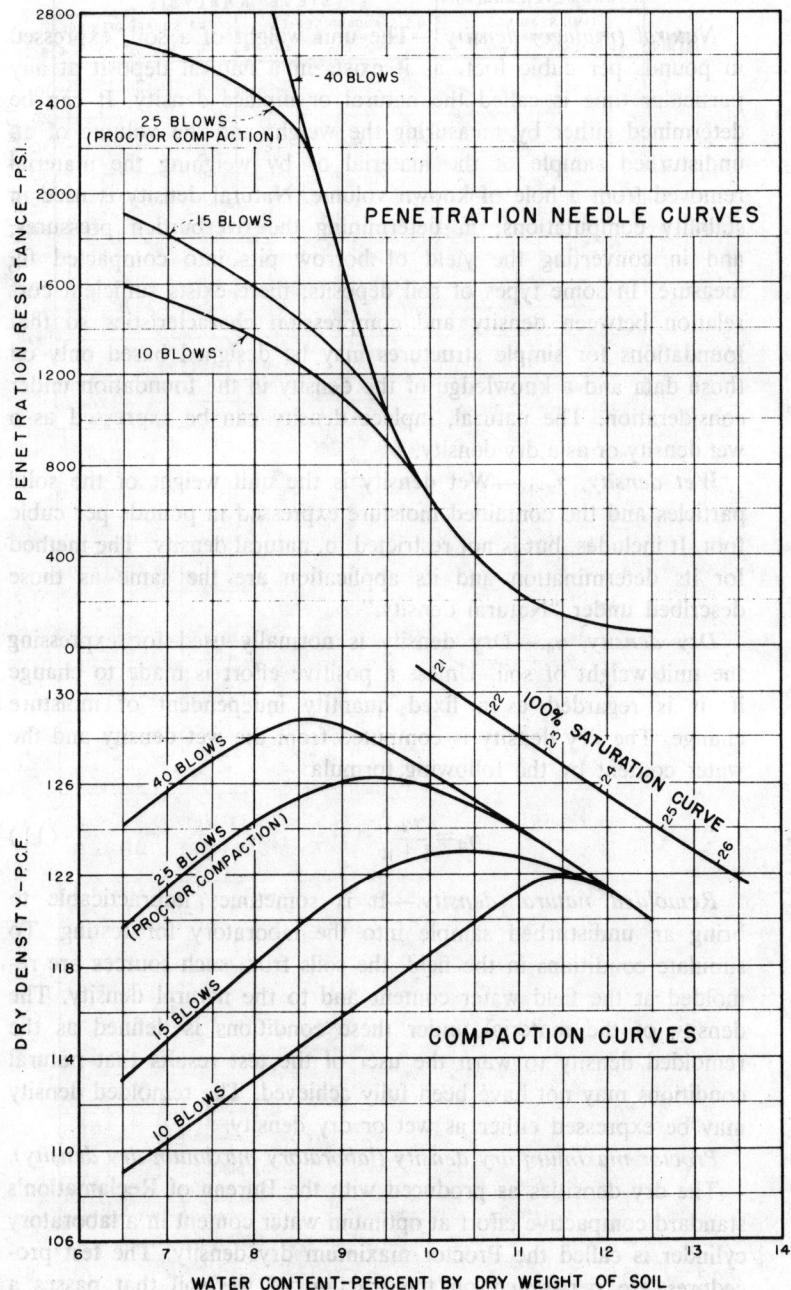

Figure 13.—Effect of compactive effort on the compaction and penetration-resistance curves. 101-D-176.

Natural (inplace) density.—The unit weight of a soil, expressed in pounds per cubic foot, as it exists in a natural deposit at any particular time is called the natural or inplace density. It can be determined either by measuring the weight and the volume of an undisturbed sample of the material or by weighing the material removed from a hole of known volume. Natural density is used in stability computations, in determining the overburden pressures, and in converting the yield of borrow pits into compacted fill measure. In some types of soil deposits, there exists sufficient correlation between density and compression characteristics so that foundations for simple structures may be designed based only on those data and a knowledge of the density in the foundation under consideration. The natural, inplace density can be expressed as a wet density or as a dry density.

Wet density, γ_{wet}.—Wet density is the unit weight of the solid particles and the contained moisture expressed in pounds per cubic foot. It includes, but is not restricted to, natural density. The method for its determination and its application are the same as those described under "Natural density."

Dry density, γ_d.—Dry density is normally used for expressing the unit weight of soil. Unless a positive effort is made to change it, it is regarded as a fixed quantity independent of moisture change. The dry density is computed from the wet density and the water content by the following formula:

$$\gamma_d = \frac{\gamma_{wet}}{1+w} \tag{11}$$

Remolded natural density.—It is sometimes impracticable to bring an undisturbed sample into the laboratory for testing. To simulate conditions in the field, the soils from such sources are remolded at the field water content and to the natural density. The density of the material under these conditions is defined as the remolded density to warn the user of the test results that natural conditions may not have been fully achieved. The remolded density may be expressed either as wet or dry density.

Proctor maximum dry density (laboratory maximum dry density).—The dry densities as produced with the Bureau of Reclamation's standard compactive effort at optimum water content in a laboratory cylinder is called the Proctor maximum dry density. The test procedures are performed on the fraction of the soil that passes a No. 4 sieve, according to procedures outlined in designation E–11 which provide a compactive effort of 12,375 foot-pounds per cubic foot of soil. Figure 12 shows the variation in laboratory maximum dry density for several types of soil. Figure 13 shows typical compaction curves with various compactive efforts expressed in terms of number of standard blows. When the Proctor test procedure was originated, it was believed that it produced the most desirable conditions in the compacted soil that could practicably be obtained. Additional experience has shown that there are materials and conditions of use wherein it is desirable to depart from this standard. As a result there are several procedures in use that differ from the one used by the Bureau, and they produce different values of optimum water contents and laboratory maximum dry densities. Instead of using a variety of standards, the procedure adopted by the Bureau is to maintain a single standard and to specify the deviation that is to be used for those specific materials in structures where the most desirable density and water content differ from the Bureau standard.

Relative density, D_d.—Soils that consist almost exclusively of coarse-grained particles—that is, sands and gravels—when compacted according to procedures outlined for determining Proctor maximum dry density, have density and moisture relationships that correlate poorly with other properties. Furthermore, the compaction curves are erratic and often do not produce a definable maximum density. For evaluating such soils, the relative density procedure, designation E–12, has been developed. It has been found that when the shear strength of these soils is correlated with changes in density in the range between minimum and maximum densities obtainable, a fairly reliable relationship exists. Furthermore, it has been found that with a reasonable amount of construction control, a given type and amount of compactive effort can be expected to produce a related relative density. In practice, a relative density of 70 percent has been found satisfactory for most conditions. While there are several different ways that the relative density concept might be expressed, the one chosen originally and that has come into general use is to express relative density in terms of void ratio by the following formula:

$$D_d = \frac{e_{max} - e}{e_{max} - e_{min}} \qquad (12)$$

where: e_{max} = void ratio of the soil in its loosest state,

e = void ratio of the soil in the state of the test,

e_{min} = void ratio of the soil in its densest state, and

D_d is usually expressed as a percentage from 0 to 100 percent.

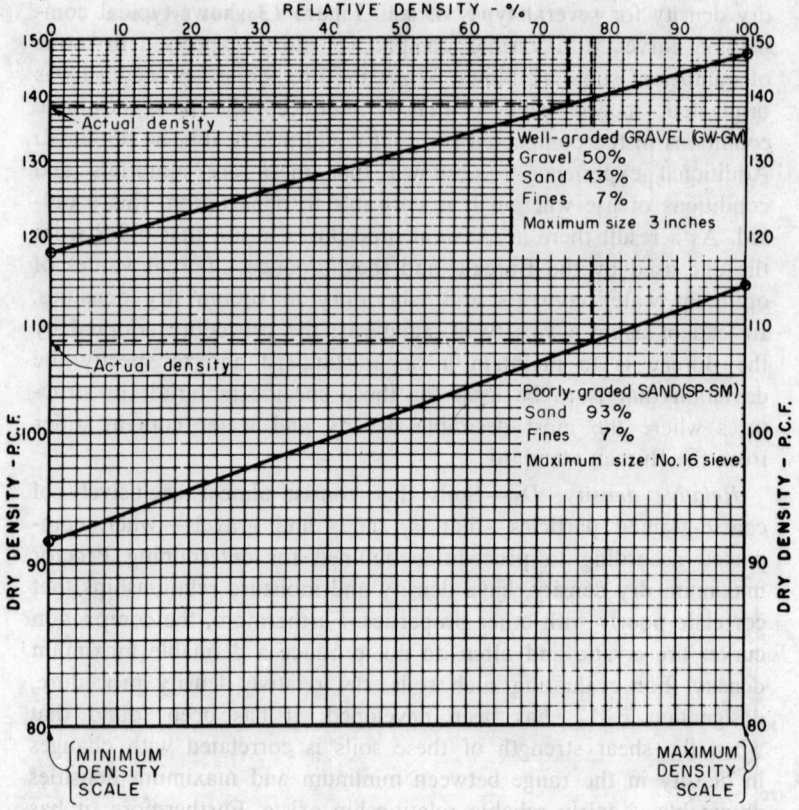

Figure 14.—Relation of relative density and dry density (scaled to plot as a straight line on form 7–1595.) 101–D–572.

Figure 14 shows a chart from which relative density may be determined if the maximum, minimum, and the actual densities of the material are known. Figure 15 shows the maximum and minimum densities for typical sand and gravel soils.

Compacted fill density.—The density of the soil as it is compacted into a manmade fill is called the compacted fill density to distinguish it from the natural density of a foundation or any of the various densities determined in the laboratory. It is usually expressed as dry density, although wet densities are also used to some extent. The compacted fill density is determined according to procedures outlined in designation E–24. It is used primarily for construction control to assure that the compacted fill is at least as dense as the conditions assumed in preparing designs.

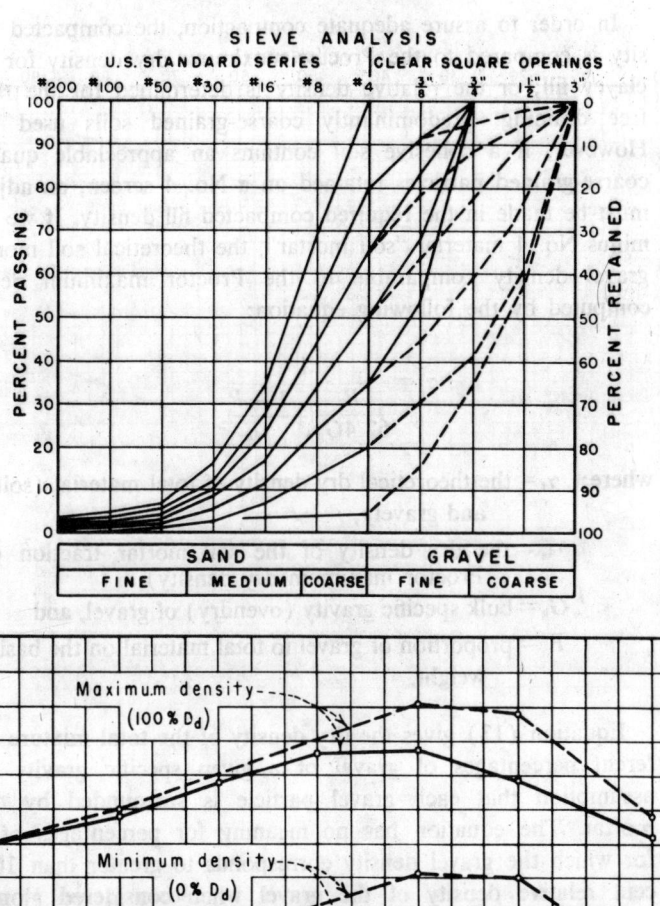

Figure 15.—Maximum and minimum densities of typical sand and gravel soils.
101-D-173.

In order to assure adequate compaction, the compacted fill density is compared to the Proctor maximum dry density for silty or clayey fill; or the relative density is determined for the relatively free draining, predominantly coarse-grained soils used for fill. However, if a cohesive soil contains an appreciable quantity of coarse-grained particles retained on a No. 4 screen, an adjustment must be made in the required compacted fill density. If we call the minus No. 4 material "soil mortar", the theoretical soil mortar and gravel density comparable to the Proctor maximum density is computed by the following equation:

$$\gamma_t = \frac{1}{\dfrac{P}{62.4G_g} + \dfrac{1-P}{\gamma_{d_s}}} \tag{13}$$

where: γ_t = the theoretical dry density of total material (soil mortar and gravel),

γ_{d_s} = the dry density of the soil mortar fraction (usually Proctor maximum dry density),

G_g = bulk specific gravity (ovendry) of gravel, and

P = proportion of gravel to total material on the basis of dry weight.

Equation (13) gives the dry density of the total mixture for different percentages of gravel of a given specific gravity on the assumption that each gravel particle is surrounded by the soil mortar. The equation has no meaning for percentages of gravel for which the gravel density corresponds to greater than 100 percent relative density of the gravel when considered alone (see fig. 16). This occurs at about 70 to 80 percent of gravel, depending on the gradation and on the specific gravity of the gravel. For gravel contents above about 30 percent, it has been found that compaction of the soil mortar in the field is less than the theoretical, although for many soils compaction is not seriously affected until the percentage of gravel rises to about 50. For these soils the usual assumptions of correlation between density and other soil properties do not hold. If such soils are to be used in fill, special tests must be made. Figure 16 illustrates the relationships of the various densities in soil mixtures.

15. Penetration Resistance.—In 1933, Mr. R. R. Proctor reported the development of a simple instrument called a soil plasticity needle which could be used to determine moisture and density relationships of soils used in construction. He noted that for a given compactive effort

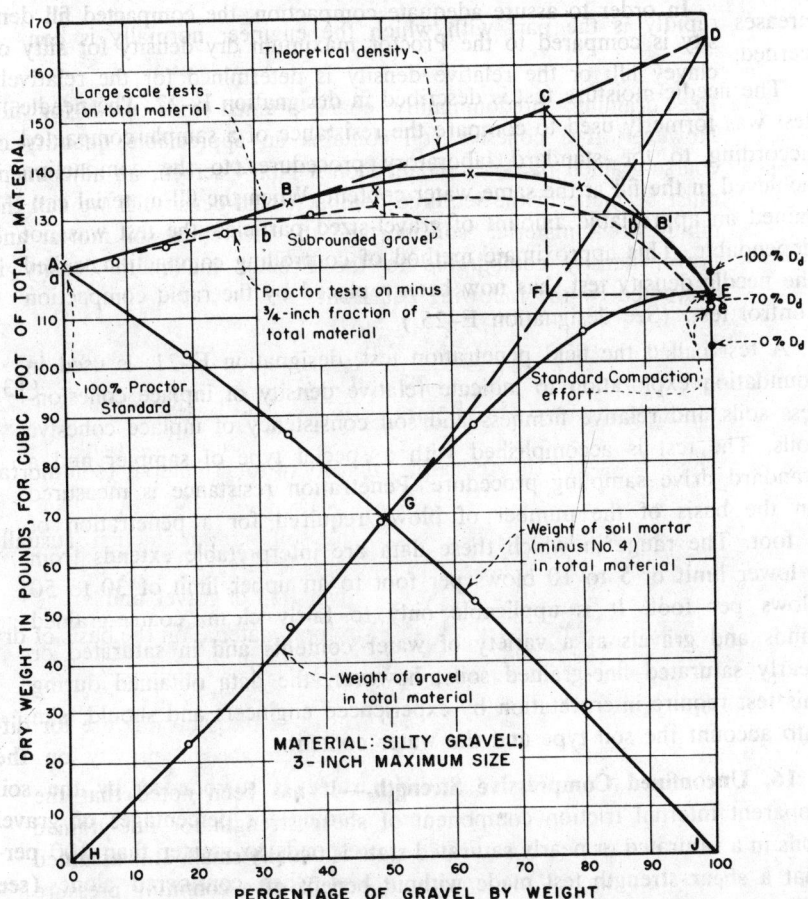

**Figure 16.—Maximum dry density versus gravel content for compacted soils.
101–D–218.**

on a given soil, a curve could be developed which related the penetration resistance of a small rod with the water content of the soil.

Typical curves of penetration resistance plotted as ordinates versus water content as abscissa are shown in figures 12 and 13. The curve starts with a rather high penetration resistance at a point on the compaction curve dry of the optimum water content. The penetration resistance then decreases rapidly and almost uniformly to a rather low value at a point wet of the optimum water content. From there on, the decrease in penetration resistance is gradual to zero near the liquid limit. The straight part of the curve where penetration resistance decreases rapidly is the part with which the engineer normally is concerned.

The needle-moisture test is described in designation E–22. The needle test was formerly used to compare the resistance of a sample compacted according to the standard laboratory procedures to the compaction achieved in the fill at the same water content. When the fill material contained an appreciable amount of gravel-sized particles, the test was not dependable. This approximate method of controlling compaction, called the needle density test, has now been replaced by the rapid compaction control test. (See designation E–25.)

A test called the field penetration test, designation E–21, is used in foundation exploration to indicate relative density of inplace cohesionless soils and relative firmness and soil consistency of inplace cohesive soils. The test is accomplished with a special type of sampler and a standard drive sampling procedure. Penetration resistance is measured on the basis of the number of blows required for a penetration of 1 foot. The range in which these data are interpretable extends from a lower limit of 5 to 10 blows per foot to an upper limit of 30 to 50 blows per foot. It is applicable only to fairly clean, coarse-grained sands and gravels at a variety of water contents and in saturated or nearly saturated fine-grained soils; however, the data obtained during this test require interpretation by experienced engineers and should take into account the soil type and the water content.

16. Unconfined Compressive Strength.—It has been noted that the apparent internal friction component of shear strength for fine-grained soils in a saturated or nearly saturated state is usually small. It is assumed that a shear strength test made without benefit of a confining pressure will provide an index of the available shear strength of a foundation soil that will be somewhat conservative. A variety of apparatus has been developed for making such tests by the various organizations specializing in evaluation of soil foundations. The apparatus ranges from simplified equipment and procedures from which the results can be considered of value only as an index test; to rather complex equipment and test procedures that provide reliable data on engineering properties. The value of unconfined compression tests is still controversial, and recommended procedures are so variable that use of such tests by the Bureau of Reclamation is limited to special problems and the results are considered to represent index properties rather than engineering properties. The tests are accomplished as special cases under designation E–17, in which the unconfined compressive strength of a soil is the strength of a specimen (maximum axial load divided by area) tested in triaxial shear with the lateral pressure equal to atmospheric or zero gage pressure.

17. Soluble Salts.—The quantity of soluble salts present in a soil may be an important factor when considering the suitability of a soil for constructing embankments. The quantity of soluble salts is determined by the procedure outlined in designation E–8. The effect of soluble salts on an earth structure is dependent on the kind of salts and their solubility characteristics; the coefficient of permeability and thus the amount of water passing through the soil; temperature; chemical characteristics of the natural water; and other factors. Therefore, the percentage of soluble salts is only an indication of the possible effects. Soluble salts are more objectionable in materials with moderate to high permeability than in soils with low permeability.

The kinds of salts present in a soil or being carried by the ground water are important when considering the type of cement to be used in constructing concrete structures in contact with the soil. Most prominent among aggressive substances which affect concrete structures are the sulfates of sodium, magnesium, and calcium. The relative degrees of attack on concrete by sulfates from soils and ground waters and other substances may be found in the Concrete Manual.[2] The kinds of salts present in the ground and ground water must be determined by chemical analysis. The presence of appreciable quantities of soluble salts indicates that engineering properties may change in the presence of percolating water.

C. Engineering Properties

18. General.—In the evaluation of a soil, the transition is gradual between those properties which serve only as a broad guide to the character of the material and the quantitative properties which define specific performance characteristics. For example, water content and density are at times used as index properties and at other times used as engineering properties. The importance of these two properties in investigation and in construction control, as indicators of the nature of the material and of the quality of compaction, has frequently resulted in the belief that low water content and high density are the only desirable characteristics to attain in soils. This is not necessarily true. These properties can be regarded only as indices of the probable engineering behavior of a soil. On the other hand both density and water content are so intimately related with the computations involved in the design of

---

[2] Concrete Manual, Bureau of Reclamation, Denver, Colo., seventh edition, revised reprint 1966.

structures of soil that they must frequently be evaluated as engineering properties.

The principal distinction between index and engineering properties is that the determination of index properties is simple and may be accomplished by personnel with comparatively little training, whereas the determination of engineering properties requires considerable knowledge and skill if reliable information is to be developed. On the other hand considerably more special knowledge and skill are required to interpret and utilize index information than are required in using information from tests obtained for engineering properties. The characteristics of soil most commonly listed in a catalog of engineering properties are shear strength, compressibility, and permeability. In addition, but less clearly defined, are such characteristics as deterioration and workability for which quantitative evaluation is still imperfectly developed.

19. Shear Strength.—(a) *General*.—Soil has very little strength compared to other materials used for building a structure. Moreover, compared to the maximum soil strength that may be found, there is a very large variation both from soil to soil and within a given soil type depending on how it was deposited or placed. By trial and error, in the course of time a series of conventional practices for considering soil strength has been developed. Originally, these practices were based on the type of structure involved, but with little regard to the kind of soil and almost no concern as to its inplace condition. Some of these practices avoid the problems associated with soil strength either by removal of the soil or use of point bearing piles, piers, or similar devices. Other construction or design practices require so little strength in a soil for the support of the structure that in most cases knowledge of the soil strength is unnecessary. On the other hand, to an increasing extent situations are being encountered where lack of knowledge of strength can lead only to extravagant costs or to failure; therefore more rational approaches are desired.

The engineering computations concerned with the strength of a soil deal primarily with its shear strength; that is, the resistance to sliding of one mass of soil against another, and rarely with the compressive or tensile strength. In 1776, Coulomb, a French scientist, observed that the shear strength of a soil was made up of two parts, one part dependent on the stress acting normal to the shear plane, the other part independent of that stress. These two parts are called the internal friction and the cohesion, respectively. The factor relating the normal component of the load to strength is designated by $\tan \phi$. The unit of cohesion is designated by c. The shear strength, s, may then be expressed as:

$$s = c + \sigma \tan \phi \qquad (14)$$

where $\sigma =$ normal stress on the sliding surface.

It was soon noticed that many soils were either dominantly cohesive or noncohesive. For either type of soil, the engineering computations could be simplified by dropping the smaller term and the resulting solution would be on the safe side. Much of soil mechanics practice is based on this simplifying assumption. The terms cohesive and noncohesive (cohesionless) soil are in common use in referring to these soils.

Occasionally the structure is so massive or the soil foundation is so weak that this simplifying assumption cannot be used. For a large earth dam this is almost always the case. In such a situation it is important to determine the values of c and tan ϕ with precision for both the proposed structure and foundation under the worst probable condition. In order that the proper values can be determined by tests, it is of the utmost importance that undisturbed samples from foundations secured during investigations be truly representative of materials and conditions. Also, material placed in these earth structures should comply with the established design limitations based on laboratory tests.

The measurement of inplace shear strength is accomplished in the field indirectly by the field penetration test, designation E–21, or directly by the vane test, designation E–20. Correlations between strengths of sandy soils and penetration have been determined. Studies are being made to determine whether a similar relationship exists for silty and clayey soils. Field penetration test results on the latter soils are considered in a relative rather than an absolute sense and are used to indicate the location of the weakest soils for more detailed testing. The vane test is used in soft saturated clays and silts.

(b) *Direct Shear.*—The measurement of shear strength in the laboratory is accomplished either by the direct shear test or by the triaxial test. The direct shear test was developed first and is the test that is still most widely used. It has the advantage of simplicity but the results are very sensitive to test procedure; hence, test results are open to question. Shear resistance determined by this method is called apparent shear resistance to warn the user of test results that all the factors influencing shear strength may not have been evaluated.

To overcome this difficulty three types of direct shear tests have generally been accepted. These tests are called the quick test (Q), the consolidated-quick test (Q$_c$), and the slow test (S). These tests provide information on the range of variation of shear resistance and for many problems are adequate for the required solutions. The quick test is made on the sample in its natural condition or initial state of compaction and the shear stresses or forces are applied so rapidly that little readjustment or drainage in the sample is possible. In the consolidated-quick test the normal load is supplied and the sample is allowed to adjust under this stress, after which the shear stress is rapidly applied without allowing further readjustment. In the slow test the sample is allowed to come to equilibrium after each load increment is applied.

(c) *Triaxial Shear.*—Since there are many factors which influence the shear strength that cannot be evaluated with the direct shear test apparatus, the triaxial shear test apparatus was developed to permit control of these factors. Used as a research tool it has been instrumental in the development of a better understanding of the mechanics of shear in soils. The same three types of tests may be made as discussed in subsection (b). Because of its flexibility, triaxial shear apparatus is being used increasingly for routine testing, but in such tests the full capability of the equipment is seldom utilized. It should be borne in mind, therefore, that the fact that triaxial testing has been used does not give assurance that the validity of the result is superior. Although the mechanics of the triaxial test are complicated and operators of the equipment require special training, this test procedure in which pore-water pressures are measured is used by the Bureau for almost all its shear testing requirements (see designation E–17). Figure 17 indicates the relation between the relative density and friction factor for remolded coarse-grained soils.

(d) *Pore-Water Pressure.*—Shear strength is dependent primarily on the normal or confining stress. Under this stress the soil grains are forced into more intimate contact and the volume of the soil mass is decreased somewhat. Since the volume of the soil grains cannot be changed appreciably, this volume change must take place primarily in the voids or pores of the soil. If these pores are completely filled with water their volume cannot be changed unless some of the water is drained from the soil mass. If drainage is prevented a stress will be developed in the pore water opposing the externally applied stress. The developed stress is called pore-water pressure. Even if the pores are only partially filled with water, pressures in the fluid combination of air and water will develop, but to a lesser degree because volume change is possible and additional stresses can be carried by the soil grains. However, there will be a difference between the pore-air pressure and the pore-water pressure resulting from the capillary suction of the water films. Therefore, consideration must be given in the analysis of whether the pore-air or the pore-water pressure is used. Since these pore pressures are opposed to the normal or confining stress, the shear resistance will be reduced whenever positive pore-water pressure is present. On the basis of this observation, for the general case Coulomb's equation must be rewritten:

$$s = c' + (\sigma - u) \tan \phi' \tag{15}$$

where u is the pore pressure. The analysis could be made considering

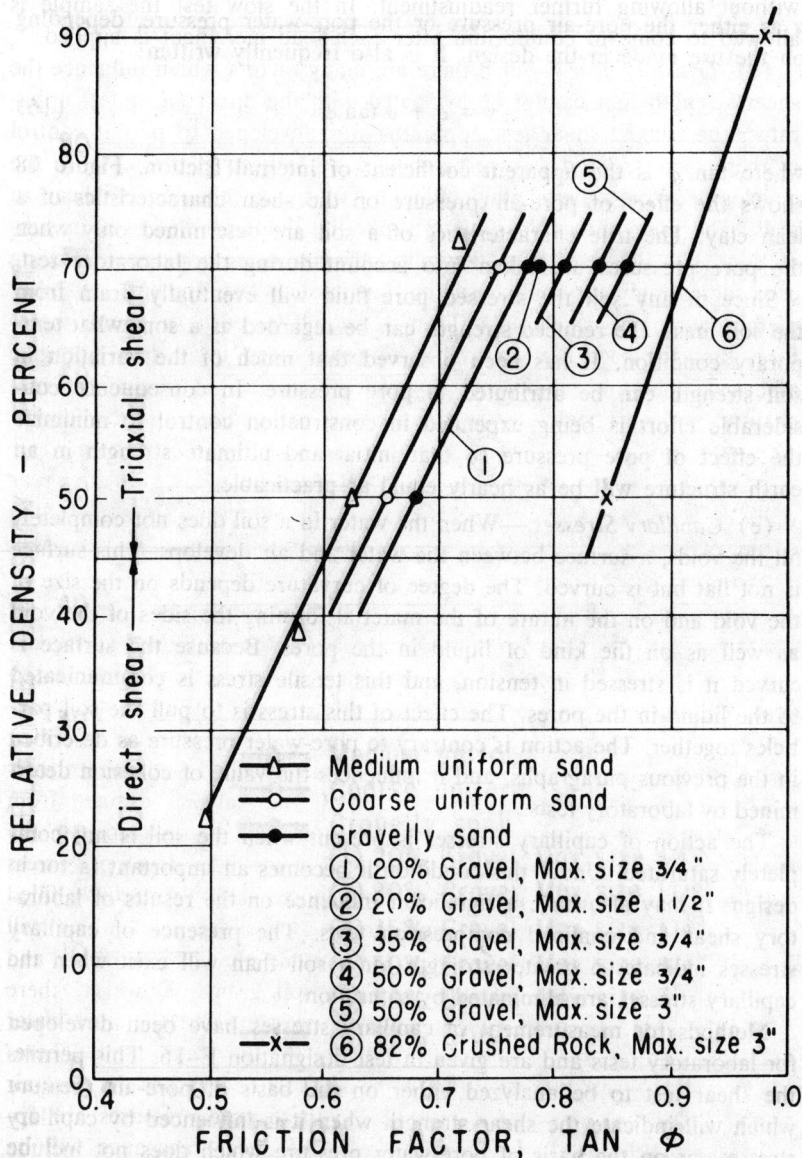

Figure 17.—Effect of relative density on the friction factor for coarse-grained soils. 101–D–573.

u as either the pore-air pressure or the pore-water pressure, depending on the use made in the design. It is also frequently written:

$$s = c + \sigma \tan \phi \qquad (16)$$

where $\tan \phi$ is the apparent coefficient of internal friction. Figure 18 shows the effect of pore-air pressure on the shear characteristics of a lean clay. The true characteristics of a soil are determined only when the pore pressures are taken into account during the laboratory test.

Since in any soil the stressed pore fluid will eventually drain from the soil mass, the reduced strength can be regarded as a somewhat temporary condition. It has been observed that much of the variation in soil strength can be attributed to pore pressure. In consequence considerable effort is being expended in construction control to minimize the effect of pore pressure so that initial and ultimate strength in an earth structure will be as nearly equal as practicable.

(e) *Capillary Stresses.*—When the water in a soil does not completely fill the voids, a surface between the water and air develops. This surface is not flat but is curved. The degree of curvature depends on the size of the void and on the nature of the material forming the sides of the void as well as on the kind of liquid in the pores. Because the surface is curved it is stressed in tension, and this tensile stress is communicated to the liquid in the pores. The effect of this stress is to pull the soil particles together. The action is contrary to pore-water pressure as described in the previous paragraphs, and it influences the value of cohesion determined by laboratory tests.

The action of capillary stresses is present when the soil is not completely saturated. Under this condition it becomes an important factor in design. It may also have a significant influence on the results of laboratory shear and confined compression tests. The presence of capillary stresses indicates a greater strength for a soil than will exist when the capillary stresses are eliminated by saturation.

Methods for measurement of capillary stresses have been developed for laboratory tests and are given in test designation E–16. This permits the shear test to be analyzed either on the basis of pore-air pressure which will indicate the shear strength when it is influenced by capillary stresses, or on the basis of pore-water pressure which does not include the influence of capillary stresses and which can represent the condition of saturation.

(f) *Sliding Resistance.*—A special type of shear strength investigation involves the shear strength between dissimilar substances, most commonly soil and concrete. To identify this type of shear it is usually called sliding resistance. The nature of the problem makes it necessary to solve

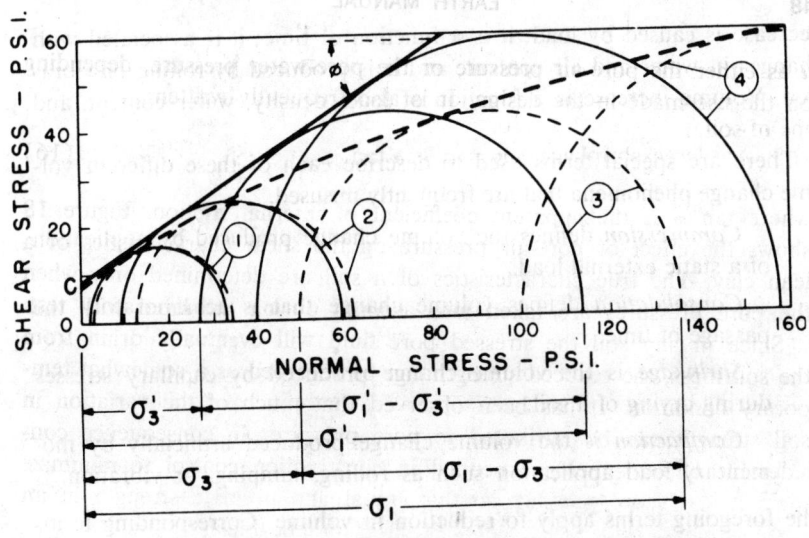

—— Stresses corrected for pore pressure
— — Applied Stresses (not corrected for pore pressure)

SPECIMEN No. _ _ _ _ _ _ _ _ _ _ _ _ _ _ _ _ _ _ 1   2   3   4
APPLIED LATERAL STRESS σ_3 (p.s.i.)   3.1   12.5   50.0 100.0
Measured pore pressure, u (p.s.i.)   1.1   3.3   22.8   59.7
Effective lateral stress, $\bar\sigma_3 = \sigma_3 - u$ (p.s.i.)   2.0   9.2   27.2   40.3
Deviator stress (p.s.i.) $\sigma_1 - \sigma_3$ 31.3   49.7   86.0 117.5
True Tan $\phi = 0.62$ and $c = 7.7$ p.s.i.

PLACEMENT DATA (Avg. of 4 tests)
Dry density, (p.c.f.)   114.1
Void ratio (e)   0.50
Water content (%)   13.5
Degree of saturation (%)   74.4

Figure 18.—Results of an undrained triaxial shear test on lean clay with and without correction for pore-air pressure. 101–D–574.

this case by direct shear test methods and the result may therefore be an apparent rather than a true shear resistance.

20. Compressibility.—(a) *General.*—Volume change in a soil mass due both to natural and artificial causes introduces problems peculiar to soils that are not encountered with other construction materials. Volume decrease is caused by load; it is a function of time; it is associated with changes in water and air content; and it is produced by rolling or vibration. Volume increase is a function of load, density, water content, and type of soil.

There are special terms used to describe each of these different volume change phenomena that are frequently misused.

Compression defines the volume change produced by application of a static external load.

Consolidation defines volume change that is achieved with the passage of time.

Shrinkage is the volume change produced by capillary stresses during drying of a soil.

Compaction is the volume change produced artificially by momentary load application such as rolling, tamping, or vibration.

The foregoing terms apply to reduction in volume. Corresponding terms that apply to increase in volume are:

Rebound as opposed to compression.

Expansion as opposed to consolidation.

Swell as opposed to shrinkage.

Loosening, scarifying, or similar terms describing the operation used in opposition to compaction.

Heave is used to describe volume change produced by frost action or expansive soils.

For the most part the phenomenon of compressibility is associated with changes in volume of the voids and only to a very limited extent with changes in the solid particles. If the voids are to a large extent filled with air, the addition of a load on the soil mass will result in compression without appreciable subsequent consolidation. On the other hand, if the voids are very nearly or completely filled with water, very little or no compression will take place immediately upon application of a load and only as the water drains from the soil mass can consolidation take place. If the water can readily drain from the soil mass, consolidation may take place within a short period of time; but if the soil is very tight and the soil mass is large, complete consolidation may require many years.

Some volume change results from particle rearrangement (in reality localized shear), from particle breakdown, and from physical or chemical absorption of moisture. Particle rearrangement is usually associated with clay soils deposited under water that in the past have been loaded only with the weight of the soil above them. Particle breakdown is most commonly found where residual soil is made from rock that has been weathered and altered in place. Most clay soils have affinity for moisture that can be removed only with considerable effort. Fortunately many of the clay minerals attain a state of saturation without very great volume change; but a few, such as the montmorillonite clays, will absorb or release large volumes of water and experience very large shrinkage and swell.

(b) *Control of Compressibility.*—By far the most frequently occurring problem with which the soils engineer must deal involves the question of compressibility. There have been some very spectacular failures due to this factor, but the most commonly observed effect is the cracking of structures. The classification system presented in part A of this chapter was developed initially to distinguish the different types of soil according to their potential compressibility because of the frequency with which failures due to this cause occur rather than because of their severity. Shear strength is affected indirectly, since the greater the compressibility of a soil the greater the potential for high pore pressures and the more likely that the apparent shear strength will be small. Figure 19 shows the variation of compressibility with soil type for compacted earth embankments.

Although the potential for a high compressibility is usually associated with the fine-grained, highly plastic soils, a number of procedures have been developed by which a soil may be treated to minimize this effect. In manmade fills, for most soil types sufficient compaction can be applied to keep the compression within a few percent. Where it can be applied, moisture control will tend to reduce future consolidation by facilitating compaction. It should be noted that if soils are compacted when very dry of optimum, the grain structure will not assume its densest form as can be readily seen from the compaction curves shown in figure 12(C). With such a condition a subsequent wetting by percolating water may result in a particle rearrangement and an accompanying sudden volume change called saturation collapse. This phenomenon may occur also in soils placed near Proctor maximum dry density but overly dry. Soils subject to shrinkage and swell can be utilized if compacted under good moisture control and loaded sufficiently with other materials to prevent swelling. Embankments utilizing swelling-type soils require flatter slopes and greater volumes than are needed when nonswelling types are involved. This frequently justifies securing materials from a distant source in preference to using a swelling-type soil found close at hand.

Compressibility in a foundation soil is difficult to control. Removal of the soil, use of piles or piers, and spread footings or mats have been applied in specific cases with success. For the most part variation in compressibility within the construction area produces greater difficulty

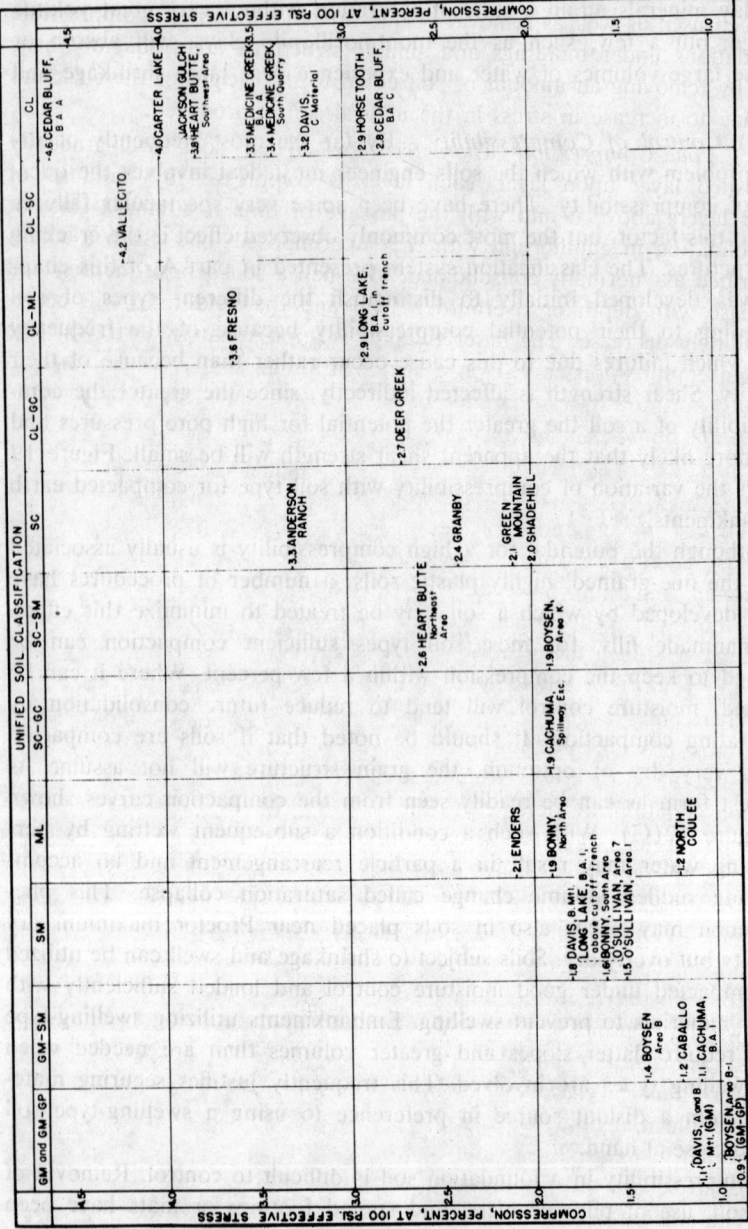

Figure 19.—Compression characteristics of various compacted embankment soils based on field measurements. 101-D-239.

than the amount of compressibility encountered. Under earth dams, drainage of wet soils, wetting of dry soils, and spread embankments have all been used as well as removal of questionable materials. Compressible foundations under buildings and similar structures are frequently handled by removing an amount of soil equal to the weight of the structure so that no increase in stress in the natural soil is produced.

(c) *Load-Compression Characteristics.*—When soils are gradually deposited layer upon layer, each layer is compressed by the load of other layers above it and with the passage of time it attains a state of consolidation in equilibrium with the superimposed load. Such a soil is described as "normally consolidated." In such a soil deposit, density will increase with depth or overburden load and the void ratio will correspondingly decrease. The latter relation is approximately a straight line (virgin compression curve) on a semilogarithmic scale (see fig. 20). If all or a portion of the superimposed load is then removed, some rebound will occur but the change in void ratio will usually be small compared to that produced by the initial compression and consolidation. Such a soil is said to be "preconsolidated." In this state if such a soil is again loaded, the change in void ratio will be small compared to that produced with similar loading increments during initial consolidation. The total decrease in void ratio, however, will be somewhat greater than that originally obtained. Where such a condition is found as a result of a glacier having overridden the area, or where extensive erosion has occurred, it is often possible to use a foundation without difficulty.

Where compressibility is potentially a problem, a foundation investigation must provide information not only as to the soil types found but also information on their present undistorted state. Indications as to the state of consolidation are obtained in the field by the use of the field penetration test, designation E–21, and the vane test, designation E–20. The process of compacting soils produces a state of overconsolidation. Tests for the compacted condition are made with the one-dimensional consolidation test, designation E–15, and also by use of the permeability and settlement tests, designations E–13 and E–14. Figure 20 shows the type of compression results obtained from tests on cohesive soil. Figure 21 shows the relation of relative density and void ratio to compressibility for a compacted, fine sand.

(d) *Load-Expansion Characteristics.*—In addition to the normal rebound phenomenon which occurs on release of a compressive load as indicated in figure 20, certain types of clay soils and clay shales exhibit expansive characteristics in the presence of water.

The amount of expansion depends on the type of clay mineral and the availability of water, and is a function of time, confining load, initial density, and initial water content. Clays of the montmorillonoid type are the

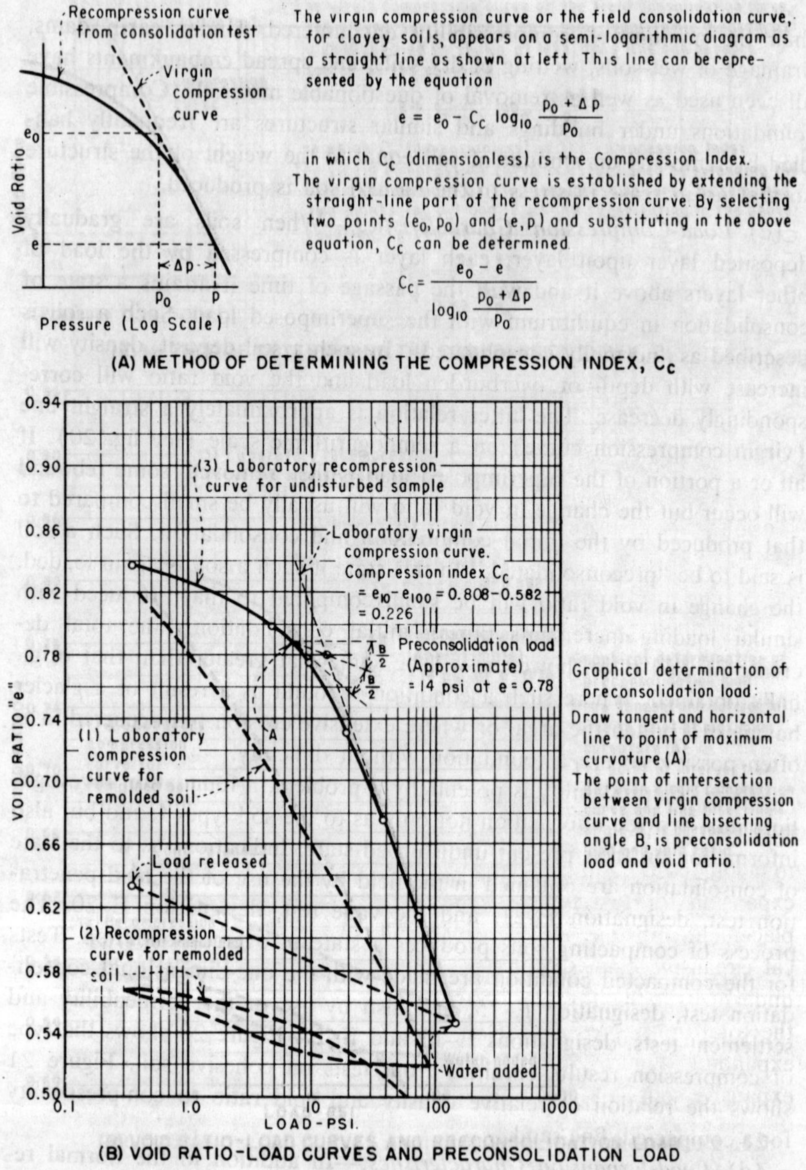

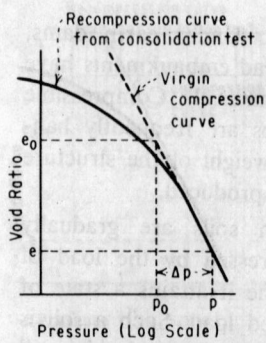

The virgin compression curve or the field consolidation curve, for clayey soils, appears on a semi-logarithmic diagram as a straight line as shown at left. This line can be represented by the equation

$$e = e_0 - C_c \log_{10} \frac{p_0 + \Delta p}{p_0}$$

in which C_c (dimensionless) is the Compression Index.
The virgin compression curve is established by extending the straight-line part of the recompression curve. By selecting two points (e_0, p_0) and (e, p) and substituting in the above equation, C_c can be determined

$$C_c = \frac{e_0 - e}{\log_{10} \frac{p_0 + \Delta p}{p_0}}$$

(A) METHOD OF DETERMINING THE COMPRESSION INDEX, C_c

(3) Laboratory recompression curve for undisturbed sample

Laboratory virgin compression curve.
Compression index C_c
= $e_{10} - e_{100}$ = 0.808 - 0.582
= 0.226

Preconsolidation load (Approximate) = 14 psi at e = 0.78

(1) Laboratory compression curve for remolded soil

Load released

(2) Recompression curve for remolded soil

Water added

Graphical determination of preconsolidation load:
Draw tangent and horizontal line to point of maximum curvature (A)
The point of intersection between virgin compression curve and line bisecting angle B, is preconsolidation load and void ratio.

(B) VOID RATIO-LOAD CURVES AND PRECONSOLIDATION LOAD

Figure 20.—Void ratio-load curve, compression index, and preconsolidation load.
101-D-171.

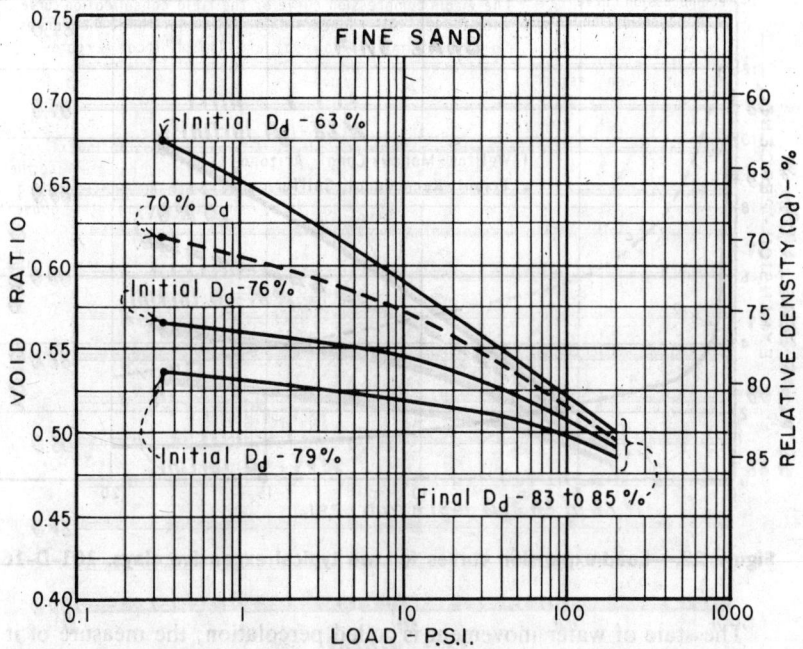

Figure 21.—Compressibility characteristics of a fine sand in relation to placement relative density. 101–D–162.

chief sources of difficulty. Since hydraulic structures always provide a source of water for expansion, these clays must be identified and treated to avoid costly failures. Figure 22 is a comparison between load and expansion for two typical expansive clays. Figure 23 shows effects of placement moisture and density on expansion characteristics for a typical expansive clay. Graph (a) shows the volume change that occurred under a light loading of 1 pound per square inch, and graph (b) shows the total uplift pressures developed when this clay was restrained from expanding. Figure 24 shows examples of canal lining failures caused by expansive soils.

21. Permeability.—(a) *Definition.*—The voids in a soil mass provide not only the mechanism for compressibility but also passages by which water may move through the soil mass. Such passages are variable in size and the paths of flow are tortuous and interconnected. If a sufficiently large number of such paths are considered as acting together, an average rate of flow for the soil mass can be determined under controlled conditions that will be representative of larger masses of the same soil under similar conditions.

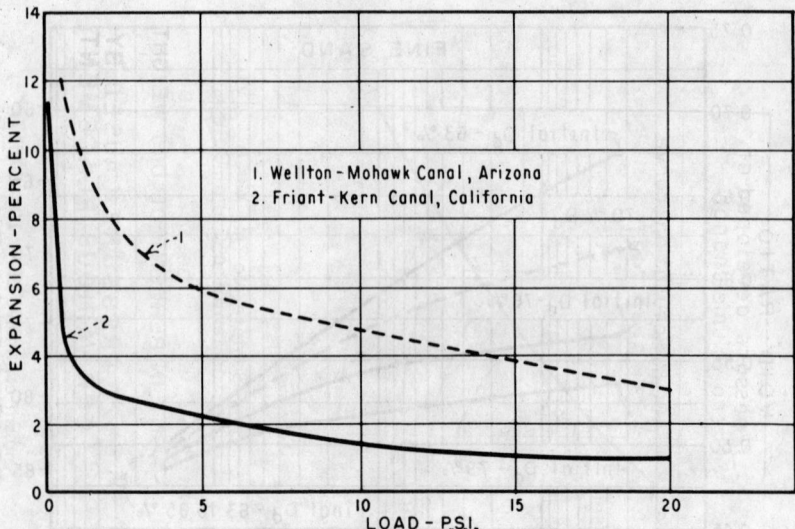

Figure 22.—Load-expansion curves for two typical expansive clays. 101-D-163.

The state of water movement is called percolation; the measure of it is called permeability; and the factor relating permeability to unit conditions of control is called the coefficient of permeability.

The coefficient of permeability, k, is defined as:

$$k = \frac{Q}{A}\frac{L}{h} = \frac{v}{i} \tag{17}$$

where: $Q =$ the quantity of water per unit of time,
           $A =$ the gross cross-sectional area through which Q flows,
           $h =$ the pressure head lost,
           $L =$ the distance through which the head is lost,
           $v =$ the discharge velocity, and
           $i =$ the hydraulic gradient, that is, the ratio of the head lost to the distance in which it is lost.

This equation is commonly known as Darcy's law. Although it is recognized that temperature and viscosity of water affect the coefficient of permeability, these factors are usually neglected in permeability determinations. There are many units of measurement in common use for expressing the coefficient of permeability. Those preferred for Bureau use are feet per year, or cubic feet per square foot per year at unit gradient. A term feet per day is used to some extent in canal design, water supply engineers favor gallons per square foot per day, and most technical literature uses centimeters per second.

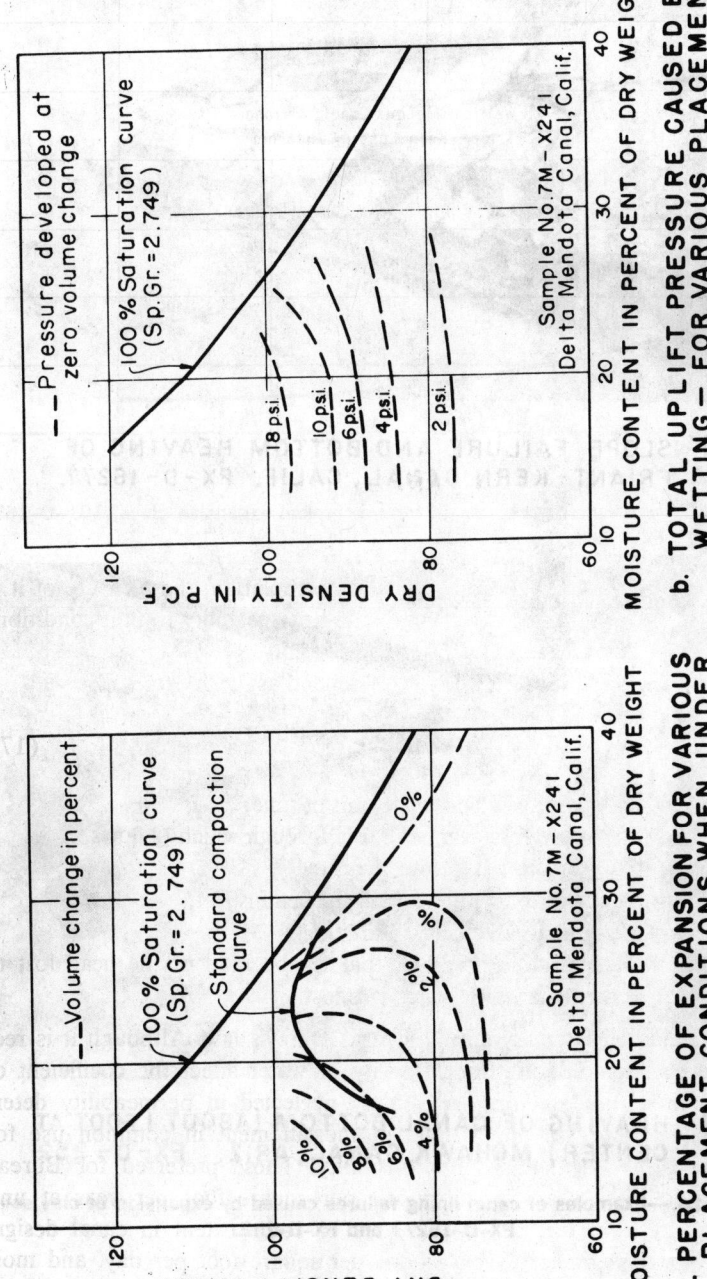

Figure 23.—Typical example of the effect of placement moisture and density on an expansive clay. 101-D-517.

(A)   SLOPE FAILURE AND BOTTOM HEAVING OF
       FRIANT-KERN CANAL, CALIF. PX-D-16277.

(B)   HEAVING OF CANAL BOTTOM (ABOUT I FOOT AT
       CENTER) MOHAWK CANAL, ARIZ.   PX-D-252

Figure 24.—Examples of canal lining failures caused by expansion of clay soils.
PX-D-16277 and PX-D-252.

(b) *Ranges of Permeability.*—The coefficient of permeability found in natural soil deposits ranges from millions of feet per year to less than one-thousandth of a foot per year. In many soil deposits, the permeability parallel to the bedding planes may be 100 or even 1,000 times larger than the permeability perpendicular to the bedding plane. An exception to this is loess in which the vertical permeability is several times greater than the horizontal permeability. Permeability in some soils is very sensitive to small changes in density, water content, or gradation. In certain ranges, a few percent variation in any one of these factors may result in a thousand percent variation in permeability. Because of the wide variation in permeability that is possible, measurement of great accuracy is not required for designs; rather, the order of magnitude of the permeability is of importance.

It has become customary to describe soils with permeabilities less than 1 foot a year as impervious, those with permeabilities between 1 and 100 feet per year as semipervious, and soils with permeabilities greater than 100 feet per year as pervious. However, soils in the pervious range as defined above have been used successfully in the impervious zones of earth dams, and impervious materials by this definition have been used successfully in the pervious zones of earth dams. In differentiating between soils for different zones of a dam, a ratio in permeability of 100 is desirable but 10 is acceptable if a greater range is not readily obtainable. Figure 25 shows the results of permeability tests on relatively clean sand-gravel mixtures.

(c) *Control of Permeability.*—The determination of permeability rates is important in water retention and water conveyance structures because water lost through pervious soils is an economic loss that must be charged to the structure, and the continuous movement of water through the soil of a structure may result in the removal of soluble solids or may result in internal erosion called piping. Piping must particularly be guarded against since it occurs gradually and is often not apparent until failure of the structure is imminent. With nonhydraulic structures, permeability is important only when work below water table is required or when changes in water table occur in the area influenced by the construction of the structure.

The control of percolating water is accomplished in a variety of ways that may conveniently be divided into three classes: (1) reduction of the coefficient of permeability, (2) reduction of the hydraulic gradient, and (3) control of the effluent. Reduction of permeability in embankments is accomplished by selection of material, compaction control, and the use of additives. Reduction of percolation through foundations is accomplished by use of cutoffs of which there are many varieties, injection of a material into the foundation by grouting methods, and densification by

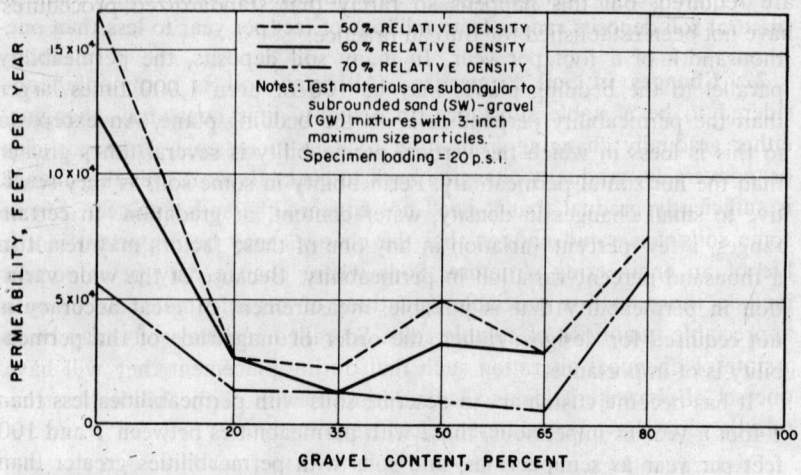

Figure 25.—Relationship of permeability to gravel content for specimens of various relative densities. Specimens are sand-gravel mixtures with 3-inch maximum size particles. 101–D–214.

shock or loading often accompanied by effecting a change in water content to hasten or make densification easier. Reduction in hydraulic gradient is accomplished either by reducing the head or increasing the length of the seepage path, usually by the use of some type of blanket. Control of effluent requires making a design in such a manner that seepage pressures are at all points kept sufficiently low so that uplift and rupture of the material above the seepage path is prevented. This is accomplished by zoning in dams and canal banks, use of filters, installation of drains, and installation of pressure relief wells. There are also a number of combination control methods in common use including use of spread embankments, impervious diaphragms, and linings. The method of control selected depends primarily on the cost of treatment compared to the benefit received both in prevention of water loss and in assurance against piping failure.

(d) *Determination of Permeability Values.*—Permeability is determined in the field by means of a variety of tests based either on forcing water into the material or removing it under controlled conditions. Designation E–18 describes water loss tests made in bore holes. Designation E–19 describes a test procedure sometimes used in canal alinement investigations. In the laboratory, permeability is usually determined according to designation E–13, E–14, or E–15. There are ranges of permeability for which these tests are not satisfactory and special procedures are required, but this happens so rarely that standardized procedures have not been established for Bureau practice.

22. Changes in Soil Properties.—Although soil is commonly considered to be a stable substance, in actuality it is constantly changing; either gradually changing from solid rock to increasingly finer particles or, conversely, gradually changing back to rock. In most soils this change is sufficiently gradual that it need not concern the soils engineer; but in some soils it is rapid enough that the changes may be important in the life of an engineering structure. Soils where change may be important include those with appreciable quantities of organic materials or with appreciable quantities of soluble salts. So-called residual soils may be in a state of chemical alteration such that during placement they will have one set of characteristics and later during the life of the hydraulic structure they may have very different characteristics.

Frequently there are foundations that in their natural state have been stable for many years and give every indication that they will remain so. Nevertheless, minor manmade developments result in failure on some of these foundations. One of the materials that is frequently responsible for such failures is called sensitive clay. This type of clay, when tested in an undisturbed condition, has substantial strength which is to a large extent lost upon being remolded. Another material is the so-called quicksand for which continued disturbance causes a moderately solid foundation to become liquid. This phenomenon is also known as spontaneous liquefaction. There is another group of soils where minor changes in moisture content result in an abrupt change in strength. In some cases these soils, such as loessial soils, have been deposited very loosely and exhibit change in strength and can subside when the moisture is increased. Swelling clays frequently exhibit this change of strength characteristic due to increase in moisture. Laterite clays are often responsible for structural failures when they dry out.

Among those soils that through desiccation and chemical action have changed to forms commonly regarded as rock, there are varieties of shale, sandstone, and limestone which, when exposed to air, can result in marked changes in character. Some shales flake off or weather rapidly and turn into soil. Some shales may dry out without any apparent effect, but if they are then rewetted they deteriorate rapidly into a mud. Some sandstones and limestones harden on exposure to air and retain their improved qualities. Other limestones and sandstones break down rapidly with fluctuating temperature and water content.

Although the condition of potential deterioration is found comparatively rarely, it is highly critical because if it is not recognized failure comes without advance warning. Engineering practices for treatment of soils which deteriorate are not well known. Consequently specialized assistance is required where engineering structures must be designed for such conditions. Where situations such as described above are suspected it is advisable to have the situation reviewed by someone who has specialized knowledge in this field.

23. Workability.—Although laboratory testing will indicate the maximum extent to which such engineering properties as strength, compressibility, and permeability of a given soil may be varied, it will seldom be practicable to reach these limits in engineering practice. From an economic standpoint the ease with which satisfactory values of engineering properties can be achieved is a very important attribute of a soil, a soil deposit, or a foundation.

The cost of procuring a unit volume of soil and placing it in a structure or for treating a unit amount of foundation varies widely not only according to soil type, but also according to type and size of structure. It is influenced also by the kind of equipment available and the state of the labor market at the moment. If the project is sufficiently large so that specialized equipment can be charged off on that one job, maximum efficiency in construction is most likely to be achieved by its use. The soils selected, however, must be workable by such methods. Where separation of oversize is not required and where mixing requirements are minimized, borrow pits that can be preprocessed to optimum water content will be preferable even though longer hauls and a somewhat larger embankment than the theoretical minimum are required for their effective use.

In practice there will be situations where separation of oversize is economical. There will be cases where mixing of two varieties of soil is worthwhile. There will be instances where extensive efforts to obtain maximum moisture control are justified. However, for the most part such operations should be avoided if possible.

There are no test procedures by which the property of workability may be given a quantitative value. Rather, it is necessary that all pertinent information concerning a soil, a borrow pit, or a foundation be tabulated so that the various design possibilities can be evaluated in the light of the current economic situation when final designs are prepared. The engineering use chart, figure 8, gives qualitative information on the workability of soils as a construction material and the relative desirability of various soil types according to structure. Borrow pits may be evaluated according to amount of work required. Because the cost of bringing in equipment is chargeable against a deposit, there is a rapid decrease in unit cost as the volume of work is increased. The change in unit cost for excavation up to about 100,000 cubic yards is quite noticeable; up to 1,000,000 cubic yards it is very gradual; and beyond that range it is nearly constant. Hauling costs are nearly constant above about 100,000 cubic yards depending only on distance. Moisture control costs depend primarily on the uniformity and slope of the borrow area and the availability of water; excavation costs are also influenced somewhat by the topography of the borrow area. Borrow pits slightly higher in elevation than the work are preferable to those below the work.

24. Frost Action.—The heaving of subgrades due to formation of ice lenses and the subsequent loss of stability on thawing is known as frost action. Water will rise by capillarity and by attraction toward the freezing zone, and forms lenses of pure ice which heave the soil. Therefore, soils most susceptible to frost action are fine-grained; that is, those in which capillarity can develop but which are sufficiently pervious to allow adequate water movement upward from below the freezing zone. The freezing of the pore water in saturated fine-grained soils will decrease the density of the mass by expansion, but will not result in an appreciable frost heave unless water movement from below can take place.

The severity of frost action depends on three factors: (1) the type of soil, (2) the availability of a source of water, and (3) the time rate of fluctuation of temperature about the freezing point. Silty soils, chalk, and some shales are frost susceptible. In the absence of a water supply, frost action will be limited to the 9 percent increase in volume of the pore water.

Of particular importance to Bureau structures are the uplifting of concrete slabs in canals, spillways, and other structures; the loosening of compacted earth linings in canals; and loss of stability in highway subgrades. Figure 26 shows ice lens formation in a chalk foundation.

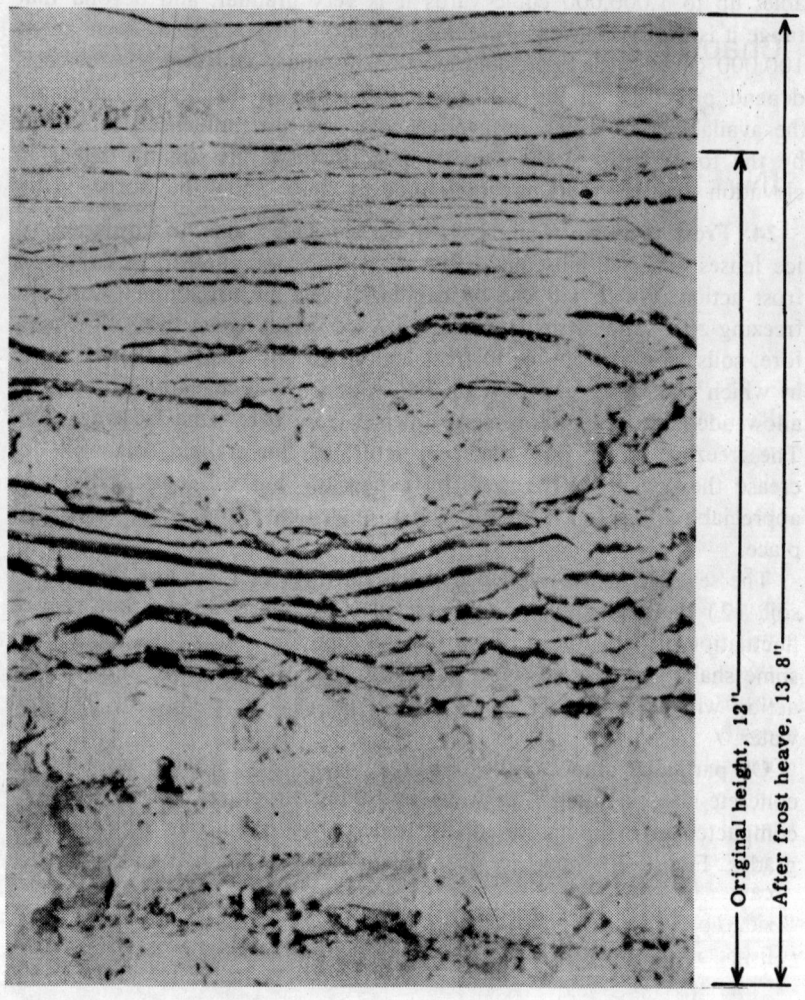

Original height, 12"

After frost heave, 13.8"

Figure 26.—Ice lens formation between chalk layers. This undisturbed sample of Niobrara chalk, Ft. Hayes member, from the spillway foundation of Kirwin Dam shows ice lenses formed after slowly freezing in the laboratory to a depth of 10 inches with free water available at the bottom. PX–D–17012.

Chapter II

INVESTIGATION

A. Stages of Investigation

25. General.—An investigation problem handled by the Bureau of Reclamation normally starts from one of two conditions: Either a supply of water exists that should be fully utilized, or a demand for water develops that needs to be satisfied. The first condition leads to a river basin report with the ultimate objective of providing a development that will utilize the available water resources to produce a maximum benefit. The second condition leads to a project plan with the objective of satisfying a demand insofar as a consideration of costs compared to benefits will permit. Inasmuch as the western part of the United States where the Bureau of Reclamation functions is an area generally deficient in water, the development of a water supply system is usually difficult and costly. Therefore, many proposals when analyzed will prove to be economically unjustified. In an attempt to satisfy a variety of conflicting interests competing for a limited water supply and to improve the economic evaluation of a project, it has become necessary to design the project facilities for a multiplicity of purposes. Because of these complicating factors, the initial planning of a project is accomplished to a considerable degree by the use of maps, statistical information, and published reports before any field investigations are required.

At the time field investigations are initiated, the investigator should have available for his use a tentative plan of development which shows the kind of facilities required, their approximate capacity, size, and general location. Often previous studies of the area have been made. Reports of such studies should be reviewed and information pertinent to the proposed investigation should be abstracted. In addition, there are generalized surveys available for many areas which may provide some information on the accessibility, geology, and soils which can be utilized as the starting point for the field investigation. Furthermore, the investigator must have at least a general knowledge of the foundation and materials requirements for the various facilities under consideration if his investigations are to be accomplished effectively and efficiently.

When developing a potential project, it is essential that the data which have been gathered be summarized from time to time and that conclusions based on all available information be reached in order to check the economic practicability. Normally, a project is developed in three stages, each identified by a word descriptive of the objective of each stage: (1) reconnaissance, (2) feasibility, and (3) specifications. Two additional stages carry a project to completion: construction, and finally operation and maintenance. It has become customary to divide the investigations into corresponding stages, each of which is characterized by a comparable degree of investigation. The degree of reliability of the data collected defining each influential factor should be of the same order, and the conclusions should be based on the probable influence each factor has on the practicability of the project.

The data secured in the reconnaissance stage are primarily descriptive. Data already available are summarized and a visual examination of the project is made. Soil properties are based largely on visual classification. The field work is usually done by a geologist or a soils engineer.

In the feasibility stage the data are primarily qualitative. Limited exploration is done to confirm the geological interpretation or to develop a new interpretation if necessary. The properties of soils are determined by index tests. The field work is done jointly by a geologist and an engineer.

In the specifications stage, the data are primarily quantitative and specific. Sufficient exploration is done to establish conditions at all critical points. Engineering properties are determined for the various soil types. The work is done primarily by an engineer with geological assistance as required.

Foundation and materials investigations during the construction stage are primarily confirmative in character. They are used to clarify those conditions which could not be resolved during the specifications stage; and to explore alternative proposals suggested as a result of having available special types of equipment or as a result of variations in weather, in labor conditions, or in general economic conditions differing from those assumed when designs were formulated. If existing explorations do not provide sufficient information to permit accurate staking of the cut-off trench excavation and other required excavation, additional exploratory drilling must be done at this stage to obtain the necessary data. For the soil characteristics, index properties or engineering properties are required, depending on the particular problem. Work is performed under the direction of the construction engineer. For earth dams, the specified borrow areas are usually thoroughly investigated prior to large scale embankment operations to determine depths of cut and moisture requirements (see sec. 45). Sometimes the only investigative work for small structures is done during construction (see sec. 75(b)).

Only in the event that the structure proves to be unsatisfactory in some way is investigation during the operation and maintenance stage required. These situations occur too rarely to warrant development of generalized procedures.

26. Reconnaissance.—(a) *Objectives.*—The data obtained during the reconnaissance stage are used in reconnaissance designs and estimates to evaluate the engineering and economic possibilities of a project and to determine whether further investigations are justified. The work is accomplished within a short period of time, from a minimum amount of data, and without delving into design details. The investigations should be no more detailed than is necessary to determine the approximate cost and physical ability of constructing the type of works proposed. Such designs and cost estimates are also used as an aid in selecting the most economical plan when several alternative possibilities exist.

The investigations during the reconnaissance stage should lead to an appraisal of the general subsurface conditions throughout the area of the project as well as an evaluation of the broad aspects of the foundation conditions of alternative sites selected for specific major or critical structures. The appraisal should define major advantages and defects of the foundations and materials deposits at the alternative sites with reasonable certainty. The most desirable sites then should be investigated further by additional surface examination, and a limited amount of subsurface exploration and testing if required, to permit the selection of the most desirable site, or sites, from the standpoint of foundation conditions and materials.

(b) *Sizes and Depths of Investigated Areas.*—Since the reconnaissance investigation is based primarily on surface indications, it may at times be necessary to go to considerable distances from the proposed structure sites in order to secure information from which subsurface conditions may be reasonably interpolated. Because underground conditions are largely hypothetical, the areas described for borrow areas should be appreciably larger than the areas ultimately required.

Descriptions of foundations for buildings and similar structures should include about 1½ times the depth of the minimum dimension. Hydraulic structures, such as dams, require a description of foundation conditions beneath their base to a depth equal to about twice the hydraulic head. Foundations for conveyance systems—that is, canals, highways, and tunnels—should be described to a sufficient depth below grade line to include foundation materials which will have a bearing on design.

27. Feasibility.—(a) *Objective.*—The objective of the feasibility stage of an investigation is to confirm or expand the work done in the reconnaissance stage in order that an adequate estimate of cost may be prepared for the project. This cost can be compared to the benefits to determine whether the project is economically justified. The purpose of a feasibility investigation is to establish a factual background for a design and estimate that serves as a basis for seeking authorization and an advance of funds for construction. There has been and will continue to be considerable variation in requirements for feasibility investigations depending on Congressional requirements and on administrative policy. There will also be considerable variation in requirements depending on local conditions.

It is occasionally permissible to waive requirements for a more extensive investigation than is provided in the reconnaissance stage for those features that contribute only a small portion of the total cost of a project. For example, a project may consist of a dam representing a large percentage of the total cost, and some canal and road construction which together with rights-of-way account for the remainder. In this case reconnaissance investigations for the canal and road might be acceptable unless unfavorable geological conditions exist that (1) might control the feasibility of the plan involving that feature, (2) might have a significant influence on project costs, or (3) unless that feature might be used as a separate part of the project such as for repayment negotiation purposes. Another project may consist of a canal whose cost is a large percentage of the total project cost and the remaining cost is for a diversion dam. In this case a more detailed foundation investigation than that provided by a reconnaissance investigation would be required for the canal but may not be required for the dam. In another example, a powerhouse might account for a large percentage of the total project cost, but since only a small percentage of the cost of the powerhouse would be affected by foundation conditions, requirements for foundation investigations beyond the reconnaissance stage could possibly be waived. Requests for waiving feasibility investigations in possibly significant cases must be submitted to the Engineering and Research Center, Denver, for review and approval.

For those structures concerned with the storage or conveyance of water, the problem of water loss should always be thoroughly investigated in the feasibility stage. Unless areas where water loss may be expected are concentrated, the cost of the required treatment will nearly always have a significant influence on the total project cost and on the overall project feature requirements. It is essential, therefore, that all proposed reservoirs be examined to determine whether water loss is a potential problem, and if so the areas of water loss should be delineated and the magnitude of the loss should be estimated. The routes of the canal system should be examined for areas where water loss may be significant. Such areas should be defined and explored to determine the volume of loss to be expected.

(b) *Organization of the Investigation.*—The increasing complexity of Reclamation projects, combined with benefit-cost ratios in the range of minimum acceptability, result in strong pressure to support project plans with cost estimates which will very closely approximate ultimate costs. This pressure in turn may be reflected in a need for making foundation and materials investigations even for minor facilities. If these are required the investigations should be made to a high degree of dependability. To produce the maximum information at the least cost, orderly procedures have been developed. These procedures and standards are described in the following paragraphs.

Even where the structure or its foundation accounts for the major part of the total cost, the conditions expected to prevail can sometimes be determined with sufficient accuracy so that a reconnaissance investigation will suffice. Since the main purpose in making a feasibility type of investigation is to provide basic data for a feasibility design from which a dependable estimate of cost and a firm plan may be determined, it is necessary to carry the geological and materials investigation only sufficiently far to assure a dependable cost estimate. Frequently, however, there are flaws in a foundation or in a material source that appreciably affect total cost, which are not readily apparent even to those with considerable experience with the type of structure involved. It is imperative, therefore, that these cases be very carefully considered before abbreviating the investigation.

A project may be so obviously feasible that its authorization may be anticipated with confidence. To minimize the time required between authorization and construction, it may then be advisable to make the investigation during the feasibility stage sufficiently detailed to provide final design data. Occasionally the cost of moving in exploratory equipment is sufficiently great that maximum economy is achieved by making a complete investigation while the equipment is there. Because of these conditions the extent of a feasibility investigation must be determined administratively project by project.

The first step in a feasibility investigation is a review of the work already done. This includes not only a review of the reconnaissance investigations, but also an examination of the reconnaissance design, if any is available, and a determination of whether the objective is still the same. The design of the structure is compared with the foundation condition, as visualized; and the materials for construction, both in quality and quantity, are compared with how they are utilized in the design.

Such an examination may disclose areas in which the investigator's observations were not understood by the designer. It may also disclose conditions that the investigator considered minor but the designer considered important or that the designer disregarded although they appeared important to the investigator. This examination will permit the investigator to locate exploratory holes so that pertinent information can be secured to clarify a questionable condition and at the same time minimize the amount of interpretation required at the proposed location of the structure.

The second step in a feasibility investigation is the preparation of a program of work. The program should show the locations of proposed test holes, the kind of exploratory procedures required, the depth to which the holes should extend, and the kind of data to be secured. The proposed test holes should be located in a systematic pattern into which future investigations can be incorporated. The program should show the order in which the explorations are to be made so that the most questionable and the most critical areas are explored first, but should remain flexible to adapt to the conditions encountered. The cost of the program should be estimated in order that those responsible for authorizing the work may be informed of the magnitude of the undertaking. Responsibility for authorizing and reviewing exploration programs is given in the Reclamation Instructions, series 110 and 130, respectively.

The third step is the performance of the exploration. This operation is usually performed by Government forces, and this method is preferred because of the greater flexibility of the operation. Explorations by contract are used when Government facilities are not available. Sampling by contract should conform to the procedures given in designation E–2. At the feasibility stage of investigation, the information obtained from each item should be as complete as time and funds permit so that the data can be used in design. A log of the test hole should be prepared on the basis of actual examination of the materials recovered from the hole, supplemented by information from the driller's report. The permeability of the various strata should be tested, and the ground-water level should be determined.

The fourth step is the preparation of a report. The report may sometimes be an amendment to the reconnaissance report. Generally, however, a new feasibility report is required. (See the Reclamation Instructions, part 118). The feasibility investigations report differs from the reconnaissance report primarily in that the descriptions of foundations and materials deposits are delimited in depths on the basis of exploratory holes rather than from surface indications, and descriptions of materials are elaborated with data from additional index and other laboratory tests.

28. Specifications.—(a) *Scope.*—The purpose of the specifications stage of an investigation is to provide data from which designs for construction of the various features of a project may be made. Since it is the general practice of the Bureau of Reclamation to secure the construction of these features by contract, it is necessary to specify the nature and volume of the work. The completion of the specifications marks the end of this stage.

The foundation and materials investigations in this stage must produce, in addition to relevant geologic data, sufficient detailed information so that quantities of materials to be moved may be established. This operation requires considerable judgment. It is necessary to secure the maximum degree of accuracy with a minimum number of measurements. This requires the location of exploratory holes in such a fashion that pertinent bedrock structures and conditions including the contours of the bedrock, or a buried horizon if that is the objective, can be determined with reasonable accuracy.

The exploratory work for the feasibility stage will usually provide sufficient information so that, with a knowledge of geologic structures, cross sections may be sketched which show up the points where additional measurements are needed. If the work done in the feasibility stage of investigation is insufficient for this purpose, the specifications stage of exploration must first be directed toward this objective.

At the specifications stage of an investigation, the size and location of the structure under consideration should be known. If this information is available, the exploratory work required can be restricted to the area to be occupied by the structure. A variety of situations may be revealed by feasibility explorations which will make it possible to limit the amount of investigations still further. For example, if feasibility explorations have shown that permeable zones or strata are not involved and geological or engineering considerations indicate that water loss is not a problem, then the number of water tests in the drill holes may be reduced. If the nature of the foundation is such that an impervious layer (usually bedrock) may be reached or that a layer of adequate strength can be reached economically, then exploratory hole operations may be limited to methods which will locate such layers of impermeability or adequate strength. Little attention need be given to the overburden provided that sufficient information on its character already exists from previous investigation.

Although it is hoped that during feasibility explorations sufficient information is developed to describe the nature of the foundation and material deposits, further explorations during the specifications stage occasionally disclose conditions that had not been anticipated. When such is the case the exploratory program will need to be modified to investigate the new condition. The individual in charge of investigations should be alert to such conditions which may not be easily recognizable. The specifications stage investigation starts where previous investigations left off. It may be necessary, therefore, to secure feasibility and specifications grade information at the same time. It may merely be required to put the finishing touches on an investigation that has to a large extent already been completed to specifications grade. On the other hand, occasionally it is necessary to complete a reconnaissance-type investigation before proceeding with detailed investigations.

(b) *Programing the Exploration.*—As for a feasibility stage investigation, a program of exploration for the specifications stage should be formulated before field work is started. Since this will be the last opportunity to consider the conditions of the site in relation to structure requirements before construction is initiated, the engineer who will be responsible for preparation of designs and specifications should review the exploratory program with a soils engineer, an engineering geologist, or both before work is initiated.

On the basis of the engineer's knowledge of structure requirements, his experience with performance of earth materials, and the indicated conditions expected to prevail at the site, he can determine where conditions are (1) obviously adequate, (2) clearly inadequate, and (3) those which are questionable. As a result he can direct the testing to questionable areas and appreciably reduce the work involved. It should be emphasized that maximum economy in exploration can be achieved only with the assistance of the engineer who will be responsible for the design work.

As the work proceeds on the exploration, the Engineering and Research Center, or the field design office in case the design work has been delegated, should be kept informed. Requirements will vary depending upon urgency; normally the geologic log of each drill hole should be forwarded as the hole is completed. For materials investigations, data accumulated should be forwarded at about weekly to 10-day intervals.

The results of a specifications stage foundation and materials investigation should be incorporated in a report or reports in similar form to that required for reconnaissance and feasibility investigations. At this stage background material can be summarized, but the additional detail work performed should be fully covered with drill logs and cross sections showing geological interpretation. The report should be prepared under the direction of the engineer in charge of the field investigations. (See the Reclamation Instructions, part 118.)

B. Principles of Investigations

29. General.—(a) *Objectives.*—The purpose of an investigation is to obtain information relating to foundation conditions and natural construction materials commensurate with the magnitude and type of structure involved and with the stage of the project. The investigation is conducted in the office, in the field, and in the laboratory. It normally follows a "learn as you go" procedure in which the characteristics of the subsurface soils and conditions are developed in progressively greater detail as the exploratory work proceeds. The data obtained must be organized to show clearly the significant features of the occurrence and properties of the soils.

The specific objectives of an investigation are to determine, as required:

(1) The location, sequence, thickness, and areal extent of each soil stratum, including a description and classification of the soils and their structure and stratification in the undisturbed state. Significant geologic features, such as concretions, and mineral and chemical constituents, should be noted.

(2) The depth to and type of bedrock, as well as the location, sequence, thickness, areal extent, attitude, depth of weathering, soundness, seams, joints, fissures, and other structural features, and description of rock in each rock stratum within the depth of influence of the structure.

(3) The characteristics of the ground water, including depth to water table, whether the water table is perched or normal, depth of and pressure in artesian zones, and quantity of soluble salts or other minerals present.

(4) The properties of the soils by one or a combination of methods commensurate with the investigation stage, the type of structure, and detailed engineering data, as follows:

By describing and identifying soils in place visually, and determining their inplace density.

By obtaining disturbed samples, describing and identifying them visually, and determining their inplace water content and their index properties. The physical properties may be estimated on the basis of their classification and the results of laboratory index tests.

By indirect methods performed in the field, such as geological interpretations, soundings, or geophysical methods, using the results of a limited amount of direct exploration and other tests to provide necessary correlations.

By observing performance of previously constructed structures built of or resting on similar soils.

By obtaining undisturbed samples, identifying them visually, describing their undisturbed state, determining inplace density and water content, and obtaining index and engineering properties by laboratory tests.

By performing tests in the field, such as standard penetration tests, pile loading tests, permeability tests, and vane shear tests.

(b) *Classification of Structure Foundations.*—The investigation requirements for foundations for structures used by the Bureau vary over a wide range and may include consideration of the foundation material for use as the foundation and also for use in the structure. As a guide to help determine the type and degree of foundation investigation required, foundations for structures can be conveniently grouped into four classes:

1. The soil or rock is of poor material which must be removed partially or entirely to provide a satisfactory foundation for the structure under consideration;

2. The soil or rock in situ will provide the structure foundation, either with or without treatment;

3. The soil or rock provides both the foundation and a major part of the structure, with the material from required foundation excavation providing the material used in the structure; and

4. Same as (3), except that substantial amounts of material are needed in addition to that available from required excavation.

For structures which rest on rock, such as large concrete dams, in addition to the investigations of the rock foundation a soil investigation is made which is concerned primarily with depth to bedrock, stability of slopes, and difficulty of excavation. The materials from excavations for structures in this class should be utilized for other purposes when practicable; hence, it is desirable during preliminary investigations to examine the soils with that possibility in mind. For example, a site considered suitable for a concrete dam will require temporary cofferdams, and consideration should be given to using materials from required excavations for that purpose.

For structures which are often founded on soil, such as buildings and canal structures, the primary objective of a soil investigation is to determine its volume change characteristics which may result in foundation settlement or heave. If heavy loading and wet soil conditions are anticipated, the shear strength should also be investigated.

For structures which are founded on soil and which utilize materials from required excavations, it is essential that materials be considered from both stability and utilization standpoints. Stability of slopes, both in cuts and fills, is a primary consideration. Compressibility varies in importance, approximately commensurate with the importance of the structure itself, having little significance where roads and laterals are concerned but major importance where paved highways and large lined canals with large structures are required. In expansive soils and in low-density soils, the probability and magnitude of uplift and subsidence must be evaluated. (Refer to sections 20 and 64.) Permeability is important on canals and laterals. Where a choice in location is possible, workability of materials is of major economic importance. For this reason, cuts into bedrock are normally avoided.

30. Sources of Map and Photo Information.—The Bureau's administrative instructions regarding maps and mapping are given in part 368, and surveying and mapping techniques and standards are given in series 550, of the Reclamation Instructions. The following subsections discuss several types of maps and air photos and the sources where information concerning them may be obtained.

(a) *Topographic Maps.*—A topographic map is indispensable in the design and construction of most civil engineering structures. Before undertaking the job of map making, a thorough search should be made for the existence of maps covering the area of the structure and the potential sources of construction materials. The U.S. Geological Survey (USGS) should be contacted for information on the availability of maps. This organization is making a series of standard topographic maps to cover the United States and Puerto Rico.

The unit of survey for the USGS maps is a quadrangle bounded by parallels of latitude and meridians of longitude. Quadrangles covering 7.5 minutes of latitude and longitude are generally published at the scale of either 1:24,000 (1 inch equals 2,000 feet) or 1:31,680 (1 inch equals ½ mile). Quadrangles covering 15 minutes of latitude and longitude are published at the scale of 1:62,500 (1 inch equals approximately 1 mile), and quadrangles covering 30 minutes of latitude and longitude are published at the scale of 1:125,000 (1 inch equals approximately 2 miles). In certain western States, a few quadrangles covering $1°$ of latitude and longitude have been published at the scale of 1:250,000 (1 inch equals approximately 4 miles). A few special maps are published at other scales. Each quadrangle is designated by the name of a city, town, or prominent natural or historical feature within it, and on the margins of the map are printed the names of adjoining quadrangle maps that have been published.

In addition to the published topographic map, information of great assistance to engineers is available from the U.S. Geological Survey for mapped areas. For example, the locations and true geodetic positions of

triangulation stations and elevations of permanent benchmarks established
by the Geological Survey are recorded. Coordinates and elevations of
First and Second Order basic control established by the Coast and
Geodetic Survey, Environmental Science Services Administration, may be
obtained from that agency. Also, map manuscripts and advance blueline
prints available at the same scale as published maps are available several
months prior to publication of the final map[1]. Figure 27 shows the status
of topographic mapping in the United States, mainly of the 7½- and
15-minute series distributed by the Geological Survey. An index map of
each individual State at the scale of 1:1,000,000 similar to figure 27 is
also available.

In the absence of topographic coverage of the area, other types of maps
may be used in the preliminary stages. Of considerable importance to
dam design are the river survey maps. These are strip maps which show
the course and fall of the stream; the configuration of the valley floor and
the adjacent slopes; and the locations of towns, scattered houses, irrigation
ditches, roads, and other cultural features. River survey maps were prepared largely in connection with the classification of public lands; hence,
most of them are of areas in the western States. If the valley is less than a
mile wide, the topography usually is shown to 100 feet or more above the
water surface; if the valley is flat and wide, topography is shown for a
strip of 1 to 2 miles wide, parallel to the river or stream. The usual scale
is 1:31,680 or 1:24,000, and the normal contour interval is 20 feet on
land and 5 feet on the water surface. Many of these maps include proposed damsites on larger scale topography and show a profile of the
stream. The standard size sheet is 22 by 28 inches.

The availability of river survey maps, other special maps and sheets,
including national parks and monuments, and a list of agents for topographic maps are indicated on the index to topographic mapping for the
various States. These indices are available from the U.S. Geological
Survey. Requests and inquiries on published maps and on the availability
of map manuscripts or other information of the area should be directed to
Distribution Section, U.S. Geological Survey, Denver Federal Center,
Denver, Colo. 80225, for maps west of the Mississippi River, or Distribution Section, U.S. Geological Survey, 1200 South Eads St., Arlington, Va.
22202, for maps east of the Mississippi River.

Topographic maps are of considerable value in the exploration of
foundations and construction materials for hydraulic structures. The locations and elevations of exploratory holes, outcrops, and erosional features
can be placed on the map, and the land forms portrayed by the contours

---

[1] See "Advance Material Available from Current Topographic Mapping, Quar-
terly Edition," available at Topographic Divisions of Regional Offices of the U.S.
Geological Survey.

indicate to some degree the type of soil and subsurface geologic conditions. Information on the origin and characteristics of some of the simpler land forms is given in section 31. In the absence of topographic map coverage or where greater detail is needed, the Bureau of Reclamation contracts with commercial firms to produce by photogrammetric methods maps of any desired scale and contour interval, and profiles or cross section data suitable for automatic data processing.

(b) *Geologic Maps.*—Considerable useful engineering information is obtainable from geologic maps. These maps identify the rock units directly underlying the project area. The characteristics of rocks are of major importance in the selection of a damsite and in the design of water retaining and conveyance structures. Many surface soils are closely related to the type of rock from which they are derived, but if the soil has been transported an appreciable distance it may overlay an entirely different rock type. When the influence of climate, relief, and geology of the area are considered, the experienced engineer can make reasonable predictions of the type of soil which will be encountered or of the association with a particular parent material. Conditions beneath the surface can often be correctly deduced by the three-dimensional information given on geologic maps. These maps are especially valuable in the areas where only limited information on soils from the agricultural standpoint is available; for example, in arid or semiarid regions where soils are thin. (See subsection (c) below.)

On geologic maps rocks are identified by name and geologic age. The smallest rock unit commonly mapped is generally a formation, but smaller subdivisions such as members or beds may be delineated. A formation is an individual bed or several beds of rock that extend over a fairly large area and that can be clearly differentiated from overlying or underlying beds because of a distinct difference in lithology, structure, or age. The areal extent of these formations is indicated on geologic maps by means of letter symbols, color, and symbolic patterns.

Letter symbols indicate the formation and geologic period. For example, "Jm" stands for the Morrison formation of the Jurassic period. Standard color and pattern conventions are followed on maps produced by the U.S. Geological Survey. Tints of yellow and orange are used for different Cenozoic rocks, tints of green for Mesozoic rocks, tints of blue and purple for Paleozoic rocks, and tints of russet and red for pre-Cambrian rocks (and for igneous rocks of different ages). The primary structural features of the rock types are depicted, as far as practicable, by conventional patterns. Variations of dot and line patterns are used for sedimentary rocks; wavy lines for metamorphic rocks; and checks, crosses or crystal-like patterns for igneous rocks. A legend of structural symbols is included on most geologic maps. One of the most important symbols is the dip-strike symbol, which indicates the direction of strike of a rock bed, fault, fold, or flow structure; the direction of dip; and the angle of dip from the horizontal in degrees.

Geologic maps often carry one or more geologic sections. The section is a graphic representation of the disposition of the various strata in depth along an arbitrary line usually marked on the map. Geologic sections are somewhat hypothetical, and must be used with caution. The vertical scale is nearly always exaggerated. Sections prepared solely from surface data may easily be erroneous; sections prepared from boring records or mining evidence are more reliable. A section compiled to show the sequence and stratigraphic relations of the rock units in one locality is called a columnar section; it shows only the succession of strata and not the structure of the beds as does the geologic section.

There are several types of geologic maps. A map showing a plan view of the bedrock from the geological standpoint in the area is a *bedrock* or *areal geologic map*. Such a map indicates the boundaries of the visible formations, and the inferred distribution of those units covered by soil or plant growth, and usually includes one or more geologic sections. Except for indicating thick deposits of alluvial, glacial, or windblown materials, areal maps do not show soil or unconsolidated mantle. In areas of complex geology where exposures of bedrock are scarce, the location of the contacts between formations is often indicated as approximate or hypothetical. *Surficial geologic maps* differentiate the unconsolidated surface materials of the area according to their geologic categories, such as stream alluvium, glacial gravel, and windblown sand. These maps indicate the areal extent, characteristics, and geologic age of the surface materials. Areal (bedrock) geologic maps of moderately deformed areas often carry enough structural symbols to provide an understanding of the structural geology of that region; in many instances generalized subsurface structure can be deduced from the distribution of the formations on the map. In highly complex areas, where a great amount of structural data is necessary for an interpretation of the geology, special *structural geologic maps* are prepared.

In addition to giving the geologic age of the mapped rocks, some maps briefly describe the rocks. Many maps, however, lack a lithologic description. The experienced geologist can make certain assumptions or generalizations from the age of the rock alone by making analogies with other areas. For more certain identification of the lithology and for details, geologic literature on the whole area must be consulted. Engineering information can be obtained from geologic maps if the user possesses a knowledge of the fundamentals of geology and an understanding of how engineers use geologic facts in design and construction. By a study of the basic geologic map, together with all the collateral geologic data that pertain to the area shown, it is possible to prepare a special map that interprets the geology in terms of construction materials. Similarly, foundation and excavation conditions, as well as surface and ground-water data, can be interpreted from geologic maps. Such information is valuable in preliminary planning activities, but is not a substitute for detailed field investigations in the feasibility and specifications stages.

The U.S. Geological Survey now publishes a series of geologic quadrangle maps which replaces the earlier folios of the Geologic Atlas of the United States, published from 1894 to 1945. The new series consists of geologic maps supplemented where possible by structural sections and other graphic means of presenting geologic data, and accompanied by a brief explanatory text to make the maps useful for general scientific, economic, and engineering purposes. Full descriptions of the areas shown on these maps and detailed interpretations of geologic history are reserved for other channels of communication, such as the bulletins and professional papers of the Geological Survey.[2] Separate maps of some quadrangles are published in the geologic quadrangle map series under such titles as "Economic Geology," "Surficial Geology," and "Engineering Geology." Each map is issued in two forms: flat for filing in large map cases, and folded for use in the field.

There are several geologic maps of special interest to planners of hydraulic structures. There is a series of maps resulting from geologic mapping and general resources investigations conducted by the Geological Survey as part of the Department of the Interior plan for study and development of the Missouri River Basin. These include maps showing construction materials and nonmetallic mineral resources, including sand and gravel deposits of several of the States in the Missouri River Basin. The Geological Survey has also published a bound set of six maps entitled "Interpreting Geologic Maps for Engineering Purposes, Holidaysburg Quadrangle, Pennsylvania," 1953, containing examples of how geologic maps are used to solve engineering problems including a problem of selection of a damsite.

Figure 28 shows the status of geologic mapping being accomplished in the United States by the U.S. Geological Survey. More detailed information about published geologic maps for individual States is given in the series of geologic map indices available from the Geological Survey. Each published geologic map is outlined on a State base map, with an explanatory key giving the source and date of publication, the author and the scale. The attention of all those engaged in searching

---

[2] Publications of the U.S. Geological Survey are available from the U.S. Government Printing Office, Superintendent of Documents, Washington, D.C. 20402.

for geological information is called to the Directory of Geological Material in North America.[3] This directory includes comprehensive lists of sources of available maps, charts, air photos, logs, cores, etc., for each State and territory of the United States, as well as for provinces in other countries of North America.[4]

(c) *Agricultural Soil Maps.*—A large portion of the United States has been surveyed by the Department of Agriculture. These investigations are surficial, extending to depths up to 6 feet, and consist of classifying soils according to color, structure, texture, physical constitution, chemical composition, biological characteristics, and morphology. The Department of Agriculture publishes reports of their surveys in which the different soils are described in detail, and their suitability for various crops is given. Included in each report is a map of the area surveyed, usually a county, showing by the pedological classification the various types of soils that occur. In addition to the county soil maps, there are many areas in which individual farms are mapped using the same system of soil classification.

Figure 29 shows the extent of published agricultural soil mapping in the United States. These surveys (if in print) are available for purchase from the Superintendent of Documents, Washington, D.C. Out-of-print maps and other unpublished surveys may be available for examination from the U.S. Department of Agriculture, county extension agents, colleges, universities, and libraries. Soil surveys using the agricultural soil classification have been made in many of the river basins in the 17 Western States for the purpose of classifying land for irrigation based on physical and chemical criteria. Inquiry should be made at the local project offices of the Bureau of Reclamation for the availability of soils data for these areas.

In order to apply agricultural soil maps to explorations of foundations and construction materials, some knowledge of the pedological system of classification is necessary. This system recognizes the fact that movement of water from the surface of the soil downward leaches inorganic colloids and soluble material from the upper portion to create a soil profile. The depth of leaching action depends on the amount of water, the permeability of the soil, and the length of time involved. This action produces distinct layers of soil. The surface layer is lacking in the fines which the subsurface layer has accumulated in addition to its original fines. The soil beneath the subsurface layer has been little

---

[3] Howell, J. V., and Levorsen, A. I., Directory of Geological Material in North America, American Geological Institute, second edition, 1957, 208 pp.

[4] Also refer to "Abstracts of North American Geology," a monthly publication of the Geological Survey, and to "Geologic and Water Supply Reports and Maps," compiled by the Geological Survey for each State.

affected by water and remains essentially unchanged. These three layers are designated from the surface downward as the A horizon, the B horizon, and the C horizon. In detailed classifications these horizons may be subdivided into A_1, A_2, etc.

The soils of the United States are first divided into main divisions depending on the cause of profile development and on its magnitude. The main soil divisions are further divided into "suborders" and then into "great soil groups" on the basis of the combined effect of climate, vegetation, and topography. Within each "great soil group" the soils are divided into *soil series,* each of which has the same age, climate, vegetation, relief, and parent material. According to this system of classification. all soil profiles of a certain soil series are similar in all respects, with the exception of a variation in the texture, or grain size, of the topsoil or A horizon. The soil series were originally named after a town, county, stream, or similar geographical source where the soil series was first identified.

The final classification unit, which is called the soil type, is made up of the soil series name plus the textural classification of the topsoil or A horizon. This textural classification is different from the Unified Soil Classification System used by the Bureau of Reclamation and other organizations for engineering purposes (part A of chapter I). Figure 30 shows the textural classification of soils of the U.S. Department of Agriculture.[5] The chart shows the terminology used for different percentages of clay (defined as particles smaller than 0.002 mm.), silt (0.002 to 0.05 mm.), and sand .(0.05 to 2.0 mm.). Note the use of the term "loam" which is defined in the chart as a mixture of sand, silt, and clay within certain percentage limits. Other terms used as adjectives to the names obtained in the triangle classification are: "gravelly" for rounded and subrounded particles from 2 mm. to 3 inches, "cherty" for gravel sizes of chert, and "stony" for sizes greater than 10 inches.

The textural classification given as part of the soil name on the agricultural soil map refers to the material in the A horizon only; hence, this is not of much value to the engineer who is interested in the entire soil profile. The combination of soil series name and textural classification to form a *soil type,* however, provides a considerable amount of significant data. For each soil series the texture, degree of compaction, presence or absence of hardpan or rock, lithology of the parent material, and chemical composition can be obtained.

---

[5] Adapted from "Supplement to Soil Classification System (7th Approximation)," Soil Conservation Service, U.S. Department of Agriculture, Second Printing, March 1967.

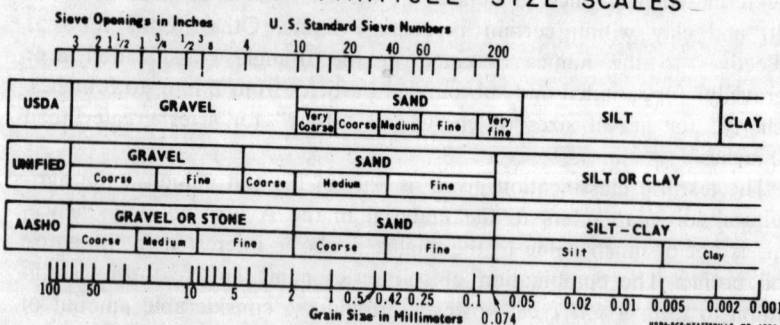

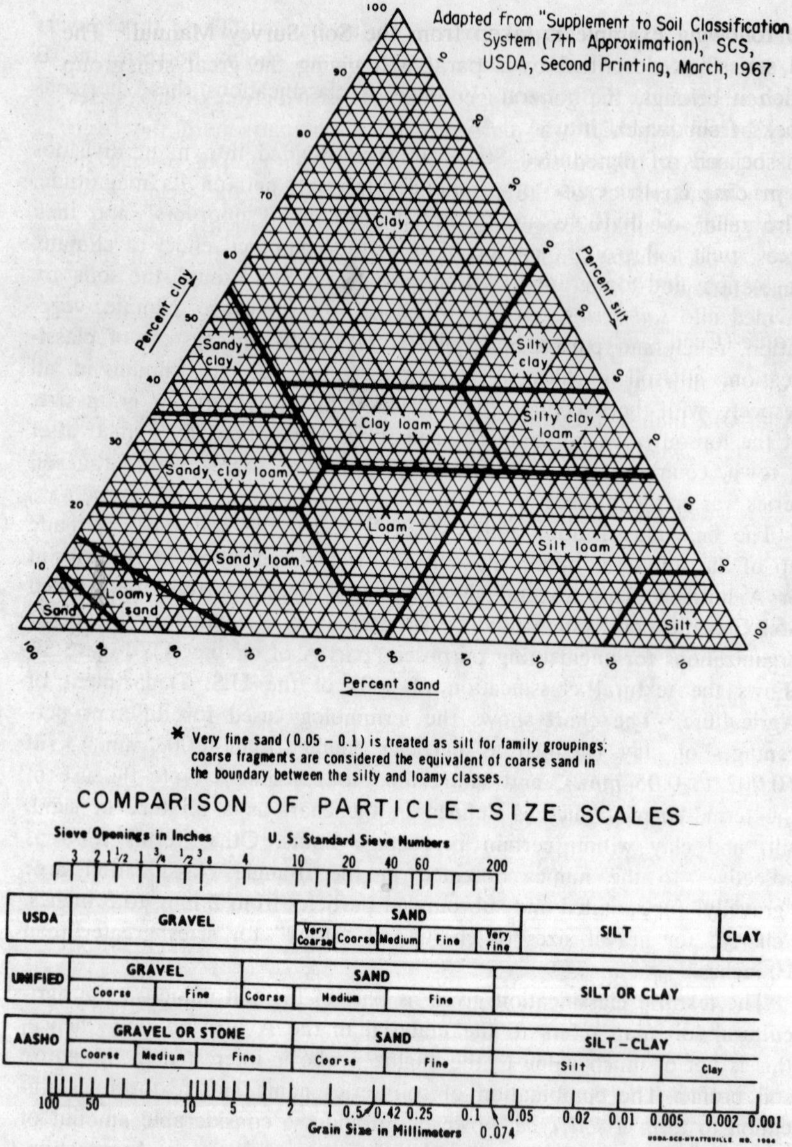

Adapted from "Supplement to Soil Classification System (7th Approximation)," SCS, USDA, Second Printing, March, 1967
Percent clay
Percent silt
Percent sand
Clay
Sandy clay
Silty clay
Clay loam
Silty clay loam
Sandy clay loam
Loam
Silt loam
Sandy loam
Loamy sand
Sand
Silt
* Very fine sand (0.05 – 0.1) is treated as silt for family groupings; coarse fragments are considered the equivalent of coarse sand in the boundary between the silty and loamy classes.
COMPARISON OF PARTICLE-SIZE SCALES
Sieve Openings in Inches
U. S. Standard Sieve Numbers
USDA
GRAVEL
SAND
Very Coarse
Coarse
Medium
Fine
Very fine
SILT
CLAY
UNIFIED
GRAVEL
Coarse
Fine
SAND
Coarse
Medium
Fine
SILT OR CLAY
AASHO
GRAVEL OR STONE
Coarse
Medium
Fine
SAND
Coarse
Fine
SILT-CLAY
Silt
Clay
Grain Size in Millimeters
USDA-SCS-HYATTSVILLE MD. 1964

Figure 30.—Soil triangle of the basic soil textural classes. (U.S. Soil Conservation Service.) 288–D–2782.

The following example is taken from the Soil Survey Manual.[6] The "Cecil series'" is described by a paragraph giving the great soil group to which it belongs, the general geographical distribution of the series, the rocks from which it was derived, and a comparison of the series with associated or related soil series. Additional paragraphs discuss the range in characteristics of the principal soil types of the Cecil series, and also relief, drainage, vegetation, land use, distribution of the series by States, type location, and remarks. A soil profile for the Cecil series is given as follows:

Soil profile (Cecil sandy loam):

A_{00} A thin layer of leaves and pine needles.

A_1 0–2 inches, brownish-gray, very friable sandy loam with fine, weak, crumb structure; strongly acid. 1 to 4 inches thick.

A_2 2–8 inches, weak-yellow to light yellowish-brown, nearly loose or very friable sandy loam; strongly acid. 4 to 10 inches thick.

B_1 8–10 inches, weak reddish-brown to strong-brown friable heavy sandy loam or light sandy clay loam with medium granular structure; strongly acid. 2 to 4 inches thick.

B_2 10–38 inches, moderate to strong reddish-brown clay that is plastic when wet, very firm when moist, and very hard when dry. The clay has a moderately blocky structure and contains some white sand grains and small mica flakes; strongly acid. 20 to 36 inches thick.

B_3 38–60 inches, light to moderately reddish-brown clay loam with mottles or splotches of yellow; firm to friable when moist. The soil contains enough small mica flakes to make it feel slick when rubbed between the fingers; it has a weak, coarse, blocky structure and is strongly acid. 10 to 30 inches thick.

C   60 inches plus, mottled or splotched light reddish-brown, yellowish-brown, light-gray, and black friable disintegrated rock material in which there is usually much mica; strongly acid. 20 to 60 inches thick.

The limitations of agricultural soil classifications for engineering purposes can be judged by the foregoing example. From the engineering point of view, information of this type is qualitative rather than quantitative, but if properly evaluated it can often be used to advantage in

---

[6] Soil Survey Manual, U.S. Department of Agriculture Handbook No. 18, 1951, p. 473.

the reconnaissance stage and in planning subsurface explorations for dams and other structures.

The agricultural soil survey report is designed to include information useful to the fa ner and to the agricultural community. However, apart from the soil maps and soil profile descriptions contained in these reports, other information of great value in planning of hydraulic structures is included. The reports discuss topography; ground surface conditions; obstructions to movement on the ground; natural vegetation; size of farms; land utilization; farm practice and cropping systems; meteorological data; drainage; flood danger; irrigation; water supply and quality; nearness to towns, roads, and railroads; electric power; and similar data.

(d) *Air Photos.*—An air photo or aerial photograph is a pictorial representation of a portion of the earth's surface taken from the air. It may be a vertical photograph in which the axis of the camera is vertical, or nearly so, or an oblique photograph where the axis of the camera is more or less inclined. High oblique photographs include the horizon; low obliques do not. The vertical photograph is commonly used as the basis for topographic mapping, agricultural soil mapping, and geological interpretations. In addition to black and white (panchromatic) aerial photography available as contact prints, it is possible to obtain under contract with commercial aerial photographers color photography, either positive transparencies or opaque prints, black and white infra-red, or color infra-red. By use of appropriate filters, photography may be obtained ranging from ultraviolet to near infra-red. Multiband cameras employing four to nine lenses and various lens, filter, and film combinations make possible photography within narrow wavelength bands throughout this range to emphasize various soil, moisture, temperature, and vegetation effects as an aid to photo interpretation. Other remote sensors are available to record and process data outside of the photographic range through thermal, infra-red, microwave, and radar wavelengths. Such phenomena as differences in the earth's gravity and magnetic properties may also be measured to afford additional interpretive tools.

Except where dense forest cover obscures large areas from view, the air photo reveals every natural and manmade detail on the surface of the ground within the resolution of the film. Relationships are exposed which could not be found on the ground under normal or routine surface investigation, no matter how careful the examination. Identification of features shown on the photo is facilitated by stereoscopic examination. The features are then interpreted for a particular purpose, such as geology, land utilization, or engineering characteristics. The experience and training of the engineer will determine his utilization of aerial photographs. Knowledge of the elements of geology and of soil science will assist him in interpreting air photos for engineering uses. Air photos are very often used for locating areas to be examined and sampled in the field and as substitutes for maps.

Virtually the entire area of the United States has been covered by aerial photography. A 28- by 42-inch index map of the United States is available from the U.S. Geological Survey, Washington, D.C. 20242. This map shows which of seven Government agencies can provide photographic prints for particular areas. When ordering photographs, specify contact prints or enlargements, glossy, matte finish, or Cronapaque; location should be given by range, township, and section, latitude and longitude, State and county, or shown on an enclosed index map of the area. Where possible, use the airphoto index to determine project symbol, film roll number, and exposure number to expedite request. Stereoscopic coverage should be requested for most uses. Aerial mosaics covering roughly 25 percent of the area of the United States are also available. A mosaic is an assemblage of individual aerial photographs fitted together systematically to form a composite view of an entire area of the earth's surface covered by the photographs. An index map showing the status of aerial mosaics of the United States, including the coverage and the agencies holding mosaic negatives, is available from the Map Information Office, U.S. Geological Survey.

Equipment is now available commercially to produce orthophotographs corrected throughout by strip photoscanning scale distortion due to topography, overprinted with contours, if desired.

Air photo interpretation of earth materials and geologic features is relatively simple and straightforward, but requires experience. The diagnostic features include terrain position, topography, drainage and erosional features, color tones, and vegetative cover. Interpretation is limited mainly to surface and near-surface conditions. There are special cases, however, where features on the photograph permit reliable predictions to be made of deep, underground conditions. Although interpretation can be rendered from any sharp photograph, the scale is a limiting factor, since small-scale photos limit the amount of detailed information that can be obtained. The scale of 1:20,000 has been found satisfactory for engineering and geologic interpretation of surface materials. Large-image photos often have application to highly detailed work, such as for reservoir clearing estimates and for geologic reconnaissance mapping of damsites.

Air photos can be used to identify certain terrain types and land forms. These topographic features are described in section 31. Stereoscopic photo inspection of an area, taking particular note of regional topography, local terrain features and drainage conditions, will usually suffice to identify the common terrain type. This permits the possible range in the soil and rock materials to be anticipated, and their characteristics to be defined within broad limits.

Geologic features that may be highly significant to the location or performance of engineering structures sometimes can be identified from air photos. In many instances these features can be more readily identified on the air photo than on the ground. It must be recognized, however, that air photo interpretation is applicable only to those features which develop recognizable surface expressions, such as drainage patterns, old river channels, and alinement of ridges or valleys. Joint systems, landslides, fault zones, lineations, folds, and other structural features sometimes are identified quickly in an aerial photo, whereas it may be difficult to find them on the ground. The importance of these items in the location of a dam and its appurtenant works is obvious. The general attitude, bedding, and jointing of exposed rock strata, as well as the presence of dikes and intrusions, often can be interpreted in air photos. Such information is valuable in appraising the possibilities of landslides into open cuts and of seepage losses in reservoirs.

Drainage patterns, particularly their type and density, provide an indication of the relative permeability of the earth materials. A dense, finely divided drainage pattern indicates an impervious soil area with high runoff and low infiltration. In contrast, the absence of a surface drainage pattern indicates a soil area with low runoff and high infiltration, provided the area is not a desert. The surface drainage pattern in areas of high water table has only limited significance as an indicator of the earth materials present. Definite alinements in the drainage pattern usually indicate control by local geologic structure.

Erosional features have significance in that they often reflect the textural characteristics of the exposed materials. Short, steep, V-shaped gullies with uniform gradients are associated with granular materials; long gullies with uniform gradients of rounded cross-sectional slopes are associated with fine-grained plastic soils. Silts and sand-clay materials usually exhibit gullies having U-shaped cross sections and compound gradients. The significance of gullies as an indicator of soil texture is modified by extreme climatic influence, such as in arid regions where "box" gullies seem to prevail, irrespective of soil texture. Regardless of the climatic influence, however, changes in the gradient or cross section of gullies or changes in the surface slope of eroded surfaces may indicate a change in the exposed soil, rock texture, or geologic structure.

Color tones (relative photographic gray values) have a general significance in that they reflect the soil moisture conditions and often reveal the relative position of the ground-water table. Light color tones are usually associated with well-drained soils, such as gravels, sands, and silts with ground-water levels well below the ground surface. Dark color tones usually indicate poorly drained organic clays and silty clays with ground-water levels near the ground surface. The significance of soil color in air photos must be appraised from the overall color pattern, since some variation may be expected in the photographic tone quality of individual air photos. It is also necessary to exclude, visually, the color tones produced by vegetative cover.

Vegetative cover is significant in that the vegetative patterns produced in the air photos often reflect the nature of soil and moisture conditions. Also, a change in vegetative pattern may indicate a change in the type or texture of the underlying bedrock. The use of vegetative patterns as an indicator of soil conditions will prove most useful in extreme climates, such as in arctic, tropical, and arid regions where the combination of soil and climate is selective of vegetative growth. In arid regions the pattern of vegetation can be used to distinguish between high and low alkali soils and between high and low ground-water levels. The effective use of vegetation as an air photo materials indicator requires a limited amount of field correlation.

Figures 31 and 32 are examples of identifiable geologic features determined from air photos by an experienced interpreter using stereoscopic procedures. Examples of typical land forms on air photos are given in section 31.

31. Surface Exploration.—(a) *General.*—A relationship between topographic features or land forms and the characteristics of the subsurface soils has been shown repeatedly. Thus, the ability to recognize terrain features on maps, on air photos, and during field reconnaissance, combined with an elementary understanding of geological processes, can be of great assistance in locating sources of construction materials and in making a general appraisal of foundation conditions.

The mechanisms which develop soil deposits are water action, ice action, and wind action for transported soils; and the mechanical-chemical action of weathering for residual soils. A soil deposit may be the product of several of these mechanisms. For the transported soils, each type of action tends to produce a group of typical land forms, modified to some extent by the nature of the parent rock. Soils found in similar locations within similar land forms usually have the same physical properties. The personnel responsible for explorations for foundations and materials for hydraulic structures should become familiar with land forms and with the soils associated with them. Such knowledge is of great assistance during the reconnaissance stage of investigations, and it is useful in controlling the extent of investigation for feasibility and specifications designs.

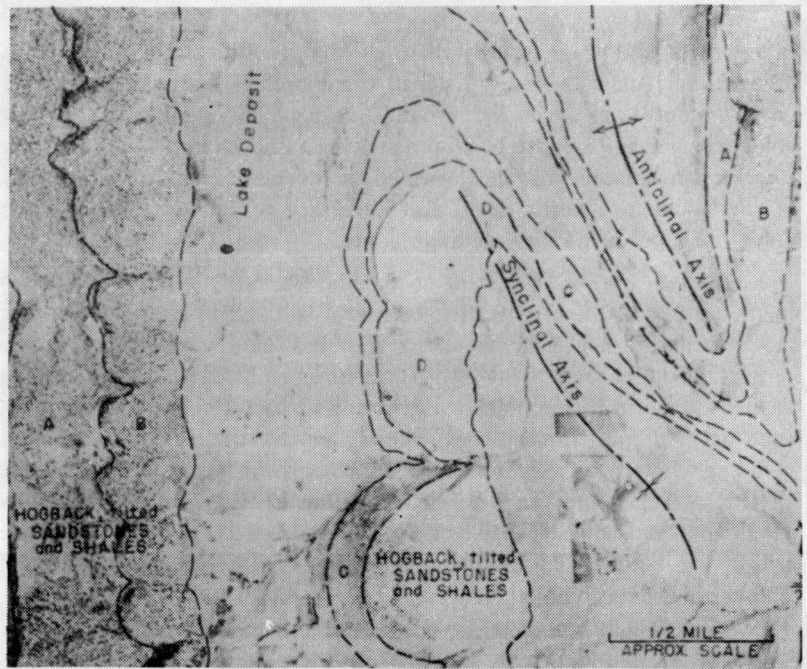

Figure 31.—Rock strata illustrating folding in sedimentary rocks. (A) Satanka formation, (B) Lyons formation, (C) Morrison formation, and (D) Lower and Middle Dakota formation. (Photo by U.S. Forest Service.) PX–D–16265.

(b) *Fluvial Soils.—*

(1) *Definition.*—Soils whose properties are affected predominantly by the action of water to which they have been subjected are designated fluvial soils. Their common characteristic is roundness of individual grains. Frequently, there is considerable sorting action, so that a deposit is likely to be stratified or lensed. Individual strata may be thin or thick, but the material in each stratum will have a small range of grain sizes. The three principal types of fluvial soils, reflecting the water velocity of deposition, are identified as torrential outwash, valley fill, and lakebeds.

(2) *Torrential outwash.*—The typical land forms of this type are alluvial cones and alluvial fans. They vary in size and character from small, steeply sloping deposits of coarse rock fragments to gently sloping plains of fine-grained alluvium several square miles in area. The deposition results from the abrupt flattening of the stream gradient that occurs at the juncture of mountainous terrain and adjacent valleys or plains. (See fig. 33.) The coarser material is deposited first; hence, it is found on the steeper slopes at the

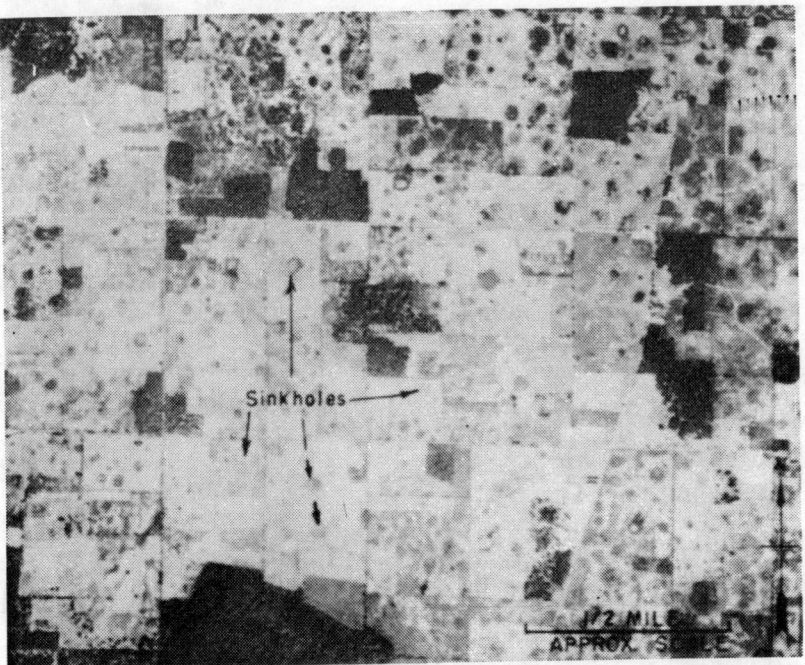

Figure 32.—Sinkhole plain indicating deep plastic soils over cavernous lime-stone, developed in humid climate. (U.S. Agricultural Stabilization and Conservation Service.) PX–D–16264.

head of the fan, while the finer material is carried to the outer edges. In arid climates where mechanical rather than chemical weathering predominates, the cones and fans are composed largely of rock fragments, gravel, sand, and silt. In humid climates where the land forms have less steep slopes, the material is expected to contain much more sand, silt, and clay.

Sands and gravels from these deposits are generally subrounded to subangular in shape, reflecting movement over relatively short distances, and the deposits have only poorly developed stratification. The torrential outwash deposits are likely sources of sand and gravel for pervious and semipervious embankment materials and for concrete aggregate. The presence of boulders is likely to limit their usefulness for some types of fill materials. The soils are typically skip-graded, resulting in a GP or SP classification. Because this type of deposit is consolidated only by its own weight, settlement should be anticipated in fine-grained soils when used as foundations for hydraulic structures. Normally, torrential outwash deposits do not provide satisfactory abutments for dams. If it is necessary to

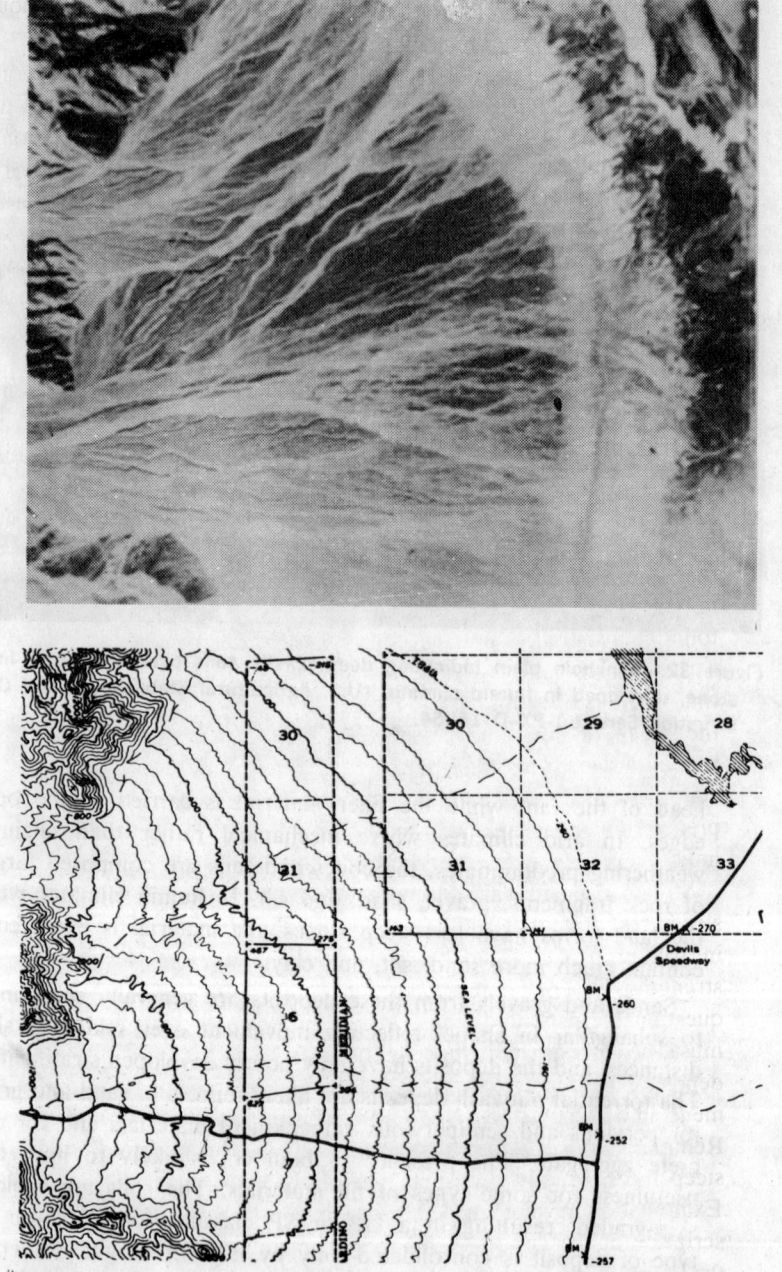

Figure 33.—Aerial view and topography of an alluvial fan, a potential source of sand and gravel. (Photo by U.S. Geological Survey.) PX–D–16262.

locate a dam in the vicinity of such a deposit, the dam should be placed along the upstream edge of the fan.

(3) *Valley fill.*—Valley fill or flood-plain deposits are generally finer, more stratified, and better sorted than are torrential outwash deposits. The degree of variation from the latter depends largely on the volume and on the gradient of the stream. The surface of these stream deposits is nearly flat. The nature of the materials in the deposit can be deduced by the characteristics of the stream. Braided streams usually indicate the presence of sand and gravel, whereas meandering streams in broad valleys are commonly associated with fine-grained soils (silts and clays).

Flood-plain deposits of sand and gravel are common sources of concrete aggregate and pervious shell materials for embankments. The soils in the various strata of river deposits may range from pervious to impervious; hence, the permeability of the resulting material sometimes can be influenced appreciably by the choice of depth of cut. Presence of a high water table is a major difficulty in the use of these deposits, especially as a source of impervious material. Also, removal of materials from the reservoir floor just upstream from a damsite may be undesirable when a positive foundation cutoff is not feasible. When considering borrowing from a river deposit downstream from a dam, it should be remembered that such operations may change the tailwater characteristics of the stream channel and that the spillway and outlet works will have to be designed for the modified channel conditions. If tailwater conditions will be affected, borrow operations must produce a predetermined channel and explorations for the specifications stage must accurately define conditions within the channel.

Stream deposits vary in competency as foundations for hydraulic structures. Potential difficulties include high water table, variation in soil properties, seepage, consolidation, and possibly low shear strengths. Except for minor structures valley fill deposits are usually questionable as foundations, hence their depths and characteristics must be investigated thoroughly. An important type of stream deposit is the terrace. It represents an earlier stage of valley development into which the river subsequently has become entrenched. Remnants of such deposits are recognized by their flat tops and steep faces, usually persistent over an extended reach of the valley. Examination of the eroded faces facilitates classification and description of the deposits, and the extent of the drainage network on the face is helpful in determining relative permeability. Free-draining material has almost no lateral erosion channels, whereas impervious clays are finely gullied laterally. Terrace sands and gravels were laid down in the geologic past. These terraces are found along streams throughout the United States and are prevalent in the glaciated regions of the northern sections. Sands and gravels in terrace deposits usually occur in layers and are well-graded. They provide excellent sources of construction materials. Figure 34 is an air photo and topographic map showing river alluvium and terrace deposits.

(4) *Lakebeds.*—Lake sediments, or lacustrine deposits, are the result of sedimentation in still water. Except near the edges of the deposits where alluvial influences are important, the materials are very likely to be fine-grained silt and clay. The stratification is frequently so fine that the materials appear to be massive in structure. Lacustrine deposits are recognizable by their flat surfaces surrounded by high ground. The materials they contain are likely to be impervious, compressible, and low in shear strength. Their principal use is for impervious cores of earthfill dams, for impervious linings for reservoirs and canals, and for low embankments. Moisture control in these soils is usually a problem, since the water content is very difficult to change.

Lake sediments generally provide poor foundations for structures. Their characteristics may be expected to be so questionable that special laboratory and field testing may be required even during the reconnaissance stage. It is desirable to obtain a specialist's advice whenever it appears necessary to locate structures on this type of foundation.

(c) *Glacial Deposits.*—

(1) *Definition.*—The results of the advances and retreats of the great North American continental ice sheets during glacial times are represented by recognizable land forms which are important sources of construction materials and which may be encountered in the foundations of hydraulic structures. Smaller scale evidences of ice action are found in high mountain valleys of the Rocky Mountains and the Sierra Nevadas; in some instances the glaciers still exist. Glacial deposits (glacial drift) are generally heterogeneous and erratic in nature; hence, they are difficult to explore economically. They contain a wide range of particle sizes, from clay or silt up to huge boulders; and the particle shapes of the coarse grains are typically subrounded or subangular, sometimes with flat faces. Deposits of the glacier proper can be distinguished from deposits formed by the glacier melt water.

(2) *Morainal deposits.*—Glacial till is that part of the glacial drift deposited directly from the ice with little or no transportation

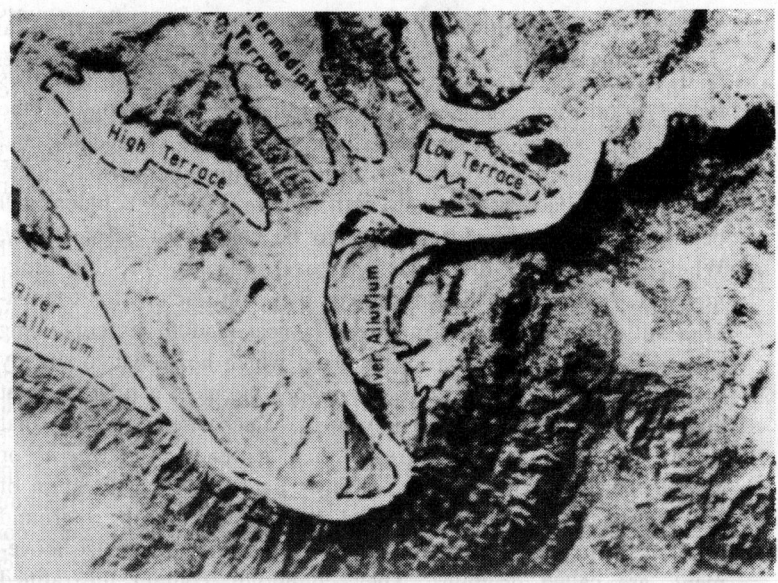

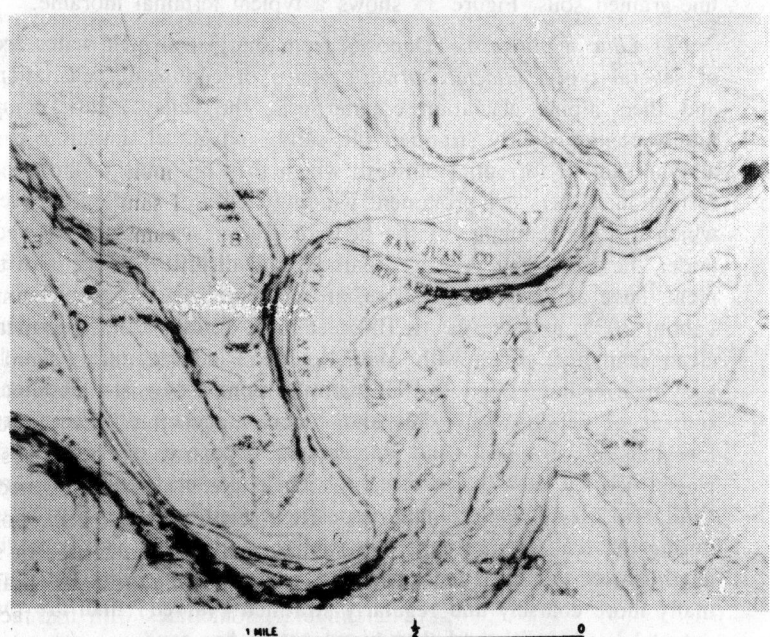

Figure 34.—Aerial view and topography of stream deposit showing river alluvium and three levels of gravel terraces. (Photo by U.S. Geological Survey.) PX–D–16259.

by water. It consists of a heterogeneous mixture of varying amounts of boulders, cobbles, gravel, and sand in a generally impervious matrix of fines. Gradation, type of rock, and mineral materials found in till vary considerably, depending on the geology of the terrain over which the ice moved and the degree of leaching and chemical weathering. Glacial tills usually produce impervious materials with satisfactory shear strength, but removal of boulders will be necessary in order for the soil to be compacted satisfactorily. Where morainal deposits have been overridden by ice, the resulting fairly high inplace density makes them satisfactory for foundations of many hydraulic structures. Typical land forms containing till are the *ground moraine* or till plain which has a flat to slightly undulating surface; the *end (or terminal) moraine,* a ridge at right angles to the direction of ice movement, which often curves so that its center is farther downstream than its ends; and *lateral* or *medial moraines* which occur as ridges parallel to the direction of ice movement. Low, cigar-shaped hills occurring on a ground moraine, with their long axis parallel to the direction of ice movement, are called drumlins. They commonly contain unstratified fine-grained soils. Figure 35 shows a typical terminal moraine.

(3) *Glacial outwash.*—Deposits from the glacial melt water are of several types. *Glacial outwash plains* of continental glaciation and their alpine glaciation counterparts, the *valley trains,* commonly contain poorly stratified silt, sand, and gravel similar to the alluvial fans of torrential outwash which they resemble in mode of formation. *Eskers* are prominent winding ridges of sand and gravel which are the remnants of the beds of glacial streams that flowed under the ice. They generally parallel the direction of ice movement, have an irregular crestline, are characterized by steep flanks (about 30°), and are 20 to 100 feet high. Eskers usually contain clean sand and gravel with some boulders and silty strata which are in irregular, poor to fair stratification. They are excellent sources of pervious materials and concrete aggregate. *Kames* are low hills of stratified sand and gravel deposited on or against glacial ice by melt-water streams. Their contents and uses are similar to eskers. Glacial lake deposits formed in temporary lakes during the Ice Áge are generally similar in character and in engineering uses to fluvial lacustrine deposits. However, they are normally more coarsely and regularly stratified (varved) than are the recent lake deposits, and they may contain fine sand.

(d) *Eolian Deposits.*—Soils deposited by the wind are known as eolian deposits. The two principal classes that are readily identifiable

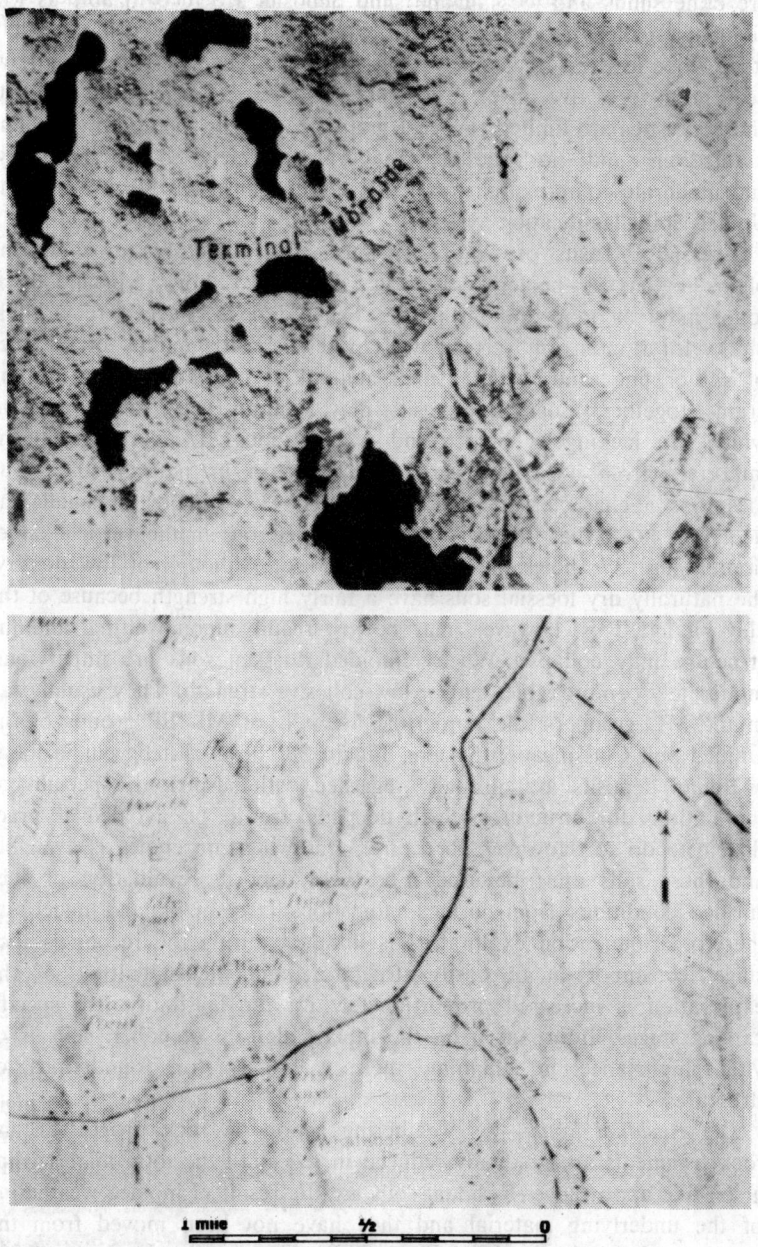

Figure 35.—Aerial view and topography of terminal moraine of continental glaci
ation. (Photo by U.S. Agricultural Stabilization and Conservation Service.
PX-D-16260.

are dune sands and loess. Dune sand deposits are recognizable as low elongated or crescent-shaped hills, with a flat slope windward and a steep slope leeward of the prevailing winds. Usually, these deposits have very little vegetative cover. The material is very rich in quartz and its characteristics are limited range of grain size usually in the range of fine to medium sand; no cohesive strength; moderately high permeability; and moderate compressibility. They generally fall in the SP group of the Unified Soil Classification System.

Loessial deposits of windblown dust cover extensive areas in the plains regions of the temperate zone. Figure 36 shows typical loessial topography by map and air photo. These deposits have a remarkable ability for standing in vertical walls. Loess consists mainly of particles of silt or fine sand, with a small amount of clay that binds the soil grains together. Loessial deposits may contain very sandy portions which are lacking in binder and are pervious. Their recognition is important from an engineering standpoint. In its natural state, true loess has a characteristic structure formed by remnants of small vertical root holes that makes it moderately pervious in the vertical direction. Figure 37 shows the structure of loess. Although of low density, the naturally dry loessial soils have a fairly high strength because of the clay binder. This, however, may be lost readily upon wetting, and the structure may collapse. When remolded, loessial soils are impervious, moderately compressible, and of low cohesive strength. They usually fall in the ML group or the borderline ML–CL or ML–SP groups of the Unified Soil Classification System. Figure 38 shows a steep cut in loess.

Eolian deposits are normally regarded with suspicion, especially as foundations for structures. Such deposits are to be avoided if practicable to do so. However, there are areas where no choice is available and these soils must be used. For these deposits, evaluation of subsurface conditions from surface indications is complex and uncertain; therefore, foundation exploration is initiated during the reconnaissance stage for important or costly structures, and the magnitude of the exploration is increased proportionately for the feasibility and specifications stages. Information on the inplace density of eolian soils is of vital importance in planning their usefulness for foundations of structures.

(e) *Residual Soils.*—As weathering action on rock progresses, the rock fragments are gradually reduced in size until the total material has a soil-like appearance. Residual soils are the result of inplace weathering of the underlying material and they have not been moved from the location where they were formed. It is sometimes difficult to define clearly the dividing line between rock and residual soil, but for engineering purposes a material may be considered soil if it can be removed

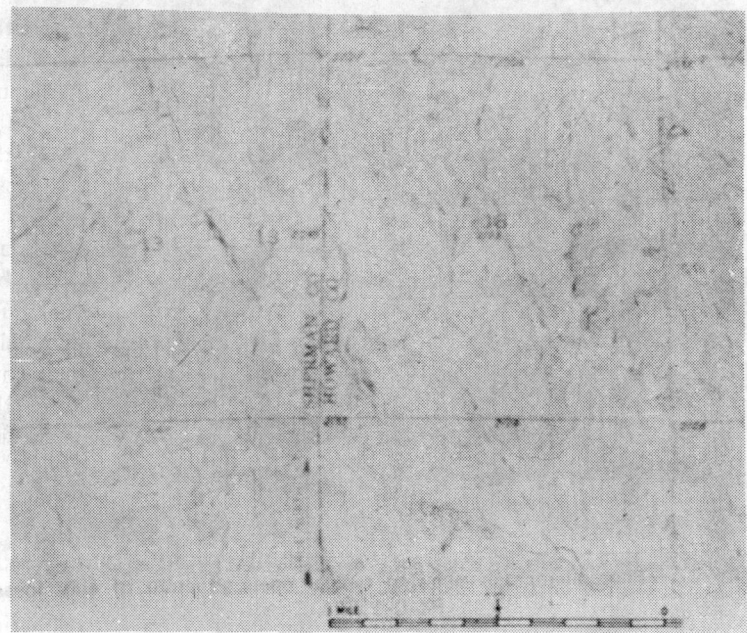

Figure 36.—Aerial view and topography of loess identified by smooth silt ridges; usually parallel, right-angle drainage patterns; and steep-sided, flat-bottomed gullies and streams. (Photo by U.S. Agricultural Stabilization and Conservation Service.) PX–D–16263.

Figure 37.—Photomicrograph showing typical open structure of silty loess. PX–D–15721.

Figure 38.—A 34-foot cut on ½:1 slope in loess formation—Franklin Canal, Nebraska. P–271–D–15713.

by commonly accepted excavating methods. Some knowledge of the engineering characteristics of residual soils can be obtained from an understanding of the bedrock from which they are derived.

It is difficult to recognize and appraise residual soils on the basis of topographic forms. Their occurrence is quite general wherever none of the other types of deposits, with their characteristic shapes, are recognizable and where the material is not clearly rock in place. Also, erosional features may be helpful in evaluating a residual deposit. Since the type of parent rock has a very pronounced influence on the character of the residual soil, the rock type should always be determined in assembling data for the appraisal of a residual deposit. The degree to which alteration has progressed largely governs the strength characteristics. Laboratory testing is required if the material appears questionable or when very large structures are planned. Identification of the clay minerals in residual soils is often necessary for an understanding of their engineering properties.

A distinguishing feature of many residual soils is that individual particles in place are angular but soft. Handling of the material during construction may appreciably reduce the grain size so that the soil as used may have entirely different characteristics than shown by standard laboratory tests of the original soil. Special laboratory testing programs are often required to determine the changes in characteristics to be expected. Full scale field test sections are sometimes appropriate before proper decisions can be made regarding utilization of such soils.

Some residual soils are entirely satisfactory for foundations or as construction materials and, in some cases, may be superior to other local soils. Others, depending on parent rock type, have such poor engineering properties that they should be avoided if possible.

32. Subsurface Exploration.—(a) *General.*—Aside from the geologic aspects, subsurface exploration is performed for three purposes: First, to determine what distinct masses of soil and rock exist in a foundation or borrow area within the area of interest; second, what the dimensions of these bodies are; and third, what their engineering properties are.

In the engineering evaluation of a foundation or borrow area, it is necessary to divide the soil structure by means of profiles or plans into a series of masses or zones within which soil properties are uniform. Materials having variable soil properties can be evaluated provided that the nature of the variation can be defined in detail. The determination of the dividing lines between what may be considered as uniform soil masses must usually be done on the basis of visual examination and requires considerable judgment. The soil classification system described in chapter I provides a satisfactory guide for the consideration of soils in a disturbed state. For the evaluation of soils in the undisturbed state, additional qualifying factors required are inplace water content and density, and stratification. Color and texture are also helpful in the delimitation of soil masses of uniform characteristics.

Occasionally, soil foundations are described in such detail that the picture presented is confused rather than clarified; however, it is better to err on the side of too much detail in foundation description since extraneous description can be eliminated but missing pertinent information cannot be added. Occasionally the only uniformity that can be found in a soil deposit is its heterogeneity. However, in many cases with careful analysis a pattern in the soil mass may be found that will assist the designer to reduce the cost of the structure involved.

The dimensions of these bodies of soil are determined by methods analogous to those used in surface surveying—that is, by making cross sections or developing the topography of the upper and lower surfaces. The preferable method used depends to some extent on the type of structure involved. Point structures or line structures (buildings or canals, pipelines, and roads) are best handled by cross sections; massive structures (dams) are best handled by topographic surfaces in addition to cross sections. Unfortunately, the problem of locating measuring points or the "breakpoints" of buried surfaces is virtually insolvable because the surface cannot be seen and the cost of covering the area with a grid of test holes is usually great. The normal solution used on an investigation starts with an estimation of the location of breakpoints based on a geologic interpretation of the subsurface. These breakpoints are then explored with test holes by successive approximations. A grid system of test holes is used only on large borrow pits and foundations for large earth dams where subsurface irregularities cannot be established by other means.

(b) *Point Structures.*—For structures such as small buildings, small pumping plants, transmission towers and bridge piers, a single test hole will often satisfy the foundation investigation requirements. Somewhat larger structures may require more holes. Where the exact location of a structure is dependent on foundation conditions, the number of holes required will increase. If for these situations two or three holes are used for preliminary exploration to establish general foundation conditions, the requirement can usually be reduced for later stages. Figure 39 shows suggested depths of preliminary exploratory holes for various point structures. Figure 40 is an example of a soil profile at a pumping plant site.

(c) *Line Structures.*—Exploration requirements for the purpose of delimiting thickness, as for the various materials in the foundations of canals, pipelines, drains, and roads, vary considerably both according to the size and importance of the structure and according to the character of the ground through which the line structure is to be located. The spacing of holes will vary, depending on the need to identify changes in subsurface conditions. Where such structures are to be located on comparatively level ground with obviously uniform soils, such as the plains areas, uplands, and beaches, relatively few holes will suffice for foundation investigation requirements; usually, holes at about 1-mile intervals for feasibility investigations and about 2,000-foot intervals for specifications stage investigations are considered maximum for canals and drains. Closer coverage is required where subsoils are likely to be erratic.

For specifications purposes on major structures, test hole requirements are quite variable; landslides, talus slopes, and outwash fans require thorough exploration. Every such geological feature should be explored with at least two holes if it affects the line structure for more than 200 feet in extent, and holes at 100-foot intervals are often necessary. Heavy cut and fill areas should also be explored with at least two test holes at roughly the ¼ and ¾ points along the length. Usually, a test hole at the point of highest fill or valley bottom is also needed. Additional off-line holes may be required for all of these features, depending

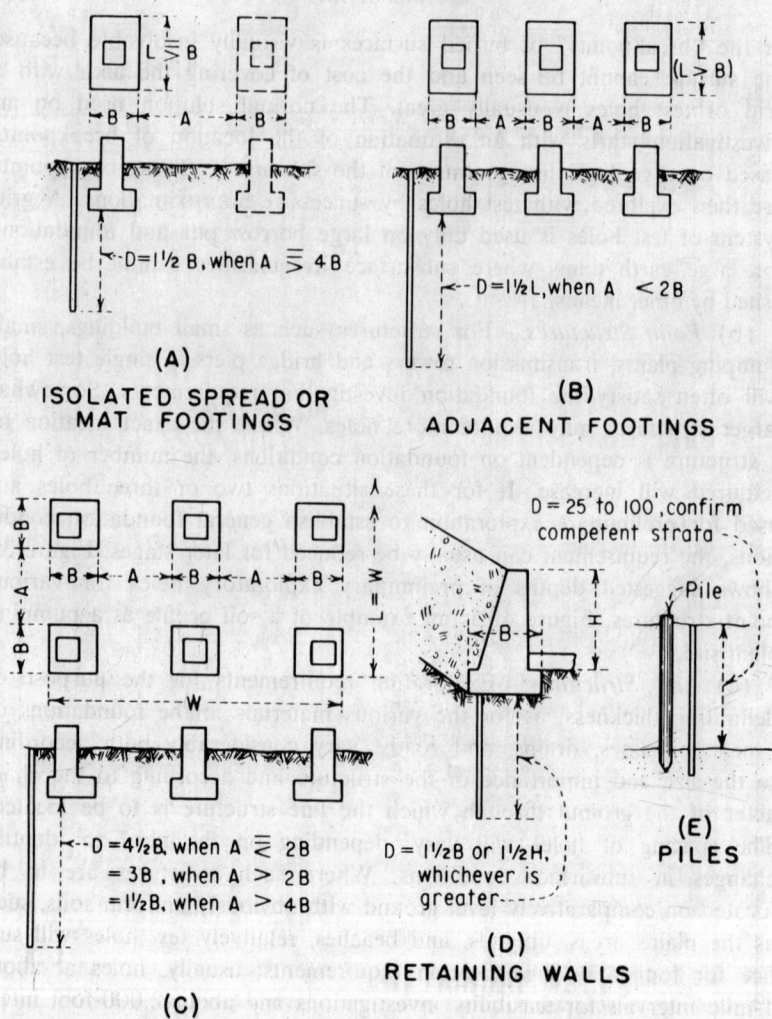

Figure 39.—Depth of preliminary exploratory holes for point structures. 101-D-211.

on the topographic, geologic, and subsurface conditions. When major items of cost are involved, the foregoing explorations may be required for feasibility estimates. Figure 41 shows suggested minimum depths of exploratory holes for major line structures. Sometimes greater depths are required to determine the character of questionable soils. Figure 42 is an example of a geologic profile along the centerline of a proposed pipeline.

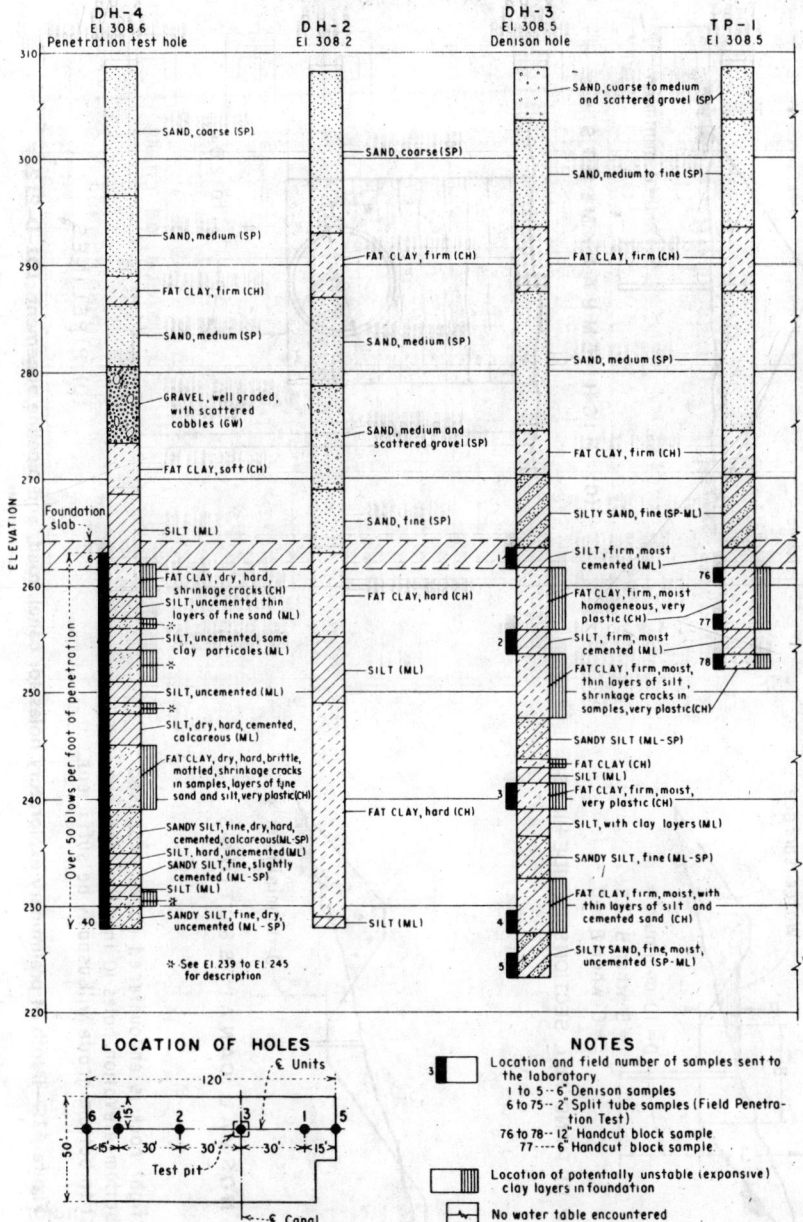

Figure 40.—Soil profile at a pumping plant. 101-D-202.

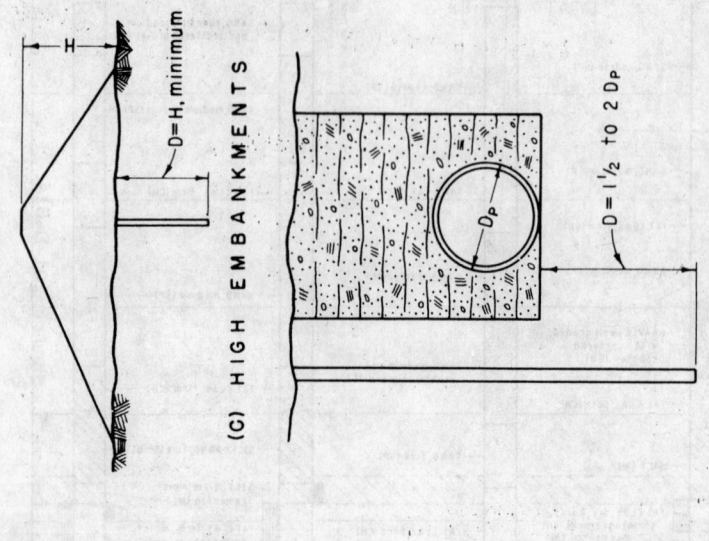

(A) DEEP CUT AND FILL SECTIONS ON SIDE HILLS

Water surface

$D=10'$ minimum *
$= B$ when $B \lessgtr C$
$= C$ when $B > C$

(B) NORMAL CANAL SECTIONS

$D=10'$ minimum *

(C) HIGH EMBANKMENTS

$D=H$, minimum

(D) PIPELINES

$D=1\frac{1}{2}$ to $2 D_P$

* If hard, tight rock is encountered above proposed
  canal bottom elevation, holes 10' into the rock, but
  at least to bottom grade will usually be sufficient.

Figure 41.—Depth of preliminary exploratory holes for canal, road, and pipeline alinement. 101-D-212.

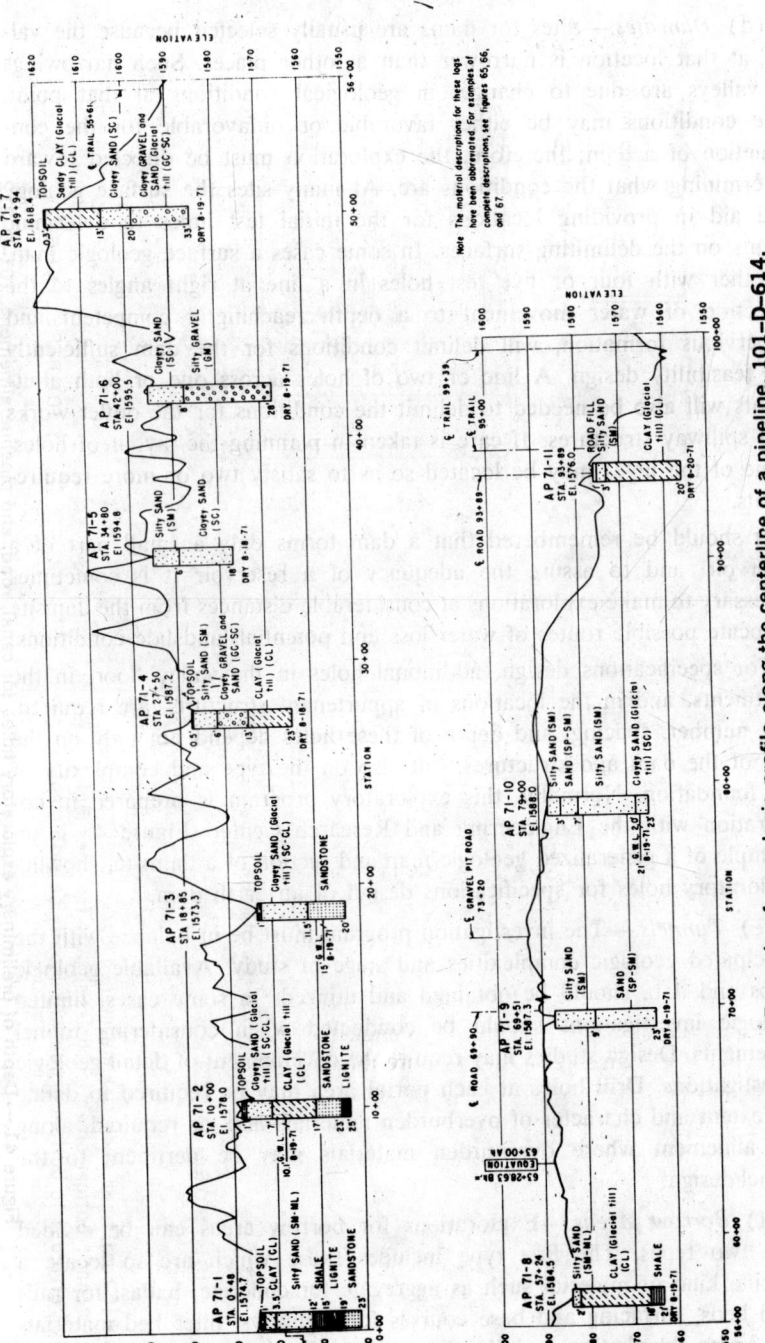

Figure 42.—Example of a geologic profile along the centerline of a pipeline. 101-D-614.

(d) *Damsites.*—Sites for dams are usually selected because the valley at that location is narrower than at other places. Such narrowings in valleys are due to changes in geological conditions at that point. The conditions may be either favorable or unfavorable for the construction of a dam; therefore, the exploration must be directed toward determining what the conditions are. At many sites the surface geology will aid in providing locations for the initial test holes to determine points on the delimiting surfaces. In some cases a surface geologic map, together with four or five test holes in a line at right angles to the direction of water movement to a depth reaching a competent and impervious formation, will delimit conditions for the dam sufficiently for feasibility design. A line or two of holes across one or both abutments will also be needed to delimit the conditions for the outlet works and spillway structures. If care is taken in planning the layout of holes, some of the holes may be located so as to satisfy two or more requirements.

It should be remembered that a dam forms only a small part of a reservoir, and to assure the adequacy of a reservoir it is sometimes necessary to make explorations at considerable distances from the damsite to locate possible routes of water loss and potential landslide conditions.

For specifications design, additional holes in the valley floor, in the abutments, and in the locations of appurtenant structures are required. The number, spacing, and depth of these holes depend not only on the size of the dam and structures, but also on the type and complexity of the foundation. Normally, this exploratory program is prepared in cooperation with the Engineering and Research Center. Figure 43 is an example of a generalized geologic map and section of a damsite, showing exploratory holes for specifications design of an earth dam.

(e) *Tunnels.*—The investigation program must be in balance with the anticipated geologic complexities and stage of study. Available geologic maps and data should be obtained and utilized. In some cases, limited geologic investigations should be conducted when considering tunnel alinements. Design studies may require the development of detail geologic investigations. Drill holes at each portal area may be required to define the extent and character of overburden materials and, as required, along the alinement where overburden materials may be pertinent to the tunnel design.

(f) *Borrow Areas.*—Explorations for borrow areas can be divided into two types. The first type includes those which are to locate a specific kind of material such as aggregate for concrete, ballast for railroad beds, surfacing and base courses for highways, filter bed materials for drains, blanketing and lining materials for canals, riprap materials

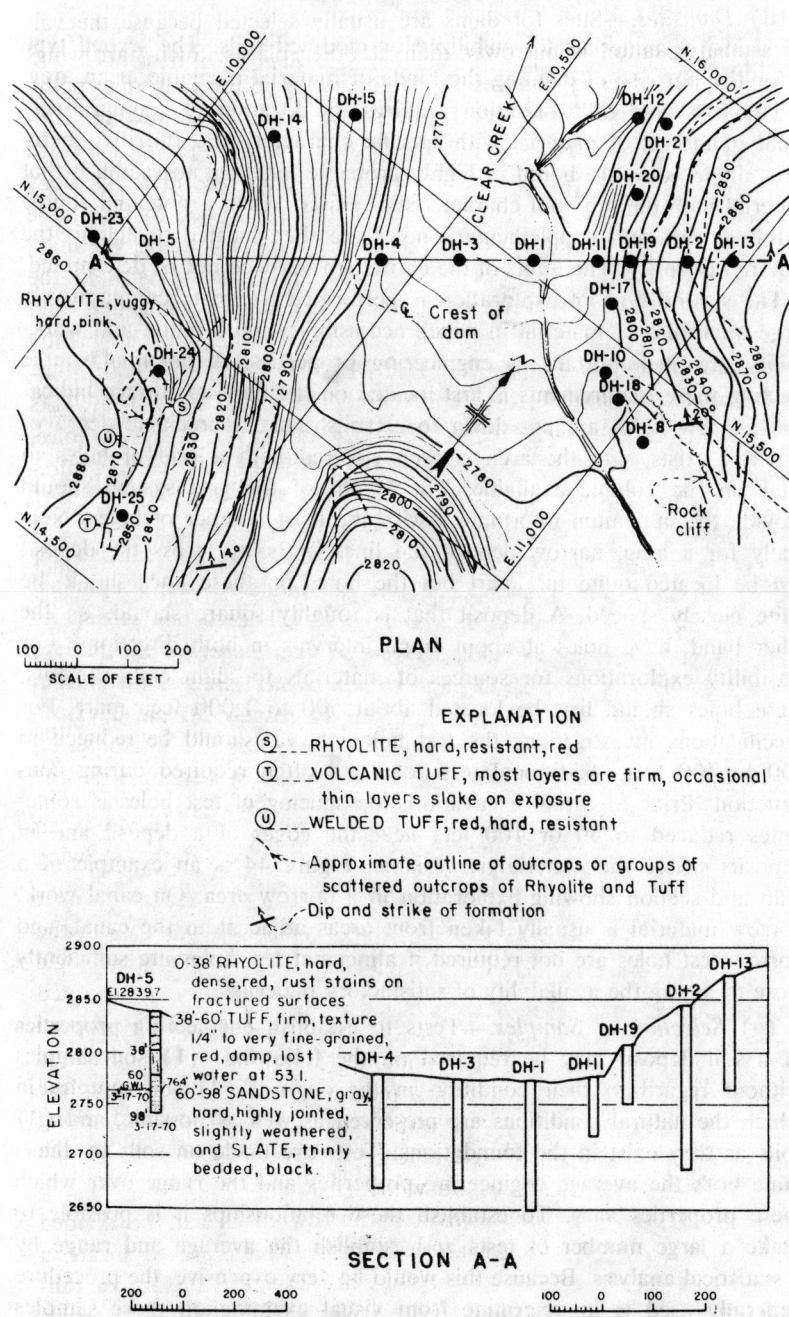

PLAN

100 0 100 200
SCALE OF FEET

EXPLANATION

S___RHYOLITE, hard, resistant, red

T___VOLCANIC TUFF, most layers are firm, occasional thin layers slake on exposure

U___WELDED TUFF, red, hard, resistant

----Approximate outline of outcrops or groups of scattered outcrops of Rhyolite and Tuff

----Dip and strike of formation

0-38' RHYOLITE, hard, dense, red, rust stains on fractured surfaces

38'-60' TUFF, firm, texture 1/4" to very fine-grained, red, damp, lost water at 53.1

60'-98' SANDSTONE, gray hard, highly jointed, slightly weathered, and SLATE, thinly bedded, black.

SECTION A-A

200 0 200 400
HORIZ. SCALE OF FEET

100 0 100 200
VERTICAL SCALE OF FEET

Figure 43.—Geologic map and section of a damsite. 101-D-229.

for dams, or materials for stabilized or modified soils. The second type is for the purpose of defining the kinds of material available in an area.

The first type of exploration requires the location of comparatively small quantities of material with specific characteristics. Initially, therefore, single holes are bored in highly probable locations to establish that material with the required characteristics exists. When a potential source is found, sufficient supplementary holes are then located to delimit the required quantity. The limits of the entire deposit need not be determined.

The second type of exploration is performed to locate comparatively large quantities of material in which accessibility, uniformity, and workability are as important as engineering properties. A potential source meeting these requirements is first located on the basis of surface indications, a few holes are put down to establish that appreciable depth of material exists, and the area is then covered with a grid of holes to establish the volume available. The layout of the grid system should provide the maximum information with the least number of holes. Normally for a long, narrow deposit the lines of holes across the deposit can be located quite far apart but the holes on these lines should be quite closely spaced. A deposit that is roughly square should, on the other hand, have holes at about equal intervals in both directions. For feasibility explorations for sources of materials for dam embankments, these holes should first be located about 500 to 1,000 feet apart. For specifications investigations the test hole interval should be reduced to 200 to 400 feet. Additional test holes are often required during construction. Prior to actual excavation, the spacing of test holes is sometimes reduced to 50 or 100 feet near the edges of a deposit and in deposits where the material is variable. Figure 44 is an example of a plan and section showing exploration in a borrow area. On canal work, borrow material is usually taken from areas adjacent to the canal, and borrow test holes are not required if alinement test holes are sufficiently close to assure the availability of satisfactory materials.

(g) *Selection of Samples.*—Tests to establish engineering properties of a soil deposit may be required on the following: (1) soil samples without regard to their condition in the deposit, (2) soil samples in which the natural conditions are preserved as well as possible, and (3) soils as they exist in the foundations. Tests are made on soils to determine both the average engineering properties and the range over which these properties vary. To establish these relationships it is possible to make a large number of tests and establish the average and range by a statistical analysis. Because this would be very expensive, the procedure generally used is to determine from visual examination those samples likely to have the poorest, median, and best properties for the engineering property considered to be critical. In the specifications stage, index tests

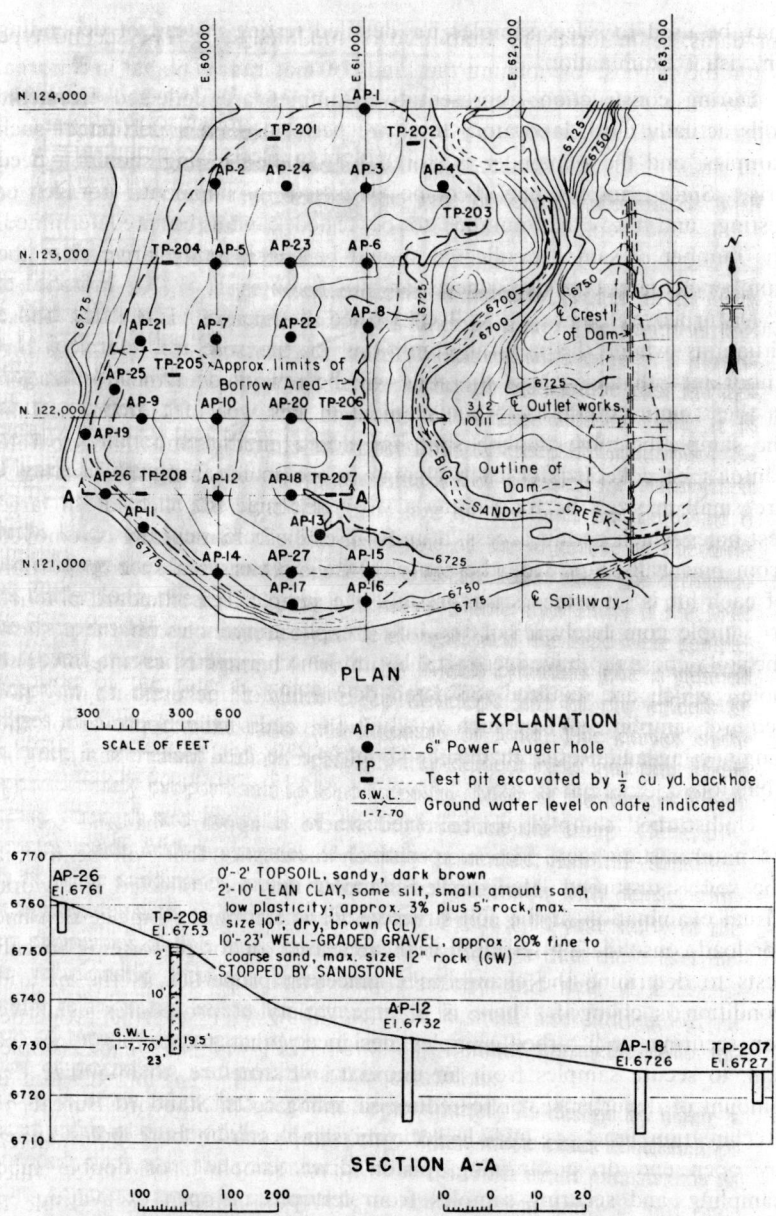

PLAN

300 0 300 600
SCALE OF FEET

EXPLANATION

AP ● ———— 6" Power Auger hole

TP ——— Test pit excavated by ½ cu. yd. backhoe

G.W.L. ——— Ground water level on date indicated
1-7-70

SECTION A-A

100 0 100 200
HORIZ. SCALE OF FEET

10 0 10 20
VERTICAL SCALE OF FEET

Figure 44.—Exploration for embankment materials—Location map and section for a typical damsite. 101–D–230.

may be used to select samples for detailed testing instead of depending on visual examination.

During construction, representative samples are collected from the soils actually used, laboratory tests are performed on a portion of such samples, and the remainder is stored for detailed testing should a need arise. Since samples may become damaged in shipment, storage, or testing, and the exact number to be tested cannot be predetermined, the number of samples collected should be considerably larger than the number actually considered necessary for testing.

Disturbed samples are collected (see designation E-1) for those situations where the natural conditions of the soil are relatively unimportant—that is, where the soils will be reworked in their utilization in the structure. The important element in this type of sampling is that the sample be uniform with sampling depth interval and that separate samples be collected for each change in the material. If the test holes are small in size, the total material from the hole is collected. In large test holes which permit access, a uniform cross section cutting is removed from one wall to provide the samples. In some cases a specific portion of each lift is set aside as a representative sample. It is standard practice to sample completely all of the initial exploratory holes. If these show the soil deposits to have recognizable uniform characteristics, intermediate holes, which are required for detail delimiting of the deposit, may not require sampling. When such a situation occurs, the approval of the Engineering and Research Center should be secured before sampling is abandoned.

Undisturbed samples are collected where it appears that the soil in its natural state may possess special characteristics that will be lost if the soil is disturbed. Undisturbed samples may be collected solely for visual examination of the soil structure, for determining inplace density, for load-consolidation testing, for shear testing, or for performing special tests to determine the change in engineering properties as the natural condition is changed. There is a large variety of procedures available for securing undisturbed samples designed either to sample types of soil, to secure samples from an unusual soil structure, to minimize the amount of disturbance, or to reduce sampling costs. Standard Bureau of Reclamation practices include securing samples from bore holes either by open end drive sampling, piston drive sampling, or double tube sampling, and securing samples from test pits or open excavation by cutting out a large block of material. The details of undisturbed sampling procedures are given in designation E-2. In situations where these procedures are unsatisfactory, special instructions and assistance will be provided by the engineering staff of the Engineering and Research Center.

(h) *Field Tests.*—Tests made on foundations in the field include inplace density and moisture tests, inplace permeability tests, penetration tests, vane tests, pile-driving tests, and pile-loading tests. Foundations for hydraulic structures (dams and canals) are tested for permeability as a standard procedure. Penetration tests are sometimes made on soil foundations and used as an index test, particularly where the bearing strength of the soil is questionable. Vane tests and pile-driving tests are made normally at the request of the Engineering and Research Center. Since the location of the test must be closely related to design requirements, the latter two tests are required mainly in connection with specifications stage investigations. In rare instances other special tests are made in the field. When such tests are required, instructions and special assistance will be provided by the Engineering and Research Center.

33. Exploration for Materials Having Specific Properties.—(a) *General.*—Frequently the immediate vicinity of an earthwork structure does not contain a sufficient variety of naturally occurring materials to permit the development of the feature except at an excessive cost. In such instances it is often economical to obtain limited quantities of materials that have specially desirable characteristics from areas at a considerable distance from the site. Such materials include impervious soils for the construction of linings or blankets; sand and gravel for concrete aggregate, filter blankets, drains, road surfacing, and, occasionally, protection from erosion; and rock fragments for riprap or rockfill, filter blankets, or concrete aggregate.

Obviously, if the required materials are found in ample quantities in the immediate vicinity of the site it is unnecessary to investigate more distant sources. If, however, there is a deficiency of impervious materials, pervious materials, or rock in the immediate area, it is not uncommon to go 25 to 50 miles for limited quantities of the deficient material and there have been cases where it was found economical to go 200 miles to secure rock for riprap. However, soil-cement or asphaltic concrete should be considered for alternate methods of slope protection if the haul distance to a suitable rock source exceeds about 20 miles. Nevertheless, the nearest rock source, or sources, must be investigated so that alternate bids can be obtained if found by the designer to be desirable.

The engineering use chart, figure 8, provides information on desirability of materials, except rock, for various uses from a quality standpoint. It is usually not economically feasible to secure any material, including rock, with ideal characteristics, and considerable judgment must be exercised by the investigator in the selection of material sources. The extent to which desirable characteristics are sought varies according to the purpose for which the material is to be used. In the utilization of materials, volume may be substituted for quality to some extent, and special processing of closer sources may be more economical than longer hauls. There must be a definite improvement in quality of the material with increasing distance from the site of utilization to justify long hauls. For distant sources, accessibility and type of transportation facilities available have an important bearing on the desirability of the source. Sometimes there are limiting conditions beyond which substitute materials no longer are economically competitive.

(b) *Impervious Materials.*—There are situations where a need for a special source of impervious materials may be necessary; namely, in the construction of canals, dams on pervious foundations, and terminal, equalizing, or regulating reservoirs. The fact that extensive permeable beds are found in the foundations for such structures is a clear-cut directive to the investigator to locate a source of impervious material. Such material should be impervious as compared to the foundation soils in order to justify its use, but highly plastic clay materials are seldom necessary or desirable. These impervious soils are applied in a blanket or lining over the pervious foundations. Hydraulic gradients through the blanket or lining will be high, so it is essential that the gradation be such that piping of the fines from the blanket or lining into the more pervious foundation materials is prevented. The material will be exposed to the water in the canal or reservoir, hence it should be capable of resisting the erosive forces of flowing water and waves. The material may be exposed to alternately wet and dry conditions and in some cases to freezing and thawing. Materials used for exposed linings or blankets should therefore be free of shrinkage and swelling characteristics. There are procedures by which these objectionable characteristics can be overcome, but, in many cases, costs are increased to a point that buried membranes or the use of additives to stabilize soils to form hard surface protective cover, or linings of asphaltic concrete, pneumatically applied mortar, compacted soil-cement, or concrete become competitive. However, blending two different soils together to form a superior blanketing material has been found to be practicable. The competitive nature of alternative procedures is such that a search for impervious material should not extend for very great distances without considering comparative costs.

(1) *Canal construction.*—On canal construction, linings are used to reduce water loss, which, in turn, is required to conserve a water supply, prevent waterlogging of adjacent lands, or reduce the size of the conveyance system. On this account, a high degree of impermeability is very desirable; however, thick linings of material of moderate impermeability have been used. Such linings can be protected from the erosion of flowing water by reducing veloci-

ties or by providing a protective cover. However, these linings are made as thin as practicable both to conserve material and minimize the necessary overexcavation; and so a procedure that has been used is to mix coarse material with the impervious soil, if such a mixture does not already exist, to produce a material that is both erosion resistant and impervious. When it is contemplated that canal velocities may be high or the natural soil is erodible, investigations for materials for canals should include a source of coarse material for blending or blanketing unless the impervious material already contains an appreciable amount of coarse particles.

In situations where canals are located on steep side hills, loss of stability due to seeping water may be critical even though water loss may not be important, and lining may be required. Regulating or equalizing reservoirs may also require hard surfaced linings or protective covers to minimize turbidity, growth of algae, and to facilitate cleaning.

(2) *Dam construction.*—On dam construction it is seldom possible to reduce water losses appreciably with partial earth blankets, but seepage gradients can be reduced in some cases to the point where piping is prevented. Materials for blankets on reservoirs do not need to be highly impervious; it is sufficient if their permeability is low compared to that of the foundation. Resistance to flowing water is usually not critical and therefore blending requirements need not be anticipated. However, the possibilities of piping the blanket into the foundation and of cracking due to the material having high shrinkage and swell characteristics are important. Investigations to locate a source of blanket material may be required for storage dams but will more often be needed in connection with investigations for diversion dams. The qualifying differences between impervious blanket material for canals and dams as described above should be noted.

(c) *Pervious Materials.*—Sand and gravel are required for concrete aggregate, for filters and drains associated with the construction of concrete structures, for blankets under riprap on earth dams, for drainage blankets under the downstream toes of earth dams, for use as transition materials to prevent piping, and for road surfacing. Except as a need for a special blending material arises, pervious material in the sense used here implies a material with substantial amounts of gravel-sized particles. The presence of fines in the material is progressively less objectionable as one proceeds through the foregoing list. In fact, for road surfacing a small amount of fines for binder is desirable, although where frost action is a factor the amount of fines must be strictly limited. Gradation, shape, and mineral composition also become progressively less important as one proceeds through the list. The distance from the site to areas in which investigations are made for locating a supply of pervious materials with special properties will vary depending on the need for such special material.

(1) *Concrete aggregates.*—Procedures for investigation of aggregates for concrete are described in the Concrete Manual [7]. It should be noted that investigations for this purpose are more exacting than for other purposes. For example, investigations for concrete aggregate can also be used for other purposes, but investigations made for other purposes generally are not adequate as concrete aggregate investigations.

(2) *Filters and drains.*—Although the quantity requirements of pervious materials for filters and drains are usually small, quality requirements are high. The principal purpose for construction which utilizes this material is the prevention of hydraulic uplift. Therefore the material must be very free draining; at the same time comparatively high heads must be dissipated without movement of either the filter material or the foundation soil. Often a single layer of material will be inadequate and a two-layer blanket should be designed. Fine sand, silt, or clay in the pervious material is objectionable, and processing by washing or screening is required to produce acceptable material from most natural deposits. Although grading requirements will be different, materials for filters are commonly secured economically from sources acceptable for concrete aggregate. However, neither particle shape nor nature of the minerals contained in the pervious material is very critical, and processed concrete aggregates rejected on these accounts can be used for the construction of drainage blankets and drains if suitable gradation is obtained.

(3) *Blankets under riprap.*—For sand and gravel blanket material under riprap, the primary requirement is coarseness. Because of this requirement, blanket material for this purpose is often secured from the rock fines developed in quarrying operations. However, if a deposit of coarse gravel can be found within a reasonable distance from the damsite, it will usually prove to be economical to develop such a source. Quantity requirements will be quite large and special processing by screening or other means will be costly. The principal purpose of this type of blanket is to prevent waves that penetrate the riprap from eroding the underlying embankment. A limited

---

[7] Concrete Manual, Bureau of Reclamation, Denver, Colo., seventh edition, revised 1966.

amount of fine material is not objectionable even though some of it will no doubt be lost through erosive wave action. The material should be durable; the material found in most gravel deposits is adequate. However, gravel deposits have been found which contained large quantities of unsuitable material. Such deposits include ancient gravel beds that have deteriorated by weathering, and talus or slopewash deposits where water action has been insufficient to remove the soft rock.

(4) *Drainage blankets.*—The materials used for drainage blankets under the downstream toes of earth dams should be pervious with respect to both the embankment and the foundation soils. The magnitude of the coefficient of permeability of such material may therefore be quite low. Particle size is relatively unimportant and the material may contain large quantities of coarse to fine sand. It should not, however, contain appreciable quantities of silt or clay. Volume of blanket material is to a large extent a substitute for quality, and if an ample supply can be found close to the point of use, a search for better materials at greater distances is seldom warranted.

(5) *Road surfacing and base course.*—Materials for road surfacing or base course are sought primarily for strength and durability. The preferred material for surfacing will consist primarily of medium to fine gravel with enough clay to bind the material together and relatively small amounts of silt and fine sand. Similar material *without* silt or clay is preferred for base courses. If the road construction to be performed by the Bureau of Reclamation is for replacement, the requirements of the local highway agency should be determined before making extended investigations for these materials.

(d) *Riprap and Rockfill.*—

(1) *General.*—Rock fragments are required in connection with earthwork structures for the protection of earth embankments or exposed excavations from the action of water, either as waves, turbulent flow, or heavy rainfall. The rock fragment protective work associated with wave action or flowing water is designated as riprap. The term rockfill is commonly applied to the more massive bodies of fill in dam embankments consisting of rock fragments which are used primarily to provide structural stability. Such rockfills may also serve as drainage blankets or blankets under riprap, and if the material used is adequate for slope protection, a separate need for riprap may be eliminated. The term rockfill is also used to describe thin blankets of rock fragments used for the protection of embankments from the erosive action of rainfall.

Material from rock sources should satisfy two main requirements: First, the rock source should produce rock fragments in suitable sizes according to required use, and secondly the rock fragments should be tough and durable enough to withstand the processes involved in procuring and placing them, and withstand normal weathering processes and other destructive forces associated with the place of use. Density is an important attribute although to some extent increase in size may be substituted for high density. There are substitutes available for riprap and rockfill blankets, such as soil-cement or asphaltic concrete for upstream slope protection, and sod cover for downstream slope protection, which must be considered when rock is not locally available.

(2) *Riprap for earth dams.*—Riprap surfaces on earth dams have to withstand very severe action from waves and the destructive forces associated with temperature changes, including freezing and thawing, heating and cooling, and wetting and drying. Securing material for these purposes is therefore most desirable. No general statement can be made as to whether one type of rock is better than another for slope protection. However, any sedimentary rock containing clay should be suspected of weakness. Laboratory tests such as the "freeze-thaw" test will disclose such weaknesses.

Frequently durability can best be judged by finding locations where the same rock is subjected to similar conditions in other reservoirs or in stream channels.

Size of rock fragments is very important; fragments up to 1 cubic yard in volume are required on dams associated with large reservoirs and on even the smallest earth dams fragment sizes up to one-half cubic yard are desirable. Figure 45 shows the results of a major blast in a riprap quarry producing rock fragments of the sizes required for a dam constructed in the western United States.

Joint spacing in a rock outcrop will help determine whether adequate size can be secured, but attention should also be given old joints which have become cemented but which would break apart during excavation. Because these mechanical weaknesses are often hard to detect, a blast test sufficient to remove 10 to 20 cubic yards is desirable. (See fig. 46.) When a bedrock exposure containing satisfactory rock that can be developed into a quarry is not available in the immediate vicinity of the damsite, it may be possible to secure riprap material by removing boulders from stream deposits or glacial till, from talus slopes (see fig. 47), and occasionally from surface deposits. The quality of many rock sources changes with depth, and the overburden on some sources rapidly becomes so thick that its removal is uneconomical. Hence, it will frequently be necessary to explore the rock sources with drill holes, depending on the geologic conditions, before establishing the deposit as an approved source.

Figure 45.—Results of a major blast in a riprap quarry. P–145–432–712.

Figure 46.—Blast test in igneous rock investigated as a source for riprap. TD–1501–CV.

Figure 47.—Talus slope of igneous rock proposed for riprap.

When it becomes necessary to seek riprap at distances greater than a few miles, it often develops that a number of sources are located that are competitive. In such situations it has been found desirable to specify quality requirements for the riprap rather than the source, so that a contractor may utilize the competitive nature of the supply to secure the material most economically. When such a situation exists the investigations should be directed toward establishing the competitive nature of several deposits. These deposits should be sampled and tested sufficiently to establish their essential characteristics. This information will be used in connection with designs to establish minimum acceptable properties for specifying quality requirements. Establishment of these minimum requirements will be influenced by initial cost, effective life of the cover, repair cost, climatic conditions, and thickness of cover; hence, the requirements cannot be established in advance of a survey that establishes the nature of the available alternatives.

(3) *Riprap for stilling basins.*—Riprap blankets are commonly used below spillway and outlet works stilling basins, and other energy dissipating structures where high-velocity turbulent flow is reduced to noneroding velocities. The quantities involved are usually comparatively small, but quality requirements may exceed those of reservoir riprap blankets. In some instances even though marginal material may be used on the dam surface, it will be necessary to

secure a high-quality riprap blanket for the protection of the related structures. Where investigations disclose only moderate quality rock in the vicinity of the site, investigations should be extended to establish the location of a high-quality source.

(4) *Riprap for canals.*—Another use for riprap is in canal construction where severe erosion of the channel would otherwise occur. These locations are usually in short reaches of canals below concrete structures, adjacent to bridge piers, and at sharp turns. Protective requirements vary over a wide range; they may be satisfied with a thin gravel cover or may require riprap equal to that required below the controlling works of dams. When these blankets are composed of riprap, they are usually 12 to 24 inches thick with corresponding rock sizes. (See sec. 74(d).)

(5) *Rockfill blankets.*—When rockfill is used to provide surface protection from rainfall, almost any rock fragments that do not break down on exposure to air or water are acceptable as rockfill. Shales and some siltstones are about the only types of rock considered unacceptable. Size is not a critical criterion except that fragments should be at least gravel size and the upper limit is controlled by the specified thickness of the blanket. Rounded gravel if readily available is frequently used for this purpose. Special investigations for this type of material are almost wholly confined to the plains areas where rock is generally absent. Substitutes include sod blankets which can be produced cheaply. Securing rockfill blanket material from appreciable distances is justified only where rock is generally absent in the immediate vicinity of the work and when aforementioned substitutes are inadequate.

(6) *Investigative procedures.*—The investigations required for rock sources for riprap are variable depending upon the stage of investigation, upon whether there is locally a choice of material, and upon the type of specifications to be prepared; that is, whether the contractor or the Government is to furnish the rock. The type of specifications to be used in turn depends on the type and choice of materials available. The complexity of the problem therefore requires specific instructions for the various situations. These instructions are included as part of designation E–39, "Investigation of Rock Sources for Riprap."

34. Materials for Stabilized and Modified Soils.—(a) *General.*—A stabilized soil is a soil whose properties are partially or completely changed by adding a dissimilar material before compacting the soil, or by injecting an additive into the soil in place. Depending upon the properties and the amount of the dissimilar material added (additive), all of the properties characteristic of the soil may be completely and permanently changed. (See secs. 54h and 80 through 84.)

A modified soil is a soil in which specific properties may be either temporarily or permanently changed. Soils are modified by adding a small amount of an additive prior to compaction or by injecting the additive into the soil in situ.

Stabilized soils are used as a substitute for riprap to protect the upstream slopes of earth dam embankments, linings for reservoirs, and temporary protection of construction during river diversion. They are also used as protective blankets and as pipe bedding (see section 77). Small quantities of additives are used to modify and improve the properties of soils used in fills, to increase erosion resistance, to reduce permeability, or to provide temporary stability during construction.

(b) *Compacted Soil-cement.*—Compacted soil-cement is a cement stabilized soil consisting of a controlled mixture of soil, cement, and water compacted to a uniform, dense mass. It is used for linings, protective blankets, and slope protection in lieu of riprap. The placement water content and density are controlled by the Proctor compaction test, designation E–11.

The most desirable soil for these purposes is a silty sand (SM) which has a good distribution of sizes with 15 to 25 percent fines and a maximum size of No. 4 sieve size to about 2 inches. Other soils may be used; however, more cement may be required to satisfy strength and durability requirements. Uniformity of the soil in texture, grading, and moisture when introduced into the mixing plant is the most important factor in assuring uniformity of the compacted soil-cement.

The soil is usually obtained from a borrow area explored in detail (see section 32f) to assure the quantity and uniformity desired. A uniform deposit is most desirable. Stratified deposits may be used provided selective excavation and processing is practical and economical compared to using other potential sources. Selective excavation and mixing during stockpiling may be necessary to provide a soil as uniform in texture, grading, and moisture content as practicable. Screening and shredding equipment may be necessary to: (1) remove undesirable organic material and oversize particles; (2) remove or reduce the size of shale, caliche, hard pan, or other particles which may not break down during normal processing; or (3) remove or reduce the size of sand, silt, and clay aggregations, called "clay balls," which tend to form if the borrow material contains lenses of silt or clay.

Considerable testing is required to determine the quantity and type of cement, the moisture limits, and compaction requirements to be specified for construction. Therefore, representative 300-pound samples of the minus 3-inch fraction of the average, finest, and coarsest material should be submitted for testing. The water proposed for mixing should be reasonably clean and free from objectionable quantities of organic matter, alkali, salts, and other impurities. Clear water that does not have a saline or brackish taste may be used; however, doubtful sources should be sampled and tested.

(c) *Plastic Soil-cement.*—Plastic soil-cement is a cement stabilized soil consisting of a mixture of soil and cement with sufficient water to form a fluid consistency which will flow easily and can be pumped without segregation. Available sands which are either well-graded or poorly-graded with less than 10 percent fines and a maximum size of about ⅜ inch are excellent soils. Mortar or concrete sand (SW) is the most desirable.

The Bureau has used plastic soil-cement for pipe bedding. Although materials from required excavation may be used if they meet the above requirements, it is usually more economical and desirable for control to locate borrow areas along the pipeline. One-hundred-pound samples representative of the materials proposed for use should be submitted for testing to determine their suitability and type and percent of cement to be used.

(d) *Asphaltic Concrete.*—Asphaltic concrete is asphalt cemented aggregate described as a controlled hot mixture of asphalt-cement and well-graded, high quality aggregate thoroughly compacted into a uniform, dense mass. It is used for slope protection on dams, linings for canals and reservoirs, and protective blankets. In general, the aggregate (coarse aggregate, minus 1½ inches to plus No. 8 sieve; fine aggregate, minus No. 8 sieve to plus No. 200 sieve) should be reasonably well-graded, durable sand and gravel particles with a maximum size ranging from the No. 4 sieve size to about 1½ inches, with nonplastic fines (filler or mineral dust, minus No. 200 sieve) ranging from about 5 to 15 percent.

A wide range in grading and type of aggregate may be used depending upon available sources by adjusting the asphalt-cement content and thickness of the final compacted layer. The aggregate may be obtained from naturally occurring materials and crushed rock, or may consist entirely of crushed rock. Selective excavation and screening may be required to assure that the aggregate does not contain organic matter, nondurable aggregations of materials, such as "chunks or clay balls", or other objectionable materials. A 100-pound sample of each fraction of the aggregate or a 200-pound sample of a pit-run material proposed for use is required for testing.

(e) *Modified Soil.*—A modified soil is a mixture of soil, water, and a small amount of an additive mixed prior to compaction or injected into the soil in situ to modify certain properties, temporarily or permanently, so that they are within tolerable limits. A modified soil usually retains most of the characteristics of the original soil since it is an aggregation of uncemented or weakly cemented particles rather than a strongly cemented mass. There is limited experience on even the most commonly used additives which include asphalt, portland cement, lime, resins, elastomers, and organic chemicals.

Research is continuing on these as well as a variety of other materials including waste products from refining and manufacturing processes to discover a material and process for improving soil properties at costs less than conventional methods of handling problem soils.

For those situations warranting the use of a modified soil, the Engineering and Research Center will issue special sampling instructions for obtaining disturbed or undisturbed samples for laboratory testing.

C. Exploratory Methods

35. General.— There are a great many methods of making exploratory holes. These may be grouped in different ways: (1) those that produce usable samples and those that do not, and (2) those that are accessible and those that are not. In the investigation of foundations or materials, the principal purpose of an exploration is to secure samples of the soil either for visual examination or for testing. Therefore, those procedures which will not produce samples should be used only where the nature of the materials to be penetrated is already known and where sufficient samples for testing have been secured. Sampling methods will vary according to the hardness of the material to be penetrated and also according to the degree of sample disturbance that is acceptable. In addition, test holes may be advanced by manual labor or by mechanical power. Exploratory holes may also be of various sizes, depending on the desirability of access, the depth to be penetrated, the size of the sample required, and the kind of material to be penetrated.

Stability of small-size holes entirely above the water table is dependent on the type of material. Holes in soil below water table, as a rule, require support by steel casing or by drilling mud with wall stabilizer. It is sometimes necessary to protect these exploratory holes with steel casing because of potential damage to the hole by the drilling operations and to prevent contamination of samples with materials from higher elevations. As part of the foundation investigation, many exploratory holes require water testing. When casing is used, specific portions of the foundation may be water tested, which simplifies the evaluation of foundations and the determination of treatment required. If water testing is a part of the requirement, drilling mud should not be used.

In soft or loose soil foundations, the in situ support of the wall of the hole may be insufficient to keep soil from flowing into the bottom of the hole. In many instances, keeping the hole filled with water will suffice to hold the materials in place. In more severe cases, a wall stabilizer, a heavy fluid, or both must be used. Information on the various stabilizers may be obtained from the Engineering and Research Center. If water testing is not required, drilling mud consisting of a mixture of commercially available bentonite and water must be used. This is a specially prepared fluid which has the required weight due to the addition of finely divided solid material. New organic chemicals, such as drilling fluid additives, are on the commercial market and may also be useful for this purpose (see designation E–2). For the most part, stability of soft foundations is the primary question with which the engineer is concerned, and securing samples for laboratory testing is the major problem of the investigator. Samples may be secured from holes supported by drilling mud or casing, either with the double-tube soil samplers, core samplers, or with drive samplers. To minimize sample disturbance, fixed-piston-type samplers are preferable for very soft or very loose soils. Figure 2–5 shows some of the samplers available for obtaining undisturbed samples.

In extending exploratory holes through hard materials where support is normally unnecessary, crushed zones or faults may be encountered from which rock fragments keep falling into the hole and tend to plug the hole or bind the drilling equipment. In such situations, a cement grout may be inserted in that area, and after the grout has set, the hole is drilled through the grout. Since these crushed zones or faults represent some of the critical conditions being investigated from an engineering standpoint, it is necessary that all possible tests, such as water tests, be made before grouting the unstable section of hole and that a full report of the conditions be made. Before a test hole is grouted, the person in charge of the explorations should be informed and his concurrence received.

All exploratory holes should be protected with suitable covers or fences to prevent foreign matter from entering the holes and to keep people and animals from falling into them. All holes should be filled or plugged after they have served their purpose.

36. Test Pits, Trenches, and Tunnels.—(a) *General.*—Open test pits, trenches, and tunnels are accessible and afford the most complete information of the ground penetrated. They also may permit examination of the surface of foundation bedrock. When the depth of overburden and ground-water conditions permit their economical use, these methods are recommended for foundation exploration in lieu of relying solely on borings. In prospecting for embankment materials or concrete aggregate containing cobbles and boulders, open pits and trenches may be the only feasible means of obtaining the required information.

(b) *Test Pits.*—Test pitting is an effective means of exploring and sampling earth foundations and construction materials. Its use facilitates inspection, sampling, and testing. The depth of a test pit is determined by investigational requirements but is usually limited to a few feet below water table. The minimum recommended cross section for a hand-dug pit is 3 by 5 feet. Dragline, backhoe, clamshell, caisson drilling or auger equipment, and bulldozer pits are usually more economical than hand-dug pits for comparatively shallow materials explorations, but are not practicable where a depth of more than about 15 feet is desired. Where the soil is hard, pneumatic paving breakers or spades operated by small trailer-mounted air compressors will facilitate progress in hand-dug pits. The use of explosives to break up large boulders is a common practice.

In hand-dug pits the materials are removed from the hole by buckets operated from a hoist or windlass which should be equipped with a ratchet device for safety. During excavation, the bottom of the hole should be kept fairly level and of full size so that each lift may represent the corresponding portion of the deposit in quantity and quality. At the surface the excavated material should be placed in an orderly manner around the pit, and marked stakes should be driven to indicate depth of pit from which the material came in order to facilitate logging and sampling. Recording the moisture condition before allowing to dry out by exposure to air should be made. All hand-dug pits should be cribbed. A convenient method of cribbing, with 3- by 6-inch lumber, is shown in figure 48. In loose materials it is advisable to keep the space between the pit walls and the cribbing at a minimum, to pack the space with hay or excelsior, and to keep the bottom of the cribbing close to the bottom of the pit.

Deep test pits should be ventilated to prevent accumulation of dead air. For this purpose connected lengths of stovepipe, starting slightly above the floor and extending about 3 feet above the mouth of the pit, have been found satisfactory. Canvas sheeting has been used to deflect wind into the hole. Test pits left open for inspection should be provided with covers or barricades for safety.

When water is encountered in the pit, a pumping system is required for further progress. Small, portable, gas-powered, self-priming, centrifugal pumps can be used; however, it is preferable to use air-powered equipment whenever possible because of the added safety against potential carbon monoxide poisoning. It is desirable that the suction hose be one-half inch larger in diameter than the discharge opening of the pump and not more than 15 feet long. This requires resetting the pump in the

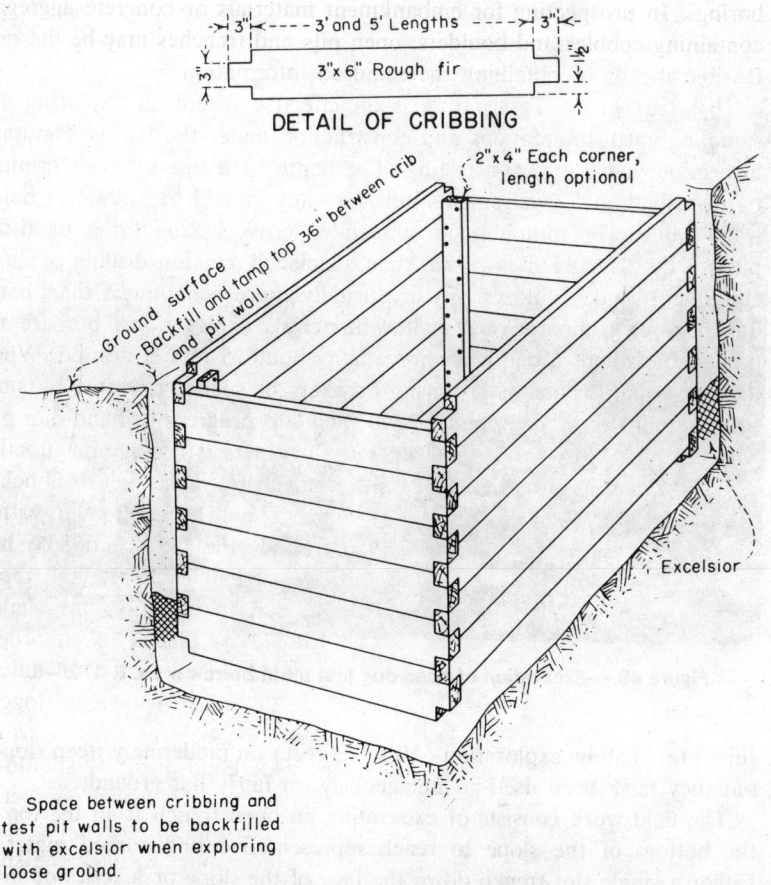

DETAIL OF CRIBBING

3' and 5' Lengths

3"

3" x 6" Rough fir

3"

2" x 4" Each corner, length optional

Ground surface

Backfill and tamp top 36" between crib and pit wall

Excelsior

Space between cribbing and test pit walls to be backfilled with excelsior when exploring loose ground.

Figure 48.—Test pit cribbing. 288–D–2624.

pit (on a frame attached to the cribbing) at intervals of about 12 feet. When an air-powered pump is not available, thus necessitating the use of the gas engine, it is necessary to pipe the exhaust gases well away from the pit when the engine is in or near the pit. When a gas engine is operating within a pit, personnel shall not be allowed in the pit for any extended period of time regardless of how well the system is vented. Unwatering test pits is usually expensive and often is unwarranted. Figures 49 and 50 show different types of test pits.

(c) *Trenches.*—Test trenches are used to provide a continuous exposure of the ground along a given line or section. In general, they serve the same purpose as the open test pits but have the added advantage of disclosing the continuity or character of particular strata. They are best

Figure 49.—Excavation of hand-dug test pit in borrow area. E–1725–6.

suited for shallow exploration (10 to 15 feet) on moderately steep slopes, but they have been used advantageously on fairly flat ground.

The field work consists of excavating an open trench from the top to the bottom of the slope to reach representative undisturbed material. Either a single slot trench down the face of the slope or a series of short trenches spaced at appropriate intervals along the slope can be excavated. Depending on the extent of the investigation required, either pick and shovel methods or bulldozers, ditching machines, backhoes, or draglines can be used. A trenching layout suitable for materials investigations is shown in figure 51. Figure 52 is a photograph of a bulldozer trench, and figure 53 shows a trench excavated by bulldozer and backhoe. Safety precautions should be used in deep trenches to prevent accidents caused by caving ground.

The profile exposed by these trenches may represent the entire depth of significant strata in an abutment of a dam; however, their shallow depth may limit investigation to only a portion of the foundation and other types of exploration may be required to explore to greater depths. As with test pitting, trenching permits visual inspection of the soil strata which facilitates logging of the profile and selection of samples. It also aids in obtaining large undisturbed samples or large disturbed individual

Figure 50.—Equipment-excavated test pit showing location of samples. LM–173.

or composite samples. Trenches in sloping ground have the advantage of being self draining.

(d) *Tunnels.*—Tunnels or drifts have been used to explore areas beneath steep slopes in or back of clifflike faces. For such purposes the exploratory tunnel or drift, is usually roughly rectangular in shape and approximately 5 feet wide and 7 feet high. The placing of lagging, when

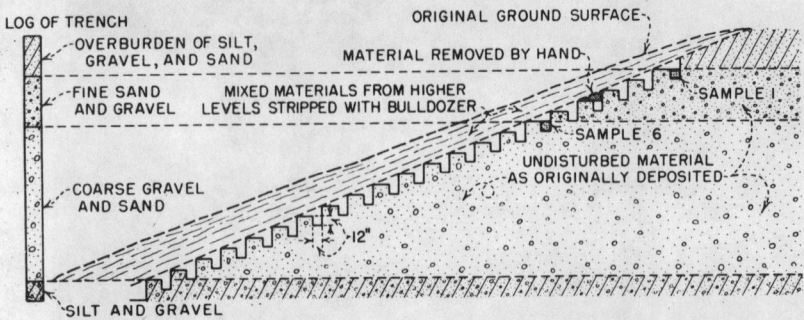

LOG OF TRENCH   ORIGINAL GROUND SURFACE
OVERBURDEN OF SILT,
GRAVEL, AND SAND   MATERIAL REMOVED BY HAND
FINE SAND   MIXED MATERIALS FROM HIGHER   SAMPLE I
AND GRAVEL   LEVELS STRIPPED WITH BULLDOZER
                                        SAMPLE 6
                              UNDISTURBED MATERIAL
COARSE GRAVEL   AS ORIGINALLY DEPOSITED
AND SAND
                         12"
SILT AND GRAVEL

**Figure 51.—Trenching, a low-cost method of obtaining soil samples.
288–D–2623.**

Figure 52.—Shallow test trench excavated by bulldozer. 222–117–37785.

required for side and roof supports, should follow excavation as closely
as practicable. Excavation of exploratory tunnels may be a slow, expensive process; consequently, this type of investigation should be utilized
only when no other method will supply the required information.

Figure 53.—Trench in steep abutment area excavated by bulldozer and backhoe.
711–400–9.

Logging and sampling of exploratory tunnels should proceed concurrently with excavation operations if possible.

If explosives have been used in the tunnel excavation, careful location selection for undisturbed samples must be made. This would include removal of all material which has been disturbed by the explosives, thereby exposing the undisturbed material.

37. Auger Borings.—Auger borings often provide the simplest method of soil investigation and sampling. They may be used for any purpose where disturbed samples are satisfactory and are valuable in advancing holes to depths at which undisturbed sampling or penetration testing is required. Depths of auger investigations are, however, limited by the ground-water table and by the amount and maximum size of gravel, cobbles, and boulders as compared with the size of equipment used. Hand-operated post hole augers 4 to 12 inches in diameter can be used for exploration to shallow depths (figs. 54 and 55). Machine-driven augers are of three types: helical augers 3 to 16 inches in diameter; disc augers up to 42 inches in diameter; and bucket augers up to 48 inches in diameter. Figure 56 illustrates the basic differences of these three types of augers. Figure 57 shows a double-tube sampler which has a helical auger on the outside barrel to remove cuttings as the inside tube slides over the cut sample, thus permitting sampling without contamination by drilling fluid. A disc auger is shown in figure 58 and figure 59 shows a bucket auger.

An auger boring is made by turning the auger the desired distance into the soil, withdrawing it, and removing the soil for examination and sampling. The auger is inserted into the hole again, and the process is repeated. Pipe casing is required in unstable soil in which the borehole

Figure 54.—Exploring a borrow area with a hand auger. PX–D–16277.

fails to stay open, and especially where the boring is extended below the ground-water level. The inside diameter of the casing must be slightly larger than the diameter of the auger used. The casing is driven to a depth not greater than the top of the next sample and is cleaned out by means of the auger. The auger can then be inserted into the borehole and turned below the bottom of the casing to obtain the sample.

Enclosed augers have been used successfully in lieu of casing (see fig. 60). The outer barrel acts as casing, is connected to the sampler on

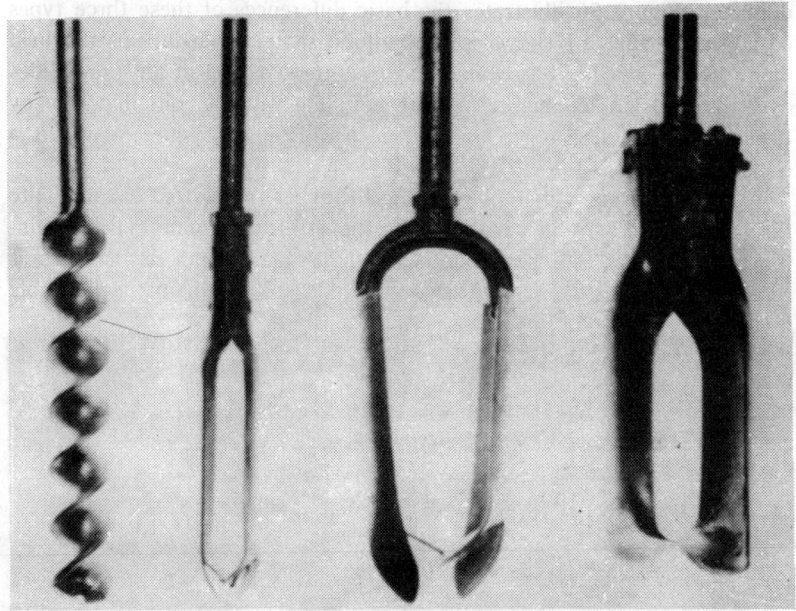

Figure 55.—Types of hand augers (2-inch helical, 2- and 6-inch Iwan, and 6-inch Fenn (adjustable)). PX–D–16998.

a swivel-type head, and remains in a stationary position as the auger is rotated.

The sampler is lowered to the bottom of the hole and auger rotation started. As the auger penetrates unsampled soil, the outer barrel is pulled down with the unit, thereby holding out any caved or foreign material. The sampled material is retained on the helical auger inside the outer barrel. After completing the sample run, the final penetration depth is carefully noted and the sampler removed from the hole. The auger is reverse rotated to eject the sampled material.

The sampler is again lowered into the hole and the bottom depth noted. Any discrepancy between the final penetration depth, determined on the first run, and the hole bottom depth would be due to caved material in the hole. This material would be recovered on the top of the sampler and must be discarded upon ejection to avoid mixing with the newly sampled material.

The soil auger can be used both for boring the hole and for bringing up disturbed samples of the soil encountered. It operates best in somewhat loose, moderately cohesive, moist soils. Holes are usually bored without the addition of water; but in hard, dry soils or in cohesionless

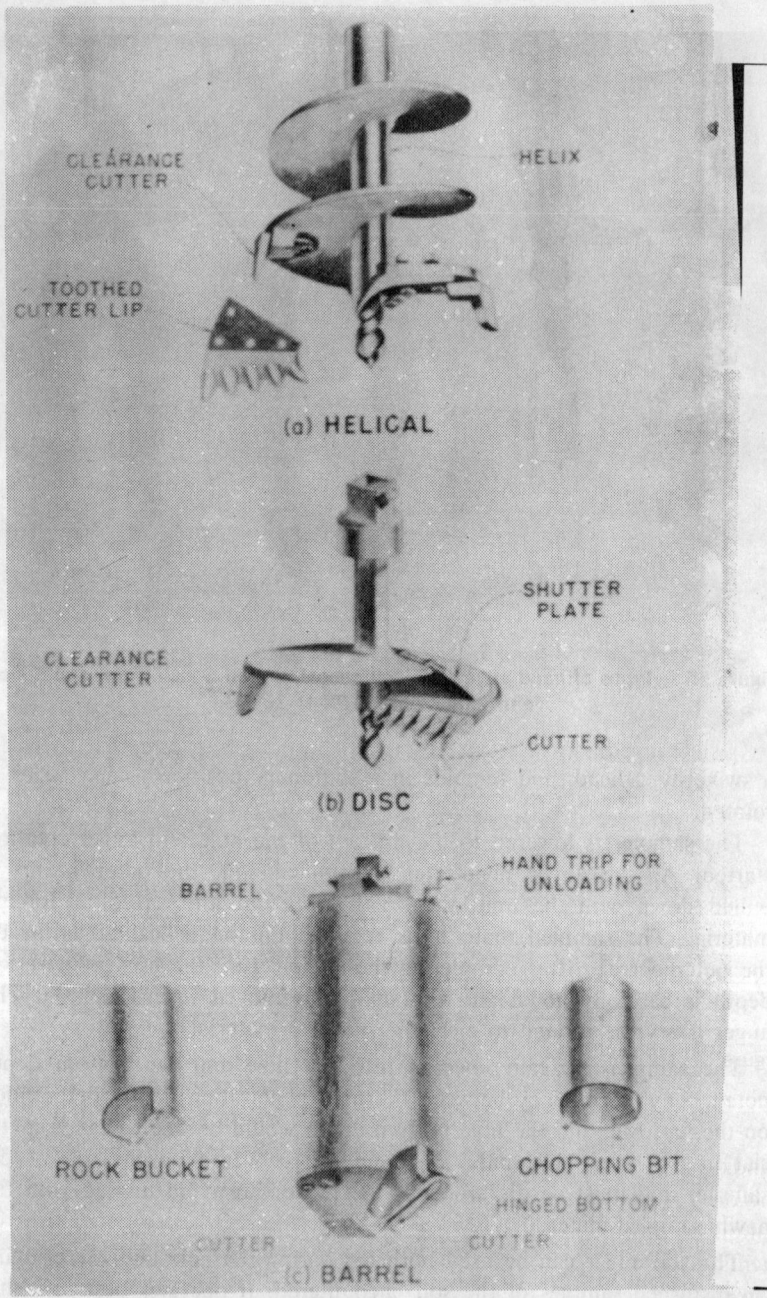

(a) HELICAL

(b) DISC

(c) BARREL

Figure 56.—Illustration of the helical, disc, and barrel types of machine-driven augers showing basic differences. (U.S. Department of the Navy.) PX–D–69900.

AUGER SAMPLER DISASSEMBLED
A - Sludge barrel
B - Auger barrel
C - Inner tube (5-inch O.D. by 24 or 36 inches long)
D - Inner head
E - Outer head

Figure 57.—Undisturbed sampling with a double-tube helical auger and description of equipment. E–2235–1NA and PX–D–48272NA.

sands the introduction of a small amount of water into the hole will aid the drilling and sample extraction. It is difficult to avoid some contamination or mixing of soil samples obtained by small augers. Rock fragments larger than about one-tenth of the diameter of the hole cannot be successfully removed by normal augering methods. Large-sized holes permit examination of the soils in place, and therefore are preferred for foundation investigation.

Figure 58.—Disc auger used to explore borrow areas of fine-grained soils. FK–391–CV.

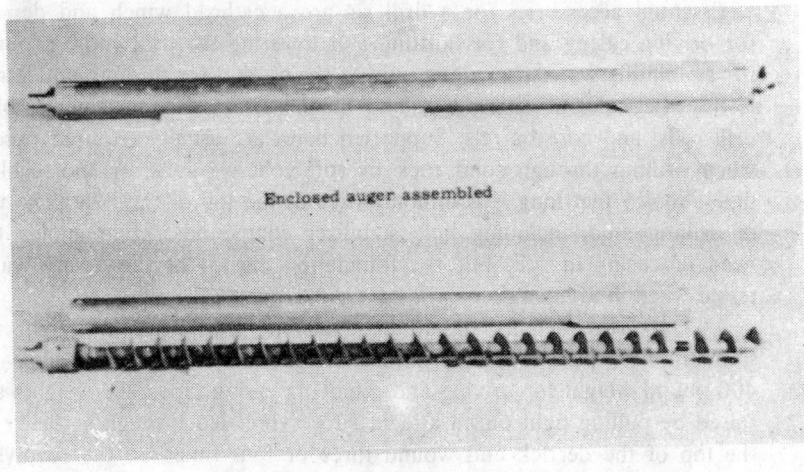

Figure 59.—Bucket auger used in exploration of a borrow area containing gravel particles. TD-763-CV.

Enclosed auger assembled

Auger disassembled

Figure 60.—Enclosed auger. 495-D-44746NA and P495-D-44745NA.

38. Rotary Drilling.—One of the most important tools for sub-surface exploration is the diamond drill—a rotary drill with a core barrel, a diamond bit, and a hydraulic or screw feed, originally developed for drilling through and sampling either hard or soft rock. Various core bit sizes are shown in designation E–2, figure 2–21. The diamond drill may actually be operated with a variety of bits, depending on the hardness of the material penetrated. Rotary drill equipment is manufactured in a wide variety of forms, which vary from highly flexible to extremely specialized equipment, from lightweight and highly mobile equipment to heavy stationary plants, and in size of hole and core range from less than 1 inch to 3 feet or more. They are capable of drilling to depths of hundreds of feet.

One of the most advanced rotary drilling developments in recent years has been the introduction of wire-line drill rod and core barrel assemblies. This is especially valuable in deep hole drilling since the method eliminates trips in and out of the hole with the coring equipment. With the wire-line technique, the core barrel becomes an integral part of the drill rod string. The drill rod serves as both a coring device and casing as it is usually not removed except while making bit changes. Core samples are retrieved by removal of the inner barrel assembly from the core barrel through the drill rod. This is accomplished by lowering an overshot or retriever, by wire-line, through the drill rod to release a locking mechanism built into the inner barrel head. The core is removed and the inner barrel returned through the drill rod and coring continued. Present day wire-line core sizes are limited to a maximum diameter of 3⅜ inches.

Essential accessories for a drill rig are a cathead winch and derrick for driving casing and for hoisting and lowering the drill rods; a pump for circulating water or drilling mud to the bit and for flushing and water testing the hole; a watermeter; and the necessary driving weights, bits, drill rods, and core barrels. Supported holes are usually required except when drilling through solid rock or stiff cohesive soils. A short collar pipe about 5 feet long is commonly used at the top of the hole. The use of drilling mud, including hole stabilizer compounds, often avoids the need of casing in soil, but the foundation cannot be effectively water tested when mud is used.

At least two weights should be available, a 140-pound weight for standard field penetration tests (see designation E–21) and a 250- to 400-pound weight for driving and removing casing pipe. The weights are raised by pulling tight on an attached rope threaded through a sheave at the top of the derrick and wound three or four times on the revolving cathead winch. Sudden loosening of the rope permits the weight to drop on the driving head attached to the casing. Various types of chopping

Figure 61.—Diamond drill rig used in foundation exploration. E–2255–4NA.

bits are used to facilitate the driving of casing through soils containing cobbles and boulders. Large boulders must be either blasted with explosives or drilled with a diamond bit or a roller rock bit. The casing is raised several feet prior to blasting. Figure 61 shows a diamond drill rig with derrick.

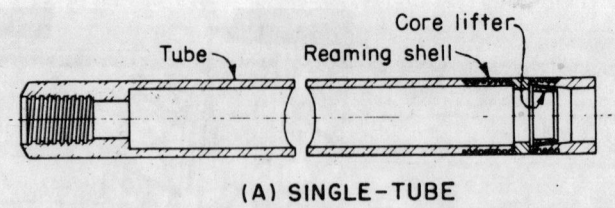

(A) SINGLE-TUBE

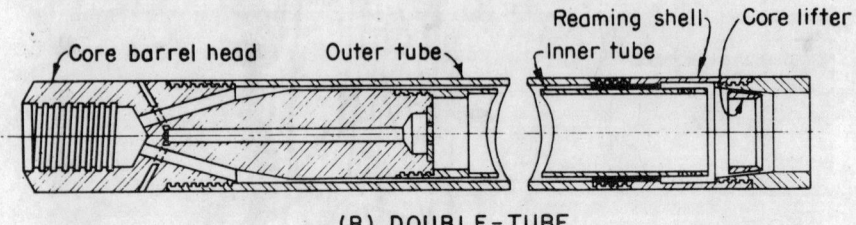

(B) DOUBLE-TUBE

Figure 62.—Core barrels used for obtaining samples of rock. 288–D–2514.

The accuracy and dependability of the records furnished by diamond drilling depend largely on the size of the core in relation to the kind of material drilled, the percentage of core recovery, the behavior during drilling, and the experience of the drill crew. Since a rock that will core well in an NX hole may break up badly in an EX hole, it is important to use the largest practicable size diameter hole and core barrel. Recovery of core is much more important than making rapid progress in drilling the hole. Portions of a core that are lost will probably represent shattered or soft, incompetent rock, whereas the recovered portions represent the best rock from which an overoptimistic evaluation of the foundation likely will be made. A reasonably large percentage of core recovery, on the other hand, will provide a more continuous section of the materials passed through. The cores provide information on the character and composition of the different formations, with evidence of the spacing and tightness of joints, seams, fissures, and other structural details. When drilling in soft materials, the water circulation must be reduced or stopped entirely, and the core recovered "dry," even though a marked delay in operations may result.

There are two principal types of core barrels, single-tube and double-tube, as shown in figure 62. The single-tube is simpler in design and consists of a core barrelhead, core barrel, and attached coring bit, which has an annular groove that permits downward passage of drilling fluid pumped through the hollow drill rod and between the core and the core barrel. This design exposes the core to drilling fluid over its entire length, which can result in serious core erosion of unconsolidated or weakly

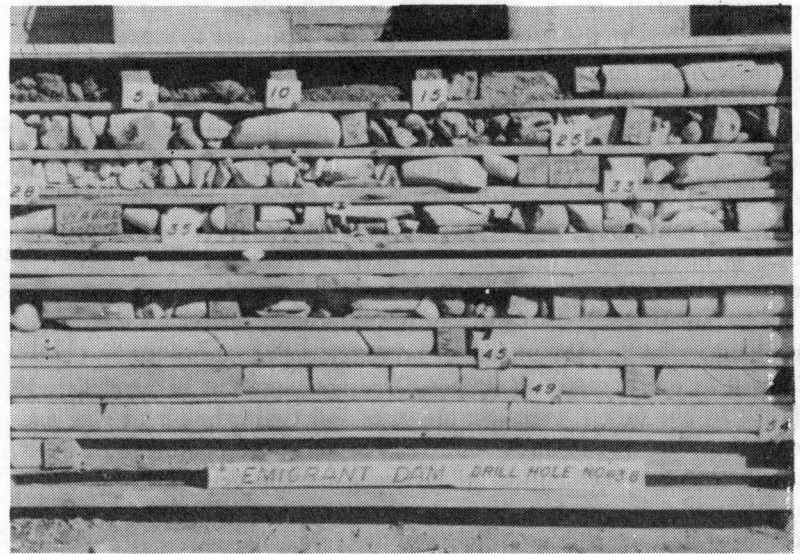

Figure 63.—Arrangement of cores in a core box to insure proper identification of samples. PX–D–16303.

cemented materials. Therefore, it is used primarily to sample hard, solid rock which requires a diamond drill bit.

In addition to an outer rotating barrel, the swivel-type, double-tube core barrel sampler provides an inner stationary barrel which protects the core from the drilling fluid and reduces the torsional forces transmitted to the core. It is used to sample soil and soft or fractured rock and may be used to obtain cores in hard, brittle, or partially cemented soils, or cores of soft, weakly cemented rocks. For these materials, hardened metal drill bits and shorter core barrels (5 feet or 2½ feet in length) are used.

The rigid-type, double-tube core barrel provides an inner barrel which rotates with the outer barrel but protects the core from the drilling fluid. The rigid-type, double-tube core barrel has a wide range of applications in soft to medium-hard formations. It is also well suited to hard, broken formations. There is a problem of core abrasion caused by the rotating inner tube and soft, easily disturbed ground cannot be cored well with the barrel.

Figure 2–20, designation E–2, shows the dimensions of standard drilling rods, casings, and cores. Figure 63 shows a standard core box and illustrates the method of placing cores in the box to insure proper identification of each core sample.

The core barrel is fitted with a drill bit and is lowered into the hole by the hollow drill rod. Circulation of the wash water should be started before the core barrel reaches the bottom of the hole to prevent cuttings or sludge from entering the core barrel at the start of coring. The optimum rotation speed of drilling varies with the type of bit used, the diameter of core barrel, and the kind of material to be cored. Excessive rotation speed will result in chattering and rapid wear of the bit and will break the core. Low speed results in less wear and tear on the bit and better cores but lower rates of progress. Approximate ranges of rotation speeds used in medium-hard rock are 300 to 1,500 revolutions per minute for diamond bits and 100 to 500 revolutions per minute for metal carbide bits.

The rate at which the coring bit is advanced depends on the firmness of the material and the amount of pressure applied downward on the bit as well as on its speed of rotation. This pressure must be carefully adjusted by the driller; excessive pressures will cause the bit to plug and may shear the core from its base. The bit pressure is controlled by a hydraulic or screw feed on the drilling machine. The weight of the column of drill rod will seldom be in excess of the optimum bit pressure for the coring of medium and hard rock, and it is frequently necessary to supply additional downward pressure.

Since the hole left in the rock is clean and the seams and fissures are not sealed off by the action of the drill, opportunity is afforded for making percolation tests to indicate the permeability of the strata and to determine probable leakage through open joints or fissures in the rock. Gravity or pressure tests are made as described in designation E–18. Large water losses or inflows into the holes during drilling are recorded as indicating, respectively, either the presence of large openings in the formation or the existence of an underground flow. Completed holes should be capped to preserve them for use in ground-water-level observations or as grout holes, or for reentry, if it is later found desirable to deepen the hole. Casing is, of course, usually required for those sections of the hole in loose material or unconsolidated subsurface soils.

Although the rotary drill is designed primarily for penetrating rock rather than soil, sample barrels and cutting bits have been developed which are satisfactory in many kinds of soil deposits. Double-tube core barrel samplers of the Denison, Denver, Pitcher, and double-tube auger types are capable of obtaining 6-inch-diameter undisturbed samples of sands, silts, or clays for laboratory testing. These samplers are described in designation E–2.

39. Drive-Tube Boring.—The drive-tube method is commonly used for foundation exploration in soil. The amount of sample disturbance, at least for visual examination, is ordinarily within tolerable limits. In its simplest form, a drive sampler may be a piece of open pipe, sharpened on the end, which is forced into the soil by hydraulic pressure, jacking, or driving; however, a relatively disturbed sample, suitable only for visual examination and identification tests, is obtained. With careful procedure and specialized equipment, samples suitable for laboratory testing can often be secured. A great many varieties of drive sampling equipment have been developed in order to improve sampling techniques for various kinds of soils. These innovations include making the walls of the pipe as thin as possible, insertion of sample holding liners, and a variety of mechanisms to aid in holding the sample in the sampler. Some types are fitted with a piston which keeps the sample chamber closed until the sampler has been pushed to the depth where sampling is required. In others, the piston is used to prevent sample disturbance and also to produce a vacuum to hold the soil in the sampler.

The most commonly used drive samplers are 2 to 5 inches in diameter and 1 to 3 feet in length. Small sizes may be used if visual examination only is desired. Samples at least 3 inches and preferably 5 inches in diameter are required for testing in the laboratory. Ordinary drive samplers are not satisfactory in extremely gravelly or rocky soils and generally samples cannot be secured from below water table using this method. The piston-type sampler is superior in the latter situation, and some special types of equipment are effective in the hands of a skillful operator. Equipment and procedures for obtaining samples with the thin-wall, open drive sampler and the piston sampler are given in designation E–2.

Drive samplers can be operated with most varieties of rotary drilling equipment. There are also available a large variety of drilling rigs which will operate only drive samplers and various kinds of nonsampling drilling equipment. However, in many situations such limited range equipment is perfectly satisfactory and specifying more elaborate equipment is not justified. In making explorations by drive-tube boring it is very desirable that a continuous sampling technique be used at least until subsurface conditions have been fully established. In this case the hole should be reamed to avoid sample contamination. The practice of advancing a hole by nonsampling methods and extracting short drive samples at 5- to 10-foot intervals is not as satisfactory as obtaining continuous samples.

Special variations of the drive sampler called the split-tube penetration sampler and the split-liner sampler are quite widely used both to secure subsurface samples and at the same time measure the inplace strength, firmness, and denseness of the foundation. Either sampler is capable of collecting a core sample about 1⅜ inches in diameter. The core barrel can be separated for examination and removal of the sample, and the resistance to penetration can be measured in terms of the number of blows of a standard weight required to produce 1 foot of penetration. In obtaining this type of information it is essential that a standardized procedure be used. In making this type of exploration, information on soil type, water content, and penetration resistance should be obtained. In a range of 5 to 50 blows per foot, fairly reliable correlation with engineering properties has been established for several varieties of material provided the water content is high. The relationship between penetration resistance and relative density for either dry or saturated sands is given in section 79(b). Procedures for the operation of the split-tube sampler are covered in designation E–21.

40. Miscellaneous Methods.—(a) *Nonsampling Borings.*—Test holes excavated merely to determine depth to some particular stratum or bedrock, or for advancing a hole to provide access to a buried layer for sampling, can be accomplished by any of the previously described methods. There are also a number of procedures in common use which will accomplish these purposes very economically. These procedures include percussion or churn drilling, wash boring, and jetting. Probing is often an economical method of establishing the depth to a firm stratum. Variations in procedure depend primarily on the nature of the soil to be penetrated, with percussion drilling being used on the hardest and most compact soils, and probing on the softest. All the operations are based on moving the tool up and down to chop away the material in the hole, using increasing amounts of water in the order listed, except probing which uses none at all.

In percussion or churn drilling, the tool is attached to the end of a cable. Water is added and the cuttings form a slurry which is removed intermittently by pumping or bailing. In wash boring and jetting, the cuttings are removed with a continuous flow of water from the top of the hole. Wash boring advances the hole by a combination of chopping and washing. Jetting depends primarily on the cutting action of a high-pressure stream of water. Care must be exercised to avoid disturbing and moistening the underlying stratum to be sampled when using these methods. Probing consists of driving or pushing a rod or pipe into the soil and measuring the effort required. Some indication of the nature of the material penetrated is secured by examination of the cuttings from the sludge or wash water, but accurate classifications require other sampling methods.

(b) *Geophysical Methods.*—The use of geophysical techniques involves taking measurements at the surface of the earth to determine subsurface conditions. Geophysical methods include seismic, electrical, magnetic, and gravity, and may be employed during any stage of the investigation: reconnaissance, feasibility, or specifications.

Use of these methods is generally in conjunction with drilling. Geophysics can be used to help determine the best location of future drill holes as well as to help extrapolate foundation conditions between existing drill holes. The cost of performing geophysical surveys is very often less expensive than the cost of drilling; therefore, a judicious use of both methods can produce the desired information at an overall lesser cost.

Although the field work required is relatively inexpensive, the interpretation of results is difficult and requires special training. In order for test results to be usable and reliable, it is necessary that a correlation be made locally with exploration made by borings. The Engineering and Research Center can provide information and assistance on the applicability of geophysical techniques to specific situations, and has available trained personnel and mobile equipment for making geophysical explorations for situations where their use is warranted.

41. Field Tests.—(a) *General.*—In addition to the quantitative data (the number of blows per foot) obtained during the standard penetration test described in designation E–21, three other field tests which are used to obtain information of the in situ ground and which are applicable in exploring foundations are: (1) permeability tests, (2) vane tests, and (3) inplace density tests. The last of these is used also in calculations to determine shrinkage or swell factors between borrow area excavation and embankment yardages.

(b) *Field Permeability Tests.*—Approximate values of permeability of individual strata penetrated by borings can be obtained by making water tests in the holes. The reliability of the values obtained depends on the homogeneity of the stratum tested and on certain restrictions of the mathematical formulas used. However, if reasonable care is exercised in adhering to the recommended procedures, useful results can be obtained during ordinary boring operations. Open end tests and packer tests in boreholes are described in designation E–18. A well permeameter test used for estimating canal seepage losses is described in designation E–19. Use of the more precise methods of determining permeability by pumping from wells with a series of observation holes to measure drawdown of the water table, or by pumping-in tests using large-diameter perforated casing, requires special instructions and advice by specialists on the staff of the Engineering and Research Center, Denver.

(c) *Field Vane Test.*—The vane method of testing to determine the inplace shear resistance of soils has been found advantageous in foundation investigations. This test is applicable primarily in soft, saturated clay soils in which the penetration resistance, as determined by the standard penetration test, is found to be very low. The vane test is much more sensitive and accurate than the penetration test and provides a direct value of shear resistance. The equipment and procedure for making the field vane test are described in designation E–20.

(d) *Inplace Density Tests.*—The sand density method is used to determine the inplace density in a foundation, a borrow area, or a compacted embankment by excavating a hole from a horizontal surface, weighing the material excavated, and determining the volume of the hole by filling it with calibrated sand. A water content determination on a sample of the excavated soil enables the dry density in the ground to be calculated. Various devices using balloons and water, or oil, have been used to measure the volume of the hole, but the sand method is most common. Designation E–24, Field Density Test Procedure, describes the sand density method.

It is often necessary to determine the inplace dry density and water content of fairly deep foundations of cohesive soils above the water table. In these cases use of the sand density method requires the excavation of a test pit to gain access to the soils to be tested. Designation E–23 describes a simple method to obtain inplace density in stages of depth in foundations and borrow areas of dry, gravel-free soils with the use of a hand auger. Undisturbed sampling methods are often used above water table and must be used below water table.

D. Recording and Reporting of Data

42. Maps.—Information that would require many pages of narrative can often be conveyed on a single sheet of paper by use of a map. Among the many varieties of mapping methods available, some suitable procedure can usually be found that will convey the necessary information clearly and easily.

In Bureau of Reclamation investigation work, three scale ranges are in common use. Maps ranging in scale between 1 and 10 miles per inch will usually prove to be suitable for showing the general area of the work; describing access and transportation facilities such as highways, railroads, rivers, and towns; and locating special types of materials deposits such as riprap or aggregates. Map scales ranging between 5,000 feet per inch and 400 feet per inch are often used for more detailed information covering the immediate vicinity of the work; the general geology of the area; the reservoir areas; the location of borrow pits; the right-of-way lines; the locations of roads, canals, and transmission lines; and similar information. To provide detailed information on a structure site, map scales ranging between 500 feet per inch and 20 feet per inch are commonly used; but long, low damsites with limited relief may be mapped at a scale of 1,000 feet per inch. Locations of small structures where local detail is important may be mapped on scales of 20 feet per inch. In selecting a scale it is desirable to keep the ratio of ground measurement to map measurement as simple as possible; for example, the added detail that can be shown using a scale of 750 feet per inch is less beneficial than the convenience of a 1,000-foot-per-inch scale. Also, the complete map should not be larger than may be conveniently spread on an ordinary table. The scale of the map must always be shown.

All detail maps should be controlled by a coordinate system or other definite means of locating points on the ground. When a coordinate system is used the grid lines should run true north and south, and east and west. If a local grid system is established, the origin of the system should be to the south and west of the area under consideration and the displacement should be predominantly in one direction so that there will be a major numerical difference between the north and east coordinates of any point. If a damsite is involved, the displacement should be sufficient so that all the work area, including borrow areas, falls within the northeast quadrant. The grid system should be referenced to public land surveys, triangulation stations, and other prominent permanent features in the area.

Variations in elevation are delineated on detail maps by the use of contours. Contour intervals required may range from 20- or 25-foot intervals to 2-foot and occasionally 1-foot intervals depending on the map scale and on the irregularity of the land surface.

In general, contours should be sufficiently close together that elevations between contours can be determined with confidence but sufficiently separated so that a contour can be followed by eye without difficulty. Elevations should preferably always be referred to sea level on the basis of the nationwide survey system of the U.S. Geological Survey or U.S. Coast and Geodetic Survey. If for reconnaissance purposes an assumed datum is used, it should be so different from the sea level datum that no confusion results.

Location and general maps are oriented with north to the top of the page. Detail maps for water conveyance or storage structures are oriented for water movement to the top or to the right of the page. Railroad and highway location maps are oriented in accordance with the practices of the organization involved. Every map should contain a north arrow.

Location maps should show all established transportation routes and communities adjacent to the area in question. Reservoir maps should show all major manmade fixed facilities including but not restricted to railroads, highways, pipelines, canals, telephone lines, and powerlines; buildings, mines, cemeteries, reservoirs, and wells. Also, the type and kind of plant cover should be delineated. Detail maps, in addition to showing the features already listed, should show rock outcrops, talus, recognizable slides, waterways, survey monuments and benchmarks, and section, township, and county lines.

For specifications a map is required showing the extent to which right-of-way will be acquired for the structure involved. Such a map should show property lines and ownership of individual areas.

43. Logging of Exploratory Holes.—(a) *Location of Holes.*—The location of exploratory holes is governed by their purpose. The initial holes drilled or excavated in an area are usually for the purpose of clarifying geological conditions, and the location is therefore governed primarily by geologic structure; the final holes are drilled or excavated primarily for engineering purposes and are located therefore on the basis of the engineering structure to be built. Holes are also drilled or excavated both to establish the form and shape of a geologic unit and to examine the character of a geologic discontinuity. Although it is desirable to locate holes so as to satisfy as many of these requirements as possible, sometimes these respective requirements are contradictory and separate holes are required. From an engineering standpoint, holes that bracket a condition are commonly the more desirable if other design requirements provide sufficient flexibility so that movement of the structure is possible to avoid an unfavorable condition. From a geological standpoint and for those engineering situations where the questionable area cannot be avoided, holes in or crossing questionable areas are preferable.

Every hole drilled or excavated should be definitely located in space; that is, it should be tied to the coordinate grid system and have a coordinate location or tied in some other satisfactory manner, such as stationing or section ties, and its elevation at the top should be established. Coordinates and elevation of an exploration hole or trench should refer to the center of the excavation. However, if it is necessary to adequately describe the materials in a trench by more than one log, as discussed in subsection (c), the coordinates and elevations of each log should be supplied. The direction of long trenches should also be indicated. Any hole drilled or excavated should be logged for the full depth of hole. If for any reason a portion cannot be logged, the interval not logged should be recorded, along with an explanation stating the reason for omission. The bearing and angle from horizontal of angle holes must be reported.

(b) *Identification of Holes.*—To assure completeness of the record and to eliminate confusion, test holes should normally be numbered in the order of excavation and the series should be continuous through the various stages of the work. If a hole is planned and programed it is preferable to maintain the hole number in the record as "not drilled" or "abandoned" with an explanatory note rather than reuse the hole number elsewhere. It is permissible, however, to move holes short distances and retain the program number where such moves are required by local conditions or by changes in engineering plans. When explorations cover several areas such as alternative sites and borrow areas, a new series for each site or borrow area should be used. The favored practice is to start numbering each new area explored at an even hundred.

Test hole numbers are prefixed with a one- or two-letter designation to describe the type of exploration. Frequent usage has established the following letter designations:

DH Drill hole

AH Auger hole (hand)

AP Auger hole (power)

TP Open pit

T   Trench

CH Churn drill hole

PR Penetration resistance holes

VT Holes in which field vane tests are made

In the above designations DH is used to include not only rotary drilling but also all the methods of hole advancement that produce core or relatively undisturbed samples, in contrast to the AH or AP holes that produce highly disturbed samples and CH holes which produce only cuttings and no samples. The CH designation should be used for all types of holes that advance by a chopping and washing action such as wash borings, jetting, or percussion drilling. The TP designation includes both hand- and machine-excavated open holes. Similarly, the T designation includes open trenches whether made by hand or machinery. A conventional designation has not been adopted for tunnels, shafts, geophysical depth determinations, or other infrequently used field test locations. For such test points the full name should be used.

Computerized logs may require a special prefix designation, regardless of type of exploration, and the Engineering and Research Center should be consulted when computerized logs are proposed for use.

(c) *Log Forms.*—A log is a written record of the data concerning materials and conditions encountered in individual test holes. It provides the fundamental facts on which all subsequent conclusions are based, such as need for additional exploration or testing, feasibility of the site, design treatment required, cost of construction, method of construction, and evaluation of structure performance. A log may represent pertinent and important information that is used over a period of years; it may be needed to delineate accurately a change of conditions with the passage of time; it may form an important part of contract documents; and it may be required as basic evidence in a court of law in case of dispute. Each log, therefore, should be factual, accurate, clear, and complete. It should not be misleading. Log-forms are used to record and provide the required information. Examples of logs for three types of exploratory holes are discussed below:

Geologic log of drill hole (fig. 64).—This form is suitable for all types of core borings which produce comparatively undisturbed samples.

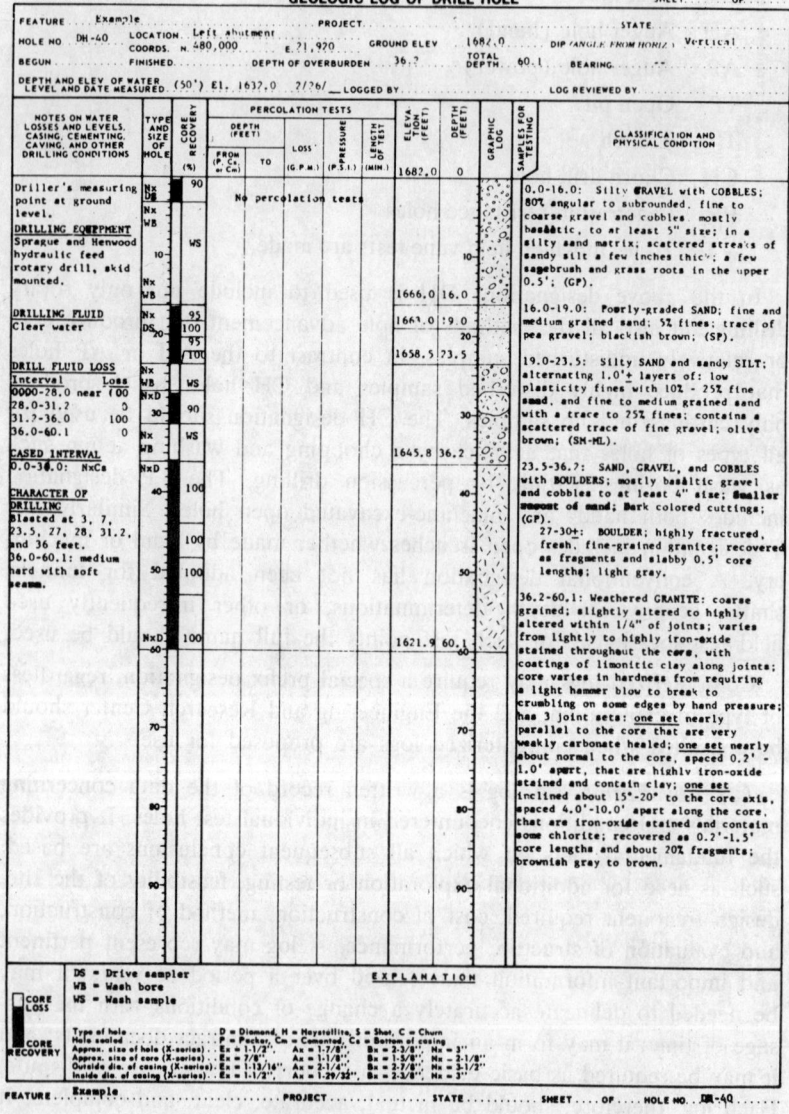

Figure 64.—Geologic log of a drill hole. 101–D–575.

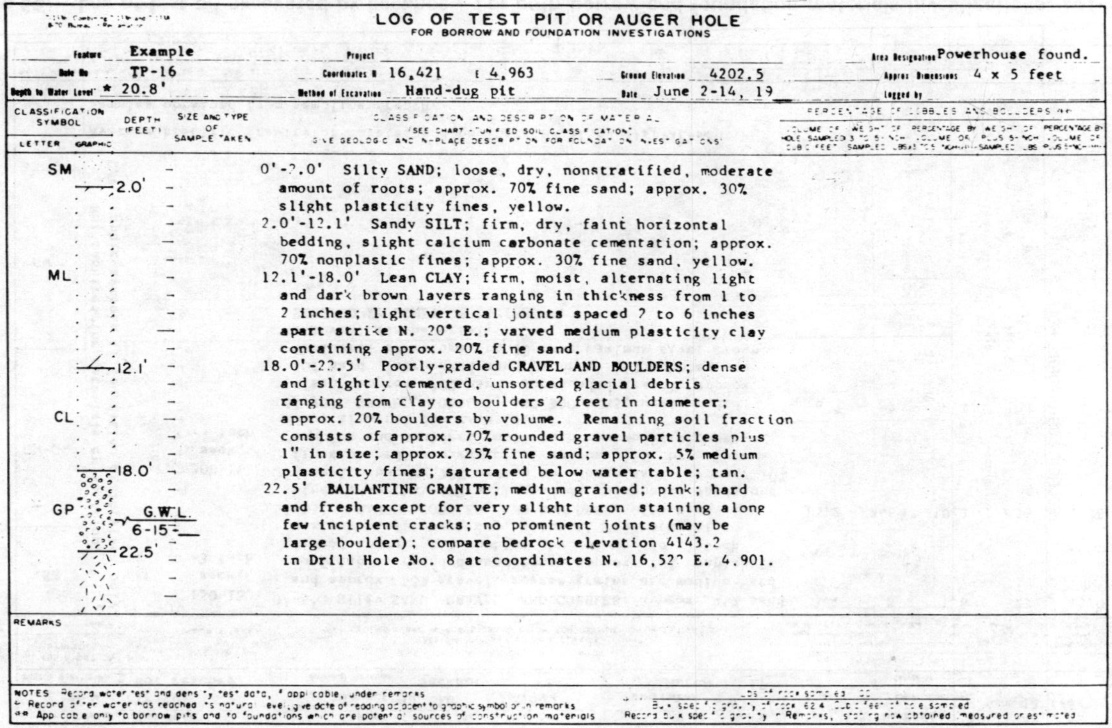

required information. Examples of logs for three types of exploratory

holes are discussed below.

LOG OF TEST PIT OR AUGER HOLE
FOR BORROW AND FOUNDATION INVESTIGATIONS

Feature: Example
Hole No: TP-16
Depth to Water Level: 20.8'
Project:
Coordinates: 16,421 (4,963
Ground Elevation: 4202.5
Method of Excavation: Hand-dug pit
Date: June 2-14, 19
Area Designation: Powerhouse found.
Approx Dimensions: 4 x 5 feet
Logged By:

CLASSIFICATION SYMBOL	DEPTH (FEET)	SIZE AND TYPE OF SAMPLE TAKEN	CLASSIFICATION AND DESCRIPTION OF MATERIALS

SM — 2.0' — 0'-2.0' Silty SAND; loose, dry, nonstratified, moderate amount of roots; approx. 70% fine sand; approx. 30% slight plasticity fines, yellow.

2.0'-12.1' Sandy SILT; firm, dry, faint horizontal bedding, slight calcium carbonate cementation; approx. 70% nonplastic fines; approx. 30% fine sand, yellow.

ML — 12.1' — 12.1'-18.0' Lean CLAY; firm, moist, alternating light and dark brown layers ranging in thickness from 1 to 2 inches; light vertical joints spaced 7 to 6 inches apart strike N. 20° E.; varved medium plasticity clay containing approx. 20% fine sand.

CL — 18.0' — 18.0'-22.5' Poorly-graded GRAVEL AND BOULDERS; dense and slightly cemented, unsorted glacial debris ranging from clay to boulders 2 feet in diameter; approx. 20% boulders by volume. Remaining soil fraction consists of approx. 70% rounded gravel particles plus 1" in size; approx. 25% fine sand; approx. 5% medium plasticity fines; saturated below water table; tan.

GP — 22.5 — 22.5' BALLANTINE GRANITE; medium grained; pink; hard and fresh except for very slight iron staining along few incipient cracks; no prominent joints (may be large boulder); compare bedrock elevation 4143.2 in Drill Hole No. 8 at coordinates N. 16,52° E. 4,901.

GWL — 6-15'

REMARKS

NOTES

Figure 65.—Log of a hand-dug test pit—For foundation investigation. 101-D-576.

LOG OF TEST PIT OR AUGER HOLE
FOR BORROW AND FOUNDATION INVESTIGATIONS

Feature: Example   Hole No: TP-101   Depth to Water Level: * Not reached

Project: _____   Coordinates: N 17,210 E 2,600   Ground Elevation: 779.4   Date: March 2, 19___

Method of Excavation: Backhoe

Area Designation: Spillway   Approx Dimensions: 4 x 13 feet   Logged by: _____

CLASSIFICATION SYMBOL (LETTER, GRAPHIC)

CLASSIFICATION AND DESCRIPTION OF MATERIAL (SEE CHART FOR UNIFIED SOIL CLASSIFICATION)
GIVE GEOLOGIC AND IN-PLACE DESCRIPTION FOR FOUNDATION INVESTIGATIONS

DEPTH (FEET), SIZE AND TYPE OF SAMPLE TAKEN

PERCENTAGE OF COBBLES AND BOULDERS

SM — 0'-5' Silty SAND, GRAVEL, AND COBBLES; approx. 45% sand and approx. 35% gravel, coarse grains are medium-hard to friable, rounded sandstone; approx. 20% nonplastic fines; approx. 15% cobbles, by volume, to 12-inch maximum size; gray; dry; alluvial fan material.
Sample: 150-lb. sack, -3 inch

GW-GC — 5'-15' GRAVEL, COBBLES, AND BOULDERS, with sand-clay binder; approx. 60% well-graded, subrounded sandstone gravel; approx. 30% medium to coarse sand; approx. 10% fines of medium plasticity; approx. 11%, by volume, cobbles and boulders to 30-inch maximum size; brown; dry; alluvial fan material.
Sample: 300-lb. sack, -3 inch

GW — 15'-25' GRAVEL, COBBLES, AND BOULDERS, well-graded, clean; approx. 70% hard, subrounded gravel; approx. 30% hard, subrounded sand; approx. 14% cobbles and boulders, by volume, to 30-inch maximum size; brown; alluvial fan material.
STOPPED BY LIMIT OF EQUIPMENT.

Not required for materials exploration.
Use graphic symbols if foundation exploration only.

Symbol	CU FT OF SAMPLE	WEIGHT (LB)	PERCENTAGE BY WEIGHT		
SM	3.1	8	1.6	73	15.0
GW-GC	15.2	246	10.3	756	10.8
GW	6.7	87	8.4	137	14.1

REMARKS: Average specific gravity of cobbles and boulders: 2.51 by displacement.
Samples obtained from sampling trench.

NOTES: ...

Figure 66.—Log of test pit excavated by backhoe—For both borrow and foundation materials investigations. 101-D-577.

LOG OF TEST PIT OR AUGER HOLE
FOR BORROW AND FOUNDATION INVESTIGATIONS

Feature: Example
Hole No.: AH-455
Depth to Water Level: *21.9 feet

Project:
Coordinates N 16,500 E 6,910
Method of Excavation: Auger hole

Area Designation: Borrow Area L-4
Approx. Dimensions: 24-inch-dia.
Ground Elevation 669.8
Date April 5-12, 19
Logged by

Depth (feet)	Size and type of sample taken	Classification and description of material	Letter symbol	40	Vol. of sample	Percentage	Percentage by weight of cobbles and boulders
0'-2'	75-lb. sack	SILT, slightly organic with some alfalfa and weed roots; nonplastic; small amount of fine sand; dark brown; dry.	ML	40	0	0	0
2'-8.5'	175-lb. sack	Lean CLAY, medium plasticity, high dry strength; approx. 25% sand and gravel to 3/4-inch size; most of gravel is shale; brown; dry.	CL	130	0	0	0
8.5'-16'		Micaceous SILT, moderate amount of very fine sand, noticeable mica flakes; very slight plasticity; tan; dry.		150	0	0	0
16'-18'	200-lb. sack	SILT, similar to material 8.5 to 16 feet, but contains about 20% shaly gravel to 1-inch size; red; dry.	ML-ML	40	0	0	0
18'-25'	90-lb. sack	GRAVEL-SAND MIXTURE WITH COBBLES, well-graded; approx. 50% gravel and 50% sand, mostly hard, sub-rounded; approx. 9% cobbles by volume to 8-inch maximum size; very small amount of nonplastic fines; black; dry above water table; river terrace gravel. STOPPED BY HARD MATERIAL.	ML	80	2430	19.1	1150	9

G.W.L.: 21.9 4/12/7?

GW-SW

Not required for materials explorations

REMARKS: Inplace density test at 8 feet: dry density 89.4 p.c.f., water content 8.9 percent.
Bulk specific gravity of cobbles and boulders: 2.55 by displacement.

NOTES: Record water test and density test data, if applicable, under remarks
* Record if water has reached its natural level; give date of reading adjacent to graphic symbol or in remarks
** Applicable only to borrow pits and to foundations which are potential sources of construction materials

Figure 67.—Log of an auger hole—For borrow materials investigation. 101-D-578.

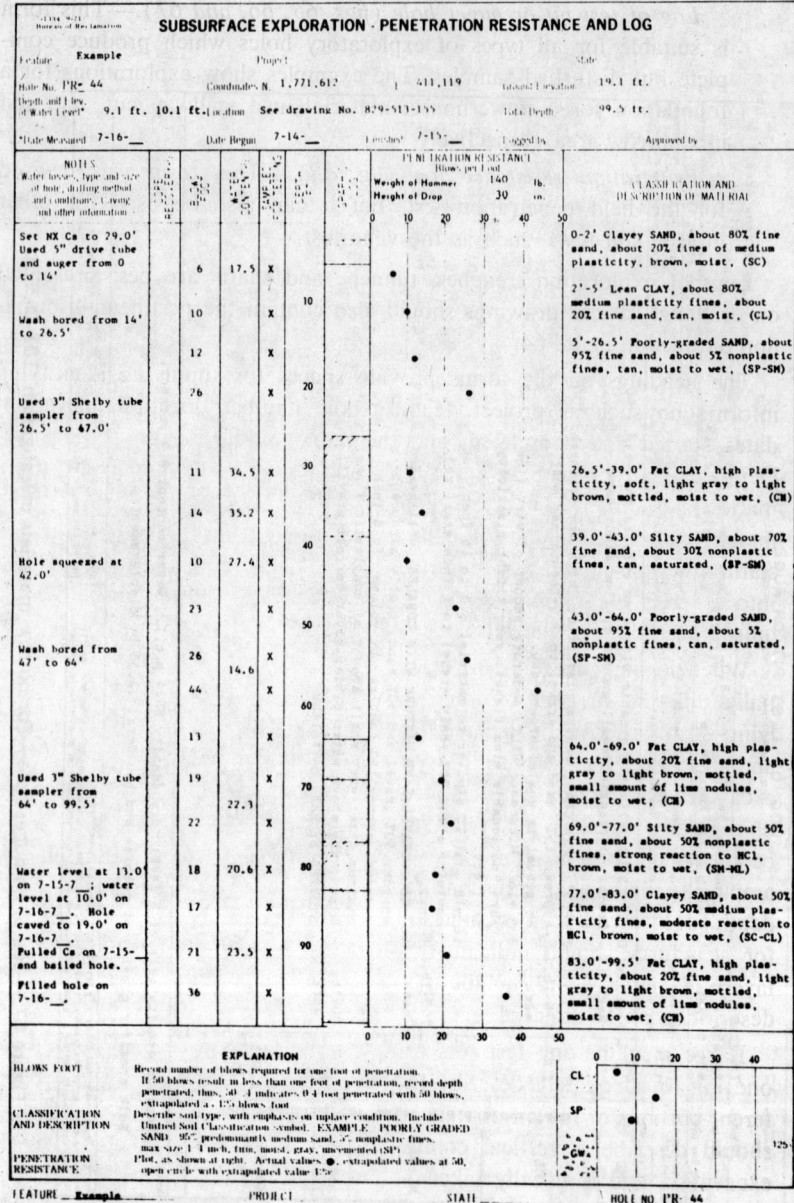

Figure 68.—Penetration resistance and drill hole data for subsurface exploration. 101-D-579.

Log of test pit or auger hole (figs. 65, 66, and 67).—This form is suitable for all types of exploratory holes which produce complete but disturbed samples. The examples show explorations for a foundation for a powerhouse, material in a spillway cut, and soil in a borrow area, respectively.

Penetration resistance log (fig. 68).—This form was devised for the field penetration test, but it can also be used for other inplace soil tests, such as the vane test.

Logs of exploration trenches, tunnels, and shafts are best presented on drawings. These drawings should also contain the pertinent information outlined in figure 64.

The headings on the forms provide spaces for supplying identifying information such as project, feature, hole number, location, elevation, dates started and completed, and the name of the person responsible. Depth to bedrock and to water table are valuable and important information and should always be reported. When this and any other information called for on the log cannot be obtained, justification in writing stating the reasons why is required. The body of the log form is divided into a series of columns covering the various kinds of information required according to the type of exploratory hole.

When logging overburden, every stratum of material that is substantially different in composition from either the overlying or the underlying strata should be located by depth interval, separately classified, and described in the body of the log. Except in holes that have been excavated for exploration of a structure foundation, thin layers or lenses of different material in a relatively uniform stratum of material should be described, but need not be separately classified on the log; for example, "A 1-inch-thick discontinuous lens of fine sand occurs at 7-foot depth." However, logs of holes excavated for exploring of foundations for structures should indicate the location by depth of all lenses and layers of material and include the classification in addition to a detailed description of the material.

Large machine-dug test pits or test trenches may require more than one log to adequately describe the variations in materials found in different portions of the pit or trench. The initial log of such pits or trenches should describe a vertical column of soil at the deepest part of the excavation and is usually taken at the center of one wall of the pit or trench. If this one log will not adequately describe the variations in the different strata exposed by the pit or trench, additional logs for other locations within the test excavation should be prepared to give a true representation of all strata encountered in the test pit or trench. In long trenches, at least one log for each 50 feet of trench wall should be prepared, regardless of the uniformity of the material or strata. A geologic section of one or both walls of long test trenches may be required to describe variations in strata and material between log locations. When more than one log is needed to describe the material in an exploratory pit or trench, coordinate location and ground surface elevation should be given for each point for which a log is prepared. A plan geologic map and geologic sections should always be prepared for test trenches which encounter bedrock in structure foundations.

A log should always contain information on the size of the hole and on the type of equipment used for boring or excavating the hole. This should include the kind of drilling bit used on drill holes, a description of the penetrating equipment or type of auger used, or method of excavating test pits. The location from which samples are collected should be indicated on the logs, and the amount of material recovered as core should be expressed as a percentage of each length of penetration of the barrel. The logs should also show the extent and the method of support used as the hole is deepened, such as size and depth of casing, location and extent of grouting if used, type of drilling mud, or type of cribbing in test pits. Caving or squeezing material should be noted on the drill hole log as this may represent a low strength stratum.

Information on the presence or absence of water levels and comments on the reliability of these data should be given on all logs. The date measurements are made should be recorded, since water levels fluctuate seasonally. Water levels should be recorded periodically from the time water is first encountered and as the test hole is deepened. Upon completion of drilling, the hole should be bailed and allowed to recover in order to obtain a true water level measurement. Perched water tables and water under artesian pressure are important to note. The extent of water-bearing members should be noted and areas where water is lost as the boring proceeds should be reported, since subsequent work on the hole may preclude duplicating such information. The log should contain information on the water tests made at intervals, as described in section 41. Since it may be desirable to obtain periodic records of water level fluctuations in drilled holes, it should be determined whether this is required before plugging and abandoning the exploratory hole.

Where cobbles and boulders are encountered in explorations for sources of embankment materials, it is important to determine their percentage by volume. The log form for test pit or auger hole (figs. 65, 66, and 67) includes a method for obtaining the percentage by volume of 3- to 5-inch rock and rock over 5 inches in size. The method involves weighing the rock, converting weight to solid volume of rock, and measuring the volume of hole containing the rock. This determination can be made either on the total volume of stratum excavated or on a representative portion of the stratum by means of a sampling trench which is described in designation E–1.

For test holes that penetrate less than 25 feet of potential borrow material, a statement should be made under "Remarks" in the log giving the reason for stopping the hole. For all other types of holes, a statement should be made at the end of the log that the work was completed as required or a statement explaining why the hole was abandoned. Material should not be described as bedrock, slide material, or similar interpretative terminology unless the exploration actually penetrated such conditions and samples were collected to substantiate these conclusions. Terminating statements similar to the following would be considered satisfactory: hole eliminated due to lack of funds; hole caved in including depth to cave; depth limited by capacity of equipment; encountered water; unable to penetrate hard material in bottom of hole.

(d) *Description of Soils.*—The person logging exploratory holes should be able to identify soils according to the Unified Soil Classification System, section 5 and designation E–5. The description of a soil in a log should include its typical name, followed by pertinent descriptive data, as listed in table 1, section 6. After the soil is described, it should be placed in the appropriate soil classification group by use of letter symbols. These group symbols represent a variety of soils having certain common characteristics; hence, by themselves they are not sufficient to describe a particular soil. Borderline classifications (two sets of symbols separated by a hyphen) should be used when the soil does not fall clearly into one of the groups but has strong characteristics of both groups.

Identification and classification of soils in logs of explorations should be based on visual examination and manual tests. Field logs should not contain refinements that can be determined only by use of laboratory equipment. Laboratory tests may be used to assist untrained personnel in verifying their field classifications.

Emphasis should be placed on the natural state of soils being investigated for foundation purposes. The use of a general type name such as loess, caliche, adobe *in addition* to the soil classification name may be helpful in identifying inplace conditions. Figure 69 shows gradation ranges used in the laboratory in identifying various types of loess in the Kansas-Nebraska area.

(e) *Description of Rock Cores.*—The basic objective of describing rock cores is to provide a concise record of the important geological and physical characteristics of the core materials.

Description of the rock core should include its typical rock name followed by data on its lithologic and structural features, physical condition including alteration, and any special geologic, mineralogic, or

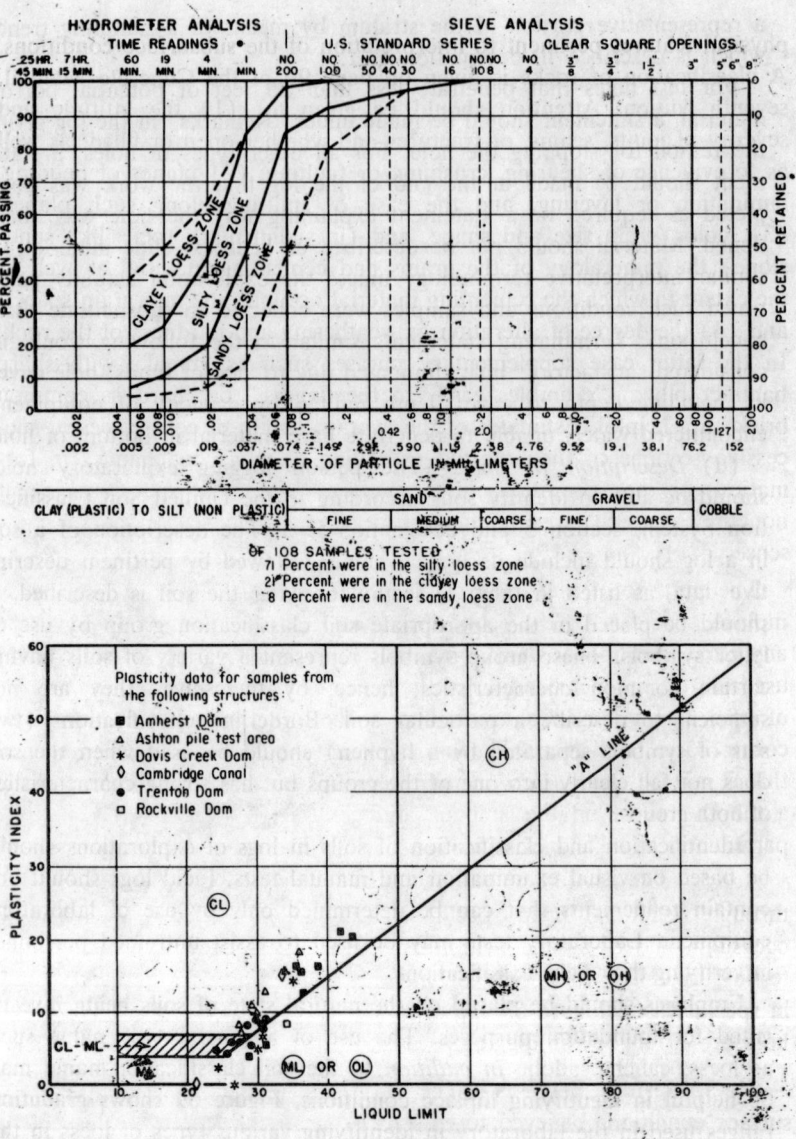

Figure 69.—Gradation and plasticity data for loessial soils in the Kansas-Nebraska area. 101–D–221.

physical features pertinent to interpretation of the subsurface conditions. A classification of rocks is given on page 96 of the Concrete Manual, seventh edition. Attention should be given to: (1) the attitude and severity of joints, seams, or fractures and whether open or filled, as well as to evidence of shearing, crushing, or faulting; (2) planes of bedding, lamination or layering, and the ease of splitting along such planes; (3), color, grain size and shape, and (in sedimentary rocks like sandstone) the mineralogy of the grains and cementing material as well as the extent to which the cementing material occupies the intergrain spaces; and (4) the degree of alteration or weathering and hardness of the rock. In the latter case supplementary phrases such as "breaks with sharp hammer blow," "crumbles easily in the fingers," or "hard as common brick" are helpful. Estimates of the average length of core pieces in successive sections of the hole aid in calling attention to changes in formations or rock conditions in the hole not otherwise recognizable but nonetheless useful in evaluating subsurface conditions in the engineering sense.

The purpose of the drilling and logging is to secure evidence of the inplace condition of the rock; therefore, care should be taken to note any core condition or damage due to the type of drill bit or core barrel used or to improper conduct of the drilling process. One common cause of core damage or breakage is due to the use of a core catcher in the coring barrel and, therefore, for most rock sampling it is recommended that it not be used unless absolutely necessary. Such factors may have a marked effect on the amount and condition of the core recovered, particularly in soft, friable, or severely fractured rock.

Adequate logs or descriptions of rock core can be prepared solely through visual or "hand specimen" examination of the core with the occasional aid of simple field tests. Detailed microscopic or laboratory testing to define rock type or mineralogy is generally necessary only in special cases. Figure 63 shows how rock cores obtained from a borehole are arranged for logging.

44. Subsurface Sections.—The use of sections to show the subsurface conditions believed to exist is both highly beneficial and potentially dangerous, since interpretation of conditions is necessarily involved. Where sections are used in contract documents, the information shown is limited to factual data such as the ground surface line and logs of drill holes located in their actual position with respect to the ground surface line. Although the choice of sections is made to simplify interpretation, actual locations of features such as bedrock, water table, etc., are not shown by continuous lines but only where they are encountered in each hole. The exceptions to these restrictions are the cross sections of exploratory trenches, tunnels, and shafts where conditions can be mapped by actual observation.

On the other hand, sections showing the subsurface conditions believed to exist are highly desirable in geological reports, materials reports, and design data for dams, canals, and other project features. The location of the sections should be chosen so as to present in the best possible way the conditions described. Cross-valley sections are generally much more informative than a series of sections parallel to the valley. Also sections should cross physical features as nearly as practicable at right angles. A clear differentiation should always be maintained between factual and interpretive data. The commonly used system which ranges between dotted and solid lines, in which dots represent purely hypothetical interpretation, a solid line represents fact, and dashed lines define the degree of reliability of intermediate data according to length of dash, is recommended. Lines of different weight should not be used for this purpose but should be reserved for use as a method of emphasis. The cross section should always show the name of the person who made the interpretation and the date the interpretation was made.

45. Sampling.—Samples of soil and of rock are collected for visual examination in order that a log of the test hole may be prepared, for preservation as representative samples in support of the descriptive log, for testing to determine index properties, and for laboratory testing to determine engineering properties.

Requirements for undisturbed samples are given in designation E–2.

When boring core holes the total material recovered as core is collected and stored in core boxes. In addition, samples of soil and rock should be collected and placed in sealed pint jars to preserve their natural water content representative of each moist or wet stratum if the method of exploration permits the collection of such samples. Samples representative of the various types of material found in the area under investigation should be collected as the work progresses. Upon completion of the investigation phase, samples may be required to be transmitted to the Engineering and Research Center. These samples should be 4 to 6 inches in length and must be representative of that material as it is found in the area, particularly as to the degree of alteration. If there is a wide variation in quality, samples representative of the best, average, and poorest of the material should be collected.

In the exploration of materials in borrow areas and in foundations where substantial quantities occur which potentially may be used in embankment construction, samples should be collected representative of each stratum in a volume sufficient to provide 75 pounds of material passing a No. 4 sieve, when practicable, to be used for testing for engineering properties. Only material larger than 3 inches should be removed from a sample and the percentage of "plus 3-inch" removed should be reported. However, in some cases larger samples are required for tests on total material. If the entire test hole appears to be in uniform material, samples from the upper, middle, and lower thirds of the test hole should be collected. Designation E–1 describes sizes and procedures for securing disturbed samples, and should be referred to for size of sample required for a particular test.

In the investigation of riprap sources, samples consisting of three or four pieces of rock totaling at least 600 pounds should be collected representative of the source. The collection of samples of concrete aggregates is covered in the Concrete Manual. The collection of samples of blanketing material, filter material, and ballast should conform to requirements for collection of borrow material for embankment construction.

From the suite of samples collected as described above (also see sec. 32(g)), samples are selected for performing index and engineering properties tests. Care should be taken to preserve sufficient samples for the support of the logs of the exploratory holes and for representative samples to be transmitted to the Engineering and Research Center laboratories in Denver for the determination of engineering properties.

Samples collected in the process of routine exploration are not as a rule satisfactory for testing associated with the determination of the properties of the soil or rock en masse in its natural condition. For this purpose, samples are collected of material unaffected by seasonal climatic influence from large-diameter boreholes (4 to 6 inches in diameter minimum), or from the bottom of open pits. Borehole samples should be 12 to 24 inches long and open-pit samples 10- to 12-inch cubes. Every effort should be made to preserve such samples in as nearly their natural condition as possible. Procedures for the collection and preservation of this type sample are described in designation E–2.

46. Reports.—The results of every investigation should be presented in a report. For the reconnaissance stage for a small structure, a letter report describing in general terms the nature of the problems associated with the investigation, the extent of the investigation, and the conclusions reached may suffice. As the investigation proceeds, additional data are collected, evaluated, and accepted or rejected. As the various stages of investigation are reached, the previously assembled material is incorporated in a progress report. During preparation of this progress report, previous reports should be examined, and those questions which have been answered noted in the report together with the solution or, if unanswered, carried over for future consideration. The final report prepared prior to preparation of specifications should either answer all the questions raised in the past or show that a positive solution cannot be reached within the scope of an investigation.

Every investigation should contain a statement of the purpose of the investigation, the stage for which the report is being prepared, the kind of structure contemplated, and the principal dimensions. The general requirements for submittal of design data for various kinds of structures and for various stages of investigation are given in the Reclamation Instructions, part 133. The following features pertaining to foundations and earthwork should be included in all reports:

Foundations.—Foundation data should reflect a recognition and consideration of the type and size of the particular engineering structure and the effect on or relationship to the structure of the significant characteristics of the foundation materials and conditions at the particular site.

The general regional geology should be described. Such a description would include the major geological features, the names of the formations found in the area, their age, their relationship to one another, and their general physical characteristics.

A description and interpretation of local geology, including the physical quality and geologic structure of the foundation strata, ground-water and seismic conditions, existing and potential slide areas, and engineering geologic interpretations appropriate to the engineering structure involved, should be given. The geologic logs of all subsurface explorations should be included in the report as well as a geologic map plotted on the topographic map of the site showing surface geology and the location of geologic sections and explorations. This map should be supplemented by geologic sections showing known and interpreted geologic conditions. Photographs of pertinent geologic and topographic features of the terrain, including aerial photographs for mosaics, if available, are valuable additions to the report.

Engineering data on overburden soils within the foundation of the proposed structure should be shown by detailed soil profiles and reported in accordance with the following:

(1) A classification in accordance with the Unified Soil Classification System of the soil in each major stratum.

(2) A description of the undisturbed state of the soil in the stratum.

(3) A delineation of the lateral extent and thickness of critical, competent, poor, or potentially unstable strata.

(4) An estimate, or a determination by tests, of the significant engineering properties of the strata, such as density, permeability, shear strength, and compressibility or expansion characteristics;

and the effect of structure load, changes in moisture, and fluctuations or permanent rise of ground water, on these properties.

(5) An estimate or a determination of the corrosive properties and sulphate content of the soil and ground water as affecting the choice of cement.

For data on bedrock, the following are required:

(1) A description of the depth to and contour of bedrock, thickness of weathered, altered, or otherwise softened zones, and other structural weaknesses and discontinuities.

(2) A delineation of structurally weak, pervious, and potentially unstable zones and strata of soft rock and/or soil.

(3) An estimate or a determination (depending on the stage) of the significant engineering properties of the bedrock, such as density, absorption, permeability, shear strength, and strain characteristics; and the effect of structure load, changes in moisture, and fluctuations or permanent rise of ground water on these properties.

Construction materials data.—An earth materials report containing an inventory of available impermeable-type soils, permeable-type soils, and rock for riprap and rockfill is required as part of the feasibility and specifications design data for earth dams, and occasionally for canals and other large structures when appreciable quantities of these materials are required. Similar reports for small quantities of special materials are sometimes needed. The principal items to be covered in the earth materials report follow:

(1) A grid map showing the topography of the deposit and of the structure site, and the intervening terrain if within a radius of 2 miles. The location of test holes and trenches should be shown, using standard symbols.

(2) Ownership of deposit.

(3) Brief description of topography and vegetation.

(4) The estimated thickness of the deposit, including variations. Drawings showing subsurface profiles along grid lines should be included.

(5) The areal extent of the deposit.

(6) The estimated quantity of the deposit.

(7) Type and thickness of overburden.

(8) The accessibility of the source.

(9) General description of the rock.

(10) The amount of jointing and thickness of bedding of rock strata.

(11) Spacing, shape, angularity, average size, and range of sizes of natural boulder deposits.

(12) Brief description of shape and angularity of rock fragments found on slopes of rock deposits and of the manner and sizes in which the rock breaks in blasting.

(13) Logs of all auger test holes and exposed faces of test trenches or pits. Figure 67 is the recommended form to be used.

(14) An estimate or determination by tests of the pertinent index and engineering properties of the soils encountered. The amount of testing should be limited in the feasibility stage but should be more detailed for specifications.

(15) Photographs, maps, and other drawings are helpful and desirable for the record of explorations.

In most cases information gathered for the earth materials report for an earth dam, which is furnished for final design and specifications, will not be of sufficient detail to permit development of a plan for utilizing available earth materials to the best advantage. As soon as funds are available for the construction of the dam, each borrow area included in the specifications should be studied thoroughly and systematically.

The primary purposes of the detail study are to determine the depth of borrow-pit cuts, the most practicable distribution of the materials to be placed in the embankment, and the need for addition or removal of moisture. In most cases it has been found desirable to add moisture to dry, impervious borrow materials prior to excavation. Studies should include an analysis of the moisture conditions in each borrow area from which plans may be developed for irrigation of the areas. If the materials in borrow areas are too wet for proper placement, plans for draining these areas may be based on results of the detailed studies. Seasonal variations of water content, variation of moisture with depth, and rate of water penetration are items requiring consideration.

The foregoing detailed investigations are also desirable for canals and structures where large quantities of required excavations and borrow are involved. In any case, the construction forces should make sufficient preconstruction explorations to know where specified types of materials are to be obtained and where all materials are to be placed.

Information on concrete aggregates should be reported in accordance with instructions given in chapter II of the Concrete Manual, seventh edition. Information on sources and character of acceptable road surfacing materials, if required, should be given in the construction materials report. Reference should be made to results of sampling and analysis of materials, including previous tests made in the Engineering and Research Center laboratories. Figures 70 and 71 are examples of forms for summarizing field and laboratory tests on embankment materials which accompany preconstruction reports.

Form 7-1465 (3-58)
BUREAU OF RECLAMATION

SUMMARY OF FIELD AND LABORATORY TESTS FOR EMBANKMENT MATERIALS REPORT
IMPERMEABLE TYPE MATERIALS

Feature _____
Project _____
Date of report _____
Borrow area _____

		FIELD						FIELD AND LABORATORY					LABORATORY											
					OVERSIZE PERCENTAGE OF VOLUME		DENSITY IN PLACE MINUS No.4 FRACTION			MECHANICAL AND PSS MINUS No.4 FRACTION % OF DRY WEIGHT		CHARACTERISTICS AT STANDARD LABORATORY COMPACTION				PERCOLATION - SETTLEMENT TEST								
																BEFORE SATURATION				SATURATED			
IDENTIFICATION OF EXPLORATION (PIECE AND NUMBER)	LOCATION OF EXPLORATION (COORDINATES)	COMPOSITE DEPTH OF SAMPLE TESTED OR DEPTH OF DENSITY TEST (FT)	FIELD IDENTIFICATION (UNIFIED SOIL CLASSIFICATION GROUP SYMBOL)	MAX PARTICLE SIZE (INCHES)	OVER 5 INCHES	OVER 3 INCHES	SPECIFIC GRAVITY PLUS No.4 FRACTION	DRY DENSITY (LBS PER CF)	WATER CONTENT % OF DRY WT	% PASSING No.4 SIEVE	% PASSING No.200 SIEVE	PROCTOR MAX DRY DENSITY LBS PER CF OF DRY WT	OPTIMUM WATER CONTENT % OF DRY WT	PROCTOR NEEDLE (LBS PER SQ IN)	SPECIFIC GRAVITY MINUS No.4 FRACTION	DRY DENSITY AS PLACED LBS PER CF OF DRY WT	WATER CONTENT AS PLACED % OF DRY WT	APPLIED PRESSURE LBS PER SQ IN	SETTLEMENT % OF PLACED SAMPLE	PROCTOR NEEDLE LBS PER SQ IN	WATER CONTENT % OF DRY WT	SETTLEMENT % OF PLACED SAMPLE	PERCOLATION RATE (FT PER YR)
(1)	(2)	(3)	(4)	(5)	(6)	(7)	(8)	(9)	(10)	(11)	(12)	(13)	(14)	(15)	(16)	(17)	(18)	(19)	(20)	(21)	(22)	(23)	(24)
TP-200	N 202,000-10,000 E	10-21.0	GW-GC	6	3.0	7.0	2.54	—	—	41	10.0	126.1	12.3	1200	2.89	123.1	10.3	20	3.1	900	14.3	3.1	0.37
AH-400	N 177,000-8,000 E	15-21.5	CL	8	0	0	—	114.0	8.2	82.3	6.0	112.2	19.1	1100	2.73	109.8	17.1	100	4.1	450	22.6	41	0.012
TP-201	N 202,000-10,500 E	5.0	GC	3	0	0	—	—	—	40.2	11.0	125.6	12.6	1100	—	—	—	—	—	—	—	—	—

NOTES

Column
(1) TP-200 and AH-400 are in different borrow areas and would ordinarily be on different sheets
(2) Depths are from ground surface. Topsoil should not be included in the sample
(4) Classification is for composite sample
(5) Maximum size of particle encountered in depths shown in column (3)
(6) Obtained by weighing rock, assuming or measuring specific gravity and measuring volume of hole or sampling trench
(8) When determined in laboratory.
(9)(10) See field density test procedure. Tests required for coarsest, finest, and average material for each proposed borrow source.
(13) Proctor compaction test
(14) Use optimum water content and maximum dry density (Columns (13) and (14))
(19) Use 100 psi for manifestly impervious soils, 20 psi for soils of doubtful imperviousness.
(20) Determined from settlement gage readings after consolidation under load is practically complete, and before saturation
(21) Average of 3 readings at end of test
(22) Obtained by drying and weighing of sample.
(23) Determined from settlement gage reading at end of test, with load (column (19)) held constant

Figure 70.—Summary of field and laboratory tests for embankment materials report—Impermeable-type materials. 101-D-580.

Figure 71.—Summary of field and laboratory tests for embankment materials report—Permeable-type materials. 101-D-581.

Chapter III

CONTROL OF EARTH CONSTRUCTION

A. Principles of Construction Control

47. Importance of Control.—The preparation of plans for any engineering structure requires the selection of materials which have adequate strength to withstand the forces to which the structure will be subjected during its life. In addition, these materials should be used in such quantities and in such a way that the cost of the structure will be as small as possible. In practice there will be a range within which materials and the method of utilization are considered to meet these criteria. No structure can be considered adequate unless the construction is maintained within these limits, regardless of the nature of the design.

In many types of engineering work, structural materials are manufactured to have certain characteristics and their method of utilization is prescribed by building codes, handbooks, and codes of practice established by various engineering organizations. However, for earth construction it is common practice to use material that is available locally rather than to specify that a particular type of material of specific properties be secured. Likewise, there is usually a variety of procedures available by which earth materials may be incorporated into a structure satisfactorily. When earth is the construction material, therefore, it becomes necessary for personnel in charge of construction control to be familiar with the design requirements and to see that the finished product meets these requirements.

The designer must provide a greater range of tolerance in earth material utilization than other construction materials require for two reasons: (1) because of the inherent variability in the properties of earth materials, both from deposit-to-deposit and within the confines of a single deposit, and (2) because of the large dimensions of many soil structures and the type of construction methods used. For maximum economy, tolerance ranges will vary according to available materials, conditions of use, and available methods of construction. A much closer relationship among the operations of inspection, design, and construction is required for earthwork than is needed in other fields of engineering. Construction control of earthwork involves not only practices similar to those normally required for structures using manufactured materials, but also the supervision and inspection normally performed at the manufacturing plant. In earthwork construction the processes by which an acceptable material is produced are performed in the field.

The requirements for knowledge and experience, for the use of judgment, and for responsibility and authority of those engaged in the administration and inspection of earthwork cannot be overemphasized. These requirements have not been met on so many occasions in the past that every authority on earthwork has commented on this problem at one time or another. For example, J. P. Justin in his book, "Earth Dam Projects", wrote:

> "An entirely safe and substantial design may be entirely ruined by careless and shoddy execution, and the failure of the structure may very possibly be the result. Careful attention to the details of construction is, therefore, fully as important as the preliminary investigation and design."

48. Organization.—The structures designed by the Bureau of Reclamation are usually built by contract. The basis of the contract is a set of plans and specifications which contains a schedule of items of work. On the basis of the information provided in the plans and specifications, the contractor proposes a set of prices for performing the items of work which, when accepted, becomes a part of the agreement between the Government and the contractor. The primary functions of the construction control organization are to assure that the structure is built according to the plans and specifications, to certify to what extent the items of work have been performed, and to determine the payment due to the contractor. Although the Government has prepared the plans and specifications defining the work to be performed, once a contract is signed the Government representatives have no more right to change the contract requirements than the contractor has to change his unit prices. Occasionally, a condition will exist in the field which is drastically different from that anticipated to exist during the preparation of the specifications. If a changed condition does exist and adjustments to the designs are necessary, an "Order for Changes" is agreed upon by both the contractor and the contracting officer.

The Government representative is known as the contracting officer, usually the project engineer, and his representative at the site of the work is called the construction engineer. The construction engineer, with the help of his organization, inspects the work, determines the quantities of work performed, and prepares vouchers for payment to the contractor for work completed. The functions of the construction control organization can be divided into three parts: administration of the contract, inspection of the work, and testing for compliance with requirements.

49. Specifications.—It is not unusual to find that differences of opinion exist between the contractor and construction engineer as to the intent and requirements of the plans and specifications, especially during the early stages of the contract. In the event that a difference of opinion arises, the construction engineer attempts to explain the requirements for the work in a form mutually acceptable to the contractor and the contracting officer. If an understanding cannot be reached, the problem is referred to the contracting officer and, as may be necessary, ultimately to the Interior Board of Contract Appeals.

The inherent variability of soil and the wide range of treatments designed to utilize the variations in soil characteristics are so great that it is impracticable to write specifications sufficiently exact to cover every condition likely to be encountered. Moreover, experiences of the contractor and the construction engineer have very likely been different, and it is only natural that each will try to interpret the requirement of the work at hand in the light of his experience.

It is important that the construction engineer understand the design criteria which are to be satisfied by the requirements of the specifications. In many cases, the Engineering and Research Center, Denver, prepares a document called "Unit Design Considerations" which explains and clarifies the more critical design criteria. A review of the specifications by the construction engineer in the Engineering and Research Center prior to issuance is another means by which the construction engineer becomes familiar with design criteria.

Specifications requirements for earthwork construction may be grouped into two types: those in which requirements are based on performance, and those in which requirements are based on procedures. This distinction must be clearly recognized. Generally when performance is the essence of the requirement, it is improper for the construction engineer to insist on any particular construction procedure or equipment being used to produce the specified result. When procedure is the essence of the requirement, the construction engineer should not require that a specific performance be achieved. The requirement for earthfill in dam embankment is based on both a minimum procedure and on a performance. If the two requirements are found to be incompatible, adjustment in the performance is provided for in order to obtain compatibility. The desirability of a performance-type requirement is recognized, but the present state of knowledge of soil behavior, the complexity of specifications required, and the extensive testing requirement makes this type of specification impracticable for some types of earthwork construction. Although it is desirable to be continually on the lookout for procedures that will produce a better product at a lesser cost, there is no point in continually experimenting with ideas that have been tried and found wanting. Therefore, all proposals for modification of specifications requirements, either of procedures or performance, should have the concurrence of the Engineering and Research Center before being put into practice.

Specifications requirements may also be divided into two groups as follows: those in which the requirement is definite, and those qualified by such phrases as "as directed by the contracting officer." The undesirability of the second type of requirement is recognized and this type is avoided whenever possible. However, there are three situations in which the "as directed" type of requirement is used: First, in the establishment of minor dimensions in areas where investigations sufficient to establish such dimensions are not justified; second, situations where any of several possible methods would be satisfactory; and third, new conditions for which requirements have not been established.

Where the "as directed" requirement refers to dimensioning, either maximum and minimum dimensions or an average dimension is given on the drawings. If maximum and minimum dimensions are stated, the usual practice is to excavate to the minimum dimension and, where visual examination indicates that the excavation is still in inferior material, to continue the excavation up to the maximum dimension, if necessary, to reach a satisfactory foundation. Where average dimensions are given, deviations up to 25 percent are considered allowable; and where dimensions are small, deviations up to 50 percent may be acceptable.

Where the "as directed" requirement is based on several possible methods being adequate, no specific method should be required; and if the quality of the work is equivalent to that normally found for such work, it should be accepted.

Where the "as directed" requirement refers to a new condition, the joint effort of the contractor and construction engineer should be utilized to establish performance limits which are practicable or to establish procedures which will produce satisfactory results where performance cannot readily be defined. Some experimentation is desirable but this aspect should be small compared to the total requirement. The construction engineer should prepare a report of his experience to the Engineering and Research Center covering each of these new conditions whenever the contracts he supervises contain such requirements. Discussions of such situations should be incorporated in the monthly report (see section 71).

50. Inspection.—The adequacy of construction is determined by visual examination, by measurements, and by testing. The extent to which each of these procedures is employed will depend on local conditions, on the importance and value of the work being inspected, and on the skill of the inspector; qualified inspectors being most important to the achievement of high quality construction. The relative amounts of each type of inspection will vary as the work progresses. During the initial construction stages the judgment of the inspector should be confirmed at frequent intervals by tests and measurements until his ability at determining adequacy by visual means is proved. Although in some cases the amount of measuring and testing can be reduced as the work progresses, it should never be eliminated.

The inspection operation determines whether requirements of the plans and specifications are being satisfied; it does not determine what these requirements should be. The inspector's primary objective should be to determine precisely the specified requirements for the work he is to inspect and to satisfy himself that the interpretation he makes is proper and just. In determining whether the work meets the requirement, he should be familiar with the way the requirement is defined. For earthwork construction, there are dimensional requirements and quality requirements. A requirement may be defined as a condition to be achieved from which certain deviations plus or minus are tolerable, or it may be defined as a limiting condition from which deviation is allowable in only one specified direction. Both methods are in common use. It should be noted that for some types of work, notably earthfill in dam embankments, that although a procedure is specified, inspection may be made on the basis of satisfying a performance requirement.

Basically, the inspection function determines only whether work is acceptable or not acceptable, but it is usually desirable to determine also the extent to which work is acceptable or unacceptable. An inspector should see that safety requirements are satisfied and that unlawful practices are not being followed by becoming familiar with the Bureau's Safety and Health Regulations for Construction, and provisions applicable to his job. He should be fully informed on the progress of the work and on the future program of work and should be alert to prevent the omission of any part of the work.

During an investigational stage on an earthwork project, tests are made to establish the characteristics of the various kinds of material available. A design is then formulated which may use any or all of the different kinds of material with or without selectivity requirements and with or without processing requirements to develop special characteristics. To assist inspection and to provide positive information that work is performed as planned, a field laboratory is established for performing the necessary tests. Depending on the nature and complexity of the work and the stringency of the performance requirements, testing may range from simple visual classification tests to some of the more complex engineering properties tests.

In order to facilitate construction control, relationships between engineering properties test results and index test results are established wherever possible. In turn, visual observation of the soil characteristics are correlated with both index and engineering properties. When there are very noticeable differences between acceptable and unacceptable work or materials, sufficient testing is required to confirm that the differences persist. As these differences become less obvious, the amount of testing should be increased. In any event, testing should be sufficient to provide adequate quality control and to furnish the necessary permanent records.

It is impracticable to test completely all the work performed. The usual procedure is to select samples of the work or materials for testing which are representative of some unit of work or material. The accuracy of such procedures depends on the relationship of sample size to size of the unit it represents, on the procedures used for sample selection, and on the frequency of sampling. For most of the earthwork construction performed for the Bureau the ratio of sample size to unit of work or material represented is very small. On this basis alone representativeness of a sample is not very reliable. To improve this situation, special sample selection procedures are used. Since the principal objective is to assure adequate work, one practice is to select samples for testing from portions of the work least likely to be adequate. If such samples show minimum acceptable performance or better, the adequacy of the work in general is almost certainly assured.

The other practice for sample selection in inspection operations is to select samples at random at the minimum recommended frequency. When this latter practice is followed, the test results will more nearly represent the average condition of the work performed, but the lower range of performance is not as accurately defined as when the biased sampling practice described above is used. In the evaluation of test results, it is important to know the method used for sample selection.

Bureau construction experience indicates that for most materials likely to be encountered with normal construction procedures and proper construction control, specified requirements will be achieved. Conditions which produce test results below the requirements should be remedied. If there is an appreciable number of borderline test results, immediate steps should be taken to ascertain the cause and to correct it. Methods for locating faulty procedures will be discussed in connection with detailed inspection procedures in the subsequent sections of this manual.

51. Field Laboratory Facilities.—The purpose of the field laboratory is to perform routine testing of construction materials. The test data serve as bases for determining and insuring compliance with the specifications, for securing the maximum benefit from the materials being used, and for providing a complete record of the materials placed.

Laboratory studies of earth materials during investigational stages are made in the Engineering and Research Center laboratories, Denver, and tests for solutions of construction problems may be performed entirely in the field laboratory or in the Engineering and Research Center laboratories, Denver, depending on the complexity of the problem. Progress reports and complete final reports of nonroutine soils investigations made in field laboratories are required to be submitted promptly to the Engineering and Research Center so that timely design decisions may be made. Reports are submitted in duplicate.

The size and type of laboratory are dependent on the magnitude of the job and on the types of structures. If it is necessary to construct a laboratory building, the requirements will usually be met by one of the designs shown in figure 72 or a suitable modification thereof. The arrangement shown for type C is appropriate for major earthwork projects. This design was used for Palisades Dam, which contains approximately 14 million cubic yards of embankment and 150,000 cubic yards of concrete. It was also used at Trinity Dam which has about 30 million cubic yards of earth embankment and about 100,000 cubic yards of concrete. The type B laboratory is applicable to projects where facilities are required for both earth and concrete testing of moderate scale. The type B drawing is a floor plan of the laboratory used at Heart Butte Dam where the volume of earthwork was 1,400,000 cubic yards and the volume of concrete was about 10,000 cubic yards. The small laboratories shown as type A, mobile or stationary, are suitable for small projects, or as a satellite laboratory on a large earthwork project or on projects divided into several work divisions. They may be truck-mounted or trailer units varying in size from 7 by 12 feet to 8 by 30 feet of floor space. An example of a mobile laboratory is shown on figure 73a. For a discussion of concrete laboratory facilities, see sections 59, 60, and 61 of the Concrete Manual, seventh edition.

It may be difficult, occasionally, to determine which type of laboratory will be suitable for a particular job. The size of the contemplated work, the concentration of the work, and the complexity of the materials to be used are the main factors in determining the size and number of laboratories. In most cases, the concrete and earth materials are combined in the same building. However, for small satellite control laboratories, separate facilities for earth control may be desirable. When the earthwork is concentrated at one location, as for a dam, the project

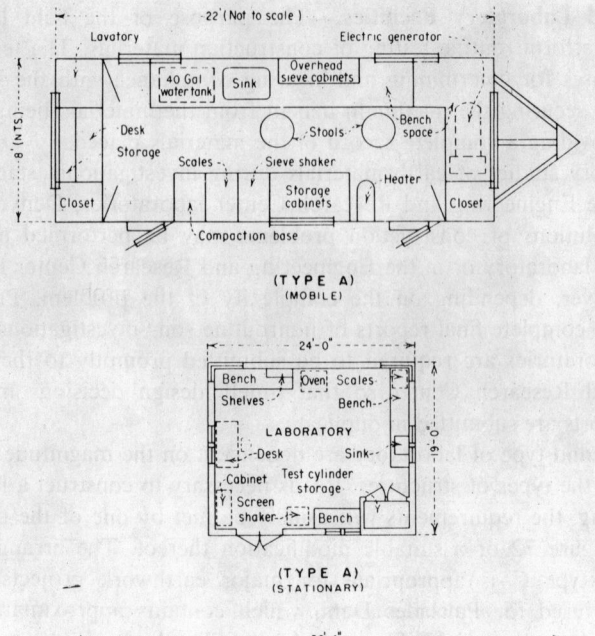

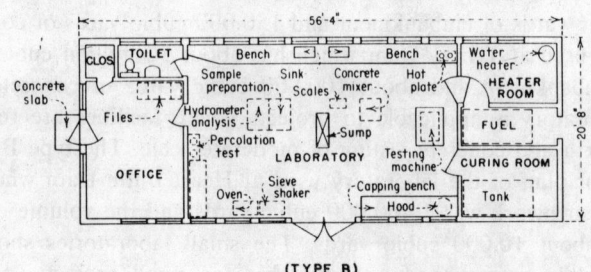

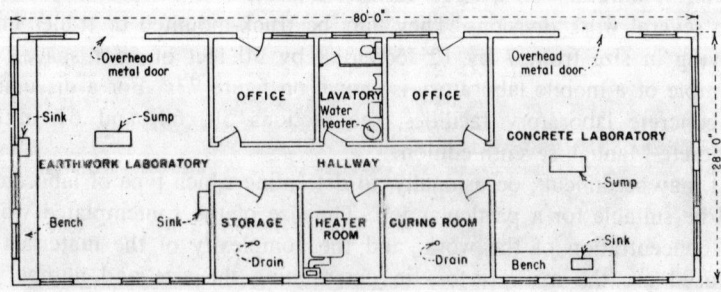

Figure 72.—Examples of floor plans for field control laboratories. 101–D–215 and 101–D–216.

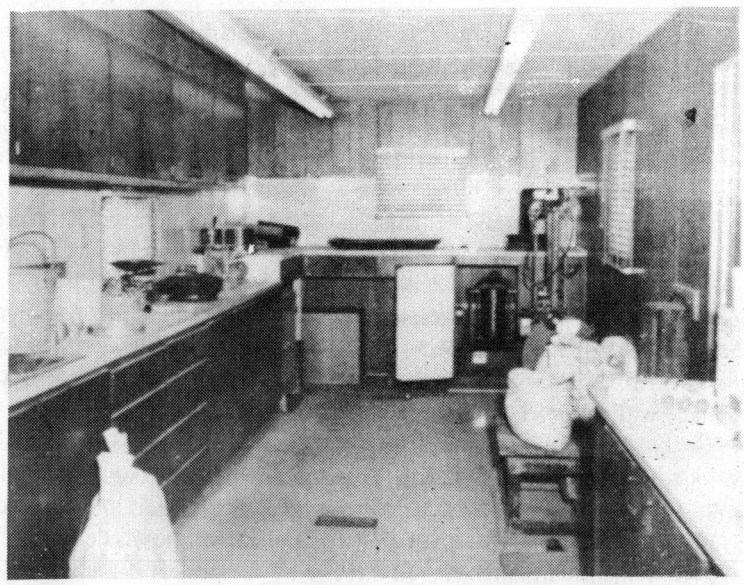

a. Semimobile trailer unit used as a field laboratory.

b. Type V laboratory—Vehicle equipped for on-site earthwork control tests.

Figure 73.—Typical field laboratories. P888–141–179NA and PX–D–16327.

laboratory can be erected near the site of the work so that the laboratory facilities are immediately available for necessary control work. When the earthwork is spread out over long distances, as in the cases of canal and road construction, testing facilities in addition to those at the main project laboratory must be provided near the work. Some projects utilize a type V laboratory; that is, a station-wagon or equivalent equipped with the necessary testing equipment, as shown in figure 73b. Other projects have used small skid-mounted buildings with about 6 by 8 feet of floor space. These buildings are pulled to new locations as the work progresses. Still other projects have utilized large portable boxes in which equipment can be stored. It is advantageous to have sheltered facilities for the testing work in very dry or rainy weather so that the soil control tests can be made without objectionable soil moisture changes.

Designation E-4 contains itemized lists of equipment recommended for each type of laboratory. These lists of equipment are considered minimum to fill the objectives of the respective field laboratories. Special problems may arise where additional equipment will be needed. Laboratory equipment is usually ordered directly by the project office. However, assistance can be obtained from the Engineering and Research Center laboratories when desired.

52. Reports.—A record should be maintained of construction operations. It is valuable in the event that repairs or modification of the structure are required at a later date. A record is necessary in the event that claims are made either by the contractor or the Government that work required or performed was not in accordance with the contract. Recorded data are beneficial in improving engineering knowledge and practices for future work. The basic documents of the construction record are the plans and specifications, modifications adopted that were considered to come within the terms of the contract, amendments made to the contract as extra work orders or orders for changes, and protests. The construction record also contains the results of tests, measurements of work performed, and contract earnings.

To assure that a proper record of construction is developed and available, various periodic reports are required. Review of these reports permits supervisory personnel to determine whether proper performance is being achieved or whether deficiencies or misunderstandings exist. On the basis of such reports, necessary corrections can be made quickly. The report of progress permits the coordination of the various operations required for servicing a contract to be performed in a timely and efficient manner.

The amount of reporting required varies according to function and degree of supervision. Reports are made of every test performed in the laboratory and in the field. Inspectors make daily reports concerning adequacy, progress, and comments on decisions. These daily reports may be of vital importance in subsequent actions. Administrative personnel make monthly reports on quantity of work performed, contract earnings, safety, employment records, and various other statistical information, as required. Intermediate and supervisory personnel summarize these basic data periodically and make frequent informal and monthly formal reports as required which include reports of all decisions made on controversial matters. The basis for reports on decisions is usually the daily entries in personal diaries maintained by supervisory personnel.

Part 175 of the Reclamation Instructions requires that every construction project shall submit a series of periodic progress reports at prescribed intervals. These reports discuss the progress of current construction activities and the structural behavior of completed or partially completed features upon which observations are being made. They also provide a continuous record of events and data for future reference. The reports required for these purposes, where applicable, are as follows:

Symbol	Title
L29	Construction Progress Report, Concrete Construction Data, and Earthwork Construction Data
L10	Foundation Grouting Report
L15	Progress Report on Earth Dam Instrumentation
L16	Final Report on Earth Dam Instrumentation
L23	Periodic Report on Earth Dam Instrumentation
L21	Technical Installation Report.

All except L29 are primarily reports for the transmittal of specific technical data and information.

These "technical" or "specialist" reports contain tabulations of test results, statements of progress and volume of work performed, any questions concerning interpretation of requirements or test results, and descriptions of abnormal conditions or methods that affect either the quality or quantity of work accomplished. Extra copies of each of these reports, including the separate technical portions of the L29, are to be transmitted to the Engineering and Research Center, Denver, for immediate review by specialists so that assistance or advice concerning the performance of the work or testing can be provided promptly, if necessary. Details for reports for earth dams are discussed in section 71 and reports for canals are discussed in section 75.

A survey of all field laboratories is made for each calendar year. A report is required from all field offices during January, which provides information on (1) the type of laboratory, (2) principal tests, (3) location, (4) operation, (5) status of buildings, (6) personnel data, (7) status of testing facilities, and (8) major equipment details. Forms for recording the required data can be obtained by request from the Engineering and Research Center, Division of General Research.

B. Earthwork

53. General.—Differences in construction procedures have led to the practice of classifying fill construction into three types: embankments, linings and blankets, and backfill. The division lines between types are usually not very distinct. Embankment construction applies primarily to laterally unsupported fills built on top of the natural ground surface; however, refill of cutoff trenches or key trenches is included as embankment construction, especially regarding construction control; lining and blanket construction applies mainly to relatively thin sheets of fill spread over an area either of natural ground, excavated surfaces, or embankment; and backfill refers to refill of holes excavated below the ground surface, or earth placed in confined spaces and against rigid structures.

Based on the amount and kind of work required, each of these groups is further divided into several types. Because of the indistinctness of definition, where two or more different types of construction are required involving separate pay items, it is customary to establish arbitrary boundaries for the different types of work. Separation may be definite, or overlapping may be involved. The lines of distinction will vary from job to job, so reference to the particular specifications involved is required in all cases.

54. Embankment.—(a) *Types of Embankment.*—The engineering properties of soil may be varied, and often improved upon, by selection, compaction, moisture control, mixing, and stabilization through the use of various admixtures. In the order stated, each of these operations is successively more expensive to perform. Although the engineering properties are improved and made more uniform thereby, some of these procedures may not be justified for all structures.

In construction, therefore, any of the following types of embankment construction may be specified: dumped fill, selected fill, equipment-compacted embankment, rolled earthfill, tractor-compacted embankment, blended earthfill, modified soil fill, or hydraulic fill.

(b) *Dumped Fill.*—The simplest construction operation of moving material from excavation and depositing it in a fill to lines and grades is called dumped fill construction. This type of embankment is used on construction of minor roads, canals, and laterals where the necessary engineering properties can be developed in the available soil without special effort. Although selection and distribution of material are not specifically required, dumped fills should be free from tree stumps, organic matter, trash, sod, peat, and similar materials. The fill should also have rocks, cobbles, and similar material distributed through the section and not nested or piled together. In order that the fill may be reasonably uniform throughout, the materials should be dumped in approximately horizontal layers. "End dumping," a process by which the material is pushed off the edge of the fill and allowed to roll down the slope, is objectionable and should be avoided wherever possible. Dumped fill is often further classified as dragline-placed fill, truck-dumped fill, and scraper-placed fill. If there is traffic over the fill during construction, either by construction equipment or otherwise, it should be routed to distribute the compaction as much as practicable.

On canal construction, there is a further requirement to place the finer and more impervious material on the water-side of the embankment. On road work the gravelly material should be on the top of the fill and large rock should be well buried. Inspection will consist of visual examination to assure that the above requirements are carried out and that dimensional requirements are attained. Laboratory testing of dumped fill construction is not required except for record purposes.

(c) *Selected Fill.*—Selected fill types in common use include selected impervious fill, selected sand and gravel fill, rockfill, and riprap. Selected fill is a dumped fill constructed of selected materials and occasionally compaction is required. This type of construction is widely used where one engineering property is required far more than others and soil with that property can be secured by selective excavation. Accordingly, selected fill may be specified in order to satisfy one of several engineering properties. The construction and inspection requirements will vary according to the engineering property being emphasized.

The selected impervious fill type of embankment has been used in canal construction. Its use is justified when selective excavation of the canal prism will provide a superior structure with little extra effort. Excavation from borrow pits and short hauls may be required, but benefits derived solely from selection usually do not warrant the cost of securing a better material from a more distant source.

Selected sand and gravel fill is specified to improve stability, to prevent frost heave, to provide an improved wearing surface on roadways, to prevent wave erosion of the underlying embankment, and to provide for removal of seepage water without piping. Benefits derived from this type of construction often warrant going a considerable distance to secure satisfactory material, and processing to improve gradation is sometimes justified. In choosing material for sand and gravel fill, the gravel is the more important component and a well-graded rather than a poorly-graded material is preferred. It is desirable to confirm visual inspection with an occasional test of gradation; sometimes this is required by the specifications.

Riprap and rockfill are used primarily for surface protection. Riprap is used to protect against the action of flowing water and waves, and to protect against rain and surface runoff. Rockfill is used to protect against rain and minor surface runoff and to provide stability to a fill structure; to some extent it is also used as a substitute for sand and gravel fill in dams. Durability and gradation are very important requirements for riprap. These are also desirable characteristics of rockfill, but the tolerance ranges are much broader for the latter. Selection is required to secure these properties and also to avoid the inclusion of large quantities of fine material. Laboratory tests are required in selecting sources. Visual examination usually suffices during construction.

(d) *Equipment-Compacted Embankment.*—There are situations where selected fill does not produce an adequate structure and the addition of compactive effort by routing of equipment will produce an acceptable fill. This kind of construction is mostly used on canal embankment and road construction. Specifications for construction of these features usually do not require a definite degree of compaction. Control of moisture in the material may or may not be required. Reference should be made to the pertinent specifications. Separate pay items may be used for the addition of water. The engineer in charge determines the amount of water to be added largely on the basis of the magnitude of the fill to be constructed and the nature of the materials being used. Visual inspection is sometimes supplemented by laboratory tests, depending on the nature of the material and the experience and ability of the inspection force.

(e) *Rolled Earthfill.*—The improvement of engineering properties to the maximum practicable extent by selection, compaction, moisture control, and special processing is generally justified in the construction of earth dams and in canal embankments or portions thereof. The limiting criterion for standard rolled earthfill construction is failure which could result in the loss of life or substantial property damage, and development of conditions of excessive water loss or maintenance. To assure development of desirable engineering properties to the maximum practicable extent, the procedures and equipment where pertinent are specified.

Selection of materials will to a large extent have been accomplished on the basis of preliminary investigations; that is, borrow areas will have been designated. It is expected that preliminary investigations will have disclosed the nature of the average materials and the probable range of variation. Field personnel in charge of construction operations should review the preconstruction investigation results, perform additional exploration and testing if necessary, and learn to recognize the kinds of material that will be acceptable for the adopted design. Since earth dams and major canal embankments are designed to accommodate materials available in the vicinity, the materials used for various purposes will differ from site to site. If, as a result of more detailed exploration or in the process of construction, materials are encountered whose characteristics differ appreciably from those anticipated, such materials should not be used without advice from the Engineering and Research Center.

In specifying compaction, requirements will include the general type of roller to be used, the thickness of lifts, and the number of passes. These requirements are based on extensive experience and statistical data and will produce a satisfactory fill when placed using good construction procedures. Specifications will also include requirements for the removal of oversize and that the material be homogeneous in texture; that is, free from lenses or pockets of material differing in gradation or classification from the average material.

To secure the maximum benefit from compaction requires that the moisture in the soil be controlled. The specifications will require that the water content be uniform throughout the layer to be compacted and that it be as close as practicable to that content which will result in the maximum densification of the material to be compacted. In general for the specified effort for earthfill in dams, this water content will be slightly less than the optimum water content as determined by the Proctor compaction test (designation E-11).

Inspection will determine whether the material is uniform and free from oversize rocks, whether compaction equipment complies with specifications and is maintained in working order, whether the thickness of lifts and the number of passes are according to specifications, and whether moisture is uniform within the layer. The inspector should determine that moisture is proper by manual tests to observe the feel and how the material compacts in the hand, and confirm this by tests using the rapid compaction control method (designation E-25). The penetration needle (designation E-22) was formerly used extensively for this purpose. Inspection will be supported with field and laboratory tests (designations E-24 and E-25) to determine the degree of compaction and the variation of water content from optimum, and designation E-13 to determine permeability.

(f) *Tractor-Compacted Embankment.*—To improve the engineering properties of strength and consolidation in comparatively permeable soils, that is, clean sands and gravels (GW, GP, SW, and SP), compaction by tractor is usually specified in the construction of earth dams. However, there is a variety of other methods of compaction that will produce satisfactory densification when impermeability is not an objective. Some of these methods are specified both for highway and airport construction. If Bureau of Reclamation forces are required to supervise the construction of highways or airports where maximum densification of sand and gravel soils is required, the specifications of the agency concerned will be followed and instructions necessary for field control will be issued.

In the event that the contractor for the construction of an earth dam should propose some other method than tractor compaction for the densification of permeable embankment materials, his proposal will be considered and, if the proposal is acceptable, appropriate specifications will be prepared which will form the basis for an order for changes. For those accepted proposals, letter instructions covering construction control will be given when required.

In the selection of material to be compacted by tractor, the emphasis will be to minimize contamination of the soil by excessive amounts of fines (minus No. 200 material). Quantities of this fine material in excess of 3 to 5 percent will in some cases tighten up the material sufficiently to make workability difficult, although permeability will remain sufficiently high for design requirements. Field inspection will, therefore, be directed toward removal of overburden fines, avoidance of silt and clay lenses or pockets, and prevention of excavation into tighter materials below the sand and gravel deposit. Thickness of placed embankment layers is commonly adjusted to avoid a requirement for removal of oversize material, and maintenance of uniformity will not require the attention that impervious fill construction requires. However, an attempt should be made to place the coarser material toward the outer slopes. Also, oversize material should be embedded in the fill. Where tractor compaction is specified, the material source will be designated, minimum size tractor described, lift thickness and number of passes enumerated, and moisture requirement defined.

Compaction by tractor depends primarily on the vibration produced by the equipment in operation. A secondary benefit results both from size and weight; for example, thicker layers can be compacted with larger equipment. Speed of operation of the tractors is considered beneficial in that the effect of increased vibration with high speeds more than compensates for any detrimental effects of short period application. It has been found that maximum compaction of sands and gravels is obtained by use of the tractors when the soil is either completely dry or thoroughly wetted. The wet condition is specified because satisfactory density is more readily obtained and field tests are more likely to be reliable with wet compaction. However, under certain conditions, dry compaction has been permitted under a change order. In these circumstances, extra compactive effort is necessary in lieu of thorough wetting. To secure a thoroughly wetted material, excavation of materials from below water table has been permitted in some instances. This procedure is successful when small-sized excavation equipment combined with a relatively slow placement rate is used. With large capacity equipment and a rapid placement rate, drainage of excess moisture may be too slow to permit satisfactory compaction.

Inspection will consist of noting that proper quality of material is used and that the specified thickness of lifts and number of passes are obtained. Water content is sufficient if free moisture appears in the crawler tracks immediately following the passage of the equipment. With proper water content the compacted fill will appear firm and solid. If the fill remains soft, moisture is too great; if the fill is fluffy, water content is insufficient. Inspection of appreciable quantities of this type of fill should be supported by relative density tests (designation E–12), and if any question exists as to the adequacy of the materials, permeability tests (designation E–14) should also be made. Figure 74 shows the equipment used in making a permeability test on sand and gravel in a field laboratory.

(g) *Blended Earthfill.*—There are some circumstances in which two materials by themselves do not have adequate engineering properties but when combined produce a satisfactory material. There are other cases in which a material with inferior properties may be combined in a mixture with another material with adequate engineering properties to increase the quantity of satisfactory material available. When the materials to be blended occur as strata, one above the other, in the same borrow pit, excavation by shovel can readily blend them together with little extra effort or expense. Construction control in this case will require maintaining the height of cut to obtain the necessary proportions of each type material and to assure that the two materials are being blended. With dragline excavation satisfactory mixing can be accomplished, but greater attention must be given to the operation since the tendency will be to remove material in horizontal cuts. Several types of excavating equipment have been developed, such as wheel excavators and belt loaders, which have excellent capabilities for blending vertical cut material.

When scraper excavation is used the difficulty of securing a satisfactory mixture is increased, and it is often necessary to provide supplemental mixing on the fill by plows, discs, rippers, or a blading operation. Excavation is performed by making a slanting cut across the different types of materials, and care must be exercised to secure the proper proportions of each type of material in each scraper load. A common practice is to spread the materials in half-lift thicknesses to improve uniformity and minimize requirements for mixing on the fill.

Considerations of cost limit the procedure for blending materials from separate sources to special situations such as blanket and lining construction. One of the materials is first stockpiled upon the other so that

Figure 74.—Large-scale permeability apparatus in a field laboratory. VP-192-R2.

excavation can be made across the two materials, or the materials are placed in thin layers on the fill and then mixed by blading, plowing, or similar procedures before being compacted. Mixing machines may be specified to obtain an intimate mixture for earth linings. Blended fills receive the same kind of attention that is required for rolled earthfill.

Material is tested in the laboratory for density, moisture content, and gradation.

(h) *Modified Soil Fill.*—A fill using a soil which has been modified by adding minor quantities of a selected admixture prior to compaction is called a modified soil fill (see secs. 34, and 80 through 84). Commonly used additives include lime, cement, asphalt, and salts. The use of modified soils should be considered in lieu of replacing poor soils with selected material from a distant source. For example, to construct a switchyard fill, the use of lime to modify the expansive characteristics of a clay soil from immediately available borrow was more economical than importing selected material from a distant source.

Conventional methods are usually used to spread and compact the modified fill. In addition, some type of equipment must be adopted to distribute and mix the additive. This equipment will require initial calibration and periodic checks during construction to insure that the proper amount of additive is being applied. Special tests of the modified soil may be required to verify the amount of additive, and visual observations should be made to insure that the additive is uniformly distributed. Construction control testing of placement moisture and density is also usually required.

Inspection requirements will be comparable to those required for concrete or the other specialized materials, and extensive testing may also be anticipated. For those situations warranting stabilized fill construction, special instructions will be issued by the Engineering and Research Center covering construction control. Considerable experimentation to develop suitable construction procedures may be required. During the period of experimentation the assistance of an engineer from the Engineering and Research Center who has special knowledge concerning the type of stabilization specified may be advantageous.

(i) *Hydraulic Fill.*—The previously described methods of embankment construction all involve the control of water content. There are some situations where placement of fill material under conditions of excess water content is required. These situations may involve the excavation and transportation of the fill material by the use of flowing water. The more common procedures involve only the placement of material in still water or the process of washing material into place or densifying it with a stream of water.

Terminology for the different types of hydraulic construction is not standardized. In general, the term "hydraulic fill" is applied to the complete operation of excavation, transportation, and placing by flowing water. When the hydraulic operation is confined to placing, it is described as semihydraulic construction. The term "puddled" is applied to special types of semihydraulic fill, and the term "sluiced fill" is used for material washed into place.

The economical utilization of hydraulic fill construction depends on (1) the availability of a material that is readily sorted by the action of flowing water into a pervious material and an impervious material, (2) a large volume operation, and (3) a source of cheap power. This combination of conditions is experienced so rarely that control procedures for hydraulic fill construction have not been developed for Bureau operations. In the event that hydraulic or semihydraulic fill construction is specified, special instructions will be issued by the engineering staff in the Engineer and Research Center.

Puddled fill construction is sometimes used where climatic conditions make moisture control operations impracticable, for backfill around pipelines and structures, and in lieu of special compaction against rough and irregular surfaces. The puddling process consists of depositing the soil in a pool of water, stirring the soil-water mixture until a fairly uniform slurry is developed, and then letting the soil settle out of the mixture. The puddling operation is used mainly for soils of low permeability, such as silts and sandy lean clays, where high density is not mandatory. Inspection is by visual examination to establish that the proper type of material is used and to assure that a thorough mixing is accomplished.

Sluiced fill construction involves working a pervious material into place by a washing operation produced by the flow of a high-velocity stream of water. Commonly, the operation is used to wash sand and gravel or rock fines into the voids of a rock mass such as a rockfill. It may also be used in lieu of tractor compaction for the consolidation of sand and gravel fill (see fig. 75). Tractor compaction is considered superior to sluiced fill construction; hence, the latter is usually restricted to areas inaccessible to tractors. Inspection will consist of visual examination to determine that appropriate materials are used and that the washing operation is properly performed; that is, that the materials are actually moved in the process. When sluiced fill construction is used in lieu of tractor compaction, relative density tests are required to assure adequate densification.

Compacted fill is sometimes specified to be consolidated in a thoroughly wet condition utilizing equipment such as pneumatic rollers, vibratory rollers, surface vibrators, or tractors. Soils placed by this method must have a relatively high permeability for proper consolidation. A density requirement is specified and is based on relative density. This type of fill is sometimes referred to as consolidated fill.

55. Linings and Blankets.—(a) *General*.—In earth construction there are many situations which require the construction of a thin layer or blanket of selected material. These situations include the layer of riprap on the upstream slopes of earth dams, the sand and gravel blanket under

Figure 75.—Sluiced backfill in cut-and-cover section of a power tunnel.
P456-108-2002.

the riprap, rockfill blankets or blankets of gravel or topsoil on the down-stream slopes of dam embankments, filter blankets under the downstream portions of dam embankments and under the floors of concrete structures, and impervious blankets on the floors of reservoirs and channel linings for canals. Base courses and surfacing on roads or highways, and ballasting for railways may also be considered to be in this category.

A common requirement of these types of blankets is proper selection; that is, locating and using materials determined to be suitable for the particular use. In part, quantity can be substituted for quality, but space requirements are often critical and the functioning and cost balance may be upset if it becomes necessary to change blanketing dimensions after construction is initiated. Sources of material with established characteristics and in ample quantity for the various blanketing requirements should always be determined in advance of construction.

A requirement of all blankets, is careful placement. Requirements may vary widely according to the type and location of the blanket placement, but in every case uniformity and thickness are very important. A discussion of types of material and general compaction requirements is given in section 75c. Blankets may be divided into four types on the basis of material used as follows: rock, sand and gravel or crushed rock, silt and clay (impervious), and topsoil.

(b) *Rock Blankets.*—The rock fragments used in rock blankets should be essentially equidimensional, angular, well-graded in size from a maximum dimension equal to the blanket thickness to about one-tenth of the blanket thickness, and sound, dense, and durable. In seeking equidimensional rock, it should be kept in mind that elongated and thin slabs of rock are undesirable, but rock fragments in which the minimum dimension is about one-third to one-fourth of the maximum dimension are not objectionable. Although angular rock is the most desirable, subrounded cobbles and boulders are not objectionable. Well-rounded cobbles and boulders, however, should be avoided except for blankets on very flat surfaces.

It is difficult to determine grading for rock fragments. Generally, blasted rock that has an appreciable quantity of fragments near the maximum size requirement will possess a satisfactory gradation. The condition to avoid is an excess of small-sized fragments. Fine material, such as rock fines, sand, and rock dust, in a volume not to exceed the volume of the voids in the larger rock, is not objectionable.

Soundness, denseness, and durability requirements may vary somewhat according to usage. For the protection of slopes downstream of stilling basins, the highest quality rock is required. The soundness requirements for riprap on dams may vary somewhat according to reservoir size and its operating characteristics; rock size can, to some extent, be substituted for density. The protection of downstream slopes of dams can be accomplished with any rock that does not break down appreciably by normal exposure to the weather. Clay shale is about the only rock that is unacceptable.

Rock blankets are normally dumped in place and roughly graded to produce the required thickness or they are built up by laying the rocks in place by hand. Hand-placed riprap is usually constructed of approximately equal-sized rock laid with the longest dimension normal to the surface. Inspection is by visual examination and measurement to establish specified thickness. Testing is not required. Figure 76 shows the placing of a 3-foot blanket of riprap over an 18-inch cobble blanket in the tailrace of a powerhouse outlet.

Figure 76.—Placing a 3-foot blanket of riprap over an 18-inch cobble blanket in the tailrace of a powerplant. P–456–108–1166.

(c) *Sand and Gravel or Crushed Rock Blankets.*—The characteristic of sand and gravel mixtures to allow the passage of water while at the same time preventing the passage of soil grains is extensively utilized in the design of water retaining structures. The properties of resistance to displacement by flowing water, resistance to wear from vehicular traffic, and the maintenance of strength and limited volume change over a large range of water contents, make sand and gravel useful in providing surface protection to canal banks, roads, and working areas in the various facilities of irrigation and power projects. The wide range in gradation possible in sand and gravel mixtures, together with the wide range in structural materials to be protected, results in a wide range of acceptability for the materials utilized for sand and gravel or crushed rock blankets.

Natural sand and gravel deposits normally contain excessive amounts of sand. However, if these materials are clean (contain less than 5 percent fines), almost any sand and gravel mixture can be used for downstream drainage blankets for earth dams by thickening the pervious blanket sufficiently so that seepage through the embankment and foundation can be carried within the blanket section.

For the pervious blankets between riprap and rolled earthfill, the requirements for the sand and gravel material become less critical as the thickness of the riprap layer increases. Generally, material from a natural deposit can be utilized if at least 50 percent of the material is in the gravel size range when riprap blankets of 3-foot normal thickness are specified. In those ranges of reservoir operation where anticipated wave action is comparatively rare, some relaxation of material requirements is also possible.

Pervious blankets under concrete structures are usually made as thin as possible. This condition makes processing of material from natural deposits almost a necessity in order to produce the specified gradation of material. Two-layer filter systems are frequently used. The filter criteria as discussed in section 79(f) will govern the selection of material for this type of construction.

Sand and gravel blankets used as base courses and surface courses on roads, highways and switchyards are normally pit-run material with thickness substituted for quality in some situations. Blankets for these purposes differ from pervious blankets for other uses in that a certain amount of clayey material to bind the blanket material together is considered desirable and is in some instances actually added to improve the quality of surface courses.

Pervious blankets used to prevent erosion in canal channels require a material predominantly in the gravel range. Appreciable quantities of sand are considered undesirable and in some instances removal of fine sand may be required. A sand layer may be required between a coarse gravel layer and the subgrade to prevent the movement of fine subgrade soils through the gravel. The materials for a pervious gravel blanket and the materials for an impervious blanket may be combined to provide an erosion-resistant, impervious blanket. Filter criteria are usually not important in impervious blankets.

Crushed rock blankets may be used for any of the above-described purposes. Being a manufactured material, the crushed rock is processed to meet specifications requirements, and inspection to secure proper material is seldom a problem.

Sand and gravel or crushed rock blankets are usually densified by travel of hauling equipment, the use of vibrating equipment, or smooth rollers. They may be placed either dry or thoroughly wetted. Inspection is commonly by visual examination, but gradation and relative density tests may in some cases be required.

(d) *Impervious Blankets and Linings.*—Impervious soils are used on the floors of reservoirs and in the channels of canals to reduce seepage. The material for these purposes must be of adequate impermeability, should be free from shrinking and swelling characteristics, should resist the erosion of flowing water, and when placed on the sides of canals and reservoirs should have good stability characteristics. Probably the best material for these purposes is a well-graded gravel with plastic fines (GW–GC) material, which offers both impermeability and excellent erosion resistance. Either a clayey gravel (GC) material or a silty gravel (GM) material is also good. Materials in other soil groups can be used in accordance with the engineering use chart, figure 8. When satisfactory soils are not available, manufacturing soil by blending operations or protecting the fine-grained soil with a blanket of sand and gravel is necessary.

Blankets and linings which will be permanently under water need not meet the erosion resistance and shrinking and swelling criteria, and blankets and linings placed on essentially horizontal surfaces need not possess high stability characteristics.

Thicknesses of impervious blankets are usually controlled by practicable placement procedures. However, with high reservoir heads thicknesses greater than the minimum required for construction operations may be needed. Also, it may be desirable to thicken impervious blankets where some swelling or shrinkage is anticipated so as to maintain an effective thickness of blanket unaffected by swelling or shrinkage cracks. For canal linings, thicknesses are usually varied depending on the design requirements; the availability of lining material may also affect the thickness and construction methods may need to be revised to attain desired engineering properties.

Construction and inspection requirements are similar to those for rolled earthfill and blended earthfill as described under embankment construction (sec. 54). Material selection requirements are high, and if it is necessary to haul materials great distances, other types of lining may be competitive.

(e) *Topsoil Blankets or Zones.*—Topsoil blankets or zones are sometimes specified for downstream slopes of dams so that the slopes can be seeded for protection of the underlying zone against erosion by wind and rain. More recently, the use of topsoil for the restoration of borrow areas, or other areas in which the existing topsoil has been destroyed or removed, is specified to fulfill ecological and environmental requirements.

Normally, stripping from required excavation for the dam and appurtenant structures and approved stripping from borrow areas are the sources of topsoil. It is usually not feasible to borrow topsoil; therefore, suitable materials from stripping which would otherwise be disposed of as waste material should be selected during excavation and, if necessary, stockpiled for later use.

For topsoil blankets, specifications normally require a 1-foot normal thickness; thus, areas to receive topsoil should be brought to within 1 foot of the prescribed final cross section at all points and finished smooth and uniform before the topsoil is applied. Topsoil should be evenly placed and spread over the graded area and compacted in two layers, each by one pass of a roller weighing not less than 50 pounds per linear inch of length of drum.

Occasionally, for larger dams requiring topsoil, it will be more advantageous and economical to specify a thin topsoil zone for downstream slope protection. One prime advantage is that excavated topsoil can be placed concurrently with the rest of the embankment; eliminating the requirement for stockpiling. When such a zone is specified, construction and inspection requirements are similar to those for rolled earthfill as described in section 54.

In most cases, seeding of topsoil on downstream slopes is required; thus, the material should be selected from required excavation or from the stockpile so as to contain the most fertile soil and shall be free from excessive quantities of large roots, brush, rocks, and other objectionable matter. Topsoil should not be placed when the subgrade is frozen or in a condition otherwise detrimental to proper grading and seeding.

56. Backfill.—(a) *General.*—The construction of earthfill in confined spaces, such as the refilling operations about concrete structures, is described as backfill. When the refilling operation is associated with embankment construction and not readily separable therefrom, the work adjacent to the concrete structure and in other confined areas may be called special compaction rather than backfill. Backfill operations may be divided into three classes, the terms used being: backfill, compacted backfill of clayey and silty soils, and compacted backfill of cohesionless free-draining soils.

For contract payment, backfill is usually listed as one pay item, and compaction is an additional pay item. Similarly, special compaction is usually paid for as a separate item in addition to being paid for as embankment construction. However, there may be exceptions to this procedure; hence, the pertinent specifications should be consulted.

(b) *Compacted Backfill.*—Compacted backfill covers two types of materials and compacting operations. The first is the compacting of clayey and silty materials of low permeability. These soils are used for backfilling around or under pipelines and structures when seepage is to be limited, or if drainage is not required. This type of backfill is normally compacted by tamping rollers when space is available or by hand or power tampers in confined areas. The soils are normally compacted at Proctor optimum moisture, and the percentage of Proctor maximum density, "D-ratio", is specified.

The second type of compacted backfill is related to the consolidating of cohesionless free-draining soils of high permeability. These soils are used around or under pipelines and structures when impermeability is not required, when free drainage is desired, or when particularly good bedding or foundations of low compressibility are desired. Like tractor-compacted fill or fills compacted by consolidating processes, the materials used must be free-draining sandy and gravelly soils. When high stability and low settlement are the design requirements, this type of backfill is often preferred over compacted backfill of clayey and silty soils because of the ease and economy of securing a backfill, particularly in confined areas. For example, when good bedding is required under and around concrete pipe, the material can easily be made to flow and can be compacted to high density in the critical wedge under the pipe and between the pipe and trench surfaces, providing proper procedures are used for wetting and vibrating the material (see sec. 77(d)). This type of fill or backfill is often specified for use under pumping plants and other structures when necessary to improve the bearing capacity of soft foundation soils. This is accomplished by removing the soft materials and replacing with selected backfill and by spreading the load to lower strata. It is also preferable as backfill adjacent to some structures when excessive surface settlements are undesirable or when space is limited.

Consolidation of the soils is usually accomplished by pneumatic rollers, vibratory rollers, tractors, surface vibrators, or internal vibrators, after the soils have been thoroughly wetted. Figure 77 shows a surface vibrator compacting a sand layer beneath a spillway slab. Figure 78 shows a typical consolidating operation on a large pipeline.

Proper selection of materials is important for successful results. Excessive amounts of fines will plug the voids between the coarser particles and will prevent drainage during the consolidating period. As a guide, table 2 provides information for preliminary selection of soils suitable for consolidating.

57. Excavation.—Materials for embankment or lining construction are obtained either from required excavations or from borrow pits. To the maximum practicable extent, materials from required excavations

Table 2.—General suitability of soils for compacted backfill by consolidating processes

Soil type	Limitations
GW, GP, SW, SP	All suitable (fines* in these soils are limited to 5 percent by definition).
Borderline GW–GM, GW–GC, GP–GM, GP–GC.	Suitable if fines are less than 8 percent.
Borderline SW–SM, SP–SM, and SP–SC.	All suitable (fines in these soils are limited to 12 percent by definition).
SM and SC	Require special consideration and testing as suitability depends upon gradation and plasticity (some SM soils with fines as high as 16 percent have been suitable).

* Fines are particles smaller than No. 200 sieve size.

Figure 77.—Vibratory consolidation of sand and gravel beneath a spillway slab. PX–D–16328.

should be incorporated in embankment or lining construction. Therefore, the same practices used in borrow-pit excavations should be followed in making required excavations when the materials can be used.

Organic material such as plant growth and decaying vegetable matter should be removed from the surfaces of borrow pits within borrow areas

Figure 78.—Consolidating backfill about a conduit by means of internal vibrators. PX–D–16909.

prior to the initiation of excavation. The depth of stripping will vary according to the nature of the cover, from 2 or 3 inches for prairie grasslands to 2 or 3 feet in valley bottoms and forested areas. It is not necessary to remove completely all material containing grass and tree roots. The fine hair roots of grass and tree roots less than one-fourth inch in diameter are normally not considered sufficiently detrimental to justify wasting material containing minor amounts of these substances. It is inadvisable to strip a greater area than will be excavated in one construction season, because of recurring weed growth. The minor amount of weed growth that may develop during the latter part of a construction season on a stripped area is usually considered unimportant and may be disregarded.

Occasionally deposits of impervious soil are covered with a layer of boulders, and often sand and gravel deposits are covered with a layer of silt and clay. Such a deposit is shown in figure 79. As a result of a thorough materials investigation, it was found most economical to use the upper clay-silt-sand portion of the deposit for the impervious zone and the lower silt-sand-gravel portion for the pervious zone of an earthfill dam,

thus eliminating the need for another borrow source and subsequent additional costs to develop. When it is not practicable to salvage these superficial layers, they are classed as stripping and wasted even though the
layers may be thicker than is normally considered applicable to stripping.
If appreciable material in this class occurs, there will be covering paragraphs in the specifications.

Figure 79.—A deposit of impervious material which overlies a deposit of pervious material. P622A–427–1652NA.

In highway and canal construction where the normal method of excavation is by scraper or dragline, there is no lower limit to the depth of
cut required. In all other excavation, cut depths less than 4 to 5 feet are
avoided for reasons of economy unless deficiency of material makes
smaller cuts necessary. Except in situations where separation between
two different types of material is desired or where controlled blending is
required, maximum depths are usually limited by the range of the excavating equipment. Where shovels are used this range will be 15 to 20
feet; with draglines or loaders 6 to 10 feet is desirable. However, the
maximum depth of cut may be limited by the presence of bedrock or
hardpan and, excepting for gravel excavation, by the location of the
water table. In stratified borrow areas, excavation must be made so that

every load delivered to the point of use contains a mixture of the full designated depth of cut. Extra mixing on the fill may be required.

C. Foundations

58. General.—If investigations could completely describe the conditions of a foundation and the designs were prepared accordingly, field inspection would be reduced to merely determining that the dimensional requirements of the designs are met. Unfortunately, even the best investigation will leave portions of a foundation unexplored. As the value and importance of the structure decrease, the gaps in an investigation usually increase. Also as the importance of a structure decreases, the quality requirements for a foundation also decrease.

Because of the infinite variety of conditions that may be encountered in foundations, it is impracticable to provide a complete set of rules that are applicable for determining their adequacy. Individual judgment based largely on experience must be used. In the development of an experience background, the first objective to be achieved is an ability to discriminate between sound and unsound foundations. In practice this means classifying a foundation as adequate, inadequate, or questionable. At first most cases will seem to fall in the questionable category, but with increased experience the number of cases in this category will decrease.

To supplement judgment, test procedures have been developed for evaluating foundation conditions. Such tests should have been performed during the investigation process, and comparison by means of index tests and visual examination between tested and untested areas will usually suffice to establish foundation adequacy. Additional tests may be required during construction.

The conditions for which adequacy of a foundation should be established include bearing capacity, stability, settlement and uplift, deterioration, and permeability. The degree to which each of these elements is important depends on the nature of the proposed construction, the character of the foundation materials, and the inplace structural characteristics of the foundation.

59. Bearing Capacity.—Foundations for rigid structures are usually evaluated on the basis of bearing capacity. Bearing capacity in this sense involves both the shear strength of the soil and its consolidating characteristics. It may be determined in the laboratory by shear and consolidation tests on undisturbed samples or approximated in the field by density tests, standard penetration, or vane shear tests.

The specifications design of rigid structures should show the type of foundation treatment required based on the investigation information available. Where foundation conditions encountered appear to be the same as those disclosed at the location of the investigations, they may be assumed to be equally competent. If foundation conditions are disclosed that are observably different from those disclosed by previous investigations, the lateral and vertical extent of this different condition should be determined. If a minor amount of additional work will result in the elimination of the unsatisfactory condition in the foundation, authority to perform such work should be requested. If a substantial amount of additional work is required, undisturbed samples should be collected and, together with a description of the situation, should be forwarded to the Engineering and Research Center, Denver, for decision.

It should be noted that a dike, pocket, or lens of harder and firmer material in a foundation can be equally as detrimental as softer deposits. It is common practice, therefore, to specify that if such material is encountered it will be overexcavated and refilled with compacted earth to provide a more uniform foundation. This requirement should include beds of indurated gravel or sand, hardpan, and the erratic deposits cemented by calcareous or siliceous material. The most desirable foundation is one in which the material has a uniform character, and a moderate amount of overexcavation to develop such a condition is usually justifiable.

60. Stability.—A foundation excavation can be divided into two parts, the base and the side slopes. Except for chutes of canals and spillways, the load of the structure rests on the base of the excavation and settlement limitations will usually prevent the development of the full shear resistance of the foundation. Consequently, stability of the base of a foundation excavation seldom presents a problem. However, foundations adjacent to bodies of water or directly above waterbearing strata should be investigated for loss of support caused by artesian pressure.

Cut slopes for most construction are determined by the staff in the Engineering and Research Center; however, the construction engineer's forces are occasionally faced with the problem of determining the proper cut slopes to use in providing stable slopes. The remainder of this discussion is, therefore, limited to the problem of stability of slopes.

The use of slopes not previously analyzed or pre-established should be discouraged because of the many variables involved in determining stable slopes, such as type of material, presence of water, depth of cut, and intended use. Both experience and analytical analyses are required for this determination. However, as a guide, from long standing usage it has become customary to make cuts in earth materials on slopes between 1½ : 1 and 2 : 1, and temporary cuts on a slope of 1 : 1. An important exception is that cut slopes in loessial soils may be ¼ : 1 or ½ : 1. Examination of cut slopes on highways, railways, and canals as one travels about the country will show, however, that there are many deviations from these criteria and that there are scattered situations where sloughing has occurred. For structures such as laterals, small canals, and minor road construction, it is generally accepted that a minor amount of sloughing is tolerable. Nevertheless, the construction engineer's forces should be on the lookout for conditions of potential instability, such as cracks forming along or at the top of the slope, sloughing along or at the base of the slope, and excessive seepage. Such observations must be reported to the Engineering and Research Center if failure would result in property damage or possible loss of life.

In the determination of stable slopes, a great deal of information can be gained from examination of previously excavated slopes in similar material in the same area. If there are no excavated slopes, examination of natural slopes is helpful. In the utilization of such information, it should be remembered that stable slopes in sand and gravel are independent of height of the cut but that safe cuts in saturated clay are dependent on height and virtually independent of slopes. Since most materials are combinations of sand, gravel, and clay, safe cuts will be controlled both by height and by slope. In evaluating natural slopes it should be kept in mind that the safety factor against sloughing is variable and that any increase in slope results in a reduction of this factor of safety. Therefore, a natural slope should be steepened as little as practicable.

The presence of water has a very marked effect on slope stability. This can be in any combination of three forms: seepage, rain, or water being conveyed or stored. Figure 80 is an example of an arc failure of a natural slope (landslide) resulting from excessive moisture entering the material. Removal of the toe material for construction of a stilling basin may have precipitated the slide. Where an excavation cuts across the ground-water table, sloughing commonly follows unless slopes of the cut are extremely flat. To overcome this condition the usual practice is to remove the ground water from the area in advance of the excavation by well points, deep wells, or pumping from sumps. Figure 81 shows the use of well points in trench excavation and figure 82 shows a more common method of removal of ground water using drainage ditches and sump pumps. Since the damage results from the steep gradient produced in the seepage flow, where pumping is not practicable the excavations may be accomplished by repeated cuts resulting in a gradual lowering as the ground water is drained away, or conversely the excavations may

Figure 80.—An example of an arc failure of a natural slope (landslide) resulting from excessive moisture entering the material. P860–427–905NA.

be performed in standing water which is then slowly lowered after the excavation is complete. The success of either of these methods is dependent on a thorough knowledge of ground-water behavior and of the local soil conditions. Neither method should be attempted without the advice of the Engineering and Research Center.

Where permanent slopes intersect the ground-water table either permanently or intermittently, it is usually necessary to provide some type of treatment to prevent slope sloughing. The required treatment involves moving the free surface of the water table back from the face of the excavation. This may be accomplished either by boring drainage holes into the face of the cut, excavating and placing drains into the cut, or protecting the surface of the cut with a free-draining material such as a gravel or rock blanket. Unless the potential seepage is very serious, the rock or gravel blanket will usually suffice. It is suggested that a blanket 1 to 2 feet thick be tried in such cases that are discovered during construction, since its cost will be relatively small. If this proves ineffective, the advice of the Engineering and Research Center should be sought. Where the material in the slopes is well-graded and free-draining, a

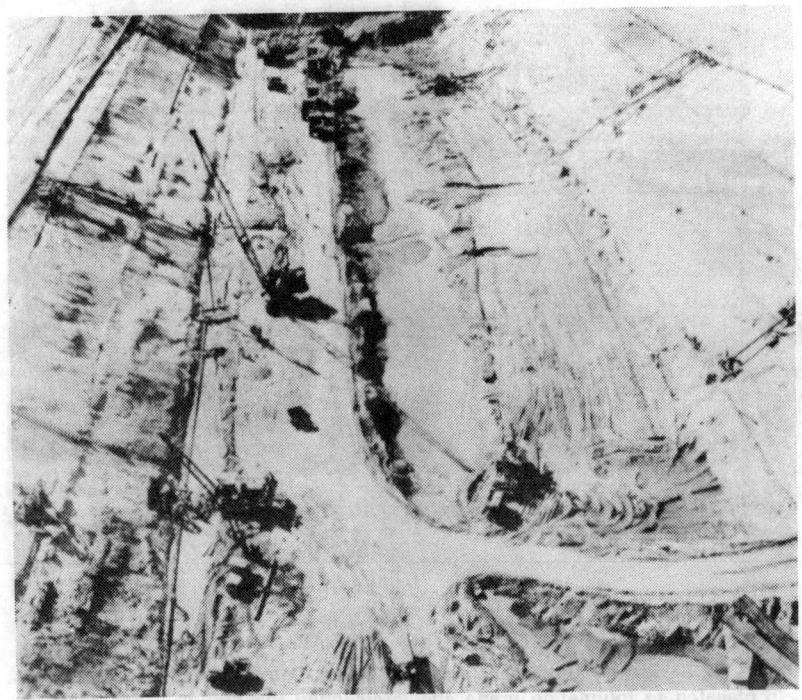

Figure 81.—View of cutoff trench at Davis Dam, Arizona-Nevada. The collecting pipe along each slope of the trench received water from the rows of well points to drain that side of the foundation. D.D. 3112.

blanket will develop naturally in many situations and no special treatment is needed.

Rainwater damage commonly takes the form of surface erosion. Preventive measures include diversion of surface runoff away from cut or fill slope faces with small ditches a short distance back from the intersection of the slope with the ground surface; surface sodding or blanketing with a thin layer of gravel or rock; constructing ditches to carry off the water laterally from the earth slopes; and steepening the slopes. However, when the rainwater can infiltrate the foundation to produce a rise in ground-water levels and produce seepage, both the sodding and slope steepening procedures may actually be detrimental and should not, therefore, be used in conjunction with a seepage condition.

Erosive action from flowing water in a canal or wave action in a reservoir is potentially the most severe type of water damage. Protection involves flattening of slopes or protection with gravel or rock blankets depending on the severity of the potential damage. There are situations

Figure 82.—View of cutoff trench at Twin Buttes Dam, Texas. Drainage ditches excavated to foundation grade and intercepting the ground water. Water is removed by sump pumps. P825-523-434A.

where sloughing of the slopes is not particularly important such as in borrow pits in the reservoir area. Such slopes should be flattened sufficiently so as not to provide a hazard; usually 1½ : 1 is satisfactory. However, 3 : 1 or flatter slopes are more desirable where the area is open to public recreation.

There are situations where selection of slopes is controlled by criteria other than stability alone. For example, some highway divisions use slopes closely approximating those found to prevail naturally as a method of drift control both of sand and of snow. In some canal work flat 2 : 1 slopes are specified instead of the more conventional slopes to minimize maintenance and to provide more stable channels. Slopes are sometimes flattened in connection with earth dams and high canal fills to distribute stresses due to embankment loads. On the other hand, where the height of cut is relatively small, vertical slopes may be used in connection with the construction of structures, perhaps without the use of intervening forms if trimming to dimensions is possible and the slopes will stand temporarily. There are also a few soil deposits, for example loess that is not wetted, which will tolerate steep slopes, whereas flatter slopes will deteriorate as a result of minor rainfall.

61. Settlement and Uplift.—Unless a structure rests on a solid rock foundation some settlement cannot be avoided. Where it is infeasible to found a structure on solid rock either by placing the structure directly on bedrock or by means of piles, piers, caissons, or walls, the structure must be capable of withstanding some vertical movement in the foundation. The usual method of reducing such movements to within acceptable limits is by spreading the base or footings of the structure so that unit pressures are sufficiently small. For irrigation structures and other structures where movement of water beneath the structure must be prevented, the spread base design is preferable although a solid foundation might be attainable with piles. The consolidating effect of the structure load will help prevent underseepage and piping. A firm uniform foundation free from loose sand, low-density silt, soft clay, or isolated rock masses should prove to be adequate for most structures with moderate loadings.

Uplift of structures may occur from several causes, such as hydrostatic forces, expansive subgrade soils upon wetting, and frost action, and structures and canal linings must be protected against excessive upward movement. For structures such as chutes and lined canals, as described in section 73, protection against hydrostatic uplift is provided. This is commonly accomplished by underdrains or drainage blankets of sand and gravel. Where possible, these drains will discharge into adjacent drainage channels or into the structures. For certain lined canals this may require flap valves on the outlets into the canal. Special care should be taken to keep the drains open and the gravel blankets free from contamination.

Expansive subgrade soils may also cause uplift upon wetting, particularly when the loadings are light, when the initial density is high, and when the initial moisture is low. Usual methods for reducing movement of rigid structures are by using long friction piles or belled caissons attached firmly to the structure, by increasing unit footing loads, or by removing expansive subgrade soils and replacing with nonexpansive soils to a depth sufficient to control movement. Flexible canal linings, such as asphalt or earth linings, are used in preference to rigid canal linings on expansive soils. In some cases, the structure load is less than that of the original earth removed and rebound may occur. This is particularly true when saturated expansive clays are involved. In some cases where structural loads are small, as in canal linings, it is possible to condition the expansive clay by prewetting the foundation and maintaining a moist condition so that future expansion will be minimal.

Frost action may also cause undesirable heaving and uplift. The susceptibility of soil to frost action depends on (1) the type of soil in the subgrade, (2) the temperature conditions, usually expressed in terms of a "freezing index" (a summation of the degrees below freezing

multiplied by the days of below-freezing weather), and (3) water supply. Uplift from the development of ice lenses in subgrade soils may
cause damage to such works as concrete canal linings, spillway apron
slabs, and other lightly loaded structures. Frost action may also be of
importance in the reduction of density in compacted earth canal linings.
All soils, except GW, GP, SW, and SP soils, in cold climates and with
available water may be susceptible to frost action. The susceptibility of
gravelly and sandy soils varies with the amount of fines. Silts and clays
with a plasticity index of less than 12, and varved clays are particularly
susceptible to frost action.

62. Deterioration.—Both earth and rock foundations should be preserved in their natural state as well as possible. A cover of soil should
be maintained over the foundation surface until final cleanup, and the
structure should be placed thereupon without delay. Drying out of the
foundation surface should be avoided. Some clays and shales will dry
and crack, then turn into a soft slurry upon rewetting when exposed to
the air. When it is impracticable to place the structure immediately after
such surfaces are exposed, a sprayed-on cover of asphalt or pneumatically
applied mortar or approved materials has been satisfactory in some
situations.

63. Permeability.—Water retention structures, particularly reservoirs
and canals, depend on the foundation for a part of the water barrier.
If access to a water-bearing stratum exists, either through the process of
excavation or due to natural conditions, some corrective action is indicated. However, it should be kept in mind that no soil formation is watertight and that some water loss is to be expected. Movement of water from
a reservoir or canal may result in piping or excessive loss. These problems
can be reduced to (1) control of the velocity of flow, and (2) control
of the volume of flow. The treatment required may or may not be the
same for both problems.

Structural damage to a foundation resulting from water movement
results from the gradual removal of soil particles either by solution or
mechanical movement of the particles. Solution is a problem only when
the foundation soils contain substantial amounts of soluble salts. Even
then the development of a detrimental condition is likely to be very slow
compared to the life of the structure, and unless the presence of soluble
material is obvious this factor should not cause the construction engineer
concern.

An advanced state of solution already existing in a foundation is a
serious problem. Such a condition involves primarily the bedrock material, the treatment of which is beyond the scope of this manual. It also
frequently introduces disturbances in the overburden which are indi-

Figure 83.—View of a typical sinkhole (pothole). P809–529–813NA.

cated by sinkholes. These sinkholes are usually small in extent and unless there is positive evidence that they are watertight, they should be stripped to sufficient depth and filled with compacted impervious soil when they are found along canal lines or in reservoir areas. A sinkhole is shown in figure 83. Similarly, if it becomes necessary in the excavation process to expose bedrock surfaces that contain solution channels, open holes, or cracks, such openings should be sealed with concrete or compacted impervious earth blankets. However, an extensive grouting program may be required to seal the foundation.

The conditions under which mechanical movement of particles due to seepage is likely to occur involve fine sands or silts and loosely compacted materials. If the seepage path is long compared to the head between intake and exit, no special remedial measures are necessary unless water conservation is important. For steep hydraulic gradients the control will take the form either of a barrier across the permeable channel, such as a cutoff trench; an impervious blanket over the intake area; or a filter blanket over the exit area. The latter method is used only when water conservation is not economically justifiable and is usually applied only after seepage and piping have been demonstrated by actual expe-

Figure 84.—Portable grout machine used to force a mixture of cement and water into holes drilled in the foundation. HH-1001.

rience. Such situations are also corrected by the use of relief wells, which may be installed prior to putting the structure in use where foundation conditions indicate a strong likelihood of a piping hazard.

The conservation of water may be required because of limited availability, because of the high unit cost of water storage and conveyance, or because of potential damage from the creation of a high water table below the structure. This evaluation is part of the design process, and the specified treatment usually will take the form of a positive cutoff to bedrock and sealing of the joints, seams, and cracks in the rock foundation with cement grout. For canals the treatment will be some form of canal lining. Figure 84 shows a portable grouting machine used to inject cement grout into a foundation.

For dams, it has been found that to secure an appreciable saving in water at least 95 percent of the pervious cross-sectional area of the foundation must be blocked. However, where a foundation consists of zones of differing permeability and one zone is more than 10 times as permeable as another, blocking of the more pervious zone is very effective in reducing water loss even though it may be only a small part of the total permeable area. Where water conservation is required, therefore, a permeable zone should be completely cut off by extending the excavation to cover the entire permeable zone.

Similarly, on canals any reach of lined section should be extended to cover the entire reach of the pervious stratum. Consideration should be given to fully lining or placing in pipe all constructed water conveyance systems. Full justification for using an unlined waterway is required. Consideration must be given to the value of water saved, operation and maintenance cost, required drainage, amount of land taken out of cultivation for canal site and seepage, right-of-way, allowable velocities, structure costs, and the cost of various types of lining correlated with the other conditions. The unit value to be placed on each item will vary with local conditions; therefore, the amount of loss which can be allowed before lining a canal may vary considerably in different areas.

In the construction of long canals, explorations are often insufficient to reveal fully all the pervious strata that will be intersected. Where such reaches are encountered that were not disclosed by explorations, an earth lining should be considered, provided that (1) suitable material is close to the work, and (2) slope and right-of-way limitations will permit. If space limitations, the unavailability of suitable earth blanket materials, or other considerations indicate a requirement for another type of or a more expensive lining, advice from the Engineering and Research Center should be obtained.

64. Inadequate Foundation Conditions.—(a) *General.*—Much of soils engineering literature is concerned with methods and procedures for overcoming the difficulties imposed by some of the foundation conditions which, for the purposes of this manual, will be classed as inadequate. In this sense, the term "inadequate" relates to obviously poor foundations which have not been tested and evaluated by a specialist.

For the majority of structures, unless the deposits of inadequate materials are very extensive, it will prove to be economical and satisfactory to avoid utilizing these materials for a foundation. The more common methods for avoiding the use of these materials include their removal and recompaction or replacement with compacted select material; full penetration with piles, piers, caissons, or walls; or displacement, that is, removal of a mass of material equal to the weight of the structure. In some instances it may be feasible to relocate the structure to provide a better foundation.

Among the materials placed in the inadequate category are materials containing organic matter such as topsoil, swamp muck, and peat; low-density materials such as loose deposits of silt or sand; talus deposits, spoil piles, and dumps; clays which are qualified as highly plastic, active, or sensitive and swelling clays; and generally soils in a soft and saturated condition.

If preconstruction investigations show that such questionable materials have been recognized and tested, it may be assumed that designs have been formulated taking the condition into account. If, however, after construction is in progress, the construction engineer feels that the condition is more serious or has not been adequately described, he should request a field review by a member of the engineering staff of the Engineering and Research Center familiar with the structure foundation requirements.

(b) *Topsoil.*—As in borrow deposits, the usual practice is to remove topsoil from all foundations. Topsoil in this sense is the surface layer of material containing decaying vegetable matter and roots. It is not necessary to remove all the soil containing fine hairlike roots but only the rather heavy root mat. On prairie soils containing light grass cover, 2 or 3 inches of stripping will usually suffice for topsoil removal. Agricultural lands are stripped to the bottom of the plowed zone, usually 6 to 9 inches. Valley bottom lands and brush-covered land commonly require stripping up to 2 or 3 feet for removal of inadequate material. Forest-covered land requires the removal of stumps resulting in stripping requirements between about 2 and 5 or more feet.

(c) *Swamp Muck.*—In this category are included marshlands, river backwaters, lakebeds, and flood-flow deposits. Foundations formed by deposits of these types are characterized as containing appreciable quantities of vegetable matter, mud, clay, or sand lenses deposited in water.

Shallow deposits should be removed; however, deposits may be deep and their depth and extent should be established before adopting a method of treatment. Relocation is recommended to avoid such deep deposits, particularly where roads, canals, pipelines, or transmission lines are involved. For foundations of dams, powerplants, and similar large structures, such materials are excavated and wasted; for bridges, pumping plants, and medium-sized structures where a water barrier is not required, fully penetrating piles are commonly used. Many times adequate foundations can be obtained by over-excavation to more firm material and the placing and compaction of a pad of select material.

Swamp muck deposits containing large amounts of organic matter which have been compressed into a fairly solid material described as peat might be expected to provide an adequate foundation in many cases. However, such deposits are ordinarily comparatively thin and can readily be removed or bypassed with footings or piles. Since chemical action and volume change are presumed to be occurring at a slow rate, it is prudent to avoid relying on such materials.

(d) *Silt and Sand.*—Although compact silt and sand deposits will generally provide satisfactory foundations, such materials can occur in a very loose state. Such loose materials are generally found as superficial deposits occurring as loess, sand dunes, sandbars in stream channels, alluvial fans, mud flows, deltas at the head of reservoirs and lakes, and generally the surface layers of soil subject to frost action.

Loose, poorly-graded sand that is saturated and located in an earthquake region is highly susceptible to the phenomenon of liquefaction (quicksand) when disturbed by a shock loading. Many slopes and bearing capacity failures have occurred during an earthquake because of liquefaction in loose sands.

Loose sands and wet silts will consolidate quite readily when loaded; however, foundations of loose sand and wet silt require no special treatment when used for flexible structures such as road fills and canal banks up to about 20 feet high. Dry, low-density silty soils, however, require removal or preirrigation before using as a foundation. Figure 85 shows the cracking and settling of a canal bank in dry low-density silt on filling the canal.

Foundations of loose materials and wet silts will require removal for fills higher than about 20 feet and earth dams; and for foundations of dry low-density silt, either removal or aiding its consolidation by ponding as shown in figure 86 is required. Consolidation of saturated sands can be facilitated by draining the water from the soil.

Rigid structures are usually not founded on loose silts and sands. These materials are either removed or consolidated, or the footings for the structures are extended to firmer material. Designs will normally show the method to be used.

It has been found that certain relationships exist between the inplace density and moisture conditions and the volume change due to settlement that may be anticipated for fine-grained soils. A relationship, in which density is expressed in terms of the ratio of inplace dry density to Proctor maximum dry density for various water contents, expressed in terms of Proctor optimum moisture, is shown on figure 87a. If the foundation density and moisture conditions can be plotted above and to the left of the limit line shown on the plot, very small additional volume change will occur on saturation and, therefore, no foundation treatment is normally required for small dams and canal embankments. If the foundation density and moisture conditions plot below and to the right of the limit line, significant volume change may occur on saturation of the foundation even under the loads of small dams and canal embankments, and foundation treatment may be required. Rigid structures should not be placed on either wet loose soils or dry loose soils subject to later wetting, as structure cracking may occur. Therefore, foundation treatment for rigid structures may be required for foundation soils which fall above the limit line.

Another criterion for predicting loose fine-grained soils needing treatment has been found useful. It involves only the inplace density and the

**Figure 85.—Cracking and settling of canal bank in dry, low-density silt.
P222–117–37478.**

liquid limit. The basis of the criteria is shown in figure 87b. The theory is simply that, if the natural dry density is low enough that the void spaces are larger than required to hold the liquid limit moisture content as shown for Case I, the soil can become wetted so that consistency upon saturation and the void space is sufficient to allow settlement. Conversely, if the natural dry density is high enough that the void spaces are less than required to hold the liquid limit moisture content as shown for Case III, the soil will not collapse upon saturation but will reach a plastic state in which there will always be particle-to-particle strength. Therefore, liquid limits and inplace dry densities which plot above the line show a critical density condition and points below the line do not. This latter condition of soils would only settle in a normal manner due to loading. This graph has additional usefulness in judging the quality of denseness of soils in the Case III category. For example, densities and liquid limits plotting close to the Case II line may not be susceptible to collapse but may be critical with respect to additional loading; whereas conditions plotting lower in the Case III area would result in more firm material. In the case of expansive soils with high liquid limit values and very high densities, plotting very low on the graph would indicate susceptibility to future expansion. Therefore, there would be a moderate. range of densities in Case III that would be most desirable.

Figure 86.—Ponding dry foundation of Trenton Dam, Nebraska, to facilitate consolidation. TD–328–701–3151.

Because of the uniformity of loessial soils in Kansas and Nebraska, which have liquid limit values between 30 and 40 percent, criteria of density and moisture conditions versus settlement on saturation can be expressed in terms of actual unit weights and water contents for these areas:

(1) Loess with densities of less than 80 pounds per cubic foot is considered loose and highly susceptible to settlement on saturation with little or no surface loading.

(2) Loess with densities from 80 to 90 pounds per cubic foot is medium-dense and is moderately susceptible to settlement on saturation when loaded.

(3) Loess with densities above 90 pounds per cubic foot is quite dense and capable of supporting ordinary structures without serious settlement, even on saturation.

(4) For earth dams and high canal embankments, a density of 85 pounds per cubic foot has been used as the division between high-density loess requiring no foundation treatment and low-density loess requiring treatment.

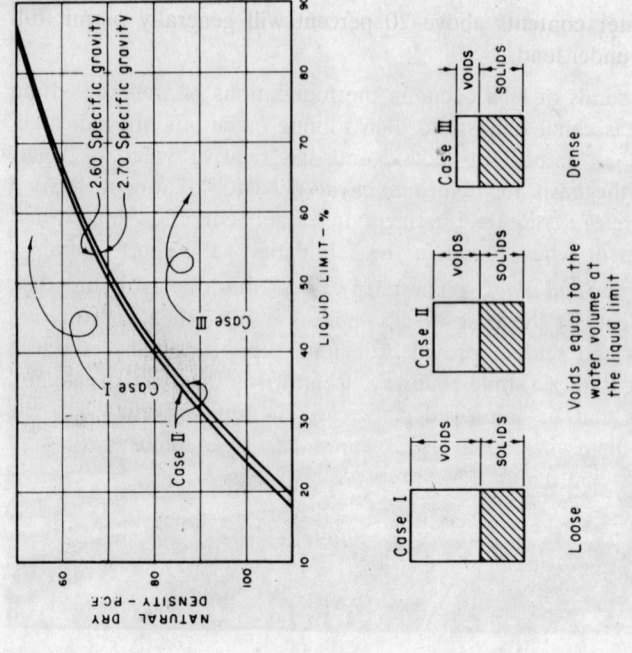

$\omega_0 - \omega =$ OPTIMUM WATER CONTENT(% BY DRY WEIGHT)–NATURAL WATER CONTENT(% BY DRY WEIGHT)

a. Considering "D-ratio," the ratio of natural (inplace) dry density to Proctor maximum dry density, and $\omega_0 - \omega$, optimum water density, and $\omega_0 - \omega$, optimum water content minus natural water content.

b. Considering natural dry density and liquid limit.

Figure 87.—Criterion for treatment of relatively dry fine-grained foundations. 101-D-550.

(5) Water contents above 20 percent will generally permit full settlement under load.

Where loose sands or silts occur in the foundations of water retention structures such as canal banks and dams, lining or cutoffs are indicated. Permeability is an important factor, and the relative value of water will usually be the basis for determining the need for treatment. A long canal may, therefore, require treatment for a soil condition that would not need treatment when found in the foundation of a short lateral.

(e) *Talus and Spoil Piles.*—The term "talus" describes soil and rock debris that collects at the base of a slope. As a rule, such deposits are loose and open and seldom provide adequate support for any structure more important than a simple roadway. Removal is the usual treatment, but some types can be improved sufficiently by sluicing fines into the voids to make them acceptable. Excavations into talus deposits are very likely to be unstable if cut slopes appreciably steeper than the slope of the talus deposits are used.

The processes used for disposal of excess or unsatisfactory materials in spoil piles are somewhat similar to the processes by which talus deposits are developed, and so such deposits have similar characteristics. Spoil piles and talus deposits should not be used for foundations of any structure except by specific directive from the Engineering and Research Center. If, during layout, structures are found to be located on such foundations, an inquiry should be made because it is likely that the designer was unaware of the foundation condition.

(f) *Clays.*—The performance of many structures founded on clays is satisfactory; yet many that have been founded on clays have failed. There does not appear to be any clear line of demarcation. Those clays that appear to be most treacherous have one or more of the following characteristics: plasticity indices as determined by Atterberg tests greater than 25 percent; colloid content greater than 20 percent; high sensitivity, that is, a piece of the natural deposit softens as it is manipulated in the hand; and shrinkage and swelling characteristics as noted by its tendency to shrink and crack as it drys out. There also appears to be a relationship between water content and the strength of clay materials. Table 3 gives criteria for identification of expansive clays.

(g) *Soft or Saturated Materials.*—Almost any soil foundation is reduced in quality when it contains large amounts of water. This is of particular importance to the Bureau of Reclamation which constructs many hydraulic structures. It should be recognized that an apparently firm dry foundation may become unstable when saturated. Soil foundations subject to saturation may be protected either by provisions which will prevent water getting to the foundation, such as drains to divert

Table 3.—Relation of soil index properties and probable volume changes for highly plastic soils

Data from index tests [1]			Estimation of probable expansion,[2] percent total volume change (dry to saturated condition)	Degree of expansion
Colloid content (percent minus 0.001 mm.)	Plasticity index	Shrinkage limit, percent		
>28	>35	<11	>30	Very high.
20–31	25–41	7–12	20–30	High.
13–23	15–28	10–16	10–20	Medium.
<15	<18	>15	<10	Low.

[1] All three index tests should be considered in estimating expansive properties.
[2] Based on a vertical loading of 1.0 p.s.i. as for concrete canal lining. For higher loadings the amount of expansion is reduced, depending on the load and on the clay characteristics.

surface runoff, or by provisions for removing water that reaches the foundation such as underdrains or drainage blankets. Structures whose foundations will be permanently under water are designed against uplift.

Because of the difficulty of preparing and inspecting foundations located beneath the water table, Bureau specifications contain provisions to require unwatering of foundations prior to excavation. However, to accomplish this for the foundation, in general, it is necessary to do some work in water such as the construction of cofferdams, pump sumps, and drains. Insofar as practicable such structures should be kept away from the planned foundation surfaces so that these surfaces will not be damaged by the construction or action of the foundation unwatering facilities. Foundations of saturated materials can be damaged by permitting an upward flow of ground water into the excavation or by permitting heavy equipment traffic in the area.

D. Rolled Earth Dams

65. Foundation Treatment.—(a) *Design Features.*—An earth dam is often the only feasible type of structure that can be built at a particular damsite because of relatively poor foundation conditions. Unconsolidated alluvial deposits consisting of layers of coarse sand and gravel, silts, or clays of depths varying from only a few feet to hundreds of feet before bedrock is reached are commonly found at damsites. Deeply weathered or faulted broken rock may also be encountered in earth dam foundations. Foundations of thick deposits of windblown silts and fine sands present especially difficult problems. Foundation design features can be placed in two broad categories: (1) control of seepage and (2) control of stability.

Cutoff trenches, cutoff walls, or combinations of trenches and walls, designed to obstruct the flow of seepage water or to lengthen the path of percolation under the dam, are most commonly used with foundations of unconsolidated permeable materials. Blankets of impervious material extending upstream from the toe of the dam on the reservoir floor and covering all or part of the abutments are sometimes used for the same purpose. The closing of seams, joints, and other openings in the bedrock is usually necessary and is accomplished by the injection of grout under pressure to form a curtain cutoff. Comparatively shallow low-pressure grouting over the foundation area, so-called "blanket" grouting, is also used under certain conditions. Cutoff features are usually located upstream from the centerline of the crest of the dam, but not so far upstream as to create a short percolation path through the embankment.

Two general types of cutoff trenches are used: (1) sloping-side trenches and (2) vertical-side trenches. The number, depth, location, and type of cutoff trenches that are provided depend on the nature of the foundation materials and on their permeability. Common practice for foundations with bedrock at a considerable depth is to excavate a single sloping-side trench located upstream from the centerline of the crest of the dam. The centerline of this trench parallels the dam centerline of the crest across the valley floor, but converges toward the centerline of the crest as it extends up the abutment in order to maintain the required embankment cover. When bedrock or an impervious stratum is relatively close to the surface, the cutoff trench may be widened to the full width of the impervious zone of the embankment and sometimes is designed to the full width of the total embankment. Two or three sloping-side trenches may be used occasionally to provide a series of cutoffs. Such trenches are excavated to bedrock when practicable, and the plane of contact sometimes is interrupted by a concrete cutoff wall. Where bedrock cannot be reached economically, the excavation is carried either to a stratum of relatively impervious material or to a depth which will provide the desired lengthening of the path of percolation. Sloping-side cutoff trenches are excavated by shovels, draglines, or scrapers, and are filled with selected impervious materials equivalent to the impervious zone in the dam embankment and which are compacted in the same manner as rolled earthfill embankment material.

Vertical-side trenches also may be used as cutoffs, and may consist of either opencut excavation by hand or trenching machine, or stoped excavation, or by slurry trench methods below water table. In some instances, the depth of a sloping-side cutoff trench may be extended by continuing the excavation as a narrow trench with vertical side walls;

this requires a portion of the fill to be specially compacted by approved tamping equipment. Stoped cutoffs, although very rarely used, are those constructed by using underground mining methods for excavation, then backfilling with concrete or impervious earth materials. They are used when construction of a cutoff would otherwise be impracticable or where the depth is such that the cost of open excavations is excessive. Stoped cutoffs are used in or across permeable strata extending below a deep overburden or where it is necessary to remove and replace breccia or debris in fault zones. The general procedure consists of excavating and concreting or backfilling in successive lifts or levels, and the methods of mining, timbering, and placing of backfill are determined by the particular conditions and problems.

Concrete cutoff walls are used infrequently. When the dam embankment is placed on bedrock, it is sometimes desirable to construct one or several concrete walls, depending on foundation conditions and on the maximum depth of water in the proposed reservoir. When more than one wall is built, the downstream wall is usually located along or near the centerline of the crest of the dam and the others are placed upstream from it. Particular geologic features may require the placing of short independent sections of wall at certain locations on the abutments or within the river channel. Concrete cutoff walls are usually required when the bedrock is hard but badly broken and irregular, or is dipping steeply, especially on the abutments, making adequate bond between embankment and foundation extremely difficult.

Sheet piling is occasionally used in combination with a cutoff trench as a comparatively economical means of increasing the depth of the cutoff, and, under certain conditions, it may be used in lieu of a cutoff trench. Interlocking steel sheet piling is the recommended type. Sheet piling cutoffs can be used in foundations of silt, sand, and fine gravel. Where stones or boulders are present or where the material is highly resistant to penetration, driving or jetting not only becomes difficult and costly, but it is highly doubtful that an effective cutoff can be obtained because of the tendency of the piling to wander and become damaged by breaks in the interlocks or tearing of the steel.

In the design of an embankment, careful consideration is given to the abutment contacts. Any continuous void spaces caused by the lack of an intimate contact between the impervious portion of the embankment and the abutments may result in dangerous seepage or even failure. Overhanging rocks and undesirable formations should be removed, and slopes should be reduced to permit embankment materials to be bonded properly to the abutments. Abutment slopes should not be steeper than 1:1 for any appreciable distance, except in special cases and then only after careful detailed studies. In thin core dams, the impervious section may be widened at the points of contact with the abutments.

Foundation grouting is a process of injecting under pressure suitable grouting materials into the underlying formations through specially drilled holes for the purpose of sealing off or filling joint seams, fissures, or other openings. Grouting may be done with neat cement, mixtures of cement and sand, or cement admixed with bentonite or other clay. In special cases, chemical or bituminous grouting may be used. The adopted program may consist of grouting through a concrete grout cap along specified lines to create a deep impermeable curtain, comparatively shallow or "blanket" grouting over portions of the foundation area, or both. Every foundation presents a separate, distinct grouting problem, depending on the composition and nature of the material and on its geologic structure as found by the explorations. Figure 88 is a geologic section of an abutment of a dam showing a grout curtain.

The path of percolation in foundations of unconsolidated permeable materials can be increased by the construction of an apron or blanket of compacted impermeable material upstream from the toe of the dam. A blanket is usually used when cutoffs to bedrock or to an impervious layer are not practicable because of excessive depth. Sometimes they are used in conjunction with partially penetrating trenches designed to increase the length of the path of percolation. The topography just upstream from the dam and the availability of impervious materials are important factors in the design of blankets. The length of upstream extension is determined by the desired reduction of the volume of seepage and of the hydraulic gradient. In zoned embankments, the blanket is usually carried below the upstream shell as a continuation of the central, impervious section. The blankets may be constructed to cover either or both abutments or particular abutment areas. The required thickness of blanket will vary with the permeability of the material of which it is constructed and with the head of water to which it will be subjected.

Methods for improving foundation stability include removal of unsuitable material; preconsolidation by unwatering, irrigation, or explosives; adequate drainage of the downstream toe by means of filters, toe drains, and relief wells to avoid piping; and construction of stabilizing fills to prevent displacement of the foundation.

The minimum requirement for foundation stripping is the removal of vegetation, sod, topsoil with high organic content, and other unsuitable material that can be removed by opencut excavation. In cases where the overburden is comparatively shallow and composed of soft clays, loose fine sands and silts, or extremely pervious sands and gravels, the entire foundation area of the dam may be stripped to bedrock.

Unwatering of the entire foundation area may be undertaken as a means of consolidating and stabilizing otherwise suitable foundation mate-

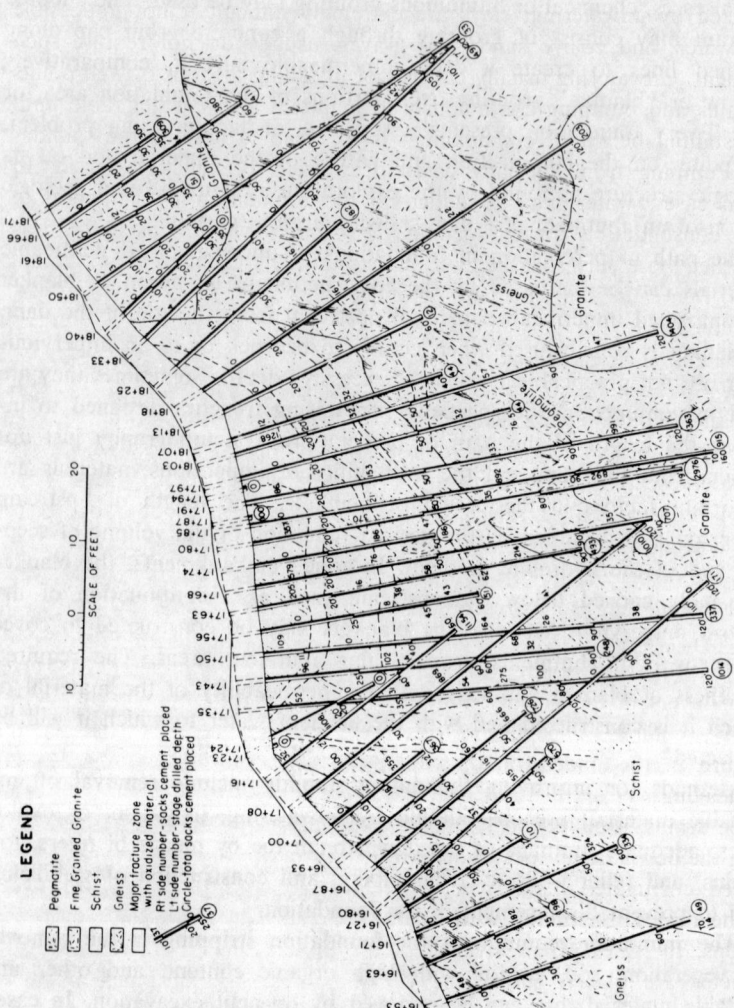

Figure 88.—Curtain grouting in right abutment of Granby Dam, Colo. 101-D-245.

rials prior to loading them with embankment, when the cost of such treatment is less than the alternative of excavating and replacing the material with selected borrow material. The purpose of unwatering is to minimize consolidation after completion of the embankment or to reduce pore-water pressures created by loading. Fine sands and silts may be stabilized by unwatering; clays are extremely difficult if not impossible to unwater; and coarse sands and gravels usually do not require such treatment. Unwatering usually consists of installing a series of well points or drains and pumping continuously to maintain a low water table until a substantial height of embankment is in place to confine the foundation. Pumping from specially cased deep wells is sometimes used. In special cases, water may be added to the foundation by irrigation to cause consolidation during loading. Dry loess (wind deposited) soils and other loose silty soils have been successfully treated by this method. Loose, fine sand deposits below ground-water table have been successfully consolidated by the detonation of buried charges of explosives.

It is impracticable to stop all seepage under the dam. Water will inevitably escape somewhere in the vicinity of the downstream toe, and when the foundation materials in that area consist of or contain appreciable quantities of fine sands, silts, or clays that may be carried away by water escaping under pressure, an inverted filter is provided. The filter is constructed of selected materials ranging from fine to coarse from bottom to top. It permits the necessary passage of water but prevents the displacement of fine soil particles. The design of a filter takes into account the grain sizes of the foundation soils and the head loss. Several layers of different sizes of sand and gravel may be required. In order to avoid buildup of high pressures beneath impervious foundation layers in the vicinity of the downstream toe, vertical relief wells may be required. The purpose of these wells is to relieve excess water pressure in the underground, and thereby prevent formation of boils and heaving.

Toe drains, usually 8 to 12 inches in diameter, are commonly installed along the downstream toes of dams. Vitrified clay, concrete tile, asbestos-bonded corrugated metal pipe, or asbestos-cement pipe may be used for the drains. Beginning with small diameters laid along the abutment sections, the lines progressively increase in size; the lines of maximum diameter are placed across the valley floor. Adverse slopes are not allowed in toe drains, regardless of size or length. Draintile or pipe is carefully laid and encased in gravel, or crushed rock, to prevent piping of surrounding material into the drain. Inspection wells may be installed in the outlet line for observation and measurement of the amount of seepage water. The function of the drains is to allow the easy escape of seepage water, thereby preventing saturation and consequent reduction of stability of the material in the downstream toe of the dam. Toe drains placed sufficiently deep in the foundation will tend to keep the ground-water level low and thus avoid the creation of boggy areas downstream from the dam.

To make foundations of low strength suitable as support for a dam, stabilizing fills for weight only are often provided at the upstream, downstream, or both toes of the dam. These fills act to restrain the tendency of the foundation materials to displace under load. The depths and extent of these fills depend on the characteristics of the foundation materials as revealed by laboratory tests on undisturbed samples. Stabilizing fills are generally contiguous and integral with the embankments.

(b) *Specifications Provisions.*—Provisions for foundation preparation and treatment are included in the section of the specifications entitled "Diversion and Care of River (Stream, or Surface Runoff) During Construction and Removal of Water from Foundations," in the section on earthwork, and in the section on embankment. Diversion and care of the river (stream, or surface runoff) during construction is an important factor in the design of an earth dam. Figure 89 shows a temporary diversion channel through an earth dam. Specifications require

Figure 89.—Temporary diversion channel through Bonny Dam, Colorado.
PX–D–16330.

that the contractor's plan for diversion should be approved by the contracting officer. Underwater excavation for the foundation of a rolled earth dam is not permitted. The contractor is required to unwater the foundation in advance of excavating an embankment cutoff trench. The unwatering is required to be accomplished in a manner that will prevent loss of fines from the foundation, will maintain stability of the excavated slopes and bottom of the cutoff trench, and will result in all construction operations being performed in the dry. Unwatering is to be continuous until the excavation has been filled, except that in filling the cutoff trenches, the ground water may be permitted to rise within 5 feet of the top of the compacted embankment after at least 10 feet of fill have been placed in the dry. This provision enables the contractors to remove the majority of the well points if they use that type of unwatering process. It is usually impracticable to prevent water from seeping out along the contact of the cutoff trench slopes with the bottom. Hence, some drains are needed to assure that the backfill is placed in the dry. Open-jointed or perforated pipe drains surrounded by gravel and leading to sumps can be used to dry up the bottom of the foundation. When these pipe drains cannot be removed, provisions must be made to grout and seal them by a system of grout connections and return pipes.

Under the section "Earthwork" there is a paragraph which provides that the area to be covered by the embankment shall be cleared of all trees, stumps, brush, rubbish, and other objectionable matter. Other paragraphs classify or define materials to be excavated and used in construction. The specifications provide that all excavations for foundations shall be performed in the dry and that all suitable excavated materials shall be used for permanent construction. The paragraph on excavation for embankment foundation provides that the area to be occupied by the embankment shall be excavated to a sufficient depth to remove all topsoil, rubbish, roots, vegetable matter of every kind, and all other perishable or objectionable material that might interfere with the proper bonding of the embankment with the foundation.

The paragraph on earthfill in the "Embankment" section of the specifications contains provisions for foundation preparation beneath the earthfill. Earth foundation surfaces are to be prepared by leveling and rolling in the same manner as specified for placing embankment materials. Rock foundation surfaces are to be cleaned of all loose and objectionable materials. Foundation surfaces of shale and other materials that tend to slake or loosen on contact with air are to be cleaned immediately prior to placing earthfill materials. All foundation surfaces are to be moist but free from standing water. Figure 90 shows a foundation being prepared by washing to remove detrimental air-slaked and loose

Figure 90.—Foundation preparation by washing to remove detrimental air-slaked and loose material at Twin Buttes Dam, Texas. P825–523–564.

material prior to placing earth and concrete. Noted should be the safety line attached to the worker and hose.

(c) *Control Techniques.*—The weak points in earth dam construction are generally found within the foundation and at the contact of the natural ground surface with placed embankment. Construction of the foundation seepage control and stability features included in the design must be carefully supervised by the inspection force in accordance with the specifications. Unwatering methods used in connection with the excavation of cutoff trenches and for stabilizing the foundation should be carefully checked to see that fine material is not being washed out of the overburden because of improper screening of drainage wells. To avoid creation of a "live" bottom due to upward flow of water, continuous pumping is required. Whenever possible, well points and sumps should be located outside the area to be excavated. Concrete footings for cutoff walls or the concrete grout cap must be on firm foundation. Blasting for the excavation of these structures should be prohibited or strictly controlled in accordance with the specifications.

Where overburden is stripped to firm formation, the foundation surface, including all pockets or depressions, should be carefully cleaned of soil or rock fragments before placing the embankment upon it. This may require handwork and compressed air cleaning. Surfaces which disintegrate rapidly on exposure should be covered immediately with embankment material or an approved protective covering.

When the foundation is earth, all organic or other unsuitable materials, such as stumps, brush, sod, and large roots, should be stripped and wasted. Stripping operations should be carefully performed to assure removal of all material that may be rendered unstable by saturation, of all material that may interfere with the creation of a proper bond between the foundation and the embankment, and of all pockets of soils significantly more compressible than the average foundation material. Test pits for further exploration should be excavated if the stripping operations indicate the presence of unstable or otherwise unsuitable material, and on completion of the pits, the Engineering and Research Center should be informed so that a final inspection may be made.

Prior to placing the first layer of embankment on an earth foundation, the foundation surface should be leveled, moistened, and compacted so that the surface material of the foundation will be as compact and well bonded with the first layer of embankment as is specified for subsequent layers of embankment. Foundation surfaces of firm formation should be moistened, but no standing water should be permitted when the first lift is placed. Sometimes the earth foundation surface requires scarification by harrows to insure proper bonding; when it can be penetrated by tamping roller feet, no additional scarification is usually necessary. Where the formation would be injured by penetration of the tamping roller feet, it is permissible to use a thicker than specified layer of earthfill for the first compacted layer. However, the first layer should never exceed 15 inches loose lift, and additional roller passes are required to insure that proper compaction is attained. Special compaction methods should be used in pockets that cannot be compacted by the specified roller instead of permitting unusually thick first lifts to obtain a uniform surface for compaction.

Although the soil should be not drier than optimum, the use of overly wet soil in the first lift against the foundation should be generally avoided; rather, the foundation should be properly moistened. On steep, irregular rock abutments, wetter-than-optimum material may be necessary or desirable in order to obtain proper bond. However, the use of such material should not be permitted without specific approval of the Engineering and Research Center. Care must be exercised when special compaction is used, to insure that bond is created between successive layers of material. This may require light scarification between lifts of tamped material. Figure 91 shows power tamping of earthfill at contact with irregular rock abutment.

66. Compacted Earthfill.—(a) *Design Considerations.*—Rolled earth dams vary considerably in the proportions and physical characteristics

Figure 91.—Power tamping of earthfill at contact with irregular rock abutment.
P860–427–358NA.

of the soils of which they are constructed. Many dams contain essentially one kind of material, hence are called homogeneous impervious structures, while others are constructed of sections of sand, gravel, rockfill, and miscellaneous soils in a variety of quantities and locations within the fill, in addition to an impervious zone of earth. This impervious zone of earthfill, usually designated zone 1 in Bureau dams, provides the water barrier and is common to all rolled earth dams.

In controlling the construction of impervious earthfill zones, the following design criteria must be met:

(1) The material must be formed into an essentially homogeneous mass free from any potential paths of percolation through the zone or along the contacts with the abutments or concrete structures.

(2) The soil mass must be sufficiently impervious to preclude excessive water loss through the dam.

(3) The material must not consolidate excessively under the weight of superimposed embankments.

(4) The soil must develop and maintain its maximum practicable shear strength.

(5) The material must not consolidate or soften excessively on saturation by water from the reservoir.

These design objectives are obtained by proper use of equipment and methods of borrow-pit conditioning, excavation, placing, and compaction.

To satisfy the design criteria of imperviousness, low consolidation, and resistance to softening, the maximum possible density is desirable. However, in order to control the development of pore pressure which would reduce shear strength and to avoid the formation of slickensided layers which may become paths of percolation, it may be desirable to place material at slightly drier than optimum water content even though it can result in a density slightly less than the Proctor maximum. The plan of control of the impervious earthfill for high dams must be a compromise that will result in a proper balance among all the design criteria.

(b) *Specifications Provisions.*—The section in the specifications entitled "Earthwork" contains a number of provisions relating to the construction of earthfill. The specifications on opencut excavation provide that all excavations for foundations shall be performed in the dry, and that all suitable excavated materials shall be used for permanent construction. Materials suitable for earthfill must be excavated separately, and in the same manner as those fill materials excavated from borrow areas. Provisions are made for stockpiling of materials.

The paragraph on borrow areas designates the borrow areas to be used for earthfill material and provides for Government control of location and extent of borrow pits within the designated borrow areas in order to obtain suitable materials. Designated borrow pits are required to be cleared and stripped of unsuitable material and are to be irrigated or drained to obtain the proper water content prior to excavation. The contracting officer designates the depth of cut in all borrow pits, and the contractor's operations must be such that the earthfill material delivered on the embankment is equivalent to a mixture of materials obtained from an approximately uniform cutting from the full height of the face of the borrow-pit excavation. When necessary, separation of oversized material from the earthfill is specified.

The section in the specifications on embankment prohibits placing frozen soil in the embankment or placing unfrozen material on frozen embankment. The maximum allowable difference in elevation between zones during construction is specified. This section provides for control by the Government of all openings through the embankment, limits the slopes of bonding surfaces, and requires careful preparation of sloping surfaces before additional material is placed. Slopes of bonding surfaces in earthfill material are usually limited to 4:1 or flatter. Each load of embankment material must be placed in the location designated by the Government.

Figure 92.—Separation plant at Meeks Cabin Dam, Wyoming, where impervious material containing oversize was screened into fractions above and below 5 inches. P145-432-468.

The paragraph on earthfill in embankments provides for placing earthfill only on unwatered and prepared foundations. All cavities within the earthfill area are to be filled. The materials to be used for earthfill are described. The requirements for obtaining a mixture of materials from designated depths of cut in the borrow area in order to produce an acceptable gradation of compacted earthfill are reiterated. In placing the earthfill, the distribution and gradation must be such as to avoid lenses, pockets, streaks, or layers which would prevent homogeneity of the fill. Within the earthfill zone the distribution of material is to be such that permeability and coarseness increase toward the outer slopes. If the earthfill material contains only occasional cobbles or rock fragments, provisions are made to remove such cobbles or rock fragments over 5 inches in size from the fill prior to rolling. When an appreciable percentage of oversize is known to exist in the source of earthfill material, a separation plant is specified (see fig. 92). When a separation plant is required, separation may be specified on either the 3-inch or 5-inch size. Placement of earthfill is to be in approximately horizontal layers not more than 6 inches in compacted thickness. Surfaces of fill that are too dry or too wet are required to be reconditioned before additional materials are placed.

Moisture control provisions of earthfill require uniformity of water content throughout the entire layer prior to and during compaction. The water content of a soil is limited to a range based on design considerations. The average water content, in general, is required to be slightly less than the Bureau laboratory standard optimum. The contractor is required to perform all operations necessary to obtain the proper water content by irrigating or draining borrow pits, performing supplementary sprinkling on the fill, and working the material in the borrow area or on the fill prior to compaction.

Tamping (sheepsfoot) rollers are required for compacting earthfill. The size, arrangement, and safety features of the drums are specified. The number, location, length, and cross-sectional area of the tamping feet are described. The weight of roller when loaded and used is required to be not less than 4,000 pounds per foot of length of drum. The contractor is required to keep the spaces between the tamping feet clear. The number of passes of the roller on each layer of material is specified. Provisions are made for the contingencies of too dry or too wet material. Where rolling with a specified roller is undesirable or impracticable, provisions are made for special compaction by tampers or other equipment and procedures which will obtain the same degree of compactness as the rolling process. Figures 93 and 94 show the operations of placing, spreading, and compacting earthfill.

The specifications provisions for control of placement water content and compaction must be implemented by establishing procedures to assure their attainment. Prior to construction of the earthfill, all the information available from field and laboratory investigations is analyzed and reviewed in the light of design considerations to arrive at a tentative plan of control. Special limiting moisture tests in the Engineering and Research Center laboratory will usually be made on representative samples of materials to be used in high earth dams to determine placement conditions that will prevent development of excessive pore-water pressures during construction on the one hand, and settlement of the loaded soil when saturated by water from the reservoir on the other hand. Based on the field and laboratory investigations and on experience with similar soils, the Engineering and Research Center will recommend placement moisture limits and minimum density for representative types of materials. These recommended limits are transmitted to the construction engineer in the "Design Considerations" or they may be included in the specifications.

(c) *Control Techniques.*—Although fill conditions are approximated by laboratory tests, it is always necessary to check the planned control against actual results obtained on the embankment and then make any required adjustments. This is done by careful observation and analysis

Figure 93.—Placing, spreading, and compacting earthfill in a relatively restricted area on Blue Mesa Dam, Colorado. P622–427–10018NA.

of the operations, control test results, and embedded instruments, especially during the early stages of the construction of the earthfill.

In structures where unusual conditions are present, the specifications may include provisions for the contractor to construct one or more test sections of earthfill material. The purpose of a test section is to deter-

Figure 94.—General view of placing, spreading, and compacting of earthfill in a relatively unrestricted area on Twin Buttes Dam, Texas, allowing for a multitude of pieces of equipment. P825-523-1637A.

mine the most practicable excavation, processing, and placing procedures for representative soils under job conditions. By varying the placement procedure within certain limits, by exercising rigid control over the relatively small volume of the section, and by keeping complete records of the tests, the most applicable procedures for the rolled earthfill portion of embankment may be determined during the initial stages of construction. Results of field density tests made on the test sections will provide the necessary information for establishing construction control procedures consistent with design requirements. Figure 95 shows a test section under construction in a borrow area at Trinity Dam.

After the proper placing procedures have been determined from the embankment test section or from the initial placing operations when no test sections are required, construction can proceed at full scale. It is essential that throughout the construction the contractor's operations adhere strictly to the limits and conditions set forth in the specifications and to the control recommendations made by the engineering staff of the Engineering Research Center. In a sense, the entire embankment is a "test section." A complete record of all operations must be kept

Figure 95.—Test section of residual soil in borrow area at Trinity Dam, California. TD–2318–CV.

so that the most efficient methods may be perfected. The relationships between laboratory compaction curves and field roller curves obtained by analysis of control test results are shown in figure 96 for three dams.

The most important variables affecting embankment construction are: selection of materials in the borrow area; distribution of materials on the embankment to obtain uniformity; placement water content, and the uniformity of this moisture throughout the spread material; water content of borrow material; methods for correcting borrow material water content if too wet or too dry; roller characteristics; number of roller passes; thickness of layers; maximum size and quantity of rock in the material;

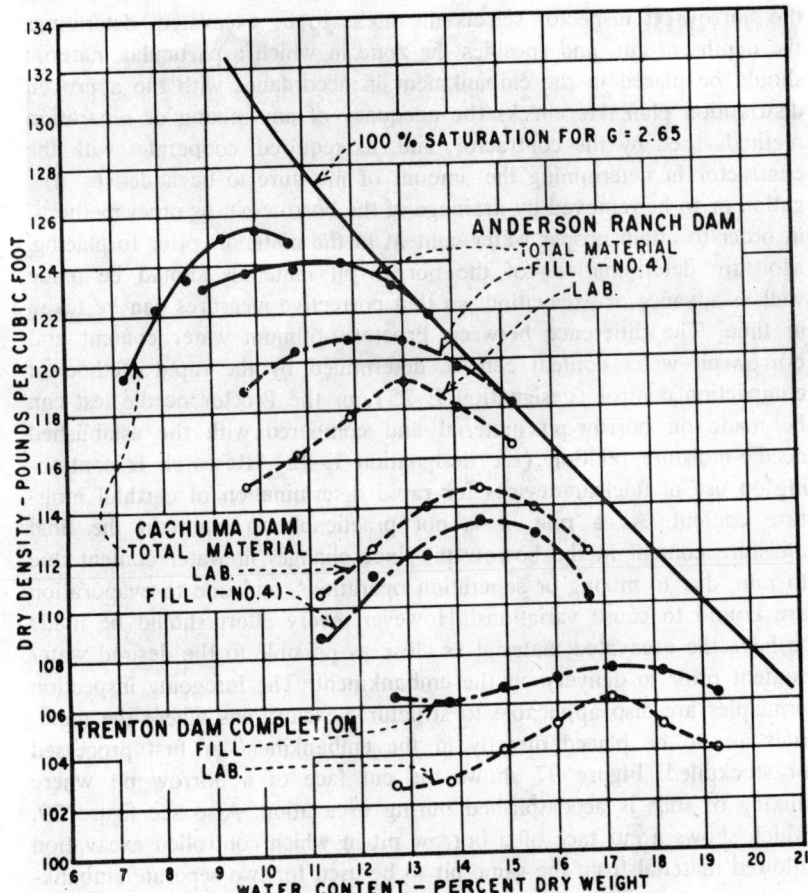

Figure 96.—Average field and laboratory compaction curves for three dam embankment soils. 101-D-248.

condition of the surface of layers after rolling; and effectiveness of power tamping in places inaccessible or undesirable for roller operation.

To maintain the control of earthfill construction, adequate inspection and laboratory personnel are essential. There should be at least one inspector on the embankment and one technician in the laboratory for each shift. Borrow operations on small jobs may be inspected by the embankment inspector. On large jobs, it is necessary to have an inspector at the borrow pit, especially when moisture control is critical.

The borrow-pit inspector controls and records all earthwork operations prior to dumping the material on the embankment. Under the supervision of the construction engineer, through the chief inspector, the borrow-pit inspector selects the areas to be excavated, determines the depths of cut, and specifies the zone in which a particular material should be placed in the embankment in accordance with the approved distribution plan. He checks the adequacy of any mixing or separation methods used by the contractor, and, as required, cooperates with the contractor in determining the amount of moisture to be added by irrigation or to be removed by drainage of the borrow pit, or other methods, in order to attain proper water content in the materials prior to placing. Moisture determinations of the borrow-pit materials should be made well in advance of excavation, so that corrective measures can be taken in time. The difference between Proctor optimum water content and borrow-pit water content can be determined by the rapid method of compaction control (designation E–25), or the Proctor needle test can be made on borrow-pit material and compared with the established needle-moisture relation (see designation E–22). Research is continuing on use of nuclear devices for rapid determination of earthfill moisture content. As a rule, it is not practicable to maintain the final moisture content in the borrow pit since changes in water content due to rain, due to mixing or separation operations, and due to evaporation are bound to cause variations. However, every effort should be made to have the excavated material as close as possible to the desired water content prior to delivery on the embankment. The foregoing inspection principles are also applicable to structural excavations where the materials are to be placed directly in the embankment or first processed or stockpiled. Figure 97 shows the cut face of a borrow pit where mixing of soils is accomplished during excavation. Also see figure 79, which shows a cut face of a borrow pit in which controlled excavation allowed material from the same pit to be used for two separate embankment zones—pervious and impervious.

The embankment inspector should be provided with means for determining the location and elevation of tests made on the embankment and for reporting the location of the contractor's operations. Horizontal control by means of coordinates or station and offset should be established. It has been found desirable to establish vertical control by benchmarks and the use of stadia rods from which the inspector can establish an elevation anywhere on the fill by using a hand level. When materials are brought on the embankment, the inspector must see that they are placed in the proper zone, as indicated by the distribution plan. Lines of demarcation can be painted on rock abutments or marked by flags if a zoned embankment is being constructed. Within a particular zone or on homogeneous embankments, the objective is to direct the placing of materials so that the most impervious soils are located in the center of the zone, and the coarser, more pervious soils are placed toward the

Figure 97.—Elevating grader equipped with vertical cutting blade for mixing upper layer of silty clay with lower sand and gravel deposit. PX–D–16336.

outer limiting slopes of zones or the outer slopes of the homogeneous embankment, so that the permeability and stability of materials will increase toward the outer slopes. In general, when materials differ in dry density but have the same percolation rate, the materials having the greater dry density should be placed, for stability purposes, in the outer portions of the zone, or dam, as the case may be.

After the materials are placed in the correct location, the embankment inspector determines whether they contain the proper amount of moisture *prior* to compaction. This is of utmost importance. The Proctor needle-moisture test (see designation E–22), the rapid compaction control method (designation E–25), or moisture determination by nuclear devices, if approved by the Engineering and Research Center, will determine whether supplementary water is necessary. Should the materials arrive on the embankment too dry, it will be necessary to condition them properly by sprinkling prior to, during, or after spreading. The contractor's operations in sprinkling and mixing the moisture with the soil will vary. However, regardless of the method used, it is of paramount importance that the *proper water content be uniformly distributed throughout the spread layer prior to rolling.*

Another important inspection task is the determination of the thickness of the compacted layer. A layer that is spread too thick will not give the desired density for given conditions of compaction. Initial placing operations or the test section will have determined the proper spread thickness of a layer that will compact to the specified thickness; this specified compacted thickness should never be exceeded. A method for determining average thickness of placed layers is to plot daily a cross section of the fill at a reference station. The inspector's report for that day will contain the number of layers placed at that section; hence, the average thickness can be determined.

The removal of oversized rock from earthfill embankment material is most efficiently done prior to delivery of the soil on the embankment when the oversized rock content is greater than about 1 percent. Smaller amounts of oversized rock can be removed by handpicking or, under favorable conditions, by various kinds of rock rakes. Oversized rock that has apparently been overlooked prior to rolling can generally be detected by the inspector during rolling by noticing a bounce when the roller passes over a hidden rock. The inspector should see that such rock is removed from the fill by the contractor.

The inspector is charged with the responsibility of seeing that the specified number of roller passes is made on each lift. An oversight in maintaining the proper number of passes may lead to a considerable drop in the desired degree of compaction. The insistence on orderliness on the fill and the establishment of routine construction operations will result in minimizing trouble from too few roller passes.

The final check on the degree of compaction is done by the field density test (designation E–24) and the rapid compaction control method (designation E–25). If the degree of compaction of the material passing the No. 4 sieve is above the minimum allowable percentage of Proctor density and the water content is within the allowable limits, then the embankment is ready for the next layer after scarifying and moistening, if necessary, to secure a good bond between the layers. The time required to obtain results on the field density test by the rapid compaction method should not exceed 1 hour.

Mechanical tamping, when used around structures, along abutments, and in areas not accessible to the rolling equipment, should be carefully watched and checked by frequent density tests. The procedure to be followed for mechanical tamping will depend greatly on the type of tamper used. Some of the factors affecting density are thickness of the layer being tamped, time of tamping, air pressure (if air tampers are used manufacturer's recommended pressures should be maintained), water content of the material, weight of tamping unit, and condition of tampers.

The embankment inspector must be alert to discover areas of low density and, when they occur, he should set out immediately to ascertain the causes and have them remedied by sprinkling, scarifying, removal, or rerolling, as required. An important function of inspection is to determine when and where to make field density tests. These tests should be made:

(1) In areas where the degree of compaction is doubtful.

(2) In areas where embankment operations are concentrated.

(3) For every 2,000 cubic yards of embankment when no doubtful or concentrated areas occur.

(4) For representative tests of every 30,000 cubic yards of earthfill placed.

(5) For "record" tests at the locations of all embedded instruments.

(6) At least one test should be made during each shift involving placing of earth materials.

Areas of doubtful density are sometimes detected by observation by the inspector. Possible locations of insufficient compaction include:

(1) The junction between areas of mechanical tamping and rolled earthfill embankment along abutments or cutoff walls.

(2) Areas where rollers turn during rolling operations.

(3) Areas where too thick a layer is being compacted.

(4) Areas where improper water content exists in a material.

(5) Areas where less than specified number of roller passes were made.

(6) Areas where dirt-clogged rollers are being used to compact the material.

(7) Areas where oversized rock which has been overlooked is contained in the fill.

(8) Areas where materials have been placed when they contained minor amounts of frost, or at nearly freezing temperatures.

(9) Areas that were compacted by rollers that have possibly lost part of their ballast.

(10) Areas containing material differing substantially from the average.

The number and location of field density tests should be such that the extent of the doubtful area is determined. All tests made in areas of doubtful density should be identified by the letter "D" on the report (see sec. 71).

When embankment operations are concentrated in a small area—that is, if many layers of material are being placed one over the other in a single day—tests should be made in this area in every third or fourth layer to assure that the desired density is being attained. All tests made in such areas should be identified by the letter "C."

If areas of doubtful compaction do not exist and no tests are required because of concentrated areas, at least one field density test should be made for each 2,000 cubic yards of compacted embankment and it should be representative of the degree of compaction being obtained. Such tests should be identified by the letter "R."

Regardless of the number and purpose of other field density tests, one test is required for each 30,000 cubic yards of earthfill placed, and the tests should be representative of the conditions of the fill. Percolation-settlement and specific gravity tests are made on the same sample (see fig. 98), and the tests should be identified by the letter "R." Additional material must be excavated adjacent to the field density test hole to make these additional tests.

"Record" tests are made at all instrument installations and are recorded separately (see sec. 70).

During a seasonal shutdown period, it is sometimes desirable to excavate one or two test pits in the compacted embankment to determine the net result of the season's operations. The effect of rolling and of superimposed loading is indicated by making field density tests as the pit is being dug. The degree of uniformity of moisture with depth is found by testing samples at various depths, and visual inspection will indicate whether successful bonding of the layers has been accomplished. If possible, photographs should be taken of the pit. Figure 99 shows a test pit in the embankment of an earth dam. If time and personnel do not permit the excavation of a test pit and the embankment is free of rock, an auger hole 5 or 10 feet deep can be used to determine average density and water content. For auger holes, the procedure to be followed is the same as for making the field density test, excepting that more material is removed and the standard sand is not recovered. If inspection of the auger hole in compacted fill shows that a uniform diameter exists for the full depth of hole, the volume of the hole may be computed without the use of standard sand. (see designation E–23).

If any unusual conditions are noted in the embankment test pit or auger hole, they should be reported to the Engineering and Research Center.

67. Compacted Pervious Fill.—(a) *Design Considerations.*—Permeable type materials are used in rolled earth dams to provide an outer shell of high strength to support the impervious core, to secure favorable hydraulic conditions of drainage, and to act as filters between materials of greatly differing grain sizes. The criteria that must be met in controlling the construction of zones of sand and gravel are:

(1) The material must be formed into a homogeneous mass free from large voids.

(2) The soil mass must be free draining.

(3) The material must not consolidate excessively under the weight of superimposed fill.

(4) The soil must have a high angle of internal friction.

7-1352 (4-70)
Bureau of Reclamation

SUMMARY OF FIELD AND LABORATORY TESTS OF COMPACTED FILL
CONTROLLED BY PROCTOR TEST

FEATURE ____
PROJECT ____
DATE OF REPORT ____
PERIOD OF REPORT ____ TO ____
ZONE ____ MATERIAL

Use separate sheets for reporting tests made in different zones

TEST NUMBER	LOCATION OF TEST ON EMBANKMENT: STATION AND OFFSET OR COORDINATES	LOCATION OF TEST ON EMBANKMENT: ELEVATION	SOURCE OF MATERIAL (BORROW AREA, REQUIRED EXCAVATION, ETC.)	METHOD OF COMPACTION (NUMBER OF TAMPER PASSES OR IF POWER TAMPED(P.T.))	FIELD DENSITY, NEEDLE, AND CYLINDER CORRELATION TESTS AT FILL MOISTURE: WET DENSITY OF EARTH AND ROCK (LB PER CU FT)	PERCENT PLUS NO. 4 BY DRY WEIGHT	WATER CONTENT OF EARTH AND ROCK (% OF DRY WT)	PROCTOR NEEDLE, COMPACTION CYLINDER (LB PER SQ IN.)	MINUS NO. 4 WATER CONTENT (% OF DRY WT)	MINUS NO. 4 DRY DENSITY (LB PER CU FT): COMPACTION CYLINDER	MINUS NO. 4 DRY DENSITY (LB PER CU FT): FILL	CHARACTERISTICS AT STANDARD LABORATORY COMPACTION: MAXIMUM DRY DENSITY (LB PER CU FT)	OPTIMUM WATER CONTENT (% OF DRY WT)	PROCTOR NEEDLE (LB PER SQ IN.)	RAPID CONTROL VALUES: OPTIMUM WATER CONTENT MINUS FILL WATER CONTENT (IN %) ($W_o - W_f$)	RATIO OF FILL DRY DENSITY TO CYLINDER DRY DENSITY AT FILL WATER CONTENT (C IN %)	RATIO OF FILL DRY DENSITY TO LABORATORY MAXIMUM DRY DENSITY (D IN %)	SPECIFIC GRAVITY: PLUS NO. 4 FRACTION	SPECIFIC GRAVITY: MINUS NO. 4 FRACTION	PERCOLATION-SETTLEMENT TEST, BEFORE SATURATION: DRY DENSITY AS PLACED (LB PER CU FT)	WATER CONTENT AS PLACED (% OF DRY WT)	APPLIED PRESSURE (LB PER SQ IN.)	PERCENT SETTLEMENT OF PLACED SAMPLE	SATURATED: PROCTOR NEEDLE (LB PER SQ IN.)	WATER CONTENT (% OF DRY WT)	PERCENT SETTLEMENT OF PLACED SAMPLE	PERCOLATION RATE (FT PER YR)
				WURD[1]	1	2	3	☒	4	5	6	7	8														
7-1-a-1-R	10+00, 10' D.S.	3892.0	Area 'A' 1st cut	12	135.5	6.4	11.1	900	13.2	117.7	118.5	117.8	13.4	600	0.2	100.6	100.6	2.63	2.70	118.5	13.2	187	4.1	1000	14.0	4.1	0.50
7-2-b-4-D	15+00, 40' U.S.	3890.5	Area 'B' 2nd cut	P.T.	131.0	5.8	10.8	800	13.0	118.1	119.0	119.2	13.2	650	0.2	100.2	99.7	—	—	—	—	—	—	—	—	—	—
(1)	(2)	(3)	(4)	(5)	(6)	(7)	(8)	(9)	(10)	(11)	(12)	(13)	(14)	(15)	(16)	(17)	(18)	(19)	(20)	(21)	(22)	(23)	(24)	(25)	(26)	(27)	(28)

←— FIELD DENSITY TEST VALUES —→

Columns (1) through (19), inclusive—Complete for each field density test.
(1) through (28), inclusive—Complete for each 30,000 C.Y. test and for each record test at internal instrument installations.
(1) -7-1-a-1-R means July 1, 1st shift, first test on shift, representative test.
-7-2-b-4-D means July 2, 2nd shift, fourth test on shift, area of doubtful compaction.
(4) -Name of source; if a source has material for more than one zone, state zone, if more than one cut, state cut.
(6) -Weigh all material from test hole to determine result.
(8) -Obtain by drying and weighing soil and rock fractions.
(9) -Average of several readings from point (1) of the rapid compaction control method.
(10) -Obtain by drying and weighing a sample.
(11)(12)(13)(14)(16)(17)(18) - Obtain by means of the rapid compaction control method, Form 7-1624
(15) -Obtain by interpolation from Proctor needle readings for rapid compaction control points.
(21)(22) - Place sample as near fill conditions as possible.
(23) -Apply a load in p.s.i. equal to the estimated stress above the sample, using the cumulative wet density of soil and rock.
(24) -Determine from gage readings after consolidation under fill load is practically complete, and before saturation.
(25) -Average of several needle readings at end of test.
(26) -Obtain by drying and weighing a sample.
(27) -Obtain from gage readings at end of test.

●RE-ROLLED
●●CHECK AFTER RE-ROLLING

[1] Applies only to punch card operation

SHEET ____ OF ____

Figure 98.—Summary of field and laboratory tests of compacted fill controlled by Proctor test. 101–D–582.

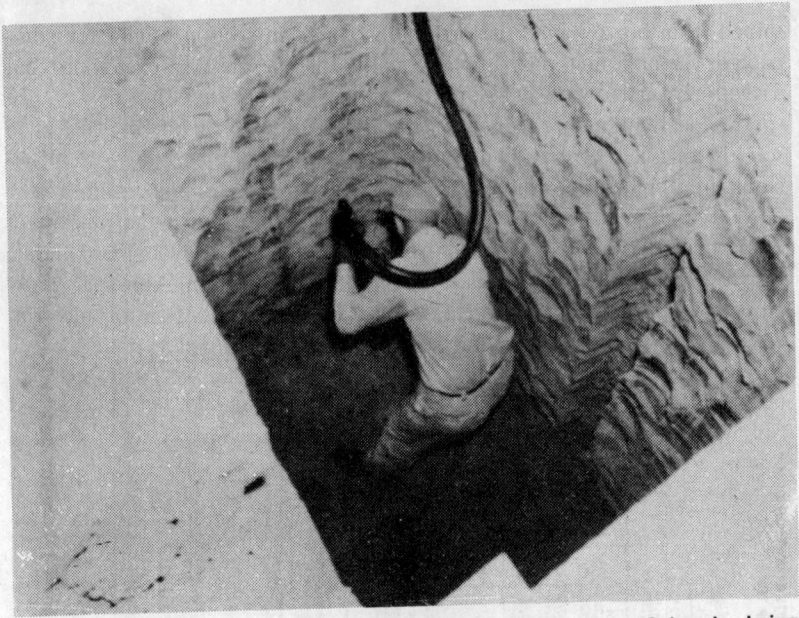

Figure 99.—Test pit in compacted embankment, Bonny Dam, Colorado, being backfilled. 794–701–351.

Homogeneity is important in order to insure distribution of the inevitable seepage from the impervious zone throughout the length of the dam, since concentration of seepage water into a few channels may induce dangerous piping. In order to meet the requirements for drainage, the ratio of permeabilities of the permeable zone to the impermeable zone should be at least 10, and preferably much larger. In addition, the permeability coefficient of the free-draining zone should be large enough to preclude development of construction pore-water pressures in this zone. The requirements of low consolidation and high shear strength for permeable zones can be obtained by proper compaction of the materials. Since there is no possibility of excessive pore-water pressure development in the pervious free-draining materials, no restriction is necessary on the density; hence, the maximum practicable compaction of such materials is desirable.

(b) *Specifications Provisions.*—The section in the specifications for an earth dam entitled "Earthwork" will contain a number of provisions applicable to the excavation and use of permeable-type soils. The specifications on opencut excavations provide that suitable materials shall be used for permanent construction. Provisions are made for temporary stockpiling of such materials. When permeable-type soils are to be obtained from borrow pits, the provisions stated in section 66(b) for compacted earthfill materials apply equally for these soils, excepting that irrigation of borrow pits is not required.

The "Embankment" section in the specifications prohibits use of frozen soil or placing unfrozen material on frozen embankment, which provisions are applicable to all kinds of soils. Provisions for maximum allowable difference in elevation between adjacent zones affect the placement of permeable-type material. Control of openings through embankments, slopes of bonding surfaces, and preparation of bonding surfaces apply equally to all zones within the embankment. Distribution of material within a permeable zone is controlled by the contracting officer.

There will be written directives in the specifications pertaining to permeable-type material used as a zone in the embankment. The material will be described and its source will be designated. The thickness of the layer will be specified, either in the form of compacted thickness or placed thickness. One method of compaction of the permeable-type material often specified requires a certain number of passes of the treads of a crawler-type tractor weighing approximately 40,000 pounds (the weight of a D-8 Caterpillar tractor with bulldozer blade). One pass of the tractor treads consists of the required number of successive tractor trips which, by means of sufficient overlap, will insure complete coverage of an entire layer by the tractor treads. Second and subsequent passes of the treads are not permitted until the previous pass has been completed as described above. Either immediately prior to or during compaction, the material is required to be thoroughly wetted.

Alternate methods of compaction are sometimes permitted by the specifications. In such cases, a minimum relative density requirement is specified.

(c) *Control Techniques.*—The workability of a permeable-type soil is reduced considerably by the inclusion of even small amounts of silt or clay; hence, every effort should be made to insure that the contractor's operations in the borrow pits and on the fill are such that contamination of the pervious soil is reduced to a minimum. As the material is brought on the fill, it must be directed to the proper zone. Within the pervious zone, individual loads should be directed so that the coarser material will be placed toward the outer slopes.

When compacted thicknesses are specified, the thickness of loose layer required which will result in the specified final thickness must be determined by the inspector during the initial stages of construction. Since the field density will be checked by relatively few tests after satisfactory placing procedures have been established, the thickness of loose layer required must be maintained within close limits throughout the job. The thickness of specified compacted layer is usually great enough to accommodate the maximum size of rock encountered in the borrow. In cases where cobbles or rock fragments greater in size than the specified thickness of layer occur, provisions will be made for special embedding, for removal to outer slopes of the pervious zone, or for removal to other zones. In order to secure the best compaction, the inspector must see that such provisions for disposal of oversized rock are followed.

After the material has been placed, spread to the desired thickness of lift, and oversized cobbles or rock fragments disposed of, the next important step is application of water. Thorough and uniform wetting of materials during or immediately prior to compaction is essential for best results. The most applicable method of adding and distributing water on the fill should be determined during the initial placement. It has been found that relaxation of the requirements of thorough wetting may result in densities far below the minimum requirement, even with great compactive effort. Different pervious materials require different amounts of water for thorough wetting and best compaction. In general, it is desirable to add as much water to the material as it will take. Too much water cannot be added to an extremely pervious soil; however, permeable-type soils containing small quantities of silt or clay may become temporarily boggy if an excessive amount of water is used. For such soils, care must be exercised in wetting the soil and the contractor's operations must be carefully controlled to avoid creation of temporary boggy conditions.

In compacting permeable-type soil by the treads of crawler-type tractors, it is desirable to have the tractor operating at the highest practicable speed during the compaction operation. High speed is conducive to greater vibration which aids in the compaction. In inspection of compaction operations using tractor treads, it is important to require the tractor to make a complete coverage of the area to be compacted prior to making the second and subsequent passes. Different widths of areas to be compacted will require different numbers of tractor trips to obtain the same number of passes of the treads. The proper number of trips should be determined in each case.

During the initial placing operations, it is recommended that relative density tests (designation E–12) and gradation analysis (designation E–6) should be made at a frequency of about one test representative of each 1,000 cubic yards placed. After placement procedures have proved satisfactory and unless significant changes in gradation occur, one relative density test for every 10,000 cubic yards of material placed will suffice. In the event of significant gradational changes in the borrow materials, increased frequency of field tests may be needed to insure satisfactory compaction of the variable materials. Form 7–1582 (fig. 100) is used to record the results of these tests.

68. Rockfill and Riprap.—Rockfill zones are used in earth dams to

Figure 100.—Summary of field and laboratory tests of compacted fill controlled by relative density test. Form 7-1582.

provide stability to the embankment and to protect exposed surfaces of the fill. Riprap is a relatively thin layer of specially selected and graded rock fragments used for protecting earth slopes from erosion by water currents and waves. Rockfill and riprap are usually not compacted, but are dumped or placed so as to obtain high shear strength by interlocking of the angular fragments. High permeability is desired in rockfills, hence the amount of fines permitted is limited. On the other hand, large unfilled voids are undesirable. The outer portion of a rockfill zone should contain the largest available rock in order to secure slope protection. The most desirable riprap surface is well keyed but rough in order to resist wave action effectively. Where very large rockfill sections are used, excessive settlement may be a problem, and sluicing may be required to compact the fill.

The specifications for earth dams containing rockfill or riprap will contain paragraphs for each of these materials. The paragraph on rockfill will describe the material and state where it is to be obtained. The rockfill will be required to be placed in horizontal layers not exceeding 3 feet thick, with the maximum sizes of rock increasing toward the outer slopes of the dam. Provisions will be made for embedding rock masses (over 1 cubic yard in volume) in the fill adjacent to the outer slope. The contractor is required to exercise care in placing rockfill adjacent to concrete. No large unfilled spaces are permitted in rockfill. Provisions are made for routing of construction traffic over the fill. Sluicing will be specified if the design requires it.

The paragraph in the specifications on riprap provides that the material will consist of hard, dense, and durable rock fragments, reasonably well graded in particular sizes. The specified sizes will depend on the reservoir fetch, the wind velocity and direction, the steepness of the slope, and the type of rock available. Graded rock up to 1 cubic yard is usually specified where wave action is severe or where marginal quality rock must be used. For less stringent conditions, rock graded up to one-half cubic yard in size is specified. No particular compaction operation is required for riprap, but the material is to be dumped and graded off so that the large rock fragments are evenly distributed and the smaller rock fragments fill spaces between larger rock fragments in a manner that will result in a uniform and closely packed layer of riprap of the required thickness. Where significant quantities of riprap are involved, for example, on most earth dams, payment will be made on a weight basis. Special care is required in placing riprap adjacent to concrete. Hand placing is required only to the extent necessary to secure the foregoing results.

Inspection may be necessary both at the rock source and on the rockfill to insure that the material is selected to avoid an excessive amount of fines in the rockfill. Breakage in quarrying, handling, and transporting should be taken into account. Placing operations must be inspected to see that segregation is avoided and that no large voids are left in the rockfill. If sluicing is required, the contractor's operations must be carefully controlled to avoid excessive wetting of the previously placed impervious zone and to insure that sufficient quantity of water is being used uniformly on the rockfill.

Inspection of riprap placement consists of visual observation of the operations and completed product to insure that a dense, rough surface of well-keyed graded rock fragments of the specified quality and sizes is obtained. Figure 101 is an example of riprap under construction.

Selection in the source should be such that the best possible product is obtained from the available rock. Methods of sorting and blasting in quarries will be subject to approval of the contracting officer. Processing may sometimes be required to eliminate excessive fines from the blasted material. Specifications also will contain provisions for control of blasting procedures.

69. Miscellaneous Fills.—Dam embankments on very plastic foundations may require toe support fills, the weight of which acts to improve stability. Excavations for foundation of the dam or for the appurtenant structures often produce material unsuitable for or in excess of the requirements for the structural zones of the dam. Such excavated material can be used for stabilizing fills at the toes of the dam. In localities where good quality riprap is very expensive, fill materials from structural excavations have been used to flatten the upstream slope of the dam to permit use of poor quality rock or, in some cases, of no rock at all. In a few cases, excess required excavation has been used in a special isolated zone in the downstream portion of the dam merely to replace material which, otherwise, would have had to be borrowed at greater expense.

The permeability of the foregoing fills is not important in the design, and they usually are not specially compacted. However, full use should be made of the compaction obtainable by routing of the hauling and placing equipment on layers of the material. Sometimes, the nature of the available materials or the design requires that some compactive effort other than routing of equipment be used. For example, sheepsfoot rolling has been used to break up fairly large pieces of soft rock to avoid excessive settlement. Compaction may be specified if the miscellaneous fill is to serve as an impervious blanket.

The specifications for miscellaneous fills will describe the materials to be used and state their sources. Thickness of layers will be specified, and the requirement of routing of construction equipment over each layer will be stated. If any moisture or special compaction requirements are needed, the specifications will so state.

a. **Riprap being placed on a gravel blanket at Bonny Dam, Colorado.**

b. **Riprap being placed on a sand, gravel, and cobble zone at Meeks Cabin Dam, Wyoming.**

Figure 101.—Examples of riprap being placed on dam embankment for upstream slope protection. PX–D–16329 and P145–432–720A.

Inspection of miscellaneous fills is entirely visual. Control tests are not ordinarily made on such fills. A few fill density tests may be required for record purposes and instruments may be inserted to obtain performance characteristics. The main problem in inspection of miscellaneous fills is to see that the specified thickness of lift is not exceeded, and to see that the hauling equipment is not channelized by a roadway, but is spread over the entire placed layer.

70. Instrument Installations.—(a) *Instruments.*—In addition to the control soil tests performed on the fill throughout construction, the behavior of many earth dams is observed during construction and during operation by means of instrumentation placed in the embankment, in the foundation, and in or on appurtenant structures. The data obtained from these instruments will indicate conditions which can affect the stability of the structure. Data from the installations and from the field control tests can be studied for comparison with design assumptions on existing structures and for determination of design practices for future construction. Such studies yield criteria which will promote better economy and greater stability in the design of earth dams.

The various installations are grouped into three categories: (1) pressure apparatus to observe pore-water pressures in the embankment and in the foundation, (2) internal and external movement devices or fixed points of reference to provide data on volume change and displacement of the embankment and foundation, and (3) vibration recording devices to indicate the result of transient stresses on the embankment.

The following types of instrumentation may be placed in, on, or adjacent to foundations, abutments, and embankments, or on concrete structures: (1) embankment and foundation piezometers or transducers; (2) porous-tube piezometers and observation wells; (3) vertical movement devices, generally at 10-foot spacing of crossarm units; (4) horizontal movement devices; (5) foundation settlement baseplates; (6) measurement points on the abutment slopes and on the completed surface of an embankment or structures for measuring settlement and deflections; (7) weirs, flumes, or other devices for measurement of seepage; and (8) accelerographs.

The types and quantity of instruments installed at a dam will vary with the design of the structure, with the purpose of the structure, with the foundation conditions, and with the material selected for the embankment. The design of the individual units for each installation has been standardized; however, the details may be modified to meet local requirements. Details of the instrument installations are described in designations E–27 through E–35.

(b) *Installation of Earth Dam Instrumentation.*—The work required for installation of the instruments is usually performed by the contractor under provisions of the specifications for construction of the dam. In special cases, however, it may be desirable for Government forces to perform all or part of the work. Generally, because of the specialized nature of the equipment, specifications for materials which become a permanent part of the installations are included in the construction specifications and purchased by the contractor subject to Government approval.

When installation of equipment is performed by either the contractor or by the Government, detailed costs should be maintained for each type of apparatus. Installation costs for all instrumentation should include costs of materials or specification pay items, estimated costs for overhead, laboratory testing, inspection, and surveying. These data should be used to determine total unit costs of the apparatus which are included in the Final Report on Earth Dam Instrumentation (L–16).

(c) *Inspection.*—The responsibility for supervising and inspecting the field installation of the behavior apparatus should be assigned to one project engineer. Additional inspectors should be assigned as required to supervise the work, to test the apparatus, and to observe and record data. Close inspection and supervision are essential to assure correct installation of equipment and proper operation of the apparatus. When materials for the installations are received at the project, each item of equipment, particularly the fabricated material, should be inspected for conformance with specifications, quantity required, and general workmanship.

The piezometer tips should be assembled and checked for loose connections. The plastic tubing should be checked with a micrometer for dimensional tolerances and for absence of foreign substances as required in the specifications. All tubing should withstand a working pressure of at least 100 pounds per square inch and should permit the free flow of water. All Bourdon-tube gages should be inspected to assure compliance with specifications regarding top connections, dial capacity, filling of Bourdon-tube element with liquid, and capping of the gage connection.

Heavier metal parts such as pipe sections for the vertical and horizontal movement devices, air trap, reading scale, and torpedo should be inspected for quantity, workmanship, and operation. The coating of protective paint, when applied, should be checked for continuity. The porous tubes for porous-tube piezometers also should be checked for possible breakage.

Materials requiring special care in handling during their installation are described in designations E–27 through E–35.

(d) *Observations.*—Periodic readings of the various installations are required as described in designations E–27 through E–34 and presented in figure 103. An analysis of these readings both during and after construction can indicate unusual behavior of the structure. Variations in readings should be described in the required reports (see sec. 71(d)). During construction, piezometer readings should be observed for the development of excessive pore-water pressures.

Internal movement readings should be watched for large vertical or horizontal movement and for heave. When the reservoir water surface is at a high level, the instruments will indicate the flow pattern through the embankment and foundation. Observations made immediately following a severe drawdown of the reservoir will provide information on pore-fluid pressures retained within and under the structure. Changes in pore-water pressures and internal or external movement will reflect stress conditions within the embankment that will influence the stability of the structure. Since behavior conditions originate during construction and vary continuously throughout operation, data from the instruments may be utilized by both construction and operating personnel, as well as by the designing engineers.

(e) *Record Tests.*—Record tests of embankment and foundation materials at internal instrument installations are necessary to obtain data on the soils adjacent to the instruments. At each unit of the internal vertical and horizontal movement devices in earthfill material, two field density tests should be made. One test sample should be obtained from the soil near the bottom of the trench excavated for the crossarm, vertical plate, or baseplate unit; the other should be obtained in the tamped backfill of the trench after it has been brought up to grade. When all or a part of the dam is founded on highly compressible materials, representative undisturbed samples should be obtained in the foundation at the locations of the vertical movement and baseplate installations unless suitable samples have been obtained previously. No tests are required when the internal movement device is placed on rock foundation unless specifically requested. Tests at the internal movement units should be designated "record rolled," "record tamped," or "undisturbed," depending on the material being tested.

One field density test is required at each embankment piezometer tip. This test should be made in the rolled fill at the tip location prior to excavating the offset trench for the tip. Where feasible, each hole drilled for foundation-type piezometer tips should be logged throughout its length, and a record sample should be obtained of the soil at the bottom of each hole where the foundation pore-water pressure will be contacted.

In addition to field density tests on embankment materials and inplace density tests on foundation materials, the following tests should be performed on the samples taken: (1) gradation analysis, (2) specific gravity, (3) liquid and plastic limits, and (4) percolation-settlement. This laboratory testing may be performed during seasonal shutdowns or at other times when the workload permits. Sufficient material must be obtained while making the density tests so that the additional tests can be made. Prior to testing, foundation samples should be sealed to prevent loss of moisture. Special care should be exercised to attain the desirable accuracy in all record tests. Embankment field moisture and density conditions must be duplicated for the percolation-settlement test and inplace conditions must be maintained for the foundation materials; hence, all stored samples should be properly marked and referenced to the pertinent density test. The results of all tests on these samples should be submitted with the Final Report on Earth Dam Instrumentation (L–16) on form 7–1352, figure 98.

For every record test made in the embankment, a 15-pound sample of the material passing the No. 4 sieve should be ovendried at 110° C. and shipped by freight for storage and special testing to the Bureau of Reclamation, Division of General Research, Engineering and Research Center, Denver Federal Center, Denver, Colo., 80225. A 6-inch-diameter by 12-inch-high metal can is a satisfactory container for the sample. All pertinent data for each density test should be submitted in duplicate and should include the identification of the instrument to which it refers, and whether the sample is "record rolled," "record tamped," or "undisturbed." Each container should be identified by painting or marking on the side of the can the following information:

(1) Name of dam.

(2) Identification symbols.

For samples taken at piezometer tips, the letter "P" and the number of the tip, as shown on the installation drawings, should be used. For example, "Heron Dam, P–14" will identify the sample. For samples taken at porous-tube piezometers, the symbol PTP should be used. For samples taken at crossarm units of a vertical movement installation, "Heron Dam, X–A–8–R" and "Heron Dam, X–A–8–T" will represent, respectively, "record rolled" and "record tamped" material adjacent to the eighth crossarm unit installed in installation "A" for the dam. Likewise, the symbols "Trinity Dam, H–B–6–1–R" and "Trinity Dam, H–B–6–1–T" will represent the samples taken from the excavation and backfill of the trench for plates from a horizontal movement device. The symbol "BP" can be used for the samples from the foundation settlement (baseplate) installations. These identifying symbols should be marked on the container in addition to identification normally required on record samples.

The record samples obtained from the location of the foundation piezometers should contain representative materials taken from the location of the foundation tip. The quantity of the sample will depend on the drilling equipment used to excavate the holes.

PROJECT _____ DAM   DATE_____

EMBANKMENT ROLLER DATA

	Roller No.	1	2	3	4	5	6
(a)	Make of roller						
(b)	Number of drums						
(c)	Length of drums						
(d)	Diameter of drums (outside)						
(e)	Knobs (k) sheepsfoot (sf), or square (sq)						
(f)	Number of horizontal rows of feet						
(g)	Number of feet per row per drum						
(h)	Total number of feet per drum (f) x (g)						
(i)	Length of feet						
(j)	Dimensions of bottom of feet						
(k)	Area of bottom of feet						
(l)	Weight of roller (empty)						
(m)	Ballast capacity (all drums)						
(n)	Weight of roller as used						
(o)	Ballast used (material)						
(p)	Weight of roller ÷ total area all feet						
(q)	Cleaners (yes or no)						
(r)	Type of frame (rigid or oscillating)						

Figure 102.—Tabulation for recording embankment roller data. 101–D–238.

71. Records and Reports.—(a) *Daily Reports.*—Daily reports must be made by the inspector covering the activities during his shift. These reports should record the progress of construction; provide pertinent information for the inspector about to go on shift, including shutdowns and orders given to the contractor; and furnish data for use in compiling the construction progress report. The form of a daily report will vary to suit the requirements of each job, but all information called for on the periodic earthwork progress report should be based on day-to-day records. It is desirable that a systematic method of identifying field density tests made on the embankment be used on each project. A suggested scheme is to designate each test by the date, shift, number on that shift, and purpose. For example, "8–2–70–a–2–D" would define a field density test made August 2, 1970, on the first shift, the second test made on that shift, and for the purpose of checking an area of doubtful compaction. The legend is as follows: a, first shift; b, second shift; c, third shift; D, doubtful area; C, concentrated area; R, representative. This designation is used on figure 98. The results of tests made daily on the

FREQUENCY OF READINGS

EARTH DAM INSTRUMENTATION

Form No.	Title on form	Progress report (L-15) During Construction		Periodic report (L-23) Operation	
		Frequency of readings		Frequency of readings	
		Construction	Shutdown	First year	Regular
7-1346	Piezometer Readings (Separate Gages)	Twice monthly	Monthly	Monthly	Monthly
7-1347	Piezometer Readings (Master Gage)	Monthly	Alternate months	Approximately 6 months after completion of dam	Annually. On same date as a set of separate gage readings
7-1600	Porous-tube Piezometer Readings	Twice monthly	Monthly	Monthly	Monthly
7-1348	Internal Vertical and Horizontal Movement Readings (Crossarm or H.M.D.)	Complete set of readings each time a unit is installed	Monthly	Complete set approximately 6 months after dam is completed	Every 2 years
7-1359	Foundation Settlement Readings (Baseplates)	Complete set of readings each time an extension is added	Monthly	Approximately 6 months after dam is completed	Every 2 years
7-1355	Measurement Points— Cumulative Settlement and Deflection Readings— Embankment	Monthly, if required, or when dam is completed	Monthly, if required	Approximately 6 months after dam is completed	Every 2 years
7-1355A	Measurement Points— Cumulative Settlement and Deflection Readings— Spillway and Outlet Works	Monthly as portions of structures are completed	Monthly	Approximately 6 months after structure is completed	Every 2 years
7-1355B	Measurement Points— Cumulative Settlement Readings—Spillway Floor Slabs	Monthly as slabs on structure are completed	Monthly, if required	Approximately 6 months after structure is completed	Every 2 years

Request standard reporting forms from Director of Design and Construction, Denver, Colorado, attention 220. Standard forms will be overprinted and transmitted when each installation is completed.

Transmit progress reports (L-15) once each month, combining readings on all installations. Include representative readings from each installation in the final report (L-16).

Obtain readings for periodic report (L-23) as scheduled for first 2 years after completion of dam. Thereafter, schedule for readings will be prepared by Denver on form DC-622. Transmit periodic readings within 1 month after date obtained.

If reservoir elevation changes more than 10 feet in a 15-day period, obtain twice-monthly separate gage piezometer readings (form 7-1346).

Submit copies of all readings and reports in duplicate.

Figure 103.—Frequency of readings—Earth dam instrumentation. 101-D-583.

embankment are reported in the progress report of embankment construction on the appropriate forms.

(b) *Periodic Progress Reports.*—Adequate technical control of construction is an important and essential function of the field forces on an earth dam project. Complete records of all data pertaining to the methods and procedures of achieving satisfactory control must be maintained as a normal part of the inspection. The report "Summary of Earthwork Construction Data" is the means by which these data are transmitted periodically to the Engineering and Research Center.

The "Summary of Earthwork Construction Data" reports are required from each project during the working season for each month that embankment placing is in progress, except that in special cases additional summaries or reports may be required by the Engineering and Research Center. The "Summary of Earthwork Construction Data" report is a part of the "Construction Progress Report (L–29), but one separate extra copy of the "Summary of Earthwork Construction Data" report is required for review by the Division of Design. A "Summary of Earthwork Construction Data" report consists of the following:

(1) *Narrative and photographic.*—The narrative shall consist of a description of earthwork operations during the reporting period. Important features of normal embankment operations to be included in the narrative of initial reports are outlined below:

Borrow-pit operations (each borrow area):
Description of material
Equipment used
Natural water content
Method of adding moisture
Depth of cut
Stratification
Mixing, separating
Transporting
Embankment operations (each zone):
Equipment used
Spreading, mixing
Method of adding moisture
Maximum size of rock fragments
Method of removing oversized rock fragments
Compaction method
Thickness of layers

In order to avoid repetitious reporting, many of the items in the suggested outline need be described in detail only once. Thereafter, only significant changes in materials or in methods and procedures should be described. Any difficulty or unusual conditions encountered should be fully described. The narrative should also clarify any results reported on the various forms which require special mention and it should report pertinent information not covered by the forms.

Photographs should be taken far enough in advance to assure no delay in submitting the reports, and should include:

a. A general view from an abutment point showing the extent of operations on the dam during the period of the report. This camera point should be located sufficiently high that the camera station can be used until the dam is completed.

b. A general view showing the extent of operations in each major source of material. Photographs of this type should be periodically submitted.

c. Closeup views of each operation, such as dumping, spreading, adding moisture, harrowing, rolling, or mechanical tamping. Photographs showing these operations need be submitted only once. Thereafter, only photographs showing different or unusual procedures should be submitted.

d. Closeup views of typical borrow-pit cut banks. These should be submitted once.

(2) *Completed form 7–1352 (fig. 98).*—This form should include results of field and laboratory tests made during the reporting period for all materials controlled by the Proctor test. All columns should be filled in for each test, except that specific gravity of the minus No. 4 material and the percolation-settlement tests need be performed only for every 30,000 cubic yards of material placed. The latter is a minimum requirement; variations in the material may require more frequent tests.

(3) *Completed form 7–1582 (fig. 100).*—This form should include results of field and laboratory tests made during the reporting period on all materials controlled by the relative density test. All columns should be filled in.

(4) *Roller data.*—Figure 102 shows a suggested tabulation of data for tamping rollers. Such data should be obtained for each roller used on the embankment to assure compliance with specifications. This information should be reported only once for each roller, except when modifications are made to the roller.

(c) *Final Embankment Construction Reports.*—On completion of embankment construction, a report summarizing the work accomplished should be prepared. In accordance with Part 175.2.1 of the Reclamation Instructions, "Final Construction Reports" are required to be prepared in duplicate and submitted to the Engineering and Research Center for all major works, including dams. A detailed summary of embankment construction need not be transmitted separately but should be included in the "Final Construction Report." A suggested outline for the embankment portion and related construction requirements of the report follows. Items in the outline not applicable to a particular project should be disregarded, and variations or additions may be made to suit individual job conditions.

(1) *General*

    a. Location and purpose of structure.

    b. Description of dam and appurtenant works, including dimensions and quantities involved.

(2) *Investigations*

    a. Foundation explorations.

        1. Description of foundation conditions.

        2. Itemized summary of foundation explorations showing number, depth, and coordinate location of drill holes, test pits, shafts, and tunnels.

        3. Description of any special tests and methods of taking undisturbed samples.

    b. Construction materials.

        1. If a preconstruction earth materials report was prepared, a general description of the materials investigation should be prepared and reference to the previously prepared report should be made. However, a complete and detailed description of additional investigations made subsequent to the preconstruction earth materials report is required.

        2. If no preconstruction earth materials report was prepared, a complete and detailed description of material investigations of borrow areas and rock sources with test results and quantities available shall be included in the "Final Construction Report."

(3) *Construction history* (Include photographs in text.)

    a. Required excavations, except borrow areas.

        1. Stripping of foundation(s)

        2. Cutoff trench or trenches

        3. Grout cap trench

        4. Miscellaneous excavations.

    All of the above shall include discussion on depth, quantities, unwatering methods, disposition of materials excavated, and type of equipment including number of pieces and rated capacity.

b. Foundation drilling and grouting.

1. General.—(To include discussion on construction of grout cap, type of equipment used and capacity, and grouting plan showing location and depth of holes.)

2. Pressure grouting—(To include discussion of grouting procedures, type of equipment and pressures used, grout take and results obtained, and any special problems and methods used to obtain satisfactory results.)

c. Borrow area operations.

1. General.—Discussion of plan and utilization of specified material sources, and any problems in borrow areas not anticipated at time of specifications issue.

2. Borrow areas.—Detailed report of initial conditions of borrow pits, methods of excavating, mixing, screening, and transporting of material including the type, size, and rated capacity of all equipment used, with special emphasis on any unusual requirements for blending or excavating procedures. Also, borrow pit yield, methods of controlling moisture, disposition of oversized material and material from stripping operations should be reported.

d. Embankment operations.

1. Zones.—Discussion to include construction methods and sequence, material source, and any variation from the approved materials distribution plan and reasons therefor. For riprap and rockfill, give gradation.

2. Placing.—Description of methods used including scarifying, spreading and leveling, rolling, power tamping, removal or reworking, and any other part of the work considered peculiar to the placing.

3. Equipment.—Discussion of type, capacity, size, number of pieces, and efficiency and suggested improvements, if any, for all equipment used in placement of embankment.

4. Testing.—Performance of field laboratory including number of testing personnel, and number and type of tests performed, shall be summarized. Discussion of control tests, such as moisture content and density, and any special tests shall include procedures in making tests and reporting methods, extent of testing, adequacy of tests, and difficulties encountered.

(4) *Earth dam instrumentation.*—Brief description of the systems installed such as piezometer tips, horizontal and vertical movement crossarms, and embankment and structure settlement measure-

ment points, and reference to the "Final Report on Earth Dam Instrumentation (L–16)" (see following section (d)).

(d) *Earth Dam Instrumentation Reports.*—A monthly report, separate from other construction reports, is required in duplicate during the construction of the dam to present the progress of the placement of the earth dam instrumentation. This report should include a short narrative on the progress of the work field drawings showing the location of the apparatus, readings made during the preceding period, and representative pictures showing current details of the work. Detailed descriptions of the data to be included in these progress reports are mentioned in separate paragraphs under each type of instrumentation in the appropriate designations. A suggested title is "Progress Report on Earth Dam Instrumentation— Month of   , 19   ,   Dam,   Project."

Within 6 months after completion of the instrument installations, a "Final Report on Earth Dam Instrumentation" is required in duplicate. This report should describe, in detail, procedures for the care, assembly, and installation of each type of apparatus. The report also should include suggestions for the improvement of the installations, the quantities of materials used, an analysis of costs incurred during construction for each type of installation (actual costs of Government forces and bid prices of contractor), and record test data for earth materials. Descriptive photographs, drawings, charts, and tables should supplement the report. A suggested title is "Final Report on Earth Dam Instrumentation—   Dam,   Project."

Following completion of the instrument installations, "Periodic Reports on Earth Dam Instrumentation (L–23)," are required during operations. These readings should be submitted in duplicate within 30 days of the date of readings. The forms required for this report and the frequency of the readings for the first 2 years of operation are shown in figure 103. The frequency of observations given therein may be extended only after permission is obtained in writing from the Engineering and Research Center, Division of Design. Form 7–1373, figure 104, may be used to transmit periodic readings from Government-operated projects.

Address all transmittals to the Bureau of Reclamation, Engineering and Research Center, Director of Design and Construction, Division of Design, Denver Federal Center, Denver, Colo. 80225.

72. Control Criteria.—The concept of limiting the moisture condition for placement of compacted earthfill in dam embankment has been used by the Bureau of Reclamation for many years. The idea of limiting the magnitude of pore-pressure buildup in cohesive soils during construction by controlling the placement moisture has been a major con-

7-1373 (8-70)
Bureau of Reclamation

INTEROFFICE TRANSMITTAL

SEND BY:			FOR:	
☐ FIRST CLASS	☐ AIR MAIL	☐ LOWEST RATE	☐ ACTION	☐ INFORMATION

T O

COMMISSIONER
☐ ATTN:

ENGINEERING AND RESEARCH CENTER
☐ ATTN:

REGIONAL DIRECTOR
☐ REGION:_ _ _ _ ATTN:

PROJECT MANAGER OR SUPERINTENDENT
☐ ATTN:

CONSTRUCTION ENGINEER
☐ ATTN:

OTHER *(Specify)*
☐

F R O M

OFFICE

CODE	NAME OR INITIAL	DATE

REMARKS: (If transmitting 2nd, 3rd, or 4th class matter, limit remarks to list of material and reference to order or request.)

Use when letter is not required and material is not self-transmitting. Use for informal questions, answers, and comments between offices which do not warrant filing of copies for record purposes.

Figure 104.—Interoffice transmittal form 7-1373.

sideration since 1946. The relationships among moisture control, pore-water pressure potential, and stability of the structure are well established and are generally recognized as a basis for an upper placement moisture limit.

In addition to this concept of an upper limit, it has been demonstrated that localized shear planes will be developed under certain conditions of placement moisture and compactive effort. This condition is also a reflection of excess pore pressure during compaction. Although there is limited research on this phenomenon for a few types of soils, insufficient data are available to define the limiting conditions under which these localized and slickensided planes will develop for soils in general. However, for Bureau procedures for compaction of earthfill, it has been shown that this moisture condition will be at or on the wet side of optimum where construction pore pressures normally increase rapidly with increased moisture content. A typical example of laboratory tests on one soil is shown in figure 105.

The extent of the development of the planes and their significance in stability of compacted embankments has not been thoroughly studied, but it is reasonable to assume that their net effect would be a significant loss of shear strength. Failures attributed in part to their development have been cited; hence, placement moisture should be so controlled as to avoid any possibility of their development.

The phenomenon of loss of strength of a dry, low-density soil as the water content is increased is also well known. This reduction in strength may result in appreciable volume change of a loaded soil, either in the foundation or in the compacted embankment. In the case of earth dams, wetting of the foundation or of the embankment following construction is not accomplished uniformly. Hence, such volume changes will usually result in differential settlement and may result in cracking of the embankment. Many failures of small earth dams are attributable to this cause. Placement moisture should be maintained in a range between the specified upper and lower moisture limits such that the average is dry of the optimum condition so that the soil will exhibit no additional consolidation on saturation, will not develop excessive pore pressures, and will not develop shear planes under the field compactive effort.

High shear strength, low consolidation, and to some extent impermeability, are all related to the dry density obtained in the fill. The design is usually based on impervious or core material placed at or slightly below the average laboratory maximum dry density of the minus No. 4 fraction. An average D-ratio (dry density of fill to Proctor maximum dry density) of near 1.00 is expected for most soils containing less than about 30 percent rock. For gravelly materials containing more than about 30 percent plus No. 4 material, a field density

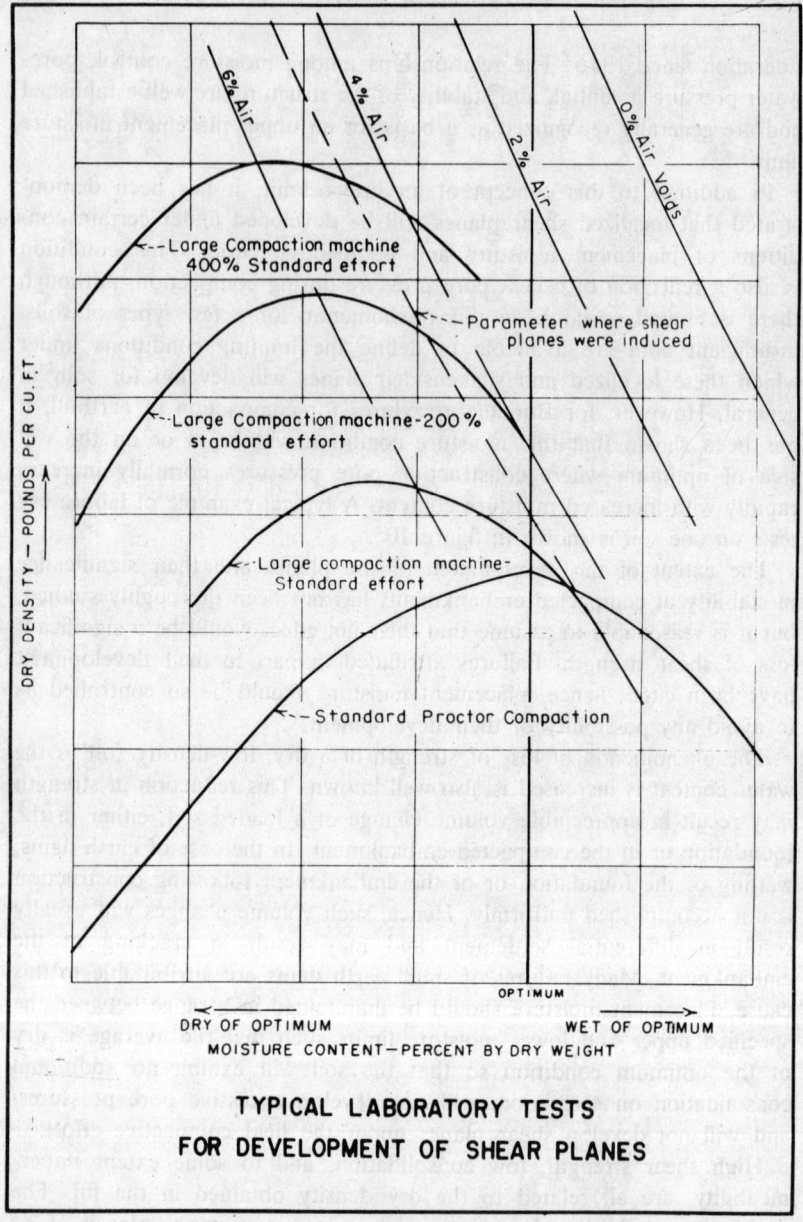

Figure 105.—Typical laboratory tests for the development of shear planes.

Table 4.—Criteria for control of compacted dam embankments

Percentages based on minus No. 4 fraction

Type of material	Percentage of + No. 4 fraction by dry weight of total material	50 feet or less in height			Greater than 50 feet high		
		Minimum acceptable density	Desirable average density	Moisture limits $w_o - w_f$	Minimum acceptable density	Desirable average density	Moisture limits $w_o - w_f$
Cohesive soils controlled by the Proctor test.	0–25	$D = 95$	$D = 98$	−2 to +2	$D = 98$	$D = 100$	2 to 0 Note [2]
	26–50	$D = 92.5$	$D = 95$		$D = 95$	$D = 98$	
	More than 50 [1]	$D = 90$	$D = 93$		$D = 93$	$D = 95$	
Cohesionless soils controlled by the relative density test.	Fine sands with 0–25.	$D_d = 75$	$D_d = 90$	Soils should be very wet.	$D_d = 75$	$D_d = 90$	Soils should be very wet.
	Medium sands with 0–25.	$D_d = 70$	$D_d = 85$		$D_d = 70$	$D_d = 85$	
	Coarse sands and gravels with 0–100.	$D_d = 65$	$D_d = 80$		$D_d = 65$	$D_d = 80$	

$w_o - w_f$ is the difference between optimum water content and fill water content in percent of dry weight of soil.
D is fill dry density divided by Proctor maximum dry density in percent.
D_d is relative density as defined in section 14, chapter I.

[1] Cohesive soils containing more than 50 percent gravel sizes should be tested for permeability of the total material if used as a water barrier.
[2] For high earth dams special instructions on placement moisture limits will ordinarily be prepared.

Example_____ DAM   ZONE __I____

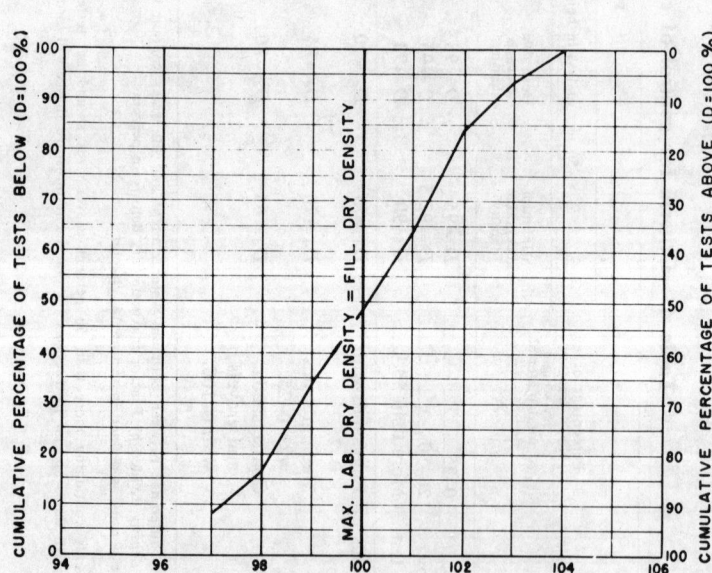

D = FILL DRY DENSITY / MAXIMUM LABORATORY DRY DENSITY (MINUS NO. 4 MATERIAL) X 100	F (PREV.)	THIS PERIOD					TO DATE		
		FREQUENCY OF OCCURRENCE		F	CUM. F	CUM. %	F	CUM. F	CUM. %
93.0 - 93.9									
94.0 - 94.9									
95.0 - 95.9									
96.0 - 96.9		IIII		4	4	8			
97.0 - 97.9		IIII		4	8	16			
98.0 - 98.9		IIII I	IIII	9	17	34			
99.0 - 99.9		IIII I	II	7	24	48			
100.0 - 100.9		IIII I	III	8	32	64			
101.0 - 101.9		IIII I	IIII I	10	42	84			
102.0 - 102.9		IIII I		5	47	94			
103.0 - 103.9		III		3	50	100			
104.0 - 104.9									
105.0 - 105.9									
TOTALS				50					

	PREV.	THIS PERIOD	TO DATE
Average max. lab. γ_D (PCF)		117.1	
Average fill γ_D (PCF)		117.1	
Mean variation from max. lab. γ_D (PCF)		0.	
Average rock content (% of plus No.4 by dry weight)		2.6	

PERIOD OF REPORT_____ TO _____
          TESTS 9-26-70-d-1R   TO   10-24-70-d-1R

$$D \text{ (IN PERCENT)} = \frac{\text{FILL DRY DENSITY}}{\text{MAXIMUM LABORATORY DRY DENSITY}} \times 100$$

Figure 106.—Statistical analysis of test results for density control of fill.
101–D–584.

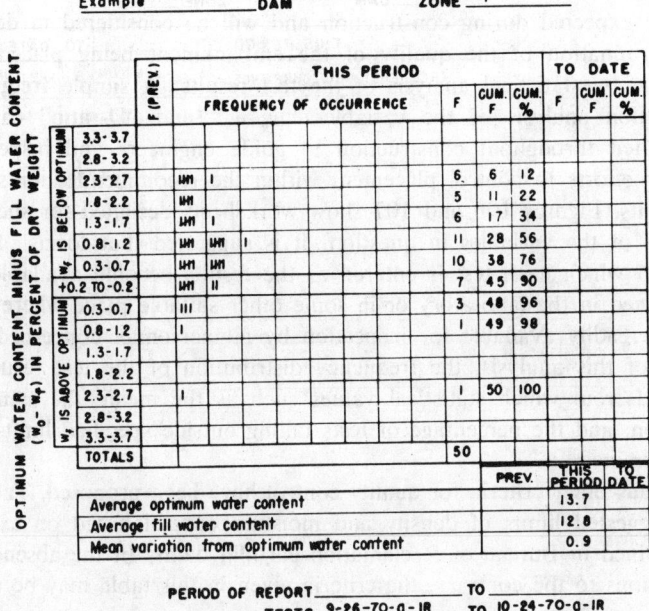

Example _____ DAM _____ ZONE __I__

	F (PREV.)	THIS PERIOD						TO DATE		
		FREQUENCY OF OCCURRENCE				F	CUM. F	CUM. %	F	CUM. F	CUM. %
3.3-3.7											
2.8-3.2											
2.3-2.7		꒐꒐꒐	I			6	6	12			
1.8-2.2		꒐꒐꒐				5	11	22			
1.3-1.7		꒐꒐꒐	I			6	17	34			
0.8-1.2		꒐꒐꒐	꒐꒐꒐	I		11	28	56			
0.3-0.7		꒐꒐꒐	꒐꒐꒐			10	38	76			
+0.2 TO -0.2		꒐꒐꒐	II			7	45	90			
0.3-0.7		III				3	48	96			
0.8-1.2		I				1	49	98			
1.3-1.7											
1.8-2.2											
2.3-2.7		I				1	50	100			
2.8-3.2											
3.3-3.7											
TOTALS						50					

OPTIMUM WATER CONTENT MINUS FILL WATER CONTENT $(w_o - w_f)$ IN PERCENT OF DRY WEIGHT — w_f IS BELOW OPTIMUM / w_f IS ABOVE OPTIMUM

	PREV.	THIS PERIOD	TO DATE
Average optimum water content		13.7	
Average fill water content		12.8	
Mean variation from optimum water content		0.9	

PERIOD OF REPORT _____ TO _____
TESTS _9-26-70-a-IR_ TO _10-24-70-a-IR_

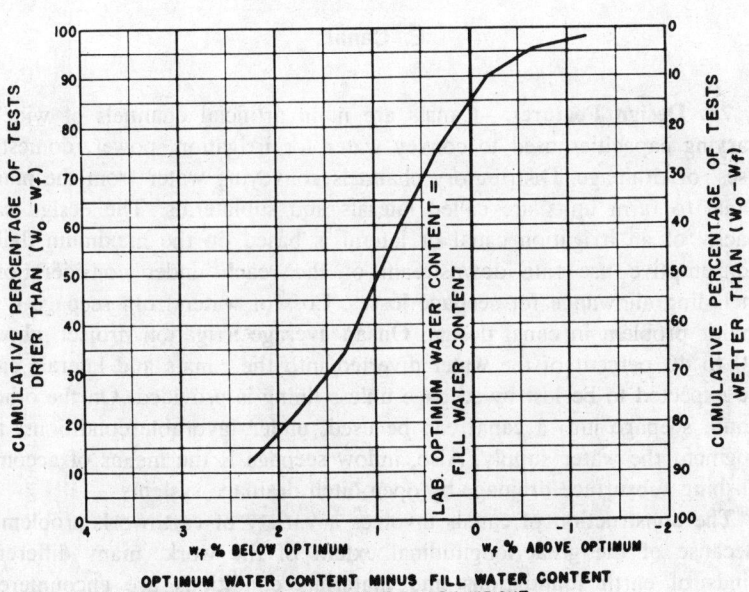

OPTIMUM WATER CONTENT MINUS FILL WATER CONTENT
$(w_o - w_f)$ IN PERCENT OF DRY WEIGHT

**Figure 107.—Statistical analysis of test results for moisture control of fill.
101-D-585.**

less than. Proctor maximum dry density of the minus No. 4 fraction
may be expected during construction and will be considered in design.

Determination of the quality of the embankment being placed can
be made by statistical analysis of the test results. A simple frequency
distribution analysis of the variables "w_o-w_f" and "D-ratio" can be
maintained throughout construction to guide engineers and inspectors
in their efforts to control placement within the recommended or specified limits. Figures 106 and 107 show worksheets designed for such an
analysis of the variables in question. It is suggested that forms of this
type, on which each test is entered as the results become available, be
maintained in the laboratory or in some other suitable place where they
will be readily available for inspection by all personnel concerned. By
means of this analysis, the frequency distribution of the test results is
obtained from which statistical values such as the mean, the standard
deviation, and the percentage of tests falling outside specified limits can
be determined.

Various other criteria for quality control have been proposed. Table 4
lists suggested limits of density and moisture control based on experience gained in Bureau of Reclamation earthfill dams. In the absence of
instructions to the contrary, the criteria given in this table may be used.

E. Canals

73. Design Features.—Canals are main artificial channels of widely
varying capacities used to convey water for irrigation, power, domestic
uses, or drainage. Distributory channels conveying water from the main
canal to farm units are called laterals and sublaterals. The design capacity of an irrigation canal or lateral is based on the maximum daily
consumptive use rate downstream of the reach under consideration,
including allowance for seepage losses. Loss of water from seepage is a
major problem in canal design. On an average irrigation project, about
20 to 40 percent of the water diverted into the canals and laterals can
be expected to be lost by seepage unless lining is provided. On the other
hand, seepage into a canal can be used, under favorable conditions, to
augment the water supply. Also, inflow seepage is the means of accom-
plishing subsurface drainage by open-ditch drainage systems.

The construction of canals involves a variety of earthwork problems.
Because of the great longitudinal extent of the work, many different
kinds of earth foundations and materials conditions are encountered
which, for maximum economy, must be treated differently. Thus, some
reaches of a canal may require nothing but uncompacted embankments

made of unselected materials excavated from the prism, while other sections may require careful control of cuts in borrow areas, strict moisture control, and compaction of earth in the form of linings or entire embankments. The following paragraphs briefly summarize the main design features related to earthwork in canal construction.

Canals are classified as unlined or lined, depending on the degree to which special provisions are made to prevent seepage. An unlined canal or lateral is an open channel excavated and shaped to the required cross section in natural earth without special treatment of the subgrade. Construction of compacted embankments at locations where the high water surface is above natural ground is considered to be a normal seepage control measure in unlined canal construction.

The cross section of the canal is selected to satisfy desirable criteria for bottom width, water depth, side slopes, freeboard, and bank dimensions, and to facilitate operation and maintenance. The ratio of bottom width to depth ranges from 2:1 for the smallest laterals to 8:1 for large canals; side slopes on unlined canals vary from $1\frac{1}{2}$:1 to 2:1, or flatter where unstable soils are encountered. The design of canals contemplates the development of a steady-state seepage condition from the normal high water surface, followed by a rapid drawdown of the water surface. The dimensions of the banks and the soils of which they are composed must assure stability under these extreme hydraulic conditions.

The water section of a canal may be entirely in cut, entirely in fill, or partially in each, depending on the economics of the route, the necessary gradient, structure design, and the requirements of safety and water distribution. If the water section is partially or entirely in fill, compacted embankments may be used to prevent excessive amounts of seepage through the fill.

The requirements for compacting an embankment under compacted earth lining depend upon: (1) height of fill, (2) earth construction materials used, and (3) method of fill construction. These requirements are reflected in the specifications.

Operation and maintenance roads of widths varying from 12 to 24 feet are usually provided on the lower bank of canals of 100 second-feet or more capacity. Where the width of the water prism exceeds 40 feet, a maintenance berm on the upper bank may be provided. In deep cuts, berms are often used to reduce bank loads and thereby prevent sloughing of earth into the canal, and to lower the elevation of the operating road for easier maintenance. They are also provided between deep cuts and the adjacent waste banks. Even where there are no regular operating roads, canal and lateral banks are made regular enough by blading to permit use of power mowers and other power equipment to control weed growth.

A lined canal or lateral is one in which the wetted perimeter of the section is lined or specially treated. Materials used to line canals include plain or reinforced concrete and pneumatically placed mortar, various types of asphalt, brick, stone, synthetic plastics, earth, and earth treated with additives (stabilized soils).

The great need for reduction of water losses in canals has stimulated a continuing search for effective low-cost linings. In addition to the savings of water that can be achieved by lining canals, other advantages include prevention of water-logging of land, lower operation and maintenance costs, less danger of canal failures, and reduced storage and diversion requirements. In the case of hard-surfaced linings, still further advantages include possibility of small canal sections and structures, less right-of-way, higher permissible velocities, and steeper gradients. Side slopes on earth-lined canals are 2:1 or flatter, and on hard-surfaced linings usually 1½:1.

Lined canals can be built in either cut or fill, or partially in each, and the economic advantages of the balanced cut and fill must be weighed against special treatment of fills to support the lining properly. The height of lining above the water surface of a canal varies, depending on the capacity, velocity, curvature, inflow of storm water; or on increases in water surface elevation caused by canal checks, checking in pumping plant forebays, bore waves, and wind action.

Various types of earth linings have been used by the Bureau to reduce water losses. The simplest method is silting, in which fine-grained soils are dispersed into the water by hydraulic or mechanical means and are deposited over the wetted perimeter of the canal. Although it is effective in reducing seepage, silting is not considered to result in a permanent seal because the silt layer is easily destroyed. Thin, loose earth blankets of fine-grained soils have sometimes been used as linings. Such linings are inexpensive and are temporarily quite effective, but they erode easily unless protected by a cover layer of gravel. Bentonite applied as a buried membrane or mixed with permeable-type soils, and buried asphalt membranes have been used; but such linings also must be protected by gravel or soil cover. When bentonite is used for the membrane, a high-swell bentonite of the sodium montmorillonite variety and without excessive moisture is required. Acceptance criteria for this material are contained in the specifications.

A wide range of impervious soils has been used successfully for compacted earth linings. The most desirable soils combine the most favorable properties of gradation, plasticity, and impermeability. The best results are obtained from gravels with sand-clay binder (GW-GC). The gravel particles provide good erosion resistance and the grading and clay provide

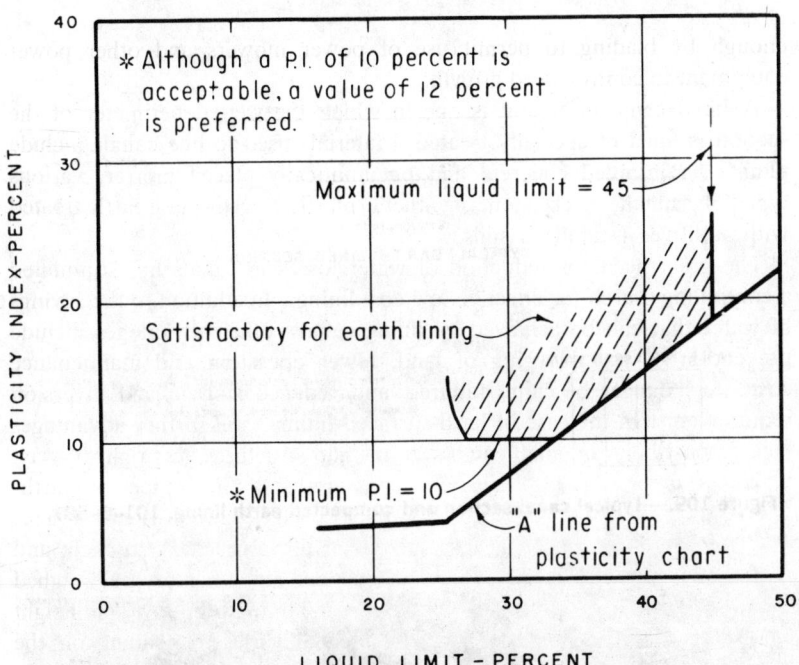

Figure 108.—Plasticity criterion for impervious, erosion-resistant compacted earth linings. 101–D–586.

imperviousness. Good stability characteristics are provided by the well-graded, sandy, gravel fraction as well as by the cohesive binder. The clayey gravels (GC) are next in quality; followed by sands with clay binder (SW-SC) and clayey sands (SC), respectively. Silty gravels (GM) have been used with success. However, the best materials of this type have just enough fines to provide imperviousness and have sufficient gravel for erosion resistance. Of the fine-grained soils, a lean clay (CL) is desirable. As a guide for selecting fine-grained soils for compacted earth linings, the chart shown in figure 108 is recommended.

Compacted earth linings 6 or 12 inches thick will usually satisfy design requirements from a seepage standpoint, but they require gravel cover for protection against erosion and are relatively expensive because of the difficulty of compacting them on side slopes. The heavy or thick compacted earth linings have proved to be the most desirable of the earth types because conventional construction equipment and methods can be used, a wider variety of soils may be used without erosion protection, the lining is not easily destroyed by cleaning operations, and excellent seepage control can be obtained. The heavy compacted earth lining is placed in

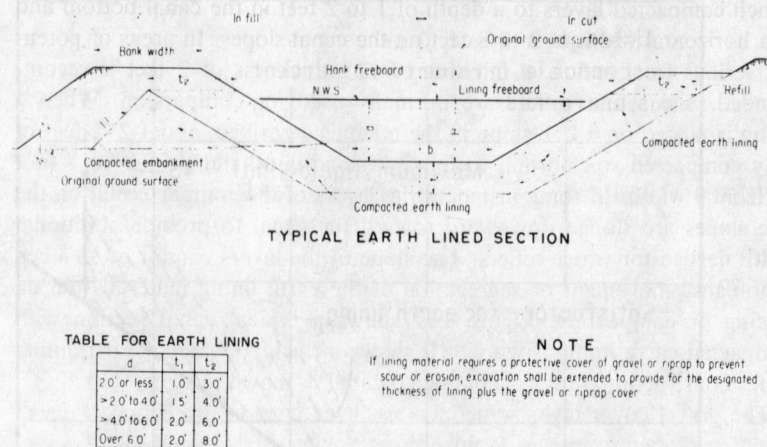

TYPICAL EARTH LINED SECTION

TABLE FOR EARTH LINING		
d	t_1	t_2
2 0' or less	1.0'	3.0'
>2.0'to 4.0'	1.5'	4.0'
>4.0'to 6.0'	2.0'	6.0'
Over 6.0'	2.0'	8.0'

NOTE

If lining material requires a protective cover of gravel or riprap to prevent scour or erosion, excavation shall be extended to provide for the designated thickness of lining plus the gravel or riprap cover

Figure 109.—Typical canal section and compacted earth lining. 101–D–587.

Figure 110.—Placing and compacting thick earth lining in Wellton-Mohawk Canal, Gila project, Arizona. P50-303-148.

6-inch compacted layers to a depth of 1 to 2 feet in the canal bottom and to a horizontal width of 3 to 8 feet on the canal slopes. In areas of potential serious frost action, a minimum lining thickness of 2 feet is recommended. Sheepsfoot rollers are normally used for compaction. When a lining is placed on a 2:1 slope in the manner described, about 2½ feet of fully compacted soil normal to the 2:1 slope is obtainable for an 8-foot horizontal width. In some instances the layers of the compacted fill on the side slopes are sloped downward toward the canal to provide additional width needed for wider rollers. The slope of the layers cannot be so steep as to cause movement or segregation of the earth lining materials during placing or compacting. Figure 109 shows a typical canal section with compacted earth lining. Figure 110 shows a thick compacted earth lining under construction, and the completed canal is shown in figure 111.

The gravel cover layer sometimes used for erosion protection of earth linings may vary from 6 to 12 inches in thickness. Pit-run sand and gravel up to 3 inches in size and containing a small amount of fines has been found satisfactory. In some instances, it has been necessary to place a 6-inch-thick sand and gravel filter layer on an open foundation structure to assure that loss of the earth lining material would not occur. Well-graded sand and gravel up to 3 inches in size are used.

Figure 111.—Completed thick earth lining in Wellton-Mohawk Canal, Gila project, Arizona. PP-50-303-685.

Riprap or coarse gravel is usually required to protect wetted earth surfaces in the vicinity of structures where high turbulence may occur. The material will vary in size and thickness, depending on the degree of protection required, but should be well graded from the smallest to the largest size.

The type of material considered suitable for covering buried membranes of asphalt or bentonite depends largely on the grade and velocity of the canal and the steepness of the side slopes. The cover materials must be stable soils when saturated. Side slopes are normally 2:1 or flatter. In some cases, earth cover (sandy, silty, or clayey soils) is adequate; in other cases where erosive forces are high, gravelly soil or gravel cover is required. Sometimes earth cover is placed directly on the membrane and a gravel cover is placed over the earth. The total thickness of the cover usually is from 6 to 12 inches. The canal subgrade must be relatively smooth as obtained by dragging, so that there are no abrupt breaks or protruding materials, and open cobble structure must be filled with a suitable finer filter material. Cover materials must be carefully placed so that puncturing or slipping of the membrane does not occur.

Stabilized soils (see secs. 34 and 80 through 84) are also used for lining canals and reservoirs. For example, compacted soil-cement and asphaltic-concrete are hard surface linings used to protect the sides and bottom of earth-lined reservoirs, to prevent erosion, to reduce turbidity of the water, and to facilitate cleaning. With additional construction experience, and information on their performance, thick linings may also prove to be sufficiently impervious to provide both seepage and erosion control.

Chemicals and other additives are also used in linings, largely on an experimental basis.

74. Specifications Provisions.—(a) *General.*—The following paragraphs summarize the various standard specifications provisions used by the Bureau for earthwork in canal construction. Their purpose is to clarify and to relate the different types of work. However, this discussion should not be considered a substitute for the published specifications for any particular job. Each set of specifications includes only those standard provisions that apply to the local situation, and in addition, modifications are made to fit each job.

(b) *Subgrades and Foundations for Embankments and Compacted Earth Lining.*—Provisions for canal excavation are included in the section of the specifications entitled "Excavation for Canal." It is usually required that the contractor excavate to sections shown on the drawings. If, because of undesirable bottom or slope conditions, changes in the sections are believed desirable by the contracting officer, such changes can be ordered by him for the removal of unsuitable material and for refilling as necessary.

Provisions for treatment of embankment foundations are included in the section of the specifications entitled "Preparation of Surfaces Under Embankments." These provisions cover such required items of work as stripping, scoring the foundations of uncompacted embankments with a plow, and scarifying and compacting the foundation surface for compacted embankments and compacted earth linings. Stripping includes the removal of all material unsuitable for embankment foundation to a depth indicated by the contracting officer.

Necessary clearing of the right-of-way, grubbing of stumps and other vegetable matter and their roots, and the burning or other disposal of the material are required of the contractor. The contractor receives no special payment for this work, but the cost is included in the unit prices bid in the schedules for excavation.

Excavation will be classified for payment as either common or rock, or will not be classified for payment. In both cases, the terms "rock" and "rock excavation" and "common" and "common excavation" are defined.

The paragraph on excavation for canals will permit the varying of slopes of excavations and embankments for canals during construction, if found desirable. Blasting is controlled. Contractor's runways must not be cut into canal slopes below the proposed water level in unlined canals or below the top of the lining in lined canals. Above the canal section, rock excavation need not be finished and will be allowed to stand at its steepest safe angle. In unlined sections of canal excavated in rock, sharp points of rock extending not more than 6 inches into the water prism are sometimes permitted. Earth slopes are to be finished neatly to lines and grades, either by cutting or by construction of earth embankments. Earth-lined sections may be excavated beyond the prescribed pay lines for earth lining, sand and gravel filter, and gravel cover, if the contractor desires; but such overexcavation is at the expense of the contractor. The excavation and trimming of concrete-lined and asphalt-membrane-lined sections in earth and in rock are described in the specifications.

The surfaces under all canal embankments, excepting rock surfaces, are required to be scored with a plow to make furrows not more than 3 feet apart and not less than 8 inches deep. For compacted embankments, the entire foundation surface is scarified or plowed to a depth of not less than 6 inches in lieu of scoring. Foundations for embankments are stripped of unsuitable material to a directed depth.

(c) *Earth Embankments and Linings.*—In specifications for construction of earth embankments and linings, general provisions are made: to adhere to the dimensions shown on the drawings; to provide allowance for settlement; to grade the tops of canal banks; to distribute gravel, cobbles, and boulders uniformly throughout the fill; and to prohibit use of frozen materials and frozen working surfaces for canal embankments and built-up canal bottoms. Provisions are made for limiting the amount of backfill or compacted backfill that can be placed about structures which are built prior to excavation of the canal and construction of the canal embankment. Use of all suitable materials from excavations for construction of fills is required. Excess material from excavation is used to strengthen canal embankments or is wasted as directed. Material from cut not needed to strengthen adjacent embankment, as well as excavated material in excess of requirements or unsuitable therefor, is deposited as directed in waste banks on Government right-of-way. Stockpiling of suitable materials for use in earth linings or earth cover for asphalt membrane linings may be required. Toes of waste piles or banks are generally set to not less than 20 feet from the edge of a canal cut or not less than 12 feet from the edge of a cut for a lateral, wasteway, or drain.

Where the canal excavation at any section does not furnish sufficient material for the required fills, the contractor is required to borrow the materials from areas designated by the Government. Adequate berms are left between embankment toes and adjacent borrow pits with $1\frac{1}{2}:1$ slopes in the pits unless otherwise specified. Borrow pits may require drainage by open ditches.

A unit of overhaul is defined as a cubic yard of excavated material hauled 1 mile in excess of the free-haul limit. Overhaul is occasionally defined in terms of station cubic yards. Unless otherwise specified, no overhaul payment is made for material paid for as backfill, riprap, gravel bedding for riprap, pit-run gravel, gravel surfacing, and compacted fill used in preparing rock foundations for concrete canal lining.

The specifications for earth canal embankments and linings require that the construction operations be such as will result in an acceptable and uniform gradation of materials to provide for impermeability and stability when compacted, and that *all* stones larger than the maximum size specified (usually 3 or 5 inches) be removed prior to compaction. These requirements are particularly important in lining construction because impermeability is controlled by a relatively thin layer of compacted soil. When blending of fine- and coarse-grained soils from different sources is required to secure the lining material desired, the specifications usually require that the fine soil be placed first, followed by the coarse soil, after which the two are blended. This procedure is used so that no coarse lenses will be left at the bottom of any finished layer. It is also required that blending be accomplished by a mixing machine built for the purpose of blending soils or by equipment producing comparable results and specifically approved for the work. The weight of the mixing equipment is important, as light equipment will rise and not accomplish proper blending for the full depth of each layer and slightly into the underlying layer. An item covering the excavation and refill of inspection trenches is often provided in the specifications. This provision is made so that proper inspection of the materials and the blending and compaction operations can be made from time to time as required to insure an acceptable lining product.

Provisions are sometimes made for compacting clayey and silty subgrade soils of low permeability for the base of a compacted earth lining in the canal bottom. This base is considered to be equivalent to about one 6-inch compacted layer in the canal bottom.

The construction of fills for canals in the forms of earth embankments, linings, backfill, and refill is covered by a number of specifications paragraphs. In order to show the relationships among the different types of fills, table 5 was prepared to summarize the pertinent provisions.

(d) *Riprap, Protective Blankets, Gravel Fills, and Gravel Subbase.*— Specifications provisions for riprap require the contractor to furnish and place riprap or coarse gravel protection at designated locations. Riprap rock can be quarried or natural boulders can be used, but the material should be hard, dense, and durable, and well graded from maximum to minimum specified sizes. Six inches of reasonably well-graded sand and gravel bedding up to 1½ inches in size are normally required under all riprap. Table 6 gives the largest and smallest permissible rock sizes for various thicknesses of riprap and coarse gravel.

Riprap need not be hand placed, but should be dumped and smoothed by adjusting the rocks to obtain a stable mass with no large unfilled voids. Rock spalls or gravel are required in amounts not in excess of the voids in the riprap. Earth, sand, or rock dust not in excess of 5 percent of the total riprap material will be permitted. Excavation outside of the normal canal prism required for placing riprap, coarse gravel, and bedding material generally is measured and paid for as excavation for structures.

Gravel fill, gravel subbase, or gravel blanket, when required, is made of pit-run gravel obtained by the contractor from approved borrow sources on Government right-of-way or from sources provided by the contractor. The material should be less than 3 inches in size, reasonably well-graded, and contain the minimum practicable amount of fines. Sometimes the gradation limits are specified. "Reasonably well-graded" means that there should be a reasonably good distribution of sizes of particles from the coarsest to the finest and without a major deficiency of any size or group of sizes. Figure 112 shows the normal limits of reasonably well-graded sand-gravel materials. Uncompacted gravel fill is placed in uniform layers to the required thickness. If compaction is specified, it is accomplished in accordance with the provision for compacting cohesionless free-draining materials.

Provisions for one-course gravel surfacing of prescribed thickness for

Table 5.—Materials and placing requirements for various fills used in canals

Type of construction and where used	Material	Placement requirements
Embankment not to be compacted. For lined and unlined canals.	Not specified.	Approximate horizontal layers. If built by excavating and hauling equipment, travel is routed over the embankment. If built by excavating machinery, thickness of layer is limited to depth of material as deposited by excavating machine. Water may be required, either added at site of excavation or by sprinkling of banks. Working may be required to secure uniformity of moisture and materials.
Embankment to be compacted. For unlined or lined canals.	Suitable material from required excavation or borrow; remove plus 5-inch material; gradation must be acceptable.	Prepare foundation; if clayey or silty soils, 6-inch compacted thickness at uniform optimum moisture; tamping rollers, 95 percent of USBR or ASTM standard density. If cohesionless and freedraining soils, horizontal layers of 6 inches compacted thickness if rollers or tampers are used; 12 inches compacted thickness if tractor treads or other heavy vibrating equipment are used; thorough wetting; 70 percent relative density.
Loose earth lining. For lined canals.	Impervious soil from canal excavation or borrow; uniform mixture required; depth of cut in borrow areas to be designated; uniform cutting of the face of the excavation; no plus 3-inch rock; may require blending.	Not specified.

Compacted earth lining. For lined canals.	Impervious soils from canal excavation or borrow; uniform mixture required; depths of cut in borrow areas to be designated; uniform cutting of the face of the excavation; no plus 3-inch rock; may require blending.	Prepare foundation; if clayey or silty soil, 6-inch compacted thickness at uniform optimum moisture; tamping rollers, 95 percent of USBR or ASTM standard density.
Refill above lining. For lined canals.	Selected from excavation or borrow.	Not specified.
Backfill. About structures and pipelines.	Approved material from required excavation or borrow.	Amount can be limited by the contracting officer; method of placement not specified, but subject to approval.
Compacted backfill. About structures and pipelines.	Selected clayey and silty material not greater than 3 inches; from required excavation or borrow. or Selected cohesionless free-draining material not greater than 3 inches; from required excavation or borrow.	Horizontal layers; compacted thickness at uniform optimum moisture shall not exceed 6 inches or two-thirds the length of roller tamping feet, whichever is the lesser; compacted thickness shall not exceed 6 inches where compaction is performed by hand or power tampers; 95 percent of USBR or ASTM standard density. Horizontal layers 6 inches compacted thickness if rollers or tampers are used; 12 inches compacted thickness if tractor treads or surface vibrators are used; and not more than the penetrating depth of the vibrator if internal vibrators are used; wetting by hoses, flooding, or jetting is required; 70 percent relative density.

Table 6.—Permissible rock sizes for various thicknesses of riprap and coarse gravel

	Nominal thickness, inches					Coarse gravel (thickness as prescribed)
	36	30	24	18	12	
Overall thickness, including bedding	42 in	36 in	30 in	24 in	18 in	6 or 12 in.
Largest permissible rock	1 cu. yd	½ cu. yd	¼ cu. yd	⅛ cu. yd	1 cu. ft	⅛ cu. ft.
Smallest permissible rock	⅒ cu. ft	⅒ cu. ft	⅒ cu. ft	⅒ cu. ft	1½ in	3/16 in.

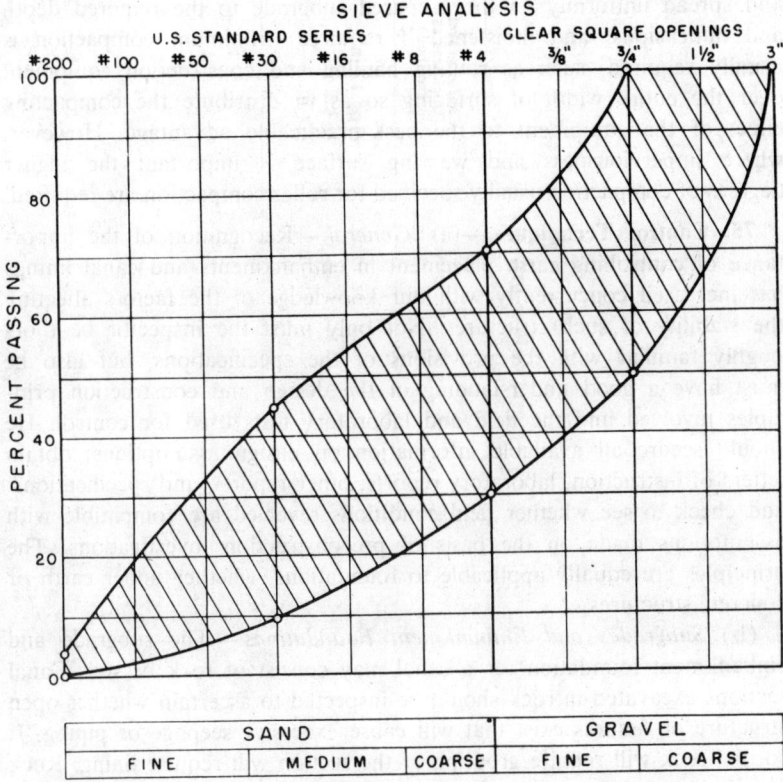

Figure 112.—Size limits for reasonably well-graded, pervious sand-gravel material. 101–D–223.

operating roads and parking areas require material consisting of sand and gravel from which all stones having a maximum dimension of more than 1½ inches have been removed. The material from gravel sources is selected to include a sufficient quantity of natural cementitious or binding material such that the surfacing will bond readily under the action of traffic. The desirable gradation conforms to ASTM Designation D-1241 for Type I, Gradation C, surface-course material. It is a well-graded material within the limits shown for reasonably well-graded gravel in figure 112, except that 8 to 15 percent fines (passing No. 200 sieve) are required to provide imperviousness and binding material. These fines are recommended to be in a limited range of plasticity such that the plasticity index (PI) is from 4 to 9 and the liquid limit (LL) is less than 35. Surfaces to receive the gravel surfacing should be bladed or dragged to secure a uniform subgrade. The surfacing material is placed and spread uniformly on the prepared subgrade to the required depth and dimensions, and moistened if required. Minimum compaction is usually required, such as routing hauling and construction equipment over the entire width of surfacing so as to distribute the compacting effect of the equipment to the best practicable advantage. However, where imperviousness and wearing surface is important the higher degrees of compaction usually specified for roller compaction are required.

75. Control Techniques.—(a) *General.*—Recognition of the importance of controlling earth placement in embankments and canal linings has increased concurrently with our knowledge of the factors affecting the stability of such structures. Not only must the inspector be thoroughly familiar with the provisions of the specifications, but also he must have a good understanding of the design and construction principles involved and the field and laboratory tests used for control. He should secure all available information on design assumptions; obtain letters of instruction, laboratory reports, other reports, and specifications; and check to see whether field conditions revealed are compatible with assumptions made on the basis of preconstruction investigations. The principles are equally applicable to foundations, whether under earth or concrete structures.

(b) *Subgrades and Embankment Foundations.*—The subgrade and embankment foundation for a canal may consist of rock or soil. Canal sections excavated in rock should be inspected to ascertain whether open structure or fissures exist that will cause excessive seepage or piping. If so, the rock will require grouting or the section will require lining. Rock foundation surfaces on which compacted earth is to be placed should be moistened before placing the first layer of earth, but standing water should not be allowed.

When canal sections are excavated in soils or embankments are constructed on soil foundations, greater care in inspection is required than for rock subgrades and foundations because of the nature of the material. In addition to noting possible seepage or piping conditions, the inspector must identify and locate conditions which could lead to high settlement and slope instability. During the investigation for the design of a line feature, it is not economically possible to predetermine all conditions which may be encountered. Therefore, during excavation or stripping operations, if conditions warrant, the inspector should call to the attention of his supervisors the necessity for additional exploratory holes; field penetration, permeability or density tests; or other sampling and tests to determine the extent of any materials of doubtful suitability for the foundation of embankments, for cut slopes, or for canal subgrades.

Some canal specifications require that the reaches of canal to be lined shall be determined immediately in advance of excavation by exploratory holes, such as bulldozer or backhoe trenches dug by the contractor. In such cases, the inspector must be familiar with the permeability characteristics of the soils encountered, making use of inplace permeability tests as required.

The inspector should examine canal subgrade soils on the slopes and bottom, with respect to probable future seepage and erosion. If high seepage losses are anticipated or if the soils are fine and lacking in cohesion and may be anticipated to erode badly under the proposed operating conditions, the inspector should call these facts to the attention of his supervisor. If on further consideration these conditions appear to be critical, advice from the engineering staff in the Engineering and Research Center should be sought. Lined sections in areas of high ground water must be protected against uplift by drainage systems, and these are normally specified. If such areas not previously known are observed during construction, the facts should be brought to the attention of the proper project supervisors, so that provisions can be made for the installation of pressure relief devices. Similarly, when high ground water is encountered near silty or fine sandy subgrade soils of concrete-lined sections in freezing climates, attention should be directed to this fact so that frost prevention measures can be instituted if necessary.

If fine-grained soils are to be placed as a sublining under concrete or membrane linings or be used for earth linings, the natural subgrade should be inspected for open voids as in cobbly soils or fissured formations through which the fine soil might erode. In such cases, the inspector should call the conditions to the attention of his supervisor so that filter blankets can be installed.

The inspector should see that all perishable matter and any soils which may become unstable upon saturation, such as highly organic soils, loose silts, very fine sands, and expansive clays, are removed or properly treated for embankments and canal linings to the extent necessary to provide a safe, stable foundation or subgrade under operating conditions. The specifications requirements for stripping and scoring earth foundations for all canal embankments, and the special treatment for preparation of the foundation surfaces of compacted embankments and earth linings, are important details that require careful visual inspection during construction. The determination of depth of stripping requires experienced judgment, since it often is a compromise between what would be desirable from a design standpoint and excessive cost. Good bond between the foundation and first layer of fill is achieved by moistening the foundation rather than using a very wet embankment layer.

Foundation control information should be included in the earthwork portion of the monthly construction progress report (L–29). If field density tests are made, the data can be reported along with the compacted

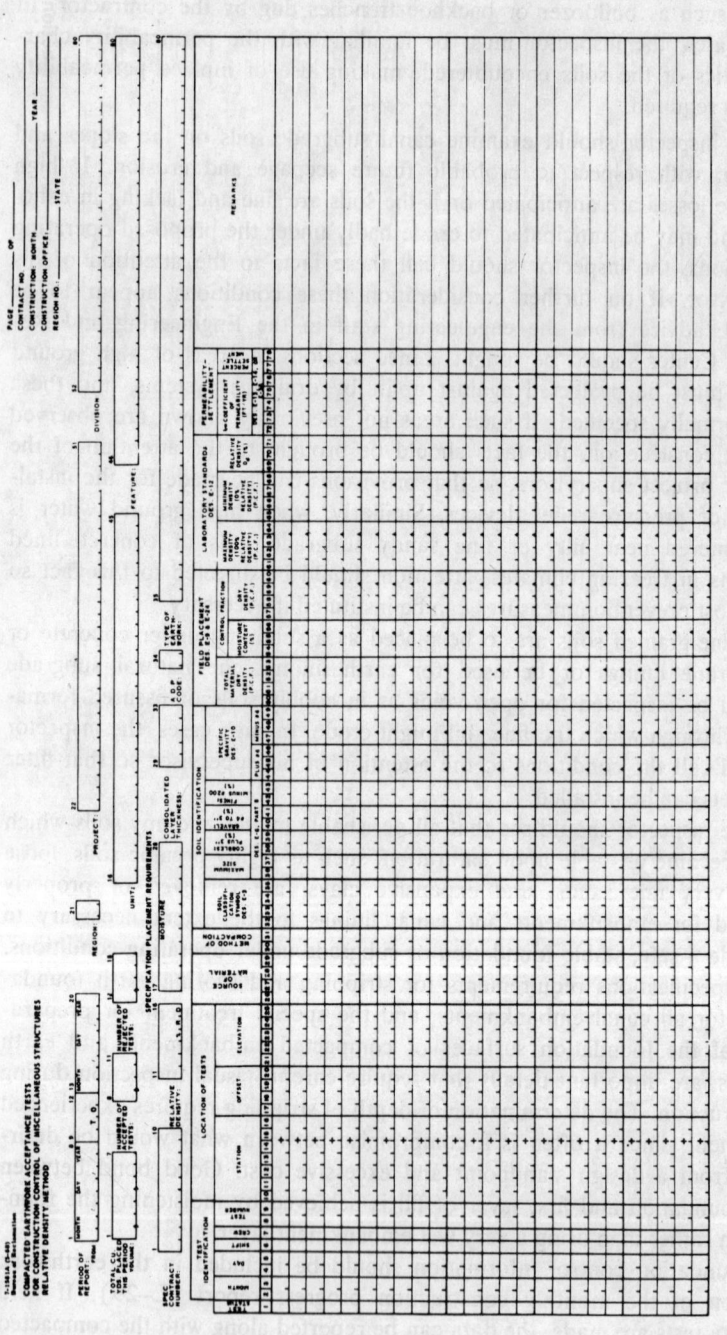

Figure 113.—Compacted earthwork acceptance tests for construction control of miscellaneous structures—Relative density method. Form 7-1581A.

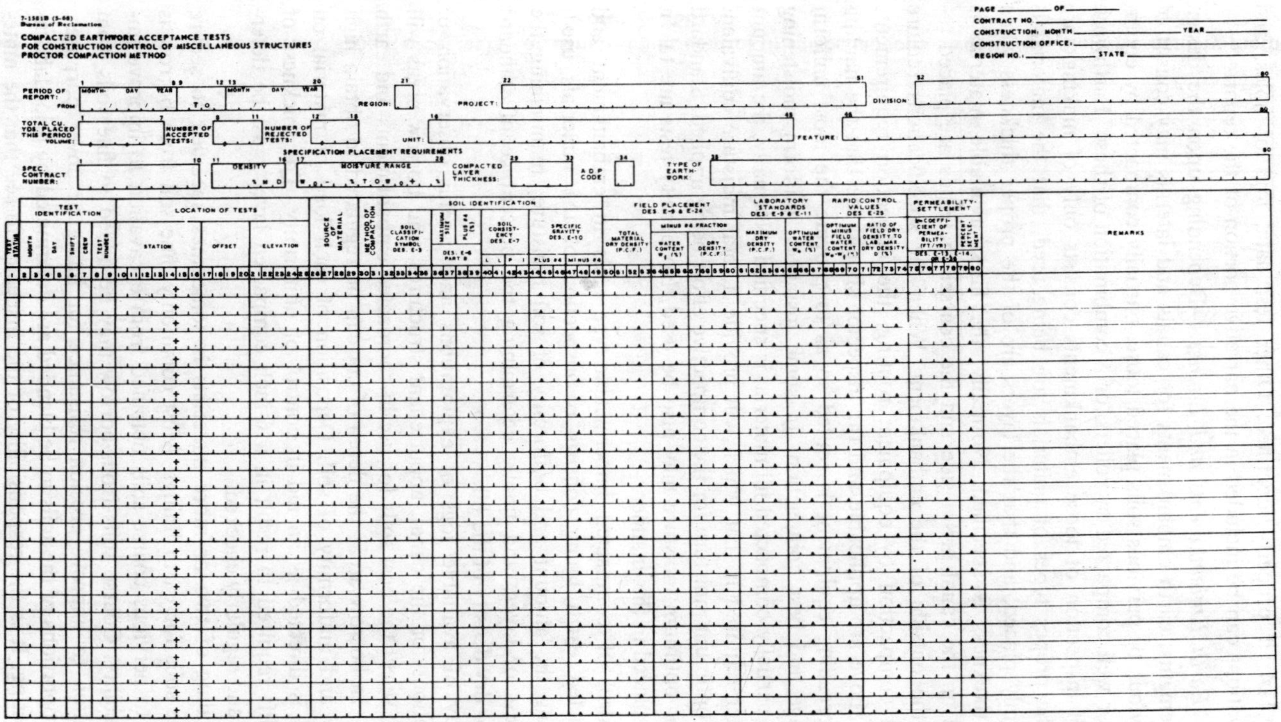

7-1581B (5-66)
Bureau of Reclamation
COMPACTED EARTHWORK ACCEPTANCE TESTS
FOR CONSTRUCTION CONTROL OF MISCELLANEOUS STRUCTURES
PROCTOR COMPACTION METHOD

PAGE ____ OF ____
CONTRACT NO. ____
CONSTRUCTION: MONTH ____ YEAR ____
CONSTRUCTION OFFICE: ____
REGION NO.: ____ STATE ____

PERIOD OF REPORT: FROM MONTH DAY YEAR TO MONTH DAY YEAR
REGION:
PROJECT:
DIVISION:
TOTAL CU. YDS. PLACED THIS PERIOD VOLUME:
NUMBER OF ACCEPTED TESTS:
NUMBER OF REJECTED TESTS:
UNIT:
FEATURE:
SPECIFICATION PLACEMENT REQUIREMENTS
SPEC. CONTRACT NUMBER:
DENSITY
MOISTURE RANGE
COMPACTED LAYER THICKNESS:
A D P CODE:
TYPE OF EARTH-WORK:

TEST IDENTIFICATION: TEST STATUS, MONTH, DAY, SHIFT, CREW, TEST NUMBER
LOCATION OF TESTS: STATION, OFFSET, ELEVATION
SOURCE OF MATERIAL
METHOD OF COMPACTION
SOIL IDENTIFICATION: SOIL CLASSIFICATION SYMBOL DES. E-3; MAXIMUM SIZE; PLUS #4 (%) DES. E-6 PART B; SOIL CONSISTENCY DES. E-7 (L L, P I); SPECIFIC GRAVITY DES. E-10 (PLUS #4, MINUS #4)
FIELD PLACEMENT DES. E-9 & E-24: TOTAL MATERIAL DRY DENSITY (P.C.F.); MINUS #4 FRACTION: WATER CONTENT w_f (%), DRY DENSITY (P.C.F.)
LABORATORY STANDARDS DES. E-9 & E-11: MAXIMUM DRY DENSITY (P.C.F.); OPTIMUM WATER CONTENT w_o (%)
RAPID CONTROL VALUES DES. E-25: OPTIMUM MINUS FIELD WATER CONTENT w_o-w_f (%); RATIO OF FIELD DRY DENSITY TO LAB. MAX. DRY DENSITY D (%)
PERMEABILITY-SETTLEMENT: COEFFICIENT OF PERMEABILITY (FT/YR) DES. E-13 OR E-36; SETTLEMENT E-14
REMARKS

Figure 114.—Compacted earthwork acceptance tests for construction control of miscellaneous structures—Proctor compaction method. Form 7-1581B.

earthwork control data on form 7–1581A, figure 113, or form 7–1581B, figure 114, if properly identified. Other test data, logs of holes, and observations can be included in the narrative portion of the report.

(c) *Earth Embankments and Linings.*—Depending upon the design requirements, earth embankments for canals and laterals may consist of impervious or pervious soils placed loose, partially compacted by equipment or well compacted by rollers, or a combination of these. Earthwork control inspection of loose embankments consists only of making sure that the proper types of materials are being used, that the material is uniformly placed, and that the layers are of the proper thickness when layer thickness is specified. Normally, the finest available material is placed on the canal prism side and the coarser materials are placed toward the outside of the embankment. When the specifications require partial compaction by equipment routing, the inspection requirements given above are applicable and the inspector should see that the hauling and spreading equipment is routed so as to provide the most uniform coverage and best compaction. In some instances, uniform moistening of the partially compacted fill material is specified. Normally, the amount of water required is that which will provide a total moisture equivalent to the optimum moisture for the compactive effort being applied. Standard Proctor optimum moisture may also be used. No density tests are taken for fill placed in this manner.

Control of compacted impermeable-type soils in embankment and compacted earth lining consists of inspection of the materials used, checking the amount and uniformity of soil moisture, maintaining the thickness of layer, and finally determining the percentage of standard Proctor dry density of the fill.

Study of available logs of explorations and careful observation of excavations for the canal, appurtenant structures, and borrow areas will help to select the best soils for use in compacted embankments and earth linings. In borrow areas the depth of cut can be regulated to obtain high quality and uniformity of soil. Deposition of each load of material on the fill is directed as an aid in control of uniformity. The thickness of loose lift required to result in a 6-inch compacted layer must be determined and regularly checked.

Blending of two or more materials is sometimes specified to secure suitable soils for earth linings. Most commonly, fine soil from borrow is added to pervious coarse soils obtained from excavation to improve impermeability. Coarse soils from borrow may be added to fine soils from excavation to improve erosion resistance for lining or cover purposes. The proportions of the soils to be blended are determined by prior laboratory testing. It is the responsibility of the inspector to see that the materials are properly proportioned and mixed. It is normally required that excavation materials to be used for one of the blended materials be completely removed from the section and be replaced in the lining layer to the depth required for proper proportioning. The correct depth of fine soil is placed first, followed by the correct depth of coarse soil. Blending then is allowed.

Blending is normally specified to be done by a machine designed for the purpose of blending soils or by other equipment obtaining comparable results. Therefore, the inspector must assure that the equipment being used provides an intimate blend to the soils for the total layer depth. This is done by frequently digging holes through the blended layer prior to compaction to observe the uniformity of the layer after blending. When soils of different colors are blended, the color of the mixture at different depths in the layer is a useful index as to the effectiveness of the mixing operation. Compaction should not be allowed to proceed until a satisfactory blend has been obtained. Inspection trenches are often included as a separate item of construction work in the specifications. These should be requested frequently at the beginning of a blended lining contract, and at lesser frequency later, so that the results of the contractor's blending and lining operations can be visually observed. The field permeability test should also be used by the inspector to check on the effectiveness of the total construction operation.

Normally, specifications paragraphs permit the contractor to use a wide variety of commercially available equipment of the sheepsfoot tamping roller type which could produce the desired kneading action during compaction, uniformity of compaction, and integration of soil layers for compacted embankments of any size. The diameter and weight of the tamping roller, number of roller passes, and length of tamping feet are not usually specified. Limitations are placed on the end area and the spacing of tamping feet to insure that the sheepsfoot roller will sink into loose soil and attain the kneading action required for a homogeneous soil mass. Inspectors should adhere to the requirement that each compacted lift be less than two-thirds the length of the tamping feet so that the roller can ride high on a compacted lift and insure the bonding of successive lifts. A determination of the height of compacted lifts is made by keeping a record of the number of lifts placed and the measured change in elevation to estimate the average thickness of compacted lifts being placed at a selected location. The above items require continued observations by the inspector. If the contractor's compacting equipment is inadequate, as evidenced by continued low or borderline densities, the contractor can be required to obtain and use suitable equipment, add ballast to existing equipment, or provide maintenance as required. Maintenance of tamping knobs is important. For instance, if the unit knob pressures become too high, as from the reduction in knob size from wearing, shearing rather than compaction will take place in soft fine-grained soils, often regardless of the number of roller passes applied.

Cobbles larger than specified, usually 3-inch or 5-inch, must be removed prior to compaction. This is especially important in all types of compacted earth linings where a relatively thin section is relied upon for seepage control. Nests of cobbles will provide an easy access for water loss and piping. Cobbles are often buried in the layer being placed and are difficult to detect or to remove. Therefore, if the percentage of oversize material is large, some means should be provided for removing this material at the point of excavation.

The inspector is responsible for controlling water content of the soil to that which is the "optimum for compaction" and assuring that the moisture is uniform throughout the layer to be compacted. Therefore, when large amounts of moisture are required, it is more efficient to add most of the water at the point of excavation with only supplemental sprinkling after the layer has been spread. Mixing after sprinkling is required to produce a uniform moisture condition throughout the layer before compaction can proceed. Inspection of moisture uniformity should be made frequently by digging holes in the loose layer just prior to compaction. Unless otherwise specified, optimum moisture requirements must be enforced for canal embankments and linings even though the required density is obtained at other moisture conditions. Adverse settlement and impermeability properties may result if the placement moisture is too low, and adverse shear properties may result if the placement moisture is too high. Therefore, the inspector must be prepared to request application or removel of moisture as required.

Control of pervious type material includes visual inspection of material for free-draining characteristics and uniformity of material. The thickness of layer is controlled, depending on the type of compaction being used. A thickness of not more than 6 inches after compaction is specified if smooth or pneumatic-tired rollers are used; not more than 12 inches after compaction if tractor treads or other surface vibratory equipment is used; or the length of the vibrating equipment if internal vibrators are used. For all types of compaction, each layer must be thoroughly wetted during the compaction operation.

The adequacy of the compaction and moisture control of impervious or pervious soils is controlled by the field density test (designation E–24) in conjunction with the rapid compaction control method (designation E–25) for clayey and silty soils, or the relative density test (designation E–12) for pervious sand and gravel soils. Unless otherwise specified, the minimum acceptable density is 95 percent of Proctor maximum density for the minus No. 4 fraction of clayey and silty soils and 70 percent relative density for the minus 3-inch fraction of pervious sand and gravel soils. With soils which are borderline, between the silty and clayey soils controlled by the Proctor test and pervious sand and gravel soils controlled by the relative density test, control is based on either 95 percent of Proctor maximum density or 70 percent relative density, whichever produces the highest unit weight. Table 2, section 56(b), is a guide for the selection of materials for which the relative density test is applicable.

Inasmuch as the adequacy of compaction is specified in terms of density achieved, the inspector is responsible for taking or arranging for the taking of sufficient tests to assure the adequacy of compaction for acceptance purposes. At the start of any work, a considerable number of tests are required to insure that the construction operations are producing the required results. After this has been established, the number of tests required are only those necessary to insure that specifications requirements are being met. Because of the widespread operations of canal work, the number of field density tests required for adequate control cannot be stated. Testing requirements should be based on area as well as on volume. The following is a guide for the minimum number of field density tests:

(1) For all types of earthwork, one test for each shift.

(2) For canal embankments, one test for each 2,000 cubic yards placed.

(3) For compacted canal linings, one test for each 1,000 cubic yards.

(4) For compacted backfill or for refill beneath structures:

    (a) Hand tamped (mechanical tamping), one test for each 200 cubic yards.

    (b) Roller or tractor tamped, one test for each 1,000 cubic yards.

(5) One complete permeability-settlement test should be made in the laboratory for each 10 density tests for compacted canal linings, impervious embankments, and impervious backfill.

When field density tests are taken in relatively narrow compacted earth linings where equipment travel is essentially along a constant route, care should be taken not to make the density tests under tractor or wheel tracks. The density at such locations may be considerably higher than the average density of the normally rolled lining.

Horizontal and vertical control must be available so that the inspector can adequately locate each field density test. Locations are usually defined in terms of station, offset from canal (or structure) centerline, and elevation above bottom grade. The inspector is responsible for locating these tests so that a complete representative record of the finished work is available. Companion Proctor tests by the rapid compaction control method for clayey and silty soils and relative density tests for free-draining sand and gravel soils should be made for each density test. A Proctor test or a relative density test is required for each control field density test to compute the percent compaction or relative density, respectively.

An additional copy of the earthwork portion of the monthly construction progress report (L–29) is required to be submitted to the Engineering and Research Center for review by the Division of General Research, for all earthwork other than earth dams. The contents of the report are given in section 71. Forms 7–1581A and 7–1581B, figures 113 and 114, are included as part of this report.

F. Pipelines

76. Design Features.—Pipelines and conduits are installed for many purposes. These include: irrigation laterals and distribution systems (including domestic and municipal and industrial deliveries), conveyance aqueducts and feeder mains, and pumping plant discharge lines. These systems may be constructed from a number of common alternative types of pipe, usually at a contractor's option, such as: concrete pressure pipe, steel pipe with a number of combinations of coatings and linings, pretensioned concrete pipe, noncircular prestressed concrete pipe, prestressed circular concrete pipe, asbestos-cement pipe, and cast iron pipe.

Other types which are being tested, but the use of which is presently in the experimental stage, include reinforced plastic mortar pipe with fiber glass reinforcement (RPM) and various types of plastic pipe in diameters 15 inches and smaller. Each of the various types of pipe has limitations of size, head class, and economic competition. While pipelines can be of any diameter, most laterals are between 12 and 96 inches in internal diameter.

Pipe trenches may vary widely in shape depending on soil conditions, type of materials, and type of pipe to be laid. Depth of the trench is determined by a minimum cover over the pipe and the outside diameter of the pipe. Pipe trenches up to 5 feet deep may be excavated with vertical sides in stable earth materials. Deeper trenches must have side slopes of ¾ : 1 or flatter or they must be shored for the safety of the men working in the trenches. Surface water, a high water table, expansive clays, and fine saturated soils are some of the natural conditions encountered which can greatly influence design of pipe trenches.

Pipe to be laid may be rigid or flexible in wall or ring strength of its transverse section. This dictates the type of bedding required and shape of the trench invert. Semi-rigid and flexible pipe may require a high degree of compaction on side walls of the pipe to a minimum height of 0.7 of its outside diameter to develop passive soil resistance and prevent excessive vertical deflection under full trench loads. Rigid-type pipe sections need only to be bedded to three-eighths of their diameter. Trench backfill loads will "arch" in deep trenches and only a part of the load will be carried by the pipe. However, this arch action is not considered to be permanent and after a long period of time the full load of the fill above will act on the pipe. This should be considered in trench design.

Trenches in wet or unsuitable materials and through irregularly excavated rock at pipeline grade must be overexcavated and replaced with a free-draining material or other selected borrow material and compacted to provide a firm pipe trench foundation. Methods of compacting earth materials in pipe trenches and the density requirements in various parts of the pipe trench are specified. In general, these specifications are similar to those provided for compacting other types of earth work. Recently installed pipelines must be protected from floating if the trench should be flooded by surface water from storms or irrigation of adjacent areas. Prompt backfilling of the trenches to a depth adequate to prevent floating is usually desirable. Very low density material at grade elevation or fine fluid material of a "quicksand" nature are conditions which may require special trench section designs and treatment or the use of a specific type of pipe, or both.

77. Specifications Provisions.—(a) *General.*—Basic pipeline trench designs are shown in figure 115. These designs set paylines for excavation and backfill in pipe trenches, as well as depth of bedding and cradle, when required. While this design is basically used for pressure pipe distribution systems, it can be applied to most other types of pipelines with minor modifications.

(b) *Pipeline Excavation.*—Pipeline excavation is governed in specifications by the paragraph "Excavation for Pipe Trenches." This paragraph describes paylines, requirements for overexcavating when trenching in unsuitable material, and replacing with selected borrow as indicated in figure 115. Safety requirements for pipe trenches require that trenches over 5 feet deep not in rock or other hard materials be protected against caving by shoring, bracing, or sloping at ¾:1 minimum from the bottom of the trench. While paylines are shown to vertical sides and at a set trench width for each pipe diameter, contractors may excavate outside of these lines if they desire to do so.

(c) *Backfill in Pipe Trenches.*—Material for backfill will normally be taken from the pipe trench excavation unless this material is unsuitable

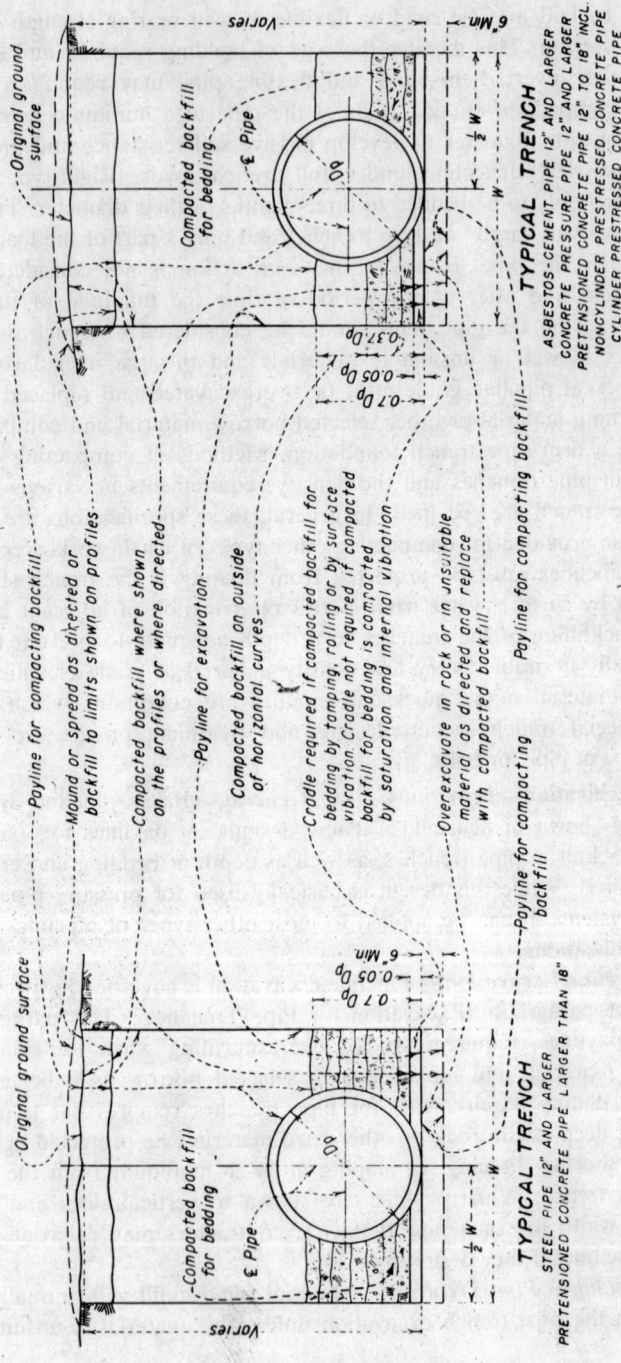

Figure 115.—Typical trench sections and compacted backfill for bedding requirements for pressure pipe. (From 40-D-6036.) (Sheet 1 of 2.) 101-D-588.

PIPE I.D. INCHES	D_p INCHES	W FEET
6 and less	I.D + 2	2.0
Over 6 thru 18	I.D + 4	0.063 (I.D + 24)
Over 18 thru 24	I.D + 4	0.063 (I.D + 40)
Over 24	1.167 I.D	0.063(I.167 I.D. + 36)

Note: .083 = Conversion factor inches to feet.

NOTES

If compacted backfill for bedding is compacted by tamping, rolling or surface vibration, the trench width, W, is the minimum width allowed.

If compacted backfill for bedding is compacted by saturation and internal vibration, the minimum trench width is $D_p + 18$".

D_p and W given in table are used for calculating pay quantities for all pipes.

I.D. is inside diameter of pipe in inches, and sizes of pipe shown refer to I.D.

Paylines for backfill will be the paylines for excavation except the volume of the pipe based on the diameter D_p will be deducted, and except where depth of backfill is limited as shown on the profiles

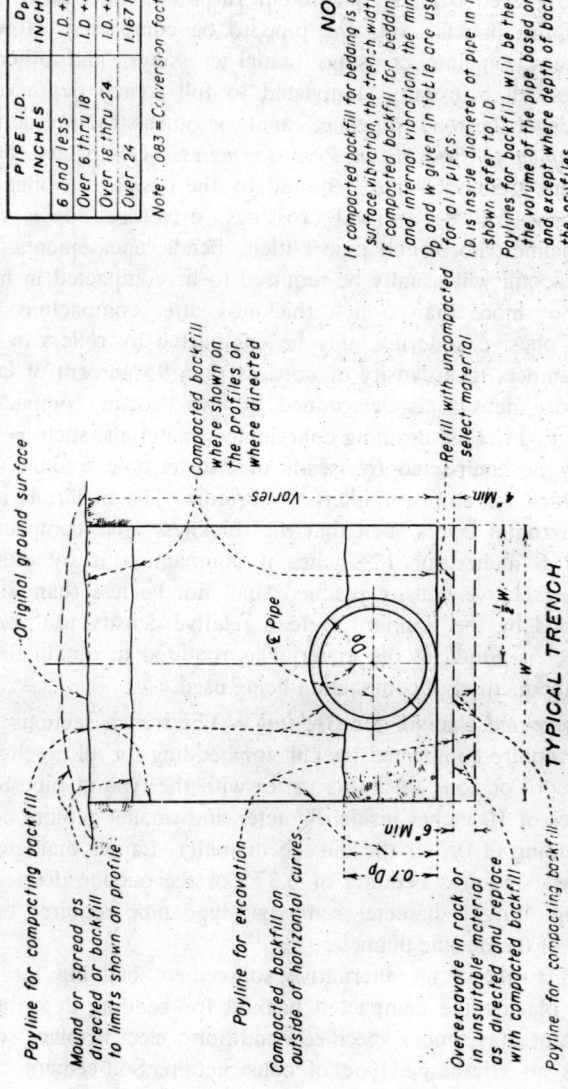

Original ground surface

Payline for compacting backfill

Mound or spread as directed or backfill to limits shown on profile

Compacted backfill where shown on the profiles or where directed

Payline for excavation

Compacted backfill on outside of horizontal curves.

Overexcavate in rock or in unsuitable material as directed and replace with compacted backfill

Payline for compacting backfill

Refill with uncompacted select material

Varies

¢ Pipe

W

½ W

4" Min

6" Min

0.7 D_p

TYPICAL TRENCH
ALL PIPE 10" AND SMALLER

Figure 115.—Typical trench sections and compacted backfill for bedding requirements for pressure pipe. (From 40-D-6036.) (Sheet 2 of 2.) 101-D-589.

or insufficient. In such cases additional backfill materials would be obtained from approved borrow pits. Backfill materials must not be dropped directly on the pipeline. Hard material larger than 3 inches in maximum dimension must not be placed within 6 inches of the pipe. For pipe with coal-tar enamel coating, hard material over 1½ inches maximum dimension must not be placed within 6 inches of the pipe, and backfill of clayey soil is usually limited to having a plasticity index of less than 22. Backfill material is placed so that the elevations of the two sides will be even, not to exceed 6 inches maximum variation. Most trench sections will not require backfill over the pipe to be compacted. However, at road crossings, pipeline crossings, canal crossings, and other similar crossings backfill is usually compacted to full trench depth or to the bottom grade of the road, pipeline, canal, or other structure as required.

(d) *Compacting Backfill in Pipe Trenches.*—Compacted backfill is placed in pipe trenches where required by the design drawings. Usually these locations will be at road crossings, driveways, open canals or laterals, pipelines, horizontal pipe outlets, bends, encasements, and collars. This backfill will usually be required to be compacted in horizontal layers of not more than 6-inch thickness after compaction for most materials. Cohesive materials may be compacted by rollers or by hand or power tampers to a density of not less than 95 percent of laboratory maximum dry density as determined by the Proctor compaction test, designation E–11. Free-draining cohesionless materials, such as sand and gravel, may be compacted by treads of crawler-type tractors, tampers, rollers, surface vibrators, or internal vibrators. The materials are deposited in horizontal layers such that the thickness after compaction does not exceed 6 inches, or 12 inches if compaction is by crawler-type tractor. The relative density reached shall not be less than 70 percent as determined by the standard Bureau relative density test, designation E–12. Water is added to the material as required to obtain the specific density by the method of compaction being used.

(e) *Compacted Backfill for Bedding.*—The trench sections shown in figure 115 require compacted backfill for bedding for all pipelines. However, the depth or zone thickness varies with the type of pipe to be laid. All pipelines of 10 inches inside diameter and smaller require compacted backfill bedding of 0.7 of the outside diameter. Larger diameter pipe of the rigid type requires bedding of 0.375 of the outside diameter of the pipe section. Larger diameter semi-rigid-type pipe requires bedding to the full 0.7 of the outside diameter.

Figure 116 shows an alternative soil-cement bedding for pipelines. Instead of placing the compacted backfill for bedding described above, the contractor may, under specified conditions, elect to place soil-cement bedding as an alternative type of construction. Soil-cement bedding is

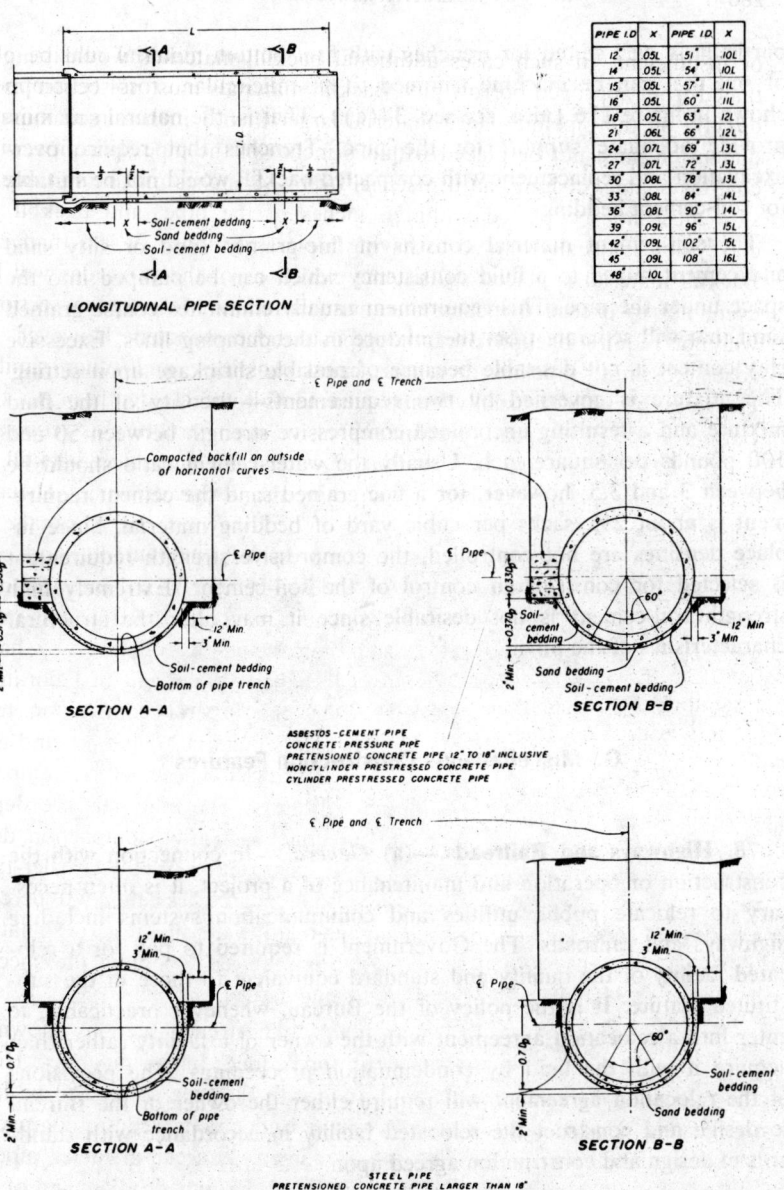

PIPE I.D.	X	PIPE I.D.	X
12"	.05L	51"	.10L
14"	.05L	54"	.10L
15"	.05L	57"	.11L
16"	.05L	60"	.11L
18"	.05L	63"	.12L
21"	.06L	66"	.12L
24"	.07L	69"	.13L
27"	.07L	72"	.13L
30"	.08L	78"	.13L
33"	.08L	84"	.14L
36"	.08L	90"	.14L
39"	.09L	96"	.15L
42"	.09L	102"	.15L
45"	.09L	108"	.16L
48"	.10L		

LONGITUDINAL PIPE SECTION

Compacted backfill on outside of horizontal curves

SECTION A-A

SECTION B-B

ASBESTOS-CEMENT PIPE
CONCRETE PRESSURE PIPE
PRETENSIONED CONCRETE PIPE 12" TO 18" INCLUSIVE
NONCYLINDER PRESTRESSED CONCRETE PIPE
CYLINDER PRESTRESSED CONCRETE PIPE

SECTION A-A

SECTION B-B

STEEL PIPE
PRETENSIONED CONCRETE PIPE LARGER THAN 18"

NOTES
Place pipe to line and grade on sand bedding as shown in Sec. B-B
Place soil-cement bedding under remainder of pipe as shown
in Sec. A-A and Sec. B-B
For D_p and trench paylines see Dwg 40-D-6036

Figure 116.—Typical trench sections and soil-cement bedding requirements for pressure pipe. (From 40-D-6185.) 101-D-590.

particularly well suited for trenches with firm bottom material consisting of soil that can be machine trimmed to a semicircular cross-section as shown in figure 116 (also, see sec. 34(c)). That is, the natural soil must provide adequate support for the pipe. Trenches that require over-excavation and replacement with compacted backfill would not be suitable for soil-cement bedding.

The soil-cement material consists of fine-grained sand or silty sand and cement mixed to a fluid consistency which can be pumped into the space under the pipe. This requirement usually eliminates coarse-grained sand that will separate from the mixture in the pumping lines. Excessive clay content is not desirable because of possible shrinkage upon setting. The mixture is governed by two requirements—viscosity of the fluid mixture and a resulting unconfined compressive strength between 50 and 100 pounds per square inch. Usually the water-cement ratio should be between 3 and 3.5; however, for a fine-grained sand the cement requirement is about 2½ sacks per cubic yard of bedding material. Since in-place densities are not controlled, the compressive strength requirement is selected for construction control of the soil-cement. Extremely high strength soil-cement is not desirable since it may alter the structural characteristics of the pipe.

G. Miscellaneous Construction Features

78. Highways and Railroads.—(a) *General.*—In connection with the construction or operation and maintenance of a project, it is often neces-sary to relocate public utilities and communication systems including highways and railroads. The Government is required to pay for a relo-cated facility of the quality and standard equivalent to those of the sub-stituted feature. It is the policy of the Bureau, whenever practicable, to enter into a relocation agreement with the owner of a facility rather than acquire it with the land by condemnation proceedings. The provisions of the relocation agreement will require either the owner or the Bureau to design and construct the relocated facility in accordance with stand-ards of design and construction agreed upon.

(b) *Design Features.*—The design of a highway or railroad consists of selecting and laying out the roadway or roadbed alinement, designing and detailing required structures, and preparing specifications for con-struction of the facility. As a basis for agreement on the design and construction of relocated highways, the Bureau will consider the Amer-ican Association of State Highway Officials' Manual, the standards of

Figure 117.—Heavy equipment constructing access road to Flaming Gorge Dam site through Dutch John Gap. P591-421-455.

the interested highway departments, and general highway construction practice; and for the design and construction of relocated and access railroads, the Bureau will consider the American Railway Engineering Association Manual, the railroad's own standards, and general railroad construction practice.

Access roads for the use of the contractor and the Government are designed and constructed by the Bureau in accordance with Bureau standards based on local proven types of road construction which will permit the use of local materials and the services of local contractors experienced with construction equipment designed for such work. Figure 117 shows construction of an access road to Flaming Gorge Dam and figure 118 shows the completed access road to Crystal Dam in Colorado. Figure 119 shows construction of the relocated railroad around Trenton Dam.

(c). *Earthwork Specifications Provisions.*—The specifications for construction of a railroad or a highway will contain a paragraph on excavation for roadway. This paragraph will require the contractor to construct the roadway to the dimensions shown on the drawings or as staked out,

**Figure 118.—Completed access road to Crystal Dam site in Colorado.
P622C–427–4210NA.**

**Figure 119.—Earthwork operations for bridge abutments of Trenton Dam rail-
road relocation. Borrow area and haul roads are shown at left. TD–268.**

and to the prescribed lines and grades. The drawings include profiles of the gradelines to either finished grade or subgrade. The term "subgrade" refers to the top of embankments and to the bottom of excavations ready to receive the roadway surface or the railroad ballast. In rock excavation, it is required that the bottom of the cut be excavated to 6 inches below the subgrade; and in common excavation, all loose rock fragments, cobbles, and boulders must be removed or excavated to a depth of not less than 6 inches below subgrade. Excavation below subgrade is required to be refilled to subgrade with material obtained from excavation for the roadway or from borrow pits, and the refilled material is required to be compacted equivalent to the compaction required for embankments. Provisions are made in the paragraph on excavation for roadway for handling of slopes shattered or loosened by blasting, and slides extending beyond the established slope lines or below subgrade. Provisions for construction of side drains and for excavation around trees, poles, or other objects which are to remain within the right-of-way for the roadway are also given in the paragraph for excavation for roadway.

In the paragraphs on preparation of surfaces under roadway embankments, provisions are included for surface treatment and for removal of unsuitable materials. The surfaces of sloping ground underneath embankments are required to be prepared to provide bond with the embankments and to prevent slipping. Where embankments are to be placed on smooth, firm surfaces and where low embankments are to be placed, the original ground surface is required to be thoroughly plowed or stepped to insure proper bonding of the new and existing material. When the foundations of the embankments are steep rock slopes, trenches are required to be blasted into the rock surface for "keying" the embankment to the foundation. The contractor is required to strip the area under an embankment of all unsuitable material to depths as directed.

The paragraph in the specifications entitled "Construction of Roadway Embankment," requires that construction be performed to the established lines and grades, increased by such heights and widths as necessary to allow for settlement. No brush, roots, sod, or other perishable or unsuitable materials are permitted to be placed in embankments. Use of materials containing excessive amounts of sand which would interfere with compaction is not permitted. Hard material or hard lumps of earth more than 12 inches in size must be broken down, or removed, and cobbles, boulders, or rock fragments more than 6 inches in maximum dimensions are not permitted in the upper 2 feet of embankment. Where there is a choice of materials, the best is to be used in the upper 1-foot of the embankment. No materials are permitted to be placed in the embankments when either the material or the surface on which it is to be placed is frozen. Good distribution of rock fragments, cobbles, and gravel within the embankment is required when such material is being used. Rockfill embankments are to be constructed so that no large unfilled spaces remain. Rock fragments are to be deposited in the outer portions of the embankments to protect the slopes against erosion, when directed.

The contractor's combined excavation and placing operations are required by the specifications to be such that the material in the embankment will be blended sufficiently to secure the best practicable degree of compaction and stability. To this end, the contracting officer may designate the locations in the embankment where the individual loads will be deposited. The required thickness of layers is 12 inches after compaction for earthfill, and not more than 24 inches after compaction for rockfill. Layers are to be constructed across the entire width of the embankments and must be built to the required slopes rather than widened with loose material dumped from the top. On sidehill fills where the width is too narrow to accommodate hauling equipment, end dumping may be used until the width of the embankment becomes great enough to permit the use of hauling equipment. The contractor is required to route his hauling equipment over the layers already in place and distribute the travel evenly over the entire width of the embankment so as to obtain the maximum amount of compaction possible.

The paragraph in the specifications entitled "Moistening and Compacting Roadway Embankments" provides for moistening and compacting where construction according to the paragraph, "Construction of Roadway Embankment", will not result in proper solidification. The paragraph states that the necessity for moistening or compacting will be determined by the contracting officer and, in general, will depend on the nature and condition of the materials available for the construction of the embankments. Where sufficient moisture is not available in the material, additional water must be provided by sprinkling as the layers of material are placed on the embankments. Moistening must be uniform, and harrowing or other working may be directed to produce the required uniformity of water content. Material containing an excess of moisture must be permitted to dry to the extent required before being compacted. The quantity of water to be used is determined by the contracting officer.

The paragraph contains provisions for rolling the embankment. The amount of rolling depends on the nature of the material to be compacted and the degree of compaction required, and will be specified for each embankment or portion of an embankment. The Bureau usually follows State highway specifications when embankments are built for a State highway department. Highway department specifications are usually of the performance type where only the end result is specified. For this type of specifications, the contractor usually selects the equipment, with approval of the contracting officer, to obtain the end result. If highway and railroad embankments are not built according to State specifications, the Bureau will build them using an end-result type of specifications requiring the fill to be compacted to 95 percent of Proctor maximum dry density.

The paragraph on borrow pits requires that additional material for construction of embankments and for excavation refills in rock cuts, where required, is to be obtained by widening the cuts and/or flattening the slopes on either or both sides of the centerline or from borrow pits. The locations of all borrow pits are designated by the contracting officer. If the additional material is taken from borrow pits adjacent to the roadway, a berm not less than 10 feet wide of original unbroken ground is required between the outside toe of the embankment and the edge of the borrow pits, with provisions for a side slope of 1½ : 1 to the bottom of the borrow pit, *provided* that, where directed on account of drainage or other requirements, the berm between the toe of the embankment and the borrow-pit slope will be not less than 40 feet wide. Provisions are made for connecting pit to pit with waterways, unless otherwise directed, and to provide ample drainage and leave no stagnant pools. Unless otherwise approved, the bottoms of pits near bridge and culvert openings are not permitted to be excavated below the surface over which the water will run to pass through such culvert or under such bridge. Provisions are made for leaving the surfaces of borrow pits in a reasonably smooth and even condition, as approved by the contracting officer.

(d) *Control Techniques.*—Many of the control techniques discussed in preceding paragraphs are applicable to highway and railroad embankments. The moisture-density relationship of soils and the related field control methods are, in fact, fundamental for any earthwork construction where compaction of the soil is required. However, the same degree of control may not be practicable nor required for highway and railroad embankments.

The control requirements will vary depending on the standards adopted by the various proprietary agencies with which the Bureau must deal. The inspector should thus first acquaint himself with the appropriate requirements and standards. He can then adapt the techniques discussed for earth dam embankments to the requirement. The principal differences in control are:

(1) Moisture, if required, will for the most part be added to material on the fill. Usually, the water content should be as near as possible to the optimum required for maximum compaction.

(2) The density requirement is usually specified in percentages of maximum laboratory density, and varies according to the maximum dry unit weight of the soil.

The frequency of testing required to maintain control on these embankments is not standardized. For each fill location, the quantity of water and the amount of rolling required to obtain results conforming to the applicable standard must be determined rapidly so that the information may be used for construction of the fill. Since construction of a fill at a given location may be a relatively short operation, speed in setting up moisture and rolling requirements is essential. The rapid method of compaction control (designation E–25) can be used for compacted highway and railroad embankments. Where compaction is required by the specifications, one field density test should be made for each 2,000 cubic yards of compacted fill.

Conformance with certain specifications provisions can be obtained by visual inspection, such as preparation of the foundation, and deposition of material. The nature of the highway or railroad work precludes, to a large extent, the selection of borrow materials permitted in earth dam work. Thickness of layers, whether they are to be compacted by roller or equipment, is normally specified, and the control of these layers is the responsibility of the inspector. Where embankment is to be compacted by equipment travel, the distribution of travel over the embankment should be directed to obtain the maximum amount of compaction.

Figure 120 shows compaction of fill over a culvert on an access road, and figure 121 shows riprap placing operations near a bridge on a relocated railroad.

79. Miscellaneous Structures.—(a) *General.*—In addition to large earth structures such as dams and embankments, discussed previously in this chapter, there are many earthwork features related to smaller structures built by the Bureau. These structures include pumping plants, powerplants, bridges, substations, warehouses, residences, and various types of canal structures such as checks, drops, turnouts, siphons, pipelines, and overchutes. Although foundation loadings for most of these structures are normally light, in the order of ½ to 1 ton per square foot, moderately heavy loadings in the order of 1 to 3 tons per square foot on soils are not uncommon for the larger units. Serious attention must be given to the adequacy of the foundation soils. There are some instances where soils which are soft, highly compressible, or expansive in a natural wet state or in a future wetted state, must be examined carefully even for the light structures. Care must also be exercised when placing compacted embankment and compacted backfill in connection with these smaller structures.

(b) *Structure Foundations on Soil or Rock.*—Provisions for structure excavation are included in the section of the specifications entitled "Excavation for Structures." The contractor is required to use methods of foundation preparation, including moistening and tamping if necessary,

Figure 120.—Backfilling operations over culvert on Flaming Gorge Dam site access road. P591–421–462.

which will provide firm foundations. Excavation is made to the lines and grades shown on the drawings or as ordered by the contracting officer as subsurface conditions become known. Overexcavation is refilled with compacted material as specified. Loosened or disturbed natural material must be compacted or be removed and replaced with suitable compacted material as required by the contracting officer. When excavation for concrete structures is in rock, the specifications sometimes require that the rock be excavated so that there will be not less than 6 inches between rock points and the bottom of the concrete slab. This space is then filled with compacted earth or gravel as specified. For other structures, concrete is often placed directly on the rock foundation which has been previously cleaned and dampened. Large unit loadings may be used in this case.

Construction of the miscellaneous structures listed in subsection (a) usually involves a variety of earthwork problems because of the variety of structure types involved and the longitudinal extent of the work, wherein a variety of soils may be encountered in foundation and materials work. For the larger structures, such as powerplants and the sizable pumping plants where loadings are moderately high and settlements must

Figure 121.—Riprap placing operations at bridge abutment at Trenton Dam railroad relocation. 328–701–3402.

be small, explorations are normally conducted before designs are prepared to determine the competency of the foundation. Measures to improve the competency of the foundation, if needed, are therefore usually known ahead of time and are included in the plans and specifications. Frequently, it has not been practicable to perform adequate predesign explorations, and provisions for drilling exploratory holes or performing tests after the excavation has been completed are contained in the specifications. It is usually not practicable to perform detailed foundation investigations for numerous small structures built by the Bureau. Regardless of the amount of prior exploration, it is the responsibility of the inspector to make certain that the soil conditions which can be observed upon completion of excavation to grade are compatible with design assumptions. If there is some doubt concerning the adequacy of the foundation, this fact should be brought to the attention of the supervisor.

There are several methods which can be employed for checking the properties of the foundation soils. When sand soils are involved, the field penetration test with the split-tube sampler (designation E–21) provides a rapid method for determining the relative density of the soil. Figure 122 shows the relationship of blow-count to relative density. Figure 122(a) is the original relationship determined from research by testing under various overburden pressures. This relationship is con-

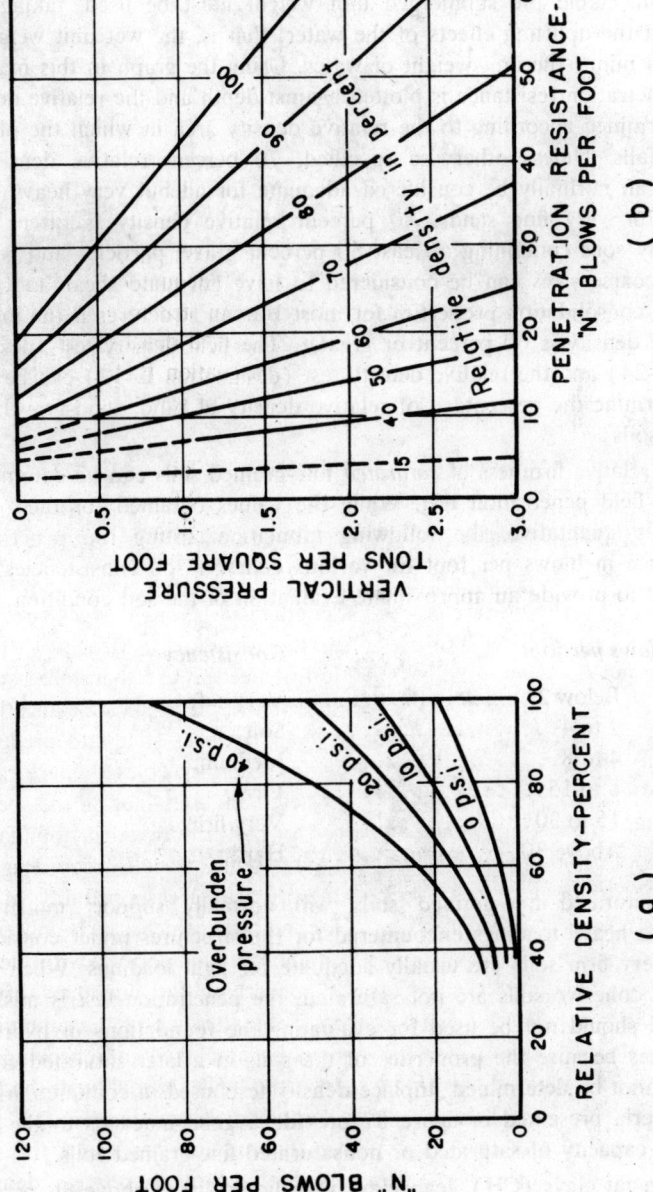

Figure 122.—Criterion for predicting relative density of sand from the penetration resistance test. 101-D-549.

verted to a more convenient form in figure 122(b) for use in analyzing data. The vertical pressure scale of the graph can be converted to a depth scale by estimating the wet unit weight of the soil. For tests below the water table the submerged unit weight must be used, taking into account the uplifting effects of the water; that is, the wet unit weight of the soil minus the unit weight of water. Using the graph in this manner, the penetration resistance is plotted against depth and the relative density is determined according to the relative density area in which the plotted point falls. Unless otherwise specified, 70 percent relative density of sands can normally be considered adequate for all but very heavy loadings. For very fine sands, 80 percent relative density is often used. Gravelly soils containing at least 50 percent gravel particles and graded up to coarse sizes can be considered to have adequate shear and satisfactory consolidation properties for most Bureau structures if the inplace relative density is 60 percent or greater. The field density test (designation E–24) and the relative density test (designation E–12) can be used to determine the percentage of relative density of sand, sand-gravel, and gravel soils.

The relative firmness of *saturated* fine-grained soils can be determined by the field penetration test. While the values obtained for these soils are only qualitative, the following tabulation, listing the penetration resistance in blows per foot for various cohesive soil consistencies, can be used to provide an approximate evaluation of the soil condition:

Blows per foot	*Consistency*
Below 2	Very soft.
2 to 4	Soft.
4 to 8	Medium.
8 to 15	Firm.
15 to 30	Very firm.
Above 30	Hard.

Hard saturated fine-grained soils will normally support moderately heavy to heavy loadings encountered for the structures under consideration. Very firm soils are usually adequate for light loadings. When fine-grained cohesive soils are not saturated, the penetration test is misleading and should not be used for evaluating the foundations of hydraulic structures because the properties of the soils in a later saturated condition cannot be determined. Inplace density tests used in conjunction with the criteria presented in figure 87 provide a good index as to the supporting capacity of saturated or nonsaturated fine-grained soils.

When fat clays (CH), lean clays with liquid limits above 40 percent, or plastic clay shales are encountered, they should be checked to determine whether they are sufficiently expansive to cause undesirable uplift of the structure. This can be accomplished by determining the gradation, plasticity index and shrinkage limit values of the soils. Any such soils falling into the "low expansion" grouping of table 3, section 64(f), are satisfactory from the expansion standpoint.

When qualitative tests indicate doubtful foundation competency, more detailed testing is necessary. This is usually accomplished by securing undisturbed soil samples and testing them in the laboratories in the Engineering and Research Center. If the size of the structure and complexity of the problem warrant, technical assistance can be requested from the Engineering and Research Center.

When the bottom grade of the structure is known to be below the water table, provisions are usually contained in the specifications for unwatering the foundation area. When water is encountered, the inspector must insure that the foundation soils are not disturbed. This may be accomplished by beginning the unwatering operation as soon as ground water is encountered or at an elevation 5 feet above bottom grade elevation. During the unwatering operation, the level of the water should be maintained continuously at least 1 foot below base grade until 48 hours after sufficient concrete has been placed to overcome uplift pressure. The unwatering may be performed by well points, trenches, or sumps, or by other methods which will prevent upward movement of water through the soils within the area of the structure foundation.

After excavation has proceeded to grade, care should be taken to insure that foundation soils are not disturbed by equipment working in the area. Excessive drying or wetting may damage certain types of soils. Shales may require a cover of asphalt, pneumatically applied mortar, earth, or a thin cover of concrete to prevent slaking.

When a structure is designed for placement on impervious soils, the inspector should assure that the material is adequately impervious and that no strata or lenses of permeable soils exist which might cause subsequent piping. Cutoff walls should be constructed to the depth shown on the drawings and into impermeable soils as required. The construction of the cutoff walls into the soil must be such that piping around the walls cannot occur. Specifications normally require either that concrete for transitions and cutoff walls should be placed against the sides of the excavation without the use of intervening forms, or that compacted backfill of silty or clayey soils should be used between the concrete and the natural ground after the forms are removed.

When the foundation area consists of more than one type of soil, care should be taken to prevent appreciable differential settlements. It is often required that pockets of soft or loose materials be removed and be replaced with compacted suitable soils to provide uniform bearing over the entire foundation area. When the foundation area is partially on soil and partially on rock, it is sometimes necessary to remove the soil to rock and refill with lean concrete. Other times it is desirable to over-excavate the rock and replace it with compacted soil. The type of treatment will depend upon the type of structure, structural loadings, and the strength and compressibility of the soil.

(c) *Pile and Caisson Foundations.*—When piles are required to provide the desired foundation bearing, the specifications will contain the requirements as to the type of piles, the cross-sectional area and in some cases the length of the pile. Frequently, it is specified that a minimum depth of penetration will be obtained. When the length of pile is not specified and unless otherwise specified, each pile is driven until the bearing power or resistance is the number of pounds shown on the drawings, but in no case shall any pile have a minimum penetration less than that shown on the drawings. Piles are of such length that after any damaged or battered portion of the head is cut off, the top of the pile will be at the specified elevation. It is required that piles be driven accurately to the required lines and depths with steam, diesel, air, or gravity hammers. Steam hammers shall have a hammer developing a total energy of not less than 6,000 foot-pounds per blow. Unless otherwise specified, the bearing power of the pile is determined by the following formulas:

$$\text{For single-acting steam or air hammers—} P = \frac{2WH(E)}{S + 0.1}$$

$$\text{For double-acting steam or air hammers—} P = \frac{2H(W + Ap)(E)}{(S + 0.3)*}$$

*The value $(S + 0.1)$ may also be specified.

$$\text{For diesel hammers—} P = \frac{2 \times 0.85 \times R}{S + 0.1}$$

$$\text{For gravity hammers—} P = \frac{2WH(E)}{S + 1.0}$$

where:

P = bearing power in pounds,
W = weight, in pounds, of striking parts of hammer,
H = fall in feet or stroke of ram in inches,
A = area of piston in square inches,
p = steam or air pressure in pounds per square inch at the hammer,
S = average penetration in inches per blow for the last 10 to 20 blows
for steam, air, or diesel hammers and 5 to 10 blows for gravity
hammers,

R = Manufacturer's certified energy rating of hammer in foot-pounds,

E = Efficiency. The efficiency factors which follow are used in the formulas for computing bearing resistance where pile-driving equipment manufacturer's substitute data are not available. Manufacturer's ratings of efficiencies or net delivered energies may be substituted or values based on actual tests.

Single-acting air or steam hammers: 0.80,

Double-acting air or steam hammers: 0.85,

Gravity hammers released by trigger: 1.00, and

Gravity hammers released by rope and winch: 0.85.

To reach the minimum penetration specified on the drawings, preboring or jetting while driving may be required. It is usually required that the number of jets and the volume of water with adequate pressure be available to freely erode material adjacent to the pile. It is suggested that the jetting system have capacity sufficient to deliver at least 100-pounds-per-square-inch pressure to two ¾-inch jet nozzles. When the minimum penetration called for is reached, the jets shall be removed and the pile finally set under normal driving by not less than 50 blows from a diesel hammer, 50 blows from a single-acting steam or air hammer, 200 blows from a double-acting hammer, or 50 blows for a gravity hammer and until the required bearing power is reached. When holes are prebored, the diameter of the prebored holes shall be slightly smaller than the diameter of the pile and the depth of the holes shall be such that the piles will reach the specified minimum penetration and will have the required bearing power.

The specifications contain detailed provisions for shoes, cutoff, boring, cutting, pile materials, preservative materials, preservative treatment, and miscellaneous accessories. Payment is normally made on the basis of total costs related to furnishing and driving piles at a unit price per linear foot. The number of linear feet on which payment is based is normally the number of linear feet of piles remaining in place in the finished structure, plus cutoff lengths up to 5 feet.

The inspector is required to assure that all piles are located properly and are driven to the specified lines, grades, and tolerances. He must assure that damage to the piles does not occur during driving and that required bearing values be obtained. Damaged piles or piles not driven to correct line and grade are ordered removed. When piles are left in place in freezing weather, the top elevation should be checked before placing of concrete. If heaving has occurred, the piles should be redriven to grade. Figures 26–4 and 26–5 (see designation E–26) are used by the inspector for recording the penetration of the pile for bearing computations.

The length of timber piles required to produce a designated bearing value can be estimated prior to driving piles by the use of data secured in the field penetration test (designation E–21) in the following empirical formula:

$$P = K_1 (d_1 N_1) + K_2 (d_2 N_2) \ldots + K_n (d_n N_n)$$

where:

P — bearing resistance in pounds,

$N_1, N_2 \ldots N_n$ = average number of blows per foot, in corresponding increments of depth $d_1, d_2, \ldots d_n$ in feet, obtained from the penetration tests, and

$K_1, K_2 \ldots K_n$ = a factor in pounds per linear foot of pile depending on the type of soil in the corresponding increment of depth.

The following values of K have been determined from field data:

Clay—100 pounds per foot
Sand—65 pounds per foot
Silt—65 pounds per foot
Loess:

    0–15 percent moisture—200 pounds per foot
    15–25 percent moisture—100 pounds per foot
    + 25 percent moisture—50 pounds per foot.

Pile-load tests are sometimes specified for determining more accurately the bearing capacity of piles. There may be instances where pile-load tests were not contemplated but are required to check or to understand load bearing values computed on the basis of driving tests. When individual pile-load tests are required and unless otherwise specified, they are performed in accordance with designation E–26 in the appendix, which is similar to ASTM Designation D 1143–69. Special pile-load tests on pile groups or lateral loading tests may be required in special cases. Special instructions are provided for these tests. When pile-load tests are performed, it is the responsibility of the inspector to see that the test piles are placed to specified line, grade, and length; that driving records are secured; that the test loadings are applied in proper manner, amount, sequence, and period of time; and that the data are secured and recorded as required. Figure 26–6 (see designation E–26) is used by the inspector for recording the pile test data.

Caisson foundations are sometimes used in lieu of pile foundations when it is necessary to transfer heavy foundation loadings to deep, competent bearing strata. Caissons may either be of constant diameter or be belled out at the bottom if higher bearing capacity is required. Belled-out piles have also been used in expansive clay foundations as anchors to prevent structure uplift. Caisson holes are usually drilled with large auger boring equipment, belled manually or by machine, and cleaned manually. Reinforcing steel and concrete are then placed as specified. It is the responsibility of the inspector to assure that all loose soil is removed from the bottom of the caisson hole, that the material at the bottom is competent for the design loadings, and that the bells, if required, are properly excavated in undisturbed material. Shoring or casing may be required to support the sides of the hole in caving soils or soils below water table. If water is encountered, the hole must be unwatered by approved methods until some specified time after the concrete is placed (usually 6 hours), unless tremie concrete is allowed in the specifications.

(d) *Transmission Tower Footings.*—There are numerous types of footings used in connection with the various steel tower transmission line structures. These can be classified in three basic types for soil foundations: (1) auger, (2) pad, and (3) pile. The auger type has an undercut and the pad type may or may not have an undercut. The specifications drawings contain information on the type of footing required or provide criteria for selecting the footing type. These criteria and the estimated number of each type of footing are usually developed from investigation data obtained for final design.

The following is an example of the criteria used to classify foundation soils for the selection of tower footing types, based on data obtained from the field penetration test (designation E–21):

Criteria for bearing classification of cohesive and noncohesive soils, except loess

Classification	Value of N (No. of blows)
Excellent	25 or over.
Good	15 to 25.
Fair	10 to 15.
Poor	Less than 10.

Criteria such as those presented above are developed on the basis of predesign foundation investigations for various tower footing types and soil conditions. They should not be used for hydraulic structures.

Provisions for footing excavation and unwatering footings are included in the sections of the specifications entitled "Excavation for Tower Footings," and "Unwatering Footing Excavations," respectively. The inspector must assure that surfaces for excavation are made to the full dimensions required and are finished to the prescribed lines and grades. Care must be taken to see that footing undercuts, bells, or bases are in undisturbed soils when specified. When soils are disturbed, it is usually required that the soil be removed and replaced with concrete. The inspector must make sure that mud or loose soil is removed from any of the footing excavations and that bells or undercuts are properly excavated before placing concrete. Unwatering is usually provided if the depth of water is greater than 6 inches. Water should not be allowed to flow upward into the excavation and thus disturb the foundation soils. It is normally required that water be kept below the foundation base during the placing of any concrete, except tremie concrete, and for not less than 6 hours after the concrete has been placed.

Requirements for the inspection of pile footings and the determination of pile bearing values from driving data or load tests are discussed in subsection (c) of this section.

(e) *Backfill.*—Backfill about structures may be specified simply as backfill in which no compaction is required, or as compacted backfill in which the compacted density is specified. The backfill may be further classified on the basis of the material required, as for example, "cohesionless free-draining materials of high permeability, commonly called pervious material," or "clayey and silty materials of low permeability, commonly called impervious material." Compacted backfill is normally specified if backfill under structures is required. Compacted backfill may be specified as compacted clayey or silty materials in which the materials are compacted by tamping rollers, power tampers, or other similar suitable equipment. Compacted backfill may also be specified as compacted cohesionless free-draining materials of high permeability in which the consolidation is accomplished by thorough wetting accompanied by operation of tractor, surface vibrator, internal vibrator, or other similar suitable equipment.

Backfill, or compacted backfill, is placed where shown on the drawings or where prescribed by the contracting officer. The type of material used, the amount, and the manner of depositing backfill are subject to the approval of the contracting officer. Insofar as possible, backfill materials are normally secured from required excavation. When suitable materials required for the specific structure are not available from required excavation, the materials are secured from approved borrow sources. The distribution of material must be uniform and such that compacted backfill is free from lenses, pockets, streaks, or other imperfections.

When silty and clayey soils are placed in compacted backfill, the required density of the soil is 95 percent of Proctor maximum (designation E–11) unless otherwise specified. It is normally required that these soils have a uniform optimum water content prior to compaction. When cohesionless, free-draining soils are placed in compacted backfill, the required density of the soil is that equivalent to 70 percent relative density (designation E–12) unless otherwise specified. Higher density requirements may be specified for fine sands when structure loadings are very high, when structure vibrations are severe, or when only very small settlements can be tolerated. The depth of layers after compaction, unless otherwise specified, is 6 inches when rollers are used, not more than 12 inches when tractors or surface vibrators are used, or a depth equal to the penetration length of the internal vibrators. It is normally required that these fine sands be thoroughly wetted during consolidation. Water jets are often used in conjunction with internal concrete vibrators. The general requirements are summarized in section 75(c).

To provide adequate protection for compacted backfill about structures when the backfill adjoins embankment, the specifications often provide that the contracting officer can direct the contractor to place a sufficient amount of embankment material over the compacted backfill within a specified length of time (often 72 hours) after compacting of the backfill has been completed. This is provided so that the backfill will not dry and shrink and cause cracking and pulling away from the structure.

Sand and gravel backfill material is often specified for placement under and about structures, under canal linings, for filters, at bridge approaches, at weep holes, or for other purposes. Various gradations of material may be specified for the different structural purposes. Laboratory tests should be made at frequent intervals to assure that the materials meet the specified gradation limits. Gradation limits for "reasonably well-graded sand and gravel" are given in section 74(d) and figure 112. Unless otherwise specified, the maximum particle size for backfill materials is 3 inches, and the amount of material finer than the No. 200 size for sand-gravel backfill is limited to 5 percent.

In general, the control techniques given in detail in sections 75(a) and (c) for compacted earth embankments and compacted linings apply to backfill. Control of compacted backfill to assure specification compliance consists of inspection of materials used in checking the amount and uniformity of soil moisture, maintaining the thickness of layers being placed, and finally, determining the percentage of Proctor maximum density or relative density obtained by the compacting or consolidating operations. It should also be determined that the soils are fully compacted or consolidated to the lines and grades specified.

(f) *Filters.*—Protective or reverse filters for canals and miscellaneous structures consist of one or more layers of free-draining sand-gravel material placed on less pervious subgrade soil to carry off seepage water and prevent erosion of the soil or damage to overlying structures from uplift pressure. The subgrade soil which is to be protected by the filter is commonly referred to as the base or the base material. A given base material can be protected by filter materials within a certain range of gradation, which gradation bears a definite relationship to that of the base material. Filters may consist of a single layer or several layers, each of a different gradation, in which case they are known as zoned filters. The filter or filter layers may be classed as uniform or graded. Uniform-grain-sized material for these filter purposes is defined as material having a narrow range of major particle sizes, such as a material having 90 percent of the particle sizes within three particle size divisions on the standard gradation chart for sieve sizes. This criterion provides a coefficient of uniformity, C_u, of about 3 to 4.

A graded filter has been defined as one in which the materials have a broad range of particle sizes. Graded materials may have concave, convex, S-shaped, or straight-line gradation curves and may be classed as poorly-graded to well-graded, depending on the shape of the gradation curve.

A material selected for the protective filter has to satisfy four main requirements as follows:

(1) The filter material should be more pervious than the base material in order that there will be no hydraulic pressure buildup to disrupt the filter and adjacent structures.

(2) The voids of the inplace filter material must be small enough to prevent base material particles from penetrating the filter and causing clogging and failure of the protective filter system.

(3) The layer of the protective filter must be sufficiently thick to provide a good distribution of all particle sizes throughout the filter and also to provide adequate insulation for the base material where frost action is involved.

(4) Filter material particles must be prevented from movement into the drainage pipes by providing sufficiently small slot openings or perforations, or additional coarser filter zone if necessary.

The gradation of materials for a filter or the various filter layers is normally given in the specifications. However, the following criteria are given as a guide for filters used in canal structures or other hydraulic structures involving high water heads where rapid dissipation of uplift pressure is desired. In the following ratios, F.M. represents the filter material and B.M. the base material.

For uniform-grain-size filters:

$$R_{50} = \frac{\text{50-percent size F.M.}}{\text{50-percent size B.M.}} = 5 \text{ to } 10$$

For graded filters of subrounded particles:

$$R_{50} = \frac{\text{50-percent size F.M.}}{\text{50-percent size B.M.}} = 12 \text{ to } 58, \text{ and}$$

$$R_{15} = \frac{\text{15-percent size F.M.}}{\text{15-percent size B.M.}} = 12 \text{ to } 40$$

For graded filters of angular particles:

$$R_{50} = \frac{\text{50-percent size F.M.}}{\text{50-percent size B.M.}} = 9 \text{ to } 30$$

$$R_{15} = \frac{\text{15-percent size F.M.}}{\text{15-percent size B.M.}} = 6 \text{ to } 18$$

When a naturally graded base material contains a significant amount of gravel but also has considerable fines, the filter material limits should sometimes be based on the minus No. 4 fraction of the base material to adequately protect the fines from washing into the filter. A general rule on this would be to use the minus No. 4 fraction of the base material as a basis for setting filter limits when the gravel content (plus No. 4) is more than 10 percent and the fines (minus No. 200) are more than 10 percent.

When the criteria cannot be satisfied by a one-layer filter with available filter material, a zoned filter can be used. In this case, the fine zone becomes the base material and the adjacent coarse zone becomes the filter material in the criteria. Other criteria are sometimes used for low-head or low-flow pressure relief systems. If it is desired to consider the use of materials falling outside the above criteria, laboratory tests should be made to confirm the suitability.

In addition to the above criteria, the following requirements for graded filters should be met:

(1) The filter material should pass the 3-inch screen for minimizing particle segregation and bridging during placement. Smaller maximum particle sizes may be specified. Also, filters must not have more than 5 percent minus No. 200 particles to prevent excessive movement of fines in the filter and into drainage pipes causing clogging.

(2) The gradation curves of the filter and the base material should approximately parallel in the range of finer sizes, because the stability and proper function of protective filters depend upon skewness of the gradation curve of the filter toward the fines, giving a support to the fines in the base material.

(3) The filter material adjacent to the drainage pipe should have sufficient coarse sizes to prevent movement of filter material

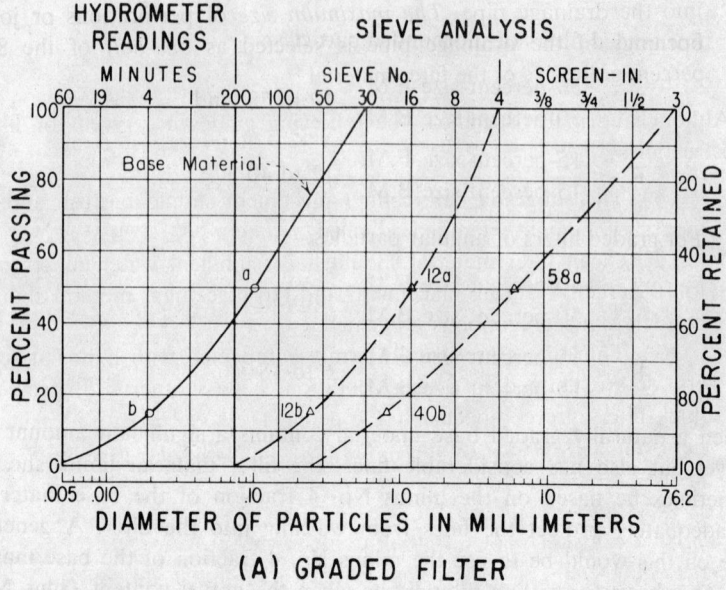

(A) GRADED FILTER

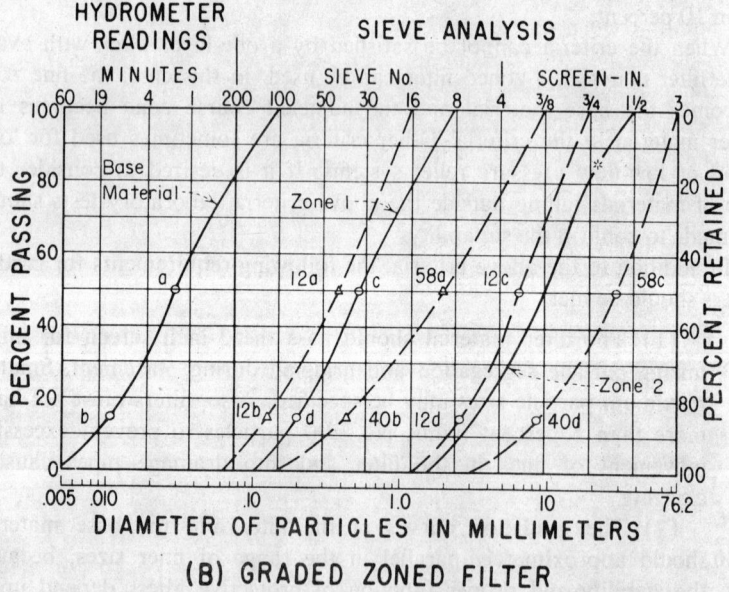

(B) GRADED ZONED FILTER

* Maximum size of opening in drain pipe $\left(\dfrac{85\% \text{ size}}{2}\right)$

Figure 123.—Examples of use of filter criteria for selecting materials.
101-D-222.

into the drainage pipe. *The maximum size* of perforations or joint openings of the drainage pipe is selected as *one-half* of the 85-percent grain size of the filter material.

Although normally specified, the following guides are given for filter construction:

(1) The subgrade before filter placement should be firm and, if necessary, be lightly tamped or rolled.

(2) Clean filter material should have sufficient water content (3 to 10 percent) during placement, and the placement method should be such that segregation is prevented.

(3) Thin filters are usually firmly compacted with light flat rollers, or are tamped to a firm condition. Unless otherwise specified, thick filters are compacted to 70 percent relative density in a manner similar to free-draining sand-gravel backfill to prevent settlement.

(4) The filter layers for coarse filter material (3-inch maximum size) are usually not less than 8 inches in thickness, and layers of finer filter material are often of 6-inch minimum thickness. However, for severe field conditions such as high head, variations in base material, or filter gradations which are near the extreme coarse limit, the minimum thickness of 8 inches may be specified. For zoned filters these minimum thicknesses may be specified and are maintained for each layer.

(5) Where drainage pipe is used in a filter system, the capacity of the pipe should be sufficient to collect the seepage water and to conduct it to a place of discharge.

(6) While the pipe is being laid, the openings are often protected from inflow of fines of the filter material by burlap or other suitable permeable material.

Figure 123 shows examples of the use of the filter criteria for selecting materials for a graded filter and a graded filter with two zones.

H. Stabilized Soils

80. General.—For stabilized soils used in fills, linings, and blankets, processing equipment and procedures are required to assure that the relatively small amount of additive is uniformly mixed. The quality and uniformity of the admixture and water are closely controlled. Thus, the uniformity of the soil and aggregate is the major factor controlling the

desired uniformity of the final product. The uniformity of the soil gradation, texture, and moisture should be controlled prior to introducing it
into the plant or adding the admixture. (Also see secs. 34 and 54h.)

Adjustments of the plant can be made during construction to allow
for the range in soil gradation permitted in the specifications or changes
in placement requirements. These adjustments are based on calibrations
obtained before construction begins.

Stabilizing soils in situ by grouting or injection methods to solve
particular foundation problems requires specialized knowledge of the
method and materials being used. The control procedures are developed
for the particular soil conditions encountered.

81. Compacted Soil-cement.—(a) *Design Considerations.*—Compacted
soil-cement is a relatively thin layer ranging from 6 inches to 2 feet in
thickness to protect earth slopes from erosion by water currents and wave
action. Figures 124 and 125 show the soil-cement facings constructed on
the upstream slopes of two Bureau structures. Compacted soil-cement
has also been used extensively for construction of road bases.

To meet design requirements for slope protection a layer must be:

    1. Formed into a homogeneous, dense, permanently cemented
mass,

    2. Free from any potential paths of percolation through it or
along contacts with earth slopes, abutments, or concrete structures,

    3. Durable and resistant to "wetting and drying" and "freezing
and thawing" action of water, and

    4. Stable with respect to the structure and of sufficient thickness
(weight) to resist displacement and uplift.

(b) *Construction Provisions.*—The section in the specifications entitled "Soil-Cement Slope Protection" specifies the type and amount of
cement, the quality and amount of water, and the borrow area for the
soil or aggregate. The permissible range of the soil and aggregate gradation is also specified. If the investigations show the deposit is variable,
selective excavation and processing may be required to produce uniformity. Oversize particles and other objectionable materials must be
removed.

A stationary mixing plant is usually required. Either batch type or
continuous feed pugmill which has a rated capacity of at least 100
cubic yards per hour is acceptable. Control over mixing time (usually
30 seconds minimum), positive interlocking of cement and soil flow, and
controls for accurately proportioning the soil, cement, and water from
suitable storage should be incorporated into the plant design. A general
view of the mixing plant used at Lubbock Regulating Reservoir is shown
in figure 126.

Figure 124.—General view of Lubbock Regulating Reservoir showing soil-cement facing 1 year after construction. P719-D-66536.

Figure 125.—General view of soil-cement facing on Merritt Dam 6 years after construction. P637-D-66532.

Trucks for transporting the soil-cement mixture should have tight, clean, smooth beds and protective covers. The spreader must produce a smooth uniform loose layer of the required width and thickness. The layers are usually placed horizontally; however, a slope as steep as 8:1 is sometimes permitted to increase the working width. The maximum time for hauling and spreading is usually specified as 30 minutes. The tractor and spreader box used for placing soil-cement on Merritt Dam is shown in figure 127.

Compaction is usually required to be completed within 60 minutes after spreading with no more than 30 minutes between operations. It is usually accomplished by several passes—at least six for a 6-inch compacted lift—of a sheepsfoot tamping roller, followed by several passes —at least four—of a pneumatic-tired roller, figure 128. The rollers should have provisions for ballast loading so that the weights can be adjusted to provide the best compaction. Roller weights have been about 2,000 pounds per foot for sheepsfoot rollers and 4,000 to 5,000 pounds per foot for pneumatic-tired rollers. The rollers may be towed or self-propelled. After compaction the compacted layer is cured by keeping the exposed surfaces continually moist using a fog spray until the overlying or adjacent layer is placed, or for a minimum of 7 days. A blanket of moist earth may be used for permanently exposed surfaces. The surface of the completed layer may require brushing just before initial set and just prior to placing an overlying layer (see figure 129).

Figure 126.—General view of soil stockpile and pugmill for mixing soil-cement— Lubbock Regulating Reservoir. P719–D–58953.

Figure 127.—Spreading soil-cement on placement area—Merritt Dam. P719–D–58952.

Figure 128.—General view of placing and rolling operations—Lubbock Regulating Reservoir. P662–525–5763.

Figure 129.—Power brooming to clean surface prior to placement of next lift— Lubbock Regulating Reservoir. P719–D–58948.

Figure 130.—Cement vane feeder and soil feed belt to pugmill—Cheney Dam. P719–D–58954.

(c) *Control Techniques.*—Compacted soil-cement is similar to compacted earthwork in that careful observations and additional control tests are required during the early stages to check planned control against results obtained under field conditions. These observations and tests are used to establish the placing conditions and procedures to be used for the remaining construction. Control should begin during excavation and stockpiling of the soil to assure that the material is within the gradation requirements and has a uniform moisture content. This is done by controlling excavation, mixing the stockpile by spreading and cross-dozing, sloping the stockpile surfaces to provide runoff without water catchments, and by sampling and testing during stockpiling to check gradation.

After the mixing plant is set up, the cement, soil, and water feeds must be calibrated individually to establish curves of setting versus quantity produced. A cement vane feeder and soil feed belt to a pugmill are shown in figure 130. This is done by timing and weighing truckloads of wet soil and smaller quantities of cement and water to check feed and meter settings. To facilitate computation of the dry soil production, the moisture content of the wet soil should be determined at the plant by a quick method, such as the carbide bomb moisture meter or hotplate method. The calibration of the cement feed is perhaps the most critical since variations in the amount of cement can affect the properties of the soil-cement the most. The cement feed is calibrated over the range of cement supply necessary by varying the speed of rotation. Since cement is a fairly uniform product and is fed under uniform conditions in a well designed plant, the cement feed is usually quite consistent. This allows for adjustments if soil feed rates change during the progress of the job. The performance and production of the plant under full load is checked by timing and weighing truckloads of the soil-cement mixture, checking mixing time, and inspecting the mixture for uniformity in texture, moisture, and distribution of cement. The soil and soil-cement mixture is sampled to determine the moisture by both quick and ovendry methods, cement content, optimum moisture, and occurrence of "clay balls" (rounded balls of fines and sand which do not break down during ordinary processing).

During construction the plant inspector should check the operation periodically. Inspection frequency will depend upon the performance of the plant and uniformity of the soil. The soil feed rate should be checked each morning by timing and weighing a truckload of wet soil. By using a quick method to determine the moisture content, the feed rate of dry soil can be computed and proper cement and water feed rates can be set. At hourly intervals, or as required, the moisture of the soil and soil-cement mixture should be determined by the quick method to provide a basis for making adjustments in the water feed if necessary. The inspector also makes a check calibration each time a record control test is performed by timing the production and weighing a truckload of soil-cement. Based on the water content and cement feed rate, the soil feed rate can be determined. The soil and soil-cement are sampled for moisture tests at the plant and record tests in the laboratory.

The placement inspector should observe the placing and compaction procedures. He should verify that the loose lift thickness, texture, and surface are uniform and that the lift is to the desired dimensions. Compaction is begun with the sheepsfoot tamping roller beginning at the outer edge. The number of passes and adequate overlap should be checked. Each pass over the entire lift should be completed before the next pass is started as the tamping feet have a tendency to follow previous tracks if successive passes are made without overlap. Because of the sandy material and moisture, the sheepsfoot roller usually does not "walk out" completely, but it should begin to "walk out" on the last passes. The compaction of each layer is completed with a pneumatic-tired roller. This is in contrast to normal earthwork where successive layers are compacted together with a sheepsfoot roller. With the time limits used on soil-cement construction, it is usually not feasible to place successive layers quickly enough to compact them together. Therefore, the entire lift must be compacted separately. The pneumatic-tired roller also begins compaction at the outer edge of the lift to minimize the amount of lateral spreading of the soil-cement. Smooth, even surfaces after compaction are desired. Rutting of the formation or crowned surfaces usually indicates a high placement water content. The thickness and width of the completed compacted lift should be checked. The placement inspector should check to be sure that the surface of the compacted layer is kept continuously moist until the overlying layer is placed, or for a minimum of 7 days. He should also observe the results of brooming (brushing) just before initial set to ascertain that the smooth compaction plane left by the roller is removed without removing excessive compacted soil-cement. The surface just prior to placing an overlying layer should be thoroughly cleaned (see figure 129). He should also check that the thickness of the equipment access ramps are adequate and are constructed to protect the edge of the top layer on which the soil-cement is being placed.

(d) *Control Testing.*—During construction, a control test is required for every 500 cubic yards placed, or a minimum of two tests per shift. The results are recorded on form 7–1737, figure 131.

The control test begins by timing the production of a truckload of soil-cement and obtaining a 10-pound sample of soil from the soil feed. The timed truckload is weighed and a 50-pound sample of soil-cement is obtained for laboratory testing. When the load is spread, the approximate

5-63
BUREAU OF RECLAMATION

SUMMARY OF FIELD AND LABORATORY TESTS OF COMPACTED SOIL-CEMENT

TOTAL CU YDS SOIL-CEMENT PLACED THIS PERIOD

FACING 24,761 (Pay Estimate 10-26 to 11-26)

OTHER ---

PERIOD OF REPORT FROM 10-31 TO 11-25

PROJECT

FEATURE

SPECIFICATIONS NO

DATE OF REPORT MONTH YEAR

Test No. (1)	Location of test: station, offset, coordinates, elev. etc. (2)	Yardage represented by field density test (3)	Source of material: borrow area, required excavation, etc. (4)	Soil classification symbol (Unified classification system) (5)	Minus No. 200 % (6)	Specific gravity (7)	Soil [illegible]	Method of compaction: SH-sheeps foot, RT-rubber tired; No. of passes, roller nos. (8)	Time: plant mixing completed (9)	Time: lab compaction completed (10)	Time: field compaction completed (11)	Time: field density test started (12)	Fill dry density: total material pcf (13)	Fill dry density: minus No. 4 pcf (14)	Laboratory standard maximum density pcf (15)	Ratio fill dry den. to lab max dry den. % (16)	Fill water content: total material % (17)	Fill water content: minus No. 4 % (18)	Laboratory optimum % (19)	Variation from optimum % (20)	Cement content, calibration from plant: by weight from batch plant (21)	By volume from batch plant (22)	By weight from chem. tests (23)	Compressive strength: 3-day (24)	7-day (25)	28-day (26)	90-day (27)
																								-	-	-	-
11-16-A-1	30+79 40'R℄	320	3277.2	SM				6SH-4RT	10:30	40	30	35	116.0	116.0	118.8	97.6	12.2	12.3	12.3	opt.	12.0		11.8	690	958	1218	-
11-16-A-2	13+36 40'R℄	310	3276.7	"				6SH-4RT	12:50	37	30	35	119.6	119.6	116.6	102.6	11.3	12.1	12.3	0.2D	12.0			-	-	-	-
11-16-A-3	13+36 43'R℄	310	3276.7	"				6SH-4RT	12:50	37	30	36	116.9	116.9	116.6	100.3	11.3	11.8	12.3	0.5D	12.0		12.1	619	828	974	-
11-17-A-1	23+50 38'R℄	480	3279.0	"	30.6			6SH-4RT	10:50	55	45	50	117.7	117.7	115.0	102.3	13.7	13.7	12.7	1.0W	12.0			631	797	-	1477
11-17-A-2	46+00 40'R℄	490	3277.6	"				6SH-4RT	13:00	30	25	30	115.8	115.8	115.8	100.0	13.1	13.0	13.0	opt.	12.0		14.5	774	1096	1461	-
11-18-A-1	7+70 38'R℄	500	3278.7	"				6SH-4RT	15:00	45	30	40	115.6	115.6	116.5	99.2	13.3	13.5	13.0	0.5W	12.0			828	1015	1445	-
11-19-A-1	43+00 37'R℄	360	3278.6	"				6SH-4RT	10:45				119.5	119.5	116.9	102.2	12.7	12.6	12.5	0.1W	11.8		13.3	893	1136	1494	-
11-19-A-2	39+50 52'R℄	260	3271.6	"				6SH-4RT					122.4	122.4	116.9	104.7	12.0	10.4	11.8	1.4D	11.8			-	-	-	-
11-19-A-3	12+10 35'R℄	260	3279.2	"				6SH-4RT	14:30	15	25	25	115.7	115.7	117.7	98.3	10.6	9.9	12.2	2.3D	11.8		12.4	630	745	1055	-
11-20-A-1	34+70 43'R℄	520	3275.6	"				6SH-4RT	10:50	65	45	45	117.4	117.4	116.3	100.9	13.5	12.8	13.4	0.6D	12.0		12.3	651	893	-	1607
11-20-A-2	34+00 39'R℄	520	3277.3	"				6SH-4RT	14:35	36	35	40	117.2	117.2	118.1	99.2	12.3	12.6	12.5	0.1W	12.0		12.7	732	776	1185	-
11-21-A-1	36+30 39'R℄	360	3277.8	"				6SH-4RT					116.3	116.3	116.7	99.7	13.2	13.0	13.0	opt.	12.0			424	610	860	-
11-21-A-2	26+00 34'R℄	360	3279.7	"				6SH-4RT	11:53	97	57	62	112.6	112.6	111.4	101.1	15.6	15.4	14.6	0.8W	12.0			623	739	1055	-
11-21-A-3	24+25 30'R℄	360	3281.3	"				6SH-4RT	14:00	35	30	32	117.9	117.9	114.8	102.7	13.2	12.9	13.8	0.9D	12.0			594	724	-	-
11-22-A-1	3+50 45'R℄	400	3276.1	"				6SH-4RT	08:40	23	23	25	117.6	117.6	116.2	101.2	13.6	13.2	12.8	0.4W	12.0			498	644	812	-
11-22-A-2	35+40 28'R℄	400	3281.8	"				6SH-4RT	11:40	50	35	38	114.8	114.8	114.9	90.9	12.1	11.9	13.0	1.1D	12.0		13.1	643	763	-	-
11-22-A-3	38+75 28'R℄	400	3282.2	"				6SH-4RT	14:45	55	30	33	114.2	114.2	115.5	98.9	13.2	12.3	12.9	0.6D	12.0		12.6	713	797	1153	-
11-23-A-1	0+50 33'R℄	580	3279.6	"	29.1	2.66		6SH-4RT	12:10	50	30	35	117.2	117.2	116.4	100.7	11.9	11.9	12.6	0.7D	12.0			705	800	1177	-
11-23-A-2	4+00 31'R℄	580	3280.6	"				6SH-4RT	14:30	40	30	35	118.4	118.4	115.7	102.3	11.7	11.8	12.4	0.6D	12.0			562	623	795	-
11-25-A-1	25+00 25'R℄	650	3282.3	"				6SH-4RT	09:00	50	40	60	114.5	114.5	116.2	98.5	12.8	13.4	13.1	0.3W	12.0			636	808	974	-
11-25-A-2	27+00 24'R℄	650	3282.8	"				6SH-4RT	14:15	70	55	60	115.0	115.0	116.2	99.0	13.4	13.0	13.5	0.5D	12.0						
				Totals for 53 tests									6209.6	6209.6	6197.3	5308.6	653.5	648.3	663.8	15.5			346.9	28,830			9885
				Average for 53 tests									117.2	117.2	116.9	100.2	12.3	12.2	12.5	0.3D			13.3	655	830	1146	1412

* Re-worked
** Check after re-working

NOTE 1: Minimum number of field density tests required: 1 test for every 500 cu yd or 2 tests per shift, whichever is greater.

Note 3: Offset refers to distance from centerline of compacted embankment.

Note 4: All soil from Caprock Sand and Gravel Company pit.

(17) Water Content as spread
(18) Water Content after compaction
(20) Variation from optimum after compaction

Figure 131.—Summary of field and laboratory tests of compacted soil-cement.—Form 7-1737. 101-D-548.

center is marked by the placement inspector. The moisture contents of the soil and soil-cement are determined by a quick method at the time of sampling, and the rest of the samples are used for laboratory tests. For the soil, the percentage of fines and the moisture content by the standard ovendrying method is determined. If cement contents are being determined by chemical titration and the soil contains a significant amount of calcium, a chemical analysis of the soil is performed. On every fourth test a complete gradation and specific gravity should be determined. For the soil-cement, the percentage of "clay balls" is calculated and the percentage of cement is determined by the chemical titration method. A three-point compaction test is performed using the rapid compaction control method. The compaction test should be performed at the same time that the layer from the sampled truck load is being compacted. This is necessary to allow for the time effects on the compaction properties of the soil-cement due to the hydration of cement. The rest of the soil-cement sample is used to prepare four compression test specimens, to be tested at 3-, 7-, 28-, and 90-day ages. The specimens are placed at the density determined by the field density test of the compacted soil-cement layer. These specimens should be formed as soon as possible after the field density has been determined.

A field density test is performed at the point marked by the placement inspector when the timed load was spread. This test should be performed as soon as compaction is completed.

Control testing includes obtaining record cores from the compacted soil-cement at least 90 days after completion. One hole should be drilled for each 5,000 cubic yards placed. The location of the core holes should be spaced to be representative of the area covered with some cores near abutments or structures. Care in drilling should be exercised to try to obtain a continuous core and comments on the bond strength should be included with the information accompanying the cores. The core holes should be carefully backfilled with cement grout and a reinforcing bar placed flush with the surface and located for future reference. The compressive strength of a section of core from each of the holes should be determined in the field laboratory and this strength compared to those of the record construction control cylinders representative of the immediate vicinity. The rest of the cores should be sent to the laboratory in the Engineering and Research Center, Denver, for durability and compression tests. A summary of the compressive strength and age of the record core and compressive strength of the specimens of material in the immediate vicinity should accompany the cores.

82. Plastic Soil-cement.—Plastic soil-cement has been used for pipe bedding in stable, firm soil foundations. The bottom of the trench is excavated to a circular section about 6 inches greater in diameter than the pipe. The pipe is then supported on sand pads about 2 inches above the bottom of the trench and the annular space around the bottom of the pipe is filled with plastic soil-cement. A typical section of pipe bedded with plastic soil-cement with a comparison to compacted backfill is shown on figure 132. (Also see sec. 77(e) and fig. 116.)

To meet design requirements, the material should have sufficient compressive strength to transfer the load directly from the pipe to the foundation. The strength required depends on the size of pipe and depth below the ground surface, but usually a minimum 7-day strength of 50 pounds per square inch is desired even if the calculated loads are less. A high strength bedding material is not desirable since it might bond to the pipe and alter its structural shape; therefore, a 7-day strength less than 300 pounds per square inch is normally specified. The bedding as placed should be in close contact and provide uniform bearing over the contact surfaces. Volumetric shrinkage from placement to cured condition should be less than 5 percent for the thin layer. This is determined by laboratory testing and most sandy materials are within this limit. After setting, the bedding material should not break down due to cycles of wetting and drying.

The materials and specific placement requirements are given in the construction specifications. The gradation of the soil including the allowable percentage of fines, percentage passing the No. 4 sieve size, maximum size of particle, and limit of organic material is usually specified. The quantity of cement, which depends upon the type of soil and strength desired, usually ranges from two to three sacks per cubic yard or about 10 percent by dry weight of soil. The water-cement ratio to form a fluid consistency usually ranges about 3:1 to 3.5:1 for most sandy soils. The water-cement ratio must be low enough to maintain the soil in suspension to form a mix that will flow or can be pumped.

Weight batching the cement and soil on a dry weight basis is preferred to the volume method. The soil moisture may be checked frequently using a carbide bomb moisture meter or the hotplate method. The fluid mixture is usually placed using a flexible hose, either by pumping or by gravity flow. During placement the mixture should be observed for uniformity, and the water content should not be excessive. The fluidity should be checked by the funnel test. A funnel 6 inches in diameter and 4½ inches high with an $11/16$-inch opening is filled with the mixed material and the time required to empty the funnel is determined. The time to empty the funnel, referred to as flow time, should range from 5 to 10 seconds for sandy soils and may be somewhat longer for clayey soils. The placement method should not erode the foundation soil. The flow should be directed between the pipe and inplace soil to place the material as near to its final position as possible. To avoid air pockets, the plastic soil-cement should

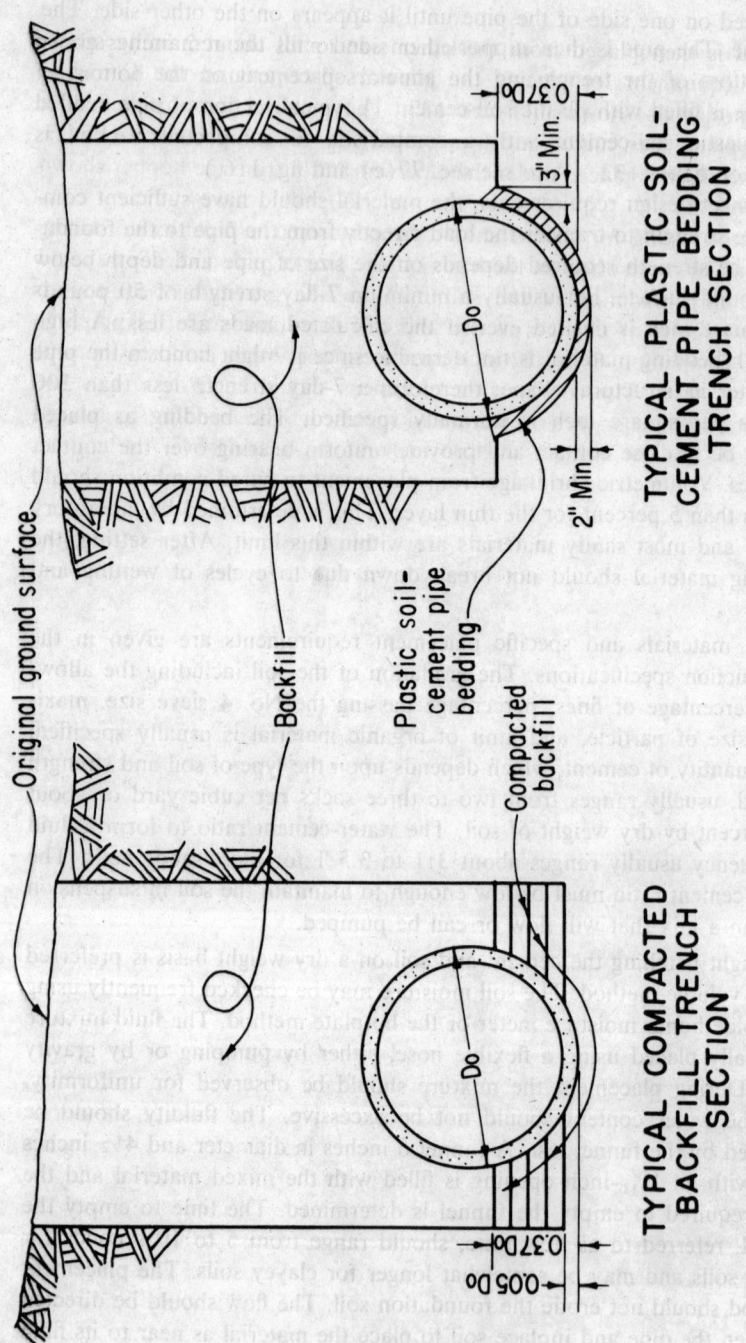

Figure 132.—A comparison of bedding methods for pipe bedded to 0.37 outside diameter—Plastic soil-cement compared to compacted backfill. 101-D-547.

be placed on one side of the pipe until it appears on the other side. The material is then placed from the other side to fill the remaining space. Special equipment for placing the plastic soil-cement on the Canadian River project is shown in figure 133. The material was batched at a central plant, then mixed and transported with a mobile concrete mixer. The material was placed under the pipe with hoses from the hopper shown in figure 133.

Control testing for record should be performed at least twice during each shift or more often on the basis of pipe length placed. Testing should include (1) gradation of the soil, (2) specific gravity as required to verify similarity or changes in the soil, (3) flow time, and (4) preparation of companion 6- by 12-inch specimens for 7-day compression and "wet-dry" tests. The compression specimens should be allowed to settle without vibration or other agitation to represent placement in the loosest condition. These specimens will normally settle before initial set and the amount of settlement should be determined before the specimens are removed from the molds.

The initial wet weight, volume of the specimen after settling, wet weight

Figure 133.—Special machine to place plastic soil-cement under pipe—Canadian River Project, Texas, P662-525-4321.

after curing, and the dry weight after testing provide data to determine the placement water content, initial dry density, and final dry density. Based on the amount of cement added or a determination of cement content, the initial water-cement ratio can be determined. These construction control test results should be reported in the monthly construction progress report.

83. Asphaltic Concrete.—(a) *General.*—Asphaltic concrete construction is used in revetments, canal and reservoir linings, thick facings for dams, resurfacing for old portland cement concrete lining, and grouting hydraulic structures. Figure 134 shows a slip-form paver placing asphaltic concrete canal lining; placing asphaltic concrete on the upstream face of a rockfill dam is shown in figure 135. Asphaltic-concrete lining can be placed successfully in small canals and laterals by slip-form pavers, provided satisfactory compaction is obtained. This lining is generally 2 inches thick. In large canals asphaltic concrete lining is generally placed to a thickness of 4 inches, and for resurfacing portland cement concrete lining asphaltic concrete is placed to a 2-inch thickness, in each case by using highway-type pavers and compactors. For the upstream facing of earth dams, as a substitute for rock riprap and the facing of rockfill dams or revetments where the asphaltic concrete acts as the impervious layer, the final normal thickness is usually 12 inches, placed in 3- to 4-inch layers, using modified highway paving equipment.

(b) *Design Considerations.*—In general, hydrostatic uplift does not cause serious harm to asphaltic-concrete linings because small bulges tend to be formed in localized areas, and if the pressure continues these bulges

Figure 134.—Slip-form paver placing asphaltic concrete canal lining.
CH-102-46.

Figure 135.—Placing asphaltic concrete on upstream face of a rockfill dam. CH–520–61.

erupt, relieving the stress and allowing the lining to settle back into place. Asphaltic concrete should be used with care for lining canals in heavy clay-type soils which are subject to saturation from runoff water from ground at a higher level. A saturated soil may exert ample pressures to cause large bulges in the canal lining. If practicable, adequate drainage protection should be provided when this condition exists. Cohesionless uniform sand subgrade may cause distress in thin linings on slopes of 1½ to 1 due to backpressure. Flatter slopes or thick linings using mass to balance pressure may be considered for use in these localities.

The decision to use an asphaltic-concrete lining involves many factors which can be evaluated adequately only through laboratory testing. Each installation must be given individual consideration relative to selection and usage of materials. Therefore, it is important that prior to construction of major lining installations, the proposed materials should be laboratory tested in order to provide data necessary for design and construction control. This information is written into the construction specifications, which not only give the minimum requirements of the asphaltic materials, but set the limits for the aggregates and establish control of the compacted density.

Hot-mix asphaltic concrete will provide a satisfactory, durable, and economical surface for dams and revetment facings, provided that the mix for each particular use is correctly designed and limitations of water velocity and turbulence, subgrade movements, and weed growth action are considered. Some design criteria are still in the process of evolution, influenced by observation of the behavior of existing work.

Additional information is given in the following Asphalt Institute publications which can be obtained from the Engineering and Research Center, or any one of the offices of The Asphalt Institute:

(1) *Mix Design Methods for Asphalt Concrete and Other Hot-Mix Types* (MS–2)

(2) *The Asphalt Handbook* (MS–4)

(3) *Asphalt in Hydraulic Structures* (MS–12).

(c) *Construction Provisions*·—To meet design requirements for water tightness and durability, asphaltic-concrete mixes for placing by slip-form are designed for an optimum degree of plasticity in the completed structure. They should have a high degree of workability (while hot) to permit placement and good compaction with slip-form equipment. Aggregate gradings should be selected to obtain maximum benefits from locally available materials. Such mixes are higher in asphalt content (7 to 10 percent) than mixes used in highway construction, and are a compromise between high plasticity to minimize cracking from subgrade movements caused by settlement or frost heave, and hardness to obtain good resistance to erosion from high velocity or turbulent water. Mix design is generally accomplished by fabricating test cylinders in accordance with ASTM Designation D–1074 and by using an immersion-compression test similar to ASTM Designation D–1075. Several additional tests have been devised to give values for a particular physical characteristic of the mix which will aid in the selection of the best percentage of each of the available aggregate materials. The following table is an example of a gradation used in canal lining construction in comparison with the current specifications:

Sieve size	Specifications	Riverton
	Percent passing	Percent passing
¾	100	100
½	85–100	87.9
No. 4	55–80	62.3
No. 10	35–60	53.4
No. 40	18–30	28.4
No. 200	5–12	9.9

A binder having a penetration of 50–60 is preferred to a relatively softer asphalt cement having penetrations of 100 to 200. This somewhat harder asphalt produces a mix that is more resistant to weed growth, is more stable on slopes and, owing to the thicker films of asphalt on the aggregate, is more durable.

In small canal and lateral lining construction, asphaltic concrete is usually placed and compacted by a slip-form paver with an adjustable, heated, and weighted ironing screed attachment. This provides compaction while the material is hot. However, provision must be made for accurately and positively adjusting the angle of attack and the total pressure of the ironing screed to obtain compaction to effect the minimum specified density of 95 percent (97 percent is desired) of laboratory compacted density. The standard laboratory density is described in the Bureau construction specifications and is obtained using ASTM Designation D–1074. Placement of thicknesses from 3 to 6 inches, as in large canals, dam facings, or revetments, usually requires use of such additional compaction equipment as vibrators of special design attached to the slip-form or rollers. In general, heavy rollers are unsuitable for compaction of the soft and highly workable mixes used for canal linings. Compaction should be performed while the mix is hot and highly workable; this gives both high density and good "structure" in the lining. Although compaction for high density and impermeability are desirable and necessary in some types of construction, in others the materials may be placed loose. The mixes are sometimes loosely spread by slip-form or may even be dumped in running water. This technique has been used extensively in river, harbor, and storm channel work for erosion control, but has not yet been used by the Bureau.

(d) *Control Testing.*—Control testing for the record should be performed at least twice during each shift or more often on the basis of the tonnage being placed. Testing should include: (1) density using the procedure outlined in Unconfined Compressive Strength Test, ASTM Designation D–1074 or Marshall Test ASTM D–1559; (2) asphalt content— Quantitative Extraction of Bitumen from Bituminous Paving Mixtures, ASTM Designation D–2172; and (3) gradation, Bureau of Reclamation Standard. The minimum compacted density of asphaltic concrete used in hydraulic construction is 95 percent of laboratory standard density with a specified 97 percent not uncommon. Complete control procedures for the mixing plant and placing equipment are available in the following Asphalt Institute publications:

(1) *Specifications and Construction Methods for Asphalt Concrete and Other Plant-Mix Types* (SS–1)

(2) *Asphalt Plant Manual* (MS–3)

(3) *Asphalt Paving Manual* (MS–8).

With respect to highway relocations, access roads, and parking areas, the Bureau generally follows the specifications used by the State in which the work is located. This is beneficial since ordinarily the State assumes maintenance of these roads after construction is completed.

84. Modified Soil.—Modified soils are considered when their use proves to be as effective and more practical, or less costly, than conventional methods of handling particular soil problems. Chemicals have been used to temporarily or permanently improve stability, increase resistance to erosion, or reduce permeability. Sands have been stabilized to permit tunneling or excavation on steep slopes. Unstable fills and foundations have been stabilized to permit underpinning operations. Erosion resistance of the sides and bottom of temporary diversion channels has been improved using cement-modified soil.

To improve foundations or soils in place at depth, admixtures are injected into soils using grouting techniques. For near surface treatment, the admixture is sometimes placed on the surface or in shallow trenches and allowed to permeate the soil by gravity flow. For construction of compacted modified soils, the admixture is mixed with the soil prior to compaction. (See secs. 34 and 54(h).)

The type and quantity of admixture and specific placement requirements are given in the construction specifications. The amount of admixture used to modify soils for construction is usually small. For example, the addition of 3 percent of lime was sufficient to modify the expansive properties of a clay soil to within tolerable limits.

For construction, an initial calibration and periodic check of the equipment and procedure for measuring and distributing the admixture is required. Observations to assure that the admixture is uniformly distributed and special tests may be required to verify the quantity of admixture in the compacted material. Conventional control tests for placement density and moisture are also required. Because of the variety of additives and soil conditions contingent upon their use, inspection procedures and control tests have not been standardized. Instructions will be issued by the Engineering and Research Center when required.

Appendix

PROCEDURES FOR SAMPLING, CLASSIFICATION, AND TESTING OF SOILS AND INSTALLATION OF INSTRUMENTS

DISTURBED SAMPLING OF SOILS

Designation E–1

1. Scope.—This designation describes methods for obtaining representative disturbed samples from natural deposits or other sources of earth materials proposed for use in foundations and embankments, canal linings, backfill, and other construction.

2. Apparatus.—The apparatus shall consist of the following (see designation E–4):

> Augers, 4 inches or larger, item 1.
> Bags, sample, item 2.
> Bucket, item 12.
> Canvas, item 16.
> Containers, sample, items 19, 20, 21 or 22.
> Mattock, item 47.
> Picks, item 63.
> Rope, item 71.
> Shovel, item 78.
> Tape measures, items 101 and 102.
> Notebook, and identification tags, preferably of the aluminum type (item 119A), or means for permanently marking containers and sacks.

3. Sample Sizes.—The quantity of soil submitted for laboratory testing is dependent on the tests to be performed. When the amount of soil needed is not otherwise specified, the sizes of disturbed samples should conform to those given in tables 1–1 and 1–2.

Table 1–1.—Samples of construction materials

Tests	Sample size	Remarks
Engineering properties	Sufficient minus 3-inch material to yield 75 pounds passing No. 4 sieve size.	Include information on percentages by volume larger than 3- and 5-inch size including maximum size. Cohesive soils—natural water content included; ship in moistureproof bags. Noncohesive soils—air-dried; ship in closely woven cloth bags.
Permeability—gravelly soils	300 pounds, minus 3-inch material	See above.
Relative density—gravelly soils	150 pounds, minus 3-inch material	See above.
Water content and identification	1 pound	Ship in sample containers, item 21.
Record samples	15 pounds, minus No. 4 sieve size	Ovendried and sealed in 6- by 12-inch cylindrical cans. Samples to be taken during construction.

4. Procedures.—Various accessible and nonaccessible methods of exploration are described in tables 1–3 and 1–4 respectively. The size and number of samples depend on the purpose of the investigation and stratification of the deposit. Each stratum over 12 inches in thickness should be sampled separately. All strata encountered to the depth of exploration should be sampled. If samples are omitted, notes on the log should give reasons why they were not taken.

(a) *Sampling in Test Pit or an Exposed Surface.*—An area of the side wall of the test pit or exposed surface is trimmed to remove all weathered or mixed soil. The trimmed surface is examined to determine the sequence, thickness, classification, and description of each stratum of material. This information is recorded on log form 7–1336, figure 66. The number of samples taken and the depth each represents is noted on the form. The sample is obtained by trenching the trimmed surface with a cut of uniform cross section and collecting the soil on a canvas spread below the trench (fig. 1–1). The minimum dimension of the sampling trench should be at least four times the diameter of the largest gravel particle in the soil. When sampling an individual soil stratum, special care is required to pre-

Individual samples are taken from each layer of soil.

Composite samples are taken from two or more layers of soil.

Figure 1–1.—Sampling trench. PX–D–4784.

Table 1–2.—Samples of foundation materials

Tests	Sample size (weight of material smaller than maximum size required)	Maximum particle size required	Application
Classification and description	1 lb / 50 lb	No. 4 / 3 in	Field log of exploration holes, describing samples in laboratory.
Gradation	1 lb / 50 lb	No. 4 / 3 in	Identification and description of soils for estimating properties of foundations.
Consistency	1 lb	No. 40	
Specific gravity	1 lb / 5 lb	No. 4 / 3 in	
Petrographic analysis	5 lb	No. 4 (and representative coarse grains).	Petrographic tests desired when expansive soils are encountered.
Free swell.			
Microscopic examination.			

Test	Sample size	Purpose
X-ray diffraction. Differential thermal. Thermo balance. Lithologic, photo-micrographic.	5 lb	Petrographic descriptions and analysis desired when unfamiliar soils are encountered.
Chemical tests. Sulfate content. Soluble salts. Corrosion. Organic content.	No. 4	Determine chemical composition for recommending type of cement or pipe coating required. Permeability and settlement of foundation influenced by soluble salts and organic matter.
Water content	No. 4 3 in. 1 lb 5 lb. (water content sample)	Determine inplace water content of foundation soils, evaluating penetration tests, comparison of inplace water content with consistency limits; determine degree of saturation and dry unit weight.
Relative density	No. 4 ¾ in. 3 in. 25 lb 100 lb. 150 lb.	Determine 0 and 100% D_d for determination of D_d of foundations.

Table 1–3.—Accessible methods

Methods	Procedure	Type of soil and inplace condition
Trenching	Excavate 3 ft. min. width by hand, dragline, power shovel, bulldozer preferable; explosives if necessary; min. bracing or slope unstable soils.	
Cuts	Same as trenching, performed on gentle to fairly steep slopes; steps up slope may be necessary.	
Test pits	Excavate rectangular hole, 3 ft. by 5 ft. min. at working level, by hand or hand-operated power tools; explosives if necessary. Cribbing required over 5 ft. depth. Log and sample as excavation progresses when sheeting, inclined poling, or notched box cribbing is required in unstable soils and for ground-water control.	Coarse-grained soils, including those containing large quantities of gravel and cobbles, and soft weathered rock; and all fine-grained soils, dense consolidated, wet or saturated or dry and hard; loose unconsolidated, wet or saturated and soft or dry and granular.
Accessible boring	Drill 28 in. min. dia. hole, using heavy power-operated disc, bucket, helical augers, single tube or core barrels in stable soils; log and sample as excavation progresses; casing required for protection during sampling and inspection.	
Accessible caissons	Same as accessible borings; casing and air pressure required in unstable soils.	
Tunnels and drifts	Excavate accessible holes, 5 by 7 ft. min. using hand or hand-operated power tools, lagging required.	
Blasting	Exposed strata using explosives and hand or power tools.	Same as above but primarily for consolidated dry soils and bedrock.

of exploration

Limitations	Use
Depth about 20 ft. or to ground water or unstable material.	Access for logging and disturbed sampling for laboratory test, for reconnaissance and feasibility design stage; and for hand-cut undisturbed sampling for final design or for field tests such as field density, permeability, full sized bearing capacity tests. Unsatisfactory in unstable cohesionless soils. Economical and best method for shallow explorations of borrow, foundation, and aggregate deposits.
Depth to 50 ft., infrequently 80 ft., or ground water if pervious strata and high flow.	Use same as above, except more expensive, used in areas of limited access by heavy equipment and for greater depths. Best method for "hand cut" undisturbed sampling except in unstable soils or below ground water. Nonaccessible methods, table 1–4, recommended for undisturbed sampling fine-grained unstable soils below water table.
Depth of 100 ft. in soil, 150 ft. in rock. Requires heavy drill rig.	Use same as above for stable soils, in place of test pits; very economical if equipment available and area is accessible.
Same as above, used only when caisson is part of construction.	Limited use, used primarily in establishing footing grade during construction for individual caissons, under very poor foundation conditions and/or under water.
Expensive, used only under special conditions.	Limited use, for final exploration of damsite abutments when other methods have disclosed questionable conditions which cannot be resolved otherwise.
Limited to exposed faces or outcrops.	Used to expose rock faces and outcrops for riprap and crushed aggregate sources and to indicate size and shape of particles which may be expected during quarrying.

Table 1–4.—Nonaccessible

Method	Procedure	Type of soil and inplace condition
Auger boring (hand).	Rotate and force auger bit into soil, withdraw and empty when full. Auger bits, 2 to 8 in., helical or post hole.	Fine-grained cohesive, fairly hard to soft; or fine-grained, non-cohesive, dense to loose, weakly cemented, dry or moist; with particles ¼ in. to 1½ in. depending upon size of auger.
Auger boring (power).	Same as above, using powered drill rigs. Auger bits, 4 to 24 in., helical, disc, or bucket. Over 28 in. considered to be accessible.	Fine-grained as above, and coarse-grained soils with particles as large as 3 in. depending upon auger.
Drive-tube boring.	Force open pipe or tube, with sharpened edges, without rotation, into soil; withdraw and remove soil. Thin- or thick-wall tubing or pipe, 2 to 8 in. dia.	Fine-grained cohesive and slightly cohesive soils such as loess, firm to soft clays, and silts.
Percussion (churn) drilling.	Chopping and cutting action by impact of heavy chisel edged bit. Water added and cuttings form slurry which is removed intermittently by pump or bailer. For holes larger than 4 in.	Coarse-grained soil containing cobbles and boulders, and hard, dense, fine-grained soils and rock.
Wash boring.	Chopping and cutting by impact and twisting action of light-weight bit, and jetting action of circulating water to remove cuttings. For holes from 2 in. to over 8 in. dia.	Fine- or coarse-grained soils, with small amounts of gravel and few cobbles; fairly hard to soft; weakly cemented to loose; above or below water table.
Jetting.	High-velocity water jet directed downward from pipe raised and lowered in short strokes; erodes soil, which is carried upward by water. For holes 2 in. to over 10 in. dia.	Fine- or coarse-grained soils; weakly cemented noncohesive, or cohesive; above or below water table.

methods of exploration

Limitation	Use
About 20 ft., 80 ft. with tripod; unsatisfactory in unstable cohesionless soils below ground water; slow in hard soils.	(1) Advance hole. (2) Data for logging. (3) Representative disturbed samples for classification, index tests, and standard properties tests. (4) Access for field penetration and permeability tests. (5) Access for undisturbed sampling.
Economical depth about 40 ft., over 100 ft. with special equipment; unsatisfactory in unstable cohesionless soils below ground water; slow in hard, dense soil.	Same as above.
About 80 ft. depending upon equipment. Not satisfactory in coarser fine-grained soils, clean sands or cohesionless soils below water table.	Same as above.
Unsatisfactory in unstable soil or fractured rock. No information for logging or samples for classification.	Used with other methods to advance hole through hard, cemented strata, coarse gravel, boulders, or other obstructions.
No information for logging or samples for classification; slow in hard or cemented layers.	(1) Used with other methods to advance hole particularly through unstable soils requiring casing. (2) Penetrate fine-grained soils to establish depth to bedrock. (3) Drill holes for ground-water observation. (4) Provide access for sampling and penetration testing of impervious soils above ground water or pervious or impervious soils below.
No information for logging or samples for classification; slow in hard cohesive soils.	Same as (1), (2), and (3) for wash boring.

Table 1–4.—Nonaccessible

Method	Procedure	Type of soil and inplace condition
Rotary drilling.	Power rotation of bit; cuttings removed by circulation of drilling mud or water; holes 1½ in. to over 10 in. dia.	Fine- or coarse-grained, compact or cemented soils and rock.
Rotary drilling.	Power rotation of bit; cuttings removed by circulation of air. Holes 2 in. to over 10 in. dia.	Fine-or coarse-grained, compact or cemented soils and rock.
Continuous sampling.	Drive-tube boring or rotary drilling (core boring) which provides samples as a result of advancing the hole.	

methods of exploration—Continued

Limitation	Use
No information for logging or samples for classification; difficult in loose, coarse-grained soil with cobbles and boulders.	(1) Advance hole. (2) Access for field penetration test (not suitable for well permeameter test or ground-water observation if drilling mud used). (3) Access for disturbed or undisturbed sampling.
Information for logging and samples for classification; unsatisfactory in loose coarse-grained soil with cobbles and boulders.	(1) Advance hole. (2) Access for field penetration test. (3) Well permeameter test. (4) Ground-water observation. (5) Access for disturbed and undisturbed sampling. (6) Advance hole to install casing for nuclear moisture-density probes.
Depends upon the method selected.	

vent inclusion of material from other strata. Where a composite sample is desired, the cross section of the sampling trench is kept constant through all the strata to be represented by the sample.

If a sample is desired from soil containing 25 percent or more of plus 3-inch material, it usually is advantageous to take representative portions of the total excavated material, such as every fifth or tenth bucketful, for the depth sampled, rather than from the sample from the side wall of the test pit or trench. When the collected sample is larger than needed for testing, the size may be reduced by first rolling and mixing the sample to obtain a uniform mixture, and then by quartering on a canvas. The mixing of the sample can successfully be achieved by two or more parties holding opposite corners of the canvas and lifting one side of the canvas at a time to roll the sample toward the opposite side. This procedure is then followed in reverse order and repeated a sufficient number of times to insure complete and uniform mixing of the sample and a near uniform gradation throughout. In quartering a sample, the material is placed in a uniform conical pile. The pile is then flattened to uniform thickness and marked into quarters by two lines intersecting at right angles at the center of the pile. Two diagonally opposite quarters are removed and the remaining material is mixed, quartered, and reduced until the desired sample weight is obtained.

(b) *Sampling from Auger Holes.*—Small auger holes cannot be sampled and logged as accurately as an open trench or test pit because the entire profile cannot be inspected and representative strata cannot be selected for sampling. Hand augers 4 inches in diameter or larger, depending upon the type of soil and inplace conditions, should be used to provide material sufficient for extensive laboratory testing. In advancing the hole, the soil removed from the hole is deposited in individual piles in an orderly depth sequence (fig. 1–2). New piles should always be started when significantly different materials are encountered. In preparing an individual sample from an auger hole, consecutive piles of similar material may be grouped to represent each stratum, by combining all materials from each pile within the stratum or depth being represented, to form the desired size of representative sample. For a composite sample, all materials from each pile representing the strata to be composited are combined. The combined samples are then reduced by mixing and quartering as described in (a) above to obtain the desired size sample.

When a large diameter (8 inches or larger) power auger is used in fine-grained soils, the flight shall be lifted at regular 9- to 12-inch depth intervals and one or more shovels full of soil from each depth piled in a depth sequence. These piles are sampled in the same manner as those from a hand auger. In soils containing large quantities of gravel, it may be necessary to pile all of the soil augered from each stratum separately and obtain a representative sample by quartering.

**SAMPLE No 1
FROM HERE**
(Similar soil)

Piles are separated when
significantly different materials
are encountered.

**SAMPLE No 2
FROM HERE**
(Similar soil)

Figure 1–2.—Auger sampling. PX–D–16331.

Sample information and a log of the hole should be recorded on log form 7–1336, figure 67.

(c) *Sampling Stockpiles and Windrows.*—As soil in stockpiles and windrows is usually segregated, sampling must be performed carefully if a representative sample is to be obtained. A representative sample from a stockpile can usually be obtained by combining small samples from several small test pits or auger holes well distributed over the pile. A windrow sample should comprise all material removed from a narrow trench cut completely across the windrow and from top to bottom. Representative soil samples from both stockpiles and windrows should initially be quite large and, after thorough mixing, should be reduced to desired size by quartering.

5. Preparation and Shipping.—(a) *Sample Containers.*—Large samples of cohesive soils and sand and gravels susceptible to breakdown shall be placed in plastic-lined bags, limiting the sample weight to between 75 and 100 pounds per bag. Nonfriable sands and gravels shall be placed in closely woven cloth bags. When samples are to be transported by public carrier, the samples shall be double sacked to decrease the probability of bag breakage. For fine-grained soil, it is generally desirable to maintain moisture conditions approximately as in situ, and plastic-lined bags (item 2, designation E–4) are recommended for this purpose.

Small disturbed samples for water content determination or inspection shall be placed within, and fill completely, moistureproof containers such as glass fruit jars or metal cans with airtight covers.

Record samples, taken during construction, shall be ovendried and placed in 6- by 12-inch cylindrical cans.

(b) *Sample Identification.*—All samples shall be plainly marked including all pertinent information given below:

(1) Field sample number.

(2) Test pit or hole number.

(3) Location: Coordinates, station, offset, section or other.

(4) Area: Name, letter or number or other identification.

(5) Depth interval sample represents.

(6) Purpose: Lining, filter, backfill, embankment, record, or other.

(7) Sack _____ of _____ (if sample placed in more than 1 sack or container).

(8) Project.

(9) Feature.

Each sample shall have duplicate identification tags, one on the outside of the container and one inside. When the soil is so wet as to cause paper or cloth tags to become illegible, an aluminum trap tag (item 119A) shall be used inside the container. Cans may be marked directly on the exterior or on labels securely fastened to the outside surface. Markings should be on the cans and not on the lids, as lids may be interchanged.

(c) *Sample Shipping.*—Samples in small containers to be shipped by common carrier shall be packed in strong boxes, preferably wood. Glass jars shall be protected with suitable packing materials to prevent breakage. Bagged samples may be shipped in the bags without further packing. A copy of the letter of transmittal shall be included with each shipment of samples. Detailed shipping instructions are given in designation E–2, paragraph 5(d).

UNDISTURBED SAMPLING OF SOILS

Designation E–2

1. Scope.—Part A of this designation describes the procedures for obtaining block or cylinder undisturbed samples by hand methods from accessible excavations, such as test pits, large auger holes, or exposed cuts; and part B describes procedures for obtaining cylindrical undisturbed samples for testing purposes from nonaccessible excavations, such as drill holes. The following drill hole samplers, their application and use are described: (1) double-tube soil samplers, (2) thin-wall open drive samplers, (3) piston samplers, and (4) core samplers.

Hand-cut samples can generally be obtained with less disturbance than samples procured by other methods, and since the excavation is accessible, representative strata can be selected prior to sampling. This method usually involves test pit excavation and is limited to relatively shallow depths, infrequently deeper than 20 to 30 feet. For greater depths and depths below the water table, the cost and difficulties of excavation, cribbing, and pumping usually make sampling by hand methods impracticable and uneconomical. For these conditions mechanical methods of undisturbed sampling are applicable. When using mechanical methods to obtain undisturbed samples, it is desirable to determine in advance the location of critical strata by means of pilot holes drilled with core samplers or small-diameter drive samplers.

It must always be kept in mind that the primary purpose is to obtain a sample which is undisturbed and of adequate size for laboratory testing. Therefore, the method and type of sampler selected must be best suited to the soil conditions, the type of soil, the water content and consistency, and the use for which the sample is being obtained. Adequate care and treatment of the sample to accomplish this primary aim must never be forfeited. A sample which has been disturbed but is submitted to the laboratory for testing as an undisturbed sample is of less value than no sample at all, since the results of tests may lead to erroneous conclusions and faulty foundation design.

Various accessible and nonaccessible methods of exploration are described in table 1–3 and table 1–4, respectively, of designation E–1. The minimum sizes of exploratory holes are given in table 2–1. Size requirements for undisturbed samples are given in table 2–2.

Table 2–1. – Minimum

Exploratory hole			Stage of exploration R—Reconnaissance F—Feasibility S –Specifications
Approx. diameter of core hole, in.	O.D. of casing bit, in.	I.D. of casing, in.	
E 1-1/2	EX 1-7/8 . . . . EW 1-7/8	EX 1-1/2 . . . . EW 1-1/2	R . . . .
A 2	AX 2-11/32 . . . . AW 2-11/32	AX 1-29/32 . . . . . . AW-1-29/32	R . . . .
B 2-1/2	BX 2-31/32 . . . . BW 2-31/32	BX 2-3/8 . . . . . . BW 2-3/8	F . . . .
N 3	NX 3-5/8 . . . . NW 3-5/8	NX 3 . . . . NW 3	R, F and S . . . . . .
H 4	HX 4-5/8 . . . . HW 4-5/8	HX 3-15/16 . . . . . HW 4	F and S . . . .
S 6	SW 6-51/64	SW 6	F and S . . . .
Z 8	ZW 8-13/16	ZW 8	F and S . . . .
10 to 28	. . . .	. . . .	S . . . .
28	. . . .	. . . .	S . . . .
36 in. and test pits	. . . .	. . . .	S . . . .
		3 by 5 ft. . . . . . . test pit	S . . . .

size of exploratory hole

Purpose	Use
Logging and sampling bedrock, 7/8-in. core.	Pilot holes in bedrock. Unsatisfactory in fractured and faulted zones.
Logging and sampling bedrock, 1-1/8-in. core.	Same as above.
Logging and sampling soils, representative disturbed samples, for classification and index tests.	Fine-grained, stable soils; max. particle size No. 4. Sample with 2-in. O.D. auger, drive-tube, or core boring methods.
(1) Logging and sampling bedrock, 1-5/8-in. core. (2) Permeability tests in bedrock.	Foundation and riprap source.
(1) Logging and sampling soils, representative disturbed samples, for classification and index tests. (2) Deep permeability tests in soil.	Fine-grained, stable, or unstable soils requiring casing; max. particle size No. 4. Sample with 2-in. O.D. auger, drive-tube, or core boring methods.
(1) Logging and sampling bedrock, 2-1/8-in. core. (2) Permeability tests in bedrock.	Foundation exploration.
(1), (2), same as 2-1/2-in. hole for soils.	Same as 2-1/2-in. hole for soils.
(3) Obtain undisturbed 2-1/8-in. core sample.	Fine-grained, firm to very compact cohesive soils; or fine-grained, cemented cohesionless soils.
(4) Field penetration test.	Sand and fine-grained soils, max. particle size No. 4.
(1), (2), same as 2-1/2 in. hole for soils. (3) Obtain representative disturbed samples for laboratory tests. (4) Well-permeameter tests.	For coarse-grained soils, max. particle size 1/2 in.
(5) Obtain undisturbed samples: 3-in. thin-wall drive 3-in. piston drive	Fine-grained, soft to firm cohesive soils; or fine-grained loose to dense cohesionless soils.
(6) Vane test using 2-, 3-, and 4-in. vanes (4-1/2-in. hole for 4-in. vane).	Inplace shear strength, firm to very soft, fine-grained soils.
(1), (2), (3), (4), same as 4-in. hole for soils.	For coarse-grained soils, max. particle size 3/4 in.
(5) Obtain undisturbed soil samples: 5-in. thin-wall drive 5-in. piston drive	Fine-grained, soft to firm cohesive soils; or fine-grained, loose to dense cohesionless soils.
(1), (2), (3), (4), same as 4-in. hole for soils.	For coarse-grained soils, max. particle size 1 in.
(5) Obtain undisturbed samples: 5-5/8-in. Denver, 5-7/8-in. Pitcher, or 5-7/8-in. Denison	Fine-grained, firm to very compact cohesive soils; or fine-grained, dense, cemented noncohesive soils.
(1) Logging and representative sampling of materials sources, and zoning borrow areas.	For fine- and coarse-grained soils, max. particle size 3 in.
(1) Accessible exploration. Logging, classification, and description inplace.	For fine- and coarse-grained soils; cribbing or casing required in unstable soils.
(1) Accessible exploration as in 28-in. hole; obtain undisturbed hand-cut samples.	For sampling fine- and coarse-grained soils; casing or cribbing required.
(2) Undisturbed hand-cut sampling.	For sampling fine- and coarse-grained soils; cribbing required.

Table 2-2. — Representative undisturbed

Size	Type of soil and maximum particle size	Test and sample length
NX cores— 2-1/8 in. .... 2-3/4 in.[1]	Fine-grained hard to firm cohesive or cemented cohesionless; max. size No. 4.	Unconfined compression, two 6-in. cores required.
		Triaxial shear, four 6-in. cores required (upon specific request only).
		Inplace density and moisture, two 6-in. cores required.
Denison— 5-7/8 in. dia. Denver— 5-5/8 in. dia. Pitcher— 5-7/8 in. dia. 4-in. dia.[1] 6-in. dia.[1]	Fine-grained hard to firm cohesive or cemented cohesionless; max. size 3/4 in.	Unconfined compression, one 6-in. core required.
		Triaxial shear, two 6-in. or one 12-in. core required.
		Permeability, one 12-in. core required.
		One-dimensional consolidation, one 6-in. core required.
		One 6-in. core required ....
		Sliding factor, two 6-in. or one 12-in. core required.
		Inplace density and moisture, one 6-in. core required.
Double-tube auger— 4-3/4 in. dia.	Fine-grained soft cohesive or slightly cohesive; max. size 3/4 in.	Same as for Denison, Denver, or Pitcher above.
Denison, Denver, Pitcher, or[1] Drive—3 in. or 5 in. Cube or cylinder— 6 in.	Fine-grained firm or soft cohesive or cemented cohesionless; max. size for 3-in. drive No. 4, for other samples 3/4 in.	Same as for Denison, Denver, or Pitcher above; for 3-in. drive, samples as required for NX cores.
Drive—3 in. or 5 in. .... Cylinder—6 in.	Fine-grained cohesionless; max. size same as above.	Inplace density and water content. Other tests same as for 5-in. drive above, providing sufficient apparent cohesion to support specimens is present.
Cube or cylinder— 12 in.	Fine- or coarse-grained; max. size 2 in.	Sample may be used for any of the tests listed for Denison, Denver, Pitcher, and Double-tube auger above.

[1] Large Diameter Design Double-Tube Core Barrels.
I.D. x O.D.—2-3/4 in. x 3-7/8 in.; 4 in. x 5-1/2 in.; 6 in. x 7-3/4 in.

sample sizes for foundation exploration

Application
Shear strength (cohesion) of fine-grained saturated soils. Note: To be satisfactory for testing, cores should be homogeneous and the inplace structure and properties should appear to be similar.
Shear strength (cohesion and friction) of fine-grained unsaturated soils.
Usually determined in connection with above tests.
Same as above.
Shear strength (cohesion and friction); same requirements as above.
Coefficient of permeability.
Load consolidation and permeability
Load expansion.
Sliding factor, between soil and concrete.
Usually determined in connection with above tests.
Same as for Denison, Denver, Pitcher, or Large Diameter Design Double-Tube Barrels.
Same as for Denison, Denver, Pitcher, NX, or Large Diameter Design Double-Tube cores.
Determination of relative density, inplace permeability by testing remolded specimen at inplace density and water content. Other application same as for Denison, Denver, Pitcher, NX, or Large Diameter Design Double-Tube cores.
Large samples are necessary when (1) it is impossible to obtain smaller undisturbed samples, or (2) a number of test specimens are required from the same stratum, or (3) the sample is to be saturated before cutting a number of test specimens, or (4) the material is heterogeneous and it is necessary to select specimens for testing, or (5) to determine the relative density of coarse-grained soils.

Part A. Hand Methods

2. Apparatus.—This equipment is not usually required for construction control; therefore, the apparatus is not listed in designation E–4 and only special or major items of equipment are listed below:

(1) Standard excavating tools as required, such as shovel, pick, trowel, large-size kitchen knife, smaller thin-blade paring knife, heavy-duty hacksaw blades, and thin piano-wire cutter for soft plastic soil.

(2) Sample containers consisting of cubical wooden boxes made to fit sample of desired size with space for packing material, or prepared metal pipe or tubes.

(3) Cheesecloth or other similar cloth wrapping material.

(4) Sealing wax[1], paintbrush, and heater. A double boiler using water and a controlled heating unit is preferable to a single container on an open fire.

(5) Sawdust or similar packing material (moistened).

(6) Shipping tags and marking crayons.

Figures 2–1, 2–2, and 2–3 show step-by-step procedures commonly used in hand-cut block sampling. Cutting and trimming block samples to desired size and shape is tedious, particularly when working with easily disturbed soft or blocky materials. The appropriate cutting tool should be used to prevent disturbance and cracking of the sample. Soft, plastic soils require thin, sharp knives; and sometimes a tight, thin piano wire is advantageous. When climatic conditions are such that quick drying occurs, moist cloths must be used or other means provided to protect the sample while it is being cut. After the sample is cut and trimmed to desired size and shape, it should be wrapped with a layer of cheesecloth and painted with warm melted wax. Following the paintbrush application of the wax, it should be rubbed with the bare hands so as to seal pores. Additional coverings of cloth and wax should be applied in the same manner, making at least three layers in all.

Other methods of wrapping and protecting samples have been suggested and used with varied success. Thin plastic wrap placed adjacent to the sample to hold natural moisture, followed by placing layers of alumi-

---

[1] Paraffin, melting point approximately 125° F., is a waxy constituent of petroleum or is produced by distilling wood, shale, or coal. Tests have shown that the microcrystalline waxes such as Petrowax A, melting point approximately 170° F., (Gulf Oil Corporation) and Eskar ML–45, melting point approximately 150° F., (Standard Oil Company of Indiana) are better than paraffin for sealing soil samples. These waxes should be applied at a temperature of about 20° F. above the melting point. If the wax is applied too cold, it will peel and crack; if too hot, it will penetrate and dry the soil.

num foil over the plastic wrap, has occasionally been used but is not considered as desirable as the wax and cloth method because it (1) has less supporting strength to the encased sample to prevent disturbance, (2) is relatively easy to rupture and tear, and (3) does not conform readily to irregular shaped surfaces, sharp corners, or small open cavities.

If a soil is easily disturbed, a firmly constructed wooden box with both ends removed should be placed over the sample before it is cut from the

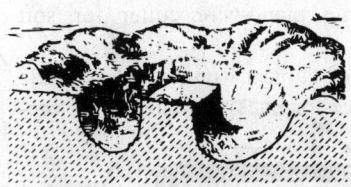

1. Smooth ground surface and mark outline of sample.
2. Carefully excavate trench around sample.

1. Carefully smooth face surface and mark outline of sample.

3. Deepen excavation and trim sides of sample to desired size with knife.

2. Carefully excavate around and in back of sample. Shape sample roughly with knife.

4. Cut sample from parent stratum, or encase sample in box before cutting if sample is easily disturbed.

3. Cut sample and carefully remove from hole, or encase sample in box before cutting if sample is easily disturbed.

(A)　　　　　　　　　　　　**(B)**

Figure 2–1.—Initial steps to obtain a hand-cut, undisturbed block sample from (A) bottom of test pit or level surface, and from (B) cutbank or side of test pit. PX–D–4788.

parent material and lifted for removal. Space between the sample and the walls should be packed with moist sawdust or similar packing material. The top cover of the box should then be placed over the packing material. After removal of the specimen, the bottom side of the specimen should be covered with the same number of wrappings of cloth and wax as the other surfaces, and the bottom of the box placed over the packing material.

In no case should hot wax be poured over the sample for sealing or packing purposes.

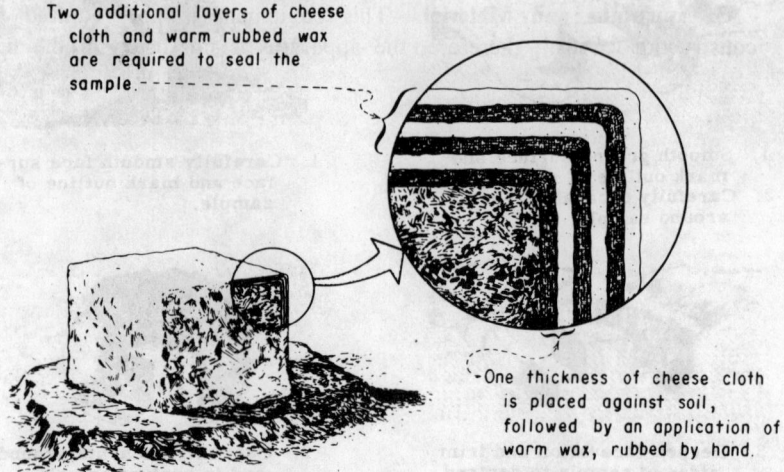

Two additional layers of cheese cloth and warm rubbed wax are required to seal the sample.

One thickness of cheese cloth is placed against soil, followed by an application of warm wax, rubbed by hand.

(A) METHOD FOR SEALING HAND-CUT UNDISTURBED SAMPLES

Fill space between sealed sample and box with moist sawdust packed to support sample.

(B) ENCASE EASILY DISTURBED SAMPLES IN BOX PRIOR TO CUTTING

Figure 2–2.—Final steps in obtaining a hand-cut, undisturbed block sample. 101–D–530.

Samples may be of various sizes, the most common being 6- or 12-inch cubes. Cylindrical samples 6 to 8 inches in diameter and 6 to 12 inches in length are frequently obtained as shown in figure 2–3. The metal cylinders may be used to confine the sample for shipping. Otherwise, the same trimming and sealing procedures discussed above apply.

All samples should be marked and prepared for shipment in accordance with paragraph 5(d) of this designation.

Part B. Mechanical Drilling Methods

3. Apparatus and Material.—This equipment is not required for construction control; therefore, the apparatus is not listed in the field

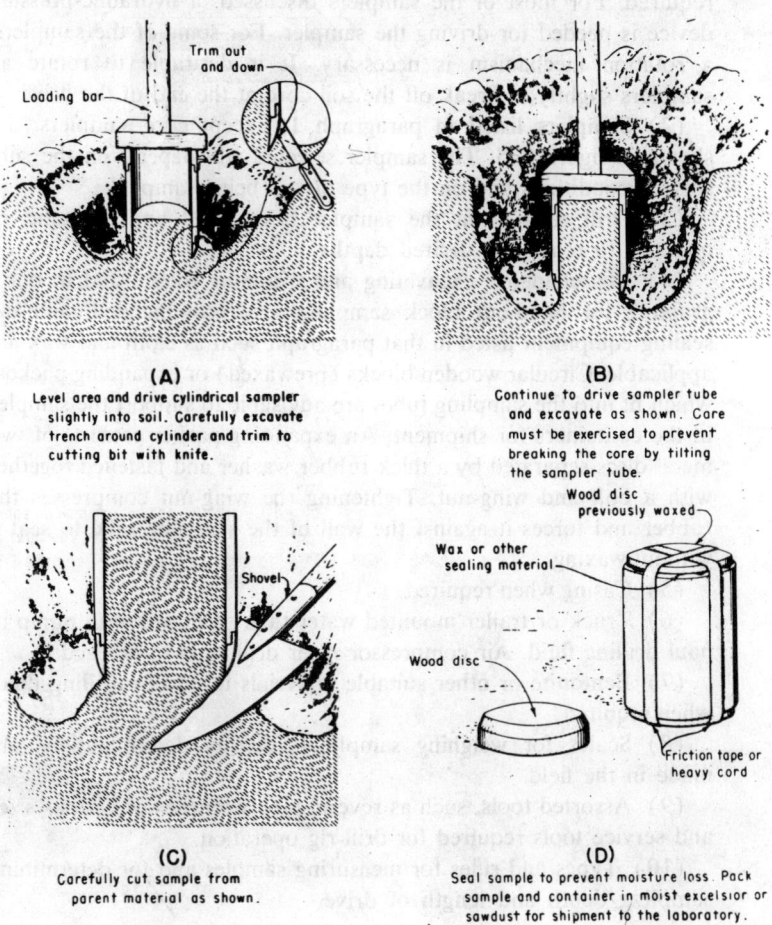

(A)
Level area and drive cylindrical sampler slightly into soil. Carefully excavate trench around cylinder and trim to cutting bit with knife.

(B)
Continue to drive sampler tube and excavate as shown. Care must be exercised to prevent breaking the core by tilting the sampler tube.

(C)
Carefully cut sample from parent material as shown.

(D)
Seal sample to prevent moisture loss. Pack sample and container in moist excelsior or sawdust for shipment to the laboratory.

Figure 2–3.—Method for obtaining a hand-cut, undisturbed cylindrical sample.
101-D-731.

laboratory equipment list, designation E–4, and only special or major items of equipment are listed below:

(1) Drilling equipment must provide a reasonably clean hole of adequate size to permit insertion of the particular sampler desired. This equipment must be such that unnecessary disturbance does not occur to the soil to be sampled. Methods such as augering, wash boring with fishtail bits, or other methods may be used to advance the holes, depending on the applicability to soil conditions. A common type of drill rig is shown in figure 2–4, which provides for mechanical preboring as well as adequate devices for operating the samplers. Rigs may be mounted on trucks, trailers, or skids, as required. For most of the samplers discussed, a hydraulic-pressure device is needed for driving the sampler. For some of the samplers, a rotation mechanism is necessary. It is desirable to rotate all samplers slightly to break off the soil core at the end of the drive.

(2) Samplers listed in paragraph 1, except core samplers, are shown in figure 2–5. The sampler selected will depend on the subsurface conditions and on the type of soil being sampled.

(3) Drill rods to fit the sampler being used and of sufficient quantity to reach the desired depths.

(4) Miscellaneous excavating and trimming tools listed in paragraph 2 for hand-cut block sampling are recommended, and the sealing equipment listed in that paragraph such as cloth and wax are applicable. Circular wooden blocks (prewaxed) or expanding packers which fit into the sampling tubes are advisable to support the samples in the containers for shipment. An expanding packer consists of two metal discs separated by a thick rubber washer and fastened together with a bolt and wing-nut. Tightening the wing-nut compresses the rubber and forces it against the wall of the sampling tube to seal it without waxing.

(5) Casing when required.

(6) Truck or trailer mounted water tank equipped with pump to haul drilling fluid. Air compressor if air drilling is to be used.

(7) Bentonite or other suitable materials to add to drilling fluid, when required.

(8) Scales for weighing sample, if density determinations are made in the field.

(9) Assorted tools, such as several pipe wrenches of various sizes and service tools required for drill-rig operation.

(10) Tapes and rules for measuring samples and for determining sampling depth and length of drive.

4. Procedure.—For all types of mechanical sampling methods it is important that the sampler be properly cared for and kept thoroughly

Figure 2-4.—Typical drill rig for mechanical sampling. E-2255-5ÑA and E-2255-6NA.

clean so that all threaded screw connections and moving parts work freely. Thin oil and lacquer coatings may be used to protect the tubes. Before sampling, the sampler should be thoroughly checked for proper operation.

For all types of samplers the hole must be prebored to the depth at which the undisturbed sample is desired. The hole must be clean and free of disturbed soil, and the material to be sampled must not be disturbed by the cleanout operation. For sampling above the ground-water table in soil which will be affected by wetting (for example, loess or fine loose sand), dry augering is an advisable method of preboring. In fine-grained soils of low permeability or in already saturated soils, rotary drilling using side discharge drag or roller rock bits is a suitable method of preboring.

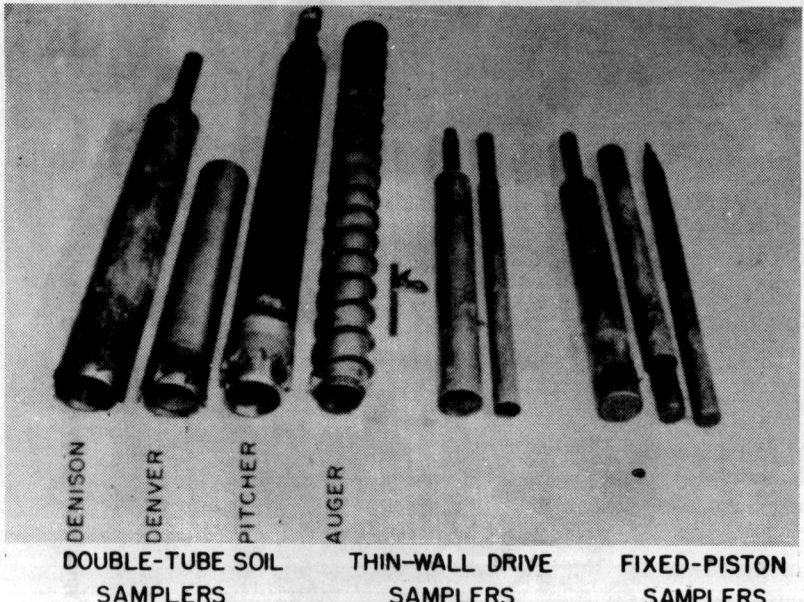

DENISON　DENVER　PITCHER　AUGER

DOUBLE-TUBE SOIL　　THIN-WALL DRIVE　　FIXED-PISTON
　　SAMPLERS　　　　　　SAMPLERS　　　　　SAMPLERS

Figure 2-5.—Samplers used for undisturbed soil sampling. E–2245–1NA.

Many soils will stand without support. However, there are others such as soft saturated soils and cohesionless soils which require support. This support can be provided by the use of drilling fluids or a casing of sufficient size to permit the insertion of the sampler. Selection of casing may be made from the information shown in figure 2–20. In some holes it may be desirable to obtain information on the location of the water table and the permeability and character of the materials. In this event casing rather than drilling fluid should be used. Where the only purpose of the hole is for securing undisturbed samples, the use of drilling fluid is recommended because it is cheaper and it aids materially in recovering the samples. Currently there are commercially available chemicals under research and development that may have usefulness as drilling fluid additives with properties of thickening and later conversion to water consistency. Where their use is permitted and appropriate they may prove beneficial.

The use of drilling fluid, commonly called drilling mud, is a technique of outstanding importance in mechanical sampling. This material has several uses; namely, assisting cutting operations by the sampler, lifting cuttings from the hole by mud circulation, supporting a hole to prevent caving and thus eliminating the necessity of casing, and assisting in holding the sample in the sampler as it is being removed. The basic features which govern the suitability of the drilling mud for these various uses are viscosity and specific gravity or weight.

The material used as drilling mud is a high-swelling bentonite (sodium montmorillonite) and water mixture. The concentration and the thoroughness of mixing of the bentonite in the water governs the viscosity of the mixture. There are available on the market commercial chemical products which have the purpose of adjusting the viscosity, either thinning or thickening the fluid and providing a stabilized filter cake along the walls of the hole which helps to give support and prevent loss of fluid. Relatively small quantities, a few ounces per barrel of water, often result in advantageous changes in the drilling mud. The use of drilling mud and these chemical additives has shown success in most applications tried. There may be occasions, however, when a heavy weight of mud is needed to support the hole. This is sometimes obtained by adding a barium compound, usually barite (barium sulphate) which is a heavy inert chemical that will increase the weight without affecting the consistency or serviceability of the mud.

The bentonite-water mixture should be sufficiently thick so that it does not penetrate into the soil sample during the cutting operation. This consistency is dependent on the material being sampled and varies from slightly thicker than water to a very thick, creamlike consistency. Thicker muds of high viscosity are required in coarser, granular soils. For supporting the drill holes and for assisting to hold the sample in the sampler, thick and heavy-weight muds are sometimes needed.

Table 2–3 summarizes these requirements. As a general rule the thinnest mud mixture meeting the job requirement is the most practical and efficient to use.

(*Notes:* Table 2–3 is presented as a guide, subject to changes according to local conditions and quality of the bentonite. Experimental trials must finally govern the mixtures required. The use of commercial chemical additives is very advantageous and is recommended for making quick

Table 2-3.—Approximate proportions of mud mixtures

Purpose of drilling mud	Approximate proportions of material per barrel of water	Viscosity [1]	Descriptive consistency
Assisting cutting operation by sampler.	10 to 30 pounds of bentonite.	Variable as needed.	Variable as needed.
For lifting cuttings from hole.	10 to 15 pounds of bentonite for fine-grained soils.	Slightly higher than water.	Thin cream.
	30 pounds of bentonite for coarse-grained soils.	About 1.3 [1] times the viscosity of water.	Very thick cream.
For supporting the drill hole.	30 pounds of bentonite and chemical additives as wall stabilizers, used as directed by the manufacturer. When barite is used about 5 pounds is recommended.	About 1.3 [1] times the viscosity of water.	Very thick cream.
For assisting to hold the sample in the sampler.	10 to 30 pounds of bentonite. Chemical additives as used above may be advantageous for this purpose. When barite is used, the amount required may vary from 0 to 10 pounds.	Slightly higher than water to 1.3 [1] times the viscosity of water.	Thin cream to very thick cream.

[1] Viscosity was measured by a Marsh funnel which is calibrated with water at 72° F. The time required for a given amount of water to flow through the funnel is considered as 1.0. The value listed above is the relative time for the same amount of mud mixture to flow through the funnel.

changes in consistency and viscosity, and improving the usability of the mud. These additives are available as thickeners, thinners, and wall builders. However, some consideration of soil and water conditions is needed since some additives are not effective in the presence of salt or lime and others which are effective may be required. Additives are continually being developed and improved, and for the latest information on commercial sources and application of chemical additives, write to the Engineering and Research Center, Division of General Research.)

(a) *Double-Tube Soil Sampler*.—As indicated in figure 2–5, there are four types of double-tube soil samplers: the Denison, Denver, Pitcher, and Double-Tube Auger samplers. These are illustrated in figures 2–6, 2–7, 2–8, and 2–9, respectively. The Denison, Denver, and Pitcher samplers are similar in design and have disposable liners for handling and shipping the soil cores. The same general rules of operation apply for each of the three (see figure 2–10). The Denver sampler was modeled after the original Denison sampler but is of smaller size and lighter weight and is equipped with smaller cutting teeth. The Denison and Pitcher samplers have liners which are 5⅞ inches inside diameter and 24 inches long. The recommended sample length is 20 inches. The Denver sampler, with overall dimensions of 6⅝ inches outside diameter by 38⅜ inches long, has a liner of 5⅝ inches inside diameter and is 28 inches long. The recommended sample length is 24 inches. The Double-Tube Auger sampler has overall dimensions of 7 inches outside diameter by 72 inches long. A standard 5-inch outside diameter by 36-inch long thin-wall drive tube serves as the inner barrel and as a handling and shipping container for the sample. The recommended sample length is 28 to 30 inches.

The Denison, Denver, and Pitcher samplers are well suited for sampling fine-grained, uncemented, or lightly cemented soils and will recover reasonably undisturbed samples if the soil is slightly cohesive and the boring is done carefully[2]. They may also be used in fairly firm to hard and brittle soils and partially cemented soils which require a cutting action rather than a simple drive penetration. They are not suitable, however, for gravelly soils, low-density cohesionless sands and silts below the ground-water table, very soft and plastic cohesive soils, or severely fissured or fractured materials.

The Double-Tube Auger sampler was developed to sample fine-grained low and medium density soils from above the ground-water table. The sampler does not require drilling fluid to remove the drill cuttings and is well suited for sampling soils that would be adversely affected by drilling fluids.

The Denison, Denver, and Pitcher samplers consist of a rotating outer barrel with cutting teeth on the bottom, a nonrotating inner barrel with a smooth cutting shoe, a spring core catcher (used only in special

---

[2] Firm or dense cohesionless sands can often be sampled if carefully handled. Loose sands may be densified during sampling operations, thus permitting samples to be obtained, but with erroneous conclusions regarding the density. When it is not known whether a sand is dense or loose, the other samplers, such as the thin-wall drive or piston samplers, are recommended. If the double-tube soil sampler is used, the recovery ratio (length of sample divided by length of drive) and results of field penetration tests will aid in recognizing changes in density.

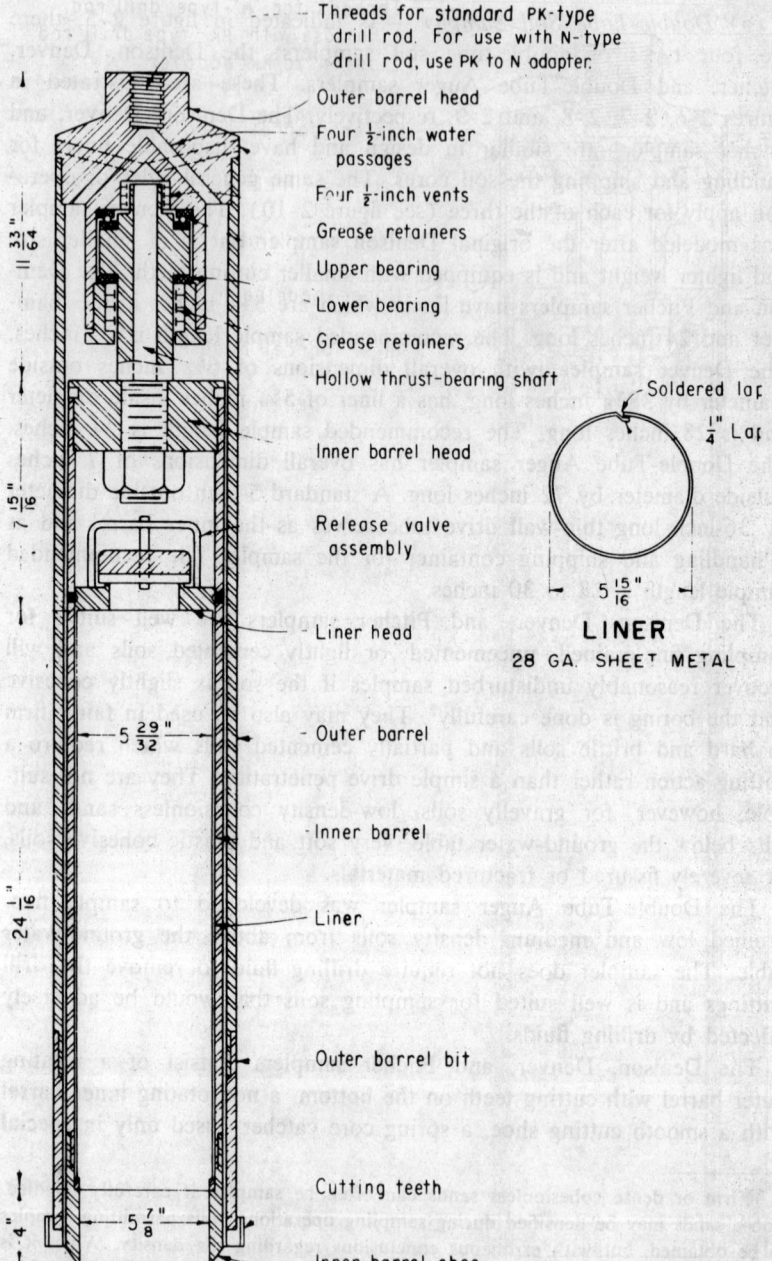

Threads for standard PK-type
drill rod. For use with N-type
drill rod, use PK to N adapter.

Outer barrel head

Four ½-inch water
passages

Four ½-inch vents

Grease retainers

Upper bearing

Lower bearing

Grease retainers

Hollow thrust-bearing shaft

Inner barrel head

Release valve
assembly

Liner head

Outer barrel

Inner barrel

Liner

Outer barrel bit

Cutting teeth

Inner barrel shoe

Soldered lap
¼" Lap

5 15/16"

LINER
28 GA. SHEET METAL

11 33/64"

7 15/16"

24 1/16"

4"

5 29/32"

5 7/8"

Figure 2-6.—Denison sampler (double-tube core barrel). 101-D-532.

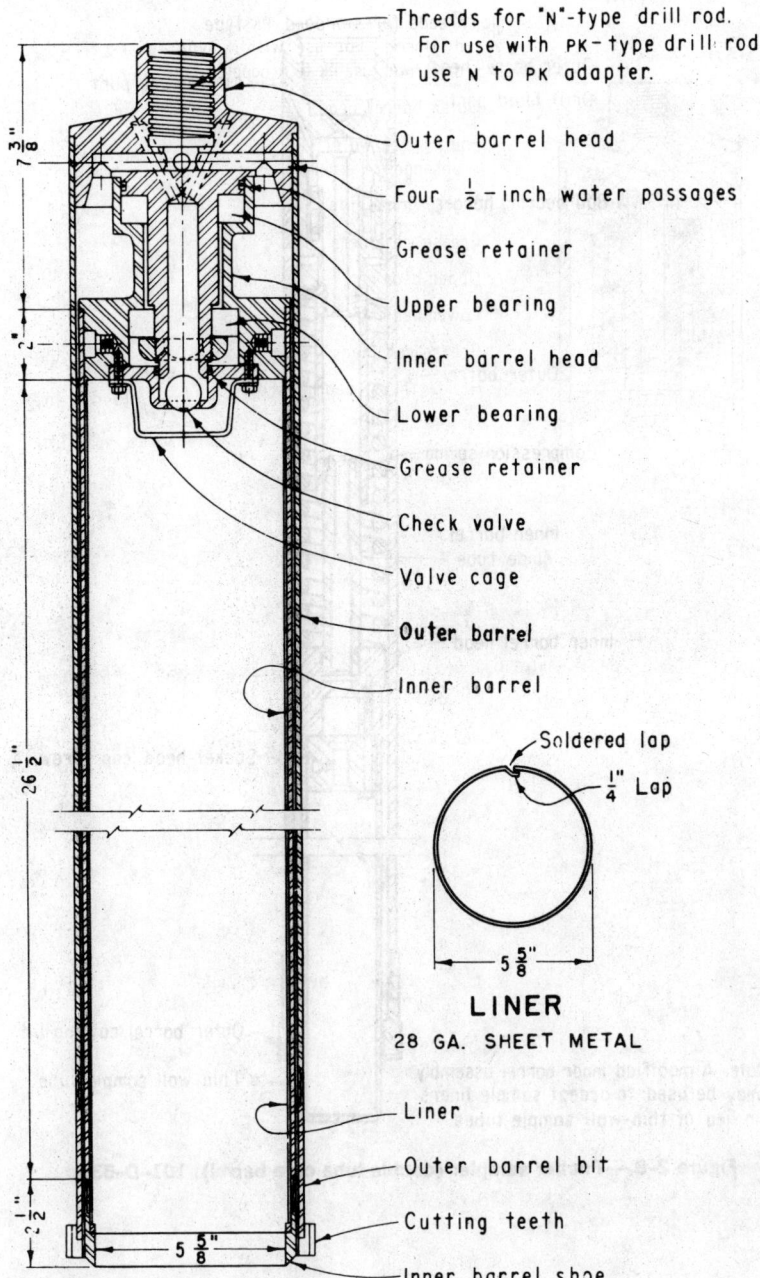

Threads for "N"-type drill rod.
For use with PK-type drill rod,
use N to PK adapter.

Outer barrel head

Four $\frac{1}{2}$-inch water passages

Grease retainer

Upper bearing

Inner barrel head

Lower bearing

Grease retainer

Check valve

Valve cage

Outer barrel

Inner barrel

$7\frac{3}{8}$"

2"

$26\frac{1}{2}$"

Soldered lap

$\frac{1}{4}$" Lap

$5\frac{5}{8}$"

LINER
28 GA. SHEET METAL

Liner

Outer barrel bit

Cutting teeth

Inner barrel shoe

$2\frac{1}{2}$"

$5\frac{5}{8}$"

Figure 2-7.—Denver sampler (double-tube core barrel). 101-D-533.

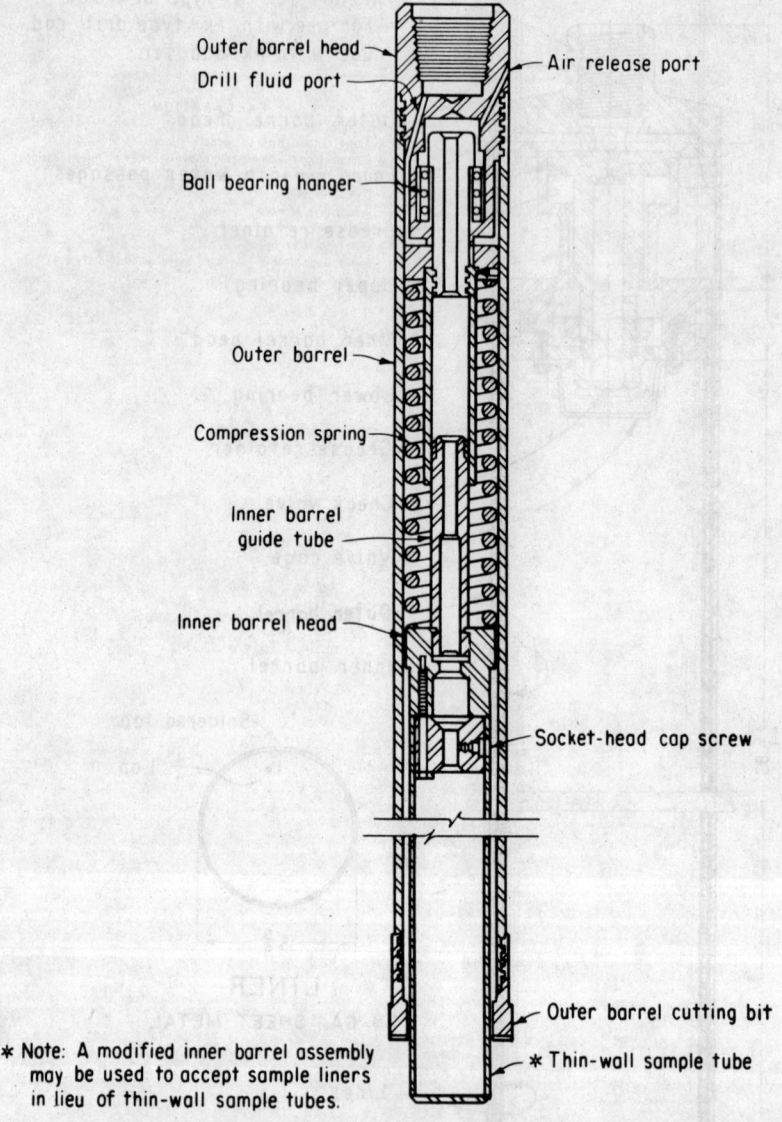

Outer barrel head

Drill fluid port

Air release port

Ball bearing hanger

Outer barrel

Compression spring

Inner barrel
guide tube

Inner barrel head

Socket-head cap screw

Outer barrel cutting bit

* Thin-wall sample tube

* Note: A modified inner barrel assembly
may be used to accept sample liners
in lieu of thin-wall sample tubes.

Figure 2–8.—Pitcher sampler (double-tube core barrel). 101–D–534.

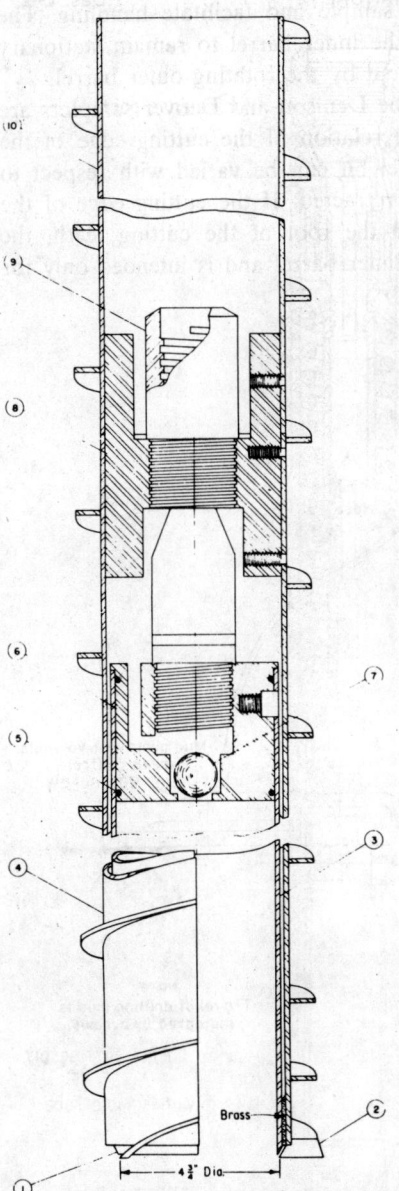

EXPLANATION

1 – Sample tube cutting bit
2 – Auger cutting bits (3 equally spaced)
3 – Sample tube (Seamless cold drawn steel, 24 or 36-inches long)
4 – Outer barrel
5 – 'O'-ring seals
6 – Inner sampler head
7 – Ball check
8 – Outer sampler head
9 – Swivel type double-tube core barrel head (old style N)
10 – Sludge barrel

Brass

← 4¾" Dia →

Figure 2–9.—Double-tube auger sampler. 101–D–404.

cases), and a liner to receive the sample and facilitate handling. The basic principle of operation is for the inner barrel to remain stationary and slide over the sample which is cut by the rotating outer barrel.

The outer barrel cutting bits of the Denison and Denver samplers are made in several lengths so that the relation of the cutting edge of the inner barrel shoe to the outer barrel bit can be varied with respect to whether the former protrudes or is retracted. If the cutting edge of the inner shoe does not extend beyond the root of the cutting teeth, the design is referred to as a retracted inner barrel and is intended only for

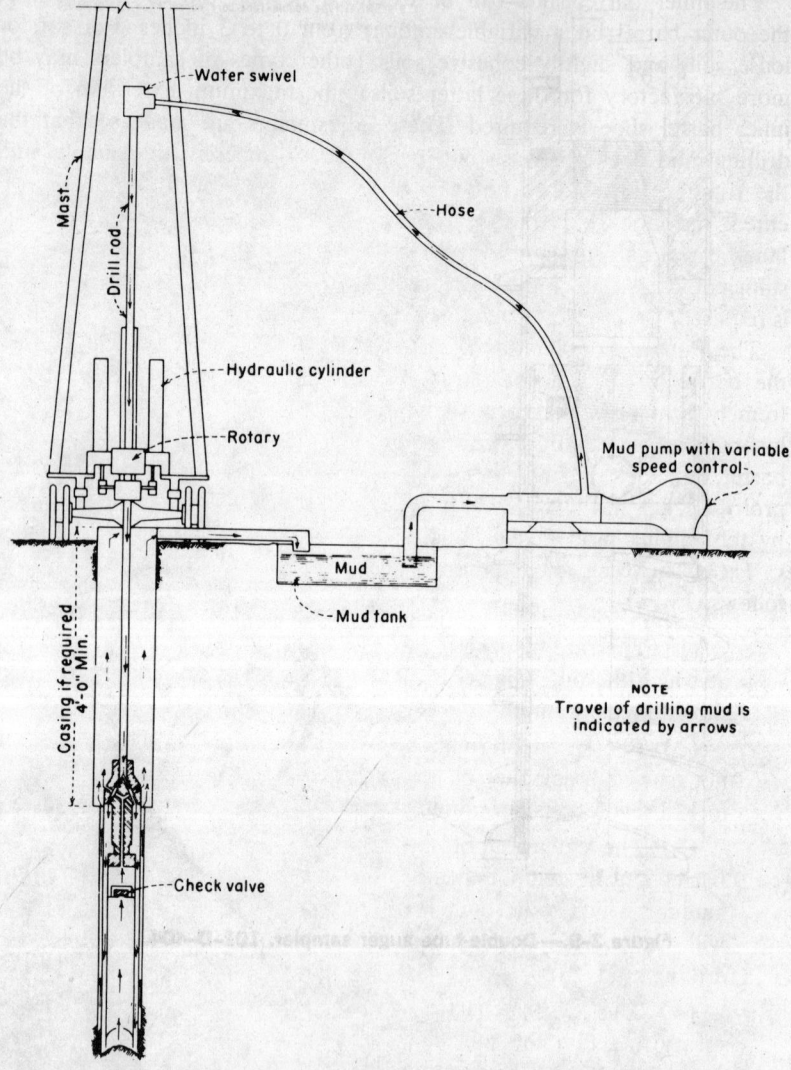

Figure 2–10.—Sampling with double-tube soil or core sampler. 101–D–332.

coring very hard soils which are not subject to erosion. If the cutting edge of the inner shoe extends beyond the root of the cutting teeth but not beyond the crest or cutting edges of the teeth so that the circulation of the drilling fluid is through the teeth, the design is referred to as a bottom-discharge bit and is used in sampling soft or broken rock and dense cemented soils which are not subject to erosion. In coring soils where additional protection of the core is needed, the inner shoe should have a sharp cutting edge and should extend beyond the cutting teeth. This design is referred to as a protruding inner barrel.

The inner barrel shoe can be extended beyond the cutting teeth of the outer barrel bit a variable amount from 0 to 3 inches. For soft or loose soils and slightly cohesive soils (other types of samplers may be more satisfactory for these latter soils), the maximum extension of the inner barrel shoe is required. These adjustments are made so that the drilling mud does not wash out, penetrate, or undercut the sample, and the rotating barrel does not tear the sample core prior to the time it enters the inner barrel. For best results, the sample should enter the inner barrel so that it fits the barrel fully, and the outer barrel should sufficiently cut the sample core so that a minimum downward pressure is required.

The Pitcher sampler has a spring loaded inner barrel which permits the barrel to protrude or retract with changes in soil firmness. In extremely firm soils the spring compresses until the cutting edge of the inner barrel shoe is flush with the crest of the cutting teeth of the outer barrel bit. In soft soils the spring extends and the inner barrel shoe protrudes below the outer barrel bit, preventing damage to the sample by the drilling fluid and drilling action.

The basic rules of operating the double-tube soil samplers are as follows:

(1) The rate of penetration should be no greater than the speed at which the outer barrel is able to cut; that is, the downward force should be a minimum.

(2) The speed of rotation should be limited to that which will not tear or break the soil sample (generally this will vary from 40 to 125 revolutions per minute).

(3) The extension of the inner barrel shoe beyond the outer barrel cutting teeth depends on the soil and should be the least amount which will result in a fully filled inner barrel and which will not cause undercutting or contamination of the sample by drilling mud.

(4) The core catcher should not be used unless absolutely necessary to retain the soil. The following procedure is recommended

to effect sample recovery without the use of the core catcher. Using the standard procedure, the sample is cored to about 1½ to 2 inches of the final desired depth; then, the drilling fluid is shut off and the coring operation completed. This will wedge the cuttings between the inner barrel shoe and the outer barrel shoe which will cause the inner barrel to rotate with the outer barrel, thus shearing the sample from the parent material. This procedure will result in some compaction of the soil in the inner barrel shoe which will form a plug and assist in sample recovery.

(5) The consistency of the drilling mud should be governed by the purposes discussed above for which it is intended. Bureau practice is to use the thinnest mud that will produce satisfactory results.

(6) The pump pressure should be the minimum amount necessary to circulate the mud freely and carry the cuttings from the hole.

(7) The total drive length should always be a few inches short of the length available for the sample to assure that the sample is not packed in the sampler. A 20-inch sample length plus the shoe length is considered the proper driving length.

After completion of the drive, the sampler is withdrawn from the hole, extreme care being taken at all times to avoid disturbance. The outer and inner barrel shoes (and core catcher if used) are removed without jarring the sample, and the soil core and the liner tube are removed. The soil at the ends is then removed to a depth of about 2 inches, to a smooth flat surface. A wooden filler block (previously waxed) is placed inside the liner tube at each end and covered with melted wax. The tube should be crimped over the wood fillers to hold them in place. The blocks should not be nailed to the container, as this causes disturbance to the sample when the container is being opened or the sample is being removed. The soil retained in the inner barrel shoe and the soil removed from the top of the core may be used for field inspection and classification.

The top and bottom of sample, hole number, project, feature, and elevations or depths between which the sample was taken, should be clearly marked on the liner before shipment to the laboratory. All samples should be marked and prepared for shipment in accordance with paragraph 5(d) of this designation.

The Double-Tube Auger sampler, figure 2–9, consists of a rotating outer auger with cutting bits at the bottom and a nonrotating inner barrel with a smooth cutting shoe. The principle of operation is for the inner barrel to remain stationary and slide over the sample which is cut by the rotating auger.

The auger has a fixed length and the relation of the inner barrel shoe to the cutting bits is varied by adjusting the distance between the inner and outer sampler heads. The cuttings from the sampling operation are carried up by the auger flights and deposited in the sludge barrel. The inner sampler barrel which is a 5-inch-outside-diameter, thin-wall, open drive sampling tube is not equipped with a core catcher and it is necessary to have available tubes with various bit clearances for sampling different soil conditions. The basic rules of operation are as follows:

(1) The speed of rotation should be about 40 revolutions per minute and the rate of penetration one to two inches per minute.

(2) The total drive length should always be a few inches short of the length available for the sample. A 30-inch sample is considered the maximum driving length.

After completion of the drive, the sampler is withdrawn from the hole, with extreme care being taken at all times to avoid disturbance. The soil core is handled and shipped in the sampling tube. Expanding packers are provided for sealing the ends of the tubes; wax is not necessary and should not be used as this makes packer removal difficult. The samples should be clearly marked and prepared for shipment as provided in paragraph 5(d) of this designation.

(b) *Thin-Wall, Open Drive Sampler.*—The thin-wall, open drive sampler is shown on figure 2–11. The details of this sampler and the principle of operation are shown in figures 2–11 and 2–12, respectively.

The thin-wall, open drive sampler is available in two sizes, 3-inch and 5-inch outside diameter, and is suitable for sampling all soils having some cohesion unless they are too hard, cemented, or too gravelly for sampler penetration. The sampler is not successful in soils which are so soft and wet that the sampling tube cannot hold them.

The basic principle of operation is to force a thin-wall cylindrical tube into the undisturbed soil in one continuous stroke without rotation. The sampling tube is cold-drawn steel tubing and may be of variable length but is commonly furnished in 36-inch lengths for Bureau work. The head fits the 3-inch-outside-diameter sampling tube, and by use of an adapter ring is also made to fit the 5-inch tube. The quality of the sample depends mainly on having as thin a tube wall as possible. The wall thickness of the tube is usually 14 gage for the 3-inch tube and ⅛ inch for the 5-inch tube. Thinner walls may be used in soft soils when less rigidity is required. The cutting edge of the tube is sharpened so that the bevel is on the outside; this edge is referred to as the bit. The inside diameter is equal to or slightly less than that of the tube, resulting in an inside clearance to facilitate entrance of the sample. Sampling tubes providing clearances of 0, ½, 1, and 1½ percent of the inside diameter are available and the bit clearance selected is that necessary to minimize drag on the inside and to assist in retaining the sample in the tube (see fig. 2–13). Cohesive soils and soils which are slightly expansive require varying amounts of clearance; whereas soils of little cohesion require little or no clearance. It is important that each drilling assembly include tubes of varying clearances so that a tube with proper clearance is available for each of the soil types encountered. Specifications for fabrication of thin-wall sampling tubes are available upon request to the Engineering and Research Center, Division of General Research.

The sampler head has vents for the escape of drilling mud and is equipped with a ball check valve to prevent entrance of the drilling mud during withdrawal and to assist in creating a partial vacuum above the

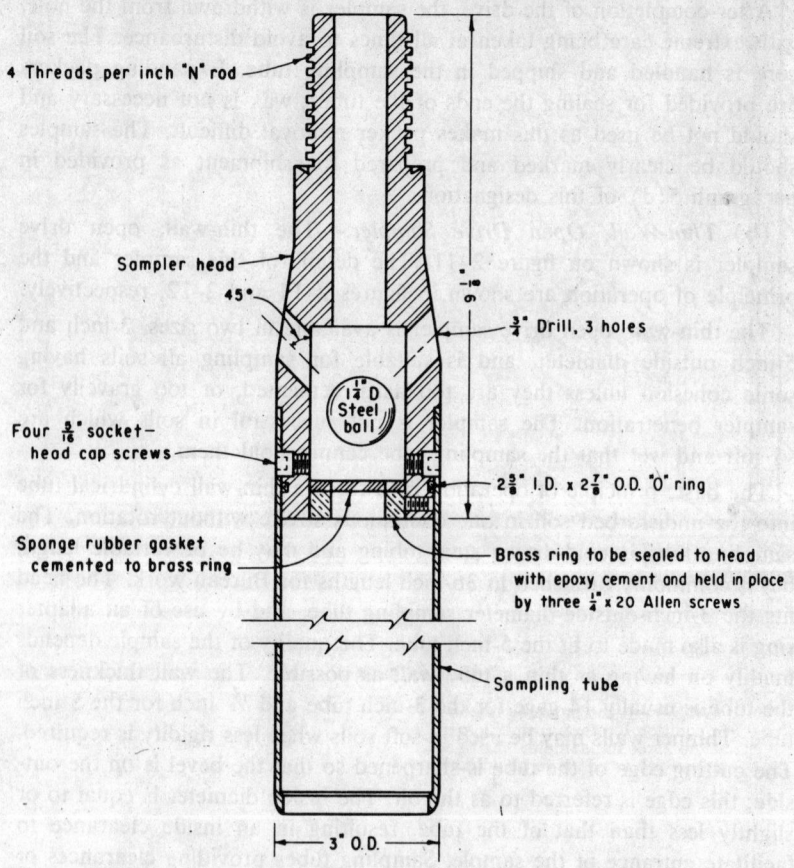

4 Threads per inch "N" rod

Sampler head

45°

¾" Drill, 3 holes

1¼" D Steel ball

Four 5/16" socket-head cap screws

2⅝" I.D. x 2⅞" O.D. "O" ring

Sponge rubber gasket cemented to brass ring

Brass ring to be sealed to head with epoxy cement and held in place by three ¼" x 20 Allen screws

Sampling tube

3" O.D.

Figure 2–11.—Thin-wall open drive sampler. 101–D–529.

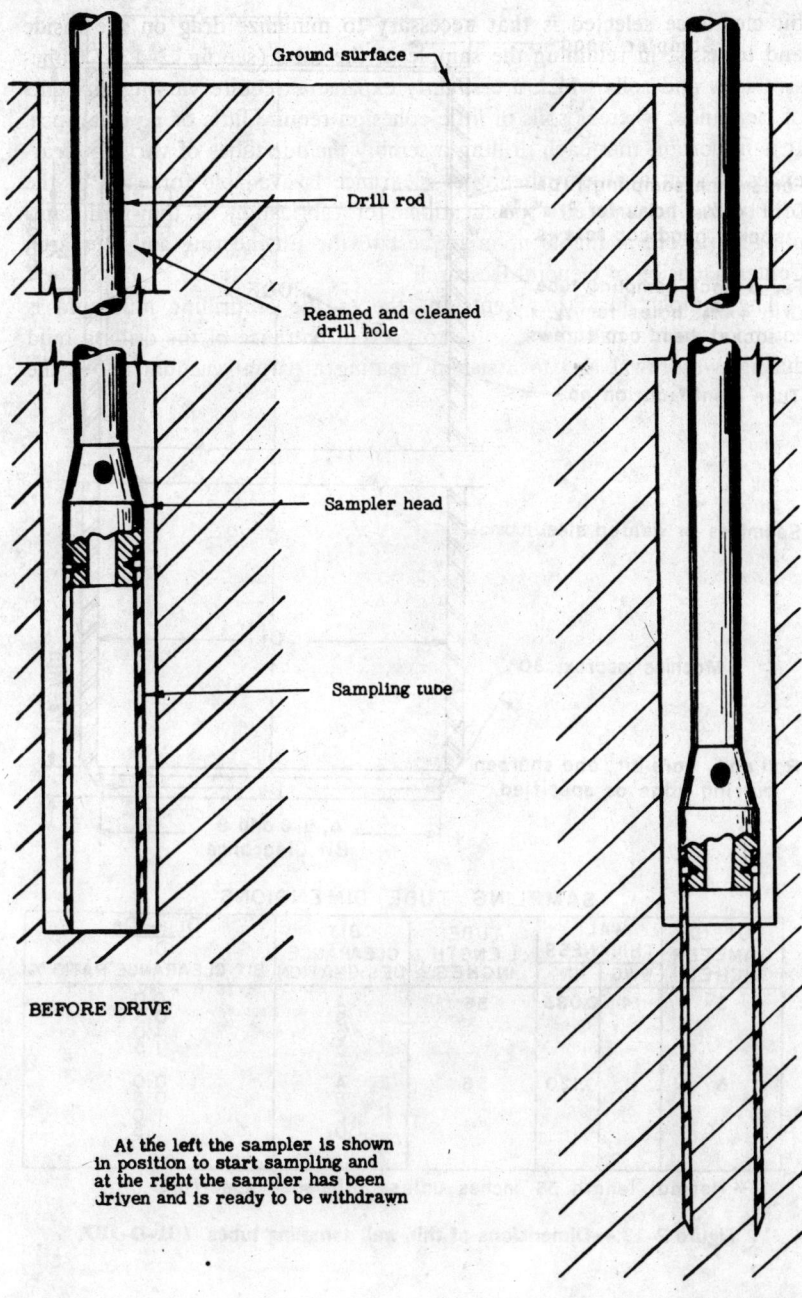

Ground surface

Drill rod

Reamed and cleaned
drill hole

Sampler head

Sampling tube

BEFORE DRIVE

At the left the sampler is shown
in position to start sampling and
at the right the sampler has been
driven and is ready to be withdrawn

AFTER DRIVE

Figure 2–12.—Sampling with thin-wall open drive sampler. 101–D–334.

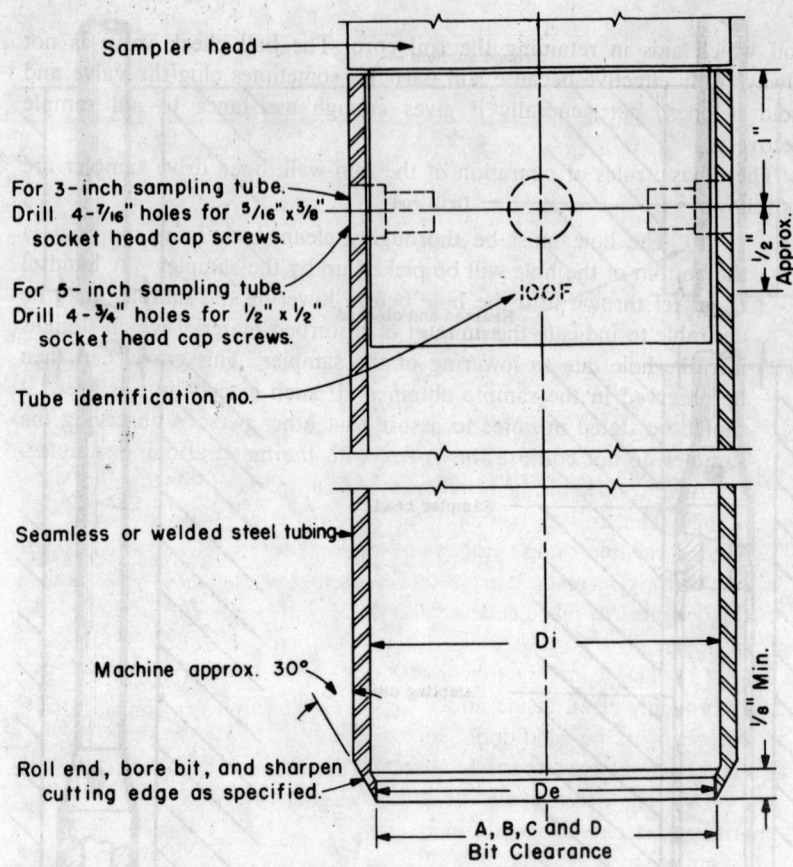

Sampler head

For 3-inch sampling tube.
Drill 4-⁷⁄₁₆" holes for ⁵⁄₁₆" x ³⁄₈"
socket head cap screws.

For 5-inch sampling tube.
Drill 4-¾" holes for ½" x ½"
socket head cap screws.

Tube identification no.

IOOF

1"

½"
Approx.

Seamless or welded steel tubing

Di

Machine approx. 30°

⅛" Min.

Roll end, bore bit, and sharpen
cutting edge as specified.

De

A, B, C and D
Bit Clearance

SAMPLING TUBE DIMENSIONS

OUTSIDE DIAMETER INCHES.	WALL THICKNESS		TUBE LENGTH INCHES *	BIT CLEARANCE DESIGNATION	$\dfrac{Di - De}{Di}$ BIT CLEARANCE RATIO %
	BWG	IN.			
3	14	0.083	36	A	0.0
				B	0.5
				C	1.0
				D	1.5
5	11	0.120	36	A	0.0
				B	0.5
				C	1.0
				D	1.5

* Normal length 36 inches unless otherwise specified.

Figure 2–13.—Dimensions of thin-wall sampling tubes. 101–D–187.

soil which aids in retaining the soil core. The ball check valve is not always fully effective because soil particles sometimes clog the valve and hold it open, but generally it gives enough assistance to aid sample recovery.

These basic rules of operation of the thin-wall, open drive sampler are as follows:

(1) The hole must be thoroughly cleaned, as loose material in the bottom of the hole will be picked up by the sampler. (A handful of gravel thrown into the hole before lowering the sampler may be desirable to indicate the amount of disturbed material which sloughs into the hole due to lowering of the sampler. This gravel can then be observed in the sample obtained. If such a technique is used, it should be stated in notes to assure that other persons observing the samples do not confuse the gravel with the material being sampled, or the gravel should be removed from the tube before shipment.)

(2) The drill rig should be provided with a hydraulic pressure device capable of exerting a driving force of 8,000 pounds. Since the drill rig serves as a reaction for driving the sampling tube, it may be desirable to place earth anchors and extra load on the rig to hold it down.

(3) The sampling tube and sampler head must be smooth and thoroughly clean inside and outside before sampling and must be in proper working condition. The tube edge must be properly sharpened and have the correct inside clearance for the soil being sampled.

(4) The drive should be made without rotation and with a continuous stroke. No additional drive should be attempted after the sampler stops.

(5) The length of drive should always be a few inches short of the length available for the sample to assure that the sample is not packed into the tube. Table 2–4 gives general recommendations for length of drive and bit clearances for various types of soils.

(6) The sampler should be rotated to break off the soil at the bottom.

(7) The sampler containing the soil sample should be carefully removed from the hole to avoid losing the sample. Disturbance by jarring must be avoided at all times.

The soil core is handled and shipped in the sampling tube. Expanding packers are provided for sealing the ends. Wax should not be used as it makes packer removal difficult. The tubes should not be crimped as suggested for the Denison liners. The samples should be clearly marked and prepared for shipment as provided in paragraph 5(d) of this designation. Sample tubes are returned to the project or regional offices from the Engineering and Research Center laboratory upon request.

Table 2–4.—General recommendations for thin-wall, open drive sampling

Soil type	Moisture condition	Soil consistency	Length of drive, inches	Bit clearance, percent	Open drive sampler recovery	Recommendations for better recovery
Gravel		(Thin-wall, open drive samplers not suitable)				
Sand	Moist	Dense	18	0 to ½	Fair to poor	
Sand	Moist	Loose	12	½	Poor	Recommend piston sampler.
Sand	Saturated	Dense	18 to 24	0	Poor	Recommend piston sampler.
Sand	Saturated	Loose	12 to 18	0	Poor	Recommend piston sampler.
Silt	Moist	Firm	18	½	Fair to good	
Silt	Moist	Soft	12 to 18	½	Fair	Recommend piston sampler.
Silt	Saturated	Firm	18 to 24	0	Fair to poor	
Silt	Saturated	Soft	12 to 18	0 to ½	Poor	Recommend piston sampler.
Clay and shale	Dry to saturated	Hard	(Thin-wall, open drive sampler not suitable)			Recommend double-tube sampler.
Clay	Moist	Firm	18	½ to 1	Good	
Clay	Moist	Soft	12 to 18	1	Fair to good	
Clay	Saturated	Firm	18 to 24	0 to 1	Good	
Clay	Saturated	Soft	18 to 24	½ to 1	Fair to poor	Recommend piston sampler.
Clay	Wet to saturated	Expansive	18 to 24	½ to 1½	Good	

(c) *Fixed-Piston Samplers.*—There are two types of fixed-piston samplers, the Hvorslev-type sampler, figures 2–14 and 2–15, and the Osterberg-type sampler, figure 2–16. These samplers are available in 3-inch-outside-diameter and 5-inch-outside-diameter sizes. With these sizes, the sampling tubes are conveniently interchangeable with the thin-wall, open drive sampler discussed in paragraph 4(b) above. The details of the fixed-piston samplers and the principles of operation are shown in figures 2–14, 2–15, 2–16, and 2–17.

The principal use of these samplers is for taking soil samples below the ground-water table. They are particularly adapted to sampling cohesionless sands and soft, wet soils that cannot be recovered by the thin-wall, open drive sampler. Although these samplers can be used for many soil conditions, they are recommended only for soils which cannot be sampled by the thin-wall, open drive sampler since they involve many working parts which complicate their operation. As in the case of the thin-wall, open drive sampler, they will not successfully sample gravelly or cemented soils or soils too hard for penetration.

The basic principle of operation for the fixed-piston samplers is the same as for the thin-wall, open drive sampler; that is, to force a thin-wall, cylindrical tube into the undisturbed soil in one continuous stroke, without rotation. The sampling tubes have been described previously in paragraph 4(b) above. However, tubes with little or no inside clearance are generally most desirable for use with the fixed-piston samplers. The sampler head of the Hvorslev-type sampler is designed for the 3-inch tube, and an adapter ring and enlarged piston assembly are available for the 5-inch tube. The Osterberg-type sampler is manufactured in the 3- and 5-inch sizes. The two sizes are available as individual samplers and the parts are not interchangeable.

Since the soil being sampled is generally saturated, water or drilling mud may be used during the drilling and sampling operation. In sampling of cohesionless sands, suitable drilling mud and proper mud techniques are often necessary for obtaining satisfactory sample recovery.

(1) *Operation of the Hvorslev-type sampler.*—The sampler is assembled with the piston locked flush with the bottom of the sampling tube, and lowered through the drilling mud to the bottom of the cleaned drill hole, drill rod being added as required. When the sampler is in place at the bottom of the hole, the piston rod, which is standard ½-inch pipe with flush couplings, is lowered through the drill rod and screwed onto the piston rod projecting at the top of the sampler. Another method which may prove more convenient is to connect the piston rod to the sampler before lowering it into the hole and add lengths of piston rod and drill rod as required. When these connections are made and the sampler shown in figure 2–14 is in position at the bottom of the hole,

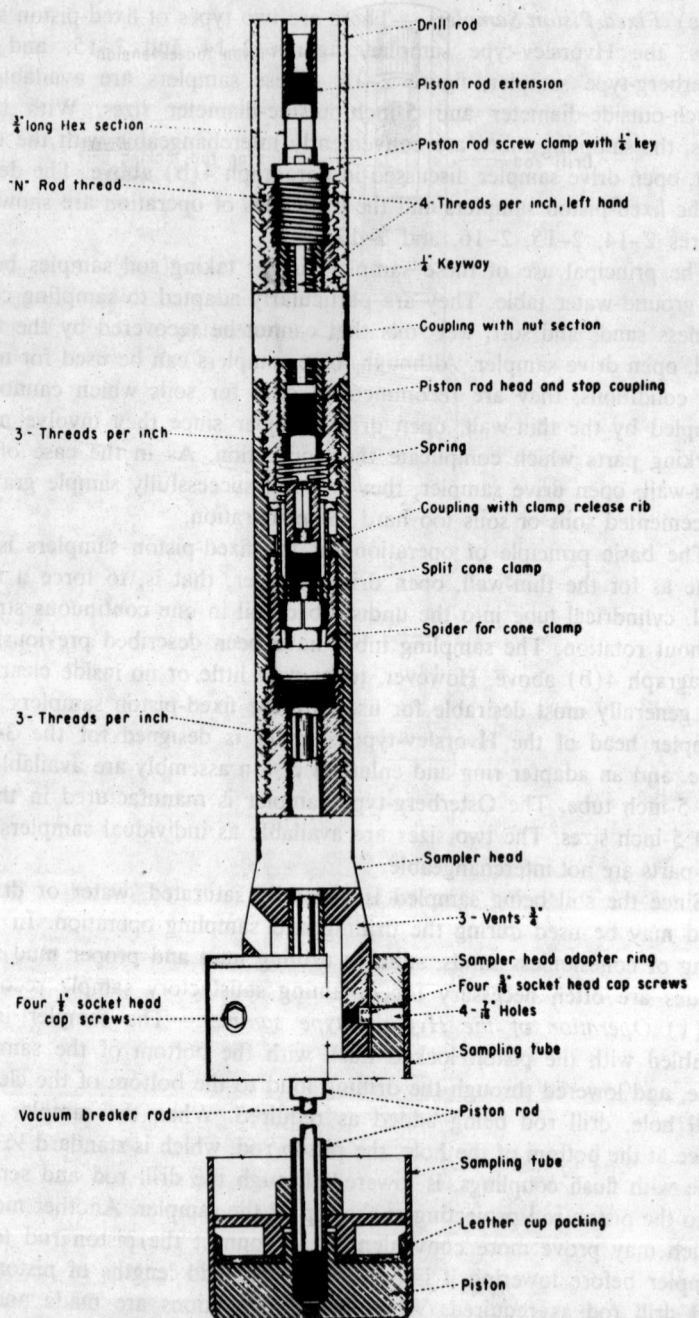

- Drill rod
- Piston rod extension
- Piston rod screw clamp with $\frac{1}{4}$ key
- "N" Rod thread
- 4 - Threads per inch, left hand
- $\frac{1}{4}$ long Hex section
- $\frac{1}{4}$ Keyway
- Coupling with nut section
- Piston rod head and stop coupling
- 3 - Threads per inch
- Spring
- Coupling with clamp release rib
- Split cone clamp
- Spider for cone clamp
- 3 - Threads per inch
- Sampler head
- 3 - Vents $\frac{3}{8}$"
- Sampler head adapter ring
- Four $\frac{3}{8}$" socket head cap screws
- 4 - $\frac{3}{8}$" Holes
- Four $\frac{1}{4}$" socket head cap screws
- Sampling tube
- Vacuum breaker rod
- Piston rod
- Sampling tube
- Leather cup packing
- Piston

Figure 2-14.—Thin-wall fixed-piston sampler (Hvorslev type). 101–D–198.

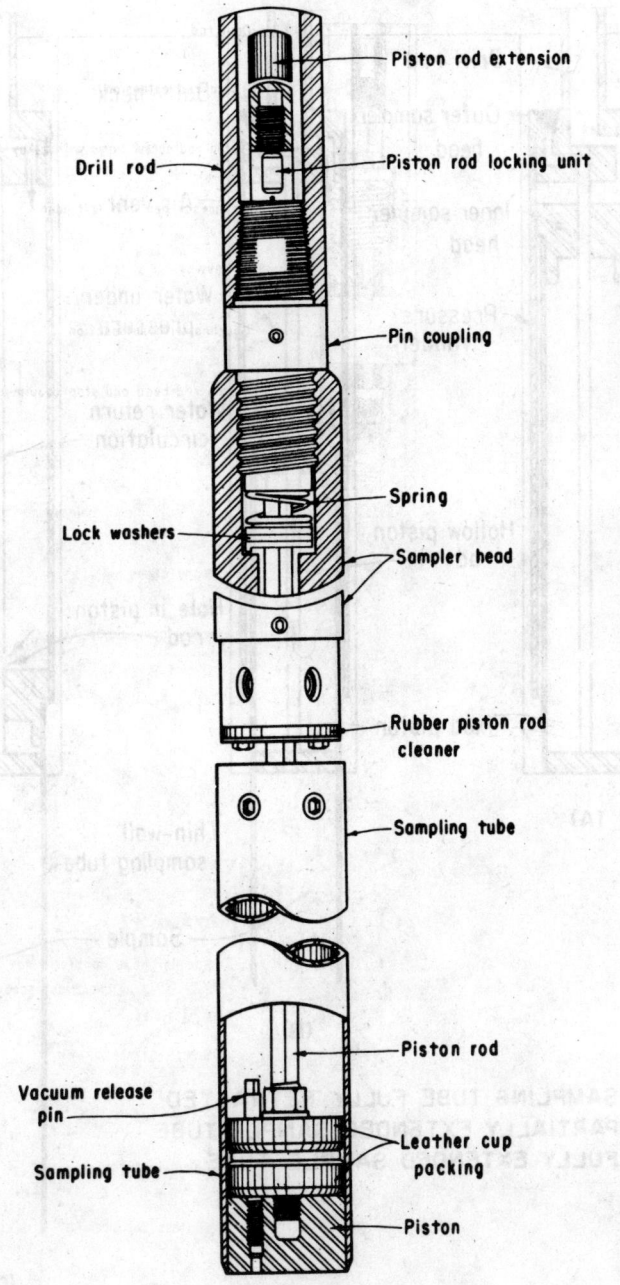

Figure 2–15.—Butter's thin-wall fixed-piston sampler (Hvorslev type). 101–D–535.

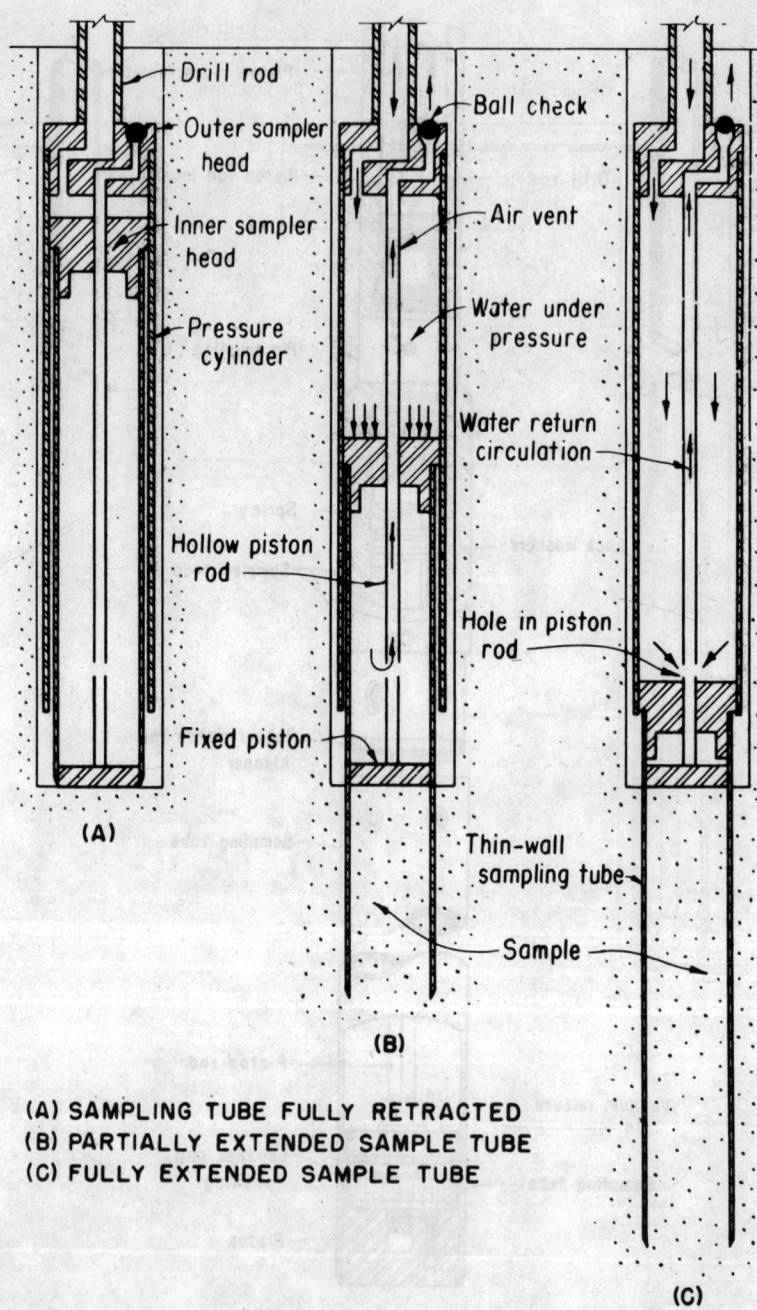

- Drill rod
- Ball check
- Outer sampler head
- Inner sampler head
- Air vent
- Pressure cylinder
- Water under pressure
- Water return circulation
- Hollow piston rod
- Hole in piston rod
- Fixed piston
- Thin-wall sampling tube
- Sample

(A)

(B)

(C)

(A) SAMPLING TUBE FULLY RETRACTED
(B) PARTIALLY EXTENDED SAMPLE TUBE
(C) FULLY EXTENDED SAMPLE TUBE

Figure 2–16.—Thin-wall fixed-piston sampler (Osterberg type). 101–D–190.

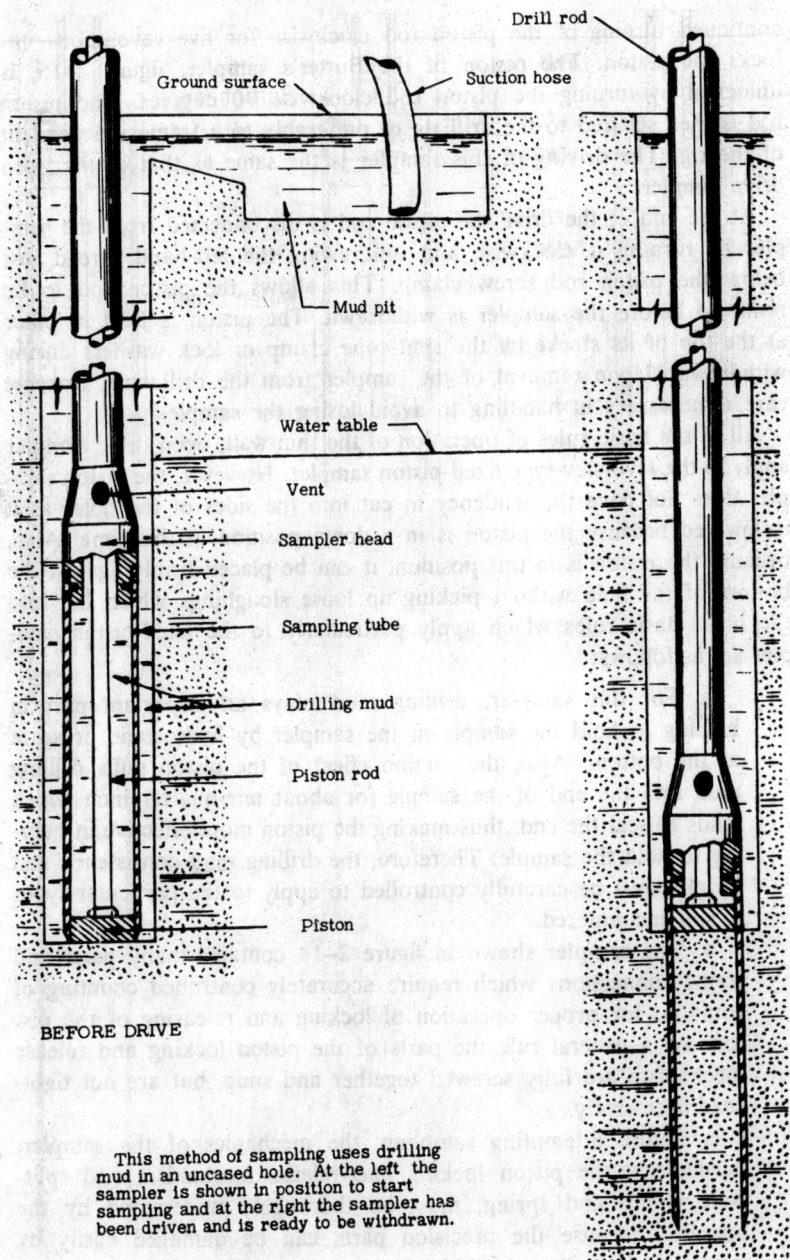

Drill rod

Ground surface

Suction hose

Mud pit

Water table

Vent

Sampler head

Sampling tube

Drilling mud

Piston rod

Piston

BEFORE DRIVE

AFTER DRIVE

This method of sampling uses drilling mud in an uncased hole. At the left the sampler is shown in position to start sampling and at the right the sampler has been driven and is ready to be withdrawn.

Figure 2-17.—Sampling with fixed-piston sampler (Hvorslev type). 101-D-345.

continued turning of the piston rod clockwise for five revolutions unlocks the piston. The piston of the Butter's sampler, figure 2–15, is
unlocked by turning the piston rod clockwise 90 degrees. The piston
rod is then secured to the drill rig or preferably to a frame independent
of the rig. The driving of this sampler is the same as that of the open
drive sampler.

At the end of the drive the piston rod is disconnected from the sampler by turning it clockwise and unscrewing the left-hand thread just
below the piston rod screw clamp. This allows the piston rod to be
removed before the sampler is withdrawn. The piston is held in place
at the top of its stroke by the split-cone clamp or lock washers during
withdrawal. Upon removal of the sampler from the drill hole, extreme
care is necessary in handling to avoid losing the sample.

All of the basic rules of operation of the thin-wall, open drive sampler
apply to the Hvorslev-type fixed-piston sampler. However, the piston sampler does not have the tendency to cut into the sides of the holes as it
is lowered because the piston is in a down position at this time. Also,
because the piston is in this position, it can be placed firmly against the
bottom of the hole without picking up loose sloughings which fall into
the hole. Basic rules which apply particularly to the fixed-piston sampler are as follows:

a. For this sampler, drilling mud plays an important part in
helping to hold the sample in the sampler by hydrostatic pressure
at the bottom. Also, the suction effect of the piston pulls drilling
mud into the end of the sample for about one-quarter inch which
tends to seal the end, thus making the piston more effective in helping to hold the sample. Therefore, the drilling mud consistency and
weight must be carefully controlled to apply to the particular types
of soil encountered.

b. The sampler shown in figure 2–14 contains many parts and
screw connections which require accurately controlled counting of
rotations for proper operation of locking and releasing of the piston. As a general rule the parts of the piston locking and release
mechanism are fully screwed together and snug, but are not tightened excessively.

c. Before attempting sampling, the mechanics of the sampler,
particularly the piston locking and release mechanism and split-
cone clamp and spring, must be thoroughly understood by the
operator because the precision parts can be damaged easily by
misuse or incorrect assembly. It is suggested that this sampler be
used only by experienced operators.

d. After the piston rod extension is screwed onto the piston rod,
continued turning rotates the piston rod screw clamp (fig. 2–14),

and five revolutions in a clockwise direction unlocks the piston. The piston of the Butter's sampler is unlocked by a clockwise 90-degree rotation of the piston rod. With this exception, procedures are the same for both Hvorslev-type samplers. This includes one extra revolution to be certain that the piston is unlocked before beginning the drive. It is good practice to assemble the sampler and check the rotations before using.

e. After the sampler and piston rod are in place, the piston rod should be secured to the drill rig or preferably to an independent frame so that it remains in a fixed position.

f. The driving of the sampler is the same as for the thin-wall, open drive sampler. The length of drive may be greater than shown in table 2-4.

g. After completion of the drive the piston rod and rod extensions are removed by continued turning clockwise until the left-hand screw (fig. 2-14) is fully released from the piston rod head. The rod extensions and the upper part of piston rod are then removed from the drill rod. The split-cone clamp or the lock washers of the Butter's sampler, holds the piston in position for removal of the sampler.

h. After rotating the sampler to break off the core, the sampler is carefully removed in the same manner as described for the thin-wall, open drive sampler. Before lifting the sampler out of the drilling mud, the hand or a block is placed over the bottom to prevent the sample from dropping out. In some instances it has been found advantageous to enlarge the drill hole to a depth of 1 foot so that a block can be slid under the end of the sampler while it is still about 1 foot down in the hole.

(2) *Operation of the Osterberg-type sampler.*—The sampler is assembled with the piston flush with the bottom of the sample tube and lowered through the drilling mud to the bottom of the cleaned drill hole, drill rod being added as required. When the sampler is in place at the bottom of the hole, the drill rod is securely clamped to the drill rig or the casing, if used, either of which will serve as the reaction for driving the sampler. By means of a high-pressure pump located at the surface, clear water is pumped through the drill rod and into the pressure cylinder above the inner sampler head (fig. 2-16). As water pressure is applied through the drill rod, the inner sampler head to which a sampling tube is attached is forced out of the pressure cylinder. When the inner sampler head has reached its full stroke and the sampling tube has penetrated its full depth into the soil, the water pressure is automatically relieved through a bypass provided in the hollow piston rod.

At the end of the drive, the sampler is rotated to break off the soil at the bottom of the tube. When the drill rod and the pressure cylinder are rotated, a friction clutch located in the inner sampler head will engage the pressure cylinder and cause the sample tube to rotate. Upon removal of the sampler from the drill hole, extreme care is necessary in handling to avoid losing the sample.

All of the basic rules of operation of the thin-wall, open drive sampler apply to the Osterberg-type fixed-piston sampler. However, the piston sampler is not susceptible to cutting into the side of the hole as it is lowered because the piston is in a down position at this time. Also, because the piston is in this position, it can be placed firmly against the bottom of the hole without picking up loose sloughings which fall into the hole. Basic rules which apply particularly to the piston sampler are as follows:

a. For this sampler, drilling mud plays an important part in helping to hold the sample in the sampler by hydrostatic pressure at the bottom. Also, the suction effect of the piston pulls drilling mud into the end of the sample for about one-quarter inch which tends to seal the end, thus making the piston more effective in helping to hold the sample. Therefore, the drilling mud consistency and weight must be carefully controlled for the particular types of soil encountered.

b. This sampler contains some moving parts and O-ring seals which must be kept clean and properly lubricated. Proper precautions must be taken to avoid rusting or damage to the inner surface of the pressure cylinder.

c. The sampler is rather difficult to assemble without causing damage to the O-ring seals; therefore, complete disassembly of the sampler should be avoided whenever possible.

d. To avoid contaminating the sampler, clean water should be used as the hydraulic medium for driving the sample tube. The use of drill rig mud pumps to furnish the driving force is not recommended unless the pumps have been thoroughly cleaned.

e. After completion of the drive the sampler should be rotated to break off the soil at the bottom of the sample tube. The rotational force should be applied in a smooth uniform manner in order to avoid damage to the friction clutch.

f. After rotating the sampler to break off the core, the sampler is carefully removed in the same manner as described for the thin-wall, open drive sampler. Before lifting the sampler out of the drilling mud, the hand or a block is placed over the bottom to prevent the sample from dropping out. In some instances it has been found advantageous to enlarge the drill hole to a depth of 1 foot so that a block can be slid under the end of the sampler while it is still about 1 foot down in the hole.

The samples should be clearly marked and prepared for shipment as required in paragraph 5(d). When cohesionless sands and saturated soft silts are sampled, often the dry density of the material is the most important consideration. In this instance the ends of the sample are cleaned out, the sample length measured, and the sample weighed in the field immediately after removal from the drill hole. After weighing, the sample may be shipped by sealing both ends of the sample tube with expanding packers. However, if only inplace density determination is required, the entire core may be removed from the sample tube and placed in an airtight can for later moisture content determination. The sample in the can, or a sacked sample of the dry material after moisture determination, should be shipped.

(d) *Core Sampler.*—Conventional rotary drilling and sampling were originally developed for drilling through both hard and soft rocks. Core barrels will obtain cores from ¾ to 6 inches in diameter. There are three principal types of core barrels: single-tube, double-tube, and triple-tube. The single-tube is the simplest in design, consisting of a core-barrel head, core barrel, and attached coring bit which cuts an annular groove that permits passage of drilling fluid pumped through the drill rod. This design exposes the core to drilling fluid over its entire length, which results in serious core erosion of unconsolidated or weakly cemented materials. Therefore, it is used primarily to sample hard, solid rock which requires the use of diamond drills and accessories. Figures 2–18 through 2–22 show dimensional standards established by the Diamond Core Drill Manufacturers Association, composed of members in the United States and Canada, for a series of nesting casings with corresponding sizes for bits and drill rods. The size combination is such that HX core-barrel bits will pass through flush coupled HX casing and will drill a hole large enough to admit flush coupled NX casing and so on to RX size. Flush joint casing is such that 6- x 7¾-inch core barrel bits will pass through ZW casing and will drill a hole large enough to admit flush jointed UW casing and so on to RW size. Figure 2–22 shows the dimensions of standard drill rods. A core box similar to that shown in figure 63 is used for storing and shipping rock cores.

The double-tube core barrel sampler, in addition to the outer rotating barrel, provides an inner stationary barrel which protects the core from the drilling fluid and reduces the torsional forces transmitted to the core. It is used to sample soft or fractured rock and may be used to obtain cores in hard, brittle, or partially cemented soils, such as shale and clay-

THREE LETTER NAMES		
FIRST LETTER	**SECOND LETTER**	**THIRD LETTER**
HOLE SIZE	GROUP	DESIGN
Casing, core barrel, diamond bit, reaming shell and drill rods designed to be used together for drilling an approximate hole size.	Key diameters standardized on an integrated group basis for progressively reducing hole size with nesting casings.	The standardization of other dimensions, including thread characteristics, to permit interchangeability of parts made by different manufacturers.

Letter	Inches	Millimeters		
R	1	25	Letters X and W are synonymous when used as the GROUP (second) letter. Any DCDMA standard tool with an X or W as the GROUP letter belongs in that DCDMA integrated group of tools designed using nesting casings and tools of sufficient strength to reach greater depths with minimum reductions in core diameter.	The DESIGN (third) letter designates the specific design of that particular tool. It does not indicate a type of design.
E	1½	40		
A	2	50		
B	2½	65		
N	3	75		
K	3½	90		
H	4	100		
P	5	125		
S	6	150		
U	7	175		
Z	8	200		

TWO LETTER NAMES	
FIRST LETTER	**SECOND LETTER**
HOLE SIZE	GROUP AND DESIGN
Approximate hole size, same as in 3-letter names.	GROUP standardization of key diameters for group integration and DESIGN standardization of other dimensions affecting interchangeability.

Figure 2-18.—Diamond Core Drill Manufacturers Association nomenclature.
288-D-2887.

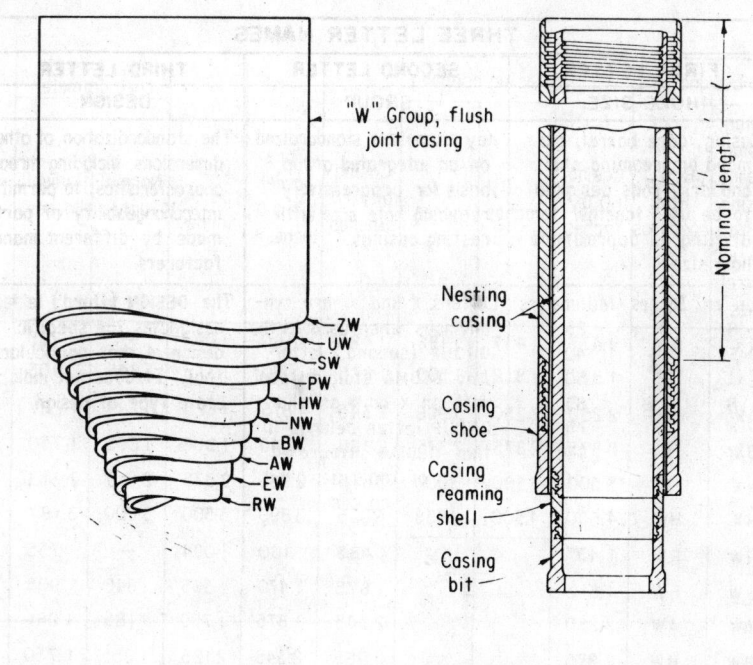

"W" Group, flush joint casing

ZW
UW
SW
PW
HW
NW
BW
AW
EW
RW

Nesting casing

Casing shoe

Casing reaming shell

Casing bit

Nominal length

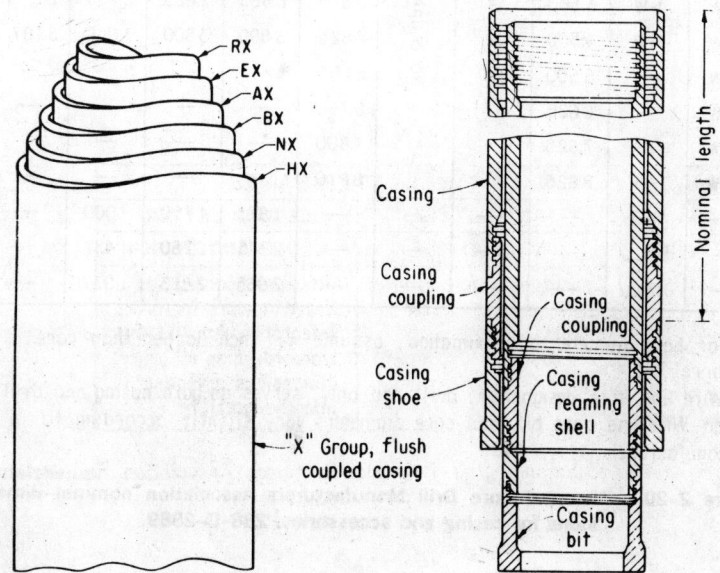

RX
EX
AX
BX
NX
HX

"X" Group, flush coupled casing

Casing

Casing coupling

Casing coupling

Casing shoe

Casing reaming shell

Casing bit

Nominal length

Note: Use of casing shoe allows nesting; use of casing bit does not.

Figure 2–19.—Core drill casing. 288-D-2888.

Size Designations (Casing; Casing coupling; Casing bits; Core barrel bits)	Rod; Rod couplings	Casing O.D., inches	Casing coupling O.D., inches	Casing coupling I.D., inches	Casing bit, O.D., inches	Core barrel bit O.D., inches*	Drill rod O.D., inches	Approximate core diameter Normal, inches	Approximate core diameter Thinwall, inches
R X	RW	1.437	1.437	1.188	1.485	1.160	1.094	—	.735
E X	E	1.812	1.812	1.500	1.875	1.470	1.313	.845	.905
A X	A	2.250	2.250	1.906	2.345	1.875	1.625	1.185	1.281
B X	B	2.875	2.875	2.375	2.965	2.345	1.906	1.655	1.750
N X	N	3.500	3.500	3.000	3.615	2.965	2.375	2.155	2.313
H X	HW	4.500	4.500	3.938	4.625	3.890	3.500	3.000	3.187
R W	RW	1.437	Flush joint	No coupling	1.485	1.160	1.094	—	.735
E W	E W	1.812			1.875	1.470	1.375	.845	.905
A W	A W	2.250			2.345	1.875	1.750	1.185	1.281
B W	B W	2.875			2.965	2.345	2.125	1.655	1.750
N W	N W	3.500			3.615	2.965	2.625	2.155	2.313
H W	H W	4.500			4.625	3.890	3.500	3.000	3.187
P W	—	5.500			5.650	—	—	—	—
S W	—	6.625			6.790	—	—	—	—
U W	—	7.625			7.800	—	—	—	—
Z W	—	8.625			8.810	—	—	—	—
—	AX 1/	—	—	—	—	1.875	1.750	1.000	—
—	BX 1/	—	—	—	—	2.345	2.250	1.437	—
—	NX 1/	—	—	—	—	2.965	2.813	1.937	—

* For hole diameter approximation, assume $\frac{1}{32}$ inch larger than core barrel bit.

1/ Wire line size designation, drill rod only, serves as both casing and drill rod. Wire line core bit, and core diameters vary slightly according to manufacturer.

Figure 2–20.—Diamond Core Drill Manufacturers Association nominal dimensions for casing and accessories. 288–D–2889.

stone, or cores of soft, partially consolidated, or weakly cemented soils. For these materials, hard metal drill bits are used.

Many of the double-tube core barrels either can be or have been modified slightly to accept a sample liner inside the inner barrel. This addition serves as a container to ship the core in and also eliminates the possibility of damage to a core upon removal from the inner barrel.

The triple-tube core barrel has been designed with a rotating outer barrel, stationary inner barrel, and a split liner inside the inner barrel to accept the core sample. Removal of the core is accomplished by exerting a hydraulic pressure over a piston located at the upper end of

Coring bit size	Nominal *		Set size *	
	O.D.	I.D.	O.D.	I.D.
RWT	$1\frac{5}{32}$	$\frac{3}{4}$	1.160	.735
EWT	$1\frac{1}{2}$	$\frac{29}{32}$	1.470	.905
EX, EXL, EWG, EWM	$1\frac{1}{2}$	$\frac{13}{16}$	1.470	.845
AWT	$1\frac{7}{8}$	$1\frac{9}{32}$	1.875	1.281
AX, AXL, AWG, AWM	$1\frac{7}{8}$	$1\frac{3}{16}$	1.875	1.185
BWT	$2\frac{3}{8}$	$1\frac{3}{4}$	2.345	1.750
BX, BXL, BWG, BWM	$2\frac{3}{8}$	$1\frac{5}{8}$	2.345	1.655
NWT	3	$2\frac{5}{16}$	2.965	2.313
NX, NXL, NWG, NWM	3	$2\frac{1}{8}$	2.965	2.155
HWT	$3\frac{29}{32}$	$3\frac{3}{16}$	3.889	3.187
HWG	$3\frac{29}{32}$	3	3.889	3.000
$2\frac{3}{4}$ x $3\frac{7}{8}$	$3\frac{7}{8}$	$2\frac{3}{4}$	3.840	2.690
4 x $5\frac{1}{2}$	$5\frac{1}{2}$	4	5.435	3.970
6 x $7\frac{3}{4}$	$7\frac{3}{4}$	6	7.655	5.970
AX Wire line 1/	$1\frac{7}{8}$	1	1.875	1.000
BX Wire line 1/	$2\frac{3}{8}$	$1\frac{7}{16}$	2.345	1.437
NX Wire line 1/	3	$1\frac{15}{16}$	2.965	1.937

* All dimensions are in inches, to convert to millimeters, multiply by 254.
1/ Wire line dimensions and designations may vary according to manufacturer.

Figure 2–21.—Diamond Core Drill Manufacturers Association commercial standards for core bit. 288–D–2891.

the split liner. Plastic or metal sample containers may be used in lieu of the split liner for shipping purposes.

Some of the latest core samplers have been designed on the order of the Pitcher double-tube soil sampler with a spring loaded retractable inner barrel. This design enables the same type of core barrel to be used for either soil or hard rock coring. The retractable inner barrel

Size designation	Remarks	Rod and coupling O.D.	Rod I.D.	Coupling I.D.	Threads per inch
E	Old standard still in use on many projects	1.313	.844	.438	3
A		1.625	1.266	.563	3
B		1.906	1.406	.625	5
N		2.375	2.000	1.000	4
RW	Present standard	1.094	.719	.406	4
EW		1.375	1.000	.437	3
AW		1.718	1.344	.625	3
BW		2.125	1.750	.750	3
NW		2.625	2.250	1.375	3
HW		3.500	3.062	2.375	3
KWY	Tapered thread (A.P.I.)	2.875	2.312	1.375	4
HWY		3.500	2.875	1.750	4
AQ	Wire line drill rods 1/ Taper thread	1.750	1.375	—	4
BQ		2.188	1.813	—	3
NQ		2.750	2.375	—	3
HQ		3.500	3.063	—	3
PQ		4.500 4.625 2/	4.063	4.063	3

1/ Wire line drill rod dimensions and designations may vary according to manufacturer.

2/ For PQ size designation, rod O.D. = 4.500 inches and coupling O.D. = 4.625 inches.

Figure 2–22.—Diamond Core Drill Manufacturers Association standard drill rod sizes. 288–D–2890.

and soil coring bits are replaced with a standard inner barrel and diamond bits for rock coring. The double-tube soil samplers previously discussed are specially designed core samplers for soils which may be used when it is required to obtain undisturbed samples of certain soils for laboratory testing. The basic rules of operation as outlined for the double-tube samplers should be followed. If soil samples believed to be satisfactory for inspection or laboratory testing have been obtained, they should be prepared for shipment, using the same precautions for sealing and marking as described for the hand-cut samples.

Core samples of soils may be used for visual classification and certain index tests. Depending upon the degree of disturbance, they may also be used for determining inplace density and water content or for other limited laboratory tests. The most desirable samples for laboratory testing are the NX to 6-inch sizes, having a length at least 2½ times the diameter.

5. Records and Field Measurements.—A complete log of the exploration hole and information for each sample should be recorded on forms 7–1336 and 7–1337, figures 64 to 67, or on other forms as required. Undisturbed soil sampling data should be recorded on form 7–1656, figure 2–23 and submitted with the samples.

(a) *Undisturbed Samples.*—Detailed notes taken for each sample should include the following as appropriate:

(1) Ground-water elevation.

(2) Method of drilling, size of hole, casing and drilling fluid used.

(3) Test pit and size.

(4) Classification and description of soil.

(5) Complete log of hole above and below the sample (unless continuous undisturbed samples are obtained).

(6) Method of cleaning hole prior to sampling.

(7) Brief statement of purpose of sample and a sketch showing the relationship of the sample to the ground surface, bottom of the structure footings, canal subgrade, or side slopes, with dimensions and elevations.

(b) *Drive Samples.*—For drive samples, the length of drive and recovery length should be recorded. If field density determinations are made, form 7–1656 (fig. 2–23) must be completed.

(c) *Undisturbed Samples for Classification, Inplace Density, and Water Content Determination.*—When samples are obtained for classification, inplace density, and water content determination, the following data and information are necessary:

(1) Size of sample (length and diameter).

(2) The wet weight of sample.

(3) The water content. This may be determined in the field, or a sealed moisture sample may be obtained for shipment to a field laboratory or to the Engineering and Research Center laboratory.

7-1656
(4-62)
Bureau of Reclamation   UNDISTURBED SOIL SAMPLING DATA

Hole No. _____ DH-19 _____ Feature ___ Example ___
Ground elev. ___ 3846.3 ___ Project _____
Water table elev. _3830.5_ Foreman _____
Location ___ N. 198,276   E. 664,508 ___

Date	10-23-70	10-23-70	10-23-70	10-23-70
Sample No.		1-PR		1
Sample depth	0.0 - 3.7	3.7 - 5.2	5.2 - 6.0	6.0 - 8.5
Sampler type	Rotary drilling	Penetration resistance	Rotary drilling	3" drive
Sampler Tube No.				709
I. D. bit (in.)				2.820
Inside area, bit (sq ft)				0.0434
Initial penetration				0.0
Drive length				2.5
Total penetration				2.5
Sample recovery length				2.3
Sample recovery (%)				92.0
Trimmed sample length				1.92
Volume of sample (cu ft)				0.083
Wt of tube & packers + soil				17.03
Wt of tube & packers				9.68
Wt of wet soil				7.35
Moisture content (%)				20.43
Dry density (pcf)				73.5
Number of blows		8		
Soil classification	Classification made from cuttings on drill bit. CL-CH; Lean to fat clay	CL-CH; Lean to fat clay, medium to high plasticity	Classification made from cuttings on drill bit. CL-CH; Lean to fat clay.	Top is CL-CH lean to fat clay, medium to high plasticity. Bottom is SC, Sandy clay, approx 65% medium plasticity fines. Good sample.
Remarks	Hole advanced with a drag bit using a mixture of bentonite and water for drilling fluid.		4½-inch	

Note: All dimensions in feet and weight in pounds unless otherwise indicated.

Figure 2-23.-—Undisturbed soil sampling data form. 101–D–336.

(d) *Marking and Chipment of Undisturbed Samples.*—Before undisturbed samples are moved from the site, they should be marked as follows: The box, or cylinder, of hand-cut samples should be marked, and the liner of the double-tube soil or core sampler, or the tube of the drive samples, should be marked rather than marking the wooden discs or expanding packers. In the case of small core samples, it may be necessary to attach a tag with the proper information:

(1) It is of extreme importance that samples be marked as to "top" and "bottom" so that they can be properly oriented.

(2) Project, feature, hole number, and field sample number.

(3) Elevation or depths between which the sample was taken.

(4) Depth to or elevation of the top of the sample.

Sealed undisturbed samples should be packed (completely surrounded) with moist sawdust or similar packing material to reduce vibrations to a minimum. Hand-cut undisturbed cylinder and block samples should be packed in box containers for shipment. Two or more double-tube soil, drive, or core samples may be packed in a box container for shipment and transported in a horizontal position. Figure 2–24 shows the required dimensions of a box container which can be used for shipping sample

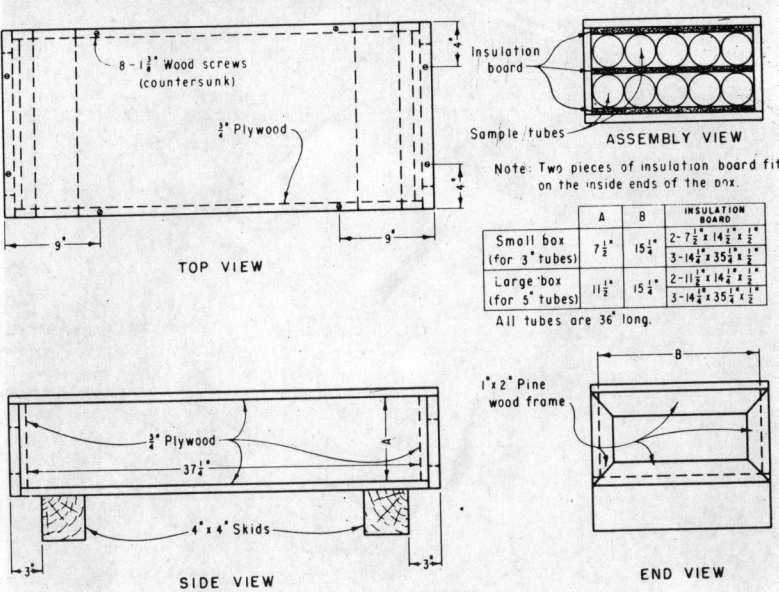

	A	B	INSULATION BOARD
Small box (for 3" tubes)	$7\frac{1}{2}$"	$15\frac{1}{4}$"	2-$7\frac{1}{2}$ x $14\frac{1}{4}$ x $\frac{1}{2}$
			3-$14\frac{1}{4}$ x $35\frac{1}{4}$ x $\frac{1}{2}$
Large box (for 5" tubes)	$11\frac{1}{2}$"	$15\frac{1}{4}$"	2-$11\frac{1}{2}$ x $14\frac{1}{4}$ x $\frac{1}{2}$
			3-$14\frac{1}{4}$ x $35\frac{1}{4}$ x $\frac{1}{2}$

All tubes are 36" long.

Figure 2–24.—Thin-wall tube box. 101–D–536.

tubes. The outside of the packing crate should be clearly marked as
follows:

>   This side up
>   Do not drop
>   Protect from freezing or overheating

Regardless of method of shipment, the samples should be sent to the
Denver Federal Center addressed as follows:

> Bureau of Reclamation, Engineering and Research Center, Division
> of General Research, Building 56, Denver Federal Center,
> Denver, Colo. 80225.

Figure 2-25 is an illustration of a box container considered as an
excellent method of packing and shipping undisturbed samples.

Shipping and sample tags, form 7-1622(3-57), are available upon
request from the above address, except use Division of Management
Support, Building 67.

(e) *Letter of Transmittal.*—A letter of transmittal should accompany
all samples. It should include:

(1) Bill of lading number and method and date of shipment.

Figure 2-25.—An excellent type of shipping container. Note the block samples
contained within separate boxes inside the shipping container. E-2179-1NA.

(2) Listing and identification of the samples giving type of sample, field sample number, hole number, elevation or depth.

(3) Purposes for which the samples were obtained.

(4) Log of exploration.

VISUAL AND LABORATORY METHODS FOR IDENTIFICATION AND CLASSIFICATION OF SOILS

Designation E–3

1. Scope.—This designation describes the methods and procedures for identifying, classifying, and describing soils in accordance with the Unified Soil Classification System.[1] The system is not limited to a particular use or geographic location. It does not conflict with other systems; in fact, the use of geologic, pedologic, textural, or local terms is encouraged as a supplement to, but not as a substitute for, the definitions, terms, and phrases established by the system and which are easy to associate with actual soils.

In this system 15 basic soil groups have been selected to define certain distinctive and peculiar engineering properties. Depending upon its basic properties, a soil is catalogued according to these groups and assigned a name and symbol; thus a soil is classified. These groups are broad; therefore, supplemental detailed word descriptions are required to point out peculiarities of a particular soil and differentiate it from others in the same group.

This system does not provide quantitative data for design purposes. It does provide qualitative information. Logs of exploration holes containing adequate soil classifications and descriptions may be used (1) in making preliminary estimates, (2) in determining the extent of additional field investigations needed for detailed design, (3) in planning an economical field testing or sampling program for laboratory testing, and (4) in extending the results of tests to additional explorations. In connection with the above, use charts have been developed to indicate the general engineering properties and potential value of the various soils

---

[1] This system based on the AC system by A. Casagrande was adopted jointly in 1952 by the Corps of Engineers and the Bureau of Reclamation. The procedure given here is adapted from a supplement to the Earth Manual published by the Bureau of Reclamation, Denver, Colo., 1953, and is similar to Technical Memorandum No. 3–357, and appendices A and B, prepared for the Office, Chief of Engineers, by Waterways Experiment Station, Vicksburg, Miss., in 1953.

for engineering uses.[2] For final detailed designs of important structures, the classification must be supplemented by laboratory tests or other quantitative data to determine the performance characteristics of the soil, such as permeability, shear strength, and compressibility under expected field conditions.

2. General Procedure.—Three steps are required to classify a soil.

(a) *First Step.*—The basic properties and characteristics of the soil components which influence the behavior of the soil as a foundation or construction material are identified. These include the sizes of particles, the amounts of the various sizes, and the influence of moisture on the characteristics of the very fine grains. Two methods are provided:

(1) The visual or field method, so called because manual (hand) tests and visual observations are employed in lieu of precise laboratory tests to define the basic soil properties. A knowledge of soil behavior and particularly an understanding of and experience in performing the gradation and soil consistency tests, upon which the hand tests and observations are based, are desirable prerequisites for competent visual classification. The visual method is used primarily in the field to classify and describe soils for logging exploration holes. This method is described in detail in Part A, Visual Method.

(2) The laboratory method, as the name implies, requires laboratory tests, specifically gradation and moisture limits, to define the basic soil properties. This method is used only when precise delineation is required, when unusual soils or conditions are encountered, or if the tests are required to supplement other laboratory tests required for design of major structures. It is also useful as an aid in teaching the visual classification method. This method is described in detail in Part B, Laboratory Method.

(b) *Second Step.*—The soil is placed into a classification group denoted by a group symbol, assigned in accordance with the criteria established by the system for the visual or laboratory method of classification.

(c) *Third Step.*—A written description of the soil is made. Regardless of the method used to identify the basic properties and characteristics, descriptive information is necessary to differentiate between soils in the same group (see pars. 8 and 13 for coarse- and fine-grained soils, respectively). The descriptive information required also depends on the purpose for which a soil is being investigated. For construction materials, as borrow for embankment, base course, backfill, or other uses, paragraph 8 or 13 applies; and for foundations for structures, the require-

---

[2] The basic principles of the system and use of the classification information are discussed in chapter I.

ments are given in paragraph 14. Examples of field classification and description are given on the classification chart, figure 3–1, and on the data forms, figure 3–2.

Part A. Visual Method

3. Apparatus.—Special apparatus or equipment is not required. However, the following items will facilitate the work:

(1) A rubber syringe or a small oil can having a capacity of approximately ½ pint.

(2) A supply of clean water.

(3) Small bottle of dilute hydrochloric acid.

(4) Classification chart, figure 3–1.

4. Procedure.—The classification of a soil by this method is based on visual observations and estimates of its behavior in a remolded state. The procedure is, in effect, a process of elimination, beginning on the left side of the classification chart, figure 3–1 (see column headed Field Identification Procedures), and working to the right until the proper group symbol is obtained. The group symbol must be supplemented by detailed word descriptions, including a description of the inplace conditions for soils to be used in situ as foundations.

By recording, briefly, the observations made in the step by step procedure given below, the information for classifying and describing the soil is obtained. The forms shown on figure 3–2 are recommended for use in the laboratory for training purposes as an aid to attaining proficiency in classification and logging procedures. However, final field classification of soils should be recorded on the form shown on figure 3–3. The classification chart and a check list of descriptive items in paragraphs 8, 13, and 14 are helpful in classifying soils in the field.

(*Note:* Many natural soils will have properties not clearly associated with any one soil group, but which are common to two or more groups. Or they may be near the borderline between two groups, either in percentages of the various sizes or in plasticity characteristics. For this substantial number of soils, borderline classifications are used; that is, an appropriate dual symbol is assigned. A dual symbol consists of the two group symbols most nearly indicating the proper soil description, connected by a hyphen as, for example GW–GC, SC–CL, ML–CL, and others.)

5. Selection and Preparation of Sample.—Select a representative sample of the soil and spread it on a flat surface or in the palm of the hand.

FIELD IDENTIFICATION PROCEDURES FOR FINE-GRAINED SOILS OR FRACTIONS

These procedures are to be performed on the minus No. 40 sieve size particles, approximately $\frac{1}{64}$ in. For field classification purposes, screening is not intended; simply remove by hand the coarse particles that interfere with the tests.

DILATANCY (Reaction to shaking)

After removing particles larger than No. 40 sieve size, prepare a pat of moist soil with a volume of about one-half cubic inch. Add enough water if necessary to make the soil soft but not sticky.

Place the pat in the open palm of one hand and shake horizontally, striking vigorously against the other hand several times. A positive reaction consists of the appearance of water on the surface of the pat which changes to a livery consistency and becomes glossy. When the sample is squeezed between the fingers, the water and gloss disappear from the surface, the pat stiffens, and finally it cracks or crumbles. The rapidity of appearance of water during shaking and of its disappearance during squeezing assist in identifying the character of the fines in a soil.

Very fine clean sands give the quickest and most distinct reaction whereas a plastic clay has no reaction. Inorganic silts, such as a typical rock flour, show a moderately quick reaction.

DRY STRENGTH (Crushing characteristics)

After removing particles larger than No. 40 sieve size, mold a pat of soil to the consistency of putty, adding water if necessary. Allow the pat to dry completely by oven, sun, or air drying, and then test its strength by breaking and crumbling between the fingers. This strength is a measure of the character and quantity of the colloidal fraction contained in the soil. The dry strength increases with increasing plasticity.

High dry strength is characteristic for clays of the CH group. A typical inorganic silt possesses only very slight dry strength. Silty fine sands and silts have about the same slight dry strength, but can be distinguished by the feel when powdering the dried specimen. Fine sand feels gritty whereas a typical silt has the smooth feel of flour.

TOUGHNESS (Consistency near plastic limit)

After removing particles larger than the No. 40 sieve size, a specimen of soil about one-half inch cube in size is molded to the consistency of putty. If too dry, water must be added and if sticky, the specimen should be spread out in a thin layer and allowed to lose some moisture by evaporation. Then the specimen is rolled out by hand on a smooth surface or between the palms into a thread about one-eighth inch in diameter. The thread is then folded and rerolled repeatedly. During this manipulation the moisture content is gradually reduced and the specimen stiffens, finally loses its plasticity, and crumbles when the plastic limit is reached.

After the thread crumbles, the pieces should be lumped together and a slight kneading action continued until the lump crumbles.

The tougher the thread near the plastic limit and the stiffer the lump when it finally crumbles, the more potent is the colloidal clay fraction in the soil. Weakness of the thread at the plastic limit and quick loss of coherence of the lump below the plastic limit indicate either inorganic clay of low plasticity, or materials such as kaolin-type clays and organic clays which occur below the A-line.

Highly organic clays have a very weak and spongy feel at the plastic limit.

(a) Estimate and record the maximum particle size in the sample.

(b) Remove all particles larger than 3 inches from the sample. Estimate the percentage and distribution, by weight (volume is satisfactory), of cobbles (particles 3 to 12 inches in diameter) and boulders (particles over 12 inches in diameter) removed, and record as descriptive information (fig. 3–2), or in the proper columns on figure 3–3. *Only that fraction of the sample smaller than 3 inches is classified.*

6. Division Between Coarse- and Fine-Grained Soils.—Classify the sample as coarse-grained or fine-grained by estimating the percent, by weight, of particles which can be individually seen by the unaided eye. Soils containing more than 50 percent individually visible particles are coarse-grained soils. Soils containing less than 50 percent individually visible particles are fine-grained soils (see fig. 3–1).

(*Note:* For classification purposes, the No. 200 sieve size (0.074 mm.) is the particle-size division between fine- and coarse-grained particles. Particles of this size are about the smallest that can be seen individually by the unaided eye.)

7. Visual Procedure for Coarse-Grained Soils.—If it has been determined that the soil is coarse-grained, the soil is further identified by estimating and recording the percentage of: (1) gravel-sized particles, size range 3 inches to the No. 4 sieve (about ¼ inch); (2) sand-sized particles, size range No. 4 sieve to No. 200 sieve; and (3) silt- and clay-sized particles, size range smaller than No. 200 sieve.

(*Note:* The fraction of a soil smaller than the No. 200 sieve size, the clay and silt fraction, is referred to as "fines.")

(a) *Gravelly Soils.*—If the percentage of gravel is greater than the sand, the soil is a GRAVEL designated by the capital letter G.

Gravel-sized particles are further divided as follows:

> Coarse gravel—3 inches to ¾ inch
> Fine gravel—¾ inch to No. 4 (about ¼ inch)

These divisions are used to describe the average size of the gravel if poorly graded.

Gravels are further identified as being CLEAN (when containing less than 5 percent fines) or DIRTY (when containing more than 12 percent fines). However, the term "dirty" is usually not used in a description; instead, the properties of the fines that make the gravel "dirty" have to be described. Gravel containing 5 to 12 percent fines are given borderline classifications, that is, dual symbols. If the soil is obviously clean, the classification will be either:

(1) GW, well-graded, if there is good representation of all particle sizes, or

7-1458 (8-71)
Bureau of Reclamation

VISUAL CLASSIFICATION OF SOIL SAMPLES

PROJECT ___ Example   FEATURE ___   TABLE NO. ___ SHEET ___ OF ___

DESCRIPTIVE CLASSIFICATION:
1. DESCRIPTIVE CLASSIFICATION
2. PARTICLE SIZE, SHAPE, AND GRADATION (UNIFORMLY, WELL, POORLY GRADED, ETC.)
3. CONSISTENCY, ELASTICITY, ETC.
4. REACTION TO SHAKING TEST, DRY STRENGTH, ETC.

SAMPLE NUMBER	HOLE NUMBER	LOCATION OR STATION	DEPTH - feet (Meters)	MAXIMUM SIZE	GRAVEL (% PLUS #4)	SAND (% #4 TO #200)	FINES (% MINUS #200)	COLOR (WET STATE)	DESCRIPTION AND SOIL CLASSIFICATION	GROUP SYMBOL
16Y-										
1	3	Borrow area A	0 - 3.0	3"	70	30	0	Gray	Well-graded GRAVEL; clean, hard, subangular gravel sizes, considerable coarse subrounded sand sizes.	GW
2	3		3.0 - 6.0	1"	60	10	30	Tan	Clayey GRAVEL; predominantly fine, hard, subrounded gravel sizes, small amount of fine sand, clay portion slightly plastic, moderate reaction to HCL.	GC
3	3		6.0 - 12.0	6"	60	30	10	Brown	Well-graded GRAVEL; fairly clean, hard, angular gravel sizes, considerable sand, clay portion moderately plastic. (approximately 15 percent oversize 3" to 6", estimates made in field), moderate reaction to HCL.	GW-GC
4	5		1.5 - 3.0	8"	--	95	5	Tan	Poorly-graded SAND; hard, subangular, no medium sand sizes very few fines (approximately 10 percent oversize 3" to 8", estimates made in field) moderate reaction to HCL.	SP
5	5		3.0 - 10.0	½"	5	70	25	Brown	Silty SAND; predominantly coarse, subangular sand sizes, contains a few angular gravel particles and considerable nonplastic fines.	SM
6	5		10.0 - 15.0	#30	--	50	50	Tan	Silty SAND; fine to medium, poorly-graded hard, micaceous slightly plastic fines.	SM-ML
7	9		0 - 2.0	#50	--	15	85	Brown	Inorganic SILT; slight plasticity, contains some fine sand no dry strength.	ML
8	10	Borrow area B	0 - 8.0	#100	--	5	95	Gray	Inorganic CLAY; high plasticity, high dry strength, contains a trace of fine sand.	CH

TABLE ___ SHEET ___ OF ___

NOTE: Numbers in parentheses are metric equivalents of numbers directly above

GPO 836-377

Figure 3-2.—Data form for visual classification of borrow area samples. (Sheet 1 of 2.) 101-D-526.

7-1469 (8-71)
Bureau of Reclamation

VISUAL CLASSIFICATION OF FOUNDATION SAMPLES

PROJECT ___ Example   FEATURE ___   TABLE NO. ___   SHEET ___ OF ___

SAMPLE NUMBER	HOLE NUMBER	LOCATION OR STATION	DEPTH - feet (Meters)	MAXIMUM SIZE	GRAVEL (% PLUS #4)	SAND (% #4 TO #200)	FINES (% MINUS #200)	COLOR (WET STATE)	DESCRIPTION - UNDISTURBED STATE	SOIL CLASSIFICATION	GROUP SYMBOL
7N-										1. DESCRIPTIVE CLASSIFICATION. 2. PARTICLE SIZE, SHAPE, GRADATION (UNIFORMLY, WELL, POORLY GRADED, ETC.) 3. CONSISTENCY, ELASTICITY, ETC.	
1	DH-1	Sta. 18+00	12.0-13.1	#200	0	5	95	Dark gray	Denison Sample — SHALE: moist, hard, laminated; has a gray 1" bentonite seam (soapy feel), no reaction to HCl, impervious. (Pierre formation).	Fat CLAY, high plasticity.	CH
2	TP-3	Borrow area D	13.2-14.2	#100	0	20	80	Brown	12-inch Cube Samples — LOESS; soft, moist, contains numerous roots and root holes, moderate reaction to HCl.	Sandy SILT; slight plasticity.	ML
3	TP-4	Sta. 3+25 30' rt. of centerline	16.3-17.3	#100	0	10	90	Tan	CLAY; homogeneous, sample appears to be disturbed, firm, fairly moist, reacts violently to HCl, impervious. (Niobrara formation).	Lean CLAY; chalky.	CL
4	TP-8	Sta. 16+00	44.3-45.3	#50	0	30	70	Blue and tan	SAND; top 6" soft, very moist, bluish gray with minor fat clay stringers, porous structure, pervious, bottom 6" firm, moist, tan.	Poorly-graded SAND; predominantly fine, slightly clayey.	SP
5	DH-5	Sta. 8+00 centerline	46.1-47.0	2"	30	60	10	Brown	3-inch Core Samples — CEMENTED SAND; hard, dense, stratified, calcareous, appears pervious. (Ogallala formation).	Well-graded SAND; rounded particles, gravelly.	SW
6	DH-9	Sta. 4+25 Spillway centerline	5.6- 8.0	--	--	--	--	Brown	VERY HARD SILTSTONE; contains calcareous lenses, impervious, (Montezuma formation).	*Bedrock is not classified as a soil or given a soil group symbol.	

NOTE: Numbers in parenthesis are metric equivalents of numbers directly above.

TABLE ___ SHEET ___ OF ___

Figure 3-2.—Data form for visual classification of foundation samples. (Sheet 2 of 2.) 101-D-527.

7-1336 (Combining 7-1336 and 7-1338)
(8-70) Bureau of Reclamation

Feature **Example**

Hole No. **TP-109**

Depth to Water Level *Not reached

LOG OF TEST PIT OR AUGER HOLE
FOR BORROW AND FOUNDATION INVESTIGATIONS

Area Requiring **Check Structure #2**

Approx. Dimensions **4 × 5 feet**

Logged by _____

Project _____

Coordinates **Station 1212+25**

Method of Excavation **Hand-dug pit**

Ground Elevation **790.5**

Date **6-1 to 6-8. 19**

CLASSIFICATION SYMBOL		DEPTH (FEET)	SIZE AND TYPE SAMPLE TAKEN	CLASSIFICATION AND DESCRIPTION OF MATERIAL (SEE CHART "UNIFIED SOIL CLASSIFICATION"; GIVE GEOLOGIC AND IN-PLACE DESCRIPTION FOR FOUNDATION INVESTIGATIONS)	PERCENTAGE OF COBBLES AND BOULDERS **		
LETTER	GRAPHIC				VOLUME OF VOLUMES 3 TO 5-INCH (CUBIC FEET)	PERCENTAGE BY WEIGHT OF PLUS 3-INCH SAMPLED (LBS) TO 5-INCH	PERCENTAGE BY VOLUME OF SAMPLED (LBS) TO 5-INCH
GW		0'-7'	150-lb. sack, -3 inch	0'-7' Well-graded GRAVEL WITH COBBLES AND BOULDERS. Clean, approx. 70% hard, subrounded gravel, coarse to fine; approx. 30% coarse and medium sand; gray. Approx. 8% cobbles and 14% boulders (by volume) to 30-inch maximum size. Inplace condition - Loose, dry, nonstratified, slightly cemented, alluvial fan material.	6.7	82	137
CH		7'-14'	150-lb. sack sample	7'-14' Fat CLAY. Approx. 90% high plasticity fines, high dry strength, high toughness; approx. 10% medium sand; brown; no reaction with HCL; maximum size, medium sand. Inplace condition - soft, wet, homogeneous.	0	0	0
SC-CL		14'-22'	150-lb. sack sample	14'-22' Clayey SAND. Approx. 50% hard, angular, coarse to fine sand, slightly micaceous; trace of gravel, maximum size, 1/2-inch; approx. 50% medium plasticity fines; yellow, moderate reaction with HCL. Inplace condition - Firm, moist, homogeneous. nonstratified.	0	0	0
Sand-stone		22'		22' SANDSTONE. Hard, highly cemented.			

Not required for materials explorations

8.4 14.1

REMARKS   Average specific gravity of cobbles and boulders - 2.51 by displacement.
Samples obtained from sampling trench.

NOTES: Record water test and density test data, if applicable, under remarks.
* Record after water has reached its natural level; give date of reading adjacent to graphic symbol or in remarks.
** Applicable only to borrow pits and to foundations which are potential sources of construction materials.

$a = \dfrac{\text{(Lbs of rock sampled) 100}}{\text{(Bulk specific gravity of rock) 62.4 (Cubic feet of hole sampled)}}$
Record bulk specific gravity in Remarks, stating how obtained (measured or estimated)

Figure 3–3.—Data form for visual classification to be used in the field. 101-D-528.

(2) GP, poorly-graded, if there is either predominant excess or absence of particle sizes within the gravel range. The letters W and P can be used in classification symbols for the coarse-grained soils only when the percentage of fines is less than 12 percent.

If the soil obviously is dirty, the classification will be either:

(3) GM if the fines have little or no plasticity (silty), or

(4) GC if the fines are of low to medium or high plasticity (clayey). (See paragraphs 9 and 19 for procedure for classifying the "fines".)

(b) *Sandy Soils.*—If the percentage of sand is greater than gravel, the soil is a SAND designated by the capital letter S.

Sand-sized particles are further divided as follows:

Coarse sand—No. 4 (about ¼ inch) to No. 10 (about 3/32 inch)

Medium sand—No. 10 (about 3/32 inch) to No. 40 (about 1/64 inch)

Fine sand—No. 40 (about 1/64 inch) to No. 200 sieve (about 3/1,000 inch)

These divisions are used to describe the average size of the sand if poorly-graded.

The same procedure is applied as for gravels, except that the word SAND replaces GRAVEL and the symbol S replaces G. Thus, the clean sands will be classified as either:

(1) SW or

(2) SP

and the dirty sands will be classified as:

(3) SM if the fines have little or no plasticity (silty), or

(4) SC if the fines are of low to medium or high plasticity (clayey).

(c) *Borderline Classifications for Coarse-Grained Soils.*—Borderline classifications can occur within the coarse-grained soil division, between soils within either the gravel grouping or the sand grouping, and between gravelly and sandy soils.

The procedure is to assume the coarser soil, when there is a choice, and complete the classification and assign the appropriate group symbol; then, beginning where the choice was made, assume the finer soil and complete the classification, assigning the second group symbol.

Borderline classifications within the separate gravel or sand groups can occur; symbols such as GW–GP, GM–GC, GW–GM, SW–SP, SM–SC, and SW–SM are common.

Borderline classifications can occur between the gravel and sand groups; symbols such as GW–SW, GP–SP, GM–SM, and GC–SC are common.

In addition to the borderline classifications within the coarse-grained division, borderline classifications also occur within the fine-grained division (par. 11 (c)).

Borderline classifications can also occur between coarse- and fine-grained soils; classifications such as SM–ML and SC–CL are common.

8. Descriptive Information for Coarse-Grained Soils.—The following information is required for a complete description of coarse-grained soils and should be recorded in the appropriate columns on figures 3–2 or 3–3. All of these descriptive data are not always needed. Judgment should be used to include pertinent information, to avoid negative information, and to eliminate repetition. However, items (1), (2), (3), (8), and (11) should always be included.

(1) Typical name.

(2) Maximum size, distribution, and approximate percentage of cobbles and boulders (particles larger than 3 inches) *in the total material.*

(3) Approximate percentage of gravel, sand, and fines in the *fraction of soil smaller than 3 inches.*

(4) For poorly-graded materials, statement of whether sand or gravel is coarse, medium, fine, or skip-graded.

(5) Shape of the grains; rounded, subrounded, angular, subangular.

(6) The surface coating, cementation, and hardness of the grains and possible breakdown when compacted.

(7) The color and organic content.

(8) Moisture conditions; dry, moist, wet, very wet (near saturation).

(9) Plasticity of fines; none, slight, medium, high plasticity.

(10) Local or geologic name.

(11) Group symbol.

9. Visual Procedure for Fine-Grained Soils.—If it has been determined that the soil is fine-grained, the soil is further identified by estimating the percentages of gravel, sand, and fines (silt- and clay-sized particles), and performing the manual identification tests for dry strength, dilatancy, and toughness. (See field identification procedures for fine-grained soils or fractions on fig. 3–1.) By comparing the results of these tests with the requirements given for the six fine-grained soil groups, the appropriate group name and symbol is assigned. The same procedures are used to identify the fine-grained fraction of coarse-grained soils to determine whether they are silty or clayey.

10. Manual Identification Tests.—The tests for identifying fine-grained soils are performed on that fraction of the soil finer than the No. 40 sieve size (about 1/64 inch).

The manual tests are considered to be performed on the "fines." The soil finer than the No. 40 includes the "fines" (minus No. 200) and fine sand (minus No. 40 to No. 200).

Select a small representative sample and remove by hand all particles larger than the No. 40 size and prepare two small specimens, each with a volume of about ½ cubic inch, by moistening until the specimens can easily be rolled into a ball. Perform the tests listed below, carefully noting the behavior of the soil pat during each test.

(*Note:* Operators with considerable experience find that it is not necessary in all cases to prepare two-pats. For example, if the soil contains dry lumps, the dry strength can be readily determined without preparing a pat for this particular purpose.)

(a) *Dilatancy (Reaction to Shaking).*—Add enough water to nearly saturate one of the soil pats. Place the pat in the open palm of one hand and shake horizontally, striking vigorously against the other hand several times. Squeeze the pat between the fingers. The appearance and disappearance of the water with shaking and squeezing is referred to as a reaction (fig. 3–4). This reaction is called (1) quick, if water appears and disappears rapidly, (2) slow, if water appears and disappears slowly, and (3) no reaction, if the water condition does not appear to change. Observe and record the type of reaction as descriptive information.

(b) *Toughness (Consistency Near Plastic Limit).*—Dry the pat used in the dilatancy test, subparagraph (a) above, by working and molding until it has the consistency of putty. The time required to dry the pat is an indication of its plasticity. Roll the pat on a smooth surface or between the palms into a thread about ⅛ inch in diameter. Fold and reroll the thread repeatedly to ⅛-inch diameter so that its water content is gradually reduced until the ⅛-inch thread just crumbles. The water content at crumbling stage is called the plastic limit, and the resistance to molding at the plastic limit is called the toughness.

After the thread crumbles, the pieces should be lumped together and a slight kneading action continued until the lump crumbles. If the lump can still be molded slightly drier than the plastic limit and if high pressure is required to roll the thread between the palms of the hands, the soil is described as possessing high toughness. Medium toughness is indicated by a medium tough thread and a lump formed of the threads slightly below the plastic limit will crumble; while slight toughness is indicated by a weak thread that breaks easily and cannot be lumped together when drier than the plastic limit. This test also provides approximate information on the plasticity index, *PI* (designation E-7), of the

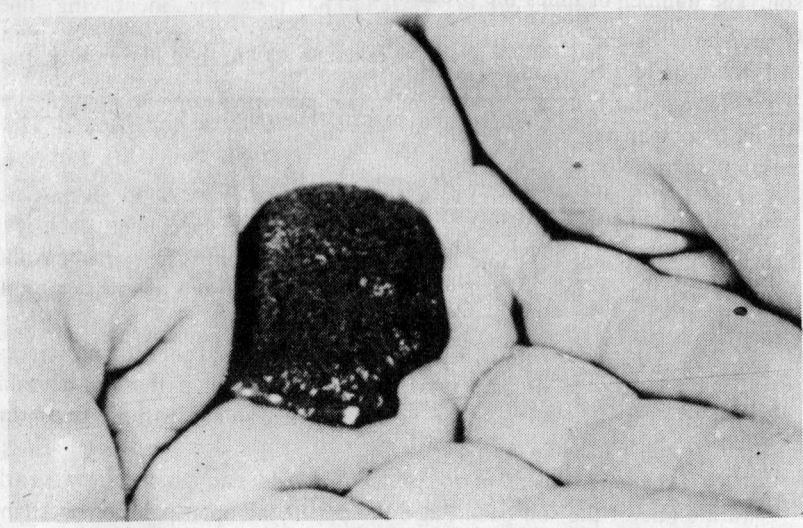

REACTION TO SHAKING

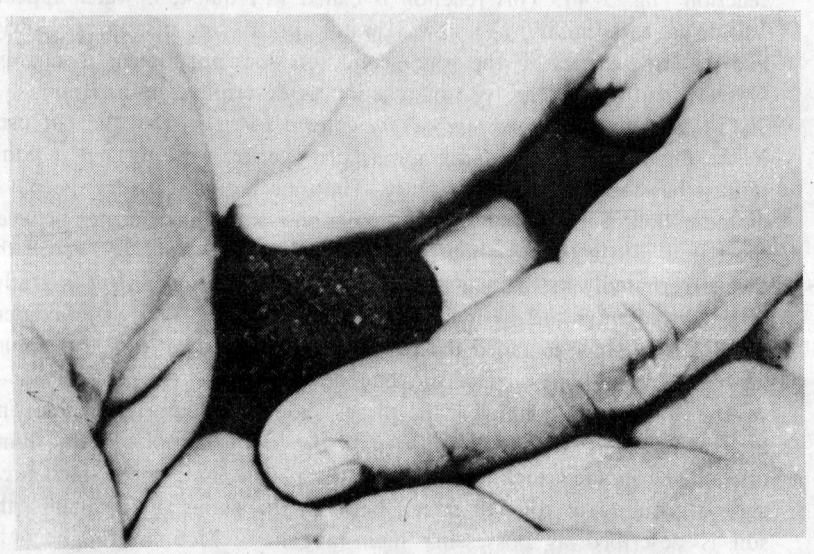

REACTION TO SQUEEZING

Figure 3-4.—Reactions of a silty soil to shaking and squeezing (dilatancy test). PX-D-16335.

soil. The number of times the procedure can be repeated is an indication of the *PI* of the material.

Highly organic clays have a very weak and spongy feel at the plastic limit. Nonplastic soils cannot be rolled into a thread of ⅛-inch diameter at any water content. Observe and record the toughness as descriptive information.

(c) *Dry Strength (Crushing Resistance).*—Completely dry one of the prepared specimens. Then measure its resistance to crumbling and powdering between the fingers. This resistance, called dry strength, is a measure of the plasticity of the soil and is influenced largely by the colloidal fraction contained. The dry strength is designated as slight if the dried pat can be easily powdered, medium if considerable finger pressure is required, and high if it cannot be powdered at all. Observe and record the dry strength as descriptive information.

(*Note:* The presence of high-strength, water-soluble cementing materials, such as calcium carbonates or iron oxides, may cause high dry strengths. Nonplastic soils, such as caliche, coral, crushed limestone, or soils containing carbonaceous agents, may have high dry strengths, but this can be detected by the effervescence caused by the application of dilute hydrochloric acid (see acid test, subpar. (e) below).)

(d) *Organic Content and Color.*—Fresh, wet, organic soils usually have a distinctive odor of decomposed organic matter. This odor can be made more noticeable by heating the wet sample. Another indication of the organic material is the distinctive dark color. Dry, inorganic clays develop an earthy odor upon moistening, which is distinctive from that of decomposed organic matter.

(e) *Other Identification Tests.*—

(1) *Acid test.*—The acid test using dilute hydrochloric acid (HCl) is primarily a test for the presence of calcium carbonate. For soils with high dry strength, a strong reaction indicates that the strength may be due to calcium carbonate as cementing agent, rather than colloidal clay. The results of this test (no reaction to HCL should be reported) should be included in the soil description. (*Note:* Dilute solution (1:3) of hydrochloric acid is one part of concentrated hydrochloric acid to three parts of distilled water. Handle with caution. Rinse with tap water if it comes in contact with skin.)

(2) *Shine.*—This is a quick supplementary procedure for determining the presence of clay. The test is performed by cutting a lump of dry or slightly moist soil with a knife. A shiny surface imparted to the soil indicates highly plastic clay, while a dull surface indicates silt or clay of slight plasticity.

(3) *Miscellaneous·*—Other criteria undoubtedly can be developed by the individual as he gains experience in classifying soils. For

example, differentiation between some of the fine-grained soils depends largely upon the experience in the "feel" of the soils. Frequent checking by laboratory tests is necessary to gain this experience.

11. Silty and Clayey Soils.—Various combinations of results of the manual identification tests indicate which grouping is proper for the soil in question.

(a) The following three groups are soils possessing *slight to medium* plasticity (symbol L):

(1) ML has little or no plasticity and may be recognized by slight dry strength, quick dilatancy, and slight toughness.

(2) CL has slight to medium plasticity and may be recognized by medium to high dry strength, very slow dilatancy, and medium toughness.

(3) OL is less plastic than the clay (CL) and may be recognized by slight to medium dry strength, medium to slow dilatancy, and slight toughness. Organic matter must be present in sufficient amount to influence the soil properties in order for a soil to be placed in this group.

(b) The following three groups are soils possessing *slight plasticity to high plasticity* (symbol H):

(1) MH is generally very absorptive. It has slight to medium plasticity and may be recognized by low dry strength, slow dilatancy, and slight to medium toughness. Some inorganic soils (such as kaolin which is a clay from a mineralogical standpoint) possessing medium dry strength and toughness will fall in this group.

(2) CH possesses high plasticity and may be recognized by high dry strength, no dilatancy, and usually high toughness.

(3) OH is less plastic than the fat clay (CH) and may be recognized by medium to high dry strength, slow dilatancy, and slight to medium toughness. Organic matter must be present in sufficient amount to influence soil properties in order for a soil to be placed in this group.

(c) *Borderline Classifications for Fine-Grained Soils.*—Borderline classifications can occur within the fine-grained soil division, between low and high liquid limit soils, and between silty and clayey soils. The procedure is comparable to that given for coarse-grained soils in paragraph 7(c); that is, first assume a coarse soil, when there is a choice, and then a finer soil and assign dual group symbols.

Borderline classifications which are common are as follows: ML–MH, CL–CH, OL–OH, CL–ML, ML–OL, CL–OL, MH–CH, MH–OH, and CH–OH.

12. Peat or Very Highly Organic Soils (Symbol Pt).—These may be readily identified by color, odor, sponginess, or fibrous texture.

13. Descriptive Information for Fine-Grained Soils.—The following information is required for a complete description of fine-grained soils and should be recorded in the appropriate columns of the log forms shown on figures 3–2 or 3–3. All of these descriptive data are not always needed. Judgment should be used to include pertinent information, to avoid negative information, and to eliminate repetition. However, items (1), (2), (6), (7), and (9) should always be included.

(1) Typical name.

(2) Maximum particle size. Distribution, and approximate percentage of cobbles and boulders (particles larger than 3 inches) *in the total material.*

(3) Approximate percentage of gravel, sand, and fines in the *fraction of soil smaller than 3 inches.*

(4) Hardness of the coarse grains, possible breakdown into smaller sizes.

(5) Color in moist condition and organic content.

(6) Moisture and conditions; dry, moist, wet, very wet (near saturation).

(7) Plasticity characteristics; none, slight, medium, high plasticity.

(8) Local or geologic name.

(9) Group symbol.

14. Descriptive Information for Foundation Soils.—The inplace condition of soils which are to be utilized as foundations for hydraulic or other structures assumes primary importance in soil classification. Logs of foundation explorations and descriptions of *undisturbed* samples, therefore, must emphasize the inplace conditions of the soil. It is necessary to present a complete word picture describing the soil as it exists in the foundation, in addition to assigning a name and proper group symbol.

Judgment should be used to include all pertinent information, to avoid negative information, and to eliminate repetition.

(a) *Coarse-Grained Soils.*—Items in table 3–1 should always be included when applicable. The information requested for each item can be recorded on the preprinted log in the approximate sequence in table 3–1. The degree of compactness and structure usually cannot be ascertained when augering; an exposed test pit or trench wall is essential for describing natural subsoil conditions. An example of a field log is shown on figure 3–3.

(b) *Fine-Grained Soils.*—Items listed in table 3–2 should always be included when applicable. The information requested for each item can be recorded on the preprinted log in the approximate sequence shown in

Table 3-1.—Check list for description of coarse-grained soils

Items of descriptive data	Typical information desired for sand and gravel
Typical name	GRAVEL; SAND; Clayey GRAVEL; Silty SAND WITH COBBLES; (Add descriptive adjectives for minor constituents—example: approximately 15 percent slight plasticity fines, medium toughness.)
Gradation	Well-graded; poorly-graded (uniformly-graded or skip-graded); (Describe range of particle sizes, such as fine to medium sand or fine to coarse gravel, or the predominant size or sizes as coarse, medium, fine sand or gravel.)
Size distribution	Approximate percent of gravel, sand, and fines in the fraction finer than 3 inches.
Plasticity of fines	None; low; medium; high.
Maximum particle size	Note percent of boulders and cobbles (by volume) as well as maximum particle size.
Mineralogy	Rock hardness for gravel and sand. Note especially presence of mica flakes, shaly particles, organic matter, or friable particles.
Grain shape	Angular; subangular; subrounded; rounded.
Color	Use one basic color, if possible.
Odor	None; earthy; organic.
Moisture condition	Dry; moist; wet; saturated.
Degree of compactness	Loose; dense.
Structure	Stratified; lensed, nonstratified; heterogeneous.
Cementation	Weak; moderate; strong. Note reaction to HCl as: none; weak; moderate; or strong.
Local or geologic name Group symbol	GP, GW, SP, SW, GM, GC, SM, SC, or the appropriate dual symbol when applicable. Should be compatible with typical name used above.

table 3-2. The items of consistency, degree of compactness, and structure usually cannot be ascertained when augering; an exposed test pit or trench wall is essential for describing natural subsoil conditions. If hard rock such as siltstone is encountered, it should not be given a soil group symbol such as ML but should be designated as siltstone on the log. An example of a field log is given on figure 3-3. The consistency of cohesive soils may be determined in place or on undisturbed samples in accordance with the identification procedure given in table 3-3.

The structural characteristics of intact soils provide important clues to their performance as foundation materials. Whenever undisturbed samples are available or when the soil profile may be inspected during sampling from a pit, the structural characteristics should be described. Stratified

Table 3–2.—Check list for description of fine-grained and partly-organic soils

Items of descriptive data	Typical information desired for silt and clay
Typical name	SILT: *Sandy SILT; CLAY; Lean or Fat CLAY; *Sandy CLAY; Silty CLAY; Organic SILT; Organic CLAY. * 25 percent or more sand must be present. "Gravelly" can be substituted for "Sandy" where applicable. Include cobbles and boulders in typical name when applicable.
Size distribution	Approximate percent of fines, sand, and gravel in fraction less than 3 inches in size. Must add to 100 percent.
Plasticity of fines	None; low; medium; high.
Dry strength	None; low; medium; high.
Dilatancy	None; very slow; slow; medium; quick.
Toughness near plastic limit	None; slight (low); medium; high.
Maximum particle size	Note percentage of cobbles and boulders (by volume) as well as maximum particle size.
Color	Use one basic color, if possible. Note presence of mottling or banding.
Odor	None; earthy; organic.
Moisture condition	Dry; moist; wet; saturated.
Consistency (see table 3–3) (for clay).	Very soft; soft; firm; hard; very hard.
Degree of compactness (for silt and fine sand).	Loose; dense.
Structure	Stratified; laminated (varved); fissured; slickensided; blocky; lensed; homogeneous. (The thickness, dip, and strike of layers should be included.)
Cementation	Weak; moderate; strong. Note reaction with HCl as: none; weak; moderate; or strong.
Local or geologic name	
Group symbol	CL, CH, ML, MH, OL, OH, Pt or the appropriate dual symbol when applicable. Should be compatible with typical name used above.

materials consist of alternating layers of varying types (or color). If the layers are less than about one-fourth inch thick, it may be described as laminated (or varved, if mostly fine-grained). Fissured materials break along definite planes of fracture with little resistance to fracturing. If the fracture planes appear polished or glossy, they should be described as slickensided. If a cohesive soil can be easily broken into small angular lumps which resist further breakdown, the structure may be described as

Table 3–3.—Identification of consistency of fine-grained soils

Consistency	Identification procedure
Very soft_____	Easily penetrate soil several inches using high thumb pressure or less.
Soft_____	Penetrate soil about 1 inch using high thumb pressure.
Firm_____	Soil indented less than ¼ inch using high thumb pressure.
Hard_____	Soil not indented using high thumb pressure. Readily indented by using thumbnail.
Very hard_____	Not indented with thumbnail.

blocky. A lensed structure is indicated by the inclusion of small pockets of different texture, such as small lenses of sand scattered through a mass of clay. The presence of special structural characteristics, such as root holes or porous openings, should also be noted. If no structural characteristics are apparent, the soil may be described as nonstratified or homogeneous.

Part B. Laboratory Method

15.—**Apparatus.**—Special apparatus is required as noted below:

(1) Equipment for performing the gradation test (see designation E–6).

(2) Equipment for performing the moisture limits tests (see designation E–7).

(3) A small bottle of dilute hydrochloric acid.

(4) Classification chart, figure 3–1.

16. **Procedure.**—The Unified Soil Classification System provides for precise delineation of the soil groups by using results of laboratory tests. For gradation and moisture limits, rather than visual estimates, see right-hand column of the classification chart, figure 3–1, entitled "Laboratory Classification Criteria." Classifying by these tests alone does not fulfill the requirements for complete classification as it does not provide an adequate description of the soil. Therefore, the descriptive information required for the visual method (pars. 8, 13, and 14) should also be included in the laboratory classification.

(a) *Preparation of Sample.*—Screen out the plus 3-inch fraction of the soil, noting the percentage.

(b) *Division between Fine- and Coarse-Grained Soils.*—Obtain the grain-size distribution of the minus 3-inch fraction by performing the laboratory gradation test. If the soil contains more than 50 percent by weight larger than the No. 200 sieve size, the soil is classified as coarse-grained; if less than 50 percent, it is classified as fine-grained.

17. Laboratory Procedure for Coarse-Grained Soils.—Coarse-grained soils are subdivided into GRAVEL and SAND (par. 7) by referring to the gradation curve instead of visually estimating the percentage of various sized particles present in the soil.

18. Gravelly or Sandy Soils.—Gravels or sands are further identified as being CLEAN or DIRTY by determining the amount of material finer than the No. 200 sieve. If less than 5 percent is finer than the No. 200 sieve, the soil will be classified as either:

(1) WELL-GRADED (GW or SW) if the coefficient of uniformity C_u is greater than 4 for gravels and 6 for sands, and the coefficient of curvature C_c is between 1 and 3; or

(2) POORLY-GRADED (GP or SP) if either one or both the C_u and C_c criteria for (1) above are not satisfied.

The coefficient of uniformity C_u and coefficient of curvature C_c are expressed as follows:

$$C_u = \frac{(D_{60})}{(D_{10})} \qquad C_c = \frac{(D_{30})^2}{(D_{10}) \times (D_{60})}$$

where:

D_{10}, D_{30}, and D_{60} are the grain-size diameters corresponding respectively to 10, 30, and 60 percent passing on the cumulative grain-size curve.

If more than 12 percent of the total soil is finer than the No. 200 sieve size, the soil will be classified as either:

(3) SILTY (GM or SM) if the results of the moisture limits tests show that the fines are silty—that is, the plot of liquid limit versus plasticity index falls below the "A" line (see plasticity chart, fig. 3–1)—or if the plasticity index is less than 4; or

(4) CLAYEY (GC or SC) if the fines are clayey—that is, the plot of liquid limit versus plasticity index falls above the "A" line and the plasticity index is greater than 7.

(a) *Borderline Classifications for Coarse-Grained Soils.*—Coarse-grained soils containing between 5 and 12 percent of fines are classified as borderline cases between the clean and the dirty gravels or sands as, for example, GW–GC or SP–SM. Similarly, borderline cases may occur in dirty gravels and dirty sands where the *PI* is between 4 and 7 as, for example, GM–GC or SM–SC. It is possible, therefore, to have a borderline case of a borderline case. The rule for correct classification in this case is to favor the nonplastic classification. For example, a gravel with 10 percent fines, a C_u of 20, a C_c of 2.0, and a PI of 6 would be classified GW–GM rather than GW–GC.

19. Laboratory Procedure for Fine-Grained Soils.—Soils containing more than 50 percent fines according to the grain-size curve are classified into one of the six fine-grained groups by the results of the moisture limits tests, as plotted on the plasticity chart, with attention being given to the organic content. Those with a liquid limit less than 50 are referred to as inorganic silts and clays of slight to medium plasticity, while those with a liquid limit greater than 50 are the elastic silts and fat clays of medium to high plasticity.

Organic silts and clays are usually distinguished from inorganic silts which have the same position on the plasticity chart by odor and color. However, when the organic content is doubtful, the material can be ovendried, remixed with water, and retested for liquid limit. The plasticity of fine-grained organic soils is greatly reduced on ovendrying owing to irreversible changes in the properties of the organic material. Ovendrying also affects the liquid limit of inorganic soils, but only to a small degree. A reduction in liquid limit after ovendrying to a value less than three-fourths of the liquid limit before ovendrying is positive identification of organic soils.

20. Subdivision of Fine-Grained Soils.—These soils are subdivided as follows:

(a) *Liquid Limit Less Than 50.*—

(1) ML has a liquid limit less than 50 and the plasticity index ranges from 0 to 22 (see area identified as ML or OL below the "A" line on the plasticity chart).

(2) CL has a liquid limit less than 50 and a plasticity index greater than 7 (see area identified as CL above the "A" line on the plasticity chart).

(3) OL contains sufficient organic material to affect the soil properties. The OL soils have liquid limits less than 50 and their plasticity indices range from 0 to 22 (see area identified as ML or OL below the "A" line on the plasticity chart).

(b) *Liquid Limit Greater Than 50.*—

(1) MH has a liquid limit greater than 50 and the plasticity index ranges from 0 to over 50 (see area identified as MH or OH below the "A" line on the plasticity chart).

(2) CH has a liquid limit greater than 50 and the plasticity index ranges from 22 to over 50 (see area identified as CH above the "A" line on the plasticity chart).

UNCLASSIFIED ROCK

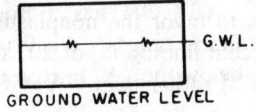

GROUND WATER LEVEL

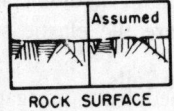

ROCK SURFACE

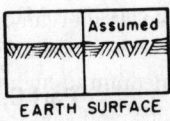

EARTH SURFACE

MISCELLANEOUS SYMBOLS

SAND

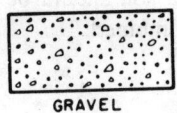

GRAVEL

SILT

CLAY

NOTE: ABOVE SYMBOLS MAY BE
COMBINED FOR BOUNDARY SOILS

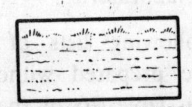

SOD, HUMUS or TOPSOIL

SOIL SYMBOLS

Figure 3-5.—Graphic symbols for soils. 103-D-345.

(3) OH has a liquid limit greater than 50 and the plasticity index ranges from 0 to over 50 (see area identified as OH or MH below the "A" line on the plasticity chart).

(c) *Borderline Classifications for Fine-Grained Soils.*—Fine-grained soils whose plot of liquid limit versus plasticity index falls on, or practically on, (1) "A" line or (2) the line LL = 50 should be assigned the appropriate borderline classification. Soils which plot above the "A" line, or practically on it, and which have a plasticity index between 4 and 7 are classified ML–CL.

Part C.　Graphic Symbols for Soils

21. Use of Symbols for Soils.—For pictorial presentation of a soil profile or log of exploration hole, graphic symbols for soils are sometimes advantageous and may be used. To avoid evaluation of a soil deposit on the basis of the graphic symbols, and for simplicity, graphic symbols are given only for the basic soil components: gravel, sand, silt, and clay (fig. 3–5). The graphic symbol for the noun in the group name assigned a soil should be used. For borderline soils the graphic symbols may be combined.

LISTS OF LABORATORY EQUIPMENT

Designation E–4

1. Scope.—This designation describes the equipment for soil materials field laboratories. The equipment requirements are divided into lists corresponding to the types of laboratories described in section 51, chapter III, and designated as follows: type A, for small-sized earthwork construction; type B, for medium-sized earthwork construction; and type C, for large centralized earthwork construction. Type V laboratory is a vehicle equipped for on-site earthwork control tests and is used as an auxiliary to type A, B, or C laboratories.

2. Equipment.—The following lists of laboratory equipment for soil materials testing were prepared without regard to the concrete control equipment listed in table 20 of the Concrete Manual, seventh edition. Therefore, there may be duplication of equipment common to both types of materials testing. Any unnecessary duplication should be eliminated when equipping a field laboratory.

Item	Quantities for laboratory				Description
	C Type	Type B	Type A	Type V	
1	2	2	1	1	Augers, hand, 4-, 6- and 8-inch diameter (expanding type or Iwan type may be used).
2	500	200	100	10	Bags, soil sample, waterproof, textile, 8-oz. burlap, 2-mil polyethylene-lined, asphalt-laminated, sewn-seam, and/or bags, textile, 7- to 8-oz. cotton with heavy filler; after-made size of either bag 17 by 30 inches.
3	1	1	1	-------	Balance, torsion, laboratory 4.5-kg. capacity, sensitive to 0.5 gram, with set of weights; or direct reading electric, 4.0-kg. capacity, sensitive to 0.1 gram, 1.0-gram graduations.
4	2	1	1	-------	Balance, laboratory, triple beam, 100-gram capacity, 0.01 gram graduations; or direct reading electric, 160-gram capacity, sensitive to 0.001 gram, 0.01-gram graduations.
5	2	1	1	-------	Balance, laboratory, 2,000-gram capacity, sensitive to 0.05 gram, graduated to 0.1 gram, with set of weights; or direct reading electric, 800-gram capacity, sensitive to 0.01 gram, 0.1-gram graduations.
6	2	2	1	-------	Base, for compaction cylinder mold, concrete filled 10-gallon carbide can.
6A				1	Base, for compaction cylinder mold, field, four flat 50-pound weights.
7	1	1	1	-------	Basket, brass wire, No. 4 mesh, approximately 8 inches in diameter by 8 inches in depth, with bail.
8	2	1	1	-------	Brush, dusting, hair-bristled.
9	2	1	1	-------	Brush, sieve, 1½-inch oval varnish.
10	2	1	1	-------	Brush, sieve, brass fine wire.
11	2	1	1	1	Brush, wire-bristled, curved handle.
12	2	1	1	-------	Bucket, metal, 4- to 5-gallon capacity.
12A	1	1	1	-------	Calibration bar, unit weight measure, metal, 3- x 12- x ⅛-inch (see fig. 12–2A).
13	1	1	1	-------	Caliper, inside micrometer, 1½- to 12-inch.
14	72	36	12	4	Cans, sample storage, round with slip-cover and side handles, 10-inch diameter by 12½-inch high.

Note. All drawings listed are available from the Engineering and Research Center.

Item	Quantities for laboratory				Description
	Type C	Type B	Type A	Type V	
15	1	1	1	--------	Can, siphon (specific gravity), fabricated from 6- by 12-inch can. (Drawing No. 101–D–97 and fig. 10–5.)
16	1	1	1	1	Canvas, lightweight, 6 by 6 feet or larger.
16A	2	2	1	--------	Carboy—5-gallon.
17	12	6	6	1	Clip board for data sheet.
18	36	36	12	--------	Clamps, rubber hose, ⅜-inch shutoff.
19	72	36	12	--------	Containers, sample, metal cans, 6- by 12-inch cylindrical.
20	400	200	200	--------	Containers, sample, paper cartons, cylindrical, low form 1-pint capacity.
21	72	36	12	--------	Containers, sample, glass jars, 1-pint size.
22	36	36	12	6	Container, sample, metal cans or glass jars, 6 or 8 ounces, with airtight fitting lids.
23	1	1	1	--------	Container, water, (specific gravity) of size to hold water for immersion of basket, wire, item 7.
23A	1	1	1	1	Crowbar, 3-foot.
24	12	8	3	--------	Cylinder, permeability, 8-inch, available commercially or may be fabricated. (Drawing No. 101–D–94 (steel) or No. 101–D–2 (aluminum).)
25	2	1	--------	--------	Cylinder, permeability, 20-inch. (Drawing No. 101–D–150 (steel) or No. 101–D–192 (aluminum).)
26	1	1	1	--------	Desiccator, for evaporating dishes, available commercially or may be fabricated. (Drawing No. 101–D–103.)
27	72	48	24	--------	Dish, evaporating, porcelain, 300-ml. capacity, approximately 4½ inches in diameter by 2 inches deep.
28	1	1	1	--------	Dish holder, for removing dishes from oven, available commercially or may be fabricated. (Drawing No. 101–D–111.)
29	24	18	12	--------	Dish, shrinkage, porcelain, or Monel milk dish, about 45 mm. diameter by 12 mm. high.
30	36	24	18	--------	Dish, weighing, aluminum, approximately 50-mm. diameter by 12-mm. high with covers.
31	36	24	12	--------	Discs, porous, carborundum, coarse grade, 7¹⁵⁄₁₆-inch diameter by 1-inch thick.

Item	Quantities for laboratory				Description
	Type C	Type B	Type A	Type V	
32	6	4			Discs, porous, carborundum, coarse grade, 19⅛-inch diameter by 2 inches thick.
32A	1	1			Extruder for compaction specimen. (Drawing No. 101–D–396 and fig. 11–6.)
33	1	1	1		Fan, electric, 12-inch.
34	12	8	4		Flask, specific gravity, 250-ml. volumetric flask, pyrex. One rubber stopper, size 0, with one hole, and a hose shutoff clamp, must be ordered separately for each flask.
35	2	1	1		Gages, dial indicator with holders for 8-inch permeability cylinder, continuous dial, 0.001-inch graduations, 1.000-inch travel, available commercially, set of two. (Drawing No. 101–D–109.)
35A	1	1			Gages, same as above except 3-inch travel, for 20-inch permeability cylinder (set of two).
35B	1	1	1		Gage, block, dial indicator gage length reference, for 8- and 20-inch permeability cylinders. (Drawing No. 101–D–515.)
36	1	1	1		Gage, ring, specimen thickness, 4- to 7-inch-outside-diameter by 3.000-inch length, made from metal pipe or tubing.
37	1	1			Gage, ring, specimen thickness, 10-inch-outside-diameter by 9.000-inch length, made from metal pipe.
38					Graduates:
	9	6	3		25 ml. graduated to 0.2 ml.
	9	6	3	1	100 ml. graduated to 1.0 ml.
	9	6	3	1	250 ml. graduated to 1.0 ml.
	9	6	3		500 ml. graduated to 1.0 ml.
	9	6	3		1,000 ml. graduated to 1.0 ml.
38A	1	1	1		Hoist, chain or electric, 1,000-pound capacity.
38B	1	1	1		Holder, dial indicator gage (fig. 12–2).
39	10	6	2		Hydrometer, range 0 to 60 grams of soil colloids per liter. (See ASTM Designation D–422 and fig. 6–1.)
40	10	6	2		Hydrometer cylinder, glass, 18- by 2½-inch diameter, graduated for 1,000 ml. (See ASTM Designation D–422.)

Item	Quantities for laboratory				Description
	Type C	Type B	Type A	Type V	
41	10	6	3	1	Jug, glass, screw top, 1-gallon capacity.
42	4	3	2	1	Knife, butcher, large.
43	1	1	1	1	Ladder, lightweight, 10-foot length (6-foot length for type V).
44	2	1	1	-------	Liquid limit device, mechanical, with grooving tool. (See ASTM Designation D–423. Motorized device shown in figs. 7–1 and 7–2 available commercially.)
45	1	1	1	-------	Loading device, hydraulic pump equal to Porto-Power Pump No. P–85, fitted with a 10,000-pound capacity gage and including flexible hose with quick couplers for fixing to rams and two ¾-inch steel balls. Two rams required: one equal to Porto-Power Ram No. RC–159 of 10-ton capacity, and one equal to Porto-Power Ram RC–210 of 50-ton capacity.
46	4	3	2	1	Mallet, medium weight, wood.
47	1	1	1	1	Mattock, lightweight.
48	1	1	1	-------	Measures, unit weight, cylindrical, metal, 1.0-, 0.5- and 0.1-cubic-foot capacity of the following dimensions and consisting of cast silicon aluminum, ASTM–SG70–A (drawing No. 101–D–181):

Volume, cu. ft.	Inside dimensions		Thickness of sides and bottom, inches
	Diameter, inches	Height, inches	
1.0	13.000	13.019	⁵⁄₁₆
0.5	11.000	9.092	⁵⁄₁₆
0.1	6.000	6.112	⁵⁄₁₆

Also, see item 69B.

Item	Type C	Type B	Type A	Type V	Description
48A	1	1	-------	-------	Mechanical compactor. (Information on approved models should be obtained from the Engineering and Research Center, Division of General Research.)
49	1	1	1	-------	Mercury, 5-pound supply.
50	1	1	1	-------	Mixer (with ⅓-horsepower, 60-cycle-per-second, 115-volt motor).
51	4	3	1	1	Mixer bowl, with collar, 3-gallon capacity.

Item	Quantities for laboratory				Description
	Type C	Type B	Type A	Type V	
51A	1	1	--------	--------	Mixer sieve, No. 4. (Drawing No. 101–D–228.)
51B	1	1	--------	--------	Mixer sieve brush. (Drawing No. 101–D–228.)
51C	3	2	1	1	Mixer paddle, hand, mixing, fabricated. (Drawing No. 101–D–88.)
51D	1	1	1	--------	Mixer paddle, mixing, type 2.
51E	1	1	1	--------	Mixer paddle, mixing, type 3. (Drawing No. 101–D–191.)
52	6	4	2	1	Mold, cylindrical, compaction, metal, $\frac{1}{20}$-cubic-foot. (Drawing No. 101–D–98 (steel) or drawing No. 101–D–100 (aluminum).)
53	1	1	1	--------	Mortar and pestle, mortar, iron, high form, 5½-inch diameter with one iron and one rubber-tipped pestle.
54	1	1	1	--------	Nozzle with hose, fine spray; can be fabricated from thick-wall, rubber tubing and ⅛-inch copper tube flattened on one end to form spray.
55	1	1	1	--------	Oven, drying, electrically heated, temperature controlled to 110° C.±3° C., minimum inside dimensions 3 feet wide by 2 feet deep by 2 feet high, available commercially or may be fabricated. (Drawings No. 101–D–113 and 101–D–114, and fig. 9–1.)

Note: Item numbers 56, 57, and 58 have been deleted.

59	36	24	12	--------	Pans, drying, small, metal, cast aluminum, approximately 8 by 8 by 2 inches deep. (Drawing No. 101–D–180.)
60	9	6	2	--------	Pans, mixing, large, metal, approximately 3 by 2 feet by 4 inches deep, may be fabricated. (Drawing No. 101–D–91.)
61	6	4	2	1	Pan, mixing, metal, approximately 16 by 16 by 2 inches, may be fabricated. (Drawing No. 101–D–105.)
62	1	1	1	--------	Pans, weighing, aluminum, approximately 100-mm. diameter by 25-mm. deep, set of eight required.
63	1	1	1	1	Picks, conventional and geologist types.

Item	Quantities for laboratory				Description
	Type C	Type B	Type A	Type V	
64	1	1	1	-------	Pouring device, two funnels of ½- and 1-inch diameter by 6-inch length, cylindrical spouts, and lipped brims for attaching to 6-inch diameter by 12-inch high metal can. (Drawing No. 101–D–340.)
* 65	3 2	2 2	1 1	1 1	Pouring devices, and sand cones, 6-, 8-, and 12-inch. (See figs. 24–1 and 24–2 and drawing No. 101–D–349 for suggested types of small devices, and drawing No. 101–D–341 for a large device.)
66	4	3	2	1	Proctor-needle tester, with six needles identified by end area in square inches as follows: 1/40, 1/20, 1/10, 1/4, 1/2, and 1. Tester and needles are available commercially or may be fabricated. (Drawing No. 101–D–84.)
67	1	1	1	-------	Pump, vacuum, or aspirator apparatus.
68	1	1	1	-------	Pycnometer, 1-pint fruit jar (edges at opening ground flat) with cap and glass plate. (Drawing No. 101–D–115 and fig. 10–5.)
69	4	3	2	-------	Rake, mixing, hand, fabricated. (Drawing No. 101–D–101.)
69A	1	1	1	-------	Reference bracket, dial indicator gage. (Drawing No. 101–D–524 and fig. 12–2B.)
69B	1	1	-------	-------	Relative density test equipment, consisting of: vibratory table, 0.1- and 0.5-cubic-foot-capacity unit weight measures, guide sleeve with three tie-down rods, one set for each size measure, surcharge weight with baseplate, one set for each size measure, and dial indicator gage holder. Vibratory table is a steel table with a cushioned steel vibrating deck about 30 by 30 inches and actuated by an electromagnetic vibrator. The vibrator should be a seminoiseless type with a net weight over 100 pounds, a frequency of 3,600 vibrations per minute, and a vibrator double ampli-

*Upper quantities—small devices. Lower quantities—large devices.

Item	Quantities for laboratory				Description
	Type C	Type B	Type A	Type V	
					tude variable between 0.002 and 0.015 inch under a 250-pound load, and be suitable for use with a 230-volt alternating current electric circuit. (Drawing No. 101–D–350 and fig. 12–1.)
70	2	1	1	--------	Rolling surface, ground glass plate or paper similar to that used for mimeographing.
71	1	1	1	--------	Rope, suitable for raising material from test holes, 100 feet, ¾-inch diameter.
72	1	1	1	--------	Sample splitter, with pans, available commercially or may be fabricated. (Drawing No. 101–D–89 and fig. 5–2.)
73	1	1	1	--------	Sampling tube, metal, about 12 inches long, tapered from 2-inch diameter end to 1-inch diameter end.
74	1	1	1	--------	Scale, portable platform or platform counter, 250-pound capacity, 0.01-pound graduations.
74A	1	1	1	--------	Scale, portable platform, 500-pound capacity, 0.5-pound graduations.
75	1	1	1	1	Scale, fan-type, 30- to 50-pound capacity, graduated in 0.01 pounds, with weights, equipped with locking device and case when used in vehicles. (Information on approved models should be obtained from the Engineering and Research Center, Division of General Research.)
76	1	1	1	1	Scoop, hand, large, available commercially or may be fabricated. (Drawing No. 101–D–93.)
77	5	4	3	1	Scoop, hand, small. (Drawing No. 101–D–104.)
78	2	1	1	1	Shovel, round point, long-handled.
79	1	1	1	--------	Shrinkage measuring apparatus, porcelain evaporating dish about 4½-inch diameter; glass dish 60-mm. diameter by 30-mm. depth with ground edge; clear acrylic plastic or glass plate, 3 by 3 by $\frac{1}{16}$ inches with three prongs. (See ASTM Designation D–427 and fig. 7–6.)
80	4	3	2	1	Sieve, 18-inch diameter, U.S. standard No. 4 with stand. (Drawings No. 101–D–234 and 101–D–339.)

Item	Quantities for laboratory				Description
	Type C	Type B	Type A	Type V	
81	2	1	1	_____	Sieves, set of hopper sieves and hopper (fig. 5–1) or 18-inch diameter hand riddles of 5-, 3-, 1½-, ¾-, and ⅜-inch, and No. 4 sizes.
82	2	1	1	_____	Sieve set, 8-inch brass sieves consisting of U.S. standard sieve series Nos. 200, 100, 50, 30, 16, 8, and 4 with pan and cover. (Tyler sieves Nos. 200, 100, 48, 28, 14, 8, and 4 are equivalent.)
83	2	1	1	_____	Sieve, 8-inch brass, U.S. standard No. 40.
84	2	1	1	_____	Sieve, washing, No. 200 mesh; may be fabricated (drawing No. 101–D–106), or 8-inch brass sieve may be used.
85	1	1	1	_____	Sieve shaker, motor-driven, electric, equipped with timer (fig. 6–2).
86	4	3	2	1	Spanner wrench for compaction cylinder mold.
87	4	2	2	_____	Spatula, 4-inch blade.
87A	1	1	1	1	Spoon, large curved.
88	_____	_____	_____	_____	Springs, loading, coil, of requirements as follows:

Spring No.	O.D. of coil (inches ±¼ inch)	Free height (inches ±¼ inch)	Static load (lb.) at deflection of 2½ inches ±¼ inch
2	4	10	1,200
3	6	10	2,400
5	6	10	7,200
6	6	10	9,000
7	6	12	15,000

The quantities for the springs are:

Item	Type C	Type B	Type A	Type V	Spring No.
	6	4	3	_____	2
	6	4	3	_____	3
	6	4	3	_____	5
	6	4	0	_____	6
	6	4	0	_____	7

(*Note:* Spring No. to be stamped on outside of coil. Spring ends to be machined or ground perpendicular to the axis. Springs to be furnished with two coats of aluminum paint.)

Item	Type C	Type B	Type A	Type V	Description
89	1	1	1	_____	Square, metal, combination, 12-inch.
90	2	1	1	_____	Stirring apparatus, electric mixing machine with special dispersion cup. (See ASTM Designation D–422 and fig. 6–1.)
91	1	1	1	1	Straightedge, metal, 6-inch length.
92	1	1	1	1	Straightedge, metal, 15-inch length.
93	1	1	_____	1	Straightedge, metal, 24-inch length.

Item	Quantities for laboratory				Description
	Type C	Type B	Type A	Type V	
94	4	3	2	1	Straightedge trimmer. T-bar, fabricated. (Drawing No. 101–D–92.)
95	12	8	3	--------	Supports, head tank, adjustable. (Drawing No. 101–D–395 and figs. 13–1 and 13–6.)
96	24	16	6	--------	T-connectors, $^3/_{16}$-inch bore ($^3/_8$-inch-inside-diameter hose T-connections).
97	6	4	2	1	Tamper, 5.5-pound, 2-inch diameter face, with height of drop guide, fabricated. (Drawing No. 101–D–87.)
97A	1	1	1	--------	Tamper, 9-pound, 4-inch diameter face, for packing material for permeability tests. (Drawing No. 101–D–516.)
98	1	1	--------	--------	Tamper, extra heavy rod weighing up to 40 pounds with about 6-inch diameter face, or suitable pneumatic tamper. (Drawing No. 101–D–338.)
99	3	2	1	--------	Tank, head, glass, set, one each of 20-, 50-, and 60-mm. diameter tanks, and two each of 30- and 40-mm. diameter tanks. (Drawing No. 101–D–85.) Rubber stoppers with one hole approximately $^1/_8$-inch, including one each of sizes 2, 10, and 12, and two each of sizes 6 and 9. Seven 3-foot lengths of 5-mm. glass tubing, 2 feet of $^1/_8$-inch plastic tubing, and hose shutoff clamps (see item 18) must be ordered separately to complete a set of head tanks.
100	--------	--------	--------	--------	Tank, head, plastic, 6-inch diameter by 40-inch length. (Figs 14–1 and 14–2.)
101	1	1	1	1	Tape measure, metallic, 50-foot length.
102	1	1	1	1	Tape measure, metallic, 6-foot length, or folding rule.
103	--------	--------	--------	--------	Templates, sand density, metal, of three sizes for 6-, 8-, and 12-inch density holes (figs. 24–1 and 24–2); one for each sand cone or pouring device— item 65.
104	9	6	3	--------	Thermometer, centigrade, 0° to 150°, 0.5° graduations.
105	12	6	6	--------	Towel.
106	4	3	2	1	Trowel, concrete finishing, 15-inch.
106A	2	1	1	--------	Tube, air dispersion, for gradation analysis (fig. 6–3).

Item	Quantities for laboratory				Description
	Type C	Type B	Type A	Type V	
107	150	100	50	-------	Tubing, rubber, thick-wall, ⅜-inch ID (feet).
108	150	100	50	-------	Tubing, rubber, thick-wall, ¼-inch ID (feet).
109	1	1	1	-------	Vibrating device, platform type (fig. 12–1); pneumatic or electric form-type or foundry-type vibrators with mounting pin for attachment to the measure are satisfactory when 0.1- and 0.5-cubic-foot small molds are used. (*Note:* Information on approved equipment should be obtained from the Engineering and Research Center, Division of General Research.)
110	9	6	2	-------	Watch, stop.
110A	1	1	1	-------	Water bath, constant temperature, specific gravity (fig. 10–1).
					Miscellaneous Supplies
110B	1	1	1	-------	Calcium chloride, for desiccator, 10-pound jar.
111	1	1	1	-------	Cheesecloth, 50-yard bolt, 36 inches wide.
112	1	1	1	-------	Dispersing agent, sodium metaphosphate, in granular or flaky form, commonly available in food stores under the trade name of Calgon.
113	-------	-------	-------	-------	Earth Manuals (copy for each laboratory employee and inspector and other persons concerned with earthwork).
114	3	2	1	-------	Forms—complete set of standard forms for soils laboratory tests, 100 copies.
115	1	1	1	-------	Hydrochloric acid (muriatic acid satisfactory), 1-pound bottle.
116	1 set	1 set	1 set	1 set	Hand tools (hammer, screwdriver, pliers, wrenches, etc.).
117	5 doz.	3 doz.	1 doz.	1 doz.	Pencils, china marking.
118	2	1	1	1	Tags, cloth shipping, plain, 5¾ by 2⅞ inches, 500 per box.
119	1	1	1	1	Tags, cloth shipping, prepared (form 7–1622), 5¾ by 2⅞ inches, 100 per package.
119A	1	1	1	1	Tags, aluminum, trap, ⅞ by 3⅛ inches, 500 per box.

Item	Quantities for laboratory				Description
	Type C	Type B	Type A	Type V	
120	2	1	1	-------	Wax, sealing, microcrystalline, 50-pound cartons.
121	-------	-------	-------	-------	Water, distilled, 10 gallons. (If not available locally, still should be provided.)
122	1	1	1	1	Funnel, sacking, 24-inch with stand. (Drawing No. 101–D–234.)
123	1	1	1	1	Gloves, rubber.
124	1	1	1	1	Tool box for hand tools.
125	-------	-------	-------	1	Slide rule, 20-inch.

PREPARATION OF SOIL SAMPLES FOR TESTING

Designation E–5

1. Scope.—This designation describes methods for preparation of samples for laboratory testing and quantitative determination of the distribution of particle sizes larger than the No. 4 sieve [1] size.

2. Apparatus.—The apparatus shall consist of the following (see designation E–4 and figs. 5–1 and 5–2):

Cans, sample storage, item 14.
Containers, sample, paper cartons, item 20.
Mallet, item 46.
Pans, mixing, large, item 60.
Sample splitter, item 72.
Scale, platform, item 74.
Sieves, item 81.
Funnel, 24-inch, sacking, with stand, item 122.

3. Procedure.—(a) Results determined from the following tests are recorded on form 7–1451 under "Sample Preparation" as illustrated in figure 5–3. After it has been determined that drying will not materially affect the soil properties, the soil sample as received from the field should be air dried in the laboratory to remove water that may interfere with sieving the material. All soil lumps shall be broken and crumbled in readiness for sieving. Considerable care must be exercised in the

---

1 According to ASTM Designations E–11–70 and E–323–70, the term "sieve" is used for woven wire cloth and perforated plates with round or square holes. Therefore, the term "sieving" in this Manual is used to describe the grain-size separation processes.

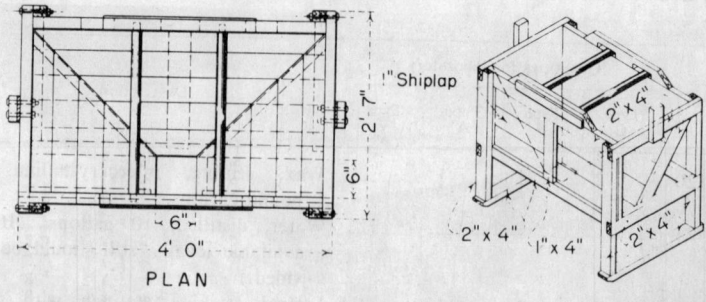

PLAN

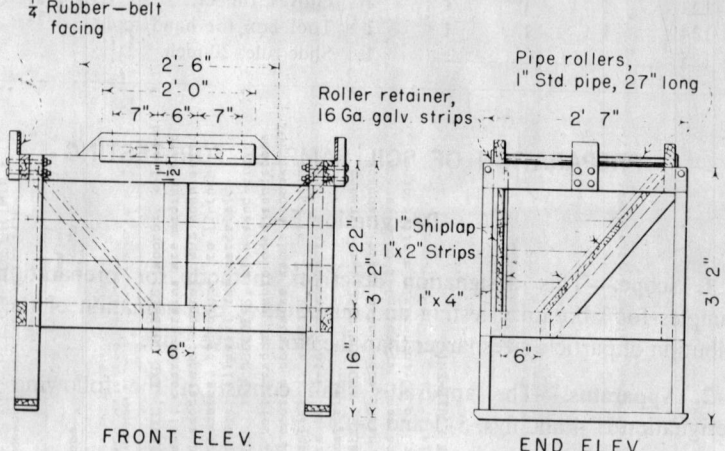

FRONT ELEV.

END ELEV.

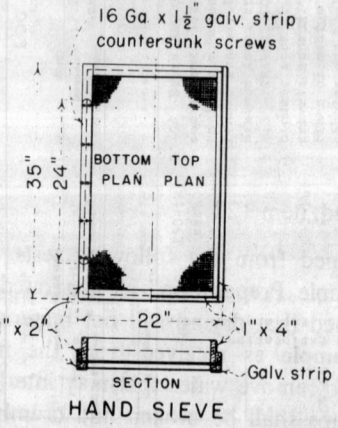

HAND SIEVE

NOTES

Inside of hopper lined with 22 ga. galv. sheet metal, lapped over top and edges.

All wood in hopper and sieves to be painted with two coats of outside gray paint.

Sieves required for soils testing to have openings of 5", 3", 1½", ¾", ⅜", and No. 4.

Sieves required for testing concrete aggregates to have openings of 6", 3", 1½", ¾", ⅜", and No. 4 unless other sizes are designated in specifications.

Where anchorage of sieves is required for nesting, lengths of 1" x 2" angle iron are attached to top of hand sieve framing.

Where applicable, test sieve openings and wire sizes should conform to dimensions and tolerances set forth in A.S.T.M. designations E–11 and E–323.

Riddles, 18-inch dia. with specified sieve openings may be substituted for hand sieves.

Figure 5–1.—Hopper and hand sieves for separating coarse aggregate and soil materials. 101-D-112.

Figure 5–2.—Sample splitter. E–2251–5NA.

handling of materials containing soft and friable rocks such as disintegrated granite, basalt, soft sandstone, soft limestone, and shale, in order to prevent undue breakdown of the particles that may result in gradation changes.

The entire sample shall be separated into sizes, using the following sieves: 5-inch, 3-inch, 1½-inch, ¾-inch, ⅜-inch, and No. 4. The weight of material retained on each sieve shall be recorded as "wet weight retained" and the material passing the No. 4 sieve recorded under "total weight passing No. 4, wet." If the sample appears to contain over 5 percent moisture, the water content of both the plus and minus No. 4 materials shall be determined and recorded (see designation E–9). Generally,

FORM 7-1451
(8-62)
BUREAU OF RECLAMATION   SAMPLE PREPARATION AND GRADATION ANALYSIS

LABORATORY SAMPLE NO. _11J-23_

FEATURE _EXAMPLE_   AREA _A_   EXC. NO. _203_   DEPTH _0.0 to 5.0'_

SAMPLE PREPARATION

PREPARED BY _____   % MOIST + NO. 4 _1.8_   WET WT. TOTAL SAMPLE _47.75_

DATE _____   % MOIST - NO. 4 _5.2_   DRY WT. TOTAL SAMPLE _45.47_

SIEVE SIZE		5"	3"	1-1/2"	3/4"	3/8"	NO. 4	TOTAL WT. PASSING NO. 4
WT. PAN + RETAINED MATERIAL								
WT. PAN								
WET WT. RETAINED		0	1.68	0.25	0.43	0.27		45.12 WET
DRY WT. RETAINED		0	1.65	0.25	0.42	0.26		42.89 DRY
DRY WT. PASSING			45.47	43.82	43.57	43.15	42.89	
% OF TOTAL PASSING			100.0	96.4	95.8	94.9	94.3 -m	

SIEVE AND HYDROMETER ANALYSIS

DISH NO. _18_   DRY WT. OF SAMPLE (W) _50.0_ gms.   FACTOR (F) = $\frac{w^1}{W} = \frac{94.3}{50.0}$ = _1.886_

DRY WT. OF SAMPLE (SIEVED) _21.4 gms._

SIEVING TIME _15 min._   DATE _____

SIEVE NO.	WEIGHT RETAINED	WEIGHT PASSING		% OF TOTAL PASSING	PARTICLE DIA (mm)	REMARKS
8	0.2	49.8	F X WEIGHT PASSING = % OF TOTAL PASSING	93.9	2.380	
16	0.4	49.6		93.5	1.190	
30	1.1	48.9		92.2	0.590	
50	3.3	46.7		88.1	0.297	
100	11.6	38.4		72.4	0.149	
200	21.4	28.6		53.9	0.074	
PAN	0.0					
TOTAL	21.4	TESTED AND COMPUTED BY _____ CHECKED BY _____ DATE _____				

HYDROMETER ANALYSIS

HYDROMETER NO. _320584_   DISPERSING AGENT _Sodium Metaphosphate_

STARTING TIME _8:00_   DATE _____   AMOUNT _125_ ml

TIME	TEMP C°	HYD READ	HYD CORR	CORR READ		% OF TOTAL PASSING	PARTICLE DIA (mm)	REMARKS
.5 MIN°					F X CORRECT READ = % OF TOTAL PASSING		0.050	
1 MIN		16.4	+3.2	19.6		37.0	0.037	
4 MIN	26.8	10.1	+3.2	13.3		25.1	0.019	
19 MIN	26.8	6.9	+3.2	10.1		19.0	0.009	
60 MIN	27.0	4.9	+3.2	8.1		15.3	0.005	
7 HR. 15 MIN°							0.002	
25 HR. 45 MIN°							0.001	

TESTED AND COMPUTED BY _____   CHECKED BY _____   DATE _____

Figure 5-3.—Sample preparation and gradation analysis data form. 101–D–329.

when samples contain less than 5 percent moisture, it is not necessary to determine water content for dry weight computations and all determinations are made on the basis of wet weight only. If the soil contains more than 20 percent gravel particles and the fines are very cohesive with considerable amounts adhering to the gravel after separation, the

7-1445
(5-56)
BUREAU OF RECLAMATION

PREPARATION OF COMPOSITE SAMPLES

COMPOSITE SAMPLE NO. 11J-X90 PROJECT EXAMPLE FEATURE _____

PIT _____ EXCAVATION 203 DEPTH 0-6' _____

COMPUTED BY _____ MIXED BY _____ DATE _____ 19___

		LABORATORY MIX				
SAMPLE NO. (1)	CLASSIFICATION (2)	DEPTH OF STRATA FT. (3)	% PASSING #4 SCREEN (4)	EFFECTIVE DEPTH FT. (5)	WEIGHT PER FT. LBS. (6)	WEIGHT MIXED LBS. (7)
11J-23	ML	5	94.3	4.72	11.49	54.23
11J-24	GM	1	50.1	0.50	11.49	5.75
11J-X90	SM	6		5.22		59.98
Col. (4) ~ from gradation curve, or sample preparation sheet						
Col. (5) ~ Col. (3) x Col. (4)						
Col. (6) ~ Total weight of composite sample desired ÷ Total of Col. (5)						
Col. (7) ~ Col. (5) x Col. (6)						

*Approximately 60 lbs. desired in this case.

FIELD MIX

MAXIMUM SIZE CONSIDERED 1½ INCHES

SAMPLE NO.	CLASSIFICATION	DEPTH OF STRATA FT.	% PASSING MAX. SIZE CONSIDERED	EFFECTIVE DEPTH FT.	WEIGHT PER FT. LBS.	WEIGHT MIXED LBS.
11J-23	ML	5	96.4	4.82	13.61	65.60
11J-24	GM	1	68.8	0.69	13.61	9.39
11J-X91	SM	6		5.51		74.99

Figure 5–4.—Composite sample preparation data form. 101–D–330.

gravel shall be washed on a No. 4 sieve, and the fines thus removed shall be combined with the minus No. 4 material.

(b) Three representative pint-sized specimens for gradation analysis, specific gravity, and soil consistency (Atterberg) tests shall be selected from the material passing the No. 4 sieve by quartering or by the use of the sample splitter illustrated in figure 5–2.

(c) The remaining material passing the No. 4 sieve and material retained on all sieves shall be stored separately in cans or sacks for other laboratory testing.

(d) Individual test pit samples from different strata may be composited to represent any desired depth for study. Samples may be composited by proportioning individual samples on the basis of percentage passing the No. 4 sieve (or on the basis of some other maximum size) and stratum depth represented in feet as shown in figure 5–4.

4. Calculations.—For samples containing more than 5 percent moisture, the dry weight for each size of plus No. 4 and for the total minus No. 4 is computed and recorded as "dry weight retained" and "total weight passing No. 4, dry," respectively. The sum of these weights is recorded as "dry weight total sample" and is also entered as "'dry weight passing" for the sieve size on which no material was retained.

When samples contain less than 5 percent moisture, dry weight determinations are eliminated and the "wet weight total sample" is entered as "dry weight passing" for the sieve size on which no material was retained.

The weight of material passing each sieve shall be computed by subtracting the amount retained from the amount passing the next larger size sieve. This weight of material passing each sieve shall be expressed as a percentage of the total sample being computed as follows:

$$\text{Percent of total passing} = \frac{\text{weight passing}}{\text{weight of total sample}} \times 100.$$

GRADATION ANALYSIS OF SOILS

Designation E–6

Part A. General Methods

1. Scope.—This part of designation E–6 describes the procedure for the quantitative determination of the distribution of particle sizes in soils. The method is similar to ASTM Designation D 422–63, the principal difference being the sieve sizes used and the separation on the No. 4 instead of No. 10 sieve for the hydrometer test.

2. Apparatus.—The apparatus shall consist of the following (see designation E–4 and figs. 6–1, 6–2, and 6–3):

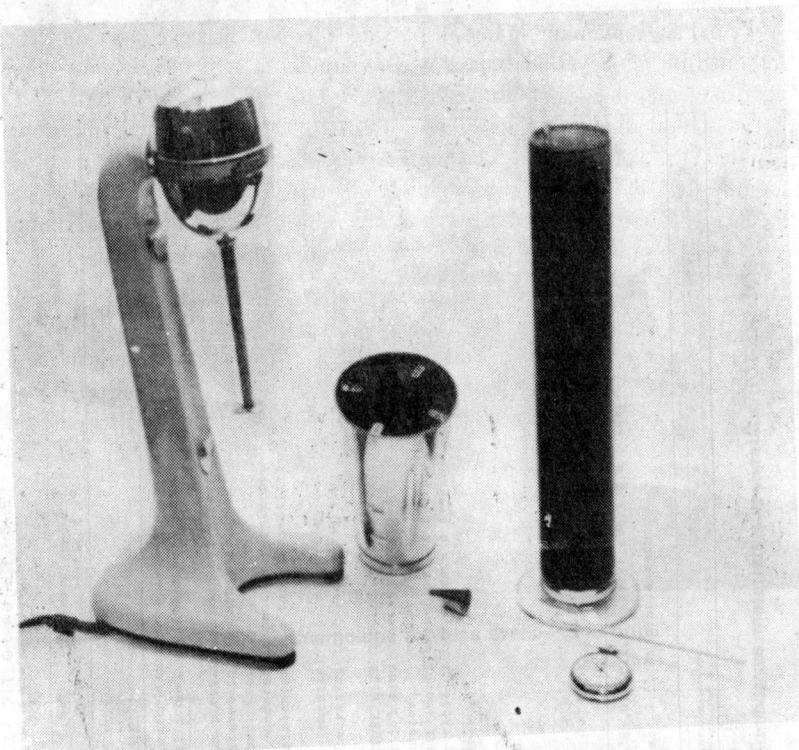

Figure 6–1.—Stirring apparatus, dispersion cup, hydrometer, and hydrometer cylinder. PX–D–16525.

Balance, laboratory, item 4.
Brushes, sieve, items 8, 9, and 10.
Desiccator, item 26.
Dishes, porcelain evaporating, item 27.
Hydrometer, item 39.
Hydrometer cylinder, item 40.
Nozzle with hose, item 54.
Pans, drying, small, item 59.
Pans, weighing, aluminum, item 62.
Sieve set, item 82.
Sieve, washing, 200 mesh, item 84.
Sieve shaker, item 85.
Stirring apparatus, item 90.
Thermometer (centigrade), item 104.
Tube, air dispersion, item 106A.
Watch, stop, item 110.

Figure 6-2.—Sieve analysis equipment. E-2251-2NA.

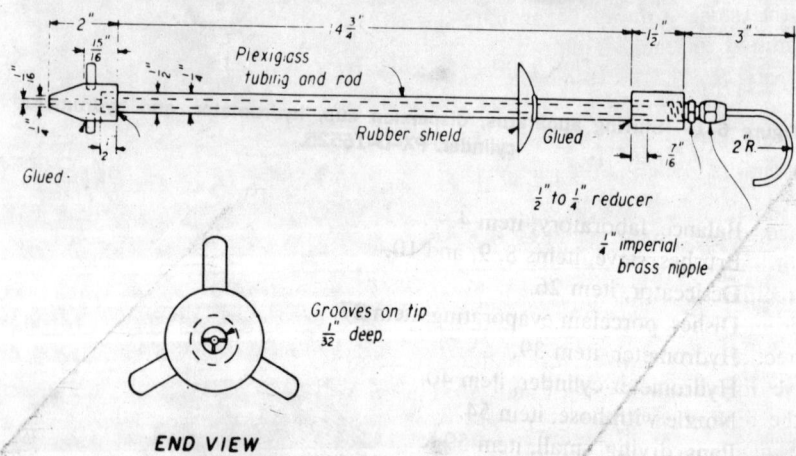

Figure 6-3.—Air dispersion tube. 101-D-523.

3. Sample.—A sample of minus No. 4 soil, prepared as described in designation E–5, for gradation analysis is dried to constant weight in an oven at 110° C. and cooled to room temperature in a desiccator.

4. Calibration.—(a) The hydrometer is calibrated to read in grams of soil in suspension per liter of mixture of soil and water at 20° C. In order to obtain test readings comparable to the calibration, corrections must be applied to the readings to compensate for errors introduced by (1) the meniscus, (2) difference in temperature, and (3) the dispersing agent (sometimes called deflocculating agent). The readings are made on the hydrometer scale at the surface of the liquid. When reading the hydrometer in soil suspensions, the surface cannot be accurately read and it is expedient to read the top of the meniscus that forms about the hydrometer stem. Therefore, a correction for the meniscus is necessary. This correction is the difference between the top and bottom readings of the meniscus in distilled water and is very near +1 for this type hydrometer. The correction is added to the readings made at the top of the meniscus.

The corrections for temperature difference and the dispersing agent can be determined by calibrating a hydrometer using two hydrometer cylinders. One cylinder contains 1 liter of distilled water and the other contains 1 liter of solution of dispersing agent consisting of 125 ml. of the stock solution of sodium metaphosphate and distilled water. Readings in each liquid are made at several temperatures between 15° and 35° C. Best results are obtained with the use of a constant-temperature water bath. Sufficient time should be allowed for the temperature of the hydrometer and liquids to equalize; usually 30 minutes are required for the hydrometer to reach equilibrium after the liquids have attained the desired temperature at which they are to be maintained. From these readings, curves may be plotted with temperature as ordinate and the hydrometer correction as abscissa. The hydrometer correction for either liquid at any temperature is equal in value to the hydrometer reading at that temperature but with the opposite sign. The corrections for temperature differences can be read directly from the curve for distilled water, and the correction for the dispersing agent is the difference in values between the two curves at any temperature. The meniscus and temperature corrections are usually constant, but calibrations must be made for different quantities and types of dispersing agent.

(b) Sodium metaphosphate (sometimes called sodium hexa-metaphosphate) appears to be the most efficient dispersing agent for most soils.[1] A 4-percent stock solution of sodium metaphosphate and distilled

---

[1] "Investigation of Dispersing Agents for Gradation Analysis of Soils," Earth Laboratory Report No. EM–617, April 10, 1961.

water is prepared by using 40 grams of sodium metaphosphate per liter of solution. The stock solution should be prepared frequently (at least once a month) or adjusted to a pH (hydrogen-ion concentration) of 8 or 9 by adding sodium carbonate. The type and amount of dispersing agent used shall be recorded on the data sheet, form 7–1451, figure 5–3 of designation E–5.

5. Procedure.—(a) *Dispersion of Soil Sample.*—From the ovendried minus No. 4 material, a representative sample of approximately 100 grams for sandy soils and approximately 50 grams for silt or clay soils shall be obtained using a sample splitter, accurately weighed and recorded, then placed in a porcelain evaporating dish. The required amount of dispersing agent (125 ml. of sodium metaphosphate stock solution) and sufficient distilled water is added to cover the soil. The mixture shall be allowed to soak for a period of at least 18 hours. It shall then be washed into the dispersion cup with distilled water and distilled water added until the cup is within 2 inches of being filled. The contents of the cup shall be mixed by the stirring apparatus for a period of 1 minute (see fig. 6–1).

(b) *Alternate Air Dispersion Method.*—The representative sample of soil, after being accurately weighed and recorded, is placed in a hydrometer cylinder. The required amount of dispersing agent (125 milliliters of sodium metaphosphate stock solution) and distilled water is added to bring the volume to about 250 milliliters. This mixture shall be allowed to soak for a period of at least 18 hours. The air dispersion tube (fig. 6–3 and item 106A of designation E–4) shall be inserted under 1- to 2-pounds-per-square-inch pressure. (*Note:* This method requires a source of air capable of producing 2 cubic feet per minute per dispersion tube. Water may condense in air lines when not in use. This must be removed either by using a water trap on the air line or by blowing the water out of the line before using for dispersion purposes.)

The contents of the hydrometer cylinder shall then be dispersed for 10 minutes with an air pressure of 15 pounds per square inch and the sides of the cylinder shall be rinsed frequently. For some samples containing coarse sand particles, it may be necessary to increase the air pressure to adequately agitate and disperse the particles.

Upon completion of dispersion, the dispersion tube while being rinsed with distilled water is withdrawn under 1- to 2-pounds-per-square-inch pressure. Distilled water is then added to the hydrometer cylinder until the mixture attains a volume of 1,000 milliliters. The remainder of the test is the same as described below.

(c) *Hydrometer Test.*—After dispersion, the mixture is transferred to a hydrometer cylinder and distilled water at room temperature is added until the mixture attains a volume of 1,000 milliliters. The contents of the hydrometer cylinder shall be thoroughly mixed for 1 minute by inverting the cylinder several times using the palm of the hand or a suitable rubber stopper over the mouth of the cylinder.

At the conclusion of mixing, the cylinder shall be quickly placed on the table and a stop watch started, and the hydrometer shall be carefully placed in the soil suspension to avoid bobbing when released. The hydrometer shall be read to the nearest 0.5 gram per liter at the top of the meniscus formed by the suspension around its stem. Readings shall be observed and recorded under "Hydrometer Reading" on form 7–1451, figure 5–3 of designation E–5, at time intervals of 1, 4, 19, and 60 minutes after the beginning of sedimentation. After the 4-minute reading, the hydrometer shall be removed and placed in distilled water, and carefully placed in the hydrometer cylinder again about 30 seconds prior to a subsequent reading. Following the 1-minute reading, the thermometer is placed in the suspension and read and recorded at the time of the hydrometer readings. Maintaining cleanliness of the hydrometer is very important, especially the reading portion of the stem. This can be accomplished by washing the hydrometer with soapy water, then rinsing it in cleaning alcohol and again in clean water.

(d) *Sieve Analysis.*—On completing the final hydrometer reading, all the material in the graduate is flushed into the 200-mesh wash sieve. The material shall then be carefully washed with tap water until the wash water is clear. That fraction retained on the No. 200 sieve shall be transferred to a suitable container, dried in an oven, and then separated into six sizes on United States standard sieves No. 8, 16, 30, 50, 100, and 200. The sieving shall be accomplished by a powered sieve shaker requiring 15 minutes of continued sieving action (fig. 6–2). After sieving, the material retained on each sieve shall be weighed; the accumulative weight of each amount is then determined and recorded under "weight retained," form 7–1451, figure 5–3 of designation E–5.

(e) *Alternate Hand Sieving Method.*—The ovendried material retained on the No. 200 sieve after washing shall be placed in the top sieve (No. 8) included in a nest of sieves containing Nos. 8, 16, 30, 50, 100, 200, and pan. A lid shall be placed on the top sieve and the complete nest of sieves shaken in a pendulum-like motion for about 30 seconds. The No. 8 sieve with lid shall then be carefully removed from the other sieves, placed on an extra pan, and vibrated vigorously in a circular motion while being tapped on the side with one hand. This shaking shall be continued until all the material finer than the No. 8 has passed the sieve. When this sieving is complete, the material retained shall be emptied into an aluminum pan and the material in the sieve pan shall be placed in the next smaller mesh sieve. The sieving procedure shall be repeated for each sieve size and the weight of material retained on each sieve determined and recorded.

6. Calculations.—The computations are divided into three parts: (1) computations of data "from preparation of sample"; (2) sieve analysis, and (3) hydrometer analysis (see form 7–1451, fig. 5–3 of designation E–5).

(a) *Preparation of Sample.*—The method and calculations required to determine the percentage and distribution of particles larger than the No. 4 sieve size are given in designation E–5.

(b) *Sieve Analysis.*—The weight of soil passing the No. 8 sieve shall be computed by subtracting the weight retained on that sieve from the dry weight of sample, *W.* Then the weight passing the No. 16 sieve shall be determined by subtracting the accumulative weight retained on the No. 16 sieve (the material retained on No. 16 sieve is combined with that retained on the No. 8 for weighing) from the weight of sample, *W.* In a similar manner, the weight passing each sieve shall be obtained by subtracting the accumulative weight retained on a particular sieve from the weight of sample, *W.* The percentage of the total sample passing a given sieve is computed using a factor as follows:

$$F = \frac{W\%}{W}$$

where: *F* is the factor,

   W% is the percentage of total sample passing the No. 4 sieve, and

   W is the dry weight of soil used in the hydrometer suspension, in grams.

The "percent of total passing" each sieve is obtained by multiplying the weight passing each sieve, "weight passing," in form 7–1451 by the factor *F.*

(c) *Hydrometer Analysis.*—The percentage of the total sample smaller than a given particle size is obtained by multiplying the corrected hydrometer reading "Corr. read." by the same factor *F* determined for the sieve analysis.

(*Note:* The results of several hundred tests on widely different soil types showed that the maximum size of particles in suspension for any given time of sedimentation varied over a comparatively narrow range. This range was found to be independent of temperature, hydrometer reading, and soil type. This study indicated that the determination of the maximum particle size in suspension, on the basis of hydrometer reading time intervals, is sufficiently accurate for analyzing soils for construction purposes. If a more accurate particle size determination is desired, the method given in the Standards of the American Society for

Testing Materials, "Grain-Size Analysis of Soils," Designation D–422–63, may be used.)

7. Plotting of Data.—The percentages of soil particles of different diameters, including the sizes larger than the No. 4 sieve, shall be plotted on semilogarithmic paper to obtain a particle size gradation curve. The percent passing (smaller) shall be plotted as ordinate on the arithmetic scale and the particle size as abscissa on the logarithmic scale. Examples of such curves are shown in figure 6–4.

8. Synthetic Analysis.—Synthetic analysis is a term used to define the method for computing the change in gradation of a soil caused by adding or removing certain fractions, or the gradation of material when two or more soils are combined whose gradations have been previously determined. This analysis is often used to study different combinations of materials prior to making composites, or to study the effects of removing or adding coarse particles.

(a) *Combining Two or More Gradation Analyses.*—From the original gradation curves determine the percentage passing each sieve size for each of the soils to be combined and record on form 7–1384, figure 6–5, in the lower right-hand corner of the divided square provided for each sieve size.

Under the column marked "strata, % of total," determine what percentage each individual sample is to be of the total combination. This may be in proportion to the depth of strata or some other desired proportion. Multiply the individual percentages recorded under the different particle sizes by the percentage each sample is of the total and record the product in the upper left-hand corner of the corresponding square. Repeat this procedure for each sample to be combined. Add the figures in the upper left-hand corner in any one vertical column together. This sum is the percent passing the size indicated by the column heading for the combined analysis.

(b) *Gradation Analysis after Removal of a Limiting Maximum Size.*—Record the original analysis in the upper left-hand corners of the square as previously described. Next, the figure "100" is entered in the lower right-hand corner of the square of the smallest particle size considered as removed, and a factor is determined as shown in figure 6–5. Multiply each of the individual percentages recorded in the upper left-hand corner by the factor and record the product in the lower right-hand corner. The figures in the lower right-hand corner represent the percent passing each size indicated by the column heading, assuming all particles larger than a given size (No. 4 sieve in the example) have been removed.

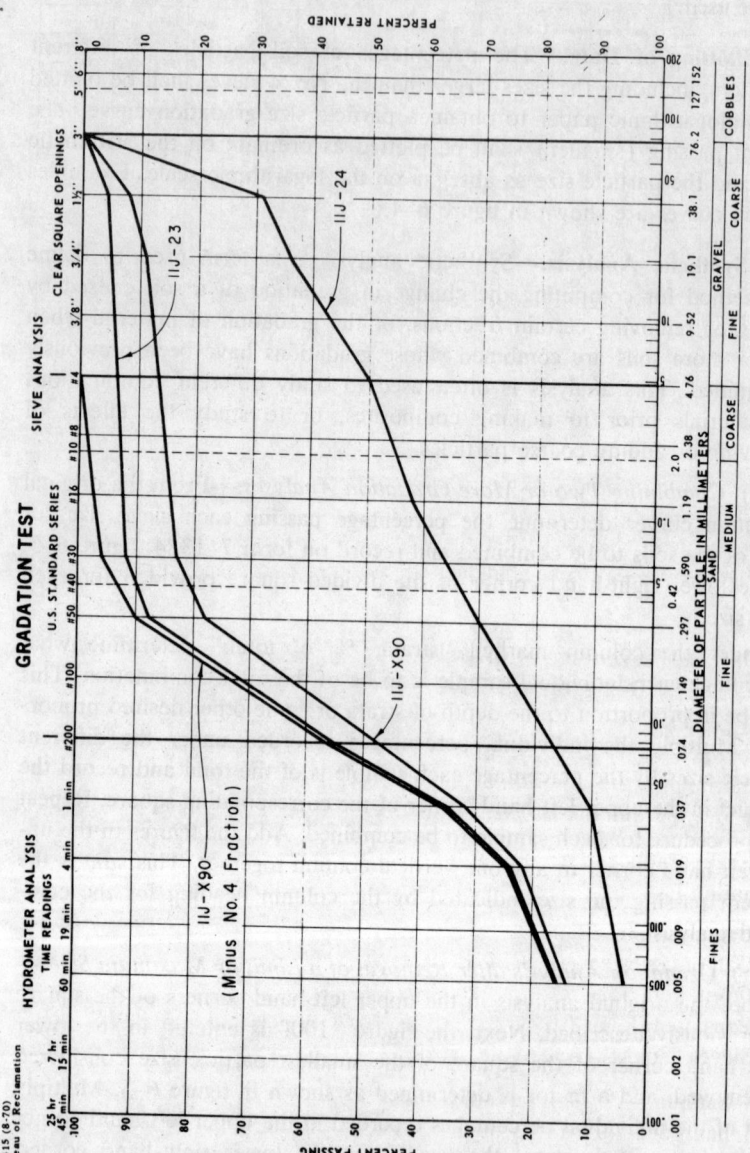

Figure 6-4.—Gradation analysis curves. 101-D-525.

SYNTHETIC GRADATION ANALYSIS COMPUTATIONS

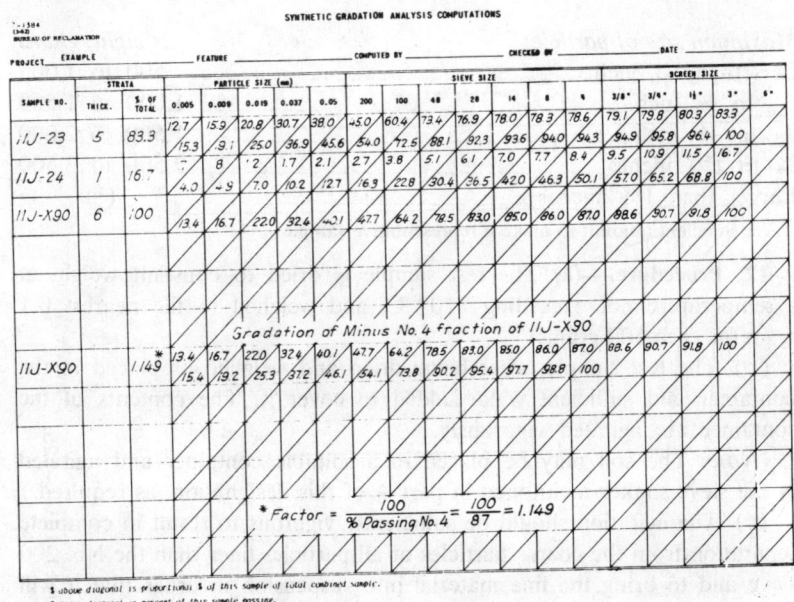

Figure 6–5.—Synthetic gradation analysis computations. 101–D–254.

Part B. Percentage of Gravel and Fines in Soil (Adapted from Designation 16, USBR Concrete Manual)

9. Scope.—This part of designation E–6 describes the procedure for determining the amount of gravel (particles retained on the No. 4 sieve) and the amount of fines (material passing the No. 200 sieve).

10. Apparatus.—The apparatus shall consist of the following (see designation E–4):

> Balance, items 3 and 5.
> Can or container, item 14 or 19.
> Oven, item 55.
> Pans, item 59 or 61.
> Sieves, Nos. 4, 16, and 200, item 82.

11. Sample.—A representative test sample is obtained from material which has been mixed thoroughly and which contains sufficient moisture to prevent segregation. The size of the sample, which depends on the maximum particle size, should be sufficient to yield not less than the appropriate weight of dried material, as shown in the following tabulation:

Maximum size of particles	Weight, grams
Less than $\frac{3}{16}$ inch_____	500 to 1,000
$\frac{3}{16}$ to $\frac{3}{8}$ inch_____	1,000 to 1,500
$\frac{3}{8}$ to $\frac{3}{4}$ inch_____	1,500 to 2,000
$\frac{3}{4}$ to $1\frac{1}{2}$ inches_____	2,500 to 3,500
Larger than $1\frac{1}{2}$ inches_____	(^{1})

^{1} A sufficient amount to make a representative sample.

12. Procedure.—(a) The test sample is dried to constant weight at a temperature not exceeding 110° C. and weighed to the nearest 0.1 gram for each 500 grams.

(b) The test sample, after being dried and weighed, is placed in the container and sufficient water added to cover it. The contents of the container are agitated vigorously.

(*Note:* The soil may be placed in a suitable container and agitated in the sieve shaker mentioned in part A of this designation, as required.)

(c) The agitation should be sufficiently vigorous to result in complete separation from the coarse particles of all particles finer than the No. 200 sieve and to bring the fine material into suspension in order that it will be removed with the wash water. Immediately after agitation, the wash water is decanted and poured over the nested sieves, arranged with the coarser sieves on top.

(*Note:* The material passing the No. 200 sieve should not be saved unless a check determination is desired (see par. 14 below).) The operation is repeated until the wash water is clear. Care must be taken to insure that all fines are washed through the No. 200 sieve.

(d) All the material retained on the nested sieves is returned to the washed sample, which is then separated into two parts by sieving on the No. 4 sieve. Each part is placed in a pan and ovendried to constant weight at a temperature not exceeding 110° C. and weighed to the nearest 0.1 gram for each 500 grams.

13. Calculations.—The results are calculated from the following formulas:

$$A = \frac{B - (C+D)}{B} \times 100$$

$$E = \frac{C}{B} \times 100$$

where: A = percentage of fines, material finer than No. 200 sieve,
    B = dry weight of test sample,
    C = dry weight of material retained on No. 4 sieve,
    D = dry weight of material passing No. 4 sieve and retained on No. 200 sieve, and
    E = percentage of gravel, material retained on No. 4 sieve.

14. Check Determinations.—When check determinations are desired for the percentage of fines, the wash water and soil passing the No. 200 sieve are saved and are either evaporated to dryness or filtered through tared filter paper, which is subsequently dried. The following formula is used:

$$A = \frac{R}{B} \times 100$$

where: R = weight of residue (material passing No. 200 sieve), and B = dry weight of test sample.

SOIL CONSISTENCY TESTS

Designation E–7

1. Scope.—This designation describes the procedures for obtaining the liquid limit, plastic limit, and shrinkage limit of soils, from which the plasticity index and shrinkage factors can be determined. These tests are also known as Atterberg Tests.

Part A. Liquid Limit of Soils (Adapted from ASTM Designation D 423–66)

2. Definition.—The liquid limit of a soil is the water content corresponding to the arbitrary limit between the liquid and plastic states of consistency of a soil. It is the water content at which a groove in a pat of soil, cut by a grooving tool of standard dimensions, will flow together for a distance of ½ inch under the impact of 25 blows in a standard liquid limit apparatus.

3. Apparatus.—The apparatus shall consist of the following (see designation E–4 and fig. 7–1):

Balance, item 4 or 5.
Dish, evaporating, porcelain, item 27.
Dishes, weighing, aluminum, item 30.
Liquid limit device, item 44.
Mortar and pestle, item 53.
Sieve, No. 40, item 83.
Spatula, item 87.

4. Calibration of the Mechanical Liquid Limit Device.—By means of the gage on the handle end of the grooving tool, and the adjustment plate, the height to which the cup is lifted shall be adjusted so that the

(A) MAJOR ITEMS FOR LIQUID LIMIT TEST

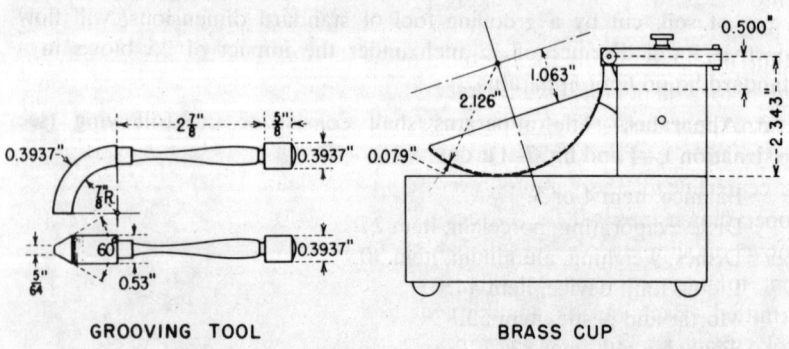

GROOVING TOOL   BRASS CUP

(B) DIMENSIONS OF GROOVING TOOL AND BRASS CUP

Figure 7–1.—Equipment for the liquid limit test. E–2249–1NA and 101–D–157.

point on the cup which comes in contact with the base is exactly 1 cm. (0.3937 in.) above the base. The adjustment plate shall then be secured by tightening the screws. The narrow edge of the grooving tool and the inside and bottom of the brass cup should be checked frequently, and they should be discarded when worn (see fig. 7–1).

5. Sample.—A representative pint sample of minus No. 4 material, obtained in accordance with the procedure given in designation E–5, shall be air-dried, and sieved by hand, using a No. 40 United States standard sieve. (*Note:* With some sensitive soils the consistency values may be affected significantly by air drying. When this is suspected, the soil should be prepared with as little loss of moisture as possible.) A mortar and rubber-tipped pestle may be used to pulverize the hard lumps, being careful not to crush the individual particles. The material passing the sieve shall be thoroughly mixed for use in the consistency tests and the material retained shall be discarded.

6. Procedure.—(a) Place a representative 100-gram sample of air-dried soil passing the No. 40 sieve in the evaporating dish and thoroughly mix with 15 to 20 milliliters of distilled water by alternately and repeatedly stirring, kneading, and chopping with a spatula. Make further additions of water in increments of 1 to 3 milliliters. Thoroughly mix each increment of water with the soil, as previously described, before adding another increment of water.

(b) When sufficient water has been thoroughly mixed with the soil to produce a consistency that will require 30 to 35 drops of the cup to cause closure, place a portion of the mixture in the cup above the spot where the cup rests on the base, and squeeze it down and spread it into the position shown in figure 7–2 with as few strokes of the spatula as possible, care being taken to prevent the entrapment of air bubbles within the mass. With the spatula, level the soil and at the same time trim it to a depth of 1 centimeter at the point of maximum thickness. Return the excess soil to the evaporating dish. Divide the soil in the cup by firm strokes of the grooving tool along the diameter through the centerline of the cam follower so that a clean, sharp groove of the proper dimensions will be formed. To avoid tearing of the sides of the groove or slipping of the soil pat on the cup, up to six strokes, from front to back or from back to front counting as one stroke, shall be permitted. Each stroke should penetrate a little deeper until the last stroke from back to front scrapes the bottom of the cup clean. Make the groove with as few strokes as possible.

(c) Lift and drop the cup by turning the crank, at the rate of two revolutions per second, until the two halves of the soil pat come in contact at the bottom of the groove along a distance of about ½ inch.

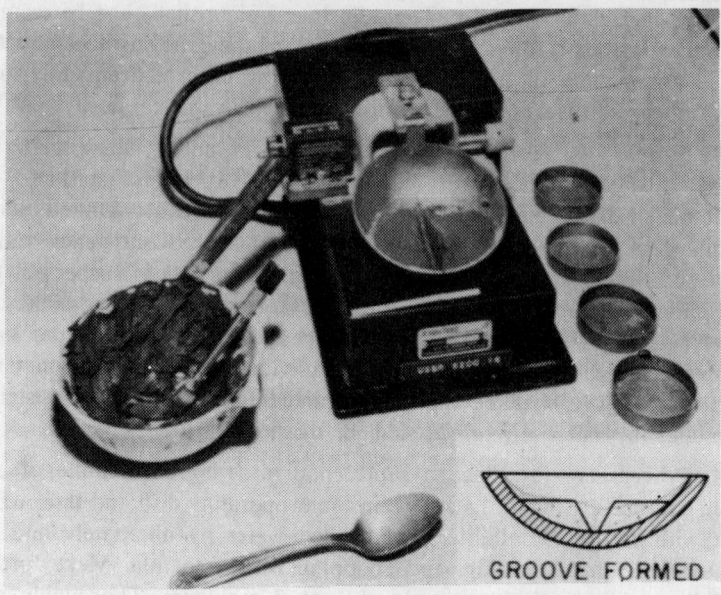

GROOVE FORMED

(A) SOIL DIVIDED BEFORE TEST

GROOVE CLOSED

(B) SOIL CLOSED AFTER TEST

Figure 7–2.—Test for liquid limit. E–2249–3NA and E–2249–4NA.

Record the number of drops required to close the groove along a distance of ½ inch.

(d) Remove a slice of soil approximately the width of the spatula, extending from edge to edge of the soil pat at right angles to the groove and including that portion of the groove in which the soil flowed together, and place in a suitable tared container for determination of the water content (see designation E-9). Weigh and record the weight. Ovendry the soil in the container to constant weight at 110° C. and reweigh as soon as it has cooled but before hygroscopic moisture can be absorbed. Record this weight. Record the loss in weight due to drying as the weight of water. The number of blows and water content data shall be recorded on form 7-1424, figure 7-3.

(e) Transfer the soil remaining in the cup to the evaporating dish. Wash and dry the cup and grooving tool and proceed with the next trial.

(f) Repeat the foregoing operations for at least two additional trials using the soil collected in the evaporating dish to which sufficient water has been added to bring the soil to a more fluid condition. The object of this procedure is to obtain samples of such consistency that the number of drops required to close the groove will be above and below 25. Generally the number of drops should be less than 35 and exceed 15. The test shall always proceed from the dryer to the wetter condition of the soil.

7. Preparation of Flow Curve.—A "flow curve" representing the relation between water contents and corresponding numbers of blows shall be plotted on a semilogarithmic graph with the water content as ordinate on the arithmetic scale, and the number of blows as abscissa on the logarithmic scale (see fig. 7-3). The flow curve is a straight line drawn as nearly as possible through the three or more plotted points.

The water content, expressed as a percentage of the weight of the ovendried soil, corresponding to the intersection of the flow curve with the 25-blow abscissa is the liquid limit of the soil.

Liquid Limit Determination by One-Point Method
(Alternate Procedure)

8. Apparatus, Calibration and Sample.—The apparatus, calibration and sample preparation are the same as specified in paragraphs 3, 4, and 5.

9. Procedure.—Follow the procedure given in paragraphs 6(a) through (e), *except* that the number of blows shall be recorded and the moisture content shall be determined for the accepted trial only.

When the number of blows required to close the groove is greater

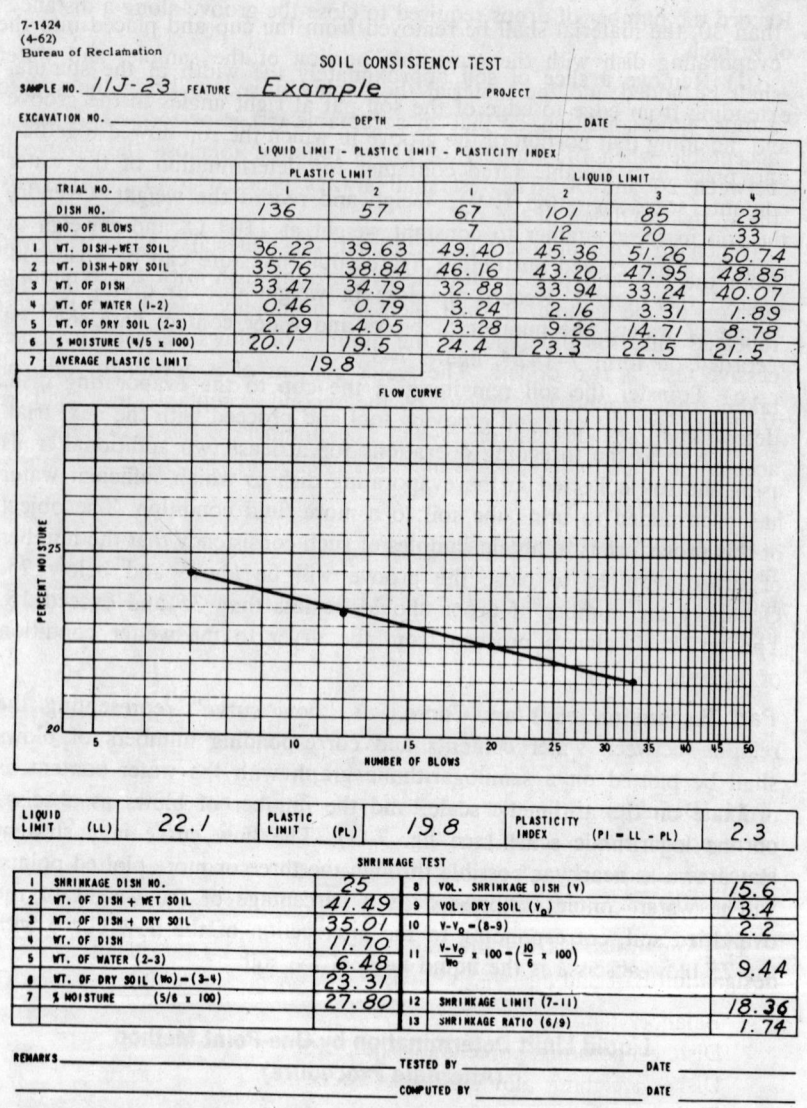

7-1424
(4-62)
Bureau of Reclamation

SOIL CONSISTENCY TEST

SAMPLE NO. *11J-23* FEATURE *Example* PROJECT _____

EXCAVATION NO. _____ DEPTH _____ DATE _____

LIQUID LIMIT - PLASTIC LIMIT - PLASTICITY INDEX

		PLASTIC LIMIT		LIQUID LIMIT			
	TRIAL NO.	1	2	1	2	3	4
	DISH NO.	136	57	67	101	85	23
	NO. OF BLOWS			7	12	20	33
1	WT. DISH + WET SOIL	36.22	39.63	49.40	45.36	51.26	50.74
2	WT. DISH + DRY SOIL	35.76	38.84	46.16	43.20	47.95	48.85
3	WT. OF DISH	33.47	34.79	32.88	33.94	33.24	40.07
4	WT. OF WATER (1-2)	0.46	0.79	3.24	2.16	3.31	1.89
5	WT. OF DRY SOIL (2-3)	2.29	4.05	13.28	9.26	14.71	8.78
6	% MOISTURE (4/5 x 100)	20.1	19.5	24.4	23.3	22.5	21.5
7	AVERAGE PLASTIC LIMIT		19.8				

FLOW CURVE

PERCENT MOISTURE — NUMBER OF BLOWS

LIQUID LIMIT (LL)	22.1	PLASTIC LIMIT (PL)	19.8	PLASTICITY INDEX (PI = LL - PL)	2.3

SHRINKAGE TEST

1	SHRINKAGE DISH NO.	25	8	VOL. SHRINKAGE DISH (V)	15.6
2	WT. OF DISH + WET SOIL	41.49	9	VOL. DRY SOIL (V_o)	13.4
3	WT. OF DISH + DRY SOIL	35.01	10	$V - V_o = (8-9)$	2.2
4	WT. OF DISH	11.70	11	$\frac{V - V_o}{W_o} \times 100 = \left(\frac{10}{6} \times 100\right)$	9.44
5	WT. OF WATER (2-3)	6.48			
6	WT. OF DRY SOIL (W_o) = (3-4)	23.31			
7	% MOISTURE (5/6 x 100)	27.80	12	SHRINKAGE LIMIT (7-11)	18.36
			13	SHRINKAGE RATIO (6/9)	1.74

REMARKS _____ TESTED BY _____ DATE _____

_____ COMPUTED BY _____ DATE _____

_____ CHECKED BY _____ DATE _____

Figure 7-3.—Data form for soil consistency tests. 101-D-255.

than 30, the material shall be removed from the cup and placed into the evaporating dish with the remaining portion of the sample, and water shall be added and the material thoroughly mixed to bring the soil to a more fluid condition, so that an acceptable test is performed. A test is acceptable when the number of blows required to close the groove is between 20 and 30. The test shall always proceed from the dryer to the wetter condition of the soil.

When an acceptable test is performed, the material shall be removed from the cup and placed into the evaporating dish with the remaining portion of the sample; it shall then be thoroughly mixed and the test repeated until the difference in the number of blows between two successive tests is two or less. The water content of a portion of the soil taken from around the groove of the second acceptable test shall be determined (see designation E–9). The number of blows for the two acceptable tests and the moisture data shall be recorded in column 1, Liquid Limit, figure 7–4.

10. Computation.—Compute the liquid limit by multiplying the moisture content of the soil, W_n, for the number of blows N, by the factor F_n corresponding to the same number of blows N as shown on figure 7–4.

Part B. Plastic Limit and Plasticity Index of Soils (Adapted from ASTM Designation D 424–54T)

11. Definition.—The plastic limit of a soil is the water content corresponding to the arbitrary limit between the plastic and semisolid states of consistency of a soil. It is the water content at which a soil will just begin to crumble when rolled into threads ⅛ inch in diameter.

12. Apparatus.—The apparatus shall consist of the following (see designation E–4 and figs. 7–5 and 7–6):

Balance, item 4.
Dish, evaporating, porcelain, item 27.
Dishes, weighing, aluminum, item 30.
Rolling surface, item 70.
Spatula, item 87.

13. Sample.—(a) If the plastic limit only is required, place a representative 15-gram sample of air-dried soil passing the No. 40 sieve in an evaporating dish and thoroughly mix with distilled water until the mass becomes sufficiently plastic to be easily shaped into a ball. Take a portion of this ball weighing about 8 grams for the test sample.

(b) If both the liquid and plastic limits are required, take a test

SOIL CONSISTENCY TEST
(ONE-POINT LIQUID LIMIT METHOD)

Sample No. **32K – 1972**   Feature **EXAMPLE**   Project _____

Air-dried ☒   Tested by _____   Date _____

Oven-dried ☐   Computed by _____   Date _____

Natural ☐   Checked by _____   Date _____

	PLASTIC LIMIT		LIQUID LIMIT	
Trial No.	1	2	1	2
Dish No.	99	145	143	
No. of blows (N)	✕	✕	21 | 22	
Wt. Dish + Wet Soil	12.10	12.24	19.60	
Wt. Dish + Dry Soil	11.05	11.16	14.32	
Wt. Dish	6.28	6.26	6.13	
Wt. Water	1.05	1.08	5.28	
Wt. Dry Soil	4.77	4.90	8.19	
% Moisture	22.01	22.04	W_n = 64.47	
Average Plastic Limit	22.0		F_n = 0.985	
			Liquid Limit 63.5	

$$LL = W_n \left(\frac{N}{25}\right)^{0.120}$$

$$F_n = \left(\frac{N}{25}\right)^{0.120}$$

$$LL = (F_n)(W_n)$$

SHRINKAGE LIMIT   10 BLOWS

1. Shrinkage Dish No.	84	
2. Wt. of Dish + Wet Soil	34.61	
3. Wt. of Dish + Dry Soil	24.98	
4. Wt. of Dish	11.56	
5. Wt. of Water (2-3)	9.63	
6. Wt. of Dry Soil (W_O)=(3-4)	13.42	
7. % Moisture (5/6 x 1.00)	71.76	
8. Vol. Shrinkage Dish (V)	14.90	
9. Vol. Dry Soil (V_O)	7.20	
10. V - V_O = (8-9)	7.70	
11. $\frac{V - V_O}{W_O} \times 100 = \left(\frac{10}{6} \times 100\right)$	57.38	
12. Shrinkage Limit (7-11)	14.38	
13. Shrinkage Ratio (6/9)	1.86	

N	F_n
20	0.974
21	0.979
㉒	0.985
23	0.990
24	0.995
25	1.000
26	1.005
27	1.009
28	1.014
29	1.018
30	1.022

LIQUID LIMIT (LL) = **63.5**

PLASTIC LIMIT (PL) = **22.0**

PLASTICITY INDEX (PI) = **41.5**

SHRINKAGE LIMIT (SL) = **14.4**

REMARKS

_____

_____

_____

_____

_____

Figure 7–4.—Soil consistency test (one-point liquid limit method). 101–D–387.

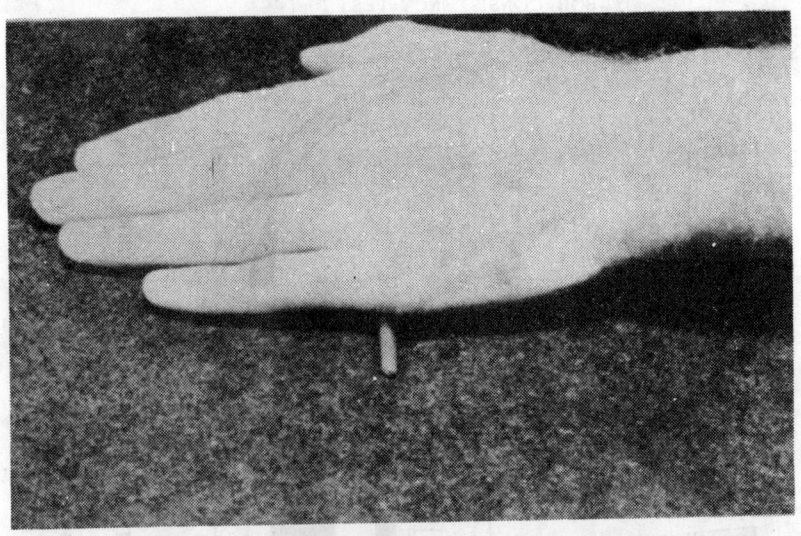

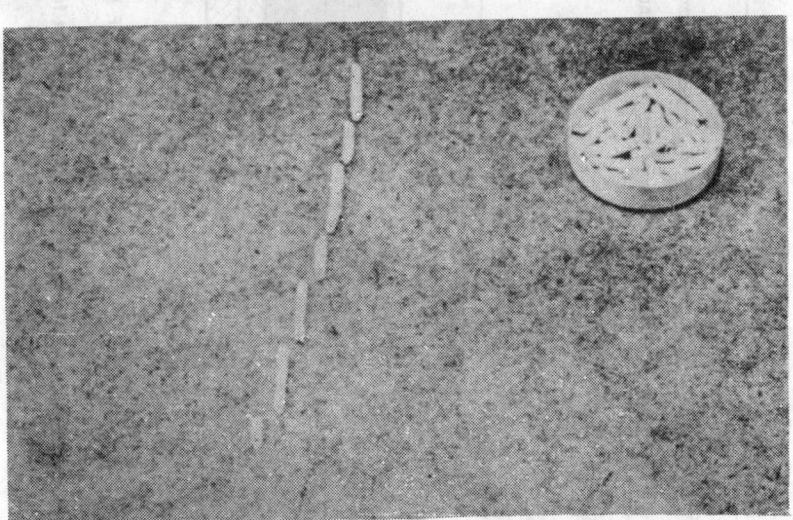

Figure 7–5.—Test for plastic limit. PX–D–16530.

sample weighing about 8 grams from the thoroughly wet and mixed portion of soil for the liquid limit test prepared in accordance with paragraph 6(a). Take the sample at any stage of the mixing process at which the mass becomes plastic enough to be easily shaped into a ball without sticking to the fingers excessively when squeezed. If the sample is taken before completion of. the liquid limit test, set it aside and allow

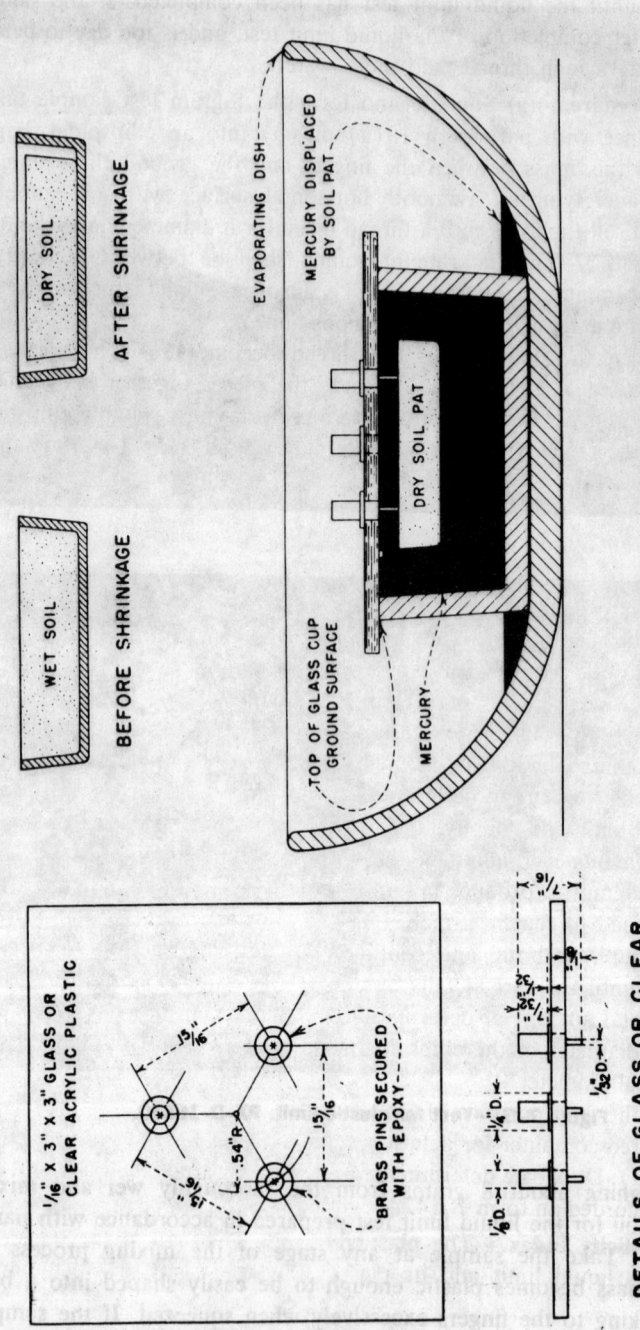

BEFORE SHRINKAGE

WET SOIL

AFTER SHRINKAGE

DRY SOIL

METHOD OF OBTAINING DISPLACED MERCURY

EVAPORATING DISH

MERCURY DISPLACED BY SOIL PAT

DRY SOIL PAT

TOP OF GLASS CUP GROUND SURFACE

MERCURY

DETAILS OF GLASS OR CLEAR ACRYLIC PLASTIC PLATE

1/16" x 3" x 3" GLASS OR CLEAR ACRYLIC PLASTIC

BRASS PINS SECURED WITH EPOXY

.54"R

.15/16"

.15/16"

.15/16"

Figure 7-6.—Apparatus for shrinkage test for soils. 101-D-159.

to season until the liquid limit test has been completed. If the sample is taken after completion of the liquid limit test, and is too dry to permit rolling to a ⅛-inch thread, add more water.

14. Procedure.—(a) Squeeze and form the 8-gram test sample taken in accordance with paragraph 13(a) or (b) into an ellipsoidal-shaped mass. Roll this mass between the fingers and the ground-glass plate or piece of paper lying on a smooth horizontal surface with just sufficient pressure to roll the mass into a thread of uniform diameter throughout its length (figure 7–5). The rate of rolling shall be between 80 and 90 strokes per minute, counting a stroke as one complete motion of the hand forward and back to the starting position.

(b) When the diameter of the thread becomes ⅛-inch, break the thread into six or eight pieces. Squeeze the pieces together between the thumbs and fingers of both hands into a uniform mass roughly ellipsoidal in shape, and reroll. Continue this alternate rolling to a ⅛-inch-diameter thread, gathering together, kneading and rerolling, until the thread crumbles under the pressure required for rolling and the soil can no longer be rolled into a thread. The crumbling may occur when the thread has a diameter greater than ⅛-inch. This shall be considered a satisfactory end point, provided the soil has been previously rolled into a ⅛-inch-diameter thread. The crumbling will manifest itself differently with the various types of soil. Some soils fall apart in numerous aggregations of particles; others may form an outside tubular layer that starts splitting at both ends. The splitting progresses toward the middle until finally the thread falls apart in many small platy particles. Fat clay soils require much pressure to deform the thread, particularly as they approach the plastic limit and, finally, the thread breaks into a series of barrel-shaped segments each about ¼ to ⅜ inch in length. At no time shall the operator attempt to produce failure at exactly ⅛-inch diameter by allowing the thread to reach ⅛-inch, then reducing the rate of rolling or the hand pressure, or both, and continuing the rolling without further deformation until the thread falls apart. It is permissible, however, to reduce the total amount of deformation for slightly plastic soils by making the initial diameter of the ellipsoidal-shaped mass nearer to the required ⅛-inch final diameter.

(c) Gather the portions of the crumbled soil together and place in a suitable tared container for determination of the water content (see designation E–9). Duplicate determinations of water content shall be made and the data recorded on form 7–1424, figure 7–3.

15. Plasticity Index.—The plasticity index of a soil is the difference between its liquid limit and its plastic limit and shall be calculated as follows:

$$\text{Plasticity index} = \text{liquid limit} - \text{plastic limit.}$$

This difference shall be reported as the plasticity index, except under the following conditions:

1. When the liquid limit or plastic limit cannot be determined using the standard testing procedures, report the plasticity index as nonplastic (NP).

2. When the soil is very sandy, the plastic limit test shall be made before the liquid limit test. If the plastic limit cannot be determined, report both the liquid limit and plastic limit as nonplastic (NP).

3. When the plastic limit is either equal to or greater than the liquid limit, report the plasticity index as nonplastic (NP).

4. When the liquid limit can be obtained, although the plasticity index is reported as nonplastic in accordance with 1, 2, and 3 above, it is of value for identifying special soil properties and should be reported along with the indicated nonplastic plasticity index.

Part C. Shrinkage Factors of Soils (Adapted from ASTM Designation D 427–61)

16. Definitions.—The shrinkage limit of a soil is the maximum water content at which a reduction in water content will not cause a decrease in the volume of the soil mass.

The shrinkage ratio of a soil is the ratio of: a given volume change, expressed as a percentage of the dry volume, to the corresponding change in water content above the shrinkage limit, expressed as a percentage of the weight of the ovendried soil.

17. Apparatus.—The apparatus shall consist of the following (see designation E–4 and fig. 7–6):

Balance, item 4.
Dish, evaporating, porcelain, item 27.
Dish, shrinkage, item 29.
Graduate, 25 ml., item 38.
Mercury, item 49.
Shrinkage measuring apparatus, item 79.
Spatula, item 87.
Straightedge, item 91.

18. Procedure.—(a) A representative 30-gram sample of the air-dried soil passing the No. 40 sieve shall be placed in an evaporating dish and thoroughly mixed with distilled water. The amount of water added shall be sufficient to fill the soil voids and to make the soil pasty enough to be readily worked into the shrinkage dish without inclusion of air bubbles. The amount of water required to furnish friable soils with the desired consistency is equal to or slightly greater than the liquid limit, and the amount necessary to furnish plastic soils with the desired consistency may exceed the liquid limit by as much as 10 percent. It has been found that an acceptable water content is one which will produce closure of the liquid limit groove in 10 blows.

(b) The inside of the shrinkage dish shall be coated with a thin layer of petroleum jelly or some other heavy lubricant to prevent adhesion of the soil to the dish. An amount of wetted soil equal to about one-third the volume of the dish shall be placed in the center of the dish, and the soil caused to flow to the edges by tapping the dish on a firm surface cushioned by several layers of blotting paper or similar material. An amount of soil, approximately equal to the first portion, shall be added and the dish tapped until the soil is thoroughly compacted and all included air has been brought to the surface. More soil shall then be added and the tapping continued until the dish is completely filled and excess soil stands out about its edge. The excess soil shall then be struck off with a straightedge, and all soil adhering to the outside shall be cleaned from the dish.

(c) The dish when filled and struck off level shall be weighed immediately and the weight recorded as the weight of dish and wet soil (see fig. 7–3). The soil pat shall be allowed to dry in air until its color changes from dark to light. It shall then be ovendried to constant weight at $110°$ C., and weighed, the weight being recorded as the weight of dish and dry soil. The weight of the empty dish shall be determined and recorded. The capacity of the dish in cubic centimeters, which is also the volume of the wet soil pat, shall be determined by measuring the volume of mercury held by the dish. Mercury shall be placed in the dish to overflowing and the excess removed by pressing a glass plate firmly over the top. (*Note:* This glass plate shall be smooth and a separate plate from that used to determine the volume of soil pat as described in the following subparagraph (d) and as shown in figure 7–6.) The volume of mercury may then be measured by means of a 25-ml. glass graduate. This volume shall be recorded as the volume of the shrinkage dish, V.

(d) The volume of the dry soil pat shall be determined by removing the pat from the shrinkage dish and immersing it in mercury contained in a glass cup in the following manner: The glass cup shall be filled to overflowing with mercury and the excess mercury shall be removed by pressing a glass plate with the three prongs (fig. 7–6) firmly over the top of the cup. Any mercury which may be adhering to the outside of the cup shall be carefully removed. The cup, filled with mercury, shall be placed in a clean evaporating dish, and the soil pat shall be placed on the surface of the mercury. The pat shall then be carefully forced under the mercury by pressing the glass plate with the three prongs firmly over the top of the cup. It is essential that no air be trapped under the soil pat. The volume of the mercury so displaced shall be measured in a glass graduate and recorded as the volume of the dry soil pat, V_o.

19. Calculations.—(a) *Water Content.*—The water content of the soil at the time it was placed in the dish expressed as a percentage of the dry weight of the soil is determined as shown on form 7-1424, figure 7-3.

(b) *Shrinkage Limit.*—The shrinkage limit, *SL,* shall be calculated from the data obtained in the volumetric shrinkage determination by the following formula:

$$SL = w - \left[\frac{V - V_o}{W_o} \times 100 \right]$$

where: SL = the shrinkage limit,
　　　w = water content of wet soil pat, expressed as a percentage of the weight of ovendried soil,
　　　V = volume of wet soil pat, in cubic centimeters,
　　　V_o = volume of dry soil pat, in cubic centimeters, and
　　　W_o = weight of ovendried soil pat, in grams.

(c) *Shrinkage Ratio.*—The shrinkage ratio, *R,* shall be calculated from the data obtained in the volumetric shrinkage determination by the following formula:

$$R = \frac{W_o}{V_o}$$

where W_o and V_o are given above.

SOLUBLE SALTS DETERMINATION OF SOILS

Designation E-8

1. Scope.—This designation describes the method for determining the amount of water soluble salts in soil.

2. Apparatus.—The apparatus for field investigation of material shall consist of the following (see designation E-4):

　　Balance, item 5.
　　Desiccator, item 26.
　　Dishes, evaporating, porcelain, item 27.
　　Graduates, 250-milliliter capacity, item 38.
　　Jug, item 41.
　　Oven, drying, item 55.

For accurate determination of amount of soluble salts as conducted in the Engineering and Research Center laboratory, the following apparatus is required (not listed in the laboratory equipment lists, designation E–4):

Steam bath evaporator for evaporating the extract.
Platinum evaporating dishes.
Balance of 100-gram capacity sensitive to 0.1 milligram.
Filtering equipment—porcelain, Gooch crucible, glass fiber disc, filtering flask, and aspirator.
Pipette, 25-milliliter capacity.
Volumetric flasks.

3. Sample.—A representative sample of soil shall be ovendried, thoroughly mixed and separated on a No. 4 sieve. The portion passing shall be reduced to approximately 500 grams by quartering or by the use of a sample splitter.

4. Procedure.—(a) A 200-gram sample of the minus No. 4 soil shall be placed in the glass jug. The remainder of the 500-gram sample shall be placed in a paper carton, properly labeled, and stored for possible future use.

(b) Two liters of distilled water shall be added to the 200-gram soil sample in the glass jug giving a soil-water ratio of 1 to 10. The soil-water mixture, referred to as a "leach," shall be agitated four or five times a day for at least 4 days.

(c) The mixture shall then be allowed to stand until the liquid becomes clear or the liquid may be filtered.

(*Note:* After settling, the mixture should be decanted, leaving the sediment in the original jar. The solution may need to be filtered several times with new filter discs before proceeding.)

(d) At least 200 cubic centimeters of the clear liquid, or extract, shall be placed in a weighed evaporating dish and evaporated, finally drying the residue to constant weight in an oven at 110° C. After cooling the dish and residue to room temperature in a desiccator, the weight of the residue shall be determined.

(*Note:* Residues over 3,000 parts per million may require drying at 180° C.)

(e) If the concentration of soluble salts in this extract as determined by paragraph 5(a) is greater than 2,000 parts per million, procedure 4 (a), (b), (c), and (d) shall be repeated except that a 70-gram soil sample and 3½ liters of distilled water (1:50 ratio) shall be used.

5. Calculations.—(a) The concentration of soluble material in the clear liquid shall be computed by the following equation:

$$C = \frac{(W)(1,000,000)}{V}$$

where: C = soluble constituents in parts per million,
W = weight of residue in grams, and
V = volume of filtrate in cubic centimeters evaporated to obtain W.

(b) The following equation shall be used to determine the percentage of soluble constituents by dry weight:

$$P = \frac{CD}{10,000}$$

where: P = percentage of soluble constituents by dry weight,
C = soluble constituents in parts per million, and
D = dilution of soil-water mixture. (Example, if soil-water ratio of mixture is 1:5, $D = 5$.)

6. Discussion.—The solubility of materials in liquids is usually expressed in terms of parts per million. For example, the solubility of calcium sulfate, $CaSO_4 x 2H_2O$ (gypsum) in water, is approximately 2,200 to 2,400 parts per million, depending upon the temperature. This means that 2,200 to 2,400 parts of gypsum by weight is the maximum amount of gypsum which can be dissolved in one million parts of water by weight. For practical use, parts per million are considered to be equivalent to milligrams per liter.

The extract from the first leach should contain all of the soluble salts present in the soil sample tested, but if gypsum is present in such amounts that the concentration of the extract exceeds 2,000 parts per million there is a possibility that all of the gypsum has not been dissolved from the sample. The simplest corrective measure for treating such samples is the preparation of a more dilute soil-water mixture to insure complete dissolving of all the gypsum present in the sample.

MOISTURE DETERMINATION OF SOILS

Designation E–9

1. Scope.—This designation describes the method for determining the moisture content (water content)[1] of soil, expressed in percent of the ovendry weight of the soil.

---

[1] The terms "water content" and "moisture content" are used interchangeably in these designations.

2. Apparatus.—The apparatus shall consist of the following (see designation E–4 and fig. 9–1):

Figure 9–1.—Drying oven. E–2250–1NA.

Sample container.—The container should be as small as practicable in keeping with the size of samples and should be of material that is resistant to corrosion and not subject to change in weight or disintegration on repeated heating and cooling. Porcelain or glass evaporating dishes, item 27, porcelain or metallic dishes, item 30, as recommended for specific tests are suitable containers.

Balance, items 3 or 4 or 5.
Desiccator, item 26.
Dish holder, item 28.
Oven, drying, item 55.

3. Sample.—A sample representative of the soil shall be obtained. The size of sample selected depends on the quantity required for good representation, which is influenced by the gradation and maximum size of particles, and on the accuracy of weighing. The following quantities are recommended for general laboratory use:

10 grams for minus No. 40 material.
200 grams for minus No. 4 material.
1,000 grams for minus ⅜-inch material.
2,000 grams for minus ¾-inch material.
3,000 grams for minus 1½-inch material.
3,000 grams or more as required to obtain representative samples of No. 4 to 3-inch material.

4. Procedure.—(a) The sample shall be placed in a suitable container and the weight of wet soil and container accurately determined. The sample and container shall then be placed in a drying oven maintained at 110° C. temperature. The drying period shall be sufficient to permit drying of the soil to constant weight. However, checking every moisture content sample to determine that it is dried to constant weight is impractical. In most cases, drying of a moisture content sample overnight (16 hours) is sufficient. When there is doubt concerning the adequacy of overnight drying, drying should be continued until the weights after two successive periods of drying indicate no change in weight. The time required may vary from a few hours for sandy soils to several days for clays. Sixteen hours is considered a minimum time. Dry soil may absorb moisture from wet samples; thus, it is best to remove the dried samples when placing wet samples in the oven.

(b) After being dried to constant weight, the sample shall be removed from the oven and cooled to room temperature in the desiccator. If containers with tight lids are used, the lids may be placed on the containers immediately upon removal from the oven and the samples cooled without using the desiccator.

(c) After cooling to room temperature, the dry weight of the sample shall be determined and recorded on forms similar to form 7-1452, figure 9-2.

7-1452
(9-62)
BUREAU OF RECLAMATION   MOISTURE DETERMINATIONS

PROJECT _Example_____   FEATURE _____   LABORATORY INDEX NO. _11J-X90_

SAMPLE NUMBER	1	2						
DISH NUMBER	15	20						
WT DISH + WET SOIL	366.1	374.6						
WT DISH + DRY SOIL	348.0	342.1						
WEIGHT OF DISH	129.4	118.0						
WEIGHT OF WATER	18.1	32.5						
WEIGHT OF DRY SOIL	218.6	224.1						
PERCENT MOISTURE	8.3	14.5						
SAMPLE NUMBER								

Figure 9-2.—Moisture determination data. 101-D-328.

5. Calculations.—The water content in percent of ovendry weight of soil shall be computed by the following equation:

$$\text{Water content, } \% = \frac{\text{weight of water evaporated}}{\text{weight of ovendry soil}} \times 100$$

SPECIFIC GRAVITY OF SOILS, AGGREGATE, AND DENSITY OF IRREGULAR BLOCKS OF SOIL

Designation E-10

1. Scope.—This designation describes methods for determining the specific gravity of fine- and coarse-grained soils. It also describes a method for determining the density (bulk specific gravity) of cores or irregular blocks of soil.

2. Definition.—The specific gravity is defined as the ratio of the weight in air of a given volume of material to the weight in air of an equal volume of distilled water at a stated temperature.

Part A. Apparent Specific Gravity of Minus No. 4 Material
by the Flask Method

3. Apparatus.—The apparatus shall consist of the following (see designation E–4 and fig. 10–1):

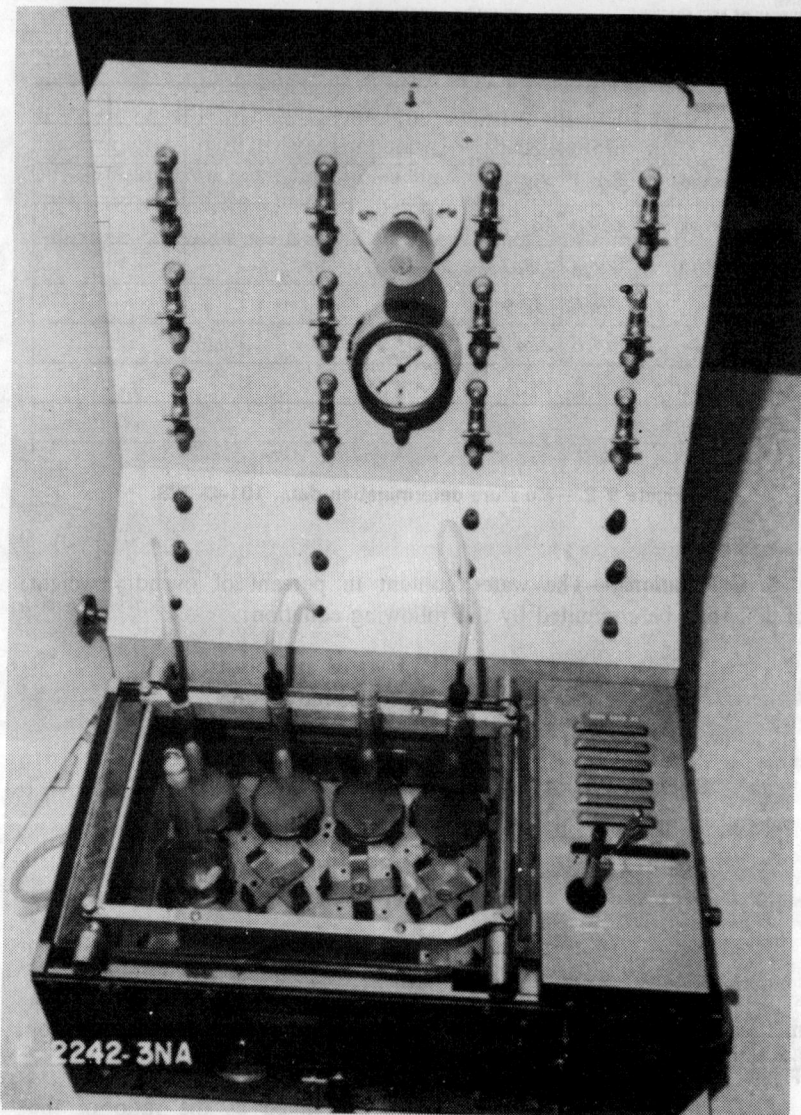

Figure 10–1.—Apparatus for determining apparent specific gravity by the flask method. E–2242–3NA.

Balance, item 5.
Flask, specific gravity, item 34.
Pump, vacuum, item 67.
Thermometer, item 104.
Tubing, rubber, item 108.

4. Calibration.—The flask shall be calibrated as follows: The weight of a clean, dry flask shall be determined to the nearest 0.1 gram and recorded. The flask shall then be filled with distilled water free of air bubbles and at room temperature until the bottom of the meniscus is at the mark on the flask neck. A vacuum may be used to remove air bubbles. The weight of the flask and water and the temperature of the water shall be determined and recorded, the weight being recorded to the nearest 0.1 gram and temperature to the nearest whole degree. This completes the calibration of the flask, and the information is recorded as items 1, 2, and 3 on form 7-1453, figure 10-2. The volume of the

7-1453
(4-62)
BUREAU OF RECLAMATION

SPECIFIC GRAVITY BY THE FLASK OR PYCNOMETER METHOD

SAMPLE NO. *11J-X90* PROJECT _Example_ FEATURE _____

COMPOSITE OF SAMPLES _11J-23 & 11J-24_

DATE _____ TESTED BY _____

(1) FLASK NO. 24	(2) WT. OF FLASK	79.88 g.
(3) WT. OF FLASK FILLED WITH WATER 328.81 g. AT TEMPERATURE		23.3 °C.
(4) COMPUTED VOLUME OF FLASK, (3-2) ÷ DENSITY OF WATER AT TEMPERATURE (3) (0.997466)		249.56 cc
(5) WT. OF SAMPLE		100.0 g.
(6) TIME UNDER VACUUM		90 min.
(7) WT. OF FLASK + SAMPLE AND WATER		391.29 g.
(8) TEMPERATURE OF WATER		23.6 °C.
(9) WT. OF FLASK AND WATER, (7) - (5)		291.29 g.
(10) WT. OF WATER IN FLASK, (9) - (2)		211.41 g.
(11) VOL. OF WATER IN FLASK, (10) ÷ DENSITY OF WATER AT TEMPERATURE (8) (0.997394)		211.96 cc
(12) VOLUME OF SOIL, (4) - (11)		37.60 cc
(13) SPECIFIC GRAVITY, (5) ÷ (12)		2.660

Figure 10-2.—Recording data and computing the apparent specific gravity of soils. 101-D-337.

flask is computed by dividing the weight of the water by the absolute density of the water (table 10–1) at the temperature recorded in item 3. A permanent record shall be made of items 1, 2, and 4, figure 10–2, for each flask used in the laboratory. It is not necessary to repeat these measurements and calculations unless the weights of the flasks have been changed. The above calibrations will suffice for field and construction control use.

5. Sample.—A sample, representative of the minus No. 4 material, shall be dried to constant weight in an oven at 110° C., and cooled to room temperature in a desiccator or a covered container.

6. Procedure.—A sample of approximately 100 grams shall be accurately weighed and carefully placed in the flask so that none of the soil is lost. Distilled water sufficient to cover the soil shall be added to the flask, the water also being used to wash the soil from the inside of the neck.

A vacuum shall then be applied to the flask with care to prevent the soil from being pulled out of the flask. Violent boiling may be decreased by reducing the vacuum or by placing the flask in a cold-water bath. The flask shall be rolled and shaken occasionally to assist in the removal of the air. As the air removal nears completion, distilled water shall be added until the water level is in the lower part of the flask neck.

When the air evacuation is complete, as indicated by no change in water level when alternately subjected to reduced and atmospheric pressure, the flask shall be filled to the calibration mark on the neck. The total weight of the flask, sample, and water shall be determined and recorded, item 7, figure 10–2. The temperature of the water in the flask shall be determined and recorded to the nearest whole degree by inserting the thermometer in the flask.

7. Calculations.—The calculations necessary to determine the specific gravity are shown in figure 10–2. The volume of water in the flask, item 11, is determined by dividing the weight of water in the flask, item 10, by the absolute density of the water from table 10–1 at the temperature recorded in item 8.

8. Calibration Method.—A procedure for calibrating a flask and preparing a calibration table which assists in computing specific gravity and saves time involved in checking results is described as follows:

The weight of the flask and distilled water is determined at a number of different temperatures, ranging from 15° to 30° C., and a calibration curve (not shown) is developed by plotting temperature in degrees centigrade against weight of flask filled with distilled water. A calibration chart, figure 10–3, is then prepared for each flask using the data from the calibration curve and the absolute density of the water in table 10–1.

Temperature degrees C.	[1]Absolute density of water in grams per cubic centimeter x 100 CC	[2]Weight of water-filled flask in grams + 100 grams	
		Flask No. 1	Flask No. 5
18.0	99.8595	424.68	418.82
.2	.8558	.67	.81
.4	.8520	.66	.79
21.0	99.7992	424.52	418.64
.2	.7948	.51	.63
.4	.7904	.50	.62
.6	.7860	.49	.61
.8	.7815	.48	.60
22.0	.7770	.47	.59
.2	.7724	.46	.58
23.0	99.7538	424.42	418.53
.2	.7490	.41	.52
.4	.7442	.40	.51
.6	.7394	.39	.50
.8	.7345	.38	.49
27.0	99.6512	424.21	418.30

[1] Values of the absolute density of water (grams per cubic centimeter) in table 10-1, multiplied by 100 cubic centimeters (the volumetric equivalent of a 100-gram sample).

[2] Weight of flask filled with distilled water at a specified temperature (obtained from calibration curve) plus 100 grams (the dry weight of the soil sample).

Note: Other sample weights may be used; however, the chart values above must be adjusted for the dry weight of sample used.

Figure 10–3.—Portions of calibration chart for specific gravity flasks for 100-gram sample. 101-D-342.

The specific gravity tests shall then be performed as described previously, using 100 grams of material. The weight of the flask, sample, and water shall be recorded as experimental weight on form 7–1589, line 6, figure 10–4, and the observed temperature shall be recorded in line 4. The chart weight, line 5, shall be obtained from the calibration chart, for the appropriate flask and temperature at which the material was tested. Line 8 is obtained from column 2 of the chart. The specific gravity is then determined by the procedures outlined on the form.

Table 10–1.—Absolute density of water in grams per cubic centimeter

Degrees C.	0	1	2	3	4	5	6	7	8	9
0	0.999841	847	854	860	866	872	878	884	889	895
1	900	905	909	914	918	923	·927	930	934	938
2	941	944	947	950	953	955	958	960	962	964
3	965	967	968	969	970	971	972	972	973	973
4	973	973	973	972	972	972	970	969	968	966
5	965	963	961	959	957	955	952	950	947	944
6	941	938	935	931	927	924	920	916	911	907
7	902	898	893	888	883	877	872	866	861	855
8	849	843	837	830	824	817	810	803	796	789
9	781	774	766	758	751	742	734	726	717	709
10	700	691	682	673	664	654	645	635	625	615
11	605	595	585	574	564	553	542	531	520	509
12	498	486	475	463	451	439	427	415	402	390
13	377	364	352	339	326	312	299	285	272	258
14	244	230	216	202	188	173	159	144	129	114
15	099	084	069	054	038	023	007	*991	*975	*959
16	0.998943	926	910	893	877	860	843	826	809	792
17	774	757	739	722	704	686	668	650	632	613
18	595	576	558	539	520	501	482	463	444	424
19	405	385	365	345	325	305	285	265	244	224
20	203	183	162	141	120	099	078	056	035	013
21	0.997992	970	948	926	904	882	860	837	815	792
22	770	747	724	701	678	655	632	608	585	561
23	538	514	490	466	442	418	394	369	345	320
24	296	271	246	221	196	171	146	120	095	069

25	044	018	*992	*967	*941	*914	*888	*862	*836	*809
26	0.996783	756	729	703	676	649	621	594	567	540
27	512	485	457	429	401	373	345	317	289	261
28	232	204	175	147	118	089	060	031	002	*973
29	0.995944	914	885	855	826	796	766	736	706	676
30	646	616	586	555	525	494	464	433	402	371

* First three significant figures shown in line below.

7-1500
(9-62)

SPECIFIC GRAVITY TEST
(Use with Calibration Chart for Specific Gravity Flasks)

PROJECT _____ *Example* _____ FEATURE _____

DATE _____ — _____ TESTED BY _____ — _____ CHECKED BY _____

1. TRIAL	*1*	*2*		
2. SAMPLE NO.	*11J-X90*	*11J-X90*		
3. FLASK NO.	*1*	*5*		
4. TEMP °C, (Experimental)	*21.4*	*21.6*		
5. CHART WEIGHT AT TEMP (4) g.	*424.500*	*418.610*		
6. EXPERIMENTAL WEIGHT, (W) g.	*386.970*	*381.140*		
7. DISPLACED WATER, (5) - (6) g.	*37.53*	*37.47*		
8. (Wt) (100), FROM CHART TEMP (4) g.	*99.7904*	*99.7860*		
9. SPECIFIC GRAVITY (8) ÷ (7)	*2.659*	*2.663*		
10. AVERAGE SPECIFIC GRAVITY	*2.661*			

Figure 10–4.—Recording and computing specific gravity by the calibration method. 101–D–343.

Part B. Apparent Specific Gravity of Minus ¾-Inch Material by Pycnometer Method

9. Apparatus.—The apparatus shall consist of the following (see designation E–4 and fig. 10–5):

Balance, item 5.
Pycnometer, item 68.
Apparatus listed for the flask method.

10. Calibration.—After the fruit jar and glass plate have been cleaned and dried, their weight shall be determined and recorded to the nearest 0.1 gram. The jar shall be filled with distilled water at room temperature, and all air bubbles eliminated by placing the cap on the jar and shaking; or applying a vacuum, if available, and then shaking. The cap shall be removed and sufficient water added until the meniscus rises above the ground edge. The excess water shall be removed by sliding the glass plate slowly across the ground surface of the jar opening. Care shall be exercised so as not to entrap any air below the plate. With the glass plate held in place, the outside of the jar and plate shall be thoroughly dried. The weight of the glass plate and the jar filled with water shall be determined and recorded. A thermometer shall then be inserted in the water, and its temperature determined and recorded to the nearest whole degree. From the weights determined and the observed tempera-

Figure 10–5.—Pycnometer (pint fruit jar with edges ground) and siphon can. PX–D–16533.

ture, the volume of the jar at room temperature shall be determined as described for the flask method. The information is recorded as shown in form 7–1453, figure 10–2, crossing out "flask" and adding "pycnometer", where applicable.

11. Sample.—A representative sample of minus ¾-inch material shall be dried to constant weight in an oven at 110° C. and cooled to room temperature in a desiccator or covered container.

12. Procedure.—A minimum of 200 grams of the ovendried soil shall be weighed to the nearest 0.1 gram and carefully placed in the fruit jar to avoid any loss of material. Sufficient distilled water shall be added to cover the sample.

A vacuum shall be applied to the jar through the plastic cap which is held in place by hand until the reduced air pressure is sufficient to hold it. The jar shall be rolled or shaken occasionally to assist in the removal of the air. When most of the entrapped air has been removed, the jar shall be filled with distilled water, and the cap replaced by sliding it horizontally across the mouth of the jar. Evacuation shall be continued until very few air bubbles appear beneath the plastic cap.

The plastic cap shall be removed and the jar filled with distilled water until the meniscus rises above the ground edge of the jar. The glass plate shall then be placed on the jar by sliding it horizontally across the mouth, removing the excess water. While the glass plate is held in place, the outside of the jar and plate shall be thoroughly dried. The weight of the jar, plate, sample, and water shall be determined and recorded (line 7, fig. 10–2). The temperature of the water in the jar shall then be determined to the nearest whole degree and recorded in line 8.

13. Calculations.—The calculations for the determination of the specific gravity by this method are shown by the step-by-step procedure in figure 10–2, and are the same as discussed for the flask method.

Part C. Specific Gravity of Gravel and Cobbles by the Displacement (Siphon) Method

14. Apparatus.—The apparatus shall consist of the following (see designation E–4 and fig. 10–5):

Balance, item 5.
Can, siphon, item 15.
Graduate, 1,000 ml., item 38.
Towel, item 105.

15. Sample.—A sample, representative of all the particle sizes in the original material and weighing about 2,000 grams, shall be obtained.

16. Procedure.—The individual particles of the sample shall be washed and allowed to remain immersed in the water at room temperature for at least 24 hours. The sample is then removed from the water and rolled in a large cloth until all visible films of water are removed, although the surfaces of the particles still appear to be damp. The larger fragments should be individually wiped, and care should be taken to avoid evaporation from the surface-dry particles. The sample, in this condition, is referred to as being saturated surface-dry. The weight of the saturated surface-dry sample shall be determined.

The siphon can shall be filled with water. The water shall be allowed to drain from the can through the siphon tube until it stops. With the stopper placed in the siphon tube   sample shall be carefully placed in the can, ensuring that all entrapped air is removed by stirring or other appropriate means.

The stopper shall be removed from the siphon tube and the water allowed to drain into the graduate cylinder, and the volume of the water displaced shall be determined and recorded.

The sample shall be removed from the can, placed in a drying oven and the ovendry weight recorded.

17. Calculations.—The specific gravity shall be computed by the the following equation:

Bulk specific gravity (saturated surface-dry) =

$$\frac{\text{weight of saturated surface-dry sample in grams}}{\text{volume of water displaced in cubic centimeters}}$$

Bulk specific gravity (ovendry) =

$$\frac{\text{weight of ovendry sample in grams}}{\text{volume of water displaced in cubic centimeters}}$$

Part D. Specific Gravity of Gravel and Cobbles and Density Determination by the Weight in Air and Water Method (Suspension Method)

18. Apparatus.—The apparatus shall consist of the following (see designation E–4):

> Balance, item 3.
> Basket, brass wire, item 7.
> Container for water, item 23.
> Towel, item 105.

19. Calibration.—The basket shall be suspended by suitable means for weighing in the container. The weight of the empty basket completely immersed but suspended in water shall then be determined. The depth of the basket below water surface should be noted and the basket immersed to the same depth when the basket contains the sample.

20. Sample.—A saturated surface-dry sample weighing at least 2,000 grams and representative of all the particle sizes present in the original material shall be prepared. (See part C of this designation for preparation of a saturated surface-dry sample.)

21. Procedure.—The saturated surface-dry sample shall be weighed and the weight recorded. The sample shall be placed in the wire basket, then suspended in water, and its weight in water determined and recorded. All entrapped air shall be removed from the particles and basket before the weight in water is determined.

The sample shall then be ovendried, and its weight in air determined and recorded.

22. Calculations.—The specific gravity shall be computed by the following equations:

$$\text{Apparent specific gravity} = \frac{A}{A - C}$$

$$\text{Bulk specific gravity (saturated surface-dry)} = \frac{B}{B - C}$$

$$\text{Bulk specific gravity (ovendry)} = \frac{A}{B - C}$$

where: A = ovendry weight of sample,
B = saturated surface-dry weight of sample, and
C = weight of sample in water.

The percentage of absorbed water can be obtained using the data above and the equation below:

$$\frac{\text{Percentage of absorption}}{\text{(saturated surface-dry basis)}} = \frac{(B - A)\ 100}{A}$$

(*Note:* The procedures described in parts C and D are normally used for determining the specific gravity of gravel particles for the field density test (designation E–24); however, saturation for 24 hours is omitted.)

Part E. Density of Cores or irregular-Shaped Blocks of Undisturbed Soil

23. Density of Soil.—The density of cores or irregular-shaped blocks of undisturbed soil may be determined by the suspended weight in air and water method.

24. Procedure.—The specimen shall be prepared by carefully trimming all rough surfaces. The specimen shall then be weighed in air and the weight recorded on form 7–1586, item 1, figure 10–6. The entire surface of the specimen is then covered with warm wax by using a paintbrush. The wax should be warmed sufficiently so that it flows when brushed on the soil, yet it should not be so hot as to penetrate the pores in the soil. Hot wax in contact with a moist sample may cause the water to vaporize, causing air bubbles under the wax. The specimen covered with wax shall be weighed in air and the weight recorded (item 4).

The specimen covered with wax is then weighed in water as discussed in paragraphs 19, 20, and 21, and the weight recorded, item 9, figure

7-1586
(4-62)
BUREAU OF RECLAMATION

DENSITY DETERMINATIONS
(Suspended Weight in Air and Water Method)

SAMPLE NO. *Example* PROJECT_____ FEATURE _____

SPECIMEN NO. _____ DATE _____ TESTED BY_____

(1) WT. SPECIMEN + CONTAINER IN AIR _____	860.45 g.
(2) WT. CONTAINER IN AIR _____	117.85 g.
(3) WT. WET SPECIMEN, (1) - (2) _____	742.60 g.
(4) WT. SPECIMEN + WAX + CONTAINER IN AIR _____	909.70 g.
(5) WT. WAX, (4) - (1) _____	49.25 g.
(6) WATER CONTENT _____	19.54 %
(7) DRY WT. SPECIMEN, (3) ÷ $\left(1 + \frac{(6)}{100}\right)$ _____	621.21 g.
(8) WT. OF CONTAINER IN WATER _____	104.45 g.
(9) WT. SPECIMEN + WAX + CONTAINER IN WATER _____	475.15 g.
(10) WT. LOSS OF SPECIMEN + WAX + CONTAINER, (4) - (9) _____	434.55 g.
(11) WT. LOSS OF CONTAINER, (2) - (8) _____	13.40 g.
(12) WT. LOSS OF WAX, (5) ÷ UNIT WT. OF WAX (see below) *(0.89)* _____	55.34 g.
(13) WT. LOSS OF WAX + CONTAINER, (11) + (12) _____	68.74 g.
(14) WT. LOSS OF SPECIMEN, (10) - (13) _____	365.81 g.
(15) DRY DENSITY OF SPECIMEN (7) x 62.29 ÷ (14) _____	105.8 lb./cu. ft.

UNIT WT. WAXES	MOISTURE DETERMINATION	
	(1) DISH NO. _____	SL-191
PARAFFIN - 0.89 g./cc	(2) WT. DISH + WET SOIL _____	407.40 g.
	(3) WT. DISH + DRY SOIL _____	366.55 g.
	(4) WT. DISH _____	157.45 g.
MICRO-CRYSTALLIN ORANGE - 0.91 g./cc	(5) WT. WATER, (2) - (3) _____	40.85 g.
	(6) WT. DRY SOIL (3) - (4) _____	209.10 g.
	(7) WATER CONTENT (5) ÷ (6) x 100 _____	19.54 %

Figure 10–6.—Recording and computing density of irregular blocks or cores of soil by the suspended weight in air and water method. 101–D–344.

10–6, being careful to eliminate air bubbles and seeing that the depth of immersion is the same as for the basket when weighed in water.

If the dry density of the soil is desired, the water content of the soil must be determined. A sample of the moist soil may be obtained by using the cuttings from the sample during the trimming operations, or using the specimen itself after the weights in air and water have been determined, making sure that all wax has been removed.

25. Calculations.—The calculations for determining the density are shown in figure 10–6.

PROCTOR COMPACTION TEST [1]
(MOISTURE-DENSITY RELATIONS OF SOIL)
Designation E–11

1. Scope.—This designation describes the method for determining the relationship between the water content of the portion of a soil passing the No. 4 sieve and resulting density (ovendry weight per cubic foot) when the soil is compacted in the laboratory as specified herein.

The Proctor maximum dry density is the greatest dry weight, in pounds per cubic foot, obtained by this procedure. The optimum water content is the water content of the soil which exists when the soil is compacted to the Proctor maximum dry density (also referred to as laboratory maximum dry density). The penetration resistance, also obtained during this test, is a measure of the firmness of the soil and is expressed in pounds per square inch.

2. Apparatus.—The apparatus shall consist of the following (see designation E–4 and figs. 11–1, 11–2, and 11–3):

> Balance, item 5.
> Base, for compaction cylinder, item 6.
> Brushes, items 8 and 11.
> Cans, item 14.
> Dishes, evaporating, porcelain, item 27.
> Extruder for compaction specimen, item 32A.
> Graduates, 100 and 500 ml., item 38.
> Knife, item 42.
> Mallet, item 46.
> Mechanical compactor, item 48A (in lieu of items 52, 97 and 86). For calibration, see designation E–37.
> Mixing tools, hand, items 51C, 69, 106.
> Mold, cylindrical compaction, $\frac{1}{20}$ cubic foot, item 52.
> Pan, mixing, item 61.
> Proctor needle tester and needles, item 66.
> Sampling tube, item 73.
> Scoops, items 76 and 77.
> Scale, item 75.
> Sieve, U.S. standard No. 4 with stand, item 80.
> Mixer with No. 4 sieve assembly, items 50, 51, 51A, 51B, 51D, and 51E (in lieu of item 80).
> Spanner wrench for compaction cylinder mold, item 86.
> Straightedge trimmer, item 94.
> Tamper, item 97.

---

[1] For compaction test for soil containing gravel, see designation E–38.

Figure 11–1.—Performing compaction test—Manual method. PX–D–16543.

Figure 11-2.—Performing compaction test—Machine method. PX-D-16544 and E-2156-5NA.

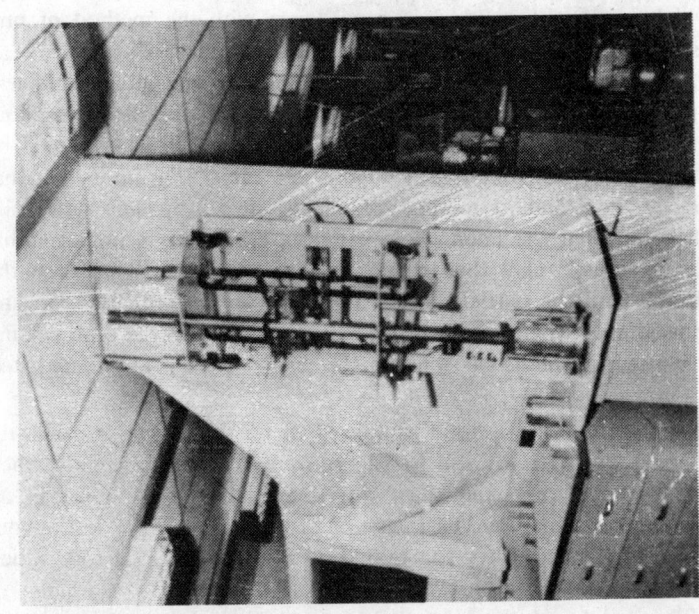

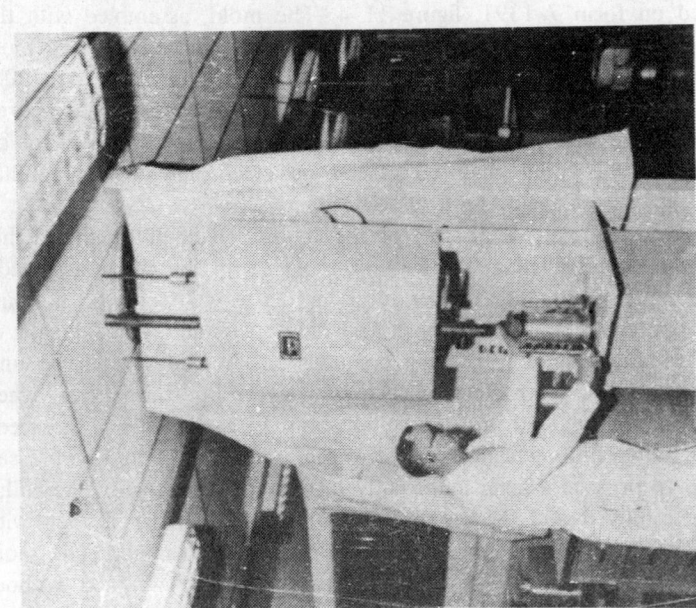

Figure 11–3.—(A) Performing compaction test in which the operator must rotate compaction cylinder, and (B) view of compactor mechanism. E-2156-7NA and E-2156-5NA.

3. Sample.—A sample weighing about 35 pounds shall be taken from the thoroughly mixed portion of the material passing the No. 4 sieve which has been prepared in accordance with the method of preparing soil samples for laboratory testing, designation E–5.

(*Note:* By reprocessing the compacted material, and adding sufficient soil to replace the moisture sample, the test can be performed on about 15 pounds of material. The Proctor maximum dry density obtained by this procedure ranges from 1 to 5 pounds per cubic foot more than that obtained by the standard method described; that is, using "new soil" for each compacted specimen. However, this procedure is not desirable when testing plastic clay soils which tend to form lumps or when the soil is friable and the particles break down during testing. Reprocessing shall be used only when a limited sample is available. If the reprocessing method is used, it should be noted in the results and on the compaction curve.)

4. Procedure.—(a) The total sample shall be thoroughly mixed with enough water to cause the soil to adhere or ball together slightly when squeezed firmly in the hand, then stored for at least 24 hours in an airtight container to permit the moisture to become uniform throughout the sample. The storage may be omitted on moist soils or on soils which readily absorb moisture.

(b) The weight of the compaction mold shall be determined and recorded on form 7–1391, figure 11–4. The mold, assembled with the collar, is then attached to the baseplate. During compaction, the mold shall rest firmly on a uniform, rigid foundation, equivalent to a 200-pound block of concrete. A carbide can filled with concrete has proved satisfactory for this purpose. When the test is performed on a fill or on the ground, the mold shall be placed on steel plates weighing 200 pounds which rest evenly on firm soil (item 6A).

(c) Approximately 7 pounds of the moist soil shall be thoroughly mixed, then compacted in the Proctor mold (with the collar attached) in three equal layers, each layer receiving 25 blows from the tamping rod dropping freely through the height-of-drop guide from a height of 18 inches above the layer being compacted. The blows shall be uniformly distributed over the surface of the layer. The third layer, when compacted, shall extend into the collar about ¼ inch, but not to exceed ½ inch. When fluffy soils are placed in the mold, they should be pressed by hand to provide a layer sufficiently firm to be compacted. The collar shall be removed and the compacted soil carefully trimmed even with the top of the mold by means of the straightedge trimmer. The mold and soil are then removed from the baseplate and weighed without trimming the soil at the bottom of the mold. The weight is recorded to the nearest 0.01 of a pound. Approved laboratory mechanical com-

7-1391
(5-63)

PROCTOR COMPACTION TEST

PROJECT **EXAMPLE** FEATURE _____ SAMPLE NO. _11J-X90_

TESTED BY _____ COMPUTED BY _____ CHECKED BY _____ DATE _____

DEGREE OF COMPACTION; PROCTOR STANDARD ☒ OR MODIFIED ☐ IF MODIFIED, SHOW RELATED INFORMATION AS FOLLOWS:

BLOWS PER LAYER _____ NO. OF LAYERS _____ HEIGHT OF DROP _____ in.

WEIGHT OF TAMPING ROD _____ lb. VOLUME OF CYLINDER _____ cu.ft.

TEST NO.	1	2	3	4	5	6	7	8
DENSITY DETERMINATIONS								
WATER ADDED (cc)	0	80	80	80	90	90		
WT CYL. + WET SOIL (lb)	10.97	11.13	11.52	11.87	11.84	11.69		
WT OF CYLINDER (lb)	5.41	5.41	5.41	5.41	5.41	5.41		
WT OF WET SOIL (lb)	5.56	5.72	6.11	6.46	6.43	6.28		
WT DENSITY (pcf)	111.2	114.4	122.2	129.2	128.6	125.6		
PENETRATION RESISTANCE DETERMINATIONS								
NEEDLE NO.								
AREA OF NEEDLE (sq.in.)	1/40	1/40	1/20	1/20	1/4	1/2		
AVERAGE READING (lb.)	70	72	95	53	58	50		
PENETRATION RESIST. (psi)	2900	2880	1900	1060	232	100		
WATER CONTENT DETERMINATIONS								
DISH NO.	233	15	44	45	55	20		
WT. DISH + WET SOIL	384.5	382.4	386.7	382.1	381.5	369.5		
WT. DISH + DRY SOIL	363.8	359.1	359.6	352.4	347.3	330.6		
WEIGHT OF DISH	121.6	129.4	133.8	140.6	138.6	118.0		
WEIGHT OF WATER	20.7	23.3	27.1	29.7	34.2	38.9		
WEIGHT OF DRY SOIL	242.2	229.7	225.8	211.8	208.7	212.7		
WATER CONT. (% DRY WT)	8.5	10.1	12.0	14.0	16.4	18.3		
DRY DENSITY (pcf.)	102.5	103.9	109.1	113.3	110.5	106.2		

QUESTIONS TO ANSWER FROM OBSERVATIONS BY OPERATORS DURING TEST

1. HOW FAST DOES SAMPLE ABSORB WATER? FAST _____ MEDIUM ✓ SLOW _____
2. IS DIFFICULTY ENCOUNTERED IN MIXING WATER WITH SOIL? No
3. ARE PENETRATION NEEDLE READINGS RELIABLE? Yes
4. AT WHAT TEST NOS. IS SAMPLE CRUMBLY? 1,2 FIRM? 3,4 SOFT? 5,6
5. WAS BLEEDING NOTICED DURING TEST? No IF SO, WHAT TEST NOS.? _____
6. AT WHAT TEST NOS. IS SAMPLE SPONGY? 5,6.
7. OTHER COMMENTS Silty sand

**Figure 11–4.—Recording data for compaction and penetration-resistance tests.
101–D–256.**

pactors mounted on a concrete base weighing at least 200 pounds may be used in lieu of the manual apparatus.

(d) The penetration-resistance test shall be performed on the compacted specimen (see fig. 11–5). The needle shall be forced about ½ inch into the compacted soil by pushing on the penetrometer barrel and the indicator clip placed against the barrel cap. Then, using the handle, the needle shall be forced into the soil an additional 2½ inches at a rate of approximately ½ inch per second. It is necessary to select a needle size that will give a reading of not less than 20 on the plunger shaft. If the compacted specimen contains hard layers that interfere with needle penetration at a fairly uniform rate, the needle shall be pushed through such layers by applying force to the penetrometer barrel, then the test continued using the handle. This method

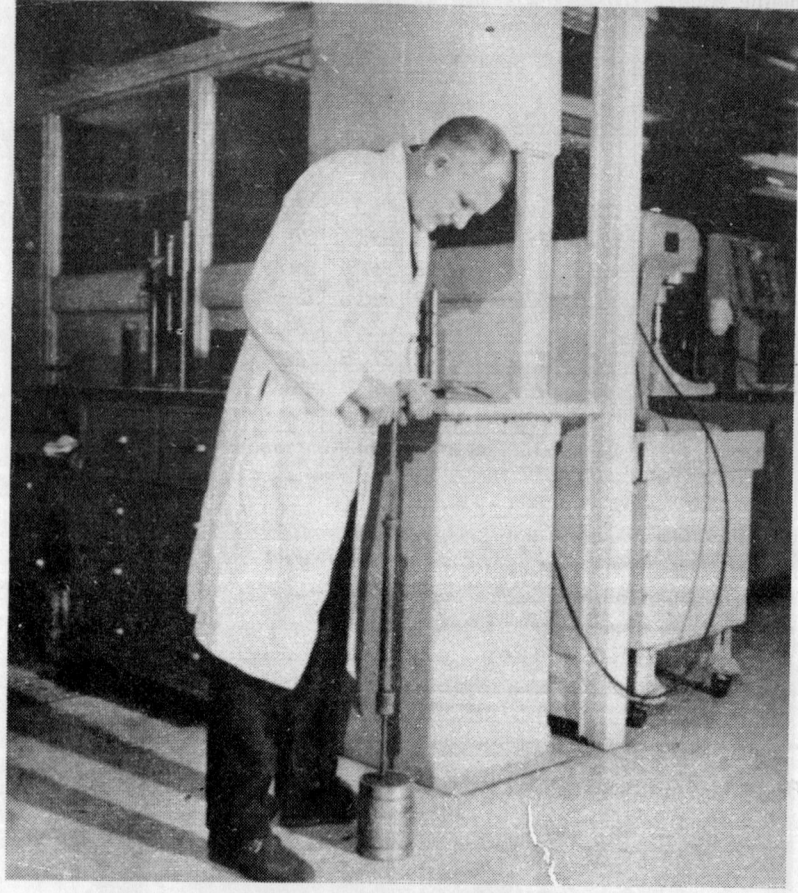

Figure 11–5.—Determining penetration resistance. E–2243–5.

gives a reading representative of the compacted soil exclusive of the hard layers. The area of the needle used and the average of three penetration-resistance readings are recorded.

(e) The compaction specimen shall be removed from the mold and a water content determination shall be made on a portion of the soil from the center of the specimen as specified in designation E–9 and the data recorded as shown in figure 11–4. An extruder such as that shown in figure 11–6 can be used to remove the specimen from the mold.

(f) At least five specimens, using new soil for each (except as provided in par. 3), shall be prepared according to procedures (a) to (e) above, with each succeeding specimen having a higher water content until the soil becomes very wet or there is a decrease in the wet weight of the compacted specimen. Five or more specimens each at increasing water contents of 1½ to 2 percent are usually required to give an adequate smooth compaction curve.

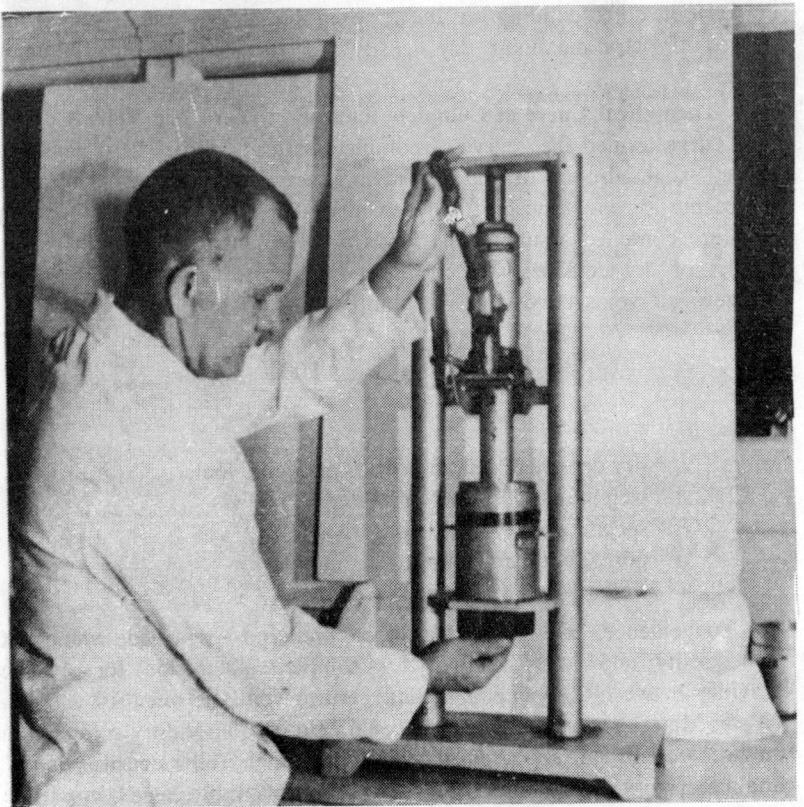

Figure 11–6.—Extruder for compaction specimens. E–2243–2NA.

5. Calculations.—The wet density in pounds per cubic foot for each specimen shall be calculated by dividing the weight of wet soil by the volume of the compaction mold (0.05 cubic foot). The dry density in pounds per cubic foot for each specimen shall be calculated by dividing the wet density by 1 + (water content as a decimal fraction). The penetration resistance in pounds per square inch for each specimen shall be computed by dividing the average penetration-resistance reading by the area of the needle used.

6. Moisture-Density Relationship.—From the data obtained in paragraph 5, dry density and penetration-resistance values shall be plotted as ordinates with corresponding water contents as abscissas (fig. 11–7). A smooth curve shall then be drawn through each set of points.

(a) *Optimum Water Content.*—From the moisture-density curve, the water content corresponding to the peak of the curve shall be termed the "optimum water content" for the standard Proctor compaction.

(b) *Proctor Maximum Dry Density.*—The dry density in pounds per cubic foot corresponding to the "optimum water content" shall be termed "Proctor maximum dry density" for the standard Proctor compaction.

7. Theoretical Curve at Complete Saturation (Zero Air Voids Curve).—A curve termed the "curve of complete saturation" or "zero air voids curve," shall also be drawn on the moisture-density sheet. This curve represents the relationship between dry densities and corresponding moisture contents, assuming the voids are completely filled with water. Values of dry densities and corresponding water contents for plotting the zero air voids curve shall be computed as follows:

$$\gamma_d = 62.4 \, G_s\left(1 - \frac{n}{100}\right)$$

$$w = \frac{62.4 \, n}{\gamma_d}$$

where: γ_d = dry density of soil, in pounds per cubic foot,
   62.4 = weight of 1 cubic foot of water, in pounds,
   G_s = specific gravity of the soil,
   n = porosity in percent, and
   w = water content in percent.

For convenience, table 11–1 has been prepared to provide values of dry density and water content for complete saturation for specific gravities between 2.45 and 2.90 and percent voids between 10 and 60. (*Note:* Materials compacted by the usual field and laboratory compaction methods contain entrapped air. Therefore, the field "roller density curves" and laboratory density test curves will approach but never cross the "zero air voids curves.")

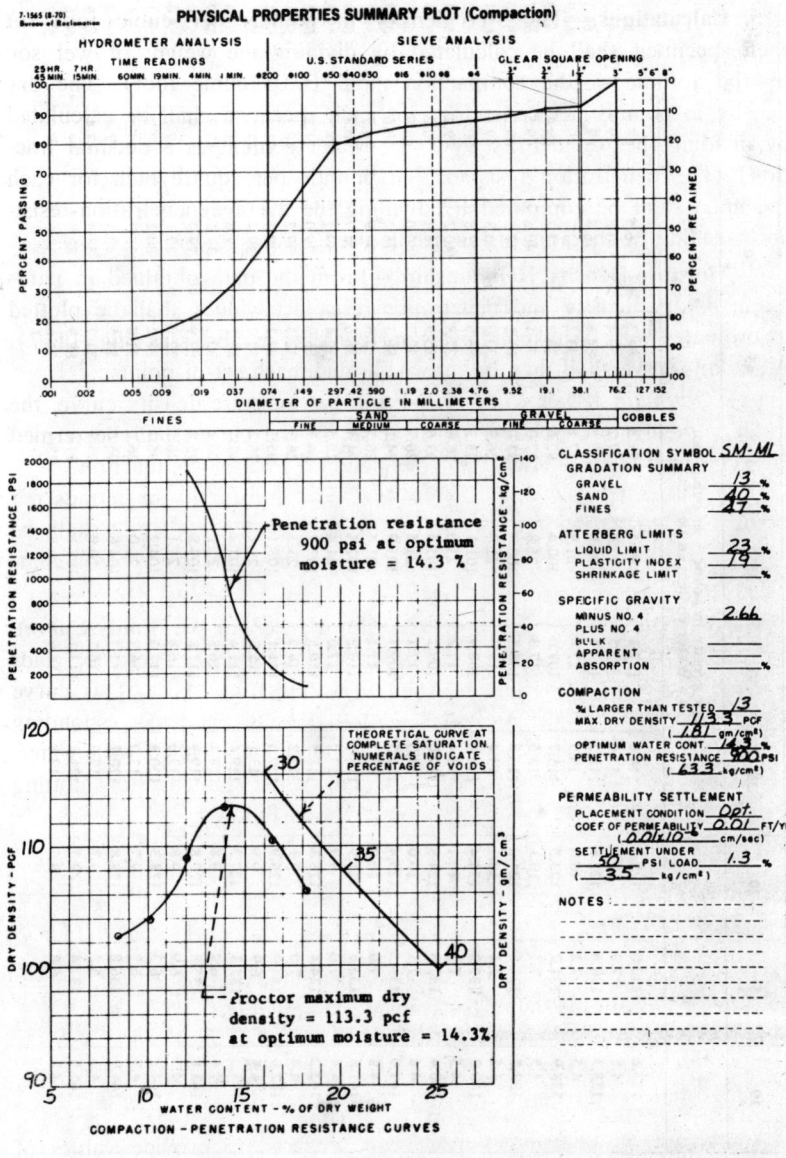

Figure 11-7.—Proctor compaction test and summary of physical properties. 101-D-591.

Table 11-1.—Points for curve of complete saturation

Dry density in pounds per cubic foot
Water content in percent

Specific gravity	Percent voids	10	15	20	25	30	35	40	45	50	55	60
2.45		137.6	130.0	122.4	114.8	107.1	99.5	91.8	84.2	76.5	68.8	61.2
		4.5	7.2	10.2	13.6	17.5	22.0	27.2	33.4	40.8	49.9	61.3
2.46		138.3	130.6	122.8	115.2	107.5	99.8	92.2	84.6	76.8	69.2	61.5
		4.5	7.2	10.2	13.5	17.4	21.9	27.1	33.2	40.7	49.6	60.9
2.47		138.9	131.2	123.5	115.7	108.0	100.4	92.6	84.9	77.2	69.4	61.7
		4.5	7.1	10.1	13.5	17.3	21.8	27.0	33.1	40.5	49.5	60.7
2.48		139.5	131.6	124.0	116.2	108.5	100.7	93.0	85.2	77.5	69.7	62.0
		4.5	7.1	10.1	13.4	17.3	21.7	26.9	33.0	40.3	49.3	60.4
2.49		140.0	132.2	124.4	116.6	108.9	101.1	93.3	85.6	77.8	70.0	62.2
		4.5	7.1	10.1	13.4	17.2	21.6	26.8	32.8	40.2	49.2	60.3
2.50		140.4	132.6	124.8	117.0	109.2	101.4	93.6	85.9	78.1	70.2	62.4
		4.4	7.1	10.0	13.3	17.1	21.5	26.7	32.7	40.0	48.9	60.0
2.51		141.0	133.2	125.4	117.5	109.7	101.9	94.1	86.3	78.2	70.6	62.7
		4.4	7.0	10.0	13.3	17.1	21.4	26.5	32.6	39.9	48.7	59.8
2.52		141.6	133.8	125.9	118.1	110.2	102.4	94.4	86.6	78.7	70.8	63.0
		4.4	7.0	9.9	13.2	17.0	21.4	26.5	32.4	39.7	48.5	59.5
2.53		142.2	134.4	126.3	118.5	110.5	102.7	94.7	86.9	79.0	71.2	63.2
		4.4	7.0	9.9	13.2	16.9	21.3	26.3	32.3	39.5	48.3	59.3
2.54		142.8	134.9	126.9	119.0	111.1	103.1	95.2	87.2	79.3	71.4	63.4
		4.4	6.9	9.9	13.1	16.9	21.2	26.2	32.2	39.4	48.1	59.1
2.55		143.2	135.3	127.3	119.4	111.4	103.4	95.5	87.5	79.6	71.6	63.6
		4.4	6.9	9.8	13.1	16.8	21.1	26.2	32.1	39.2	47.9	58.9
2.56		143.8	135.8	127.8	119.8	111.8	103.8	95.8	87.9	79.9	71.9	63.9
		4.3	6.9	9.8	13.0	16.8	21.1	26.1	32.0	39.1	47.7	58.6
2.57		144.3	136.3	128.3	120.3	112.3	104.2	96.2	88.2	80.2	72.2	64.1
		4.3	6.9	9.7	13.0	16.7	21.0	26.0	31.8	38.9	47.5	58.4
2.58		144.9	136.8	128.8	120.7	112.7	104.6	96.6	88.5	80.5	72.4	64.4
		4.3	6.8	9.7	12.9	16.7	21.0	25.9	31.7	38.8	47.4	58.1
2.59		145.5	137.4	129.3	121.2	113.1	105.1	97.0	88.9	80.8	72.7	64.6
		4.3	6.8	9.7	12.9	16.6	20.8	25.6	31.6	38.6	47.2	57.9

Des.											
2.60	146.0	137.9	129.8	121.7	113.6	105.5	97.3	89.2	81.1	73.0	64.9
	4.3	6.8	9.6	12.9	16.5	20.7	25.7	31.5	38.5	47.0	57.7
2.61	146.6	138.4	130.3	122.2	114.0	105.9	97.7	89.6	81.4	73.3	65.1
	4.3	6.7	9.6	12.8	16.4	20.6	25.6	31.3	38.3	46.8	57.5
2.62	147.2	138.9	130.8	122.6	114.5	106.3	98.1	89.9	81.8	73.6	65.4
	4.2	6.7	9.6	12.7	16.4	20.6	25.5	31.2	38.2	46.6	57.3
2.63	147.7	139.5	131.3	123.1	114.9	106.7	98.5	90.3	82.1	73.8	65.6
	4.2	6.7	9.5	12.7	16.3	20.5	25.4	31.1	38.0	46.5	57.1
2.64	148.3	140.0	131.8	123.6	115.3	107.1	98.9	90.6	82.4	74.1	65.9
	4.2	6.7	9.5	12.6	16.3	20.4	25.3	31.0	37.9	46.3	56.8
2.65	148.8	140.6	132.3	124.0	115.8	107.5	99.2	91.0	82.7	74.4	66.1
	4.2	6.6	9.5	12.6	16.2	20.3	25.2	30.8	37.7	46.1	56.6
2.66	149.4	141.1	132.8	124.5	116.2	107.9	99.6	91.3	83.0	74.7	66.4
	4.2	6.6	9.4	12.5	16.2	20.3	25.1	30.8	37.6	45.9	56.4
2.67	150.0	141.6	133.3	125.0	116.6	108.3	100.0	91.6	83.3	75.0	66.6
	4.2	6.6	9.4	12.5	16.1	20.3	25.0	30.7	37.5	45.7	56.2
2.68	150.5	142.2	133.8	125.4	117.1	108.7	100.3	92.0	83.6	75.3	66.9
	4.2	6.6	9.4	12.5	16.1	20.2	24.9	30.5	37.3	45.6	56.0
2.69	151.1	142.7	134.3	125.9	117.5	109.1	100.7	92.3	83.9	75.5	67.1
	4.1	6.6	9.3	12.4	16.0	20.1	24.8	30.4	37.2	45.5	55.8
2.70	151.6	143.2	134.8	126.4	117.9	109.5	101.1	92.7	84.2	75.8	67.4
	4.1	6.6	9.3	12.4	16.0	20.0	24.7	30.3	37.0	45.3	55.6
2.71	152.2	143.7	135.3	126.8	118.4	109.9	101.5	93.0	84.6	76.1	67.6
	4.1	6.5	9.3	12.3	15.9	19.9	24.7	30.3	36.9	45.1	55.4
2.72	152.8	144.3	135.8	127.3	118.8	110.3	101.8	93.4	84.9	76.4	67.9
	4.1	6.5	9.2	12.3	15.8	19.9	24.6	30.2	36.8	44.9	55.2
2.73	153.3	144.8	136.3	127.8	119.3	110.7	102.2	93.7	85.2	76.7	68.1
	4.1	6.5	9.2	12.2	15.8	19.8	24.6	30.1	36.6	44.7	55.0
2.74	153.9	145.3	136.8	128.2	119.7	111.1	102.6	94.0	85.5	76.9	68.4
	4.1	6.4	9.2	12.2	15.7	19.7	24.4	29.9	36.5	44.6	54.8
2.75	154.4	145.9	137.3	128.7	120.1	111.5	103.0	94.4	85.8	77.2	68.6
	4.0	6.4	9.1	12.1	15.7	19.7	24.4	29.9	36.4	44.4	54.6
2.76	155.0	146.5	137.8	129.1	120.6	111.9	103.3	94.7	86.2	77.5	68.9
	4.0	6.4	9.1	12.1	15.6	19.6	24.3	29.8	36.2	44.3	54.4
2.77	155.6	147.0	138.3	129.6	121.0	112.4	103.7	95.1	86.5	77.8	69.2
	4.0	6.4	9.0	12.0	15.5	19.5	24.2	29.7	36.2	44.2	54.2
2.78	156.1	147.5	138.8	130.1	121.4	112.8	104.1	95.4	86.8	78.1	69.4
	4.0	6.4	9.0	12.0	15.4	19.4	24.0	29.5	36.0	44.0	54.0
2.79	156.8	148.1	139.3	130.6	121.9	113.2	104.5	95.8	87.1	78.4	69.7
	4.0	6.3	9.0	12.0	15.3	19.3	23.8	29.3	35.9	43.8	53.8

Table 11–1.—Points for curve of complete saturation—Continued

Percent voids	10	15	20	25	30	35	40	45	50	55	60
Specific gravity	— Dry density in pounds per cubic foot / Water content in percent										
2.80	157.4	148.6	139.8	131.0	122.3	113.6	104.8	96.1	87.5	78.7	70.0
	4.0	6.3	8.9	11.9	15.3	19.2	23.8	29.2	35.7	43.6	53.5
2.81	157.8	149.0	140.0	131.5	122.7	114.0	105.2	96.4	87.7	78.9	70.1
	4.0	6.3	8.9	11.9	15.2	19.2	23.7	29.1	35.6	43.4	53.4
2.82	158.4	149.6	140.8	132.0	123.2	114.4	105.6	96.8	88.0	79.2	70.4
	3.9	6.3	8.9	11.8	15.2	19.1	23.6	29.0	35.5	43.3	53.2
2.83	159.0	150.2	141.3	132.4	123.6	114.8	106.0	97.1	88.3	79.5	70.7
	3.9	6.2	8.8	11.8	15.1	19.0	23.6	28.9	35.3	43.2	53.0
2.84	159.6	150.7	141.8	132.9	124.1	115.2	106.3	97.5	88.7	79.8	70.9
	3.9	6.2	8.8	11.7	15.1	18.9	23.5	28.8	35.2	43.0	52.8
2.85	160.1	151.2	142.2	133.4	124.5	115.6	106.7	97.9	89.0	80.1	71.2
	3.9	6.2	8.8	11.7	15.0	18.9	23.4	28.7	35.1	42.9	52.6
2.86	160.7	151.8	142.8	133.9	124.9	116.0	107.1	98.2	89.3	80.3	71.4
	3.9	6.2	8.7	11.6	15.0	18.8	23.3	28.6	35.0	42.8	52.5
2.87	161.1	152.3	143.3	134.3	125.4	116.4	107.5	98.5	89.6	80.7	71.7
	3.9	6.2	8.7	11.6	14.9	18.8	23.2	28.5	34.8	42.5	52.3
2.88	161.8	152.8	143.8	134.8	125.8	116.8	107.8	98.8	89.9	80.9	71.9
	3.9	6.1	8.7	11.6	14.9	18.7	23.2	28.4	34.7	42.4	52.1
2.89	162.4	153.4	144.2	135.2	126.2	117.2	108.2	99.2	90.2	81.2	72.2
	3.8	6.1	8.7	11.5	14.9	18.6	23.0	28.3	34.6	42.3	51.9
2.90	162.9	153.9	144.8	135.7	126.7	117.6	108.6	99.5	90.5	81.5	72.4
	3.8	6.1	8.6	11.5	14.8	18.6	23.0	28.2	34.5	42.1	51.7

RELATIVE DENSITY OF COHESIONLESS SOILS

Designation E–12

1. Scope.—This designation describes the method for obtaining the relative density of cohesionless free-draining soils.[1]

2. Definition.—Relative density is defined as the state of compactness of a soil with respect to the loosest and most compact states at which it can be placed by the laboratory procedures described below. It is expressed as the ratio, in percent, of the difference between the void ratio in the loosest state and the void ratio of the soil in place, to the difference between void ratios in the loosest and most compact states.

3. Apparatus.—The apparatus shall consist of the following (see designation E–4 and figures 12–1 through 12–3):

(1) Relative density test equipment, consisting of: vibratory table; 0.1- and 0.5-cubic-foot-capacity unit weight measures; guide sleeve with three tie-down rods, one set for each size measure; surcharge weight with baseplate, one set for each size measure; and dial indicator gage holder. Vibratory table is a steel table with a cushioned steel vibrating deck about 30 by 30 inches and actuated by an electromagnetic vibrator. The vibrator should be a seminoiseless type with a net weight over 100 pounds, a frequency of 3,600 vibrations per minute, and a vibrator double amplitude variable between 0.002 and 0.015 inch under a 250-pound load, and be suitable for use with a 230-volt alternating current electric circuit, item 69B [2]. (Drawing No. 101–D–350 and fig. 12–1.)

(2) Gage, dial indicator, 2-inch travel, item 35.

(3) Reference bracket, dial indicator gage, item 69A.

(4) Pouring device, item 64.

(5) Straightedge, metal, item 92.

(6) Calibration bar, unit weight measure, metal, item 12A.

(7) Scale, 30- to 50-pound capacity, item 75.

(8) Scale, 250-pound capacity, item 74.

(9) Stopwatch, item 110.

(10) Pans, mixing, items 60 and 61.

(11) Scoop, item 76.

---

[1] The soils engineer should be familiar with the types of materials for which the relative density or Proctor compaction tests are applicable (see sections 54, 56, and 75).

[2] Item numbers refer to those in the laboratory equipment list, designation E–4

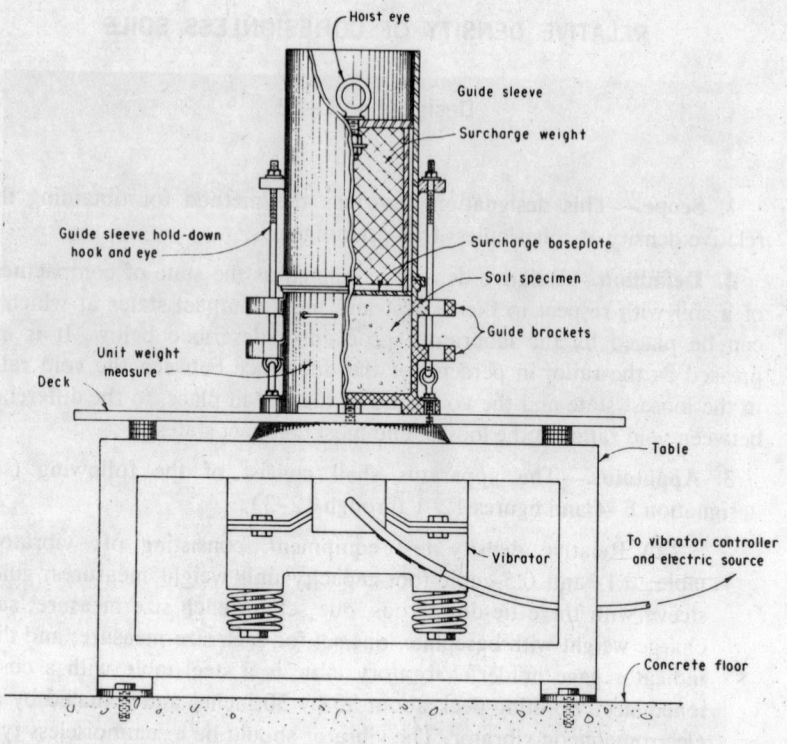

Figure 12–1.—Vibratory table. (From 101–D–350.) 101–D–592.

(12) Brush, dusting, item 8.

(13) Hoist (at least 300-pound capacity), item 38A.

4. Calibration.—The calibration sheet, figure 12–4 (form 7–1708), is provided for recording and computing data for the determination of the volume of the unit weight measure, the volume constant, and the specimen height at zero dial reading.

(a) *Volume by Direct Measurement.*—Determine the volume of the unit weight measure by direct measurement. The calculated volume should be recorded to the nearest 0.0001 cubic foot for the 0.1-cubic-foot measure and to the nearest 0.001 cubic foot for the 0.5-cubic-foot measure. Divide the volume of the measure by the average inside height to obtain the volume constant. Instruments such as inside micrometers and cylinder gages can be used to obtain the necessary measurements. All linear measurements for this and other purposes in this test should be made to the nearest 0.001 inch.

Figure 12–2.——Relative density test equipment——Dial indicator reading and check. E-2248-3NA and E-2248-4NA.

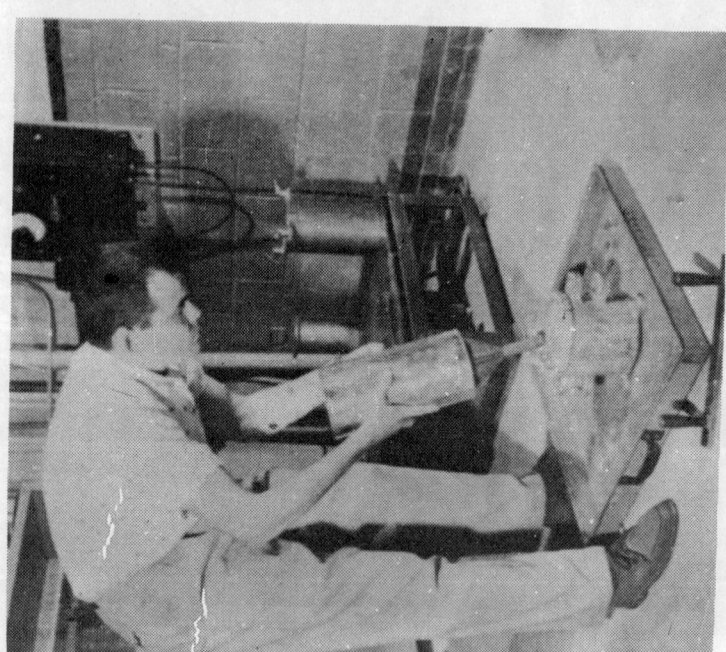

Figure 12-3.—Relative density test equipment. PX-D-27358 and PX-D-31798.

7-1708 (8-70)
Bureau of Reclamation

RELATIVE DENSITY TEST · CALIBRATION SHEET

MEASURE NO. _____ CAPACITY __0.10__ cu ft   DIAL INDICATOR NO. _____ CONDUCTED BY __Example__

	TRIAL NO.	DATE OF CALIBRATION			
DETERMINATION OF MEASURE VOLUME		Date			
INSIDE DIAMETER OF MEASURE	1	6.037			
	2	6.046			
	3	6.023			
(1) AVERAGE INSIDE DIAMETER	IN.	6.035			
INSIDE HEIGHT OF MEASURE	1	6.072			
	2	6.076			
	3	6.082			
(2) AVERAGE INSIDE HEIGHT	IN.	6.077			
*(3) VOLUME OF MEASURE $\frac{.7854 \times (1)^2 \times (2)}{1728 \text{ CU IN}}$ CU FT	CU FT	0.1006			
*(4) VOLUME CONSTANT $\frac{(3)}{(2)}$	CU FT/IN.	0.01655			

VOLUME CHECK

(5) WEIGHT OF MEASURE+GLASS PLATE+WATER	LBS	16.32			
(6) WEIGHT OF MEASURE+GLASS PLATE	LBS	10.06			
(7) WEIGHT OF WATER (5)-(6)	LBS	6.26			
(8) TEMPERATURE OF WATER	°C	17.9			
**(9) SPECIFIC WEIGHT OF WATER	PCF	62.34			
(10) VOLUME OF MEASURE $\frac{(7)}{(9)}$	CU FT	0.1004			

DETERMINATION OF DIAL READING CONSTANT

DIAL INDICATOR READINGS, LEFT SIDE	1	1.639			
	2	1.639			
	3	1.639			
RIGHT SIDE	1	1.636			
	2	1.635			
	3	1.635			
(11) AVERAGE DIAL INDICATOR READING	IN.	1.637			
(12) STRAIGHTEDGE THICKNESS	IN.	0.216			
SURCHARGE BASEPLATE THICKNESS	1	0.508			
	2	0.508			
	3	0.513			
	4	0.513			
(13) AVERAGE BASEPLATE THICKNESS	IN.	0.510			
*(14) SPECIMEN HEIGHT AT ZERO DIAL (2)-[(11-12+13]	IN.	4.146			
(15) REFERENCE BRACKET READING	IN.	0.360			

DETERMINATION OF WEIGHT OF MEASURE

*(16) WEIGHT OF MEASURE	LBS	6.85			

*VALUES USED ON MINIMUM AND MAXIMUM DENSITY TEST DATA, FORM 7-1709
**TO OBTAIN THE SPECIFIC WEIGHT OF WATER (PCF) AT THE TEST TEMPERATURE (°C) DETERMINE THE
ABSOLUTE DENSITY OF WATER (GM/CC) AT THE TEST TEMPERATURE. TABLE 10-1 DES E-10. AND MULTIPLY
BY 62.43

Figure 12–4.—Relative density test method using a vibratory table—Calibration sheet. 101–D–593.

(b) *Volume by Water.*—Check the volume of the unit weight measure by determining the weight of water required to fill the measure. Fill the measure with water until a meniscus forms at the top. Slide a glass plate over the top surface of the measure in such a manner that the measure remains completely filled (A thin film of cup grease on the top surface of the measure will help in obtaining a good contact between the glass plate and measure.) Determine the weight and temperature of the water in the measure and calculate the volume of water as described on form 7–1708, figure 12–4. All weight determinations for this check and other purposes in this test should be made to the nearest 0.01 pound. The 30- to 50-pound scale, item 75, should be used for weighings involving the 0.1-cubic-foot measure, and the 250-pound scale, item 74, should be used for the 0.5-cubic-foot measure.

(c) *Specimen Height at Zero Dial.*—With a micrometer, measure the thickness of the calibration bar at a point approximately 1 inch from one end. Mark the end measured. Record this information on form 7–1708. Place the calibration bar across the top of the unit weight measure along the axis of the guide brackets. The measured end of the calibration bar should be flush with the outer edge of the top of the unit weight measure. Insert the dial (gage) holder in the guide bracket on the unit weight measure with the dial gage stem on the top of the calibration bar and on the axis of the guide brackets as shown in figure 12–2A. The dial gage holder should be placed such that matchmarks on the guide brackets and holder are alined. Obtain three dial indicator readings and repeat the procedure for the opposite gage bracket. Measure the thickness of the surcharge baseplate at four equally spaced points approximately ½ inch in from the edge. Record the readings on form 7–1708 and calculate the "specimen height at zero dial" (line 14). This value is constant for a particular measure, surcharge baseplate, and dial indicator gage combination. Place the dial indicator gage in the reference bracket (see fig. 12–2B) and record the reading. Thereafter, before using the dial indicator gage, place it in the reference bracket and adjust it to the reference bracket reading.

5. **Sample.**—Select a representative sample of soil. The weight of sample required is determined by the maximum particle size as given in table 12–1. (*Note:* For field density control use only the minus 3-inch fraction of the soil removed from the field density hole for the corresponding relative density test.)

6. **Procedure.**—The relative density test record sheet, form 7–1709 (fig. 12–5), is provided for recording data and computing the minimum density, the wet and dry maximum density, and the relative density.

7-1709 (8-70)
Bureau of Reclamation

RELATIVE DENSITY TEST
MINIMUM AND MAXIMUM DENSITY TEST DATA

TEST OR SAMPLE NO. Example   LOCATION _____   MAX SIZE _____ DATE _____

PROJECT _____ OFFSET _____ % GRAVEL _____ TESTED BY _____

FEATURE _____ ELEV _____ % SAND _____ COMPUTED BY _____

SOURCE OF MATERIAL _____ ZONE _____ % FINES _____ CHECKED BY _____

                 MAX SIZE TESTED _____

MINIMUM DENSITY DETERMINATION (0% RELATIVE DENSITY)		1	2
(1) WT OF SOIL+MEASURE	LBS	15.74	15.73
*(2) WT OF MEASURE	LBS	6.85	6.85
(3) WT OF SOIL (1) - (2)	LBS	8.89	8.88
*(4) VOLUME OF MEASURE	CU FT	0.1006	0.1006
(5) MINIMUM DENSITY $\frac{(3)}{(4)}$	PCF	88.4	88.3

MINIMUM DENSITY TEST SPECIMEN PLACED BY

(CHECK ONE) POURING SPOUT ✓ SCOOP __ SHOVEL __

WATER CONTENT (MAXIMUM DENSITY SPECIMEN)		2
SPECIMEN NUMBER		A
PAN NUMBER		
(6) WT OF PAN+WET SOIL	LBS	13.65
(7) WT OF PAN+DRY SOIL	LBS	12.57
(8) WT OF PAN	LBS	2.95
(9) WT OF WATER (6) - (7)	LBS	1.08
(10) WT OF DRY SOIL (7) - (8)	LBS	9.62
(11) WATER CONTENT $\frac{(9)}{(10)}$ x 100	%	11.2

MAXIMUM DENSITY DETERMINATION (100% RELATIVE DENSITY)		DRY METHOD		WET METHOD	
SPECIMEN NUMBER		1		2	
(12) LEFT DIAL READING	IN.	0.599		0.844	
(13) RIGHT DIAL READING	IN.	0.630		0.804	
(14) AVERAGE DIAL READING $\frac{(12)+(13)}{2}$	IN.	0.614		0.824	
*(15) SPECIMEN HEIGHT AT ZERO DIAL	IN.	4.146		4.146	
(16) SPECIMEN HEIGHT (14)+(15)	IN.	4.760		4.970	
*(17) VOLUME CONSTANT	CU FT/IN.	0.01655		0.01655	
(18) SPECIMEN VOLUME (16)x(17)	CU FT	0.0788		0.0823	
(19) WT OF MEASURE+SOIL [DRY OR WET]	LBS	15.74		17.55	
*(20) WT OF MEASURE	LBS	6.85		6.85	
(21) WT OF DRY SOIL (19) - (20)	LBS	8.89			
(22) WT OF WET SOIL (19) - (20)	LBS			10.70	
(23) WT OF DRY SOIL $\frac{(22)}{1+\frac{(11)}{100}}$	LBS			9.62	
(24) MAXIMUM DENSITY $\frac{(21)}{(18)}$ OR $\frac{(23)}{(18)}$	PCF	112.8		116.9	

MINIMUM DENSITY _____ 88.3 _____ PCF

MAXIMUM DENSITY _____ 116.9 _____ PCF

RELATIVE DENSITY _____ 74.9 _____ %

_____ % RELATIVE DENSITY** _____ PCF

RELATIVE DENSITY COMPUTATION		
SPECIMEN NUMBER		
(25) INPLACE DENSITY	PCF	108.1
(26) MAXIMUM DENSITY (24)	PCF	116.9
(27) MINIMUM DENSITY (5)	PCF	88.3
(28) (26) x [(25) - (27)]		2,315
(29) (25) x [(26) - (27)]		3,092
(30) RELATIVE DENSITY $\frac{(28)}{(29)}$ x 100	%	74.9

REMARKS: Include items such as a brief description of material, any difficulties in conducting tests, loss of fines in wet method, segregation, etc.

*VALUES OBTAINED FROM CALIBRATION SHEET FORM NO. 7-1708.

**PERCENTAGE OF RELATIVE DENSITY GIVEN IN SPECIFICATIONS.

Figure 12-5.—Relative density test method using a vibratory table—Computation sheet. 101-D-594.

(a) *Minimum Density.*—The minimum density is at maximum void ratio; this is zero relative density.

(1) Use ovendried material. Select the pouring device and measure according to the maximum particle size as given in table 12–1.

Table 12–1.—Sample and equipment requirements

Maximum size soil particle (inches)	Sample required (pounds)	Pouring device used in minimum density test	Size of unit weight measure (cubic feet)
3	100	Shovel or extra large scoop	0.5
1½	25	Scoop	0.1
¾	25	Scoop	0.1
⅜	25	1-inch spout	0.1
No. 4 sieve	25	½-inch spout	0.1

(2) Soils containing ⅜-inch-maximum size or smaller particles.—Place soil as loosely as possible in the measure using the pouring device as shown in figure 12–3A. Maintain a steady stream of soil from the spout and adjust the height of the spout so that the free fall of the soil is about 1 inch. At the same time move the pouring device in a spiral-like motion from the outside toward the center to form a soil layer of uniform thickness without segregation. Fill the measure approximately 1 inch above the top.

(3) Soils containing particles larger than ⅜-inch-maximum size.—Place the soil by means of a large scoop (or shovel) held as close as possible to, but just above, the soil surface. The soil should slide rather than fall onto the previously placed material. If necessary, hold large particles back by hand to prevent them from rolling off the scoop. Fill the measure approximately 1 inch above the top.

(4) Screed off the excess soil level with the top by making one continuous pass with a metal straightedge. If all excess material is not removed, make an additional continuous pass. During the pouring and leveling, great care should be exercised to avoid jarring the measure. When screeding soil containing particles larger than ⅜-inch-maximum size, any slight projections of the larger particles above the top of the measure should approximately balance the larger voids in the surface below the top of the measure. For such soils it may be necessary to use the fingers in addition to the straightedge to level the surface.

(5) Weigh the measure with and without soil and record the results on lines (1) and (2) of form 7–1709. (*Note*: When an appreciable number of tests are conducted, the measure does not have to be weighed each time it is used, but should be checked periodically.)

(b) *Maximum Density: Dry Method.*—The maximum density is at minimum void ratio; this is 100-percent relative density.

(1) Mix the ovendried soil sample to provide an even distribution of particle sizes with as little segregation as possible.

(2) Fill the measure with soil by the same procedure used for the minimum density test. Normally, the measure filled with soil for the minimum density determination may be used for the maximum density test without refilling the measure.

(3) Attach the measure to the deck of the vibratory table and place the surcharge baseplate on the soil surface. Position the guide sleeve and fasten in place with the tie-down rods. Lower the surcharge weight onto the surcharge baseplate; a hoist will be required to handle the surcharge weight used in the 0.5-cubic-foot measure. The completely assembled apparatus is shown in figure 12–1.

(4) Vibrate the loaded specimens for 8 minutes with the vibrator control set to obtain maximum amplitude.

(5) Remove the surcharge weight and guide sleeve and obtain dial indicator gage readings on the two opposite sides of the surcharge baseplate, figure 12–3B. Record the results on lines (12) and (13), form 7–1709. Weigh the measure with and without soil and record on lines (19) and (20), form 7–1709.

(c) *Maximum Density: Wet Method.*—The maximum density is at minimum void ratio; this is 100-percent relative density.

(1) The wet method can be conducted on ovendried soil to which sufficient water is added or, if preferred, on wet soil. If water is added to dry soil, allow a minimum soaking period of one-half hour. The amount of water to be added to the soil can be estimated by a computation of the void ratio at the expected maximum density or by experimentation with the soil.

(2) Attach the measure to the vibratory table and slowly fill with wet soil by means of a scoop or shovel. During filling a small amount of free water should accumulate on the soil surface. During and just after filling the measure (filling of the measure should take 5 to 6 minutes), vibrate the soil for a total of 6 minutes. The amplitude of the vibrator should be reduced as much as necessary

to avoid excessive boiling and fluffing of the soil. During the final minute of vibration, remove any water appearing above the soil surface.

(3) Assemble the apparatus as in the dry method.

(4) Vibrate the loaded specimen for 8 minutes with the vibrator control set to obtain *maximum* amplitude.

(5) Remove the surcharge weight and guide sleeve and obtain dial indicator gage readings on the two opposite sides of the surcharge baseplate, figure 12–3B. Record the results on lines (12) and (13), form 7–1709. Weigh the measure with and without soil and record on lines (19) and (20), form 7–1709.

(6) Determine the water content of the complete soil sample, lines (6) to (11), form 7–1709.

(*Note:* At the beginning of a new job, or when a radical change of materials occurs, perform both the wet and dry maximum density test methods to determine which method results in the higher density. The dry method is preferred since results can be obtained in a shorter period of time. However, if the wet method produces higher densities, in excess of 1 percent, use that method.)

7. Calculations.—The procedures for performing the calculations required for the relative density test are given in form 7–1709, figure 12–5.

(a) *Minimum Density* ($\gamma_{d_{min}}$).—Calculate the minimum density as follows:

$$\gamma_{d_{min}} = \frac{\text{weight of dry soil (3)}^3}{\text{volume of measure (4)}}$$

(b) *Maximum Density* ($\gamma_{d_{max}}$).—Calculate the maximum density as follows:

$$\gamma_{d_{max}} = \frac{\text{weight of dry soil (21 or 23)}}{\text{volume of soil (18)}}$$

(c) *Inplace Density* (γ_d).—This is determined by a field density test (designation E–24) in a compacted fill or a natural deposit.

(d) *Relative Density* (D_d).—Calculate the relative density, in percent, as follows:

$$D_d = \frac{\gamma_{d_{max}} (\gamma_d - \gamma_{d_{min}})}{\gamma_d (\gamma_{d_{max}} - \gamma_{d_{min}})} \times 100$$

Or, in terms of void ratio,

$$D_d = \frac{e_{max} - e}{e_{max} - e_{min}} \times 100$$

---

[3] Numbers in parentheses refer to lines on form 7–1709, figure 12–5.

where:

$$e = \text{void ratio} = \frac{\text{volume of voids}}{\text{volume of solid particles}},$$

$e_{max} = $ void ratio in loosest soil state, and

$e_{min} = $ void ratio in most compact soil state.

(e) *Additional Calculations.*—Calculate the inplace, or placement, density when the maximum, minimum, and relative density are given, as follows:

$$\gamma_d = \frac{(\gamma_{d_{max}})(\gamma_{d_{min}})}{\gamma_{d_{max}} - \frac{D_d}{100}(\gamma_{d_{max}} - \gamma_{d_{min}})}$$

8. Plotting.—Form 7–1595, figure 12–6, is provided for plotting the results of the relative density test. This figure is scaled so that the relationship between relative density and dry density can be determined from a straight line drawn between the minimum density, plotted on the left scale, and the maximum density, plotted on the right scale.

9. Alternative Method.—The relative density procedure described above is recommended for all laboratories when control of appreciable quantities of compacted cohesionless sand and gravel soils is required. However, if tests must be conducted where adequate electrical power is not available, or if the number of tests to be conducted is small, the following alternative method can be used. Information on approved equipment should be obtained from the Engineering and Research Center, Division of General Research.

(a) *Apparatus.*—The apparatus shall consist of the following (see designation E–4):

(1) Vibrating device, item 109.
(2) Measures, unit weight, cylindrical, metal, 0.1- and 0.5-cubic-foot capacity, item 69B (measures may be modified so that the vibrating device can be attached).
(3) Pouring device, item 64.
(4) Straightedge, metal, item 92.
(5) Scale, 30- to 50-pound capacity, item 75.
(6) Scale, 250-pound capacity, item 74.
(7) Pans, mixing, items 60 and 61.
(8) Scoop, item 76.

(b) *Calibration.*—Determine the volume of the unit weight measure as described in paragraph 4, except the volume constant is not obtained.

(c) *Sample.*—See paragraph 5.

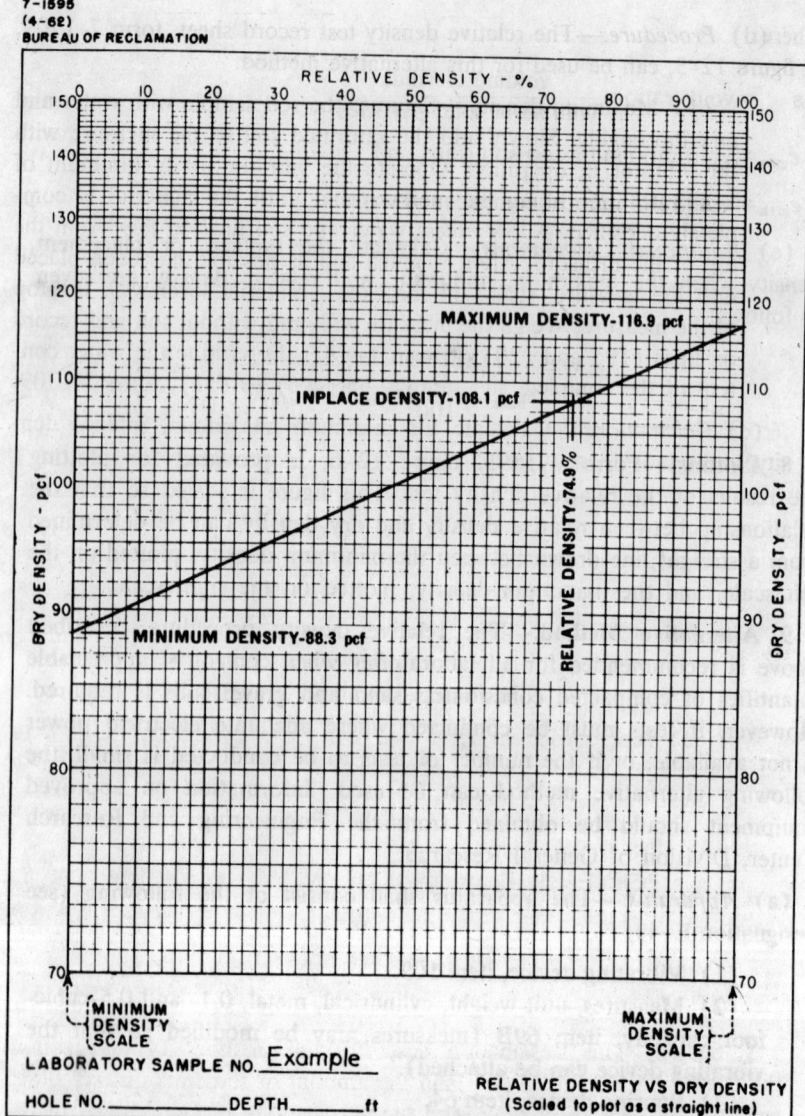

Figure 12–6.—Plotting results of relative density test. 101-D-595.

(d) *Procedure.*—The relative density test record sheet, form 7–1709, figure 12–5, can be used for this alternative method.

(1) **Maximum density.**—Saturate the soil sample with water and mix thoroughly. Slowly add the saturated soil to the measure with the vibrator in operation. Operate the vibrator for a minimum of 1 minute after filling the measure and until the material is completely densified. Allow excess water and soil fines to flow from the top of the measure. After as much soil as possible has been placed into the measure, strike off the excess material level with the top of the measure. Weigh the measure with and without soil and record on lines (19) and (20) of form 7–1709. Determine the water content of the complete soil sample, lines (6) to (11), form 7–1709.

(e) *Calculations.*—Calculate the minimum, maximum, inplace density, and relative density as follows:

(1) **Minimum density** ($\gamma_{d_{min}}$),

$$\gamma_{d_{min}} = \frac{\text{weight of dry soil (3)}^4}{\text{volume of measure (4)}}$$

(2) **Maximum density** ($\gamma_{d_{max}}$),

$$\gamma_{d_{max}} = \frac{\text{weight of dry soil (23)}}{\text{volume of measure (18)}}$$

(3) **Inplace density** (γ_d) and relative density (D_d) should be calculated as described in paragraphs 7(c) and 7(d), respectively.

PERMEABILITY AND SETTLEMENT OF SOILS

Designation E–13

1. Scope.—This designation describes the method for determining the coefficient of permeability and the amount of settlement of remolded soils passing the No. 4 sieve, and the permeability of fine-grained undisturbed soils.

2. Apparatus.—The apparatus shall consist of the following (see designation E–4 and figs. 13–1 and 13–2):

Cans, item 14.
Clamps, item 18.

---

4 Numbers in parentheses refer to lines on form 7–1709, figure 12–5.

Cylinder, permeability, 8-inch, item 24.

Discs, porous, item 31.

Gages, dial indicator, with holders, item 35.

Gage, block, dial indicator reference, item 35B.

Gage, ring specimen, thickness, item 36.

Loading device, item 45.

Pan, mixing, item 61.

Sampling tube, item 73.

Scale, item 75.

Scoops, items 76 and 77.

Springs, loading, item 88.

Square, combination, item 89.

Straightedge, metal, item 92.

Supports, head tank, item 95.

T-connectors, item 96.

Tamper, 9-pound, item 97A.

Tanks, head, item 99.

Tubing, item 107.

3. Calibration of Constant Head Tanks.—Each constant head tank shall be calibrated to determine the number of cubic centimeters per inch. The head tank is filled with water, then the air intake tube is placed in the head tank and the water level lowered to the bottom of the air intake tube. The water shall be drained slowly from the bottom of the tube into a graduated cylinder or container suitable for weighing, allowing air to enter the tank through the air intake tube as in normal operation.

The water level shall be lowered to near the bottom graduation of the tank and the volume, or weight in grams, and the temperature of the water recorded. A temperature correction (table 10–1, designation E–10) shall be made when the volume is determined by weighing. From these data the volume in cubic centimeters of the calibrated portion of the head tank shall be determined. The calibration factor, F, in cubic centimeters per inch shall be determined by dividing the volume in cubic centimeters by the length in inches for the portion calibrated, which is usually 30 inches. Once this factor has been determined, it need not be redetermined unless the air intake tube is replaced.

4. Sample.—A 15-pound sample shall be taken from the thoroughly mixed portion of the material passing the No. 4 sieve, which has been obtained in accordance with the method for "Preparation of Soil Samples for Testing," designation E–5.

5. Procedure.—(a) The water content of the 15-pound sample shall

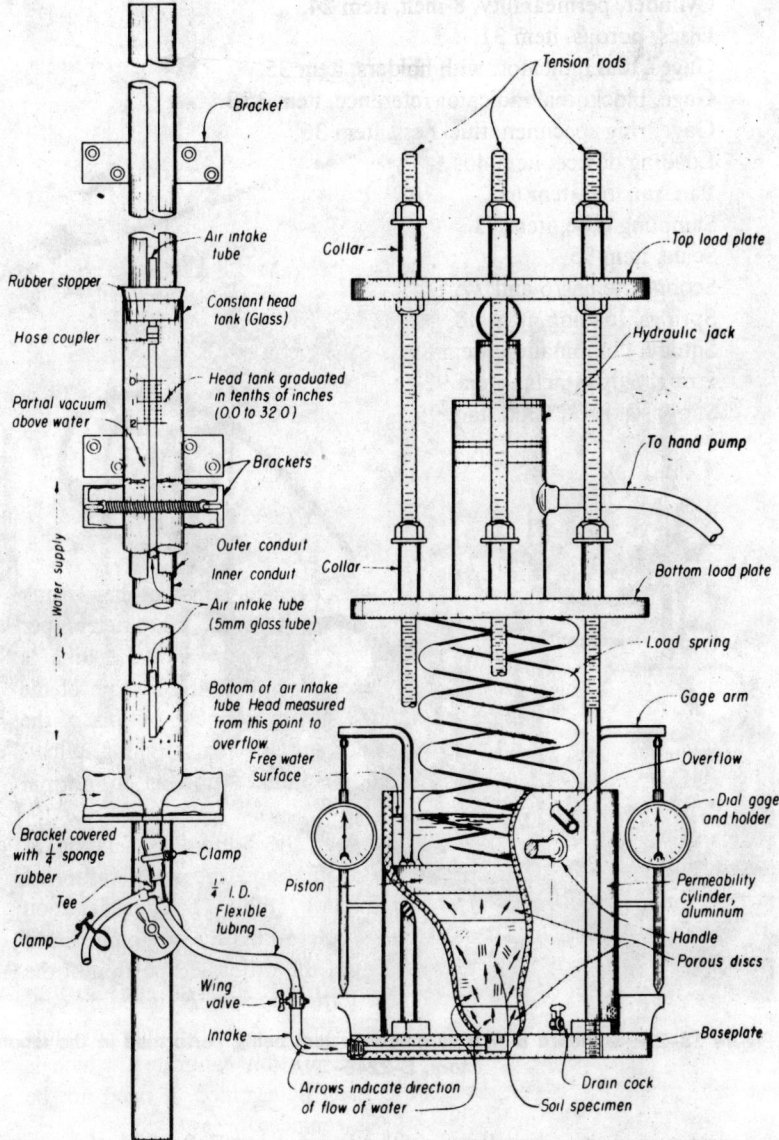

Figure 13–1.—Standard 8-inch permeability apparatus. 101–D–110.

Figure 13–2.—Standard 8-inch permeability test being performed in the laboratory. E–2246–6NA.

be determined in accordance with designation E–9, and the sample placed in an airtight storage can. The quantity of water to be added to the stored sample to give the desired water content shall be computed using the determined initial water content and the wet weight of the stored sample (form 7–1454, cols. 1 to 6, fig. 13–3). This computed quantity of water shall be spread evenly over the sample, and after thoroughly mixing, the material shall again be placed in the storage can.

7-1454
(3-62)

PREPARATION OF MATERIAL FOR REMOLDED SPECIMEN

PROJECT _Example_____ FEATURE_____ SAMPLE NO. _IIJ-X90_

COMPUTED BY _____ CHECKED BY _____ DATE _____

SAMPLE NO.	PREPARATION OF SAMPLE							DATA FOR PACKING				
	WATER CONTENT %			WEIGHT (POUNDS)			CC.	P.C.F.	%	P.C.F.	LB.	LB.
	WATER CONTENT IN CAN	WATER CONTENT DESIRED	WATER CONTENT DIFF	WET WT SAMPLE IN CAN	DRY WT SAMPLE IN CAN	WT OF WATER TO ADD	VOL OF WATER TO ADD	DESIRED DRY DENSITY	WATER CONTENT OBTAINED	WET DENSITY	WET WEIGHT REQUIRED	WET WT OF EACH LAYER
	(1)	(2)	(3)	(4)	(5)	(6)		(7)	(8)	(9)	(10)	(11)
IIJ-X90	8.3	14.3	6.3*	15.69	14.49	.91	413	113.3	14.5	129.73	11.34	3.78
	OVEN DETERMINATION	FROM COMPACTION CURVE	COL.(2) − COL.(1) + EST EVAP	TOTAL WEIGHT − WEIGHT CAN	$\frac{COL.(4)}{1 + \frac{COL.(1)}{100}}$	$COL.(5) \times \frac{COL.(3)}{100}$	COL.(6) x 453.6	FROM COMPACTION CURVE	OVEN DET NOT TO VARY OVER 0.5% FROM COL.(2)	$COL.(7)\left[1 + \frac{COL.(8)}{100}\right]$	COL.(9) x 0.0873	$\frac{COL.(10)}{NO\ OF\ LAYERS}$

* Estimated evaporation loss is 0.3%
** For specimen 3-inches in height.

Figure 13-3.—Recording preparation of sample data for permeability test. 101-D-261.

The water content of the sample shall again be determined (designation E-9) and the entire process repeated until the actual water content is within 0.5 percent of that desired.

(b) The porous discs shall be placed in a pan of water and then removed and the free water allowed to drain for 10 to 15 minutes. A porous disc shall be placed in the bottom of the permeability cylinder. The 3-inch thickness ring gage shall be placed on the bottom porous disc, the top porous disc placed on the ring gage, and the piston placed over the top disc so that the dial indicator gage arms are directly above the dial indicator supports. The positions of the piston and cylinder are matchmarked so they may be identically reassembled. The coil spring shall be placed on top of the piston to hold it firmly in place. Dial gage readings (A and B) shall be taken simultaneously and recorded as "Ring" readings on form 7-1455, figure 13-4. The total of these two readings shall be recorded as "Dial readings A + B, w/ring." The coil spring, piston, top disc, and ring thickness gage shall then be removed prior to placing the specimen. The drain cock shall be open.

(c) The soil shall be compacted into the cylinder in three 1-inch

7-1455
(5-62)
BUREAU OF RECLAMATION

PERMEABILITY–SETTLEMENT TEST, DATA SHEET I

PROJECT _Example_   FEATURE _____   SAMPLE NO. _11J-X90_

TESTED BY _____ COMPUTED BY _____   CHECKED _____   DATE _____

CYLINDER NO _14_

PLACEMENT DATA

GPO 856150

WT CONTAINER + WET SOIL	21.42 LB	FINAL WT CONTAINER	
WT 1ST LAYER	3.78 LB	+ WET SOIL	10.08 LB
WT CONTAINER + WET SOIL	17.64 LB	WT WET SOIL	11.34 LB
WT 2 ND LAYER	3.78 LB	WET DENSITY	129.73 PCF
WT CONTAINER + WET SOIL	13.86 LB	PLACEMENT WATER CONT	14.5 %
WT 3 RD LAYER	3.78 LB	PLACEMENT DRY DENSITY*	113.3 PCF
		DESIRED WATER CONT	14.3 %

	SAMPLE AREA	50.27 SQ IN
	INTENSITY OF LOAD	20 PSI
	OR	22 FEET OF FILL
	LOAD ON SPECIMEN	1005 LBS
	JACK NO	845
	GAGE READING	340

	TOTAL	1/2 TOTAL		SPECIMEN THICKNESS (INCHES)
DIAL READING A+B, W/RING	0.718	0.3590		
DIAL READING A+B, NO LOAD	0.717	0.3585		
DIFFERENCE		0.0005	3.0000 ⊕ 0.0005 · 3.0005 INITIAL	
DIAL READING A B, W/L, WET			3.0005 ⊖ 0.044 = 2.9565 (L) CONSOL, WET	

SETTLEMENT DATA

DATE 19	HOUR OBSERVED	LOAD LB/SQ FT	TRAVEL				SETTLEMENT	
			DIAL AT A		DIAL AT B		AVERAGE AMOUNT	PERCENT
			READING	AMOUNT	READING	AMOUNT		
(1)	(2)	(3)	(4)	(5)	(6)	(7)	(8)	(9)
9-7	7:40	Ring	0.377		0.341			
9-7	8:10	No Load	0.291	0.000	0.426	0.000	0.000	0.000
	8:20	2880	0.316	0.025	0.471	0.045	0.035	1.166
	9:20	"	0.320	0.029	0.483	0.057	0.043	1.433
	10:20	"	0.322	0.031	0.489	0.063	0.047	1.566
9-8	8:00	w/load	0.322	0.031	0.489	0.063	0.047	1.566
9-10	8:30	w/L, Wet	0.317	0.026	0.488	0.062	0.044	1.466
9-11	8:30	"	0.317	0.026	0.487	0.061	0.044	1.466
9-12	8:00	"	0.316	0.025	0.488	0.062	0.044	1.466
9-13	8:00	"	0.315	0.024	0.486	0.060	0.042	1.400
9-16	7:45	"	0.312	0.021	0.485	0.059	0.040	1.333
9-17	7:45	"	0.310	0.019	0.485	0.059	0.039	1.300
9-18	7:45	"	0.310	0.019	0.485	0.059	0.039	1.300
9-19	8:00	"	0.310	0.019	0.485	0.059	0.039	1.300
9-20	8:00	"	0.310	0.019	0.485	0.059	0.039	1.300 ▲
9-20	10:00	No Load	0.311	0.020	0.482	0.056	0.038	1.266
	REBOUND							

AFTER PERCOLATION DATA

WATER CONTENT DETERMINATION	
PAN NUMBER	4
WT PAN + WET SOIL	15.03
WT PAN + DRY SOIL	13.55
WT PAN	3.74
WT WATER	1.48
WT DRY SOIL	9.81
WATER CONTENT	15.1
WATER CONTENT	15.1 %

CONSOLIDATED DRY DENSITY

$$\frac{PLACEMENT\ DRY\ DENSITY}{100 - \%\ SETTLEMENT} \cdot = \frac{113.3}{100 - 1.30} ▲$$

= 114.8 PCF

NEEDLE READ. 47   NEEDLE AREA 40   PEN. RES 1880

PENETRATION RESISTANCE 1880 PSI   CONSOLIDATED DRY DENSITY 114.8 PCF.

* MAXIMUM STANDARD DRY DENSITY OR OTHER AS SPECIFIED

REMARKS: _____

Figure 13–4.—Recording compaction and settlement test data. 101-D-539.

layers. The weight of wet soil required for each layer shall be computed using the volume of the cylinder for the 1-inch depth, the desired dry density, and the moisture content of the stored sample. This information shall be recorded on form 7–1454. (See columns 7 through 11, fig. 13–3.) The soil shall be placed at the Proctor maximum dry density and optimum water content as determined by the Proctor compaction test, designation E–11. (*Note:* Other densities and water contents shall be specified as required to simulate particular field placement conditions.)

The amount of soil required for each 1-inch layer recorded on form 7–1455 is placed in the cylinder, carefully leveled while in a loose state, and compacted with the 9-pound tamper forced downward by hand from the top of the permeability cylinder. Compaction shall be continued until each soil layer is within $\frac{1}{64}$ inch of the required 1-inch thickness. The depth of the layer is determined with a metal straightedge and combination square. The surface of each layer shall be scarified to a depth of $\frac{1}{8}$ to $\frac{1}{4}$ inch before placing the next layer.

The third layer, after being compacted and leveled, is scarified, the porous disc and piston placed on top, and the piston tapped into final position with the handle of the tamper. The coil spring shall be placed on top of the piston to hold it firmly in place. The dial gage readings (A and B) shall be taken simultaneously and recorded as "no load" readings. The total of these two shall be recorded as "Dial reading A + B, no load." The final specimen thickness indicated by the dial gages shall be within ± 0.005 inch of that indicated by the 3-inch thickness gage.

(d) The assembly of the permeability apparatus shall be completed as shown in figure 13–1, making sure that the tension rods are securely screwed into the base of the permeability cylinder.

(e) The load to be applied to the soil specimen shall be as follows:

(1) For investigation of materials:

Earth dams:

For soils that are manifestly impervious, the normal load shall be 100 pounds per square inch.

For soils that are of doubtful imperviousness, the normal load shall be 20 pounds per square inch.

Canals:

For canal embankments the normal load shall be 20 pounds per square inch; and for canal linings, the normal load shall be zero if tested as shown in figure 13–6, or 1 pound per square inch if tested in a standard 8-inch permeability cylinder.

Other purposes:

The intensity of load shall be equivalent to the expected height of fill expressed as pounds per square inch.

(2) For control testing:

Earth dams:

The load applied to the specimen in the 8-inch-diameter cylinder shall be equivalent to the weight of fill, calculated as follows:

$$L = H\gamma_w (0.3491)$$

where: L = load on specimen, in pounds,

H = height of fill above, in feet, and

γ_w = wet density of soil, in pounds per cubic foot.

Canals or other purposes:

The load shall be equivalent to that imposed by the structure.

(f) The load shall be applied with a hydraulic jack and hand pump, and the nuts tightened fingertight on the bottom load plate. Immediately after the load is applied, dial gage readings (A and B) shall be taken and recorded. The load shall be checked every hour for the first 2 hours, and daily thereafter, to insure that the load maintained by the spring remains constant. Dial gage readings (A and B) shall be taken and recorded until the initial consolidation is less than 0.002 inch during a 16-hour period. The final reading is identified as "w/Load".

(g) After initial consolidation, water under a low head shall be permitted to flow upward through the specimen, using a constant head tank.

(*Note:* The size of head tank is chosen according to the anticipated permeability of the specimen, so that a full tank of water will flow through the specimen in approximately a 24-hour period under the head selected for the test.)

The head tank shall be filled from the bottom by forcing water through the tubing connecting the head tank to the permeameter intake. It is necessary that the stopper which holds the air intake tube be released before filling. After filling the head tank, the tubing shall be clamped and the lower end slipped over the intake of the permeability cylinder. With the drain cock opposite the bottom porous disc open, the head tank stopper shall be carefully and tightly inserted. Then the hose clamp on the connecting flexible tubing shall be removed, allowing water to fill the tubing and flow through the bottom porous disc. The cylinder shall be tipped so that air bubbles can escape through the drain cock. The flow of water shall be continued until all evidence of air in the system has disappeared. The drain cock shall be closed to permit water at a low head to permeate slowly upward through the specimen. The head to be used depends upon the porosity of the material as placed; it should range from 1 inch or less for specimens having high permeability, to 1 foot or more for those having low permeability. This slow permeation shall be continued until the specimen is thoroughly wetted, that is, until the top porous disc and piston are completely covered with water, or for a period of 3 days, whichever occurs first.

(*Note:* It is possible that a specimen may leak or "pipe" up the side of the cylinder and show water above the piston much earlier than the apparent permeability of the specimen would indicate. If this happens, the test shall be stopped and another specimen prepared. When testing granular cohesionless soils, special precautions should be observed. For fine sands, a water-pump grease should be applied to the cylinder wall to prevent piping between the specimen and the wall. For the coarser sands, ¼-inch-thick sponge rubber cemented to the cylinder wall has been found to be satisfactory. This will require a reduction in the diameter of the permeability cylinder for computations of soil density; a ¼-inch sponge rubber will usually compress to about ⅛ inch. The head for pervious soils should be as low as possible, usually below the critical gradient. The critical gradient is reached when the head of water on the specimen is approximately equal to the thickness of the specimen.)

(h) After the specimen is thoroughly wetted or has been wetted for 3 days, dial gage readings (A and B) shall be taken, recorded and identified as "w/L, wet".

(i) Water shall then be poured into the top of the cylinder until it flows out of the overflow. The connecting tubing between the tank and the cylinder shall then be clamped, and the head tank adjusted to the desired height and refilled in the same manner as in the initial filling, except that water shall be introduced through the bleeder tee connection just below the bottom of the head tank.

After the head tank has been completely filled and the stopper replaced, the water level in the head tank shall be lowered to the 0.0 mark or below by carefully opening and closing the clamp on the bleeder connection. The time shall be noted, the exact head tank reading observed and recorded on form 7-1394, figure 13-5, and the clamp between the head tank and the specimen released, thus starting the permeability test.

(*Note:* When readings cannot be made daily, e.g., on weekends or holidays, the head should be reduced so that the permeability test is not interrupted.) Care should be taken to see that the water in the head tank is maintained at the same temperature as the specimen, or preferably at a slightly higher temperature, about 5 degrees above that of the specimen. When water which is colder than the specimen is used, it has been found that the air dissolved in the water will come out of solution and become trapped in the voids of the specimen as air bubbles, thus resulting in an erroneously low permeability.

7-1394
(1-64)
Bureau of Reclamation

PERMEABILITY-SETTLEMENT TEST DATA SHEET 2

PROJECT _Example_   FEATURE _____   SAMPLE NO. _11J-X90_

TESTED BY ____ : ____ COMPUTED BY _____ CHECKED BY _____ DATE _____

CYLINDER NO. _14_

TIME INSTALLED		HEAD TANK		HEAD - (H_{HG}) (SEE REMARKS COLUMN)	CONSOLIDATION THICKNESS - (L') (DATA SHEET NO 1)	CONSTANT $C = 0.0738 \frac{L'}{H_{HG}}$	CONSTANT CF
DATE 19 —	HOUR	NO.	CALIBRATION FACTOR F	FEET	INCHES		
9-10	8:24	208	6.06	5.00	2.9565	0.0436	0.2642
9-13	8:24	208	6.06	1.67	2.9565	0.1307	0.7920
9-16	8:00	208	6.06	5.00	2.9565	0.0436	0.2642

TIME OF OBSERVATION		ELAPSED TIME (t')	HEAD TANK READING	DIFFERENCE IN READING	R/t'	COEF. OF PERM. ($K = CF\frac{R}{t'}$)	REMARKS (RECORD HEAD IN FEET, CHANGES IN HEAD, AND OTHER OBSERVATIONS)
19 —	HOUR	HOURS	INCHES	INCHES		FEET/YEAR	
9-10	8:24		1.00				Head at 5.00' – test
9-11	8:30	24.10	24.80	23.80	0.99	0.26	started.
	8:45		1.60				
9-12	8:00	23.25	26.60	25.00	1.08	0.29	
	8:12		1.50				
9-13	8:00	23.80	22.60	21.10	0.89	0.24	
	8:24		1.40				Head lowered to 1.67'
9-16	7:45	71.35	23.00	21.60	0.30	0.24	
	8:00		1.90				Head raised to 5.00'
9-17	7:45	23.75	22.00	20.10	0.85	0.22	⎫
	8:00		1.80				⎪
9-18	7:45	23.75	21.70	19.90	0.84	0.22	⎪
	8:00		1.00				⎬ k(ave) = 0.22
9-19	8:00	24.00	21.30	20.30	0.85	0.22	⎪
	8:12		1.00				⎪
9-20	8:00	23.80	21.70	20.70	0.87	0.23	⎭

CYLINDER LOCATION NO

Figure 13-5.—Recording permeability test data. 101-D-540.

If a head tank is operating properly, the air intake tube will remain empty. A rise in water level in the air intake tube usually indicates leakage between the large stopper and head tank or air intake tube that must be corrected. Occasionally, a slight rise in water level in the air intake tube may be caused by changes in temperature or atmospheric pressure. Observations should not be recorded unless the water level is very near the bottom of the air intake tube.

(j) The test shall be continued for a minimum of 10 days and daily settlement and permeability readings taken until the permeability rate becomes practically constant.

(k) Final dial gage readings shall be taken. The connecting tubing shall be clamped and removed from the permeability cylinder intake. The consolidating load on the specimen shall be released by means of the hydraulic jack until the load is zero. After allowing sufficient time for complete expansion of the specimen, the dial gages shall again be read and the data recorded as "no load" readings.

(1) Surplus water on top of the specimen shall be poured off; and the load plates, tension rods, spring, piston, and top porous disc shall be removed. Penetration resistance needle readings shall be taken on the specimen and recorded. A representative portion from the center of the specimen shall be removed for water content determination.

6. Calculations.—(a) The initial specimen thickness in inches shall be determined by computing the difference between one-half the total "Dial reading A + B, w/ring" and one-half the total "Dial reading A + B, no load," form 7–1455, figure 13–4. The difference shall be added to or subtracted from 3 inches.

The consolidated thickness, L', in inches shall be computed by adding to or subtracting from the initial specimen thickness the maximum "average amount" of settlement, column 8, figure 13–4, which occurred due to load and wetting (dial gage readings identified as "w/L, wet").

(b) The amount of settlement, column 5, figure 13–4, shall be computed by subtracting the "no load" reading, column 4, from subsequent readings in column 4; the same procedure is used for columns 7 and 6. The average values of columns 5 and 7 are recorded in column 8.

(c) The settlement expressed as a percentage of the initial thickness of the specimen, column 9, figure 13–4, is determined by the following equation:

$$\text{Settlement, \%} = \frac{\text{settlement, inches (column 8)}}{\text{initial specimen thickness, inches}} \times 100 \qquad (1)$$

(d) The consolidated dry density corresponding to the particular settlement is calculated by the following equation:

Consolidated dry density, in pounds per cubic foot (p.c.f.)

$$\frac{\text{placement dry density, p.c.f.}}{\dfrac{(100 - \text{settlement}, \%)}{100}} \qquad (2)$$

(e) The coefficient of permeability shall be calculated as follows:

$$k = \frac{V}{At} \div \frac{H_{wc}}{L} \text{ (for constant head) or,} \qquad (3)$$

$$k = \frac{V L}{At \, H_{wc}} \qquad (4)$$

where: k = coefficient of permeability, in feet per year,

V = volume of discharge, cubic feet in time t,

L = thickness (or length) of specimen, in feet,

A = area of specimen, in square feet,

t = elapsed time during measurement, in years, and

H_{wc} = constant head = difference between headwater and tailwater levels, in feet.

In order to simplify computations, the general equation given above has been partially solved and constants introduced as follows:

$$V, \text{ cubic feet} = \frac{RF}{28,320}$$

where R is the difference in reading of head tank levels measured in inches, and F is the calibration factor for the head tank, expressed in cubic centimeters per inch.

$$L, \text{ feet} = \frac{L'}{12}$$

where L' is the consolidated thickness of specimen measured in inches.

$$t, \text{ years} = \frac{t'}{8,760}$$

where t' is the elapsed time between head tank readings expressed in hours.

A, square feet = 0.3491 for an 8-inch cylinder.

Substituting these constants in equation (4) above,

$$k = 0.0738 \frac{RF \, L'}{t' \, H_{wc}} \qquad (5)$$

Since the value $\dfrac{L'}{H_{wc}}$ is constant for any particular head, let

$$C = 0.0738 \frac{L'}{H_{wc}} \qquad (9)$$

Then

$$k = CF\frac{R}{t} \tag{7}$$

Equations (6) and (7) are shown on figure 13–5 and are used to compute the coefficient of permeability in feet per year.

(*Note:* This method for determining the permeability is based on the use of a constant head. If a "falling head" is used for special tests, the rate may be calculated on the basis of a formula presented in designation E–15.)

7. Undisturbed Soils.—The permeability and settlement characteristics of an undisturbed soil are usually determined in the laboratories of the Engineering and Research Center by the procedures outlined in designation E–15. The standard 8-inch-diameter permeability apparatus may be used to determine the permeability of undisturbed soil by the following procedure.

The undisturbed specimen shall be trimmed into a cylinder approximately 7½ inches in diameter by 3 inches thick, the diameter and thickness being measured to a tolerance of ± ⅟₃₂ inch. A porous disc shall be placed in the bottom of the permeability cylinder, and the specimen centered in the cylinder on the porous disc. The annular space between the specimen and the cylinder shall be filled with a mixture of 10 percent dry powdered bentonite and 90 percent fine, dry sand by weight, and the mixture consolidated using a small tamping rod. The bentonite mixture swells upon wetting and forms a very effective seal. The specimen shall then be covered with a fine mesh screen, and a thin layer of uniformly-graded fine sand placed over the screen to form a seating surface for the top porous disc. The remainder of the test procedure is identical with that previously outlined for remolded soil samples, except the computations for the coefficient of permeability should be based on the size of specimen tested.

8. Nonloaded Soil Specimens.—The apparatus shown in figure 13–6 was developed to determine the permeability of disturbed or undisturbed soil specimens, with no overburden load other than the head of water. The test is used for canal lining soils. The apparatus is designed to allow water to flow downward through the specimen under any desired head.

The procedure for preparing the soil, placing it in the cylinder, and computing the coefficient of permeability is similar to the previously discussed standard permeability test. Since this is a permeability test without load, the equipment is not adapted to settlement measurements. The refinements discussed in the standard permeability test for controlling the placement density cannot be used. Instead, the soil is com-

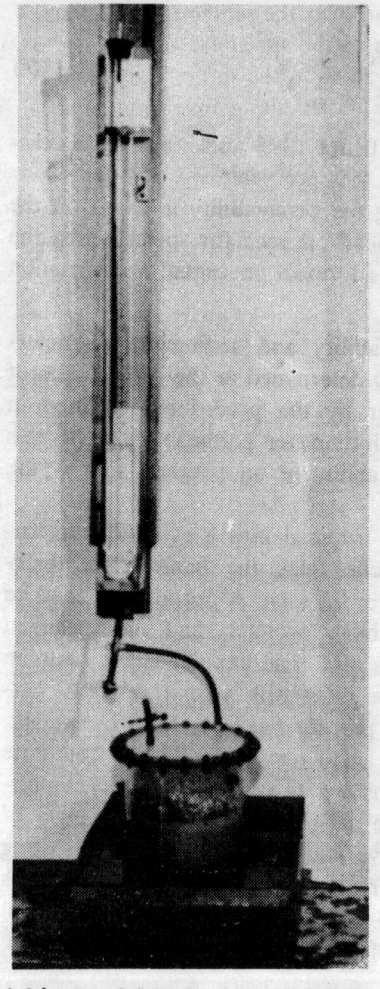

(A) PLASTIC CYLINDER (B) ALUMINUM CYLINDER

Figure 13–6.—Permeability cylinders for nonloaded specimens. PX–D–16548 and PX–D–16549.

pacted in three layers approximately 1 inch thick, as determined by using a combination square, the final result being, as nearly as possible, a specimen 3 inches in depth at the required density.

The soil is compacted, or placed in the cylinder, on a wetted porous disc. After the soil is compacted or an undisturbed specimen sealed in the cylinder, a fine mesh screen, with openings about the same as the maximum particle size, is placed on the soil, and a layer of uniformly-graded coarse sand or pea gravel is spread over the screen to a thickness of 1 to 4 inches. To prevent disturbance of the specimen when water is introduced into the cylinder, air is removed from the top of the chamber through the air release valve during the initial filling operation and as frequently thereafter as required.

PERMEABILITY AND SETTLEMENT OF SOIL CONTAINING GRAVEL

Designation E–14

1. Scope.—This designation describes the method for determining the permeability and settlement of a soil or the fraction of a soil having particles as large as 3 inches in diameter.

2. Apparatus.—The apparatus shall consist of the following (see designation E–4 and figs. 14–1 and 14–2):

> Cylinder, permeability, 20 inch, item 25.
> Disc, porous, item 32.
> Gages, dial indicator with holders, item 35A.
> Gage, block, dial indicator reference, item 35B.
> Gage, ring specimen, thickness, item 37.
> Loading device, item 45.
> Scales, item 74A.
> Springs, loading, item 88.
> Square, combination, item 89.
> Straightedge, item 93.
> Tamper, large, item 98.
> Tank, head, item 100.

3. Calibration.—The head tank factor, F, shall be determined by the method given in designation E–13, and the dimensions of the permeability cylinder and the constant C determined before starting the test.

4. Sample.—Approximately 200 pounds of minus 3-inch material shall be separated on the No. 4 sieve. The gravel particles are saturated in water and surface dried. The soil fraction is thoroughly mixed with the proper amount of water to give the optimum water content as determined by the compaction test, designation E–11. Other soil water contents may be used as specified. The soil and gravel fractions shall then be thoroughly mixed. The water content of the mixture shall be determined in accordance with designation E–9. The amount of material

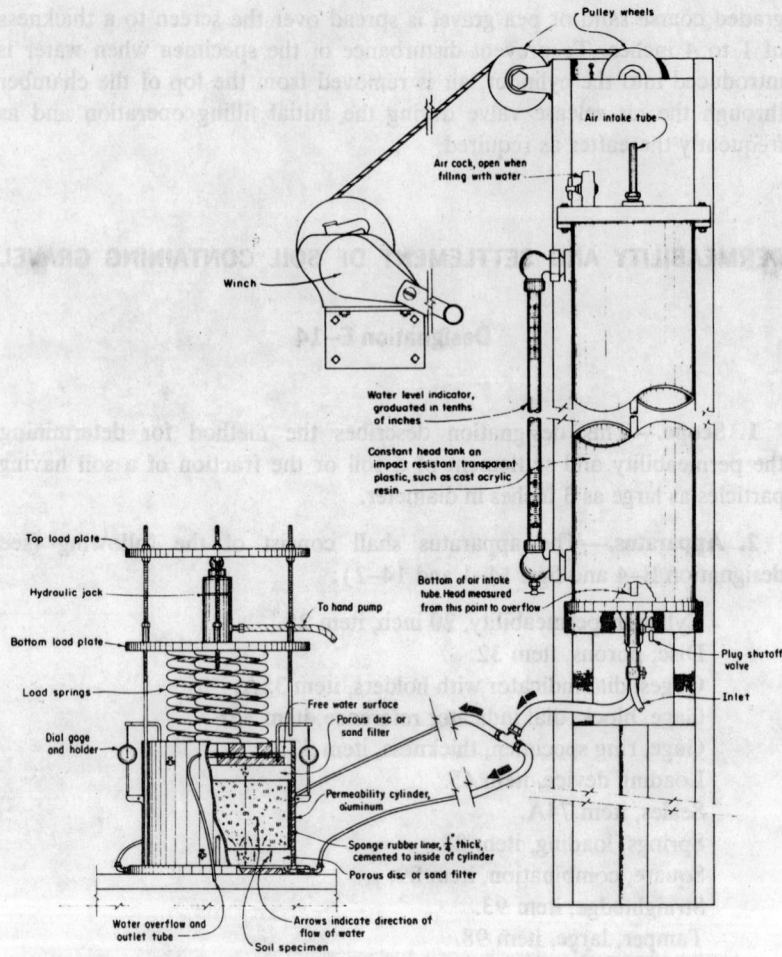

Figure 14–1.—Standard 20-inch permeability apparatus. 101–D–178.

calculated to obtain the desired wet weight of the specimen shall be weighed for placing in the container.

5. Procedure.—(a) The specimen shall be placed at the density and water content specified. The theoretical density or some percentage of the theoretical density computed on the basis of the maximum standard laboratory density of the minus No. 4 fraction and the quantity and bulk specific gravity of the gravel, is usually specified.[1] The inplace density

---

[1] See Compaction Test for Soil Containing Gravel, designation E–38.

Figure 14–2.—Standard 20-inch permeability test being performed in the laboratory. E–2246–9NA.

determined by measurements on a fill or foundation may also be specified for comparative laboratory and field tests.

(b) The bottom porous disc shall be saturated and drained and placed in the bottom of the permeability cylinder. (*Note:* If the permeability of the material to be tested is judged to be near or greater than that of the porous discs, the discs are replaced with a more pervious sand.) The 9-inch thickness gage shall be placed on the bottom porous disc, the top porous disc placed on the thickness gage, and the piston placed on the top disc so that the dial gage arms are directly over the lower bearing points for the gages. Dial gage readings shall be taken on each side of the cylinder and recorded as "Ring gage" readings, form 7–1455, designation E–13.

(*Note:* Forms 7–1454, 7–1455, and 7–1394 (figs. 13–3, 13–4, and 13–5 of designation E–13) shall be used to record the data; the constant *C*, form 7–1394, must be computed for the specimen in the 20-inch cylinder.)

The piston, top porous disc, and ring gage are then removed prior to placement of the soil specimen.

(c) One-third of the amount of minus 3-inch soil calculated for the entire specimen shall be placed in the cylinder. The layer of soil shall be compacted to a 3-inch depth with a hand or mechanical tamper to obtain the desired density for this layer. The uniformity of the depth of the layer shall be checked with the straightedge and combination square, and irregularities shall be adjusted to within $\frac{1}{16}$ inch of the 3-inch depth. The top of this layer shall be scarified with a sharp-pointed tool such as a knife or screwdriver to a depth of approximately ¼ inch. Two additional layers of soil shall be placed in a like manner to produce a 9-inch soil specimen. The depth of the specimen shall be uniform and adjusted within $\frac{1}{16}$ inch. The surface shall be scarified to a depth of ¼ inch and the top porous disc placed to insure that it is well seated and in continuous contact with the soil. The piston shall then be placed, and dial-gage readings taken on each side of the cylinder and recorded.

(d) The coil spring or springs, load plate, and tension rods shall be assembled as shown in figure 14–1, making sure that the tension rods are completely screwed into the base of the permeability cylinder.

(e) The desired load shall be applied with a hydraulic jack setup as shown in figure 14–1.

(f) The remainder of the test is similar to the standard permeability test procedure, designation E–13.

6. Calculations.—(a) The coefficient of permeability (feet per year) shall be calculated using the general equation (4) given in designation E–13.

(b) The settlement is expressed as a percentage of the initial thickness of specimen and is calculated by equation (1) in designation E–13.

(c) The consolidated density corresponding to a particular settlement, expressed as a percentage, is calculated by equation (2) in designation E–13.

ONE-DIMENSIONAL CONSOLIDATION OF SOILS

Designation E–15

1. Scope.—This designation describes a method for determining the rate and magnitude of consolidation of undisturbed and remolded soil, when restrained laterally and loaded and drained axially. In addition, the coefficient of permeability and any additional consolidation caused by saturating the soil while under load may be determined.[1] An adaption may be made so that the soil specimen container is sealed, permitting measurement of pore-fluid pressure.

2. Apparatus.—This test is not required for construction control; therefore, the apparatus is not listed in the field laboratory equipment list, designated E–4, and only special or major items of equipment are listed below:

(1) Loading device.—A suitable device for applying vertical load to the specimen. A pneumatically loaded consolidometer of 10,000 pounds maximum capacity is shown in figure 15–1.

(2) Specimen container.—A specimen container consisting essentially of a brass ring, 4¼ inches inside diameter by 1¼ inches deep or 2 inches inside diameter by ¾ inches deep, extension collar, porous plates, and dial gage (figs. 15–2 and 15–3).

(3) Stand, cutting bits and tools.—A stand with rotating platform, and guide. Cutting bits to fit specimen ring, knives, spatulas and other miscellaneous equipment for preparing undisturbed specimens, (fig. 15–4).

(4) Tamper and equipment for preparing remolded specimens.

(5) Other equipment for preparation, storage, and moisture determination.

3. Sample.—A sample having minimum dimensions of 6 inches (a 6-inch cube or a 6-inch length of core 6 inches in diameter) is required for an undisturbed specimen. At least 2 pounds and preferably 5 pounds of soil passing the No. 4 sieve, obtained in accordance with the method for "Preparation of Soil Samples for Testing," designation E–5, is required for a disturbed specimen.

---

[1] By varying the procedure to simulate specific conditions the one-dimensional consolidation apparatus is used to determine the expansive characteristics of soil and to establish placement moisture control limits for construction of high embankments.

Figure 15–1.—Pneumatically loaded one-dimensional consolidometer—10,000 pounds maximum capacity. E–2244–2NA.

Figure 15–2.—Specimen container for one-dimensional consolidometer. PX–D–16551.

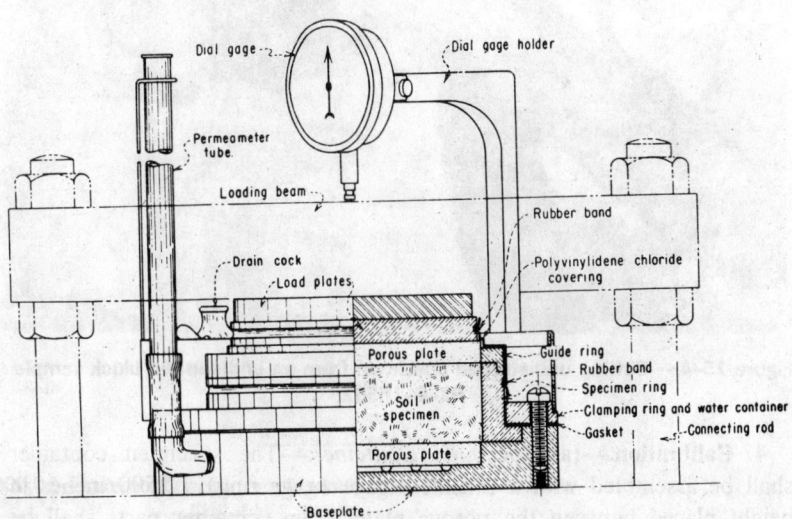

Figure 15–3.—Drawing of one-dimensional consolidometer specimen container. XD–3904.

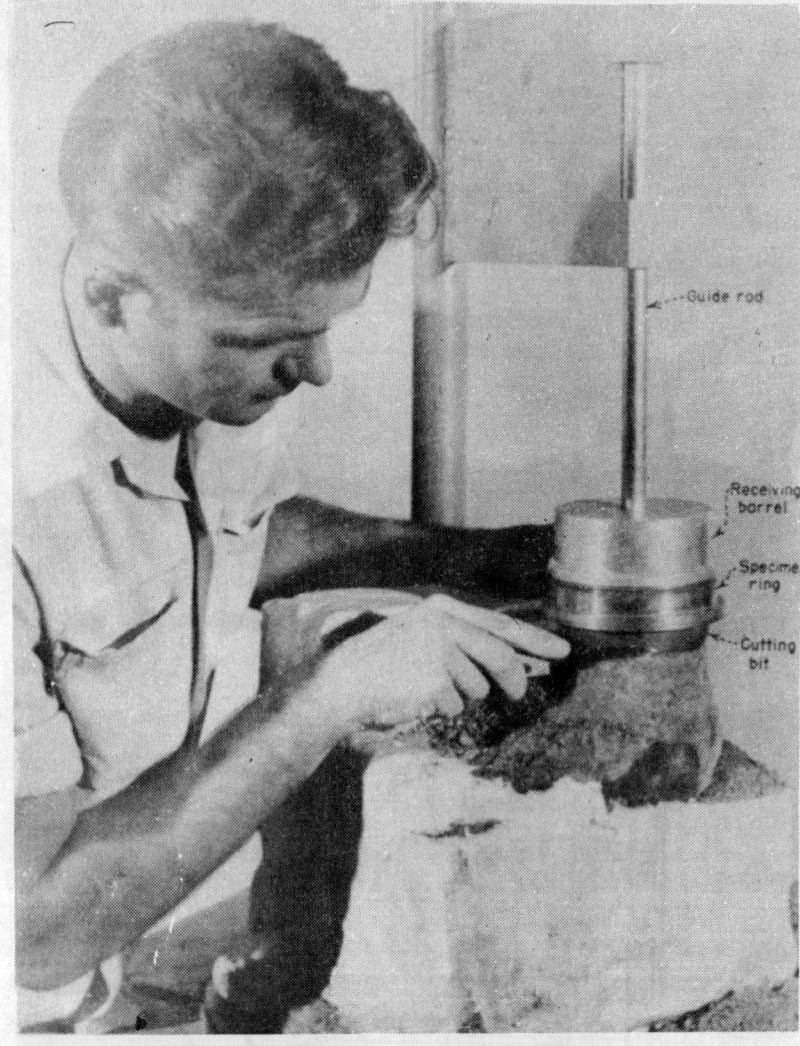

**Figure 15–4.—Cutting undisturbed specimen from an undisturbed block sample.
PX–D–16552.**

4. Calibration.—(a) *Specimen Container.*—The specimen container shall be assembled with a thickness gage (gage ring) 1.2500 inches in height placed between the porous plates. The container parts shall be match-marked to facilitate their placement in the same positions during each assembly. The assembly shall be placed in the loading device and a load equal to 1.0-pound-per-square-inch pressure applied. With this load applied, the dial gage shall be read and recorded as r_1 on figure 15–5.

ONE-DIMENSIONAL CONSOLIDATION TEST
PLACEMENT DATA, SHEET I

PROJECT _Example_   FEATURE_____   SAMPLE NO. _5L-58_

SPECIMEN NO. ___1___   SPECIMEN TYPE _Undisturbed_   CONTAINER NO. _2_   LOADING UNIT NO. _2_

TESTED BY_____   COMPUTED BY_____   CHECKED BY _____   DATE _____

CONTAINER: HEIGHT 1.25 In. DIAMETER 4.25 In. AREA 14.186 In.^{2} = 91.52 cm.^{2}

SPECIFIC GRAVITY	TYPE OF SOIL		
2.67 g	_CLAY (CL) some organic matter, very moist, black color_		
WEIGHT OF SPECIMEN CONTAINER AND SPECIMEN _(Complete assembly)_		16.5	POUNDS
WEIGHT OF SPECIMEN RING, COVER PLATES, AND WET SOIL		1622.5	GRAMS
WEIGHT OF SPECIMEN RING AND COVER PLATES		1072.8	GRAMS
WEIGHT OF WET SOIL		549.7	GRAMS
r_1 = DIAL READING WITH GAGE RING IN PLACE _(with 1.0 p.s.i. applied)_		0.2715	INCH
r_2 = DIAL READING WITH NO LOAD ON SPECIMEN _(used only in Expansion Test)_			INCH
r_3 = DIAL READING WITH 0.35 p.s.i. ON SPECIMEN		0.2705	INCH
h_1 = INITIAL SPECIMEN THICKNESS = 1.25 + (r_3 - r_1)= 1.25 +(2705-2715)		1.2490	INCH
h_0 = HEIGHT OF SOLIDS = $\dfrac{\text{WT DRY SPECIMEN IN GRAMS}}{(\text{AREA cm.}^2)(2.54)(G)} = \dfrac{422.10}{(91.52)^2(2.67)(2.54)}$		0.6801	INCH

MOISTURE DETERMINATION	MOISTURE SPECIMEN BEFORE TEST	WHOLE SPECIMEN AFTER TEST		
		*	**	
DATE	12-17	1-2		
CONTAINER NO.	221	1	72	
WET WEIGHT OF SPECIMEN + CONTAINER, GRAMS	421.9	625.35		
DRY WEIGHT OF SPECIMEN + CONTAINER, GRAMS	359.2	539.90	217.00	
WEIGHT OF CONTAINER, GRAMS	151.8	180.10	154.70	
WEIGHT OF WATER, GRAMS	62.7	85.45		
DRY WEIGHT OF SPECIMEN, GRAMS	207.4	359.80 + 63.30	= 422.10	
MOISTURE CONTENT, PERCENT OF DRY WEIGHT	30.23 w_i	23.75 w_s		

*THIS COLUMN FOR DETERMINING w_s.
**THIS COLUMN FOR DETERMINING DRY WEIGHT OF ALL MATERIAL ADHERING TO SPECIMEN RING, TRIMMED FROM SPECIMEN, AND WASHED OUT OF POROUS PLATES.

REMARKS:_____
_____
_____
_____
_____

Figure 15-5.—Specimen placement data for one-dimensional consolidation test.
101-D-263.

This reading represents a specimen height of exactly 1.2500 inches. The weight of the specimen ring and nonabsorptive glass or plastic cover plates shall be determined and recorded.

5. Procedure.—(a) *Preparation of Undisturbed Specimen.*—The cutting bit shall be attached to the bottom of the consolidometer specimen ring, assembled with the receiving barrel, and placed on the undisturbed soil sample in alinement with the guide rod on the cutting platform as shown in figure 15–4. The excess soil shall be trimmed with a knife, close to the outer edge of the cutting bit, leaving very little material to be shaved off by the cutting bit as it passes down over the specimen. Care shall be exercised in trimming in order to minimize disturbance of the specimen. When a sufficient length of specimen has been cut to protrude above the specimen ring, the soil shall be trimmed flush with the specimen ring top and bottom, using the straightedge trimmer, and covered with cover plates to retain the moisture. The water content, w_1, on figure 15–5 shall be determined on a portion of the soil trimmings, care being exercised to prevent moisture loss.

(b) *Preparation of Remolded Specimen.*—The specimen ring with the extension collar attached shall be assembled in the specimen container using a clamping ring without a water ring. A porous plate shall be placed in the baseplate, and an acetate film placed over the plate to prevent loss of soil into the plate. The soil sample shall be thoroughly mixed at the desired water content, w_i, and a quantity weighed which will give the desired density when compacted to a thickness of 1¼ inches in the 4¼-inch diameter ring. The weighed soil shall be placed in the specimen ring and compacted in three equal layers with the tamper until the specimen thickness is exactly 1¼ inches. The surface shall be carefully smoothed and covered with a cover plate until the specimen is to be placed in the loading device.

(c) *Testing.*—The specimen ring and compacted specimen are removed from the baseplate and the celluloid (acetate) disc is removed and replaced with a cover plate. The wet weight of the specimen is then obtained by weighing the specimen, specimen ring, and cover plates, and recorded. The specimen container including the soil specimen shall be assembled as shown in figure 15–3 and placed in the loading device. If the specimen is not to be saturated at the beginning of the test, a thin polyvinylidene chloride covering is placed around the specimen ring and load plate enclosing the top porous plate to prevent the loss of moisture from the soil. The polyvinylidene chloride covering shall remain on the specimen container until the specimen is to be saturated. The seating load of 1.0 pound per square inch shall be applied to the specimen and the dial reading recorded as r_3.

For this particular loading method, the load (see right-hand column of fig. 15–6) for producing the desired pressure on the specimen is computed as follows:

$$\text{Scale load} = A(1.05P) + W$$

where: A = area of specimen in square inches,
 P = desired pressure on specimen in p.s.i., and
 W = weight of soil and specimen container in pounds.

In this computation an allowance is made for friction loss equal to 5 percent of the specimen load. The specimen load shall be applied in increments of 12.5, 25, 50, and 100 percent of the maximum desired load. A greater number of increments, particularly for smaller loads, should be used when it is desired to obtain greater detail in the test curve. The increments selected are such that each succeeding load is double that of the previous load. Dial readings shall be taken and recorded for time intervals of 4, 10, 20 seconds, etc., up to 24 hours (see fig. 15–6). If the settlement is not essentially complete at the end of the 24-hour period, additional readings shall be made at 24-hour intervals until the settlement is essentially complete. The next load increment shall then be applied.

ONE-DIMENSIONAL CONSOLIDATION TEST
TIME-CONSOLIDATION DATA, SHEET 3

SAMPLE NO. 5L - 58   SPEC. NO. 1

PRESSURE	TIME ELAPSE	0:00	4	10	20	40	1:20	1:40	3:20	6:40	13:20	16:40	33:20	1:06:40	2:13:20	2:46:40	5:33:20	26:00:00		LOAD
1.0	DATE TIME	12-17 3:00																12-18 8:30		22
	% CONS. DIAL	2705																0.18 2682		
19.0	DATE TIME	12-18 8:30					8:31		8:33									12-19 8:30		300
	% CONS. DIAL	0.18 2682	3.12 2315	3.31 2291	3.64 2250	4.08 2196	4.51 2142	—	5.07 2072	5.35 2037	5.52 2015	—	5.70 1993	5.80 1981	5.87 1972	—	5.88 1970	6.06 1948		
37.5	DATE TIME	12-19 8:30																12-20 8:30		575
	% CONS. DIAL	6.06 1948	6.41 1904	6.57 1885	6.74 1863	6.92 1841	7.15 1812	—	7.53 1765	7.79 1732	7.96 1711	—	8.17 1685	8.29 1670	8.41 1655	—	8.53 1639	8.74 1613		
75.0	DATE TIME	12-20 8:30																12-23 8:30		1134
	% CONS. DIAL	8.74 1613	9.23 1552	9.39 1532	9.61 1505	9.82 1748	10.19 1432	—	10.50 1393	10.99 1332	11.23 1302	—	11.44 1276	11.59 1258	11.70 1244	—	11.84 1226	12.15 1188		
150.0	DATE TIME	12-26 8:30																12-26 8:30		2252
	% CONS. DIAL	12.15 1188	12.67 1122	12.83 1103	13.05 1075	13.35 1038	13.76 0986	—	14.28 0922		14.84 0852		15.05 0825	15.16 0812	15.25 0800		15.40 0782	15.80 0732		

PERMEABILITY COMPUTATIONS

HEAD (H₁)	DATE	TIME	TUBE READ	R	t'	c	k
16.75	12-23	2:10 3:50	0.0 0.044	0.44	1.66	0.09	0054
		4:00 8:20	0.0 3.50	3.50	16.33	0.87	0053
12-24		8:20 10:40	0.0 0.50	0.50	2.33	0.12	0052

$k = 0.053$ ft/yr   Avg.

150.0 DATE TIME 12-26 8:30 ... 12-30 8:30   LOAD 2254
% CONS. DIAL 15.80 732 ... 15.80 0732

1.0 DATE TIME 12-30 8:30 ... 12-31 8:30   LOAD 24
% CONS. DIAL 0732 ... 11.00 1331

PRESSURE IN POUNDS PER SQUARE INCH — LOAD ON BEAM IN POUNDS

Figure 15–6.—Recording time-consolidation data for one-dimensional consolidation test. 101-D-264.

When consolidation is complete under maximum loading, the permeability of the soil shall be determined. Air shall be forced from the lower porous plate and system by filling the permeameter tube and allowing water to saturate the porous plate and drain through the opened drain cock. With the drain cock in the closed position, the specimen shall then be saturated under a low head for about 24 hours or until water flows from the top of the specimen container. After the specimen is saturated, the container surrounding the specimen shall be filled with water. When required, the specimen may be saturated prior to applying the loadings. Either the constant head method, designation E–13, or the falling head method may be used to determine the coefficient of permeability.

If the falling head method is used, the data may be recorded on the data sheet shown in figure 15–6. The initial head H_i, date, time and corresponding tube reading are observed and recorded during the test. The initial tube reading and initial height of head, H_i, must be the same as used in calculations for constructing a chart as described in paragraph 6(f). If the soil is impervious several readings at 24-hour intervals may be satisfactory. If the soil is pervious and the permeameter tube empties in less than 8 hours, the time required to empty the tube shall be determined at least three times.

The load, except the 1.0-pound-per-square-inch seating load on the soil specimen, shall be released on completion of the permeability test period. The water pan shall be full during the unloading and expansion of the specimen. During the expansion phase of the test the load shall be observed frequently to maintain the desired load on the specimen. The dial gage shall be read from time to time during a 24-hour period or until the specimen reaches equilibrium.

The soil specimen shall be removed from the specimen ring after the water has been siphoned from the water pan. The water content of the soil specimen, w_e, shall be determined by drying the entire specimen, except that any material adhering to the specimen ring, any material trimmed from the specimen to remove free water, and any material washed out of the porous plates shall be dried separately and added to the dry weight of the moisture specimen after the moisture computation is made (see fig. 15–5).

6. Calculations.—(a) *Placement Data.*—The weight of wet soil, initial specimen thickness, height of solids, and water contents w_i and w_e, are computed as indicated in figure 15–5.

(b) *Time-Consolidation Data.*—The percent consolidation of the soil specimen based on the initial specimen thickness, for each dial gage reading of each load increment, as indicated in figure 15–6, shall be computed by the following equation:

$$\% \text{ consolidation} = \frac{\Delta h}{h_1} \times 100$$

where: $h_1 =$ initial specimen thickness, $1.25 + (r_3 - r_1)$, and
$\Delta h =$ change in specimen height, r_3 minus dial reading at any given time.

(c) *Load-Consolidation Data.*—The computations indicated in figure 15–6 shall be performed for each load increment, and the information summarized as on figure 15–7.

(d) *After-Test Data.*—The after-test data, consisting of the dry weight of the entire specimen, its water content, and the degree of saturation for maximum load and expanded conditions, shall be computed as indicated on figures 15–5 and 15–7.

(e) *Permeability.*—The permeability may be determined by using a constant head, in which case the computations may be simplified by the factor method described in designation E–13.

The permeability may also be determined by using a falling head; in this case, the following chart method will simplify computations.

(f) *Permeability Computation Details.*—The basic formula for computing the permeability with falling head supply is:

$$k = \frac{A_p L_s}{A_s} \times \frac{1}{t} \log_e \frac{H_i}{H_f}$$

where $k =$ permeability, feet per year,
$A_p =$ area of standpipe supplying water, square inches,
$A_s =$ area of soil specimen, square inches,
$L_s =$ length of specimen, feet,
$H_i =$ initial head, inches (difference between headwater and tailwater),
$H_f =$ final head, inches, and
$t =$ elapsed time, years.

For the apparatus used in the laboratories of the Engineering and Research Center, the following values have been established:

$A_p = 0.1075$ square inch,
$A_s = 14.186$ square inches,
$L_s = \dfrac{1.25}{12} = 0.10417$ foot,
$H_i = 16.75$ inches,
$H_f = H_i - 0.5677 R$ ($R =$ difference in tube readings), and
$t = t'/8760$ ($t' =$ elapsed time in hours).
Substituting these values in the above equation:

$$k = \frac{0.1075 \times 0.10417}{14.186} \times \frac{1}{t'/8760} \log_e \frac{16.75}{16.75 - 0.5677 R}$$

ONE-DIMENSIONAL CONSOLIDATION TEST
LOAD-CONSOLIDATION DATA, SHEET 2

SAMPLE NO. _5L-58_ SPECIMEN NO. _1_

PRESSURE (p.s.i.)	FINAL DIAL	Δ DIAL	h= h₁ - Δ	VOIDS h - ho	VOID RATIO e = (h - ho)/ho	DRY DENSITY (p.c.f.)	% CON-SOLIDATION	REMARKS
1.0	2705		1.2490	.5689	.8365	90.75	0.0	Seating load
19.0	1948	0757	1.1733	.4932	.7252	96.61	6.06	
37.5	1613	1092	1.1398	.4597	.6759	99.45	8.74	
75.0	1188	1517	1.0973	.4172	.6134	103.30	12.15	
150.0	0732	1973	1.0517	.3716	.5464	107.78	15.80	Water added at end of period
150.0	0732	1973	1.0517	.3716	.5464	107.78	15.80	
1.0	1331	1374	1.1116	.4315	.6345	101.97	11.00	Expanded under water

	INITIAL CONDITIONS	EXPANDED CONDITIONS	MAX. LOAD CONDITIONS	
DRY DENSITY, p.c.f.	90.75	101.97	107.78	
TOTAL HEIGHT, in., = h	1.2490	1.1116	1.0517	
HEIGHT OF SOLIDS, in., = h₀	.6801	.6801	.6801	
HEIGHT OF WATER, in., = h_w	.5489	.4313	.3714	
HEIGHT OF AIR, in., = h_a	.0200	.0002	.0002	*
MOISTURE CONTENT, PERCENT OF DRY WEIGHT	30.23 w_i	23.75 w_e	20.45	**
DEGREE OF SATURATION, %	96.48	99.95	99.95	

D = DRY DENSITY = $\dfrac{\text{WT. DRY SPEC., IN GRAMS}}{(3.724)(h)}$ = $\dfrac{422.10}{(3.724)(h)}$ _____ p.c.f.

h_w = HEIGHT OF WATER = $\dfrac{(\text{WT. DRY SPEC., IN GRAMS})(W_i \text{ or } W_e)}{(\text{AREA cm.}^2)(2.54)}$ $\dfrac{(422.10)(W)}{(91.52)(2.54)}$ _____ inches

h_a = HEIGHT OF AIR = $h - h_o - h_w$ = _____ inches

DEGREE OF SATURATION, % = $\dfrac{h_w \times 100}{h - h_o}$ or $\dfrac{(G)(w)}{e}$

*AIR HEIGHT ASSUMED TO REMAIN CONSTANT AFTER SATURATION OF SPECIMEN

**COMPUTED BY - $\dfrac{(h_w)(\text{AREA, cms.})(2.54)(100)}{\text{WT. DRY SPEC. IN GRAMS}}$

Figure 15-7.—Summary of load data for one-dimensional consolidation test.
101-D-267.

In this equation if R is constant, all values except t' are constant. If this constant is represented by C, the equation becomes:

$$k = \frac{C}{t'} \quad \text{where:} \quad C = 6.9 \ \log_e \frac{H_i}{H_f}$$

A chart is prepared by computing values of C for values of R and plotting as shown in figure 15–8. This chart and the equation from which it was derived are only applicable to the particular equipment for which it was computed.

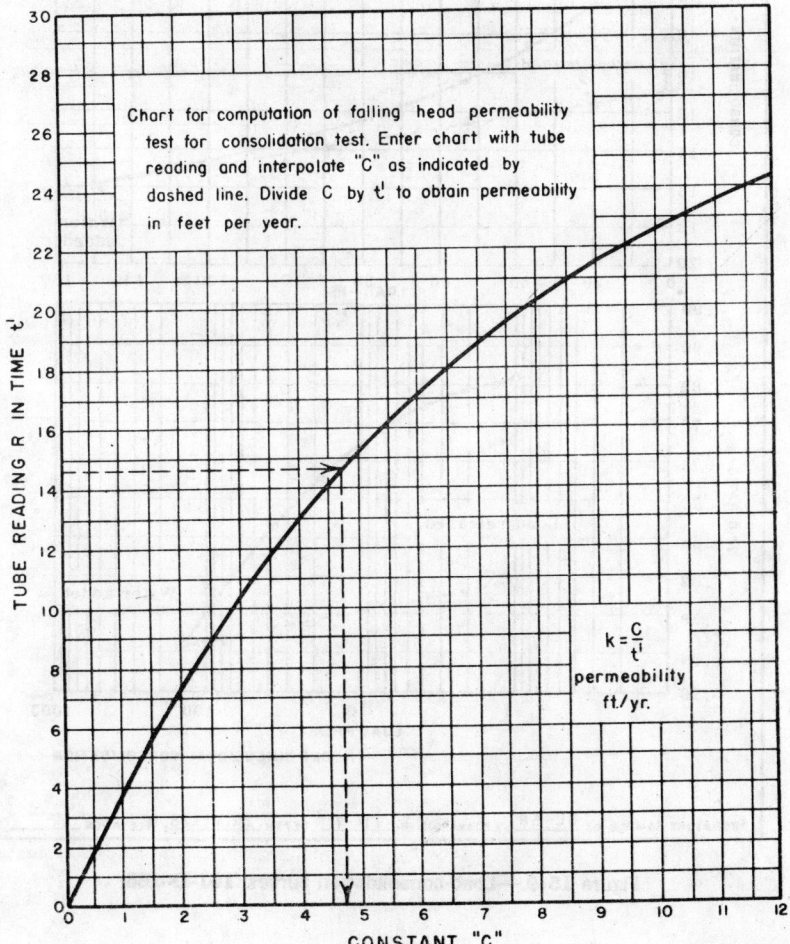

Chart for computation of falling head permeability test for consolidation test. Enter chart with tube reading and interpolate "C" as indicated by dashed line. Divide C by t' to obtain permeability in feet per year.

$$k = \frac{C}{t'}$$
permeability
ft./yr.

TUBE READING R IN TIME t'

CONSTANT "C"

Figure 15–8.—Chart for simplifying computation of permeability by falling-head method. 101–D–251.

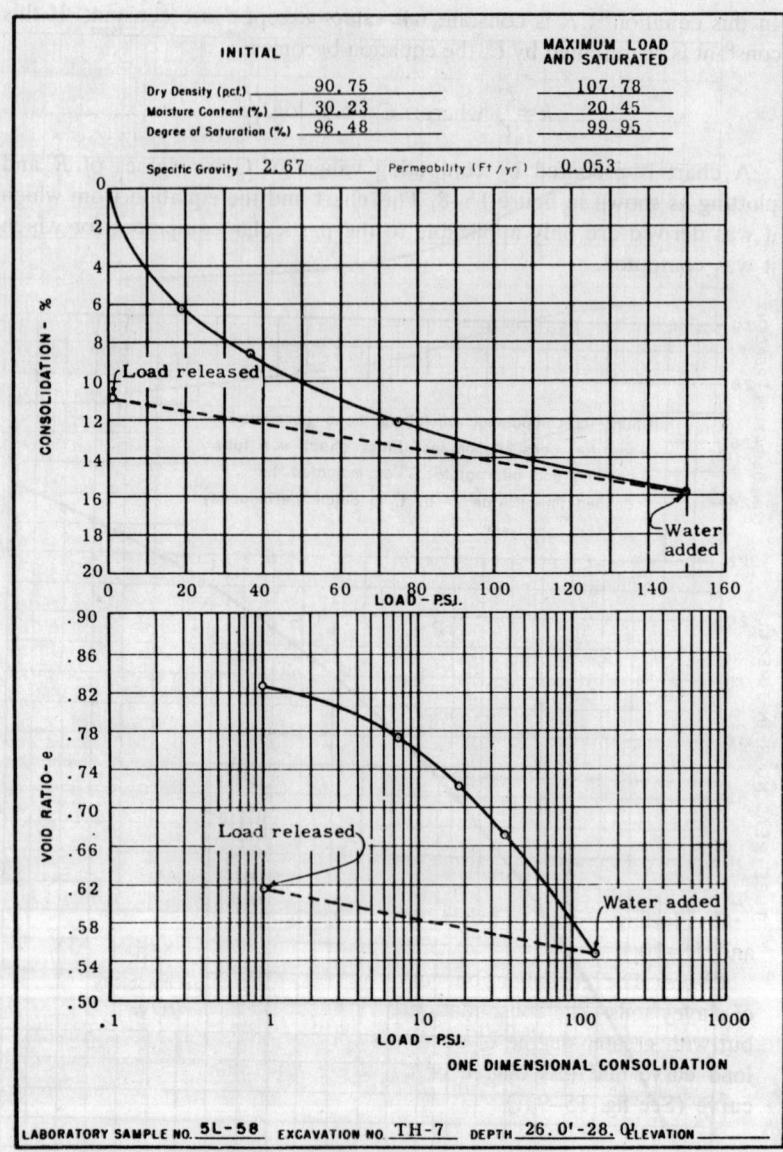

INITIAL | MAXIMUM LOAD AND SATURATED

Dry Density (pcf.) 90.75 — 107.78
Moisture Content (%) 30.23 — 20.45
Degree of Saturation (%) 96.48 — 99.95

Specific Gravity 2.67 Permeability (Ft / yr) 0.053

ONE DIMENSIONAL CONSOLIDATION

LABORATORY SAMPLE NO. 5L–58 EXCAVATION NO TH-7 DEPTH 26.0'–28.0' ELEVATION

Figure 15–9.—Load-consolidation curves. 101–D–266.

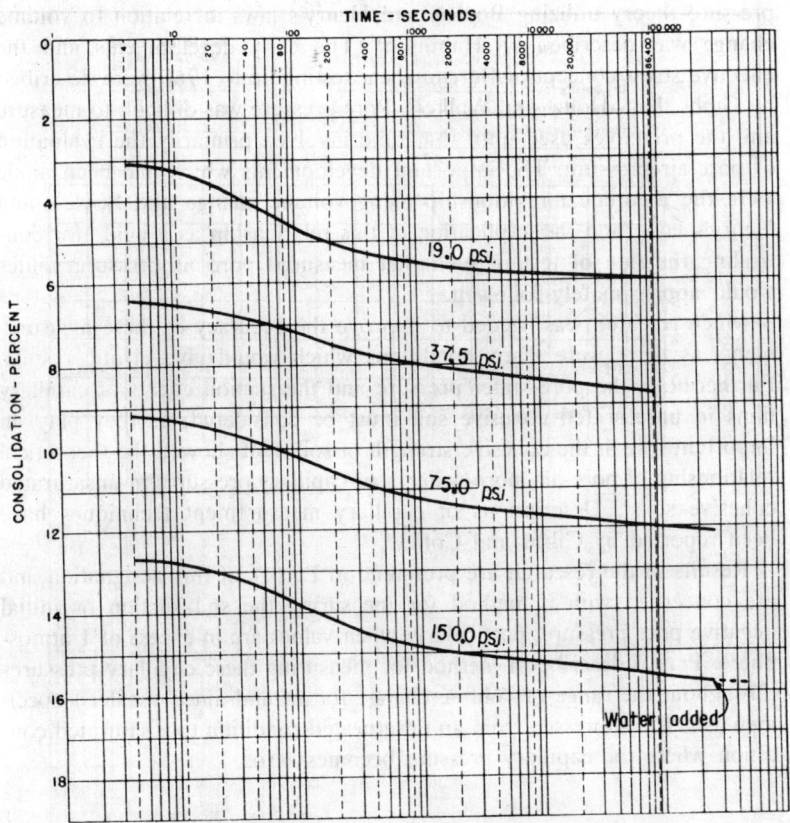

Figure 15–10.—Time-consolidation curves. 101–D–265.

7. Plotting.—The test data shall be plotted as shown in figures 15–9 and 15–10.

(*Note:* The consolidation curves are typical for undisturbed samples of clayey soils. The consolidation curves for remolded samples are similar, but with greater degree of curvature for the percent consolidation versus load curve and less degree of curvature for the void ratio versus load curve (See fig. 15–9).)

MEASUREMENT OF CAPILLARY PRESSURES IN SOILS

Designation E–16

1. Introduction.—Since about 1937, the Bureau of Reclamation has measured pore pressure in sealed (undrained) specimens. The pore-pressure theory utilizing Boyle's and Henry's laws in relation to volume change was described by Hamilton[1]. The early developments and the effective stress concept of Bureau shear testing up to 1960 were described by Gibbs, Hilf, Holtz, and Walker[2]. Pore pressure was difficult to measure and the principles used until that time involved primarily the evaluation of pore-air pressure. The important developments which had been made were the theoretical relationship using volume change and Boyle's and Henry's laws, and the application of this relationship as a guide for controlling the rate of testing such that measured pore-air pressure values would approximately follow it.

Much research was needed to improve the accuracy of these measurements as there were many conditions which would give erratic results. Furthermore, the pore-water pressure and the suction effects of capillary films in unsaturated cohesive soil must be considered, as they play an important role in the cohesive strength of soil. Hilf showed the theoretical relationship of pore-air, pore-water, and capillary pressures in unsaturated cohesive soils[3]. Descriptions of capillary measurement techniques have been reported by Gibbs and Coffey[4].

Results of the research are presented in Part A of this designation and are concerned with a method for measuring the soil suction or initial negative pore pressure, particularly when values are in excess of 1 atmosphere. Part B describes a method for measuring these capillary pressures throughout the range of volume change for an undrained (sealed) specimen as it is compressed from an unsaturated condition to a saturated condition where the capillary pressure becomes zero.

Part A. Test Procedure for Negative Pore Pressure of Soil by the Exposed End Plate Method

2. Scope.—This designation describes a laboratory procedure for measuring the negative pore pressure (capillary pressure) of unconfined unsaturated specimens of fine-grained soils. The procedure is varied for

[1] Hamilton, L. W., "The Effect of Internal Hydrostatic Pressure on the Shearing Strength of Soils," Proc. ASTM, Vol. 39, 1939, p. 1,100.

[2] Gibbs, H. J., Hilf, J. W., Holtz, W. G., and Walker, F. C., "Shear Strength of Cohesive Soils," ASCE Research Conference on Shear Strength of Cohesive Soils, Boulder, Colo., June 1960, pp. 33–162.

[3] Hilf, J. W., "An Investigation of Pore-water Pressure in Compacted Cohesive Soils," (Doctoral Thesis, University of Colorado), Technical Memorandum No. 654, Bureau of Reclamation, Denver, Colo., October 1956.

[4] Gibbs, H. J., and Coffey, C. T., "Techniques for Pore Pressure Measurements and Shear Testing of Soils," Proc. Seventh International Conference on Soil Mechanics and Foundation Engineering, Mexico, 1969, pp. 151–157.

soils having negative pore pressures greater than 5 atmospheres. The test is intended to be made on specimens that are prepared for further testing by other methods such as triaxial shear or consolidation. The pore pressure measured can be used in the analysis of results from these other tests.

3. Definitions.—According to the pore pressure theory discussed in this designation, the pressure of the pore water (u_w) is the sum of the pore air pressure (u_a) and the capillary pressure (u_c). If the pore air is at atmospheric pressure, it is considered for the purpose of this test to be at zero and the capillary pressure is then the same as that of the pore-water fluid. Since the specimen used in this test is not compressed, the capillary pressure measured is called the "initial capillary pressure." This is in contrast to capillary pressures measured in enclosed specimens where soil volume changes occur as they are compressed with loading.

4. Apparatus.—This test is not required for construction control; therefore, the apparatus is not listed in the field laboratory equipment list, designation E–4, and only special or major items of equipment are listed below:

(1) *General.*—The apparatus (figure 16–1) consists essentially of a ceramic end plate for a soil specimen in an air pressure chamber. There are sources of air for the chamber, and water for the end plate. In addition, there are provisions for measuring air and water pressures.

(2) *Air Chamber.*—Air pressure of 300 pounds per square inch (psi) is needed and the chamber should be designated to withstand this pressure. There is a pan for water around the baseplate of the chamber.

(3) *End Plate.*—The ceramic stone in the end plate should have a bubbling pressure (pressure required to force air through a saturated stone) greater than the capillary pressure of the soil being tested. (The bubbling pressure is sometimes called the air entry value.) The stone diameter should be at least one-fourth inch larger than the diameter of the soil specimen.

(4) *Measuring System.*—Air and water pressures are measured by pressure transducers connected to a multichannel electrical pressure recorder. The transducers should be designed for a working pressure of 300 pounds per square inch and the accuracy and linearity should be within 1 percent of full scale.

(5) *Head Tank.*—A head tank for water provides a means of completely filling the voids in the ceramic stone and the connecting tubing to the pressure-measuring device.

(6) *Tubing and Valves.*—It is important that the change in water volume between valve A on the head tank and the end plate connection, including the branch to the pressure-measuring device, be as small as possible. Therefore, valve A should be selected to have the smallest pos-

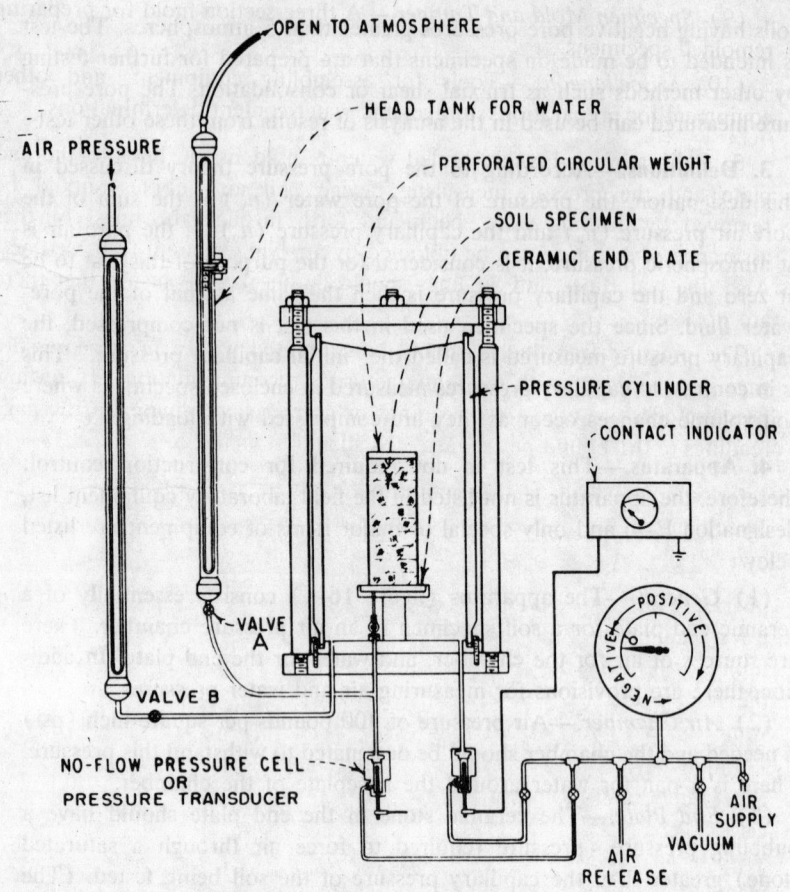

Figure 16–1.—Test chamber for measuring negative pore pressure by the exposed end-plate method. 101–D–551.

sible volume change from open to closed positions. Also, the tubing should have a wall thickness sufficient to hold the volume change to a maximum of 0.25 cubic centimeter (cc) for a 300-pound-per-square-inch chamber pressure.

(7) *Volume Determination Apparatus.*—A water container sufficiently large to accommodate the test specimens and a balance for weighing specimens in air and water.

(8) *Stands and Cutting Discs and Tools.*—A stand with rotating platform and guide, cutting discs, knives, spatulas, and other miscellaneous equipment for preparing undisturbed specimens.

(9) *Specimen Mold and Tamper.*—A three-section mold for preparing remolded specimens.

(10) *Miscellaneous.*—Tools for assembling equipment, and other equipment for preparation, storage, and water content determinations.

5. **Calibration.**—Freshly boiled water is used in the lines to the end plate and the pressure-measuring device. All air bubbles should be removed by flushing water back and forth through the lines. Further assurance of de-airing can be obtained by applying vacuum to these lines. Also, the waterlines and air lines should be checked to insure that there are no leaks.

The pressure transducers should be calibrated with a master pressure gage in accordance with the manufacturer's recommendations. If no-flow pressure cells are used, zero readings or corrections to indicate true gage readings to 0.1 pound per square inch should be made.

6. **Sample.**—This test is nondestructive and is intended to be made on soil specimens that will be used for further testing by other procedures.

7. **Procedure for Soils with Negative Pressures Less than 5 Atmospheres.**—Saturate the end plate by submerging it in water and boiling it for a minimum of 30 minutes with the connecting valve in the end plate tubing open. Cool the saturated end plate to room temperature by placing the boiling container in a cold water bath. Close the connecting valve while the end plate is still submerged.

Fill the water pan surrounding the chamber baseplate with water at room temperature to cover the outlet to the pressure transducer. Place the ceramic end plate on the chamber base and connect the valve under water. With water covering the valve, turn it to the open position. During this installation, keep the ceramic stone wet, with free water standing on its surface to prevent the development of menisci.

To check the system for saturation and sensitivity to negative pressure measurement, wipe the ceramic stone with a towel to remove any free water. Blow air over the surface of the ceramic stone to accelerate evaporation and cause a negative pressure to develop in the stone. Read the value of the negative pressure developed on the pressure transducer. Do not carry this check to pressures exceeding minus 8.0 pounds per square inch (about two-thirds of the prevailing atmospheric pressure); cavitation in the measuring system will take place at a water tension of about 1 atmosphere. As an additional operation in checking the system, release the menisci developed on the surface of the ceramic stone by opening valve A to the tank. The release of the menisci in the ceramic end plate is indicated by a zero reading on the pressure-measurement device. This procedure insures that the stone is resaturated.

Drain the water pan surrounding the chamber baseplate and place a wet sponge over the chamber pressure inlet. This will maintain a high humidity inside the air chamber and protect the specimen from drying during the test. Until the test is started and the chamber pressure applied, occasionally open tank valve A to maintain saturation of the end plate. In case any free water does accumulate on the end plate, remove it with a moist towel or cloth before placing the soil specimen in the center of the end plate (figure 16–2). During this installation, keep the pressure in the end plate near zero by occasionally opening and quickly closing the tank valve A. This condition must be maintained until the pressure chamber has been sealed by placement of the top cover plate.

Figure 16–2.—Soil specimen on exposed ceramic end plate. PX–D–64721 and E–2035–5.

Install the pressure chamber cylinder and securely fasten the top cover plate with the tie bolts. A final saturation of the end plate is made by opening and quickly closing the head tank valve A. The negative pore pressure of the soil will immediately act on the water in the ceramic end plate and develop menisci on the surface. To counteract this, air pressure must be applied as needed in the chamber.

The negative pressure of the soil is the algebraic difference between the applied chamber pressure and the ceramic end plate pressure. The applied chamber pressure must be enough so that the measured pressure on the end plate will not exceed minus 8.0 pounds per square inch. This is much less than the suction pressure at which the water column in the measuring system could cavitate.

Increase the applied chamber pressure in 5-pounds-per-square-inch increments and make these changes when or before the measuring system reaches a value of minus 5 pounds per square inch. This increment is recommended to avoid an excessive reduction of pressure in the measuring system and thus prevent dissolved air from coming out of solution. The data sheet example shown on figure 16-3 is a record of a negative pore-pressure test. After each pressure application, the pressure in the ceramic end plate is returned to zero by opening and quickly closing valve A. This procedure is repeated until the maximum value of negative pressure is reached as shown on figures 16-3 and 16-4. The total applied chamber pressure required to protect the measuring system from cavitation should not exceed the pressure required to reach equilibrium with the soil. Do not decrease the chamber pressure at any time during the test because this may cause air to come out of solution and collect in the measuring system.

At the end of the test, open valve A to insure saturation of the ceramic end plate. Then open valve B and reduce the chamber pressure to zero. Disassemble the pressure chamber and remove the soil specimen from the ceramic end plate. Obtain a wet weight of the soil specimen and carefully store it in a moistureproof container for further testing. Remove the ceramic end plate by disconnecting the cock union then immediately closing valve A.

8. Procedure for Soils with Negative Pressures Greater than 5 Atmospheres.—For soils having negative pressures greater than 5 atmospheres, it is necessary to vary the test procedure so that the test may be completed in an 8-hour work day. Follow the procedure described in section 7 until the soil specimen is placed on the ceramic end plate. Adjust the height of the water supply in the measuring system to the same height as the surface on the ceramic end plate. Open valve A and install the chamber cylinder and top cover plate.

With valve A open, apply 10-pounds-per-square-inch air pressure to the chamber. The 10-pounds-per-square-inch chamber pressure will be supported by menisci being formed within each pore of the ceramic end plate due to drainage of water from the measuring system. The end plate measuring system will then be at atmospheric pressure (zero for this test). Close valve A and if the negative pressure of the soil is considerably greater than the chamber pressure the soil will quickly act on the ceramic end plate, reducing the pressure in the measuring system to a negative

EL-589 (7-71)
Bureau of Reclamation

INITIAL NEGATIVE PORE PRESSURE DATA SHEET

Sheet __1__ of __2__

Project __Example__ Feature _____ Sample No. __22L-1__

Specimen No. __3__ Recorded by _____ Date _____

Water Content (%) __24.9__ Initial Dry Density __90.2__

End Plate Correction (A) __0.0__ Chamber Pressure Correction (B) __+0.5__

Date (1)	Time actual (2)	Temp °F. (3)	Bottom end plate pressure (psi) (4)	Chamber pressure (psi) (5)	Corrected bottom end plate pressure (4)+(A) (psi) (6)	Corrected chamber pressure (5)+(B) (psi) (7)	Negative pore pressure (7)–(6) (psi) (8)	Accumulated time (hours and minutes) (9)
			0 = 0	0 = -0.5				
8/19	7:30	73.0	0.0	-0.5	0.0	0.0	0.0	0.0
	7:40	73.0	-5.0	-0.5	-5.0	0.0	+5.0	0:10
Applied 5 psi to chamber, reduced end plate pressure to 0 psi								
	7:41	73.0	0.0	+4.5	0.0	+5.0	+5.0	0:11
	7:55	73.0	-5.0	+4.5	-5.0	+5.0	+10.0	0:25
Applied 10 psi to chamber, reduced end plate pressure to 0 psi								
	7:57	73.0	-0.2	+9.5	-0.2	+10.0	+10.2	0:27
	8:13	73.0	-5.0	+9.5	-5.0	+10.0	+15.0	0:43
Applied 15 psi to chamber, reduced end plate pressure to 0 psi								
	8:15	73.0	-0.1	+14.5	-0.1	+15.0	+15.1	0:45
	8:35	73.0	-5.0	+14.5	-5.0	+15.0	+20.0	1:05
Applied 20 psi to chamber, reduced end plate pressure to 0 psi								
Note:	Applied pressures between 20 and 50 psi not shown.							
Applied 50 psi to chamber, reduced end plate pressure to 0 psi								
	1:46	73.5	+0.1	+49.5	+0.1	+50.0	+49.9	6:16
	2:00	73.5	-1.0	+49.5	-1.0	+50.0	+51.0	6:30
	2:30	73.5	-3.3	+49.5	-3.3	+50.0	+53.3	7:00
	3:00	73.5	-4.6	+49.5	-4.6	+50.0	+54.6	7:30

Figure 16–3.—Data tabulated from a pore pressure test with an initial capillary pressure less than 5 atmospheres negative. (Sheet 1 of 2.) 101–D–552.

EL-589 (7-71)
Bureau of Reclamation

INITIAL NEGATIVE PORE PRESSURE DATA SHEET

Sheet _2_ of _2_

Project ___Example___ Feature _____ Sample No. _22L-1_

Specimen No. _3_ Recorded by _____ Date _____

Water Content (%) ___24.9___ Initial Dry Density ___90.2___

End Plate Correction (A) ___0.0___   Chamber Pressure Correction (B) ___+0.5___

Date (1)	Time actual (2)	Temp °F (3)	Bottom end plate (psi) (4)	Chamber pressure (psi) (5)	Corrected bottom end plate pressure (4)+(A) (psi) (6)	Corrected chamber pressure (5)+(B) (psi) (7)	Negative pore pressure (7)–(6) (psi) (8)	Accumulated time (hours and minutes) (9)
	3:15	73.5	-5.0	+49.5	-5.0	+50.0	+55.0	7:45
Applied 55 psi to chamber, reduced end plate pressure to 0 psi								
	3:16	73.5	+0.1	+54.5	+0.1	+55.0	+54.9	7:46
	4:00	73.5	-2.1	+54.5	-2.1	+55.0	+57.1	8:30
	6:00	73.0	-2.1	+54.5	-2.1	+55.0	+57.1	10:30
	9:30	73.0	-1.9	+54.6	-1.9	+55.1	+57.0	14:00
8/20	7:30	72.5	+3.0	+54.6	+3.0	+55.1	+52.1	24:00
Completion of test								

Figure 16–3.—Data tabulated from a pore pressure test with an initial capillary pressure less than 5 atmospheres negative. (Sheet 2 of 2.) 101–D–553.

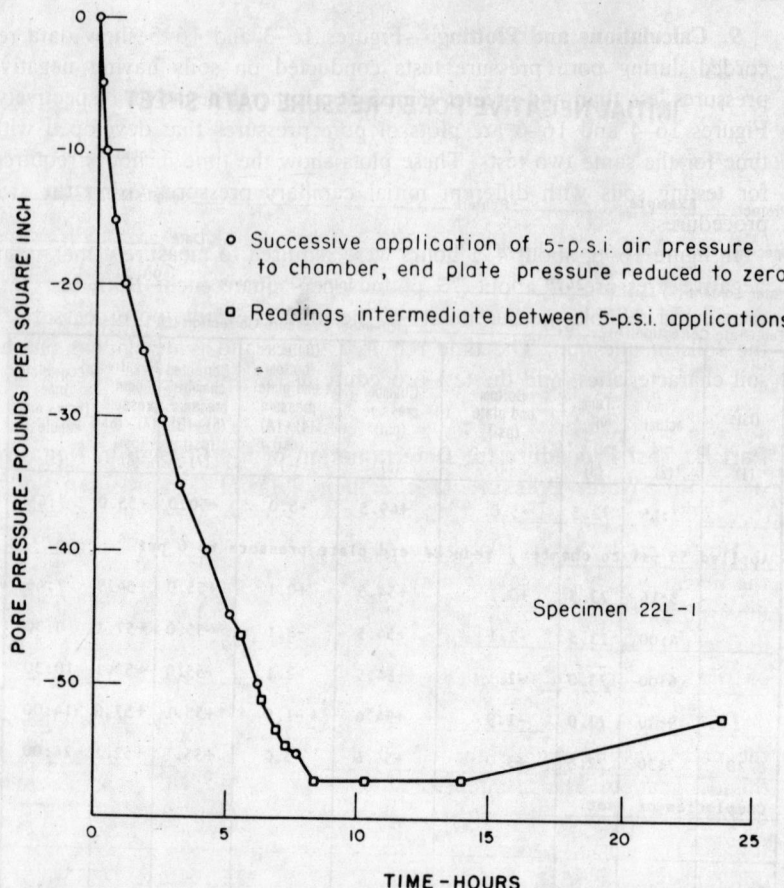

Figure 16-4.—Pore pressure development for a soil with an initial capillary pressure less than 5 atmospheres negative. 101-D-554.

value. Open valve A and apply an additional increment of 10-pounds-per-square-inch air pressure to the chamber. Close valve A and note the pressure in the ceramic end plate. Repeat this procedure until the reaction time of the soil on the closed end plate system shows a marked decrease. Complete the test using the procedure described in the last three paragraphs of section 7.

During this test, do not allow the ceramic end plate pressure to exceed plus 5 pounds per square inch as greater positive pressures could consolidate the soil specimen.

An example of the test data obtained during this test procedure is shown in figure 16-5 and a graphic presentation of the computed data is given in figure 16-6.

9. Calculations and Plotting.—Figures 16–3 and 16–5 show data recorded during pore pressure tests conducted on soils having negative pressures less than and greater than 5 atmospheres negative, respectively. Figures 16–4 and 16–6 are plots of pore pressures that developed with time for the same two tests. These plots show the time in hours required for testing soils with different initial capillary pressures using the two procedures.

In figure 16–6, about 4½ hours were required to measure a maximum negative pressure of about 73 pounds per square inch. Both tests are considered reliable measurements of the initial negative pore pressure of the soils in question. The time required varies and is dependent on the soil characteristics, and the test procedure used.

Part B. Test Procedure for Determination of the Change in Negative Pore-Water Pressure of a Soil by Triaxial Compression

10. Scope.—This test procedure describes a method of determining the negative pore pressure (capillary pressure) of unsaturated soil starting from an initial condition and continuing throughout volume change during undrained compression. The test data will be used in the analysis of triaxial shear of unsaturated soil.

11. Definitions.—The negative pore pressure of a soil is defined as the pore pressure due to the effect of capillary forces, as discussed in this designation. The resultant pore-water pressure of a soil is the result of the pore-air pressure in the soil pores and the effect of the capillary forces. The negative pore pressure in a soil may be evaluated as the difference between the pore-air pressure and the resultant pore-water pressure. At any given soil condition, the negative pore pressure could be given by a general formula:

$$u_c = -(u_a - u_w)$$

where: u_c is the negative pore pressure or capillary pressure

u_a is the pore-air pressure

u_w is the resultant pore-water pressure.

Therefore, the results obtained by this test procedure provide an evaluation of the negative pore pressure or capillary pressure of a soil at an initial condition, and show how it progressively changes through decreases in volume and increased saturation by triaxial compression under undrained conditions.

12. Apparatus.—The major test apparatus as shown in figure 16–7 consists of the following:

INITIAL NEGATIVE PORE PRESSURE DATA SHEET

Sheet **1** of **2**

Project **Example** _____ Feature _____ Sample No. **43M-14**

Specimen No. **6** _____ Recorded by _____ Date _____

Water Content (%) **16.2** _____ Initial Dry Density **105.1** _____

End Plate Correction (A) **0.0** _____ Chamber Pressure Correction (B) **+0.5**

Date (1)	Time actual (2)	Temp °F (3)	Bottom end plate (psi) (4)	Chamber pressure (psi) (5)	Corrected bottom end plate pressure (4)+(A) (psi) (6)	Corrected chamber pressure (5)+(B) (psi) (7)	Negative pore pressure (7)–(6) (psi) (8)	Accumulated time (hours and minutes) (9)
			0 = 0.0	0 = -0.5				
8/24	9:10	73.0	0.0	-0.5	0.0	0.0	0.0	0:0
	9:11	73.0	-1.5	-0.5	-1.5	0.0	+1.5	0:01
	Applied 10 psi to chamber							
	9:12	73.0	0.0	+9.5	0.0	+10.0	+10.0	0:02
	9:12	73.0	-1.0	+9.5	-1.0	+10.0	+11.0	0:02
	Applied 20 psi to chamber							
	9:13	73.0	0.0	+19.5	0.0	+20.0	+20.0	0:03
	9:16	73.0	-1.5	+19.5	-1.5	+20.0	+21.5	0:06
	Applied 30 psi to chamber							
	9:19	73.0	0.0	+29.5	0.0	+30.0	+30.0	0:09
	9:20	73.0	-2.0	+29.5	-2.0	+30.0	+32.0	0:10
	Applied 40 psi to chamber							
	9:21	73.0	0.0	+39.5	0.0	+40.0	+40.0	0:11
	9:25	73.0	-4.0	+39.5	-4.0	+40.0	+44.0	0:15
	Applied 50 psi to chamber							
	9:26	73.0	0.0	+49.5	0.0	+50.0	+50.0	0:16
	9:29	73.0	-3.0	+49.5	-3.0	+50.0	+53.0	0:19
	Applied 55 psi to chamber							

Figure 16–5.—Data tabulated from a pore pressure test with an initial capillary pressure greater than 5 atmospheres negative. (Sheet 1 of 2.) 101–D–611.

EL-589 (7-71)
Bureau of Reclamation

INITIAL NEGATIVE PORE PRESSURE DATA SHEET

Sheet __2__ of __2__

Project __Example__ _____ Feature _____ Sample No. __43M-14__

Specimen No. ____6____ Recorded by _____ Date _____

Water Content (%) __16.2_____ Initial Dry Density __105.1_____

End Plate Correction (A) ___0.0____ Chamber Pressure Correction (B) __+0.5____

Date (1)	Time actual (2)	Temp °F (3)	Bottom end plate (psi) (4)	Chamber pressure (psi) (5)	Corrected bottom end plate pressure (4)+(A) (psi) (6)	Corrected chamber pressure (5)+(B) (psi) (7)	Negative pore pressure (7)−(6) (psi) (8)	Accumulated time (hours and minutes) (9)
Note:	Applied pressures between 55 and 70 psi not shown.							
	Applied 70.0 psi to chamber, reduced end plate pressure to 0 psi							
	11:32	73.0	0.0	+69.5	0.0	+70.0	+70.0	2:25
	12:25	73.0	-2.3	+69.5	-2.3	+70.0	+72.3	3:15
	12:40	73.0	-2.5	+69.5	-2.5	+70.0	+72.5	3:30
	1:10	73.0	-2.6	+69.5	-2.6	+70.0	+72.6	4:00
	1:50	73.0	-2.9	+69.5	-2.9	+70.0	+72.9	4:40
	2:25	73.0	-2.9	+69.5	-2.9	+70.0	+72.9	5:15
	3:10	73.0	-2.5	+69.5	-2.5	+70.0	+72.5	6:00
	4:00	73.0	-2.3	+69.5	-2.3	+70.0	+72.3	6:50
8/25	7:15	72.0	+8.6	+69.5	+8.6	+70.0	+61.4	22:05
Completion of test								

Figure 16–5.—Data tabulated from a pore pressure test with an initial capillary pressure greater than 5 atmospheres negative. (Sheet 2 of 2.) 101–D–612.

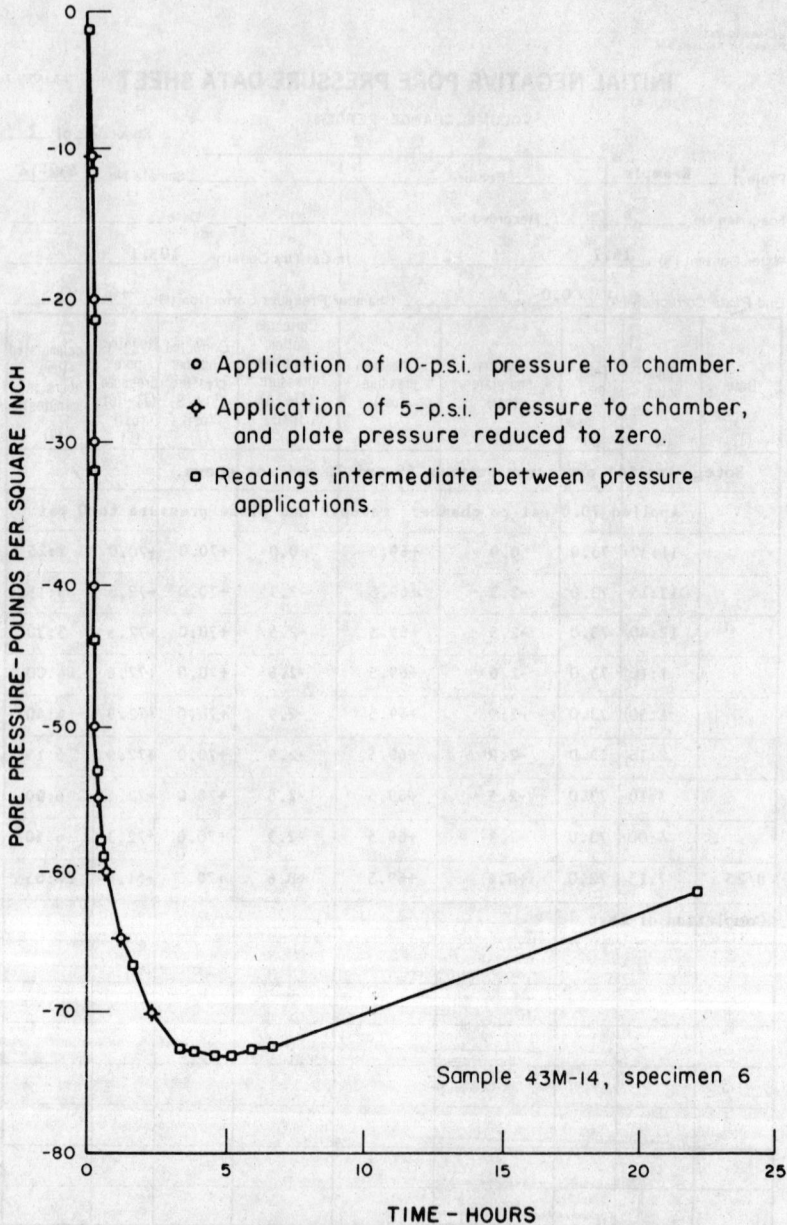

Figure 16–6.—Pore pressure development for a soil with an initial capillary pressure greater than 5 atmospheres negative. 101–D–555.

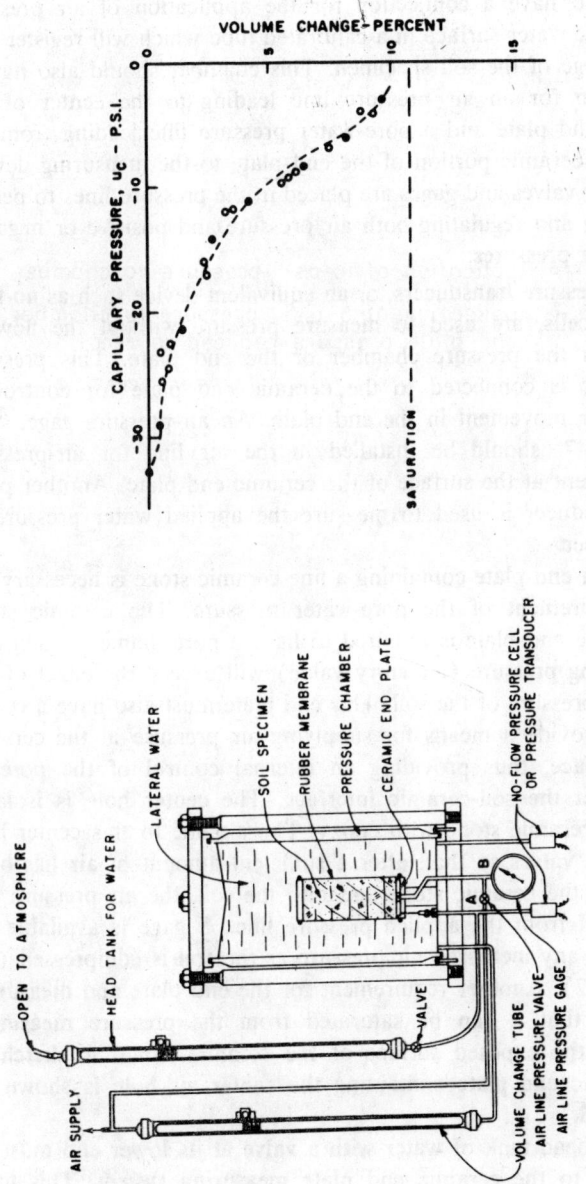

(a) PORE PRESSURE APPARATUS AS OPERATED WITH AIR
    EXPOSURE AT THE CENTER OF THE END PLATE

(b) EXAMPLE OF TEST RESULTS FOR u_c USING
    FINAL TEST PROCEDURE

Figure 16–7.—Pore pressure test for u_c during undrained compression. 101-D-556.

(1) A pressure chamber capable of containing a sealed soil specimen under an all-around water pressure of 300 pounds per square inch or higher, as required for special soil characteristics. The chamber should have a connection for the application of air pressure against the water surface in a calibrated tube which will register volume change of the soil specimen. This chamber should also have a connection for an air pressure line leading to the center of the ceramic end plate and a pore-water pressure line leading from the saturated ceramic portion of the end plate to the measuring device. Necessary valves and gages are placed in the pressure lines to permit measuring and regulating both air pressure and positive or negative pore-water pressures.

(2) Pressure transducers, or an equivalent device such as no-flow pressure cells, are used to measure pressure without the flow of fluid from the pressure chamber or the end plate. This pressure transducer is connected to the ceramic end plate for control of pore-water movement in the end plate. An air-pressure gage, "B", figure 16-7, should be installed in the air line for air-pressure measurement at the surface of the ceramic end plate. Another pressure transducer is used to measure the applied water pressure in the chamber.

(3) An end plate containing a fine ceramic stone is necessary for the measurement of the pore-water pressure. The ceramic stone used in the end plate is required to have a pore diameter such that its bubbling pressure (air entry value) will exceed the effect of the capillary pressure of the soil. This end plate must also have a center hole to provide a means for supplying air pressure at the ceramic stone surface, thus providing an internal control of the pore-air pressure at the soil-ceramic interface. The center hole is isolated from the ceramic stone with epoxy. The air line to this center hole must have valves so that, after a sufficient amount of air has been applied at the ceramic stone and into the soil, the air pressure can be shut off from the applied pressure line. A gage is available for measuring any increased air pressure as the soil is compressed (see figure 16-7). Another requirement for the end plate and measuring system is that it can be saturated from the pressure measuring device to the exposed surface of the ceramic stone. A sketch of the ceramic end plate containing the center air hole is shown by figure 16-8.

(4) A head tank of water with a valve at its lower end must be connected to the ceramic end plate measuring system. This head of water must be available to keep the end plate saturated during specimen installation and during sealing of the chamber until a

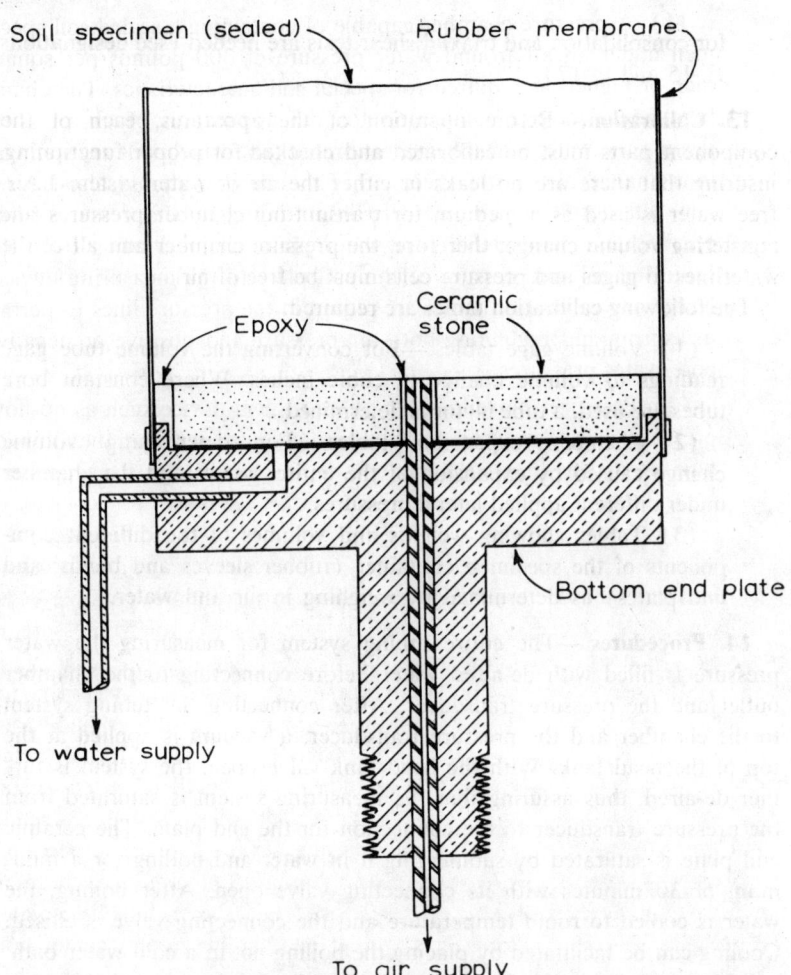

Figure 16-8.—End plate with small air inlet at the center. 101-D-557.

sufficient back pressure can be applied at the soil-ceramic interface to protect the measuring system from cavitation. The valve should be of a type requiring a minimum volume change from an open to a closed position, such as an appropriate plug valve.

(5) An electronic multichannel recorder is used for obtaining the readouts of pressure transducer values. A pressure gage panel may be used to obtain the pressure readings when no-flow pressure cells are used.

(6) The usual tools and equipment used for sample preparation

for consolidation and triaxial shear tests are needed (see designations E–15 and E–17).

13. Calibration.—Before operation of the apparatus, each of the component parts must be calibrated and checked for proper functioning, insuring that there are no leaks in either the air or water systems. Air-free water is used as a medium for transmitting chamber pressures and registering volume change; therefore, the pressure chamber and all of the waterlines to gages and pressure cells must be free of air.

The following calibration tables are required:

(1) Volume gage tables.—For converting the volume tube gage readings to volume change in cubic inches. Where constant bore tubes are used, a tube factor is determined.

(2) Chamber expansion tables.—For determining volume change caused by expansion of the entire system and the chamber under different applied lateral pressures.

(3) Tables showing weights and volumes of the different components of the specimen container (rubber sleeves and bands, and end plates) as determined by weighing in air and water.

14. Procedures.—The entire tubing system for measuring the water pressure is filled with de-aired water before connecting to the chamber outlet and the pressure transducer. After connecting the tubing system to the chamber and the pressure transducer, a vacuum is applied at the top of the head tank. With the head tank valve open, the system is further de-aired, thus assuring that the measuring system is saturated from the pressure transducer to the connection for the end plate. The ceramic end plate is saturated by submerging it in water and boiling for a minimum of 30 minutes with its connecting valve open. After boiling, the water is cooled to room temperature and the connecting valve is closed. Cooling can be facilitated by placing the boiling pot in a cold water bath.

After preparation of the soil specimen, the specimen is placed in a sealed container for a minimum of 24 hours. Immediately before testing, the soil specimen is enclosed in a rubber membrane, sealed on both ends by means of blank end plates, and an initial volume is obtained for use in making the volume change computations during compression.

While under water, the valve connection of the ceramic end plate is placed in an open position, the covering cap is closed, and a nut containing a small reservoir of water is placed over the connection. By these means, the ceramic end plate has access to a supply of water to maintain saturation after placement of the soil specimen on the end plate and until final hookup is made in the compression apparatus. The end plate is then removed from the water bath; the center air line is blown free of water. The end plate is placed on an assembly base and the free water is removed from the surface of the ceramic stone. The soil specimen with the membrane is then placed on the end plate and sealed. The soil specimen and end plate are now ready to be placed in the compression chamber apparatus. The chamber connection for the pore-water pressure measurement is made under water in a pool provided to cover the chamber base. The center air line for the back pressure measurement is connected to a position above the water to ensure no water being present in the air line. From the time the end plate with the soil specimen is connected to the pressure chamber base, the end plate must be kept continuously saturated by occasionally opening and quickly closing the head tank valve which is connected to the end plate water system. The chamber cylinder is then installed and filled with de-aired water and the top cover plate is fastened securely with the tie bolts provided. A setting is made on the volume change tube for zero volume change.

The negative pore-water pressure of the soil will immediately begin formation of menisci on the surface of the ceramic stone. To counteract the tension of the water in the measuring system, air pressure must be applied through the center of the end plate for a back pressure on the ceramic stone surface. This is done to provide an internal control of the pore-water pressure at the soil-ceramic interface similar to the external control given by the "Exposed End Plate Method," part A.

During the no-compression-load portion of the test, the air pressure is applied through the center of the end plate simultaneously with the chamber pressure. The air pressure is increased in increments, to protect the measuring system from cavitation, until the initial capillary pressure of the soil can be obtained.

At equilibrium the negative pore-water pressure of the soil is measured as the sum of the applied back pressure and the remaining water tension in the ceramic end plate. It is desired to control the speed of this initial portion of the test such that the volume change of the soil is kept insignificant. If volume change does occur, it will be determined through the volume-change measuring system.

The next step in the procedure is to close the valve to the center air line. Subsequent increases of chamber pressure then become an ambient compression load. The internal air pressure which develops from the compression at the soil-ceramic interface is measured by the pressure gage (B) between the closed valve (A) and the soil-ceramic interface. The volume change of the soil specimen is observed by measuring the water going into the pressure chamber as the specimen compresses. When equilibrium is reached for each increment of increased chamber pressure, u_c is the algebraic difference of the air pressure and the measured pore-water pressure.

The compression loadings are continued until the resultant pore-water pressure at the ceramic end plate reaches a value of minus 8 pounds per square inch, or they could continue until saturation of the soil is reached. The pore pressure is observed at various time intervals for each compression loading until stable values of volume change, air pressure, and pore-water pressure are reached. An example of the test procedure, computations, and presentation of the data are given below.

15. Data and Computations.—The specimen data are recorded on form EL–560, figure 16–9. Instructions for completing the computations are shown on the form. The data obtained during a typical test are shown on the data sheet, figure 16–10, columns (1), (2), (3), (4), (6), (11), and (12). The test results are also computed and tabulated on this form. The column numbers which follow are the column numbers on this form:

Column (5)—Volume change correction due to lateral pressure, obtained from calibration tables, considered positive.

Column (7)—Volume tube difference. The total difference in volume tube reading from initial volume tube setting resulting from compression or expansion of the soil specimen during testing.

Column (8)—Volume tube change. Equals volume tube difference, column (7), converted to cubic inches from tube calibration, considered negative for compression.

Column (9)—Total volume change. Algebraic sum of columns (5) and (8).

Column (10)—Percent volume change. Computed by dividing the value in column (9) by the initial specimen volume, line 9, form EL-560, figure 16–9, and multiplying by 100. The initial specimen volume is used until the air line valve is closed; at this point, the specimen volume is considered to be that volume at the time of closing of the valve.

Column (13)—Negative pressure of the soil is the algebraic difference of the air line pressure gage reading, column (12), and the bottom pore pressure reading, column (11).

Columns (14) and (15)—Conversion of percent volume change to void ratio. Computed as shown in columns (14) and (15).

Column (16)—Degree of saturation. Computed by the formula as given by column (16).

Columns (14), (15), and (16) are computed after the test is completed for additional data needed to plot appropriate curves.

A graphic presentation of the computed data is shown in figure 16–7(b). This shows the equilibrium negative pressure for the initial soil condition and progressing through each successive compression load. The equilibrium pore-water pressure relationship to the volume change is

EL-560
(5-58)
Bureau of Reclamation

SPECIMEN PLACEMENT – DATA SHEET 1
(FOR TRIAXIAL SHEAR AND THREE DIMENSIONAL TESTS)

SHEET __1__ OF __1__

COMPACTION

SAMPLE NO. __43M-14__   _____ % MD   FEATURE __Example__

SPECIMEN NO. __7__   __105.3__ ' pcf   DATE _____

CONTAINER NO. __1__   __17.2__ % MOISTURE

	INITIAL	FINAL	WETTED	
(1) WEIGHT OF SPECIMEN + CONTAINER IN AIR	664.22	665.50	_____	(e)
(2) WEIGHT OF CONTAINER IN AIR	155.79	157.07	_____	(e)

	INITIAL	FINAL
RUBBER	15.52	16.7
END PLATE		

	INITIAL	FINAL	WETTED	
(3) WET WEIGHT OF SPECIMEN = (1) - (2)	508.43	508.43	_____	(e)
(4) DRY WEIGHT OF SPECIMEN = 1 + $\frac{(3)}{\frac{(11)}{100}}$	432.01	_____	_____	(e)
(5) WEIGHT OF SPECIMEN + CONTAINER IN WATER	327.62	341.10	_____	(e)
(6) WT. LOSS OF SPEC. + CONT. = (1) - (5)	336.60	_____	_____	(e)
(7) WEIGHT LOSS OF CONTAINER	78.57	_____	_____	(e)
(8) WEIGHT LOSS OF SPECIMEN = (6) - (7)	258.03	_____	_____	(e)
(9) SPECIMEN VOLUME = $\frac{(8)}{16.350}$	15.782	_____	_____	(cu. in.)
(10) DRY DENSITY = $\frac{(4) \times 62.29}{(8)}$	104.29	_____	_____	(pcf)
(11) MOISTURE, % DRY WT. = $\left[\frac{(3)}{(4)} - 1\right] \times 100$	17.69	17.69	_____	(%)
(12) ABS. SOIL DENS. = $Sp. G \times 62.43$	174.80	_____	_____	(pcf)
(13) SOIL VOLUME = $\frac{(10)}{(12)} \times 100$	59.66	_____	_____	(%)
(14) WATER VOLUME = $\frac{(10) \times (11)}{62.29}$	29.62	_____	_____	(%)
(15) FREE AIR VOL. = 100 - $[(13) + (14)]$	10.72	_____	_____	(%)
(16) VOID RATIO = $\frac{100}{(13)} - 1$	0.6762	_____	_____	(%)
(17) DEGREE OF SATURATION = $\frac{(14) \times 100}{100 - (13)}$	73.4	_____	_____	(%)

		WEIGHED	MEASURED	
CONSOLIDATED DENSITY _____ pcf	(18) VOLUME CHANGE	_____	_____	(cu. in.)
FAILURE DENSITY _____ pcf	(19) DRAINAGE	_____	_____	(cu. in.)

PREPARED BY _____ COMPUTED BY _____ CHECKED BY _____

Figure 16–9.—Sample data sheet for triaxial shear and three-dimensional speci-
men placement. 101-D-558.

Sample No. 43M-14 - Specimen No. 7

(1) Date	(2) Time	(3) Room Temp. °F	(4) Lateral pressure - psi	(5) Lateral pressure correction	(6) Volume tube reading	(7) Volume tube difference	(8) Volume tube change (7)×0.4407	(9) Total volume change (8)−(5)	(10) % Volume change	(11) Bottom pore pressure-psi	(12) P_b - gage reading - psi	(13) Net negative pressure - uc (12)−(11)	(14) $\frac{(10)}{(1+e_i)} \times 100$	(15) Void ratio $e_i - (14)$	(16) Saturation-% $S = wG_s/e$
9-14	8:20	73.0	0 = 0.0	--	4.00	--			--	0 = 0.0	0 = 0.0	0.0		0.6762	73.2
	8:22	73.0	0.0	--	4.00	--			--	0.0	0.0	-5.0		.6762	73.2
Applied 5.0-psi lateral and back pressure, opened and closed valve														
	8:23	73.0	5.0	0.0220	4.10	0.10	0.0441	0.0221	0.10	0.0	5.0	-5.0	0.0017	.6745	73.4
	8:25	73.0	5.0	0.0220	4.10	0.10	0.0441	0.0221	0.10	-5.0	5.0	-10.0			
Applied 10.0-psi lateral and back pressure, opened and closed valve														
	8:26	73.0	10.0	.0420	4.06	0.06	.0264	+.0156	+0.10	0.0	10.0	-10.0	.0017	.6729	72.7
	8:30	73.0	10.0	.0420	4.06	0.06	.0264	+.0156	+0.10	-5.0	10.0	-15.0			
Applied 15.0-psi lateral and back pressure, opened and closed valve														
	8:31	73.0	15.0	.0610	4.15	0.15	.0661	.0051	0.03	0.0	15.0	-15.0	.0005	.6757	73.3
	8:34	73.0	15.0	.0610	4.15	0.15	.0661	.0051	0.03	-5.0	15.0	-20.0			
Applied 20.0-psi lateral and back pressure, opened and closed valve														
	8:36	73.0	20.0	.0660	4.20	0.20	.0881	.0221	0.10	0.0	20.0	-20.0	.0017	.6759	73.3
	8:46	73.0	20.0	.0660	4.20	0.20	.0881	.0221	0.10	-5.0	20.0	-25.0			
Applied 25.0-psi lateral and back pressure, opened and closed valve														
	8:49	73.0	25.0	.0750	4.25	0.25	.1102	.0352	0.22	0.0	25.0	-25.0	.0060	.6702	73.9
	9:50	73.0	25.0	.0750	4.25	0.25	.1102	.0352	0.22	-1.6	25.0	-26.6			
	10:20	73.0	25.0	.0750	4.25	0.25	.1102	.0352	0.22	-2.6	25.0	-27.6			
	10:30	73.0	25.0	.0750	4.25	0.25	.1102	.0352	0.22	-3.4	25.0	-28.4			
	11:25	73.0	25.0	.0750	4.30	0.30	.1322	.0572	0.36	-7.5	25.0	-32.5			
	12:15	73.0	25.0	.0750	4.30	0.30	.1322	.0572	0.36	-7.5	25.0	-32.5			

Figure 16-10.—Sample data and computation sheet. (Sheet 1 of 3.) 101-D-559.

Sample No. 43M-14 - Specimen No. 7

(1) Date	(2) Time	(3) Room Temp.°F	(4) Lateral pressure - psi	(5) Lateral pressure correction	(6) Volume tube reading	(7) Volume tube difference	(8) Volume tube change (7)×0.4407	(9) Total volume change (8)−(5)	(10) % Volume change	(11) Bottom pore pressure-psi	(12) P_b - gage reading - psi	(13) Net negative pressure - uc (12)−(11) Percent	(14) $\frac{100}{(10)}\times(1+e_i)$ volume change	(15) Void ratio e_i − (14)	(16) Saturation-% S = wGs/e
Closed back pressure valve. Volume tube drained to starting compression point. started from zero; new initial void ratio e_1 = .6702	12:16	72.0	25.0	0.0750	5.00				0.0	-7.5	25.0	-32.5	0.0060	0.6702	73.9
Open and closed valve, applied 28.1-psi lateral pressure	12:17	72.0	28.1	.0050	5.18	0.18	0.0793	0.0743	0.47	+4.0	27.0	-23.0			
	12:30	72.0	28.1	.0050	5.18	0.18	.0793	.0743	0.47	+1.5	27.0	-25.5			
	12:40	72.0	28.1	.0050	5.18	0.18	.0793	.0743	0.47	-0.6	27.0	-27.6			
9-14	1:00	72.0	28.1	.0050	5.18	0.18	.0793	.0743	0.47	-1.0	27.0	-28.0			
	1:25	72.0	28.1	.0050	5.18				0.47	-1.5	27.0	-28.5			
	2:05	72.0	28.1	.0050	5.18				0.47	-2.0	27.0	-29.0			
	3:00	72.0	28.1	.0050	5.18	0.18	.0793	.0743	0.47	-2.5	26.5	-29.0	.0079	.6683	74.1
Opened and closed valve, applied 34.3-psi lateral pressure	3:01	72.0	34.3	.0150	5.36	0.36	.1586	.1436	0.91	+2.5	27.0	-24.5			
	3:16	72.0	34.3	.0150	5.37	0.37	.1631	.1481	0.94	3.2	27.5	-24.3			
	3:50	72.0	34.3	.0150	5.37	0.37	.1631	.1481	0.94	5.0	27.5	-22.5			
9-15	7:30	72.0	34.3	.0150	5.55	0.55	.2424	.2274	1.45	0.2	21.5	-21.3			
	8:00	72.0	34.3	.0150	5.55	0.55	.2424	.2274	1.45	0.2	21.5	-21.3	.0243	.6519	76.0
Applied 46.8-psi lateral pressure	8:03	72.0	46.8	.0370	5.83	0.83	.3658	.3288	2.09	6.3	22.5	-16.2			
	8:15	72.0	46.8		5.86	0.86	.3790	.3420	2.17	6.3	23.0	-16.7			
	9:30	73.0	46.8		5.86	0.86	.3790	.3420	2.17	7.0	23.0	-16.0			
	12:17	74.0	46.8		5.83	0.83	.3658	.3288	2.09	6.0	22.0	-16.0			
9-16	3:00	73.0	46.8		5.83					4.9	21.0	-16.1			
	7:30	72.0	46.8		6.05	1.05	.4627	.4257	2.71	1.8	16.0	-14.2			
	9:50	73.0	46.8	.0370	6.05	1.05	.4627	.4257	2.71	1.8	16.0	-14.2	.0454	.6308	78.5

Figure 16–10.—Sample data and computation sheet. (Sheet 2 of 3.) 101–D–560.

Sample No. 43M14 — Specimen No. 7

(1) Date	(2) Time	(3) Room Temp.°F	(4) Lateral pressure-psi	(5) Lateral pressure correction	(6) Volume tube reading	(7) Volume tube difference	(8) Volume tube change (7)x0.4407	(9) Total volume change (8)−(5)	(10) Volume change %	(11) Bottom pore pressure-psi	(12) P_b – gage reading - psi	(13) Net negative pressure – uc (12)−(11)	(14) $(1+e_i)\times\frac{(10)}{100}$	(15) Void ratio e_i − (14)	(16) Saturation-% $S = wG_s/e$
Applied 59.3-psi lateral pressure				0.0570											
	9:55	73.0	59.3		6.30	1.30	.5729	0.5159	3.28	4.5	16.5	−12.0	0.0681	0.6081	81.5
	10:10	73.0	59.3		6.36	1.36	.5994	.5424	3.45	5.1	17.0	−11.9			
	12:20	73.0	59.3		6.39	1.39	.6126	.5556	3.53	5.5	17.0	−11.5			
9-17	3:00	73.0	59.3		6.39	1.39	.6126	.5556	3.53	5.3	17.0	−11.8			
	8:05	72.0	59.3		6.58	1.58	.6963	.6393	4.06	2.0	13.0	−11.0			
	8:45	72.0	59.3	.0570	6.58	1.58	.6963	.6393	4.06	2.0	13.0	−11.0			
Applied 84.3-psi lateral pressure				.0890											
	8:38	72.0	84.3		7.18	2.18	.9607	.8717	5.54	8.2	18.0	− 9.8	.1061	.5701	86.9
	9:10	72.0	84.3		7.30	2.30	1.0136	.9246	5.88	9.7	19.0	− 9.3			
9-17	11:10	72.0	84.3		7.35	2.35	1.0356	.9466	6.02	10.0	19.0	− 9.0			
	3:00	72.0	84.3		7.36	2.36	1.0401	.9511	6.05	10.0	19.0	− 9.0			
9-18	8:45	70.0	84.3		7.46	2.46	1.0841	.9951	6.33	6.4	15.0	− 8.6			
	9:20	70.0	84.3	.0890	7.46	2.46	1.0841	.9951	6.33	6.4	15.0	− 8.6			
Applied 109.3-psi lateral pressure				.1310											
	9:24	71.0	109.3		7.94	2.94	1.2957	1.1647	7.40	14.0	21.0	− 7.0	.1368	.5394	91.8
	9:30	71.0	109.3		8.00	3.00	1.3221	1.1911	7.57	15.0	22.0	− 7.0			
9-20	7:25	72.0	109.3		8.21	3.21	1.4146	1.2816	8.15	8.8	15.0	− 6.2			
	8:15	72.0	109.3		8.21	3.21	1.4146	1.2816	8.15	8.8	15.0	− 6.2			
	12:05	72.0	109.3	.1310	8.21	3.21	1.4146	1.2836	8.16	8.8	15.0	− 6.2			
Applied 134.3-psi lateral pressure				.1690											
	12:06	73.0	134.3		8.51	3.51	1.5469	1.3779	8.76	14.0	18.0	− 4.0	.1629	.5133	96.5
	1:10	73.0	134.3		8.65	3.65	1.6086	1.4396	9.15	17.0	21.5	− 4.5			
	2:25	73.0	134.3		8.65	3.68	1.6218	1.4528	9.24	18.0	22.0	− 4.0			
9-21	7:30	71.0	134.3		8.85	3.85	1.6967	1.5277	9.71	16.0	19.0	− 3.0			
	8:30	72.0	134.3		8.85	3.85	1.6967	1.5277	9.71	16.0	19.0	− 3.0			
	9:50	73.0	134.3	.1690	8.85	3.85	1.6967	1.5277	9.72	16.0	19.0	− 3.0			
Completion of test															

Figure 16-10.—Sample data and computation sheet. (Sheet 3 of 3.) 101-D-561.

represented by the solid black points. The hollow points represent the progress of the test before equilibrium is reached.

TRIAXIAL SHEAR OF SOILS
Designation E–17

1. Scope.—This designation describes a method for determining the shear strength of soils, expressed as the coefficient of internal friction and cohesion. The unconfined compressive strength test of a soil is considered a part of this designation as it is the strength of a specimen (maximum axial load divided by area) tested in triaxial shear with the lateral pressure equal to atmospheric or zero gage pressure.

Triaxial shear testing involves the evaluation of pore pressure for an effective stress analysis. This analysis has been based on the pore-air pressure and the theoretical relationship with volume change according to Boyle's Law for compression of air combined with Henry's Law for air solubility in water. More recently consideration has been given to capillary (suction) pressures in partially saturated soils. Research on measurements to include all of these pore pressure effects has been made in the Bureau of Reclamation and a workable test procedure has been developed[1]. The measurement called capillary stress in this designation and designation E–16 includes all of the effects that make up the suction. The consideration of capillary stress will cause the pore-water pressure to be different than the pore-air pressure. Therefore, if the effective stress is analyzed on the basis of deducting the pore-air pressure from the applied stresses, the strength is evaluated at the void ratio and degree of saturation of the soil in the test specimen which should represent a significant condition of the structure, such as the compacted condition of the embankment or the inplace condition of the foundation. The strength so evaluated would include effects of capillary stress. However, if the stress is analyzed on the basis of pore-water pressure being deducted from the applied stresses, the result would be a very basic interpretation of strength that would not include the capillary pressures as effective stress. It is believed that there is a distinct relationship between pore pressure and volume change which the observed data of any compression test should follow if the pore pressure is initially equal-

---

[1] Gibbs, H. J., and Coffey, C. T., "Techniques for Pore Pressure Measurements and Shear Testing of Soil," Proc. Seventh International Conference on Soil Mechanics and Foundation Engineering, Mexico, vol. 1, 1969, p. 151. (Also published as "Application of Pore Pressure Measurements to Shear Strength of Cohesive Soils," Report No. E.M.–761, Department of the Interior, Bureau of Reclamation, June 1969.)

ized throughout the test specimen. Also, it is considered that during tests on undrained (sealed) specimens the pore pressure will regain equalization quite rapidly because there is not a major flow of fluids within the test specimen. Thus, for practical testing procedures this means that sealed tests may be performed in a short time such as a normal work day and end plates may be utilized for measuring pore pressure during testing.

2. Pore-Pressure Analysis.—In a sealed unsaturated specimen the voids contain partly water and partly air. The air can develop pressure, u_a, as it compresses and likewise the water will develop pressure, u_w, as additional load is applied. However, the water pressure is always lower (see fig. 17–1a and equation 1) and may even be held negative as a result of the capillary suction of the water films in the small soil pores. The fluid pressures must be in equilibrium or there will be flow until they are in equilibrium, that is:

$$u_c = -(u_a - u_w) \tag{1}$$

where: u_c = the portion of fluid pressure that assimilates capillary stress and is always negative. However, it approaches zero as compression increases enough to cause saturation.

The theoretical evaluation of pore-air pressure, u_a, as shown by Hamilton[2] considers the air volume to be the volume of free air plus the dissolved air according to Henry's Law. Then, using the principle of Boyle's Law at constant temperature:

$$P_a(V_{ao} + hV_w) = (P_a + u_a)\ (V_{ac} + hV_w),\ \text{and solving for}\ u_a,$$

$$u_a = P_a \frac{V_{ao} + hV_w}{V_{ac} + hV_w} - 1 \tag{2}$$

where: P_a = atmospheric pressure,
       $h = 0.02$, the coefficient of air solubility in water by volume according to Henry's Law,
       V_{ao} = initial volume of free air in percent of initial volume,
       V_{ac} = final volume of free air in percent of initial volume, and
       V_w = volume of water in percent of initial volume.

Hilf[3] suggested a more convenient form of this equation by using percent volume change Δ_v, which is equal to $(V_{ao} - V_{ac})$.

---

[2] Hamilton, L. W., "The Effect of Internal Hydrostatic Pressure on the Shearing Strength of Soils," Proc. ASTM, vol. 39, p. 1,100, 1939.

[3] Hilf, J. W., "An Investigation of Pore-Water Pressure in Compacted Cohesive Soils." (Doctoral Thesis, University of Colorado), Technical Memorandum No. 654, Bureau of Reclamation, Denver, Colo., October 1956

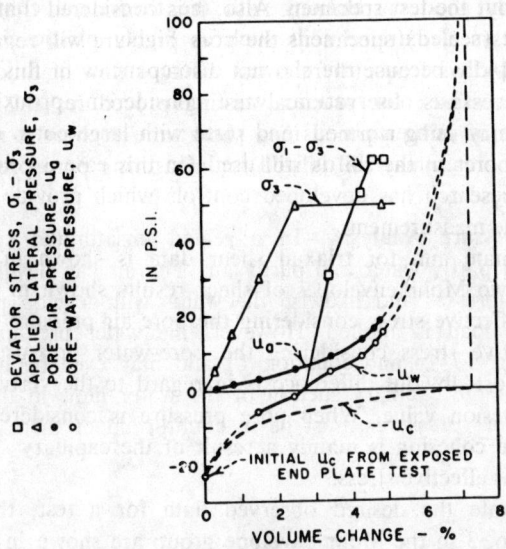

(a) Data for specimen 3

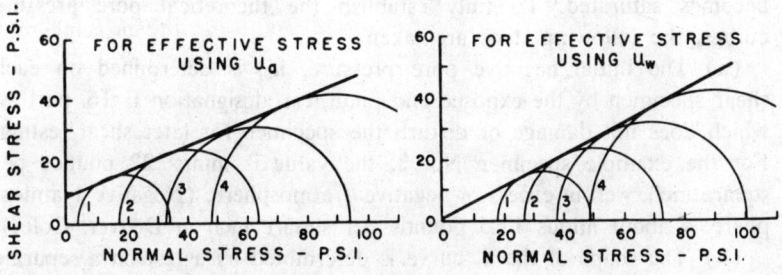

(b) Shear envelopes for the two pore-
pressure values (u_a and u_w).

Figure 17–1.—Triaxial shear stress test data. 101–D–562.

$$u_a = \frac{P_a \, \Delta_v}{V_{ao} + h V_w - \Delta_v} \qquad (3)$$

The early test measurements of pore pressure after about 1939 had been essentially that of pore-air pressure. The end plates contained drilled perforations that were obviously quite large in comparison to the soil pores. These large pores were incapable of transmitting the effects of the capillary menisci. Although tests on moderate soil types showed fairly good correlations with the theoretical computations for pore-air pressure, extreme soils such as highly plastic clays or highly consolidated clays frequently showed discrepancies in this measurement and pore pressure observations were considered approximate. Today the principle of using a porous end plate with large pores relative to the size of the pores in the soil is still used (in this case a coarse ceramic), but recent research has developed controls which provide the precision desired in the measurement.

The ultimate aim for triaxial shear data is shown in figure 17–1. There are two Mohr envelopes of shear results shown in figure 17–1b; one is for effective stress considering the pore-air pressure and the other is for effective stress considering the pore-water pressure. As can be seen, the most obvious difference is in regard to the vertical axis intercept or cohesion value. When pore pressure is considered as pore-air pressure, the cohesion is mainly a result of the capillary stress which is considered as effective stress.

To illustrate the desired observed data for a test, the results for specimen No. 3 in the Mohr envelope group are shown in figure 17–1a. All pressure observations are plotted against volume change. Firstly, the theoretical curves are established. A sealed specimen is limited to a maximum volume change equal to the volume of air voids and this is shown by the vertical dashed line. At this volume change the specimen becomes saturated. To fully establish the theoretical pore pressure curves, the following steps are taken:

(a) The initial negative pore pressure, u_c, is determined on each shear specimen by the exposed end plate test, designation E–16, part A, which does not damage or disturb the specimen for later shear testing. For the example specimen No. 3, the value is minus 22 pounds per square inch, well in excess of negative 1 atmosphere. (Negative 1 atmosphere is about minus 12.5 pounds per square inch in Denver, Colo.).

(b) The shape of the u_c curve is determined by a test on a separate specimen prepared as a duplicate of the regular shear specimen. This test is described in designation E–16, part B.

(c) The theoretical pore-air pressure line, u_a, is computed using equation 3.

(d) The theoretical pore-water-pressure line, u_w, is computed using equation 1, in which the u_c values have previously been obtained from a confined compression test.

These theoretical curves are guides for analyzing the observed data, and controlling the rate of testing in triaxial shear. Test procedures for measuring soil suction, also referred to as capillary pressure and as negative pore pressure, by the exposed end plate method and the change in this negative pore pressure of partially saturated soils with volume change have been presented in detail under parts A and B of designation E–16.

3. Pore-air-pressure Measurement.—The correctness of the pore-air-pressure measurement is dependent on the "breakline" or separation of the menisci between the water of the soil and the water of the end plate. If this does not occur, it is visualized that several sources of discrepancies in observations could result. For example, an intermixing of air and water between the end plate and the soil could cause an unsettled flow condition, resulting in a false reading, or a change of actual moisture in the soil could occur. A high degree of saturation could develop near the end plate, causing excessively high pore pressure as loading compresses the specimen against the end plate. On the other hand, there could be a high suction in the fine-grained soil pulling water from the end plate, resulting in a reduction of the air pressure and larger air voids at the contact that would cause lower than expected pore pressure measurements. Although many moderately acting soils have shown fair results, the above examples are speculative conditions in the rather difficult soils, which indicate the need of more precise control of this test measurement. The usual end plate method of test is illustrated diagrammatically in figure 17–2. There are two important features of this measuring procedure:

(a) The water control burette, which is a small bore size, is maintained at an exact level which the operator establishes to give a very small suction on the end plate surface and, thus, a very slight menisci curvature amounting to a negative pressure of about 0.5 pound per square inch. This would be the zero reading from which all remaining air-pressure readings are measured. This level is maintained precisely with pressure or suction by an air control injector as required, because any movement up or down would indicate water either being taken out of or being forced into the end plate, respectively.

(b) To ascertain further a "breakline" between the water of the soil and that of the end plate, a fine (No. 200 mesh) screen which has been sprayed with silicone mold release to minimize its effect on the surface water tension is placed between the end plate and the soil as shown in figure 17–2.

4. Summary of the Triaxial Shear Testing Method.—The triaxial shear test chamber in figure 17–3 incorporates the previously mentioned testing features. The bottom end plate contains a coarse ceramic stone and is fitted with the silicone treated No. 200 mesh screen and water burette controlled at a precise water level. The top end plate contains a fine ceramic stone capable of registering suction pressure. However, this end plate cannot begin its measurement until the pore-water suction

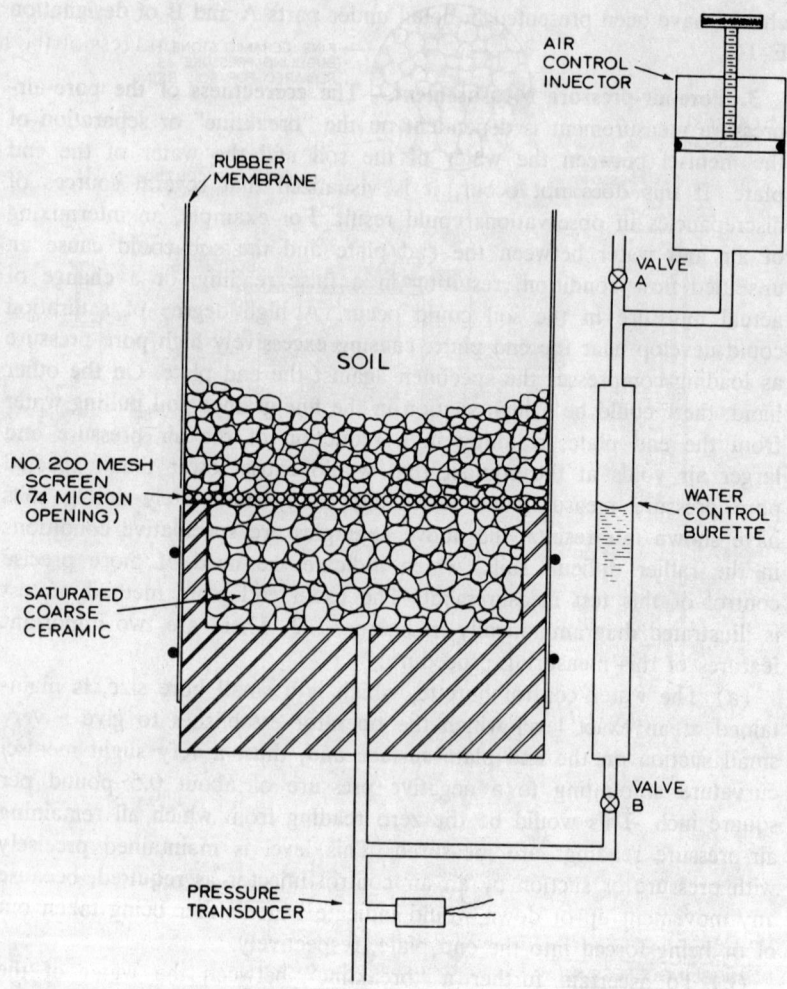

Figure 17–2.—Pore-air pressure measuring system with positive separation between soil and ceramic end plate. 101–D–563.

is less than 1 atmosphere. In the meantime, to protect the end plate from cavitation when the pore-water suction exceeds 1 atmosphere, an end plate control injector is provided in the waterline so that the operator can flood the end plate enough to keep the suction from causing cavitation. When compression is enough to cause the pore-water suction to be within the limitation of less than 1 atmosphere, direct pressure readings can be made for the remainder of the test. Referring back to

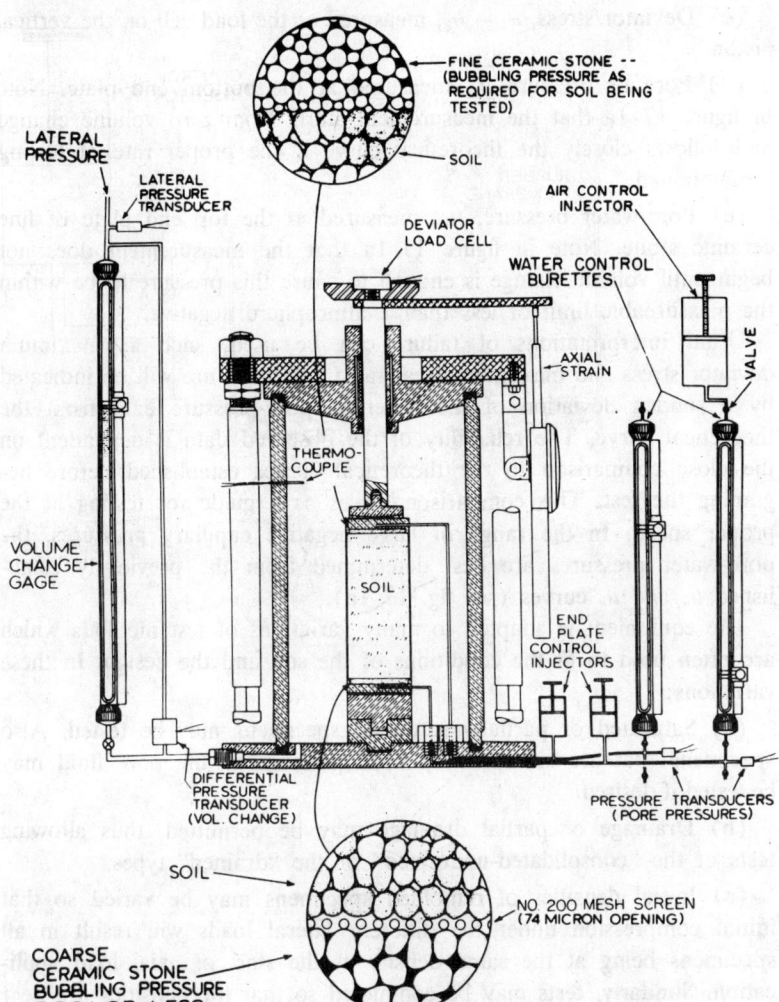

Figure 17–3.—Triaxial shear test chamber. 101–D–564.

the example data in figure 17–1 and apparatus in figure 17–3, the following data are observed:

(a) Volume change by observing the water level in the volume-change gage tube or the value shown by a differential pressure transducer.

(b) Lateral pressure, σ_3, applied to the top of the volume-change tube.

(c) Deviator stress, $\sigma_1 - \sigma_3$, measured at the load cell on the vertical piston.

(d) Pore-air pressure u_a, measured at the bottom end plate. Note in figure 17–1a that the measurement starts from zero volume change and follows closely the theoretical curve if the proper rate of testing is maintained.

(e) Pore-water pressure, u_w, measured at the top end plate of fine ceramic stone. Note in figure 17–1a that the measurement does not begin until volume change is enough to cause this pressure to be within the measureable limit of less than 1 atmosphere negative.

Usual interpretations of failure can be used, such as maximum deviator stress and maximum stress ratio. Often failure will be indicated by a sudden deviation of the observed pore-pressure data from the theoretical curve. The reliability of the observed data is dependent on the close comparison to the theoretical curves established before beginning the test. This comparison serves as a guide for testing at the proper speed. In the range of large negative capillary pressures, the pore-water pressures are best determined from the previously established u_a and u_c curves (see fig. 17–1a).

The equipment is adapted to many variations of test methods which are often used to fit the conditions of the soil and the design. In these variations:

(a) Saturated or partially saturated specimens may be tested. Also specimens that are saturated by back pressure of the pore fluid may be tested if desired.

(b) Drainage or partial drainage may be permitted, thus allowing tests of the "consolidated-undrained" or the "drained" types.

(c) Initial densities of remolded specimens may be varied so that initial compression under the different lateral loads will result in all specimens being at the same density at the start of axial load application. Similarly, tests may be conducted so that the densities are near the same at failure.

(d) Tests may be conducted with the *effective* lateral pressure held constant [4]. This is done by continually adjusting the applied lateral pressure by an amount to compensate for the pore-air pressure change so that the difference in applied lateral pressure and pore pressure is held constant.

(e) Very special adaptations of the triaxial test include the following:

---

[4] This method of test is described by J. W. Hilf and H. J. Gibbs in "Triaxial Shear Tests Holding Effective Lateral Stress Constant," Proc. Fourth International Conference on Soil Mechanics and Foundation Engineering, London, 1957, vol. 1, p. 156.

Figure 17–4.—Dynamic triaxial testing equipment for testing soils finer than the No. 4 sieve size under pulsating vertical loads. E–2150.

(1) Repetitive loading.—Repetitive vertical loads are applied during the triaxial test to simulate the effects of dynamic or pulsating loads, such as earthquakes (see fig. 17–4). Available results indicate the dynamic soil strength is sometimes different and for low density soil may be substantially lower than that obtained from the more normal static tests. Dense, cohesive, fine-grained soil generally shows a higher strength under dynamic loading while loose, cohesionless sand tends to liquify and exhibit lower strength.

(2) Zero lateral strain tests.—The advent of broader use of finite element analysis has increased the demand for tests yielding modulus of deformation and Poisson's Ratio values. A triaxial test in which the lateral strain is held to no change supplies these values. This test procedure is called the K_o-test in some texts and is described in detail in paragraph 10.

(f) Pore-fluid pressures may be measured at either end plate and inserts have occasionally been placed within the body of the specimen for additional measurements. Specimen drainage is through the end plates when required. The active elements in the end plates are ceramic stones of varying porosities. Successful pore-fluid measurement or drainage requires careful and complete de-airing of the pore-pressure measuring system.

The particular method of testing used is described when presenting the results.

5. Apparatus.—The triaxial shear test is not required for construction control; therefore, the apparatus is not listed in the field laboratory equipment list, designation E–4, and only special or major items of equipment are listed below:

Figure 17–5.—Small triaxial shear machine for testing 1⅜- by 3-inch specimens. E–2211–4NA.

(1) *Shear machine.*—Many triaxial shear machines are available in the Engineering and Research Center laboratory in Denver for testing soil specimens ranging in size from 1⅜ inches in diameter by 3 inches long to 9 inches in diameter by 22½ inches long (see figs. 17–5 through 17–9). A multi-channel electronic recorder (fig. 17–7) is available for recording the values sensed by appropriate devices of pore pressures, strains, loads, volume change, and temperature. The apparatus can be used for measuring and recording pressures of six to eight parameters within 20 seconds, or to cycle continuously. The data may be recorded manually or automatically on magnetic tape. Computer programs for data

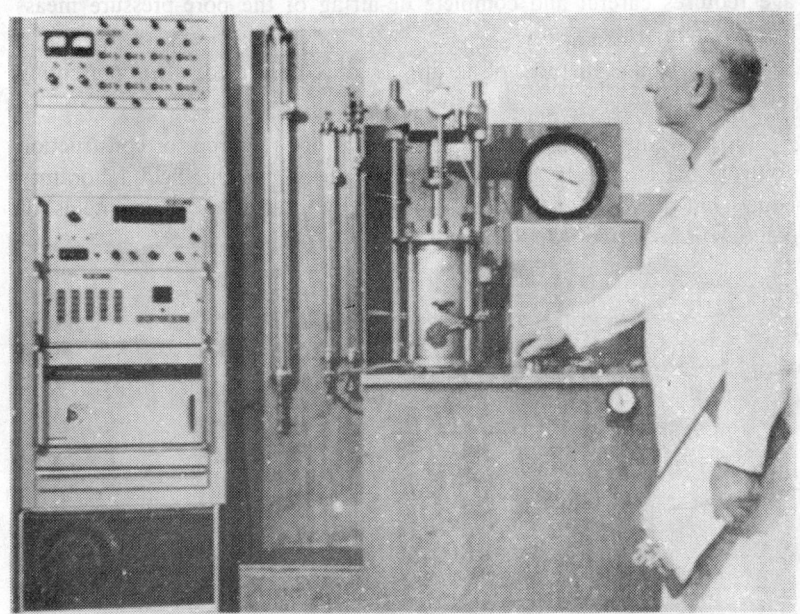

Figure 17–6.—Triaxial shear machine for testing 2- by 5-inch specimens.
E–2282–9NA.

processing are available. The equipment and testing procedures are similar for all machines.

(2) Specimen volume determination apparatus.—Water container, sufficiently large to accommodate the test specimens, and balance and wire carriage for weighing specimens in air and water.

(3) Stand, cutting bits, and tools.—A stand with rotating platform and guide; cutting bits, knives, spatulas, and other miscellaneous equipment for preparing undisturbed specimens.

(4) Specimen molds and tampers.—A three-section mold for preparing remolded specimens.

(5) Miscellaneous.—Machine tools for assembling equipment; and other equipment for preparation, storage, and water content determination.

6. Calibration.—Before using the triaxial shear machine, each of the component parts must be calibrated and checked for proper functioning, insuring that there are no leaks in either the air or water system. Air-free water is used as the medium for transmitting pressures and

Figure 17-7.—High-pressure triaxial shear machine for testing 2- by 5-inch specimens, shown with a data logging system at left of photograph. E-2247-1NA.

showing volume changes; therefore, the pressure chamber and all the waterlines to gages and transducers must be free of air.

The following calibrations are required:

(1) Volume gage tables.—For converting volume change gage or differential pressure transducer readings to volume change in cubic inches.

(2) System expansion calibration.—For determining volume change caused by expansion of the entire system and cylinder under differently applied lateral pressures.

Figure 17-8.—Equipment for testing 6- by 12-inch specimens using chamber pressures up to 1,000 pounds per square inch. PX–D–57614NA.

(3) Deviator load cell.—A calibration table for determining loadings on the specimen from deviator load cell readings. As the universal testing machine has a self-contained calibrated load measuring system, the axial load readings are provided directly so no calibration table is necessary when testing the larger size specimens.

(4) Tables showing the weights and volumes of the different components of the specimen container (rubber sleeves, bands or O-rings, end plates with tubes, porous ceramic stones, and valves filled with water) as determined by weighing in air and then water.

7. **Sample.**—At least three specimens are required for one complete triaxial shear test. Undisturbed soil samples having minimum dimensions of 6 inches are usually required for preparing four test specimens each of 1⅜-inch diameter by 3-inch length or 2-inch diameter by 5-inch length [5]. For preparing four remolded test specimens, 10 pounds of soil passing the No. 4 sieve is required for the 1⅜-inch diameter specimens

---

[5] There is little or no advantage gained in testing larger undisturbed specimens of fine-grained materials. Because of the difficulties in handling and obtaining identical undisturbed specimens that are uniform in density, water content, and structure, the larger specimens are seldom used.

Figure 17–9.—Equipment for triaxial testing of 6- by 12-inch and 9- by 22½-inch specimens up to 250 pounds per square inch chamber pressure. E-2204–11NA.

and 30 pounds of soil for the 2-inch-diameter specimens [6]. For each 6-inch specimen, 50 pounds of soil passing the ¾-inch sieve is required; and for each 9-inch specimen, 150 pounds of soil passing the 3-inch sieve is required. The tabulation below may be used as a guide for selection of specimen size based on the type of soil [7].

8. Procedure.—The procedure and equipment for testing the most-frequently used undisturbed and remolded specimens are described in detail. With minor variations, the operation of equipment and specimen preparation are similar for all machines and other specimen sizes.

(a) *Preparation of Undisturbed Specimen.*—Undisturbed specimens, most frequently the 1⅜-inch size, are cut from undisturbed samples as

---

[6] Because of difficulty in obtaining the precise measurements required to provide the desired degree of accuracy on small quantities of soil, the 1⅜-inch specimen is seldom used for remolded specimens except for special tests wherein special remolding practices are employed.

[7] "First Progress Report of Research on Triaxial Shear Testing of Gravelly Soils," Bureau of Reclamation, Earth Laboratory Report No. EM–350, July 17, 1953. (Also, published paper, "Triaxial Shear Tests on Pervious Gravelly Soils," by W. G. Holtz and H. J. Gibbs, Proc. ASCE Journal, Soil Mechanics and Foundation Division, paper 867, vol. 82, January 1956.

Specimen size, inches	Undisturbed or remolded	Type of soil	Maximum particle size or range of sizes
1⅜ by 3	Undisturbed	Fine-grained	Fine sand
2 by 5	Undisturbed or remolded	Fine- and coarse-grained	No. 4 sieve
6 by 15	Remolded	Coarse-grained	No. 4 to ¾ inch
9 by 22.5	Remolded	Very coarse-grained	¾ to 3 inches

shown in figure 15–4, designation E–15. The excess soil shall be trimmed from the specimen with a knife close to the edge of the cutting bit, leaving very little material to be shaved off by the cutting bit as it passes down over the specimen. Care shall be exercised in trimming the specimen in order to minimize the disturbance of the specimen. When a sufficient length has been cut for the test specimen, it shall be placed in the three-section specimen mold and trimmed with the straightedge trimmer to the 3-inch length. After the specimen has been trimmed to length, a plain end plate shall be placed on top and bottom to protect the edges, and a rubber sleeve shall be placed around the specimen and end plates, using a vacuum expander to enlarge the rubber sleeve. The specimen shall then be weighed in air and in water for weight and volume determinations.

(b) *Preparation of Remolded Specimen.*—The specimen mold (2-inch size selected for description) shall be assembled with an acetate film lining (or treated paper for large specimen) around a plain bottom end plate and clamped to the baseplate. The prepared soil sample at the desired water content shall be thoroughly mixed and a quantity weighed which will give the desired density when compacted in the specimen mold to a depth of 5 inches. The weighed soil shall then be divided into 10 equal parts. Each part shall be placed in the mold and compacted with a ¾-inch-diameter tamper. Following compaction of the specimen, the extension collar shall be carefully removed and the top of the specimen trimmed flush with the top of the mold by means of the straightedge trimmer. The mold and lining should be carefully removed. The rubber sleeve and top and bottom end plates shall be placed on the specimen in the same manner as described for the undisturbed specimen.

(c) *Weight and Volume Determinations.*—The complete specimen assembly (specimen, rubber sleeve, end plates, and bands or O-rings) and wire carriage shall be weighed in air and the weight recorded. Then the assembly and carriage shall be ʷuspended from the balance in a container of water at room temperature to a depth equal to the initial head on the specimen when placed in the triaxial machine. All air bubbles shall be removed from the surfaces of the specimen assembly (usually by a small jet of water from a syringe). Its weight shall then be determined and recorded [8].

(d) *Assembly of Triaxial Machine and Specimen.*—The pore-fluid measuring systems and the volume-change measuring system are thoroughly de-aired using procedures essentially the same as described in designation E–16. The end plates are installed in the machine before the test specimen is placed between them. Then, the rubber membrane is placed over the specimen and the bands or O-rings applied to seal the specimen container. The rubber gasket in the baseplate shall be reset under water to remove any air that might have become trapped. The pressure cylinder shall be placed on the baseplate and filled with air-free water at room temperature. In filling the cylinder, care shall be exercised to insure that no air remains in the cylinder. The cylinder cap shall be set in place with the piston in position to allow a calibrated clearance between the specimen end plate and cap when clamped tight. As the clamp is tightened, the excess water shall be drained from the volume-change gage.

(e) *Volume Change under Lateral Pressure.*—Readings on all gages, pressure cells, and pressure transducers shall be made and recorded before the lateral pressure is applied. The desired lateral pressure shall be applied and the specimen allowed to compress for 15 minutes or longer as required before the application of the axial load. During this period, readings of volume change and pore pressure shall be made and recorded at 1, 5, and 15 minutes after application of the lateral pressure. Following this period, the piston shall be carefully lowered until the specimen with end plates assembled, base, and piston are barely touching as in figure 17–3. This contact will be indicated by a slight increase in the deviator load cell reading. While moving the piston, the piston friction (under that particular lateral load) shall be observed on the deviator load cell and shall be used to correct subsequent deviator load cell readings. Readings of deviator load cell, axial strain, volume-change, and pore pressure shall be made and recorded. Water shall be drained from the volume-change gage to allow adequate volume for piston displacement during the test.

(f) *Axial Loading.*—The desired rate (0.001 to 0.025 inch per minute for most soils) of axial deformation shall be maintained and the

---

[8] Weight of the specimen in water is recorded immediately if no temperature differential exists between the specimen and the water. A 15-minute immersion is recommended if there is a significant temperature difference before weights are recorded.

axial loading of the specimen controlled manually to conform to the rate indicated by a plot of the pore-air pressure and volume-change data similar to that in figure 17–1a. When clays are tested, very slow rates of deformation are necessary to allow equalization of pore pressure throughout the specimen and at the ends where the pore pressure is measured. Readings of the deviator load cell, axial strain, volume change, and pore pressure shall be made and recorded at regular increments of deformation (usually 0.1 or 0.05 inch). Failure may be indicated by a deviator load cell reading which is less, equal to, or even slightly greater than the preceding reading. The pore pressure versus volume-change data plot may serve as a supplemental indication of failure as shown by a deviation from the computed values. A few additional readings are usually made beyond failure.

(g) *Disassembling.*—Following failure of the specimen, the load piston shall be repositioned to permit expansion of the specimen without restraint. The lateral pressure shall be reduced to atmospheric and the specimen allowed to expand until its volume becomes constant. Final readings of deviator load cell, strain, volume change, and pore pressure shall be made and recorded. The specimen assembly shall be removed from the machine and determinations of final weight and volume made and recorded. The entire soil specimen shall be ovendried for the determination of water content.

9. Failure Analysis and Plotting.—The shear values (tan ϕ and c) are determined from the data at the point of specimen failure. Failure has been defined as that stress condition which exists when the ratio of the major principal effective stress $\bar{\sigma}_1$, to the minor principal effective stress $\bar{\sigma}_3$, is a maximum[9]. However, the most recent work places considerable emphasis on the failure which occurs when the pore pressure is near a maximum value. In the method of test in which the lateral effective stress is held constant throughout the test, the maximum principal stress ratio, $\bar{\sigma}_1/\bar{\sigma}_3$, always occurs at the same point as the maximum deviator stress, $\bar{\sigma}_1 - \bar{\sigma}_3$. These two failure criteria are also the same when complete drainage is obtained during a drained test when pore-water pressure is zero.

The shear values are determined from a plot of Mohr's stress circles constructed for failure conditions for three or more specimens of a given sample tested under different lateral pressures (see fig. 17–10). The diameter of the stress circle shall be the deviator stress, $\bar{\sigma}_1 - \bar{\sigma}_3$ at failure, and the center of the circle shall be located a distance from

---

[9] This is called the maximum stress ratio criterion of failure. For reference, see W. G. Holtz, "The Use of the Maximum Principal Stress Ratio as the Failure Criterion in Evaluating Triaxial Shear Tests on Earth Materials", Proc. ASTM, vol. 47, p. 1,067, 1947.

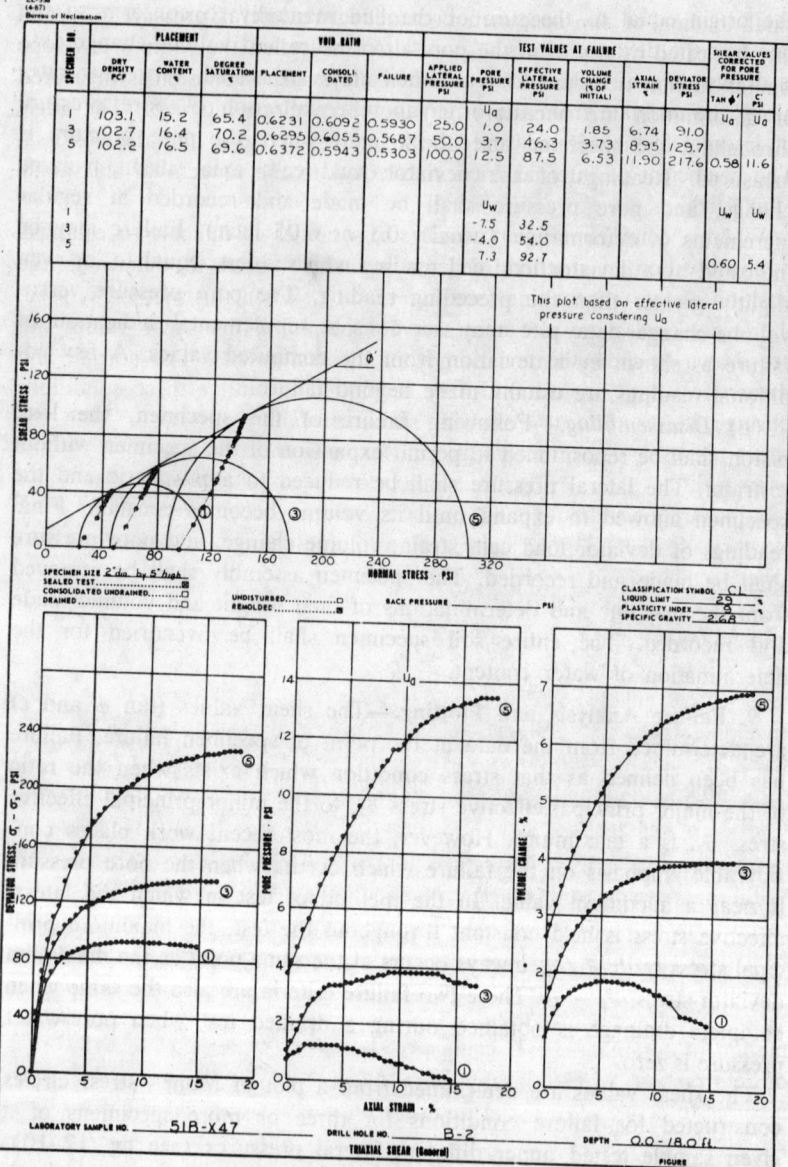

Figure 17–10.—The effect of capillary (suction) pressure on the cohesion and φ values. 101-D-565.

the origin equal to the sum of the effective lateral pressure, $\bar{\sigma}_3$, and one-half the deviator stress, $\bar{\sigma}_1 - \bar{\sigma}_3$. A line called the Mohr's envelope is drawn tangent to the three or more circles. The cohesion is measured along the ordinate from the origin to the envelope intercept. The angle of internal friction is measured between the envelope and the abscissa. The coefficient of internal friction is equal to the tangent of this angle ($\tan \phi$).

When a common tangent to all stress circles is difficult to establish by observation, a method of least squares is used to determine the position of the tangent line (see fig. 17–11). When this method is used, the stress circles are not necessary for obtaining the shear values but are drawn to show the consistency of the test results. The normal and shear stress data or vectors [10] may be plotted to outline these conditions during the test and to assist in specimen failure selection. Both the stress circles and the vectors are shown in figure 17–10.

In addition to the Mohr's envelope, curves may be drawn to show the relationship of volume change, pore pressure, and deviator stress to axial strain. All of these curves for one sample may be plotted on one sheet (see fig. 17–10).

As has been discussed in paragraph 2 of this designation and as shown in figure 17–1, it is possible to obtain two Mohr envelopes of shear results. One is for effective stress considering pore-air pressure and the other is for effective stress considering pore-water pressure. The Mohr envelope with vectors shown in figure 17–10 was plotted for effective stress considering pore-air pressure. The pore-water pressures and corresponding $\tan \phi$ and cohesion values are also shown in the table at the top of the figure.

10. Triaxial Shear Test with Zero Lateral Strain—(K_o-Test).— With the development of electronic computers it has become feasible to perform analyses of stresses in soil masses using powerful numerical analytical techniques such as the finite element method. In order to use this analytical method two of the soil parameters required are Poisson's ratio, μ, and a modulus value, E. Since there is an unrecoverable deformation in obtaining the E value for a soil, it is referred to as modulus of deformation. A method of obtaining the two parameters, μ, and E, for a soil is to perform a triaxial test with no lateral strain, commonly called a K_o-test.

(a) *Theory and Basic Equations.*—The theories used when performing the K_o-test are based on the assumptions that the soil is both isotropic

---

[10] These vector curves are discussed by A. Casagrande and S. D. Wilson in "Prestress Induced in Consolidated-quick Triaxial Tests", Proc. Third International Conference on Soil Mechanics and Foundation Engineering, Switzerland, vol. 1, pp. 106–110, 1953.

EL–563
(7–64)
BUREAU OF RECLAMATION

TRIAXIAL SHEAR TEST - COMPUTATION SHEET 4
(LEAST SQUARES METHOD)

PROJECT _____ Example _____   FEATURE _____

SAMPLE NO. _____   DATE _____

SPEC. NO.	TEST NO.	σ_1	$(\sigma_1)^2$	σ_3	$(\sigma_3)^2$	*$\sigma_1+\sigma_3$	$(\sigma_1\,\sigma_3)$	$\sigma_1(\sigma_1+\sigma_3)$	$\sigma_3(\sigma_1+\sigma_3)$
1	1	115.0	13225.00	24.0	576.00	139.0	2760.00	15985.00	3336.00
3	2	176.0	30976.00	46.3	2143.69	222.3	8148.80	39124.80	10292.49
5	3	305.1	93086.01	87.5	7656.25	392.6	26696.25	119782.26	34352.50
Σ	$\eta=3$	596.1 (1)	137287.01 (2)	157.8 (3)	10375.94 (4)	753.9 (5)	37605.05 (6)	174892.06 (7)	47980.99 (8)

$$[\eta \times (2)] - (1)^2$$
$$= \underline{411861.03} - 355335.21$$
$$= \underline{56525.82}$$
$$(9)$$

$$A^2 = \frac{(9)}{(10)} = \frac{56525.82}{6226.98} = \underline{9.078}$$

$$A = \underline{3.013} \quad \sqrt{A} = \underline{1.736} \quad 2\sqrt{A} = \underline{3.472}$$

$$\text{COHESION} = \frac{(1)-[A \times (3)]}{\eta \, 2\sqrt{A}} = \frac{596.1 - 475.451}{3 \times 3.472} = \frac{120.649}{10.416}$$

$$= \underline{11.58}$$

$$\text{TAN}\phi = \frac{A-1}{2\sqrt{A}} = \frac{2.013}{3.472} = \underline{0.58}$$

$$\phi = \underline{30.1^\circ}$$

$$[\eta \times (4)] - (3)^2$$
$$= \underline{31127.82} - 24900.84$$
$$= \underline{6226.98}$$
$$(10)$$

$$[\eta \times (6)] - [(1) \times (3)]$$
$$= \underline{112815.15} - 94064.58 = \underline{18750.57}$$
$$(11)$$

$$[\eta \times (7)] - [(1) \times (5)]$$
$$= \underline{524676.18} - 449399.79 = \underline{75276.39}$$
$$(12)$$

$$[\eta \times (8)] - [(3) \times (5)]$$
$$= \underline{143942.97} - 118965.42 = \underline{24977.55}$$
$$(13)$$

$$(12) - (11) = (9)$$
$$= \underline{75276.39} - 18750.57 = \underline{56525.82}$$

$$(13) - (11) = (10)$$
$$= \underline{24977.55} - 18750.57 = \underline{6226.98}$$

*ALL COMPUTATIONS TO THE RIGHT OF THE DOUBLE LINE ARE MADE FOR CHECKING RESULTS OBTAINED FOR (9) AND (10).

COMPUTED BY _____ CHECKED BY _____

Figure 17–11.—Least squares method of computations for coefficient of internal friction and cohesion. 101–D–566.

and homogeneous and that the ratio between a linear stress, σ, and the corresponding linear strain, ε, is a useable relationship in the form of a modulus.

$$\frac{\sigma}{\varepsilon} = E \tag{4}$$

The positive vertical strain produced by a vertical pressure is associated with a negative horizontal strain, ε_1. The absolute value of the ratio between the strains is called Poisson's ratio, μ.

$$\mu = \frac{\varepsilon_1}{\varepsilon} = \frac{\varepsilon_1}{\sigma} E \tag{5}$$

In a K_o-test there is to be no lateral strain, so in an element of soil which is laterally confined and compressed by a vertical pressure, σ_y, there will be compression stresses σ_x and σ_z which act on the sides of the element (see fig. 17–12).

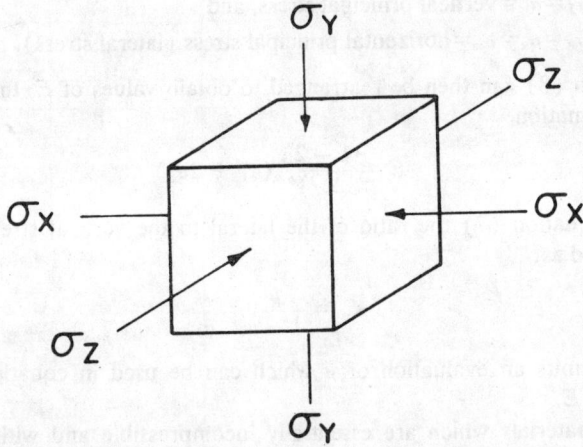

Figure 17–12.—An element of soil which is laterally confined (no lateral strain) and compressed by a vertical pressure, σ_y, which results in compression stresses, σ_x, and σ_z, which act on the sides of the element. 101–D–567.

The stress acting on the side of an element will cause strain in the direction of the stress as well as strains in the two directions perpendicular to the stress. The strain in the perpendicular directions is equal to the strain in the direction of stress times Poisson's ratio, μ. Since lateral displacement is not permitted in a K_o-test, the stresses in the x (or z) direction are in equilibrium so that:

$$\sigma_x - \mu\sigma_y - \mu\sigma_z = 0 \tag{6}$$

The K_o-test is performed on a cylindrical specimen which is symmetrical; therefore, $\sigma_x = \sigma_z$. From equation (6), the ratio K_o of the lateral stresses σ_x or σ_z to the vertical stress σ_y is:

$$\frac{\sigma_x}{\sigma_y} = \frac{\sigma_z}{\sigma_y} = \frac{\mu}{1 - \mu} = K_o \tag{7}$$

The volume change per unit of volume of a soil specimen resulting from these applied stresses is equal to the sum of the volume change per unit of volume produced by each stress individually. Since the specimen on which K_o-tests are performed are cylindrical, $\sigma_x = \sigma_z$, and the state of stress has circular symmetry about the σ_y axis. The volume change Δ_v/V per unit of volume for such a configuration is:

$$\Delta_v/V = \frac{1 - 2\mu}{E} (\sigma_y + 2\sigma_x) \tag{8}$$

A more common notation of these stresses might be:

$\sigma_1 = \sigma_y =$ vertical principal stress, and

$\sigma_3 = \sigma_x = \sigma_z =$ horizontal principal stress (lateral stress).

Equation (8) can then be rearranged to obtain values of E, the modulus of deformation:

$$E = \frac{1 - 2\mu}{\Delta_v/V} (\sigma_1 + 2\sigma_3) \tag{9}$$

From equation (7) the ratio of the lateral to the vertical stress may be expressed as:

$$K_o = \frac{\mu}{1 - \mu} = \frac{\sigma_3}{\sigma_1} \tag{10}$$

This permits an evaluation of μ which can be used in equation (9) to evaluate E.

For materials which are essentially incompressible and without shear strength, the value of μ is 0.5 and the corresponding value of K_o is 1.0. For soils, K_o normally corresponds to the coefficient of earth pressure at rest in the theory of earth pressure.

(b) *Determination of the Constants, E and μ.*—Triaxial tests with no lateral strain may be performed on undrained or consolidated undrained specimens which are partly saturated and develop shear strength. Pore-pressure measurements are made so that effective stresses may be evaluated. Since no flow of pore water occurs, equilibrium pressures are rapidly established in the pore spaces, even in very fine-grained materials of low permeability. The range of effective stress change is limited to that causing a volume change equal to the initial free-air volume.

It is also possible to perform drained tests on saturated, partly satu-

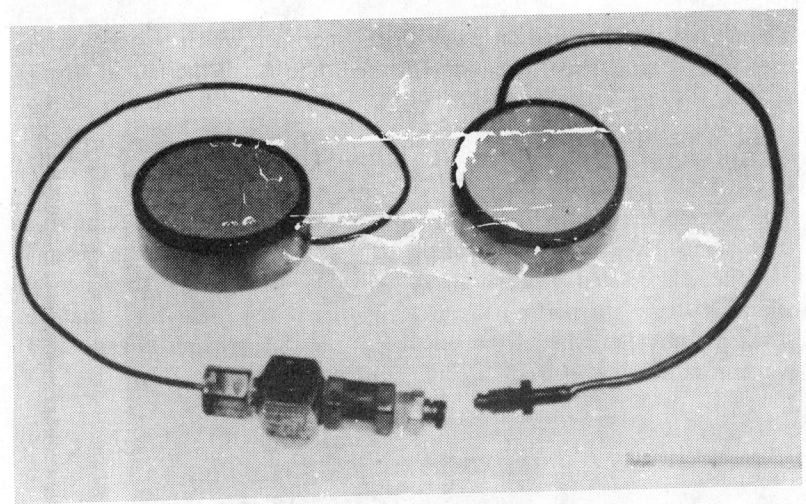

Figure 17–13.—Specimen end plates. E–2282–9.

rated, or dry specimens. Since the application of lateral pressure is controlled manually, it generally would not be practical to perform a drained test on soil of low permeability because of the time required to assure complete pore-pressure dissipation. A drained test would, however, be performed on a sand or gravel material in a reasonable length of time.

Preparation of the specimen and assembly of the equipment is performed in exactly the same manner as that for a standard triaxial shear test previously discussed. Pore-pressure measurements are made through saturated ceramic stones in the end plates and volume measurements are determined through the use of a volume-change gage read directly or by special arrangement of a differential pressure transducer, as described for the standard triaxial shear test and shown in figure 17–3.

The top and bottom end plates are shown in figure 17–13. The top end plate has a fine ceramic stone insert and the bottom a coarse ceramic stone. A soil specimen with end plates in place is shown in figure 17–14A, with a soil specimen sealed in a rubber membrane in figure 17–14B.

The rubber-sheathed specimen is placed in a standard triaxial shear pressure chamber and surrounded by air-free water. The chamber is then sealed and the axial (vertical) loading piston is brought just to contact with the specimen as in figure 17–3. The initial pore pressure, volume change, and load cell readings are recorded and the test is begun by slowly loading the axial piston. As the specimen is loaded axially the cell pressure, (σ_3), is increased to maintain zero lateral strain. The required increase in σ_3 is computed so that volume strain is equal to only that in the vertical direction. Thus, volume

(B) SOIL SPECIMEN IN RUBBER MEMBRANE

(A) SOIL SPECIMEN WITH END PLATES

Figure 17-14.—Soil specimen with (A) end plates in place, and (B) enclosed with a rubber membrane. E-2282-5NA and E-2282-4NA.

Figure 17–15.—Soil specimen with girth gage. E–2282–11.

change, or strain, only occurs in the vertical or axial direction so that volume change per unit volume and axial strain per unit length are actually the same value. The lateral strain may be monitored by either the volume-change gage or the girth gage. When using the volume-change gage for control, it is necessary to know the volume change due to the axial load piston movement and also to system expansion under varying lateral pressure. When the girth gage is used for lateral strain control, it is read directly and the setup shown in figures 17–15, 17–16, and 17–17 would be used. If the specimen expands or contracts laterally during testing, it immediately registers through the electronic strain device and the lateral pressure is adjusted accordingly. Readings of axial strain, pore pressures, lateral pressure, load, and volume change are observed and recorded as the test progresses. The test may only be continued until the volume change is equal to the initial volume of air in the specimen. As complete saturation is approached, the lateral pressure, σ_3, necessary to prevent lateral deformation increases very rapidly (see fig. 17–18). The axial loading is stopped at this point and the test is terminated. The lateral, or cell, pressure is then slowly released and the axial piston

**Figure 17–16.—Triaxial apparatus with cylinder, cap, and piston in place.
E-2282-12NA.**

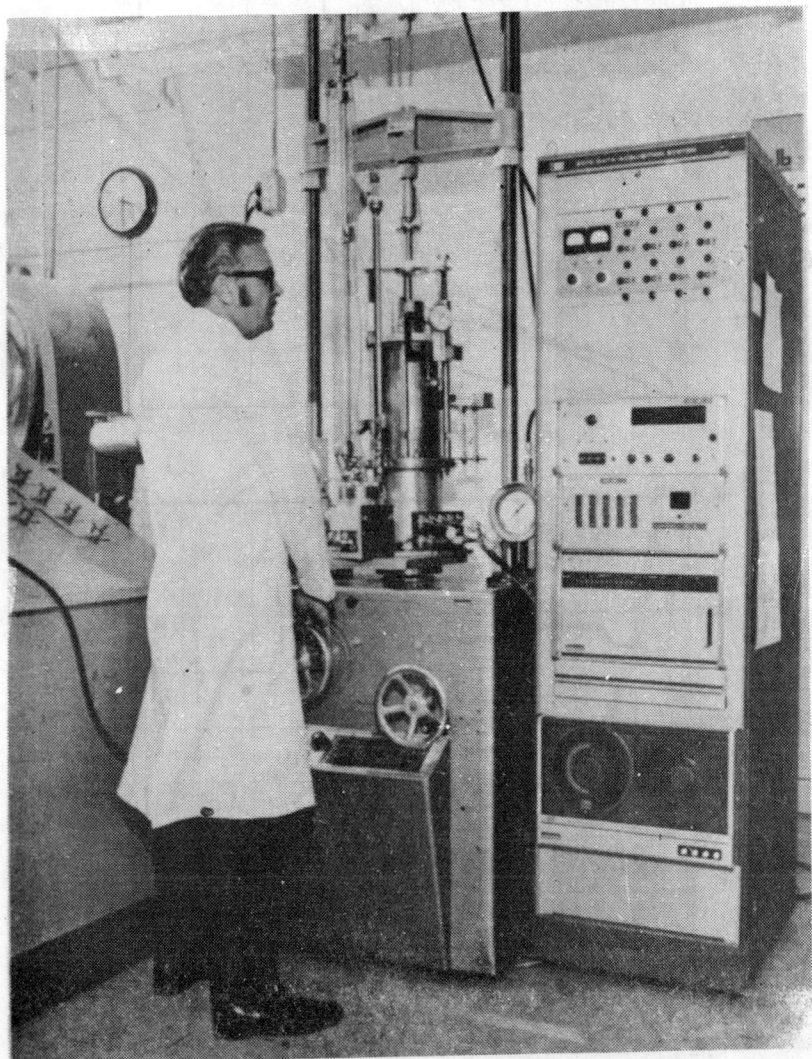

Figure 17–17.—Triaxial machine with girth gage. E–2253–1.

is lifted slightly to allow observation of the amount of rebound of the specimen. After the specimen is removed from the pressure cell, it is again weighed to determine its final volume and a final moisture content determination is also made. Results of a typical K_o-test are plotted in figure 17–18.

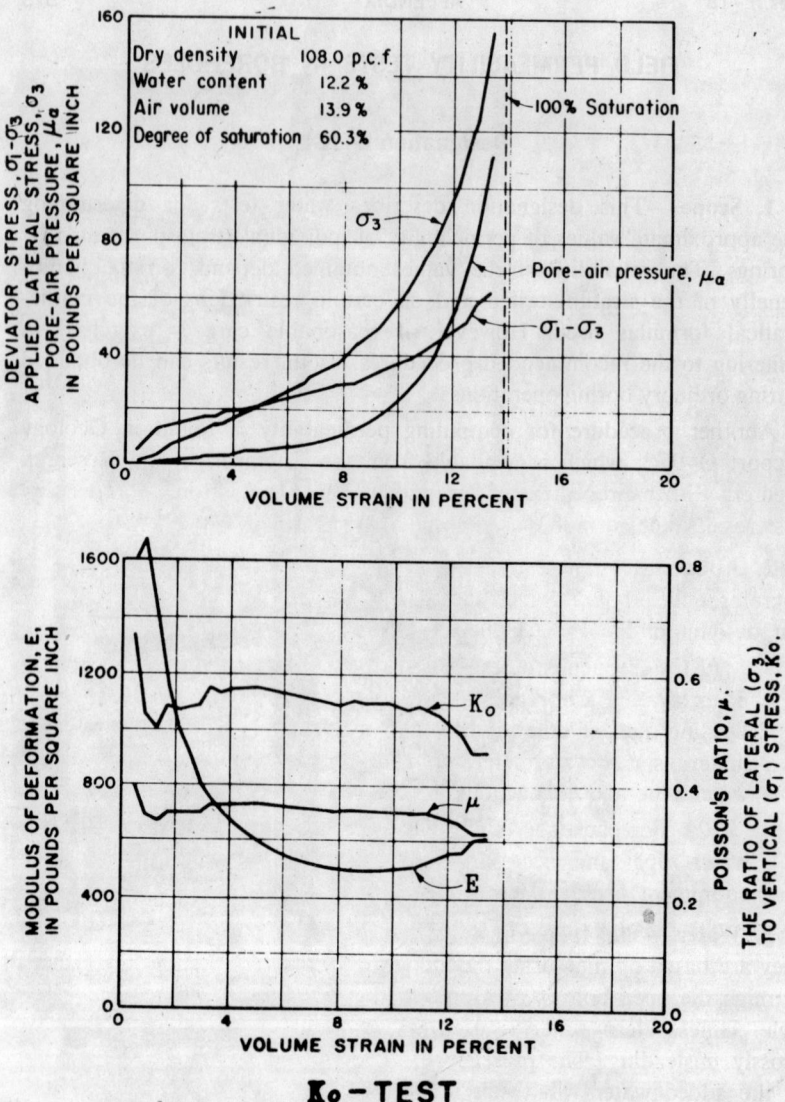

Figure 17–18.—Results of a typical Kₒ-test. 101–D–568.

The K_o-test has been successfully performed on many types of materials, ranging from clay to gravel of 3-inch maximum size. The specimen for the various types of material range in size from 1⅜ inches in diameter by 3 inches high to those 9 inches in diameter by 22½ inches high. The test is normally performed on a strain controlled basis and the rate of strain is dependent upon the soil type and specimen size.

FIELD PERMEABILITY TESTS IN BOREHOLES

Designation E–18

1. Scope.—This designation describes water tests for determining the approximate values of permeability of individual strata penetrated by borings. The reliability of the values obtained depends on the homogeneity of the stratum tested and on certain restrictions of the mathematical formulas used. However, if reasonable care is exercised in adhering to the recommended procedures, useful results can be obtained during ordinary boring operations.

Another procedure for computing permeability is found in Geology Report G–97 [1], which is available from the Engineering and Research Center. Either procedure is acceptable. When submitting permeability test results, the formulas used should be included with the data.

2. Apparatus.—These tests are not required for construction control; therefore, the apparatus is not listed in the field laboratory equipment list, designation E–4.

(1) For open-end tests (fig. 18–1), a drill rig or other means of excavating a borehole and driving pipe casing is needed. (*Note: Drilling mud or other additives must not be used in unconsolidated materials.*) A watermeter, pressure gage, pump, and the necessary water pipe and connections are also required.

(2) For packer tests (fig. 18–2), a supply of packers, perforated water pipe, and necessary fittings are needed in addition to the equipment listed under (1) above.

3. Water.—The following tests are of the pumping-in type, that is, they are based on measuring the amount of water accepted by the ground through the open bottom of a pipe or through an uncased section of the hole. Unless clear water is used, these tests are invalid and may be grossly misleading. The presence of even small amounts of silt or clay in the added water will result in plugging of the test section and give permeability results that are too low. By means of a settling tank or a filter, efforts should be made to assure that only clear water is used. It is desirable for the temperature of the added water to be higher than ground-water temperature, so as to preclude the creation of air bubbles in the ground which may greatly reduce the acceptance of water.

4. Open-End Tests.—(a) *Procedure.*—Figures 18–1 (A) and (B) show tests made through the open end of a pipe casing which has been

---

[1] "Permeability Tests Using Drill Holes and Wells", Geology Report No. G–97, Research and Geology Division, Bureau of Reclamation, January 3, 1951.

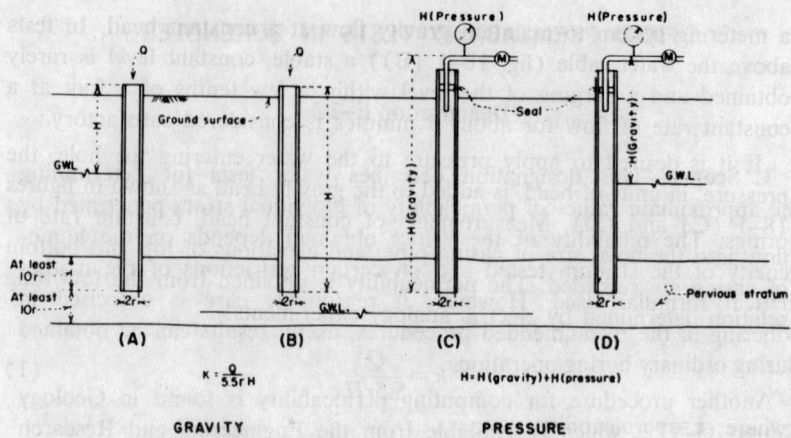

Figure 18–1.—An open-end pipe test for soil permeability which can be made
in the field. PX–D–16267.

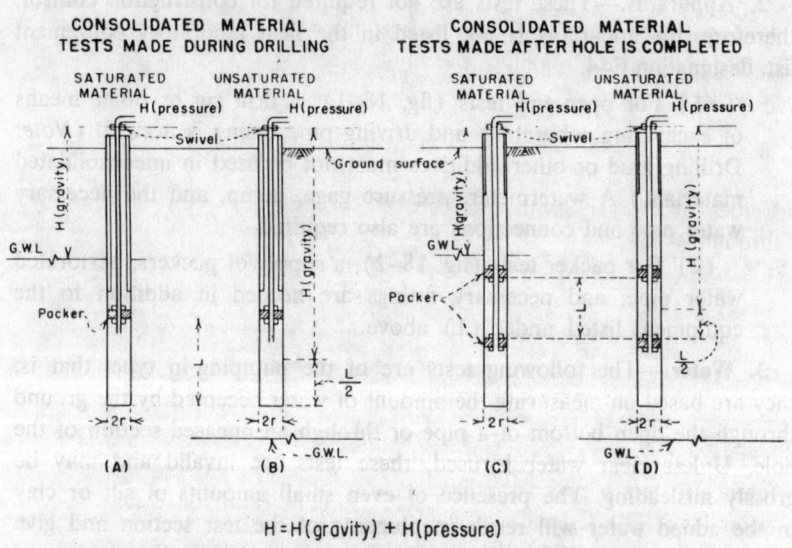

Figure 18–2.—The packer test for soil permeability. PX–D–4785.

sunk to the desired depth and which has been carefully cleaned out
just to the bottom of the casing. When the hole extends below the ground-
water table, it is recommended that the hole be kept filled with water
during cleaning and especially during withdrawal of tools to avoid
squeezing of soil into the bottom of the pipe. After the hole is cleaned
to the proper depth, the test is begun by adding clear water through

a metering system to maintain gravity flow at a constant head. In tests above the water table (fig. 18–1 (B)) a stable, constant level is rarely obtained and a surging of the level within a few tenths of a foot at a constant rate of flow for about 5 minutes is considered satisfactory.

If it is desired to apply pressure to the water entering the hole, the pressure, in units of head, is added to the gravity head as shown in figures 18–1 (C) and (D). Measurements of constant head, constant rate of flow into the hole, size of casing pipe, and elevations of top and bottom of casing are recorded. The permeability is obtained from the following relation determined by electric analogy experiments:

$$k = \frac{Q}{5.5rH} \qquad (1)$$

where: k = permeability,

Q = constant rate of flow into the hole,
r = internal radius of casing, and
H = differential head of water.

Any consistent set of units may be used. For convenience, if k is measured in feet per year, Q in gallons per minute, and H in feet, equation (1) can be written:

$$k = C_1 \frac{Q}{H}$$

Values of C_1 vary with the size of casing as follows (see figs. 2–18 through 2–22):

Size of casing	EX	AX	BX	NX
C_1	204,000	160,000	129,000	102,000

The value of H for gravity tests made below water table is the difference in feet between the level of water in the casing and the ground-water level. For tests above water table, H is the depth of water in the hole. For pressure tests the applied pressure in feet of water (1 pound per square inch = 2.31 feet) is added to the gravity head to obtain H.

(b) *Example for Condition Shown in Figure 18–1 (A)*.—
Given: NX casing

Q = 10.1 gallons per minute
H = 21.4 feet

$$k = C_1 \frac{Q}{H} = \frac{(102,000)\,(10.1)}{21.4} = 48,100 \text{ feet per year.}$$

(c) *Example for Condition Shown in Figure 18–1 (D).—*

Given: NX casing

$Q = 7$ gallons per minute
H (gravity) $= 24.6$ feet
H (pressure) $= 5$ p.s.i. $= 5 \times 2.31 = 11.6$ feet of water.

Then: $H = 24.6 + 11.6 = 36.2$ feet

$$k = C_1 \frac{Q}{H} = \frac{(102,000)\ (7)}{36.2} = 19,700 \text{ feet per year.}$$

5. Packer Tests.—(a) *Procedure.*—Figure 18–2 shows a permeability test made in a portion of a drill hole below the casing. This test can be made both above and below the water table provided the hole will remain open. It is commonly used for pressure testing of bedrock using packers, but it can be used in unconsolidated materials where a top packer is placed just inside the casing. When the packer is placed inside the casing, measures must be taken to properly seal the annular space between the casing and drill hole wall to prevent water under pressure from escaping.

The formulas for this test are:

$$k = \frac{Q}{2\pi LH}\ \log_e \frac{L}{r}\ ;\ L \geqq 10r \tag{2}$$

$$k = \frac{Q}{2\pi LH}\ \sinh^{-1} \frac{L}{2r},\ 10r > L \geqq r \tag{3}$$

where: $k =$ permeability,
$Q =$ constant rate of flow into the hole,
$L =$ length of the portion of the hole tested,
$H =$ differential head of water,
$r =$ radius of hole tested,
$\log_e =$ natural logarithm, and
$\sinh^{-1} =$ inverse hyperbolic sine.

These formulas have best validity when the thickness of the stratum tested is at least $5L$, and they are considered to be more accurate for tests below ground-water table than above it.

For convenience, the formulas can be written:

$$k = C_p \frac{Q}{H}$$

where k is in feet per year, Q is in gallons per minute, and H is the head of water in feet acting on the test length. Where the test length is below the water table, H is the distance in feet from the water table to the water swivel (see fig. 18–2) plus applied pressure in units of feet of water. Where the test length is above the water table, H is the distance in feet from the center of the length tested to the swivel plus the applied pressure in units of feet of water. For gravity tests (no applied pressure) measurements for H are made to the water level inside the casing (usually the level of the ground).

Values of C_p are given in the following table for various lengths of test sections and hole diameters:

Length of test section in feet, L	C_p values			
	Diameter of test hole			
	EX	AX	BX	NX
1	31,000	28,500	25,800	23,300
2	19,400	18,100	16,800	15,500
3	14,400	13,600	12,700	11,800
4	11,600	11,000	10,300	9,700
5	9,800	9,300	8,800	8,200
6	8,500	8,100	7,600	7,200
7	7,500	7,200	6,800	6,400
8	6,800	6,500	6,100	5,800
9	6,200	5,900	5,600	5,300
10	5,700	5,400	5,200	4,900
15	4,100	3,900	3,700	3,600
20	3,200	3,100	3,000	2,800

The usual procedure is to drill the hole, remove the core barrel or other tool, seat the packer, make the test, remove the packer, drill the hole deeper, set the packer again to test the newly drilled section, and repeat the test (see fig. 18–2 (A)). If the hole stands without casing, a common procedure is to drill it to final depth, fill with water, surge it, and bail it out. Then set two packers on pipe or drill stem as shown in figures 18–2 (C) and (D). The length of packer when expanded should be at least five times the diameter of the hole. The bottom of the pipe holding the packer must be plugged and its perforated portion must be between the packers. In testing between two packers, it is desirable to start from the bottom of the hole and work upward.

(b) Example for Condition Shown in Figure 18–2 (A).—

Given: NX casing set to depth of 5 feet
$Q = 2.2$ gallons per minute
$L = 1$ foot
H (gravity) = distance from ground-water level to swivel
= 3.5 feet

$$H \text{ (pressure)} = 5 \text{ p.s.i.} \times 2.31 = 11.6 \text{ feet of water}$$
$$H = H \text{ (gravity)} + H \text{ (pressure)} = 15.1 \text{ feet.}$$

From table, $C_p = 23,300$

$$K = C_p \frac{Q}{H} = \frac{(23,300)(2.2)}{15.1} = 3,400 \text{ feet per year.}$$

FIELD PERMEABILITY TEST (WELL PERMEAMETER METHOD)

Designation E–19

1. Scope.—This designation describes a test method for determining the permeability of a soil in place. The method consists of measuring the rate at which water flows outward from an uncased well under constant head. It is particularly useful for estimating canal seepage prior to construction for predetermining the need for canal lining.

2. Apparatus.—This test is not required for construction control; therefore, the apparatus is not listed in the field laboratory equipment list, designation E–4. The apparatus for a single test shall consist of the following (figs. 19–1 and 19–2):

(1) Augers.—Lightweight power auger or hand auger suitable for excavating permeability wells (fig. 19–3).

(2) Reservoir.—A 50-gallon calibrated metal drum with gage tube.

(3) Valve.—A bob-float valve with operating arm (see fig. 19–4 for size).

(4) Float.—A wooden float with brass stem.

(5) Casing.—A galvanized iron casing for float, 3½ inches in diameter by 12 inches in height.

(6) Counterweights.—Brass counterweights.

3. Materials.—(a) *Density Sand.*—A pervious coarse sand (or fine gravel) shall be used for backfilling test wells used in the well permeameter test. Clean washed sand of No. 4 to No. 8 size or gravel of ⅜-inch to No. 4 size is recommended for this purpose.

(b) *Water.*—The water to be used for well permeameter tests shall be clean. Small amounts of sediment or other suspended matter in the water will become deposited in the soil adjacent to the well and greatly reduce the flow. The water should preferably be from the same source as that expected to be used in the canal. In some soils, a base exchange reaction will take place between a particular water and soil which may increase or decrease soil permeability; therefore, using water from a

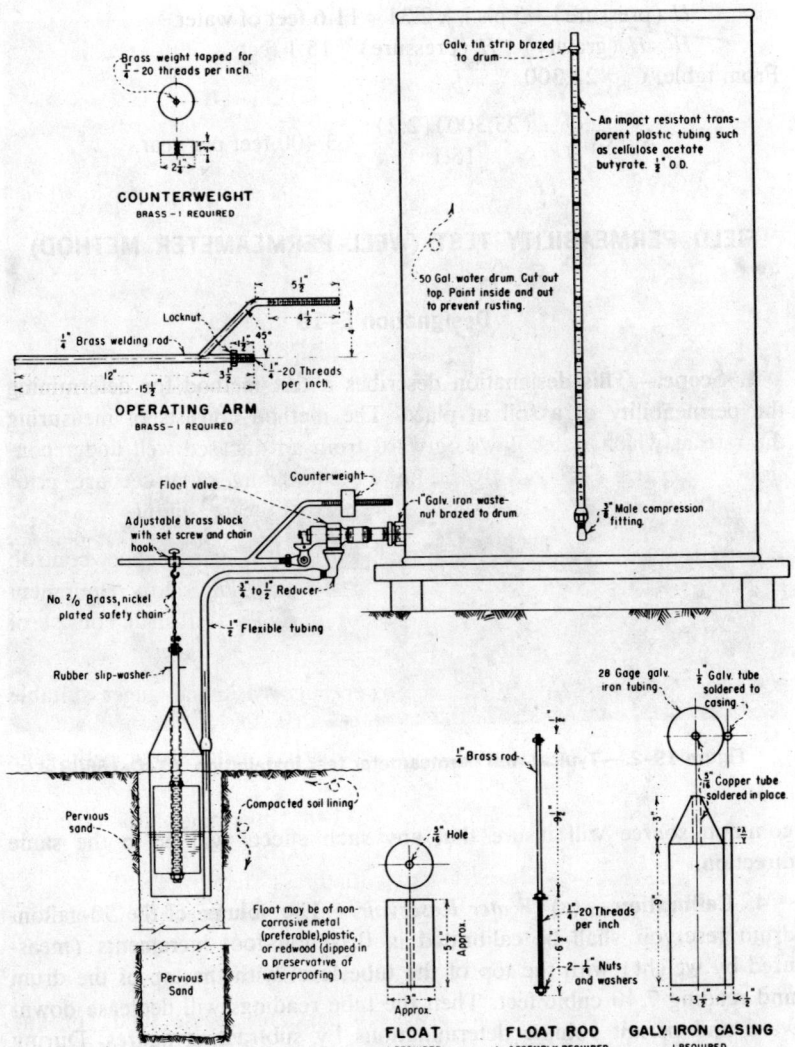

WELL PERMEAMETER EQUIPMENT
FOR
ESTIMATION OF CANAL SEEPAGE LOSSES

Figure 19–1.—Drawing of well permeameter test apparatus. 101–D–38.

Figure 19–2.—Typical well permeameter test installation. PX–D–16608.

common source will insure that any such effects will be in the same direction.

4. Calibration.—(a) *Water Reservoir.*—The volume of the 50-gallon-drum reservoir shall be calibrated in 0.1-cubic-foot increments (measured by weight) with the top of the tube level with the top of the drum and reading 7.40 cubic feet. Then the tube readings will decrease downward and permit volume determinations by subtracting figures. During calibration, after each increment of water is drawn off, a mark is placed on the plastic tubing. (*Note:* An ink with an acetate base will make a permanent mark on the plastic tubing. India ink can be used if the surface of the plastic tube is first rubbed with emery cloth or steel wool. The tube with ink marks shall be given a clear lacquer coating to restore transparency and preserve the ink marks.)

(b) *Density Sand.*—The density of the pervious sand used for filling the test wells shall be determined prior to use (see designation E–24). The density shall be determined by pouring the sand into a pipe with dimensions approximately those of the test well to be used. The pouring

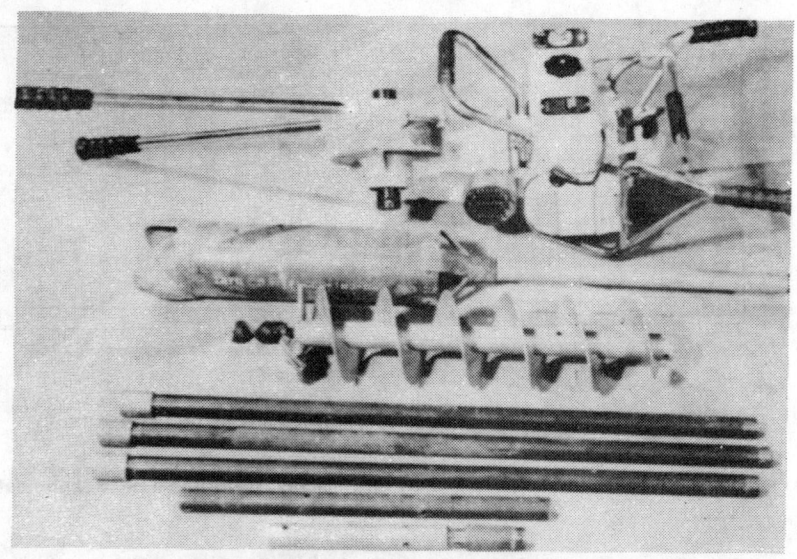

(A) POWER AUGER EQUIPMENT

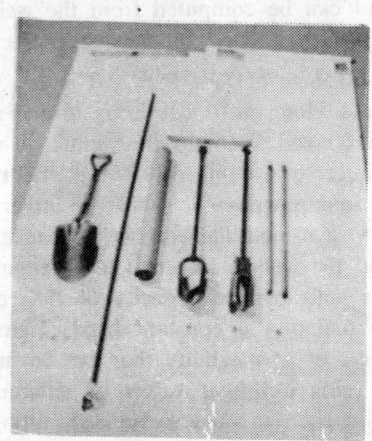

(B) HAND AUGER EQUIPMENT

1. Shovel for starting well and leveling area of reservoir
2. Spudding bar for dislodging rock in well
3. Light metal casing for casing soils
4. Iwan-type auger
5. Seymour expanding-type auger
6. Pipe extensions for auger

Figure 19-3.—Hand and power augers for excavating test wells. PX-D-16610 and PX-D-16695.

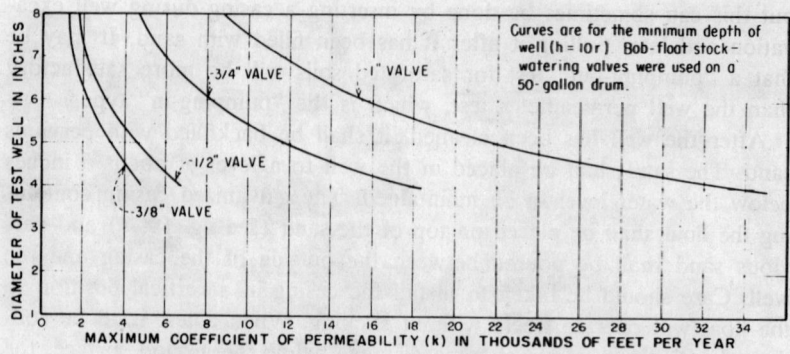

VALVE SIZE REQUIRED TO OBTAIN A GIVEN COEFFICIENT OF PERMEABILITY
FOR TEST WELLS FROM FOUR TO EIGHT INCHES IN DIAMETER

Figure 19–4.—Maximum permeability coefficients measurable with bob-float valves of the sizes shown. 101–D–80.

height shall be approximately that to be used in the well. The calibrated density of the sand can be computed from the weight used to fill the pipe, the depth of sand, and the volume of the pipe.

5. Procedure.—(a) *Size of Test Well.*—The test well may be of any desired dimensions so long as it conforms to the rule that the depth should be between 10 and 50 times the radius. A well excavated with a 4-inch-diameter auger to a depth of 2 feet is about the smallest practicable size, and for some purposes it should be larger. A 6-inch-diameter well is preferable to a 4-inch-diameter well because the volume of soil being tested around the well is larger. The maximum well size will be limited in pervious soils by the capacity of the equipment to supply sufficient water to maintain a constant head. Figure 19–4 shows the maximum coefficients of permeability that can be measured in wells of various diameters using bob-float valves of different sizes. This is of assistance in selecting the size valve to be used, although a ¾-inch valve is often used for general purposes. Also, the plot is of assistance in estimating minimum permeability of soil where the capacity of the valve is exceeded.

(b) *Preparation of Test Well.*—Wells for permeameter tests shall be prepared with care in order to cause as little disturbance to the surrounding soil as possible. They may be excavated with a hand or power auger. After the well is excavated, the sides and bottom shall be lightly brushed or shaved where necessary to remove any accumulation or compaction of the soil, and the loose soil shall be removed from the well bottom. (*Note:* It is often difficult to auger a well below the water table, but this can sometimes be done by inserting a casing during well excavation and later pulling it after it has been filled with sand. It may be that a "pumping-out" test for saturated soils will be more satisfactory than the well permeameter test, which is the "pumping-in" type.)

After the well has been cleaned, it shall be backfilled with pervious sand. The sand shall be placed in the well to a level of about 6 inches below the water level to be maintained. The galvanized casing containing the float shall be placed on top of the sand (see fig. 19–1) and pervious sand shall be poured between the outside of the casing and the well. Care should be taken to install the casing in a vertical position so the float will operate freely without sticking. When a test is to be conducted with the water level some distance below the ground surface, the casing can be lowered on the float by the light chain to the top of the backfill sand, and a little sand dropped around the casing to hold it in place during the test. The rubber slipwasher on the float stem is to prevent falling sand particles from becoming lodged between the float stem and casing guide. The sand around the casing need not be weighed, as it is not considered in computations for well radius. Depth measurements in the test well can be conveniently made from a common baseline formed by stretching a string across the tops of two stakes driven on either side of the well.

(c) *Sand for Test Well.*—The sand (or fine gravel) placed in the test well serves two purposes: First, it serves in place of a casing to support the sides of the well against sloughing during saturation of the soil; and secondly, it provides a means of indirectly measuring the average radius of the well which is necessary for permeability computations but which is otherwise rather difficult to measure accurately. For determining the average radius of the well, the weight of sand used in the well, and depth of sand are recorded. The volume is then computed from the sand density previously determined. The radius is determined from the volume and depth of the cylindrical well filled with sand. A form (7–1429) for recording soil classifications, well dimensions, and the sand weights necessary for computing the well radius is shown in figure 19–5.

(d) *Setting up Test Equipment.*—The permeability equipment should be set up as shown in figures 19–1 and 19–2. The reservoir should be set on a platform or cribbing at a convenient height. The ½-inch tube on the side of the casing can be used as a thermometer well, or the flexible water hose from the float valve can be connected to it. The length of the light chain from the float stem to the valve operating arm should be adjusted and the counterweight positioned to balance the float when it is in water.

(e) *Water Temperature.*—Because of the wide variation in tempera-

7-1429
(7-54)
BUREAU OF RECLAMATION

WELL PERMEAMETER TEST
DETERMINATION OF TEST WELL DIMENSIONS

TEST NO. *Example* GROUND ELEVATION **868.2** DATE **July 1, 1971** MADE BY **W.P.T. ~ C.A.P.**

TEST LOCATION **Canal & Sta. 846 + 25 Muddy River Canal**

Central Arizona Project

TEST LIMITS: STATION **844 + 00** TO STATION **848 + 50**   CANAL DATA: SIDE SLOPES **1½ : 1**

BOTTOM WIDTH (ft) **26.0**   WATER DEPTH (ft) **6.2**   DISCHARGE (cu ft/sec) **1050**

OBSERVATION HOLE

STRATA DEPTH (ft) FROM	TO	SOIL CLASSIFICATION
O	1.5	SILTY CLAY, topsoil (roots present); about 85% fines of medium plasticity; about 15% fine sand; maximum size, fine sand; dark gray; moist; (CL)
1.5	5.8	CLAYEY SILT, about 95% fines of slight plasticity; about 5% fine sand; maximum size, fine sand; brown; firm and dry inplace; (ML)
5.8	12.7	SILTY SAND, gravelly; about 60% coarse to fine sand; about 20% hard, angular gravel; about 20% nonplastic fines; maximum size, ½"; brown; compact and moist inplace; (SM)

1. DEPTH (ft) TO WATER TABLE (FROM GROUND SURFACE) **12.3**

WELL DIMENSIONS (DEPTHS FROM STRING BASELINE)

2. DEPTH (ft) TO GROUND SURFACE **0.70**
3. DEPTH (ft) TO BOTTOM OF WELL **4.01**
4. DEPTH (ft) TO TOP OF SAND **1.23**
5. DEPTH (ft) OF SAND (3) - (4) **2.78**
6. DEPTH (ft) TO WATER SURFACE IN WELL **0.92**
7. DEPTH (ft) OF WATER IN WELL h = (3) - (6) **3.09**

DETERMINATION OF WELL RADIUS

8. DENSITY (pcf) OF STANDARD SAND **87.40**
9. WEIGHT (lb) OF SAND + CONTAINER BEFORE FILLING WELL **75.00**
10. WEIGHT (lb) OF SAND + CONTAINER AFTER FILLING WELL **6.30**
11. WEIGHT (lb) OF SAND USED (9) - (10) **68.70**
12. VOLUME (cu ft) OF WELL $\frac{(11)}{(8)}$ **0.786**
13. RADIUS (ft) OF WELL $r = \sqrt{\frac{(12)}{(5)\pi}}$ **0.30**

Figure 19-5.—Data on soils and well for well permeameter test. 101-D-279.

tures in the field and the change in water viscosity due to temperature, it is necessary to record water temperature during the test and to correct the coefficient of permeability to a 20° C. standard. This correction is included in the nomographs of figures 19-6 and 19-7 for computing coefficients of permeability. The temperature of the water in, or the ground around, the test well should be taken. The temperature in a well where the water level is some distance below the ground surface can be obtained with reasonable accuracy by lowering a thermometer into the well on the top of the backfill sand, leaving it there for 5 minutes and reading it quickly upon removal.

It is desirable, if possible, to have the water introduced into the soil at a temperature somewhat above that of the soil. This will result in a decreasing temperature gradient as the water flows through the soil and will tend to prevent the clogging of voids in the soil with bubbles of air coming out of solution. The presence of air may unduly decrease the flow of water through the soil.

(f) *Records of Discharge and Time.*—The field permeability test is conducted by recording the gage tube readings on the 50-gallon reservoir at timed intervals, on the form shown in figure 19-8. From these data, it is possible to plot a curve showing the accumulative discharge against time and to compute discharge rates for any time period (see fig. 19-9). In general, the dry soil at the start of the test absorbs water at a comparatively high rate; but, as the soil below the test well becomes saturated, the rate decreases to a point where it is practically constant. When this occurs, as evidenced by the plotted points on the curve falling on practically a straight line for several hours, the slope of the line gives the rate of flow to be used in computations of coefficient of permeability and the test may be discontinued. If gage readings are made at equal time intervals, it is not necessary to plot the curve because a straight portion on the curve will be indicated by equal readings. Usually the steady-state flow causing a straight line on the curve will occur within an 8-hour period for soils with moderate to high permeability. If this condition does not occur within that time, the minimum time as discussed below shall be used as a guide to determine test duration.

(g) *Test Duration.*—The test should be run long enough to develop a saturated envelope in the soil but not long enough to build up the water table or produce an excessively large saturated envelope which will cause erroneous results. Thus, there is introduced a concept of minimum and maximum time limits within which the test results are valid.

(1) *Minimum time.*—The minimum time for the duration of the test is the time required to discharge the minimum volume (cubic feet) of water into the soil to form a saturated envelope of hemi-

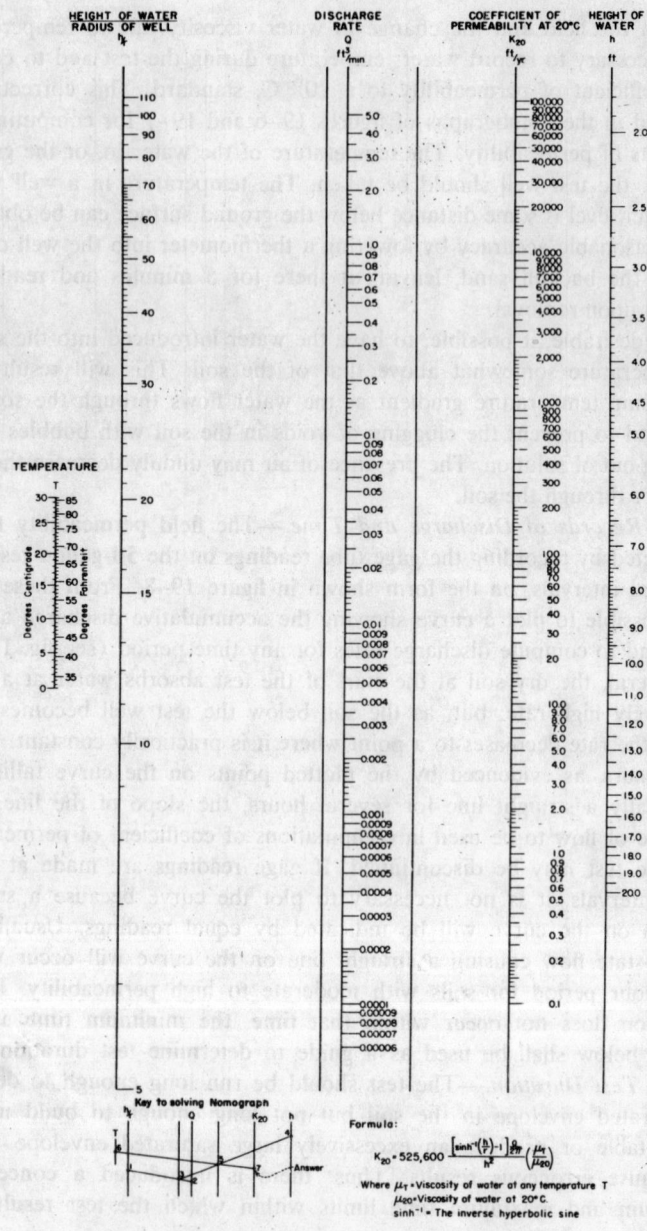

Figure 19–6.—Nomograph for determining coefficient of permeability from well test with a low water table condition. 28?-D–2397.

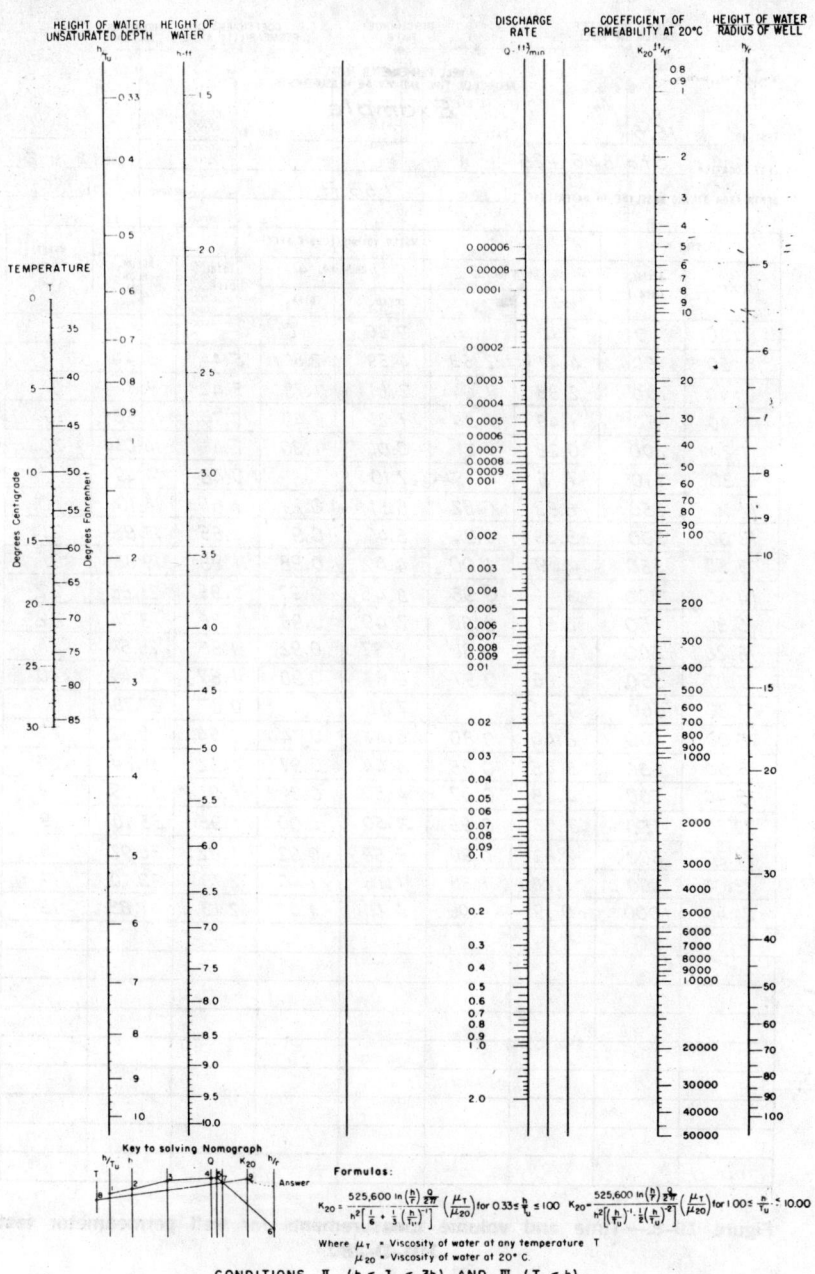

Figure 19-7.—Nomograph for determining coefficient of permeability from well test with a high water-table condition. 288-2398.

Z-1428
(7-54)
BUREAU OF RECLAMATION

WELL PERMEAMETER TEST
RECORD OF TIME AND VOLUME MEASUREMENTS

Example

TEST NO. ___1616___   DATE _____   MADE BY _____

TEST LOCATION ___Sta. 846 + 25___   SHEET _2_ OF _2_

DEPTH FROM STRING BASELINE TO WATER LEVEL ___1.53 ft.___   GROUND TEMP. (T) _18°C_

| TIME | | WATER VOLUME (CUBIC FEET) | | | | | | WATER |
| CLOCK | ACCUM.- (MIN.) | DRUM NO. 3 | | DRUM NO. 4 | | TOTAL DIFF. | ACCUM. FLOW (Q) | TEMP. (Tw) |
		READ.	DIFF.	READ.	DIFF.			
8:00	0	7:10		7.20				
8:50	50	4.47	2.63	4.39	2.81	5.44	5.44	19
9:40	100	2.83	1.64	2.61	1.78	3.42	8.86	19
10:30	150	1.49	1.34	1.21	1.40	2.74	11.60	19
11:20	200	0.28	1.21	0.01	1.20	2.41	14.01	20
11:30	210	7.15		7.10		0.48	14.49	
12:10	250	6.33	0.82	6.31	0.79	1.61	16.10	20
13:00	300	5.39	0.94	5.40	0.91	1.85	17.95	21
13.50	350	4.39	1.00	4.42	0.98	1.98	19.93	21
14.40	400	3.41	0.98	3.45	0.97	1.95	21.88	22
15.30	450	2.51	0.90	2.49	0.96	1.86	23.74	22
16.20	500	1.63	0.88	1.57	0.92	1.80	25.54	21
17.10	550	0.66	0.97	0.67	0.90	1.87	27.41	20
17.20	560	7.20		7.15		0.37	27.78	
18.00	600	6.40	0.80	6.41	0.74	1.54	29.32	20
18.50	650	5.45	0.95	5.44	0.97	1.92	31.24	20
19.40	700	4.48	0.97	4.50	0.94	1.91	33.15	19
20.30	750	3.53	0.95	3.50	1.00	1.95	35.10	19
21.20	800	2.63	0.90	2.58	0.92	1.82	36.92	18
23.00	900	1.25	1.38	1.18	1.40	2.78	39.70	17
24.40	1000	0.19	1.06	0.11	1.07	2.13	41.83	15

Figure 19–8.—Time and volume measurements for well permeameter test.
101-D-280.

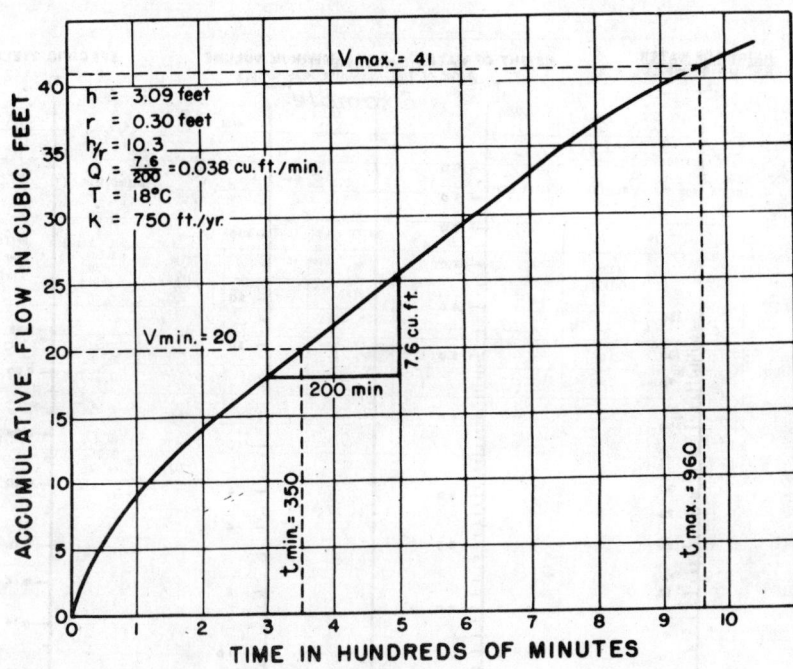

Figure 19–9.—Example of discharge-time curve for well permeameter test. 101–D–281.

spherical shape with a radius B. This is determined by the formula:

$$V_{min} = 2.09 Y_s \left[h \sqrt{\frac{2}{\sinh^{-1}\left(\frac{h}{r}\right) - 1}} \right]^3 \tag{1}$$

where: V_{min} = minimum volume, cubic feet,
  Y_s = specific yield of the soil,
  h = depth of water in well, feet, and
  r = well radius, feet.

(*Note:* the bracketed quantity is the theoretical determination for radius B.)

This equation can be solved conveniently and the minimum volume determined by the nomograph [1] of figure 19–10. Specific yield of soil is defined as the fraction of total saturated volume of the soil mass

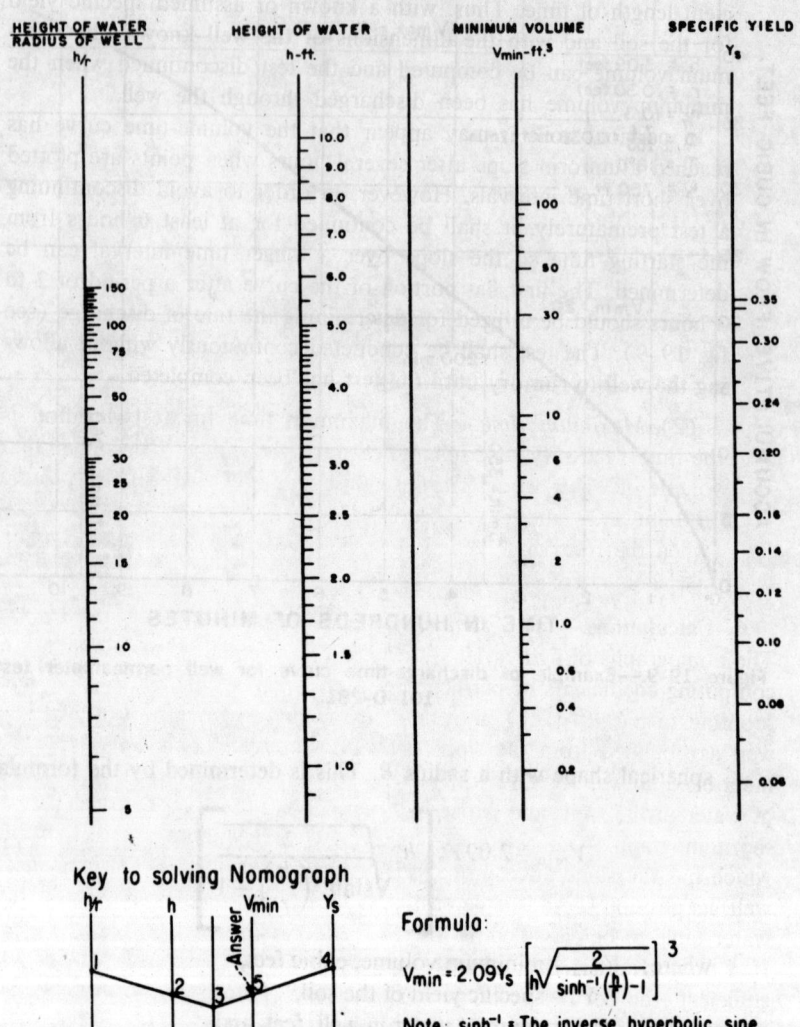

Figure 19–10.—Nomograph for computing minimum volume of water to be discharged during well permeameter test to determine test duration. 288-D-2396.

recoverable as water by natural drainage. For common soils, the specific yield value varies from 0.10 for fine-grained soils to 0.35 for coarse-grained soils. When the specific yield is unknown, the value of 0.35 should be used to give a conservative value for minimum volume and to make sure that the test has been run for a sufficient length of time. Thus, with a known or assumed specific yield for the soil and with the dimensions of the well known, the minimum volume can be computed and the test discontinued when the minimum volume has been discharged through the well.

In pervious soils it may appear that the volume-time curve has reached a uniform slope after several hours when points are plotted over short time intervals. However, in order to avoid discontinuing a test prematurely, it shall be continued for at least 6 hours from the starting time so the slope over a longer time interval can be determined. The first flat portion of the curve after a period of 2 to 3 hours should be utilized for determining the rate of discharge (see fig. 19–9). The test shall be conducted continuously without allowing the well to run dry until the test has been completed.

(2) *Maximum time.*—The maximum time for test duration is the time necessary to discharge through the test well the maximum volume of water as determined by equation (1) above, substituting 15.0 for 2.09 and in this case using an assumed minimum value (when the true value is unknown) of 0.10 for specific yield, or

$$V_{max} = 2.05 \, V_{min} \tag{2}$$

6. Calculations.—(a) *Computation of Coefficient of Permeability.*— The nomographs of figures 19–6 and 19–7 have been developed to aid in computing coefficients of permeability for the well permeameter test (see footnote to par. 5(g)(1) above). The rate of water flow from the test well as obtained from the slope of the accumulative time-volume curve mentioned above, together with the effective radius of the well, the height of water in the well, and water or ground temperature, is needed to use the nomograph. Also, when the water table, or an impervious soil layer which has the same effect in reducing seepage, is relatively near the test well, its position should be determined. This determination will enable the water table to be classified as low or high, as illustrated in figure 19–11. If a nomograph is not available or the range of the nomograph is not sufficient, the coefficient of permeability for various conditions may be computed as follows:

(1) *Low water table.*—When the distance from the water surface in the test well to the ground-water table (or an impervious soil layer which is considered for test purposes to be equivalent to a water table) is greater than three times the depth of water in the well, a low water table condition exists as illustrated by condition I, figure 19–11. For the determination of the coefficient of permeability under such a condition, equation (3) given below should be used.

(2) *High water table.*—When the distance from the water surface in the test well to the ground-water table (or an impervious layer) is

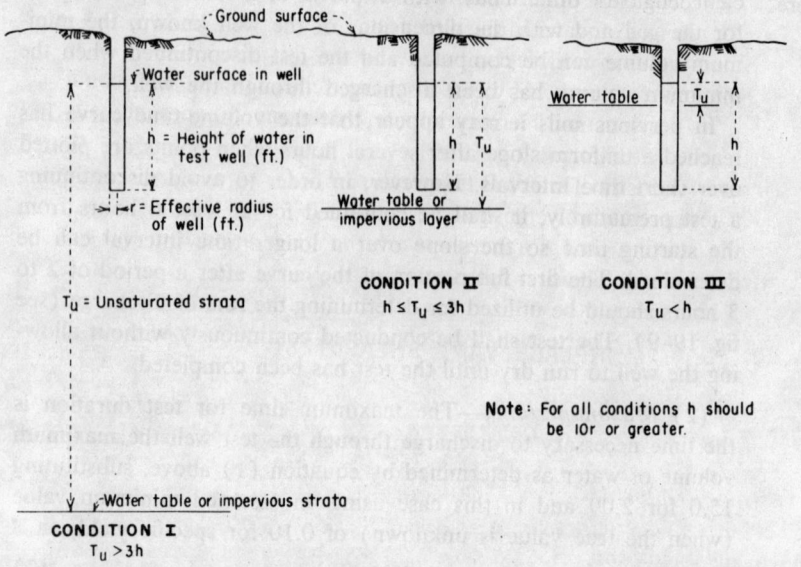

Figure 19–11.—Relationship between depth of water in test well and distance to water table in well permeameter test. 101–D–282.

less than three times the depth of water in the well, a high water table condition exists as illustrated by conditions II and III. Condition II shows a high water table condition with the water table below the well bottom, and for this condition equation (4) given below should be used.

Condition III shows a high water table condition with the water table above the well bottom. For this condition equation (5) given below should be used.

$$k_{20} = 525,600 \frac{\left[\sinh^{-1}\left(\dfrac{h}{r}\right) - 1\right]}{h^2} \frac{Q}{2\pi} \left(\frac{\mu_T}{\mu_{20}}\right) \tag{3}$$

$$k_{20} = \frac{525,600 \log_e\left(\dfrac{h}{r}\right)\dfrac{Q}{2\pi}}{h^2\left[\dfrac{1}{6} + \dfrac{1}{3}\left(\dfrac{h}{T_u}\right)^{-1}\right]} \left(\frac{\mu_T}{\mu_{20}}\right) \tag{4}$$

$$k_{20} = \frac{525,600 \log_e\left(\dfrac{h}{r}\right)\dfrac{Q}{2\pi}}{h^2\left[\left(\dfrac{h}{T_u}\right)^{-1} - \dfrac{1}{2}\left(\dfrac{h}{T_u}\right)^{-2}\right]} \left(\frac{\mu_T}{\mu_{20}}\right) \tag{5}$$

where: k_{20} = coefficient of permeability, feet per year,

          h = height of water in the well, feet,

          r = radius of well, feet,

          Q = discharge rate of water from the well for steady state condition, cubic feet per minute, (determined experimentally, see example, fig. 19–9),

          μ_T = viscosity of water at temperature T,

          μ_{20} = viscosity of water at 20° C., and

          T_u = unsaturated distance between the water surface in the well and the water table, feet.

INPLACE VANE SHEAR TEST

Designation E–20

1. Scope.—This designation describes a method for determining the inplace shear resistance of soft saturated fine-grained soils. The test is used to determine subsurface conditions with respect to shear resistance for the design of foundations and natural soil slopes.

2. Definition.—The vane shear test basically consists of placing a four-bladed vane in the undisturbed soil and rotating it from the surface to determine the torsional force required for failing a cylindrical surface sheared by the vane, which is converted to the shear resistance of the cylindrical surface.

3. Apparatus.—This test is not required for construction control; therefore, the apparatus is not listed in the field laboratory equipment list, designation E–4, and only special or major items are listed below. (See figs. 20–1, 20–2, and 20–3.) The equipment can be obtained for project use from the Engineering and Research Center, Division of General Research.

     (1) Vane shear test apparatus.—The apparatus is illustrated in figure 20–1. It consists of several principal parts as discussed below:

The torque applicator assembly contains a sturdy ring, 5 inches outside diameter, 1 inch high, and 0.57 inch thick, for applying a balanced moment to the vane rod without the necessity of a thrust bearing. The ring has a section cut from it and it will deform as torque is applied. The deformation is indicated by a dial indicator gage and the assembly is calibrated for foot-pounds of torque. A 10-inch gearwheel, which holds the torque ring, is geared to a crank handle which is operated at controlled speed. This gearwheel is marked in degrees and has a pointer for registering rotation.

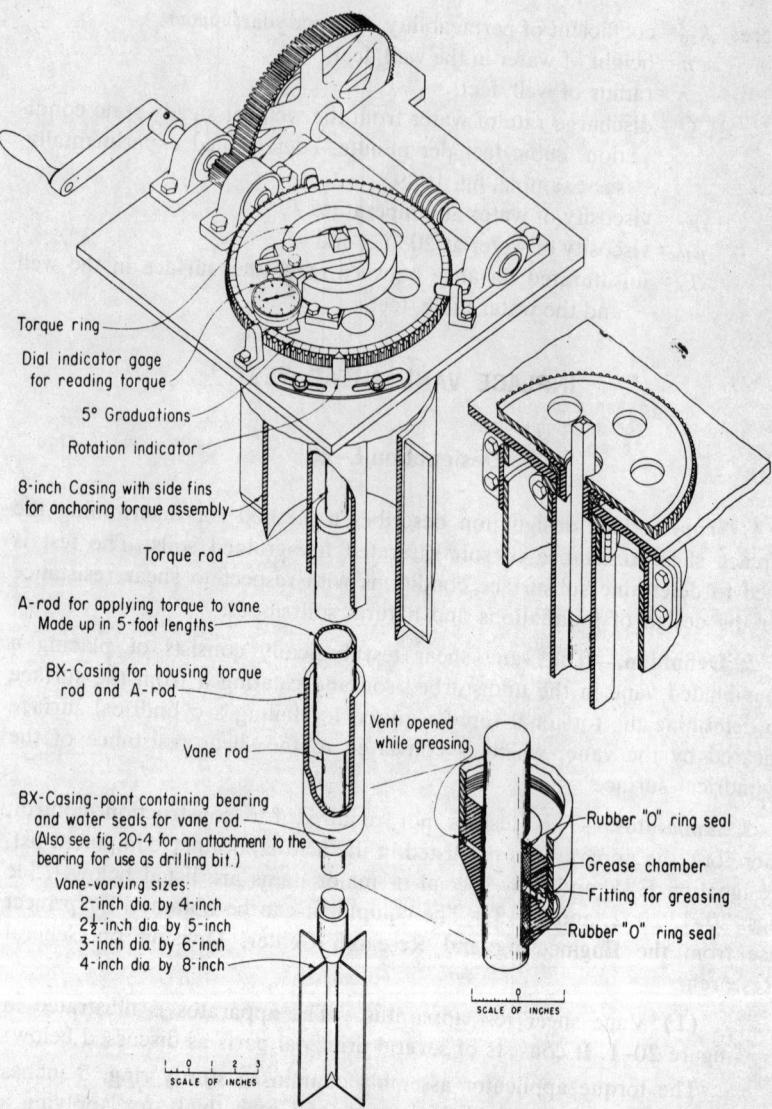

Torque ring

Dial indicator gage
for reading torque

5° Graduations

Rotation indicator

8-inch Casing with side fins
for anchoring torque assembly

Torque rod

A-rod for applying torque to vane.
Made up in 5-foot lengths

BX-Casing for housing torque
rod and A-rod

Vane rod

Vent opened
while greasing

BX-Casing-point containing bearing
and water seals for vane rod.
(Also see fig. 20-4 for an attachment to the
bearing for use as drilling bit.)

Vane-varying sizes:
2-inch dia. by 4-inch
2½-inch dia. by 5-inch
3-inch dia. by 6-inch
4-inch dia. by 8-inch

Rubber "O" ring seal

Grease chamber

Fitting for greasing

Rubber "O" ring seal

SCALE OF INCHES

SCALE OF INCHES

Figure 20-1.—Inplace vane shear test apparatus. 101-D-185.

A standard 8-inch pipe casing 2 feet in length, with side fins for embedding 1 foot into the surface of the ground, is used to anchor the torque applicator. Extensions in length can be used to vary the level of the instrument and thus the depth of vane placement, or to raise the instrument above water level when necessary.

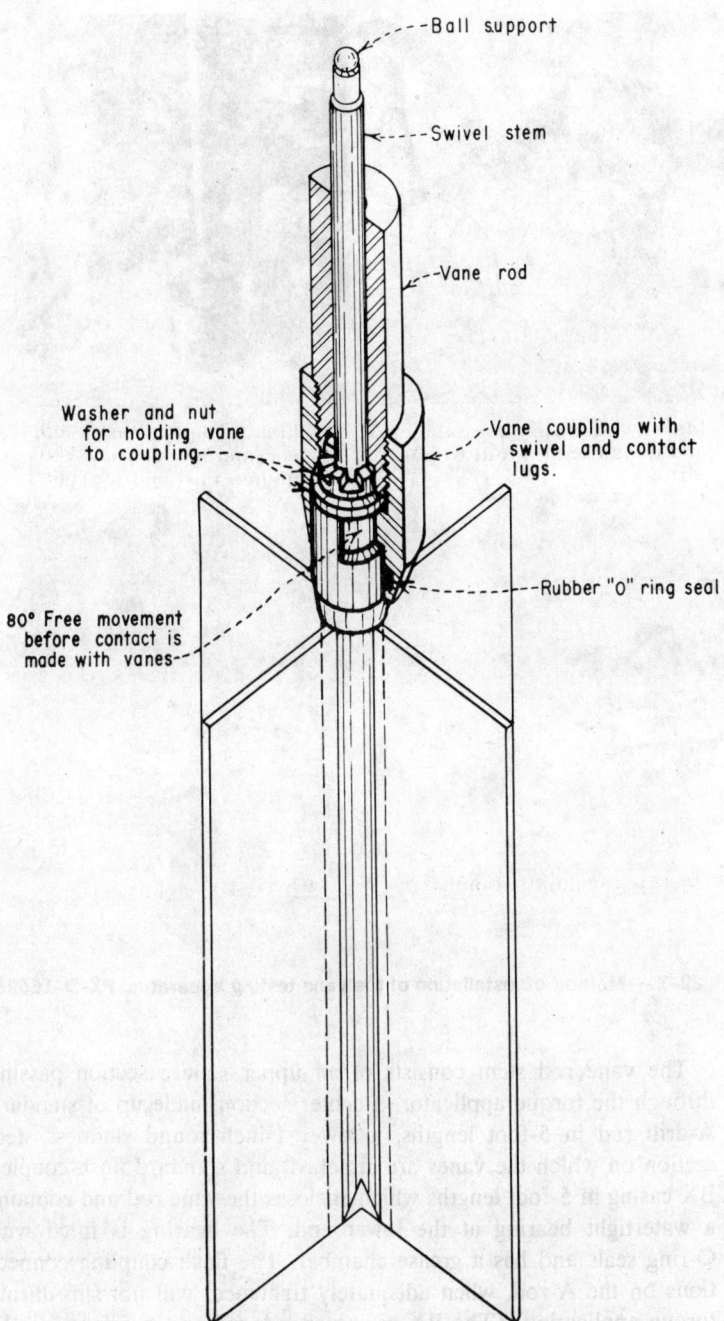

Ball support

Swivel stem

Vane rod

Washer and nut
for holding vane
to coupling:

Vane coupling with
swivel and contact
lugs.

Rubber "o" ring seal

80° Free movement
before contact is
made with vanes

Figure 20–2.—Modified vane for friction determination. 101–D–199.

(A) Install 8-inch anchor casing and press vane stem to position

(B) Install upper square vane rod and press vanes 30 inches into undisturbed soil

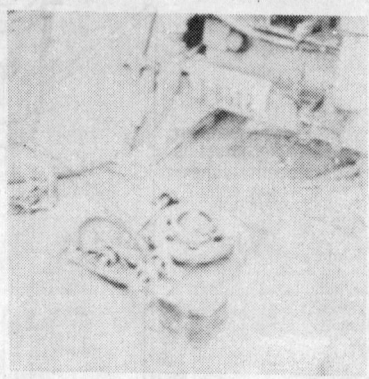

(C) Install torque assembly

(D) Test is ready to begin

Figure 20-3.—Method of installation of the vane testing apparatus. PX-D-16696.

The vane rod stem consists of an upper square section passing through the torque applicator, a center section made up of standard A-drill rod in 5-foot lengths, a lower 1-inch round stainless steel section on which the vanes are attached, and standard flush-coupled BX casing in 5-foot lengths which encloses the vane rod and contains a watertight bearing at the lower end. The bearing is fitted with O-ring seals and has a grease chamber. The flush coupling connections on the A-rod, when adequately tightened, will not slip during torque application. (The BX casing and A-rod are in 5-foot lengths to permit testing at intervals of 5-foot depth. A special short piece of BX casing and accompanying A-rod are sometimes used to vary the testing depth.)

Vanes which attach to the end of the vane rod are of four sizes: 4 inches diameter by 8 inches long, 3 inches by 6 inches, 2½ inches by 5 inches, and 2 inches by 4 inches. The modified vane illustrated in figure 20–2 is constructed to allow a definite amount of free rotation to determine the friction correction for each test prior to applying force to the vane. A rigid-type vane is available for special uses but requires separate determinations of friction using a blank stem or assumed friction corrections in order to interpret results.

(2) Drill rig.—A drill rig and accessory equipment is generally needed for placing the vane stem to testing position and pulling it after the tests. In stiff or firm soils predrilling is often necessary. Hand placement methods are possible for shallow depths in very soft soils.

(3) Stopwatch with sweep second-hand.

(4) Miscellaneous pipe wrenches, hoisting plugs, greasing equipment.

4. Procedure of Installation.—The instrument shall be installed as illustrated briefly in figure 20–3 and according to the following steps:

(a) The 8-inch anchor casing is installed with side fins about 1 foot into the soil by driving or pressing or by inserting in a prepared hole. The top level may be varied a few inches, or extension lengths may be installed to permit the vanes to be located at a desired depth. Soil in the center of the casing is removed.

(b) The vane rod stem is assembled to contain the vane on the lower end, the lower watertight bearing on the 1⅛-inch round rod, and the first sections of BX casing and A-rod. All vane rod and A-rod couplings must be tightened with forces in excess of 200 foot-pounds (the capacity of the instrument) to prevent slippage during the test.

(c) With the vane held firmly in the up position so that the vane coupling touches the BX bearing point, the assembled vane rod stem is installed to a depth 30 inches above the elevation at which the shear strength is desired. In soft soil, the vane stem can be pressed for short distances into the ground without boring a hole. Where overlying strata are relatively firm, a predrilled hole is made. Drilling mud as described in designation E–2, or 4- to 6-inch casing may be needed to support the hole above the elevation of placement.

In 1959, a feature was added to the original equipment to permit using the vane rod stem as a drilling tool during placement. This apparatus is shown in figure 20–4. A hard-faced cutting bit is attached to a special lower bearing which contains ports for passage of the drilling fluid. The drilling fluid is circulated through the BX casing and lower bearing and

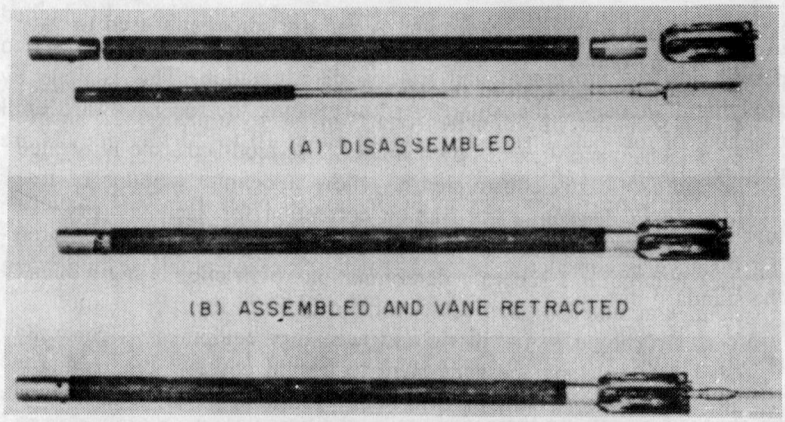

(A) DISASSEMBLED

(B) ASSEMBLED AND VANE RETRACTED

(C) ASSEMBLED AND VANE EXTENDED

Figure 20–4.—Rotary cutting bit for use with the vane shear test apparatus. PX–D–14465, PX–D–14466, and PX–D–14467.

is discharged from the bottom of the cutting bit. During drilling, the vane is held up and inside of the cutting bit. The coupling device for holding the vane rod in the up position is attached to the top of the BX casing and contains screws which are tightened against the A-rod. This coupling also serves as a connection to standard fittings for the drilling fluid hose and the hoisting plug. The entire vane assembly is used as a rotary drilling tool and can be used to advance the hole to within about 3 to 9 inches of the desired placement depth of the bottom of the bit. In very soft soil, the stem can be used as a wash boring tool and placed without using rotary action; this method would be useful when a rotary drill rig is unavailable. Care must be taken to stop the flow of drilling fluid just prior to reaching the desired drilling depth to prevent excessive washing. The cutting bit should be pressed to the desired depth into the bottom of the hole.

The desirable procedure is to press the vane stem to the desired depth with a moderate steady force of a drill rig. (Hammering or excessive forces of up and down working are not permitted and generally cause damage to the vane.) It is important to hold the vane in the up position during lowering in the hole and pressing to greater depth so as to accomplish a vertically plumb placement and reduce any tendency to veer off center.

The BX casing is placed so that its top is 1⅜ to 2 inches above the 8-inch casing, and the first section of A-rod is of such length that it extends a similar distance above the BX casing. This gives wrench gripping space for installing the instrument and adding additional standard lengths to BX casing and A-rod for subsequent testing depths.

(d) The upper square section of the vane rod is attached and tightened in excess of 200 foot-pounds. At this time, the modified vane is set to permit the free movement and friction determination. This is done by applying a wrench to the square bar and turning the rod backward until touching of the vane lugs is felt. (Only about 10 to 20 foot-pounds are required.)

(e) The vane rod is pressed 30 inches to place the vane in undisturbed soil and at the location planned for testing. (The minimum pressing distance is considered to be 18 inches for acceptable testing when the standard distance of 30 inches is not applicable.)

(f) The torque applicator assembly is placed over the square rod, resting on the BX casing, and the BX clamp is tightened. The vane rod must be free in the square hole for the zero reading and can be checked by rattling the rod. Finally, the baseplate is securely clamped to the 8-inch anchor casing so that it is sufficiently tight to resist the maximum torsional force of the instrument. The test is ready to begin.

5. Procedure of Testing.—The test is made in the following steps to obtain the data, an example of which is shown in figure 20–5 and the graph of figure 20–6. (Keypunch forms and instructions for completing the forms are available from the Engineering and Research Center.)

The initial readings of the dial indicator gage and protractor are made after checking the rod for looseness (zero torque) and taking up slack. The crank is turned at a uniform rate of 12 turns per minute which is conveniently paced at one revolution for each 5 seconds on a stopwatch. This corresponds to 0.1° rotation of the vane rod per second (standard rate). The dial indicator gage readings are made at intervals of 5° rotation.

The first part of the test is the friction correction determination. Six observations are made during the first 30° of rotation while using the standard rate. Following this, a more rapid turning without observations can be used to complete the remaining portion of the free movement. The operator must use care in observing, by the dial indicator gage, when the contact with the vane is reached so that the standard controlled speed is again used during the application of force to the vanes.

The second part of the test is the determination of the undisturbed strength. While turning the crank at the standard rate, dial indicator gage readings are made at 5° intervals and at the angle when the maximum dial indicator gage reading is observed. The test is then continued for five additional readings at 5° intervals.

The third part of the test is the determination of remolded strength after failure. The vane is rotated without observations through an angle

VANE TEST DATA
For In-Place Shear
of Soils

Sheet ___ of ___

Feature _Example_   Ground el _4200.9_   Hole No. _DH-202-VT_

Project _____   El water table _4199.6_   Location _N-381,152_ _E-1,825,984_

State _____   Date of test _____   Depth at which vane

Foreman _____   Size of vane _3"_   test is made _60'_

Remarks _Test was satisfactory, no difficulty encountered._

_____

_____

Protractor reading (degrees)	Strain gage reading (0.0001")	Protractor reading (degrees)	Strain gage reading (0.0001")	Protractor reading (degrees)	Strain gage reading (0.0001")
Initial 80	Reading 1216	185	2970		
85	1360	190	2860		
90	1360	195	2758		
95	1363	200	2685		
100	1367	Remolded 360			
105	1369	205	2220		
110	1373	210	2245		
		215	2244		
146	Start	220	2235		
150	1690	225	2225		
155	2040	No load	1287		
160	2370				
165	2650				
170	2900				
175	3110				
178	3164				
180	3100				

Figure 20–5.—Vane test data sheet. 101-D-283.

EL-590 (8-70)
Bureau of Reclamation

VANE SHEAR TEST PLOT SHEET

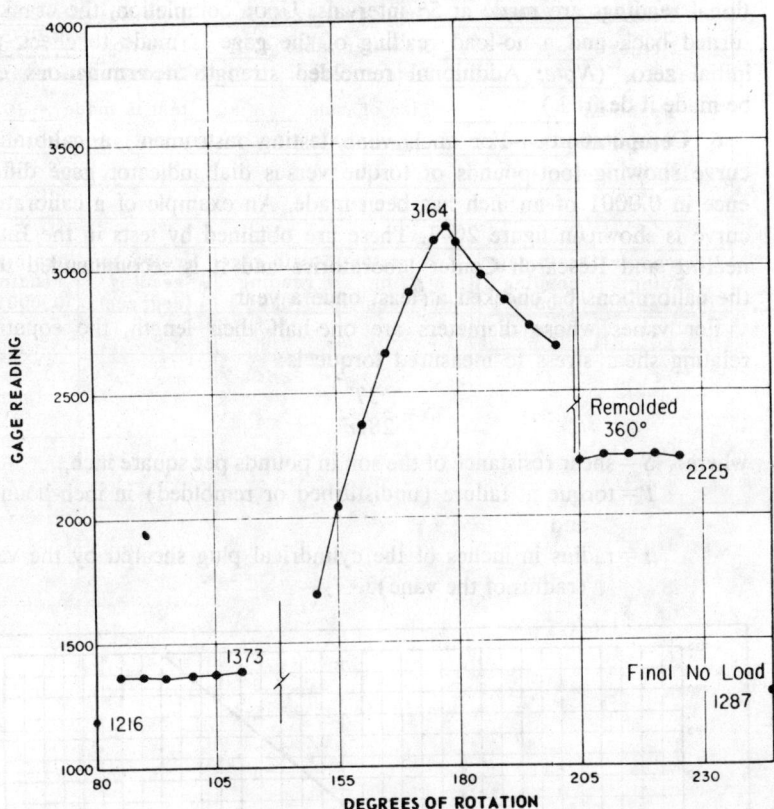

	Undisturbed	Remolded	
		360 °	°
Gage Reading	3164	2225	
Friction Gage Reading	1373	1373	
Difference	1791	852	
Torque - ft-lb	156	77	
(m-kg)	(21.6)	(10.6) ( ) ( )
Shear - psi	18.9	9.3	
(kg/cm²)	(1.3)	(0.7) ( ) ( )

Vane Size __3 x 6__ in. Hole No. __DH – 202 VT__ Depth __60__ ft
(__7.6 x 15.4__ cm) (__18.3__ m)

Figure 20–6.—Plot of vane test results. 101–D–596.

of 360 degrees, or additional rotations may be requested. After this rotation, testing at standard controlled rate is again resumed. Five additional readings are made at 5° intervals. Upon completion, the crank is turned back and a no-load reading of the gage is made to check the initial zero. (*Note:* Additional remolded strength determinations can be made if desired.)

6. Computations.—For each vane testing instrument, a calibration curve showing foot-pounds of torque versus dial indicator gage difference in 0.0001 of an inch has been made. An example of a calibration curve is shown in figure 20–7. These are obtained by tests in the Engineering and Research Center laboratories and it is recommended that the calibrations be checked at least once a year.

For vanes whose diameters are one-half their length, the equation relating shear stress to measured torque is:

$$S = \frac{3T}{28\pi r^3} \qquad (1)$$

where: S = shear resistance of the soil in pounds per square inch,

        T = torque at failure (undisturbed or remolded) in inch-pounds, and

        r = radius in inches of the cylindrical plug sheared by the vane (radius of the vane).

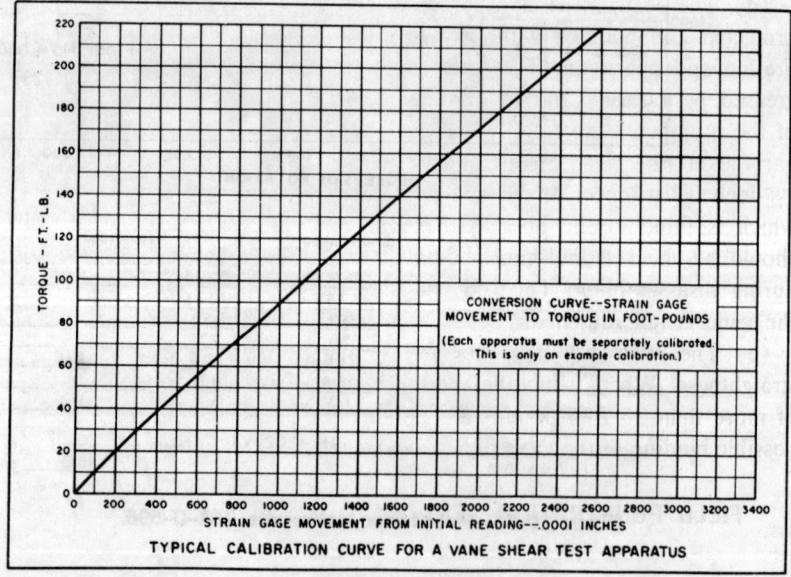

TYPICAL CALIBRATION CURVE FOR A VANE SHEAR TEST APPARATUS

Figure 20–7.—Typical calibration curve for a vane shear test apparatus.
101–D–284.

Typical results are shown in figure 20–6. The shear resistance has been determined by the use of the calibration curve in figure 20–7 and the above equation.

7. Summary Comments.—The operator should be aware of certain limitations and precautions in care and operation of the equipment.

(a) The torque applicator will permit observations from 0 to 200 foot-pounds, and this limitation must not be exceeded.

(b) The distance of pressing the vane into undisturbed soil is 30 inches. This allows about 3 inches clearance of the lower A-rod reducer above the BX bearing point, and this reducer must not be allowed to ride on the bearing during testing.

(c) The vane size selected should be the largest size suitable for the general soil conditions. The 4-inch-diameter vane is limited to obtaining shear values from 0 to 10 pounds per square inch and is recommended for very soft saturated soils. The 3-inch-diameter vane is limited to obtaining shear values from 0 to 24 pounds per square inch and is recommended for soft to moderately firm saturated soils. The 2½-inch-diameter vane is limited to obtaining shear values from 0 to 40 pounds per square inch and is recommended for moderately firm saturated soils. The 2-inch-diameter vane is used infrequently but is intended for firm saturated soils.

(d) The instrument is a precision testing apparatus and must be protected and handled with care in order to maintain the proper calibration and operation. The water seals in the lower bearing should be greased periodically and at least once for every drill hole. The swivel of the modified vane is fully enclosed when assembled, but the internal parts should be clean of dirt and well greased. Attention is called to the $\frac{3}{16}$-inch ball support of the modified vane swivel, shown in figure 20–2, which is fitted in the hole in the vane rod. The presence of the ball should be checked and care should be taken so as not to lose this ball during disassembling. The rod end should be covered by a cap when the vane is not attached.

(e) The vanes and the vane rod should be continually checked for straightness. When using the modified vanes, rod friction observations of more than 25 foot-pounds are indicators of off-center movement and possible binding of the vane rod.

FIELD PENETRATION TEST WITH SPLIT-TUBE SAMPLER

Designation E–21

1. Scope.—This designation describes a procedure to obtain a record of the resistance of subsoils to the penetration of a standard sampler and to obtain representative disturbed samples of the soil for identification purposes. The test and identification information are used to outline subsurface conditions with respect to bearing capacity for foundation design.

2. Definition.—The penetration resistance is expressed as the number of blows, of a 140-pound hammer freely dropping 30 inches, required to force the sampler 1 foot into the soil.

3. Apparatus.—This test is not required for construction control; therefore, the apparatus is not listed in the field laboratory equipment list, designation E–4, and only special or major items of equipment are listed below (see figs. 21–1 through 21–6):

(1) Boring equipment.—Methods of hole advancement include conventional rotary or wire line drilling; and standard, hollow-stem, or hand-type augers as required to provide a reasonably clean hole, without unnecessary disturbance at the bottom of the hole or soil to be sampled. Holes for penetration testing shall be at least BX casing size (2 $^{15}/_{16}$-inch OD). Wash borings using bottom-discharge fishtail bits or jetting through an open tube are not recommended, as the soil to be sampled may be disturbed and the moisture may affect the strength of the material being tested. (Side discharge bits are permissible below ground water.) The drilling equipment or suitable auxiliary equipment shall be such as to permit the driving of the sampler to obtain the sample and penetration-resistance record according to the procedure described below.

(2) Sampler.—Penetration samplers as shown in figures 21–5 and 21–6. The sampler shall be clean and lightly coated with oil at the beginning of each test. The outer wall, inner wall or liner, and cutting bit shall be smooth and free from scars made by tools and rocks.

(3) Drive hammer assembly.—Drive hammer assembly consisting of a 140-pound weight, guide pipe sufficiently long to allow a 30-inch free fall, and a jar coupling (connection between the guide pipe and sampler rod string for the hammer to strike). Various types of drive hammers are available commercially including the safety type, figure 21–1; open type, figures 21–2 and 21–3; and the automatic-trip type, figure 21–4, which is desirable to have better control of height of drop.

(4) Rod.—AW- or B-rod is considered standard and recommended for all operations. Most commercial samplers are available with AW-rod connections.

(5) Casing.—Drill hole casing as required.

(6) Miscellaneous.—Airtight sample containers, labels, field log

Figure 21–1.—Standard drilling equipment used to perform penetration tests. PX–D–34356 (E–1995–13).

sheets (form 7–1334, fig. 68), and other necessary tools and supplies.

4. Procedure.—(a) Using the equipment listed above, the hole shall be cleaned to the sampling elevation by means of drilling bits, augers, or other equipment insuring that the material to be tested and sampled is not disturbed by the drilling and cleanout operation. If an obstruction is encountered, such as a boulder or rock, the obstruction shall be removed by a chopping bit or other means, or shall be drilled through.

Figure 21–2.—A portable power-operated rig used to perform penetration tests.
PX–D–16752.

Figure 21–3.—A portable hand-operated rig used to perform penetration tests in
shallow holes. PX–D–16697.

Figure 21–4.—Automatic-trip drive hammer. PX–D–17150D.

In no case shall the sampler be used as a chopping bit. Penetration resistance testing or sampling shall be at 5-foot depth intervals and at each change of material unless other intervals are specified. Where casing is used, the casing shall not be driven into the layer to be tested (and sampled) in advance of the penetration and sampling operation.

(b) The standard sampler is attached to the drill rod (AW or B), each rod joint is securely tightened to prevent jarring loose during the driving operation, and then is lowered to the bottom of the hole. If either hollow-stem auger or wire line methods are used in the drilling operation, the sampler may be lowered to the bottom of the hole through the auger or wire line rod. Both methods will serve as hole casing and the operation is simplified by eliminating the drilling assembly removal for sampler insertion. The jar coupling, guide pipe, and drive hammer are then added to the sampler rod string and the hammer is allowed to fall on the jar coupling until the sampler has penetrated one-half foot into the soil.

The penetration test is then started, and the number of blows (30-inch drops of the 140-pound drive weight) required to drive the sampler an additional full 1 foot is recorded as the penetration resistance. Extreme care must be exercised in obtaining an accurate 30-inch free fall condition during the test. This is especially critical when using either the safety or the open type hammer as the drop height and hammer fall are controlled by the operator. Failure to completely release the hoist rope

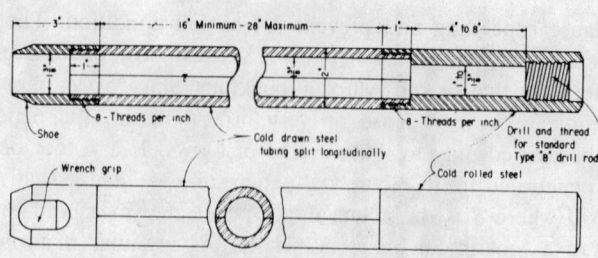

STANDARD SPLIT-TUBE SAMPLER

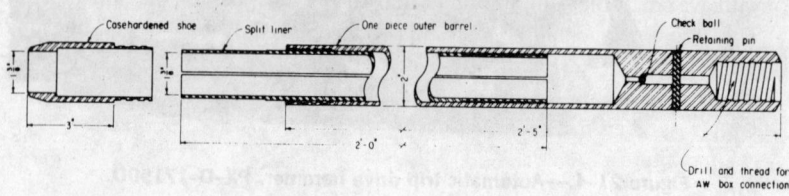

2-INCH O.D. x 1⅜-INCH I.D. PENETRATION RESISTANCE SAMPLER

Figure 21–5.—Standard split-tube and split-liner penetration resistance samplers. 101–D–616.

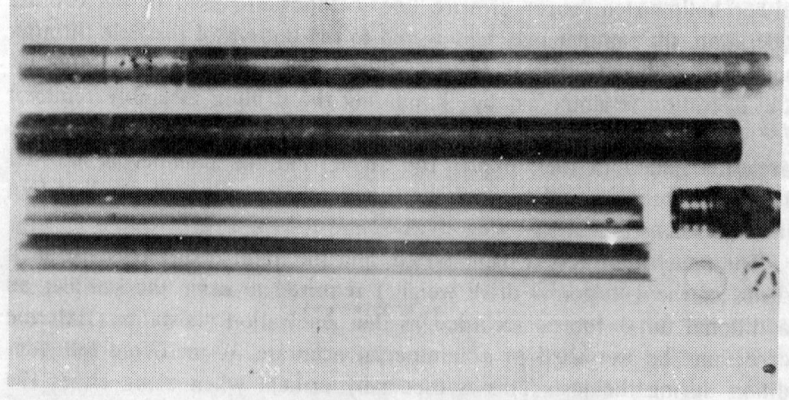

Figure 21–6.—Split-liner sampler. E–2245–3NA.

will cause rope drag and prevent a free fall condition. The rod, above the hole collar, should be held in a vertical position to prevent energy displacement due to rod whip or buckling. Since 50 or more blows per foot indicate a very dense or very firm material, the penetration test shall be carried only to this number of blows. If a 1 foot depth has not been penetrated in 50 blows, the penetration resistance shall be recorded as $50/d$ where d is the actual depth penetrated measured to the nearest tenth of a foot in 50 blows. When testing granular material below the water table, the water level in the bore hole should be maintained at, or above, the ground-water level.

(c) Immediately after each penetration test, duplicate samples of a representative portion of the soil core from the bottom 1-foot drive shall be placed in airtight containers. The sample container shall be sealed with wax or other suitable material, to prevent loss of the soil moisture. The containers shall be marked to indicate the sample number, date, project, feature, structure, hole number, location, depth or elevation at which the sample was taken, penetration resistance record, amount of sample recovered, and classification of soil. If more than one type of soil is encountered in the 1-foot drive, separate containers shall be prepared for each type of soil. The moisture content of one set of samples should be determined in the field and the data recorded on the field log, form 7-1334. The duplicate set shall be stored in suitable boxes for shipment to the laboratories in the Engineering and Research Center, Denver. If facilities are not available in the field for making moisture determinations, both sets of samples shall be shipped to Denver. Samples and containers shall be protected from freezing and breakage.

(d) Complete ground-water information shall be obtained, including ground-water level and elevations at which water was lost or water under pressure was encountered. Where hollow-stem auger, wire line drill rod, or casing is used, ground-water levels shall be measured before and after the specific type of casing is pulled. In sands, the water level shall be determined at least 30 minutes after the boring is completed, and in silts, after 24 hours. In clays it may be difficult to obtain an accurate measurement unless pervious seams are present. However, water levels in clays shall be taken after 24 hours. If ground water is not encountered, such information should also be reported.

(e) All data obtained in the borings shall be recorded on the field log, form 7-1334, and should include all information listed on the form as shown on the example, figure 68. The penetration data shall also be plotted on the form as shown. Comments on reliability of results, use of equipment, and any other information should be noted if pertinent. All samples of strata should be classified, with emphasis on natural or inplace condition of the soils.

NEEDLE-MOISTURE DETERMINATION OF SOILS

Designation E-22

1. Scope.—This designation describes a rapid method for determining the approximate amount of moisture in a soil, by the Proctor needle, provided the needle-moisture relationship for the soil has been determined previously (designation E-11).

2. Apparatus.—The apparatus shall consist of the following (see designation E-4):

> Mallet, item 46.
> Mold, cylindrical compaction, metal, $\frac{1}{20}$ cubic foot, item 52.
> Proctor needle tester, item 66.
> Scoops, items 76 and 77.
> Shovel, item 78.
> Sieve, U.S. standard No. 4, item 80.
> Straightedge trimmer, item 94.
> Tamper, item 97.

3. Sample.—A representative soil sample weighing about 25 pounds shall be selected from a spread layer or other prepared soil before compaction or from the face of the excavation in the borrow pit. The sample shall be sieved in order to provide approximately 10 pounds of soil passing the No. 4 sieve. A space on the ground can be smoothed to catch the material passing the No. 4 sieve.

4. Procedure.—A portion of the soil passing the No. 4 sieve shall be compacted in the mold and the penetration resistance of the compacted specimen determined as provided in designation E-11.

5. Computations.—The penetration resistance in pounds per square inch shall be computed by dividing the average scale reading by the area, in square inches, of the needle used.

6. Report.—The results shall be reported on the form shown on figure 22-1.

FIELD DENSITY OF DRY, GRAVEL-FREE SOILS

Designation E-23

1. Scope.—This designation describes a method for determining the inplace density of dry, gravel-free soils. It is often necessary to determine the natural dry density and water content of fairly deep foundations of

PROCTOR NEEDLE TESTS

DATE _____ DAM

| LOCATION | | NEEDLE MOISTURE BEFORE ROLLING |
STATION AND OFFSET OR COORDINATES	ELEVATION	LBS. PER IN.^{2}

Acceptable Range: from_____psi, to_____psi.

Figure 22-1.—Form for reporting results of Proctor needle test. 101-D-597.

cohesive soils above the water table. The sand density method, given in designation E–24, requires a test pit or large-diameter auger hole to gain access to the foundation. The following simple method has been used successfully to obtain inplace density in stages of depth in foundations and borrow areas of relatively dry, gravel-free soils with the use of a hand auger.

2. Apparatus.—This test is not required for construction control; therefore, the apparatus is not listed in the field laboratory equipment list, designation E–4. The major equipment items are listed below:

(1) *Platform.*—A platform should be built on which the investigator can stand without bearing on the soil within 2 feet of the hole he is to auger. A system of cribbing covered by decking with an opening about 12 inches in diameter in the center will suffice.

(2) *Auger.*—An 8-inch-diameter hand auger.

(3) Measuring tape or rule, 0.01 of a foot graduations.

(4) Inside caliper, 8-inch.

(5) Field weighing scale of 150- or 200-pound capacity.

(6) A tripod and hoist may be used to facilitate augering for deep holes.

3. Procedure.—The procedure is to start a hole with an 8-inch-diameter hand auger, penetrating the soil for a distance of between 6 inches and 1 foot, depending on the probable depth of stripping. The soil removed is discarded, and the depth from the surface of the ground to the apex of the cone at the bottom of the hole is measured to within 0.01 foot. The hole is then excavated with the auger to a depth of 3 feet or to any apparent change in soil structure, whichever occurs first; and the soil removed is placed on a clean canvas and a sample obtained for water content. The depth from the bottom of the hole to the top of the ground is carefully measured again to the nearest 0.01 foot, and the diameter of the hole is measured at about 1 foot below the surface of the ground, using the inside caliper.

4. Computations.—The volume of the hole sampled is the difference between the two depths measured, multiplied by the area of the hole as computed from the measured diameter. Thus, the wet density and the dry density can be determined for each tested depth below stripping. The tests can be continued to the limit of the practicable hand-auger depth, which is about 20 feet for an 8-inch-diameter auger. This depth can be extended by use of a tripod to aid in removing the auger from the hole. Since about 1 cubic foot of material is extracted for a 3-foot depth, an accuracy of weighing to about 1 pound is satisfactory.

This method is applicable only to relatively dry cohesive soils. If the hole tends to cave, this method obviously fails, and test pits or undisturbed sampling devices must be used.

FIELD DENSITY TEST PROCEDURE

Designation E–24

.1. Scope.—This designation describes a method for determining the inplace density of soils using sand for determining the volume of the test hole. The consecutive steps to be followed are given in figure 24–4.

2. Apparatus.—The apparatus shall consist of the following (see designation E–4 and figs. 24–1 and 24–2):

Augers, 4-, 6-, and 8-inch diameter, item 1.
Cans, sample storage, item 14.
Measures, unit weight, cylindrical, metal, item 48.
Picks, item 63.
Pouring devices, and sand cones, 6-, 8-, and 12-inch, item 65.
Proctor needle tester with needles, item 66.
Scale, item 74.
Scoops, items 76 and 77.
Shovel, item 78.
Sieve, item 80.
Straightedges, items 92 and 93.
Templates, item 103.
Trowel, item 106.
Sand.—Approximately 100 pounds of clean, air-dry, uniformly graded sand passing the No. 16 sieve and retained on the No. 30 sieve. Clean "blow sand" or clean dune sand is often suitable for the purpose, as well as the sand furnished by the commercial suppliers for foundries or fracturing sand used by the oil well drilling industry.

(*Note:* When large test holes are used in gravelly soils and there are appreciable voids between the particles, coarse sand having rounded particles and passing the No. 4 and retained on the No. 8 sieve is recommended.)

3. Calibration of Density Sand.—The sand shall be calibrated before using in the density tests to determine its weight per unit volume when poured. The sand-pouring or sand-cone device shall be filled with clean, dry sand and weighed.

The poured density of the sand is affected by atmospheric moisture and changes in relative humidity. If tests are made at infrequent inter-

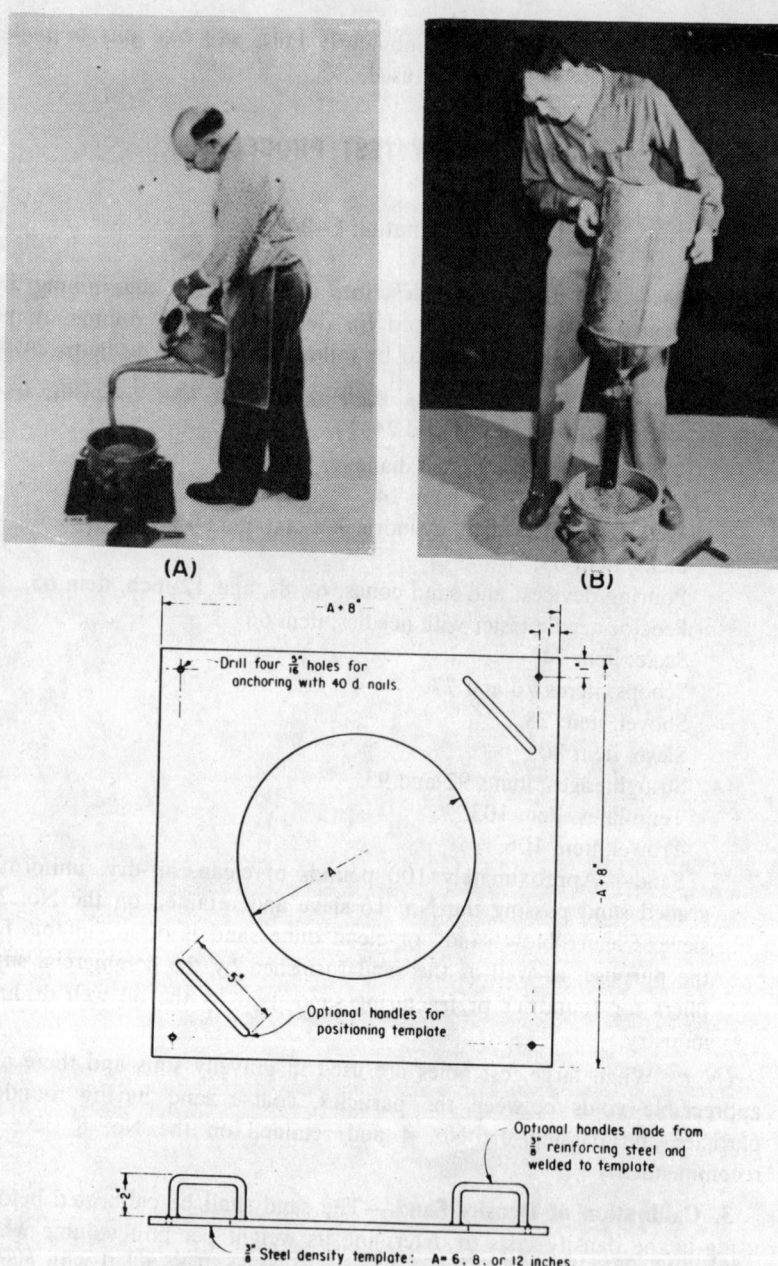

(A)

(B)

(C)

Figure 24-1.—Sand pouring devices (A) and (B), and template (C) for inplace soil density tests. E-1908-3 (PX-D-16698), E-1908-1 (PX-D-16699), and 101-D-615 (template).

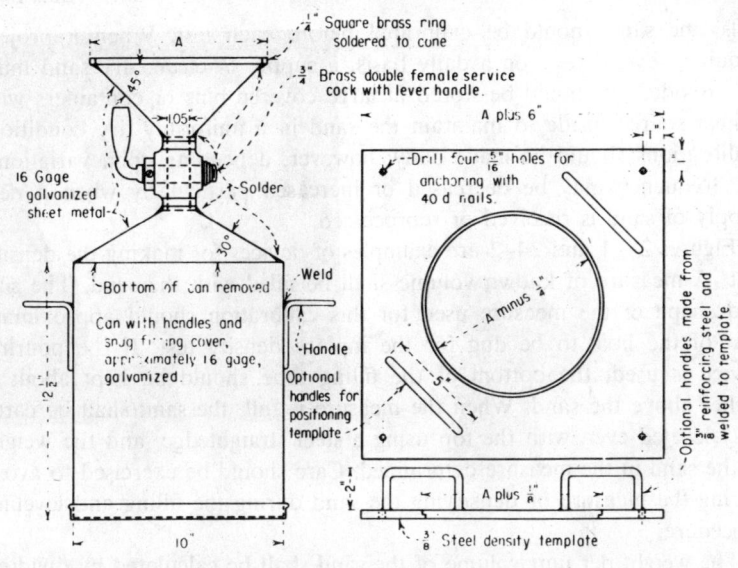

DIMENSIONS

DENSITY HOLE		CONE AND TEMPLATE
DIAMETER INCHES	DEPTH INCHES	A INCHES
6	9	6
8	12-14	8
10-12	12-14 *	12

* Conical shaped hole

(A) SAND CONE DEVICE (FOR INPLACE SOIL DENSITY TEST)

(B) CONE DEVICE DESIGNED FOR USE ON SLOPES (REMOVABLE STANDARD 6-BY 12-INCH METAL CAN USED FOR SAND CONTAINER)

**Figure 24–2.—Cone-type sand pouring device for inplace soil density tests.
101–D–349 and PX–D–16700.**

vals, the sand should be calibrated before each test. When a project requires several tests on a daily basis, a supply of clean, dry sand must be provided. It should be stored in large covered bins or containers with a heat source inside to maintain the sand in a uniformly dry condition. Calibrations should be made daily; however, depending upon variations, the frequency may be decreased or increased particularly when a new supply of sand is received or reprocessed.

Figures 24–1 and 24–2 are examples of devices for making the density test. A measure of known volume shall be filled with the sand. The size and shape of the measure used for this calibration should approximate that of the hole to be dug for the inplace density test. If the pouring device is used, the bottom of the filling tube should be kept about 2 inches above the sand. When the measure is full, the sand shall be carefully leveled even with the top using a steel straightedge, and the weight of the sand in the measure determined. Care should be exercised to avoid jarring the measure or densifying the sand during the filling and leveling procedure.

The weight per unit volume of the sand shall be calculated by dividing the weight of sand in pounds by the volume of the measure in cubic feet. Several trials shall be made until consistent results with a variation of less than ½ percent have been obtained (line 6, form 7–1425, fig. 24–3).

The weight of sand required to fill the hole in the template shall be determined by placing the template on a smooth surface covered with paper, then carefully filling the hole with density sand. Several trials shall be made and an average value obtained (line 4, fig. 24–3). Care shall be exercised to avoid densifying the sand when leveling to the top of the template.

When using the sand-cone device, the weight of sand required to fill the cone and template shall be determined. The cone device is filled with sand and weighed. The template is placed on a smooth, flat surface. The cone is placed in position, and the valve is opened. After the cone and template are filled with sand, the valve is closed, and the cone device and sand remaining in the device are weighed. Several trials shall be made and the average difference in weight recorded (line 4, fig. 24–3).

4. Procedure.—(a) At the location to be tested, all loose soil shall be removed from an area 18 to 24 inches square, and the exposed area shall be leveled until a firm, smooth surface is obtained. The operator should not step on or near the area selected for testing. A working platform, supported at least 3 feet from the edge of the test hole, should be provided when testing materials which may flow or deform during the test.

(b) The size of density hole or pit shall be selected in accordance with the following criteria:

Form 7-1425(11-58)
Bureau of Reclamation

FIELD DENSITY TEST RECORD
(Including Rapid Compaction Control)

TEST NO. *Example* FEATURE_____ TESTED BY_____

LOCATION_____ OFFSET____ ELEV____ ZONE____ COMPUTED BY_____

SOURCE OF MATERIAL _____ DATE OF TEST_____

(1) Wt. Sand & Can, No., 1 ,	94.1 lbs.	(22)	Rock Water Content, $\frac{[(15)-(21)]}{(21)} \times 100$	1.1	%
(2) Wt. Sand Residue & Can	16.3 lbs.				
(3) Wt. Sand used, [(1)-(2)]	77.8 lbs.	(23)	Wt. Wet Soil, [(10)-(15)]	65.1	lbs.
(4) Wt. Sand in plate, No. 4,	11.0 lbs.	(24)	Wet Density of Soil, $\frac{(23)}{[(7)-(17)]}$	137.1	p.c.f.
(5) Wt. Sand in Hole, [(3)-(4)]	66.8 lbs.	(25)	Wt Dry Soil, $\frac{(23)}{1+W_f}$	55.8	lbs.
(6) Sand Calibration	84.4 p.c.f.	(26)	Wt Dry Soil & Rock, [(25)+(21)]	102.7	lbs.
(7) Vol. of Hole, $\frac{(5)}{(6)}$	0.791 ft³	(27)	Percentage of Rock, $\frac{(21)}{(26)} \times 100$	45.7	%
(8) Wt. Wet Soil, Rock, & Can	115.7 lbs.	(28)	Water Content, Soil & Rock, $\frac{[(10)-(26)]}{(26)} \times 100$	9.5	%
(9) Wt. of Can, No. 2,	3.2 lbs.				
(10) Wt. Wet Soil & Rock, (8)-(9)	112.5 lbs.	(29)	Fill Cyl. Needle.[b]	860	p.s.i.
(11) Wet Density Soil & Rock $\frac{(10)}{(7)}$	142.2 p.c.f.	(30)	Needle at opt.[b]		p.s.i.
			RAPID CONTROL VALUES		
(12) Dry Density Soil & Rock, $\frac{(11)}{1+(28)}$	129.9 p.c.f.	(31)	D =		%
		(32)	C =		%
		(33)	w_o-w_f =		%
(13) Wt. Wet Rock[o] & Pan	50.0 lbs.	(34)	Fill Water Content, w_f	16.7	%
(14) Wt. of Pan	2.6 lbs.	(35)	Fill Dry Density, No. 4, $\frac{(24)}{1+W_f}$	117.5	p.c.f.
(15) Wt. Wet Rock,[o] [(13)-(14)]	47.4 lbs.	(36)	Max. Lab. Dry Density		p.c.f.
(16) Wt. Rock in Water	27.7 lbs.	(37)	Cylinder Dry Density		p.c.f.
(17) Volume of Rock[o] $\frac{(15)-(16)}{62.4}$ *	0.316 ft³	(38)	Opt. Water Content, W_o		%

Method	Passes		
(18) Sp. G. of Rock,[o] $\frac{(15)}{[(15)-(16)]}$: 2.41	Tamping roller	✓	Canal Lining ✓
(19) Wt. Dry Rock & Pan 49.5 lbs.	Tractor treads		Embankment
(20) Wt. Pan 2.6 lbs.	Equip. Tamp.		Str. B'kfill
	Power Tamp.		Unified Soil Class **GC**
(21) Wt. Oven Dry Rock 46.9 lbs. [(19)-(20)]	Hand Tamp.		
	REMARKS_____		

* Or (17)$_A$ by measuring the water displaced from a siphon can; then:
$$(18)_A = \frac{(15)}{(17) \times 62.4}$$
(a) Wet surface dried condition.
(b) If obtained for moisture control by needle moisture test.

Figure 24-3.—Field density test record (including rapid compaction control).
101-D-226.

(1) For soils with little or no gravel:

Canal lining, embankments, structure backfill, foundation, or subgrade—6-inch diameter, 9-inch depth, cylindrical hole.

Earth dam embankment—8-inch diameter, 12- to 14-inch depth, cylindrical hole.

(2) For soils containing appreciable gravel:

10- to 12-inch diameter, 12- to 14-inch depth, conical hole.

(3) For cohesionless soils containing appreciable cobbles:

Small test pit at least 2 by 3 feet in size and 1½ to 2 feet deep.

(c) The template with the proper size hole shall be selected and placed firmly on the test area. An excavation slightly smaller than the hole in the template shall be dug to a depth of approximately 6 inches. This may be done with hand tools, or a soil auger may be used when gravel is not present. While excavating, extreme care shall be exercised to avoid deforming the hole; the movement of heavy equipment in the immediate test area should not be permitted.

All material taken from the excavations described in (1) and (2) above shall be placed in an airtight container for subsequent weighing in the field or laboratory. For (3) above, only a representative sample need be protected against moisture loss. To avoid undue loss of moisture, the cover shall be kept on the container at all times when soil is not being placed in it, and in hot, dry climates, shade for the test area and a damp cloth over the container should be provided. A sealed plastic bag can be used inside the container to hold the soil removed and prevent moisture loss.

The excavation shall be continued to the required depth and the hole carefully trimmed by hand to the required size and shape to remove any material that has been compacted or loosened by the auger or hand tools. The weight of the wet soil and container shall be determined to 0.01 pound and recorded (line 8, fig. 24–3).

(d) The volume of the hole (line 7, fig. 24–3) shall be determined by carefully filling the hole with calibrated density sand using a sand-pouring device or sand-cone device. The weight of the container filled with sand shall be determined to 0.01 pound and recorded (line 1, fig. 24–3). The sand shall be poured using the same procedure as for calibrating the sand in the laboratory. When the pouring device is used, the sand shall be carefully leveled at the top of the template with the steel straightedge and all excess sand returned to the pouring container and weighed (line 2, fig. 24–3).

When the sand-cone device is used, the cone shall be placed on the template, the valve opened, and the sand allowed to fill the hole and

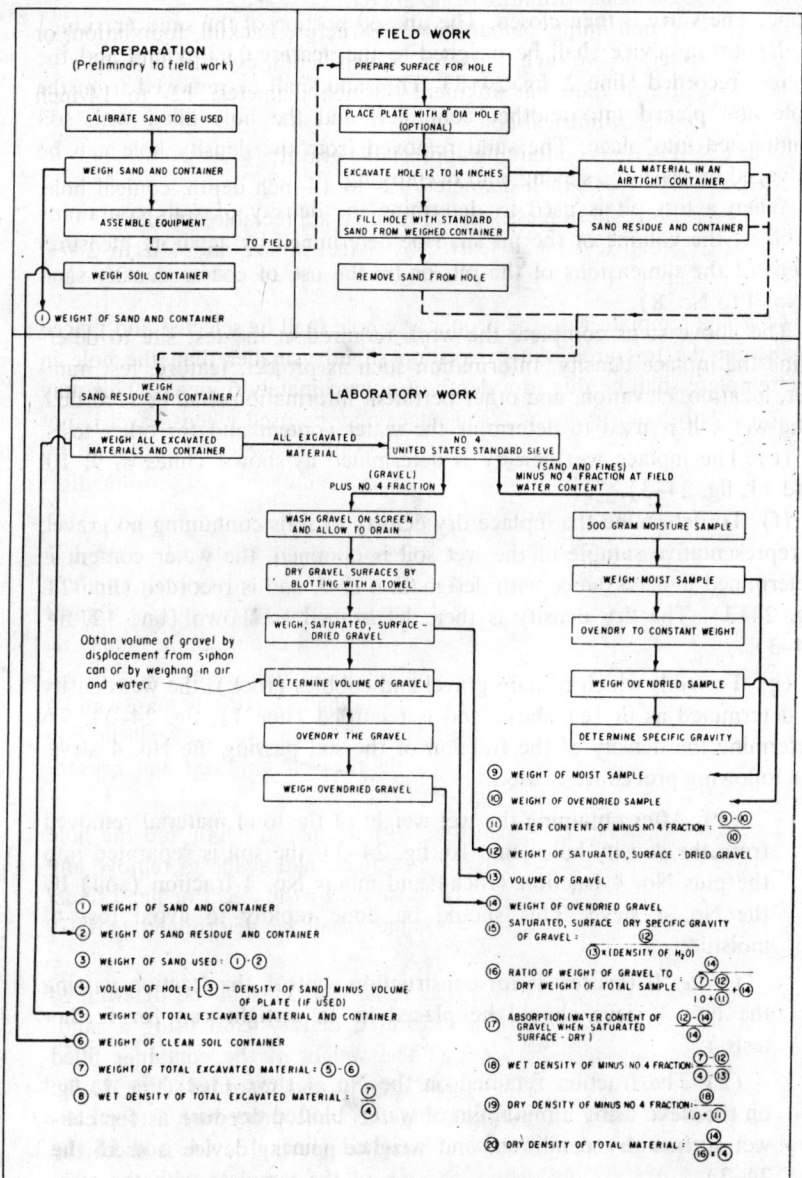

Figure 24-4.—Procedure for inplace density test. 101-D-285.

cone. The valve is then closed. The unused portion of the sand (residue) and pouring device shall be weighed to the nearest 0.01 pound and the weight recorded (line 2, fig. 24–3). The sand shall be removed from the hole and placed into another container, and the hole filled with soil compacted into place. The sand removed from the density hole can be salvaged by washing, sieving, and ovendrying.

When a test pit is used to determine the density of soils containing cobbles, the volume of the pit shall be determined by accurate measurement of the dimensions of the pit, or by the use of coarse density sand (No. 4 to No. 8).

The above steps complete the work required at the test site to determine the inplace density. Information such as project, feature, test number, location, elevation, and other pertinent information shall be recorded. The wet soil is used to determine the water content and for other tests.

(e) The inplace wet density is determined as shown (lines 8, 9, 10, and 11, fig. 24–3).

(f) To determine the inplace dry density of soils containing no gravel, a representative sample of the wet soil is obtained, the water content is determined in accordance with designation E–9, and is recorded (line 28, fig. 24–3). The dry density is then determined as shown (line 12, fig. 24–3).

(g) For soils which contain gravel and cobbles (rock), the wet density is determined as in (e) above and is recorded (line 11, fig. 24–3). To determine the density of the fraction of the soil passing the No. 4 sieve, the following procedure is used:

(1) After obtaining the wet weight of the total material removed from the density hole (line 10, fig. 24–3), the soil is separated into the plus No. 4 fraction (rock) and minus No. 4 fraction (soil) by the No. 4 sieve. This should be done rapidly to avoid loss of moisture.

(*Note:* If this test is for construction control, the fraction passing the No. 4 sieve should be placed in an airtight can for further tests.)

(2) The fraction retained on the No. 4 sieve (rock) is washed on the sieve using a minimum of water, blotted dry with a towel to a wet surface-dry condition, and weighed (lines 13, 14, and 15, fig. 24–3).

(3) The volume of the rock, in a wet surface-dry condition, is then determined by displacement of water from a siphon can from which the overflow can be accurately measured, or by weighing in air and in water (line 17, fig. 24–3). The specific gravity of rock (line 18) may then be computed.

(*Note:* For construction control, the volume of rock need not be measured every time a test is made. After several tests have shown that the specific gravity of the rock from a particular source is virtually constant, the specific gravity can be assumed and the volume computed by obtaining the weight of the rock in a wet, surface-dry condition (line 15) and dividing the weight by the assumed unit weight of rock (specific gravity times 62.4 pounds per cubic foot.))

(4) The wet rock is placed in an oven and the oven-dry weight and water content determined (lines 19, 20, 21, and 22, fig. 24–3).

(5) The wet weight and wet density of the soil fraction passing the No. 4 sieve are determined as shown (lines 23 and 24, fig. 24–3).

(6) The water content of the soil fraction passing the No. 4 sieve is determined as in designation E–9, and recorded (line 34, fig. 24–3), and dry weight of soil computed (line 25).

(7) The dry density of the soil fraction passing the No. 4 sieve is computed by dividing the wet density of soil (line 24) by the quantity, 1 plus water content expressed as a decimal, and recording on line 35.

(8) The water content of total material (soil and rock) is the difference between total wet weight (line 10) and the sum of the dry weight of soil and rock (line 26) divided by the latter.

(9) The percentage of rock in the material on a dry weight basis is computed as shown (line 27, fig. 24–3).

(10) The dry density of the total material (soil and rock) (line 12, fig. 24–3) is computed by dividing the wet density of soil and rock (line 11) by 1 plus the water content, expressed as a decimal, of the soil and rock.

RAPID COMPACTION CONTROL

Designation E–25

1. Scope.—The field density test discussed in designation E–24 determines the dry density and the water content of the compacted fill. For control purposes, these values must be compared with the laboratory maximum dry density and the optimum water content. The rapid control procedure described herein yields the exact ratio of fill dry density to laboratory maximum dry density and a very close approximation of the difference between optimum water content and fill water content of a field density sample, without requiring determinations of water contents. Thus, compaction control can be effected within 1 hour from the time the field test is made. Only one water content—the fill water content—is measured, and after it is available (usually the following day) the values of field dry density, cylinder dry density at fill water content, laboratory maximum dry density, and optimum water content are determined for record purposes. The theoretical basis for this method was described by Hilf.[1]

Succeeding parts of this designation list the laboratory equipment, describe required laboratory procedures, explain the rapid control technique, and present the forms necessary for recording the data. The rapid control procedure is also used for comparing the inplace density and moisture conditions in borrow areas and foundations with standard laboratory values.

2. Apparatus.—The items of equipment required for the rapid compaction control method are listed below and are shown in figure 25-1. Each item is listed according to its number on the photograph, followed by its name and then by the item number given in the equipment list in designation E-4. The asterisks indicate mechanical equipment not required when handmixing methods are used.

    1. Scale, fan type, 30-pound capacity, graduated to 0.01 pound (with locking device), item 75.

    2. Scale weights, item 75.

    3. Needles for Proctor-needle tester, item 66.

    4. Proctor-needle tester (penetrometer), item 66.

    5. Mallet, wooden, item 46.

    6. Tamper (5.5-pound compaction hammer with 18-inch guide), item 97.

    7. Knife, large butcher, item 42.

    8. Brush, wire, item 11.

    9. Scoop, small hand, item 77.

    10. Pan, large mixing, item 61.

    11. Clipboard for data sheets, item 17.

    12. *Fan, electric, item 33.

    13. Scoop, large hand, item 76.

    14. Glass graduate, 100-milliliter capacity, item 38.

    15. *Mixer paddle, type 3, item 51E.

    16. *Mixer bowl, with collar, 3-gallon capacity, item 51.

    17. *Mixer, with ⅓-horsepower, 115-volt, 60-cycle-per-second electric motor, item 50.

    18. Towel, item 105.

---

[1] Hilf, J. W., "A Rapid Method of Construction Control for Embankments of Cohesive Soils," Engineering Monograph No. 26, Department of the Interior, Bureau of Reclamation, September 1961.

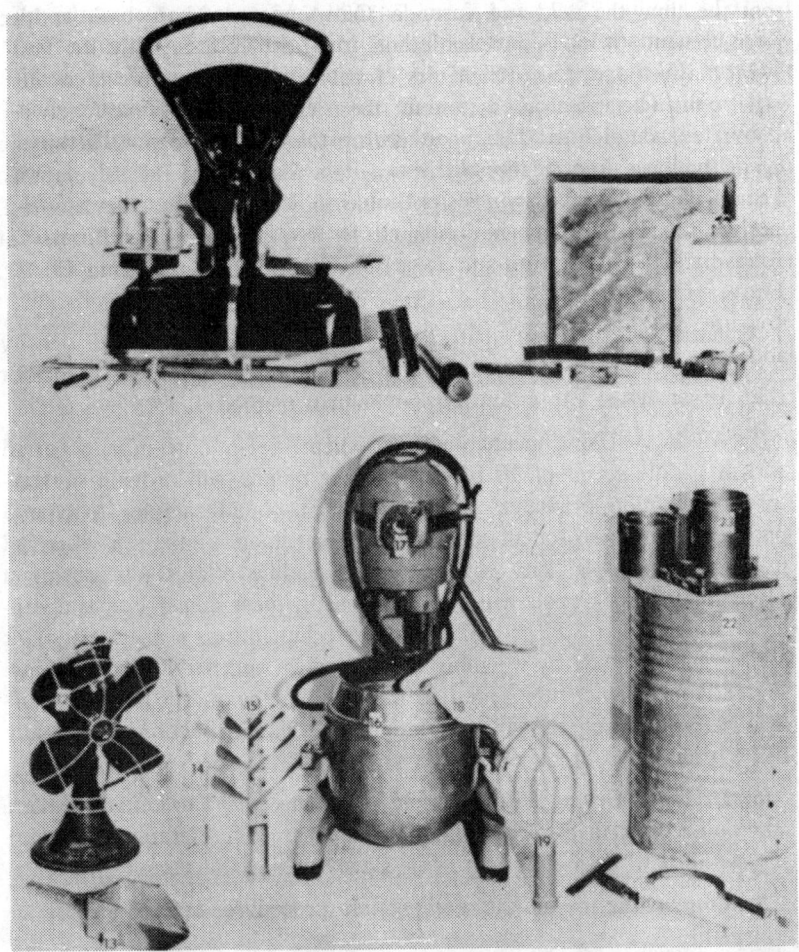

Figure 25-1.—Laboratory equipment required for rapid compaction control. E-1861-2.

19. *Mixer paddle, type 2, item 51D.

20. Straightedge trimmer (T-bar), item 94.

21. Spanner wrench for compaction cylinder mold, item 86.

22. Base, compaction cylinder mold, laboratory (concrete-filled carbide can), item 6.

23. Mold, compaction cylinder, $\frac{1}{20}$-cubic-foot, with baseplate and collar, item 52.

Note: Mechanical compactor, item 48A, may be used in lieu of items 52, 86, and 97.

3. Forms.—The forms needed to obtain rapid control are form 7–1624, figure 25–2, and form 7–1624A, figure 25–3, both entitled "Rapid Compaction Control Method for 7.50 Pounds of Moist Soil". Figures 25–6 and 25–7 are tables of coordinates of the maximum density point. The values used to enter these tables are differences between converted wet densities obtained during the test. Form 7–1911, figure 25–8, may be used to record the test data and outline the calculations. Forms and full-size drawings applicable to Rapid Compaction Control, designation E–25, may be obtained by request from the Bureau of Reclamation, Engineering and Research Center, Denver Federal Center, Denver, Colo. 80225.

4. Sample.—Sufficient material is obtained from the field density test (designation E–24) to provide 7.50 pounds of minus No. 4 soil for each point in the rapid compaction control method.

5. Mixing.—Using mechanical equipment, a 1-minute mixing period with a paddle speed of 90 revolutions per minute will provide satisfactory distribution of water through the 7.50-pound sample. A type 3 paddle may be used for all soils; however, satisfactory mixing is obtained in soils of low plasticity with a type 2 paddle. Water loss during a 1-minute mixing period will be negligible in most cases. For temperatures greater than 80° F. and low relative humidities, a check of water loss should be made by weighing the soil, bowl, and paddle after mixing. Water is most efficiently added to the sample by pouring it on the surface before starting the mixer. Handmixing should be done as rapidly and thoroughly as practicable.

6. Drying.—Drying a soil sample, when necessary to remove water, may be accomplished fairly rapidly by blowing air (preferably warm air) across the bowl or mixing pan while slowly mixing the sample.

7. Compacting.—For the compaction procedure, refer to designation E–11.

8. Rapid Method.—(a) Compact a sample of minus No. 4 fraction of the soil by the standard method of laboratory compaction at fill water content. Record the data on form 7–1911, figure 25–8, and plot the resulting cylinder wet density as point (1) on the 0-percent vertical line shown on the graph of form 7–1624, figure 25–2, or of form 7–1624A, figure 25–3. The latter form is used for density values below 110 pounds per cubic foot.

(b) To obtain point (2), weigh out 7.50 pounds of soil at fill water content. Add 68 cubic centimeters = 0.15 pound (2 percent) water; mix and compact into a cylinder to determine the wet density. Convert the wet density to "wet density at fill water content (called the converted wet density)" by dividing the wet density by 1.02. Plot the converted wet density on the +2-percent vertical line. The division can be done graphically by using the diagonal lines on figure 25–2 or figure 25–3. Instructions for the graphical procedure are given on figure 25–4.

(c) To obtain point (3) on the plot, the procedure varies depending on the relative positions of points (1) and (2). If point (2) has a *greater* converted wet density than point (1), to 7.50 pounds of soil at fill water content, add 136 cubic centimeters = 0.30 pound (4 percent) of water, mix, and compact into a cylinder to determine the wet density. Convert wet density to wet density at fill water content by dividing the wet density by 1.04. Plot the converted wet density on the +4 percent vertical line. The division can also be done graphically using the diagonal lines.

If the converted wet density of point (2) is *less* than point (1), permit 7.50 pounds of soil at fill water content to dry without loss of soil, then weigh. (*Note:* If the converted wet density of point (2) is less than point (1) but is within 3 pounds per cubic foot, the requirement for drying may be eliminated. Point (3) is found by adding 1 percent of water as explained in paragraph 9 of this designation.) The table on the right-hand portion of form 7–1624, figure 25–2, gives the percentage of water loss corresponding to the weight of the dried soil. Compact the dried soil into a cylinder and determine its wet density. Convert this wet density to wet density at fill water content by dividing the wet density by the quantity (1 + the negative percentage of water lost, as indicated by the table). Plot point (3) on the vertical line corresponding to the correct negative percentage, interpolating as necessary. If desired, the graphical method employing the diagonal lines may be used to obtain converted wet density.

(d) Three points, (1), (2), and (3), are sufficient to determine the curve of wet density at fill water content if the wet densities at fill water content of both the left and right points are less than the center point. If not, a fourth point is necessary except in the special procedure described in paragraph 9 below. Find the maximum density point of the converted wet density curve from the coordinates obtained from figure 25–6. Note that the table can be used only if points (1), (2), and (3) are equally spaced horizontally in 2-percent increments.

The maximum density point of the converted wet density curve can also be found by the graphical parabola method shown on figure 25–5. The graphical parabola method must be used if the points are not equally spaced in 2-percent increments. The maximum density point can be found by sketching the converted wet density curve if the number and location of points permit accuracy by sketching.

(e) To obtain *D*, the ratio of fill dry density to laboratory maximum dry density, divide the value of fill wet density (item 24, figure 24–3,

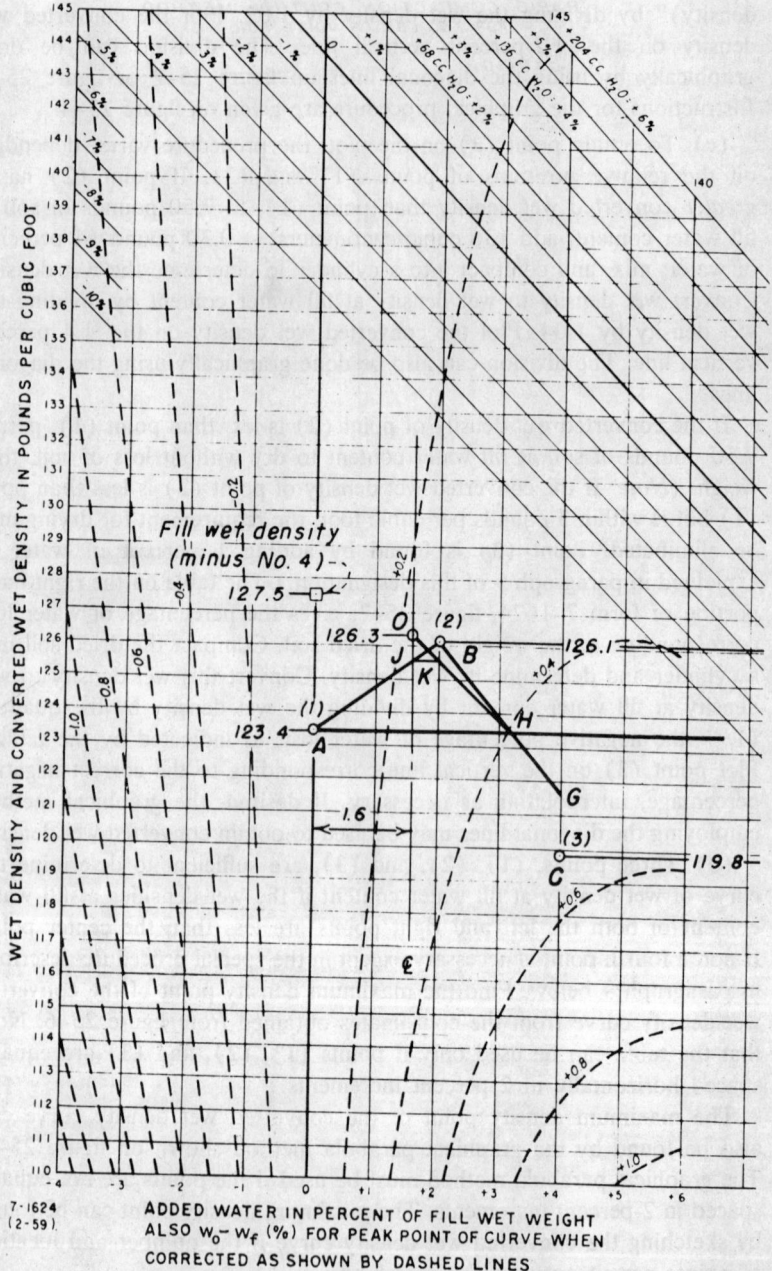

Figure 25-2.—Rapid compaction control method for 7.50 pounds of moist soil (normal density range). (Sheet 1 of 2.) 101-D-324.

RAPID COMPACTION CONTROL METHOD
FOR 7.50 LBS. OF MOIST SOIL

Example

FEATURE .... DATE .... TEST NO. ....

DRIED WEIGHT	%	DRIED WEIGHT	%
7.49	-0.1	7.32	-2.4
7.48	-0.3	7.31	-2.5
7.47	-0.4	7.30	-2.7
7.46	-0.5	7.29	-2.8
7.45	-0.7	7.28	-2.9
7.44	-0.8	7.27	-3.1
7.43	-0.9	7.26	-3.2
7.42	-1.1	7.25	-3.3
7.41	-1.2	7.24	-3.5
7.40	-1.3	7.23	-3.6
7.39	-1.5	7.22	-3.7
7.38	-1.6	7.21	-3.9
7.37	-1.7	7.20	-4.0
7.36	-1.9	7.19	-4.1
7.35	-2.0	7.18	-4.3
7.34	-2.1	7.17	-4.4
7.33	-2.3	7.16	-4.5

$D = 101.0$ %

$C = 103.3$ %

$w_0 - w_f = +1.8$ %

Fill water content, w_f : 15.0 %

Fill dry density of - No 4 : 110.9 $\#/ft^3$

Lab max. dry density : 109.8 $\#/ft^3$

Cylinder dry density : 107.3 $\#/ft^3$

Optimum water content, w_0 : 16.8 %

Figure 25-2.—Rapid compaction control method for 7.50 pounds of moist soil (normal density range). (Sheet 2 of 2.) 101–D–324.

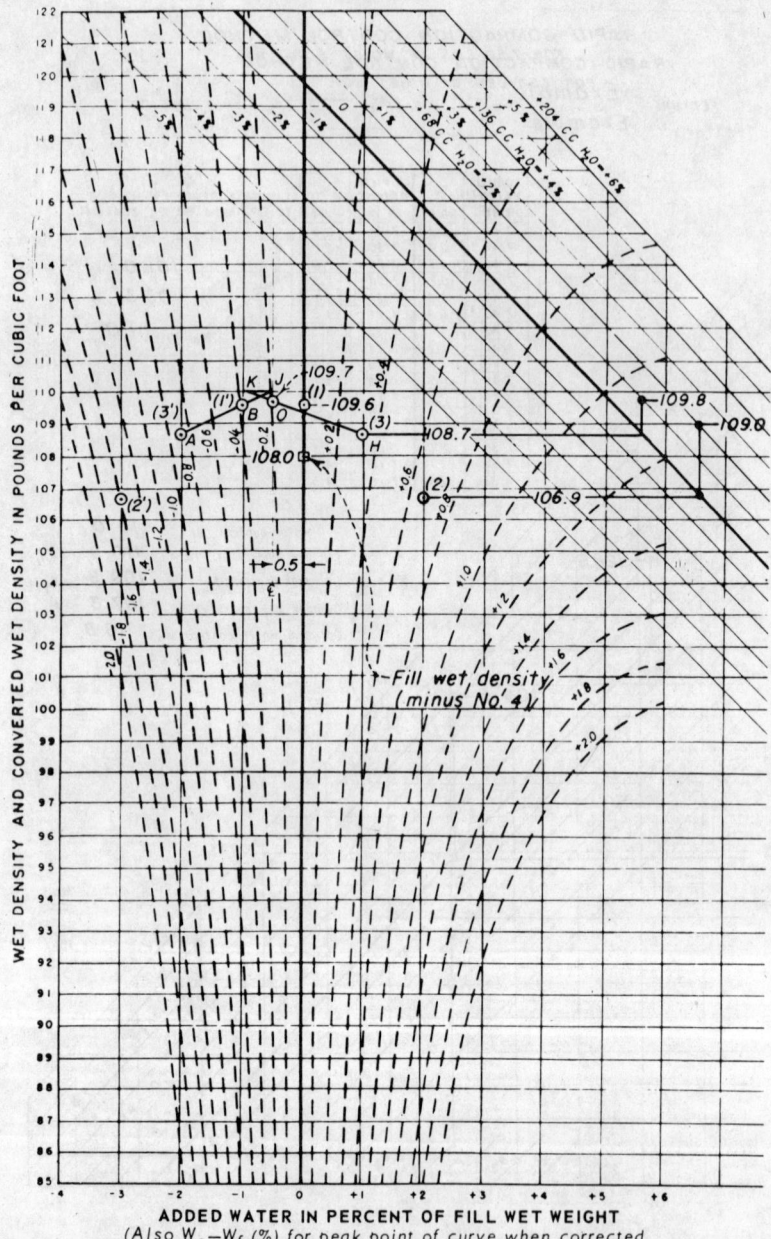

Figure 25–3.—Rapid compaction control method for 7.50 pounds of moist soil (low density range). (Sheet 1 of 2.) 101–D–325.

RAPID COMPACTION CONTROL METHOD
FOR 7.50 LBS. OF MOIST SOIL

7-1624A
(12-59)

FEATURE ..... *Example* ..... DATE ..... ——— ..... TEST NO. ——

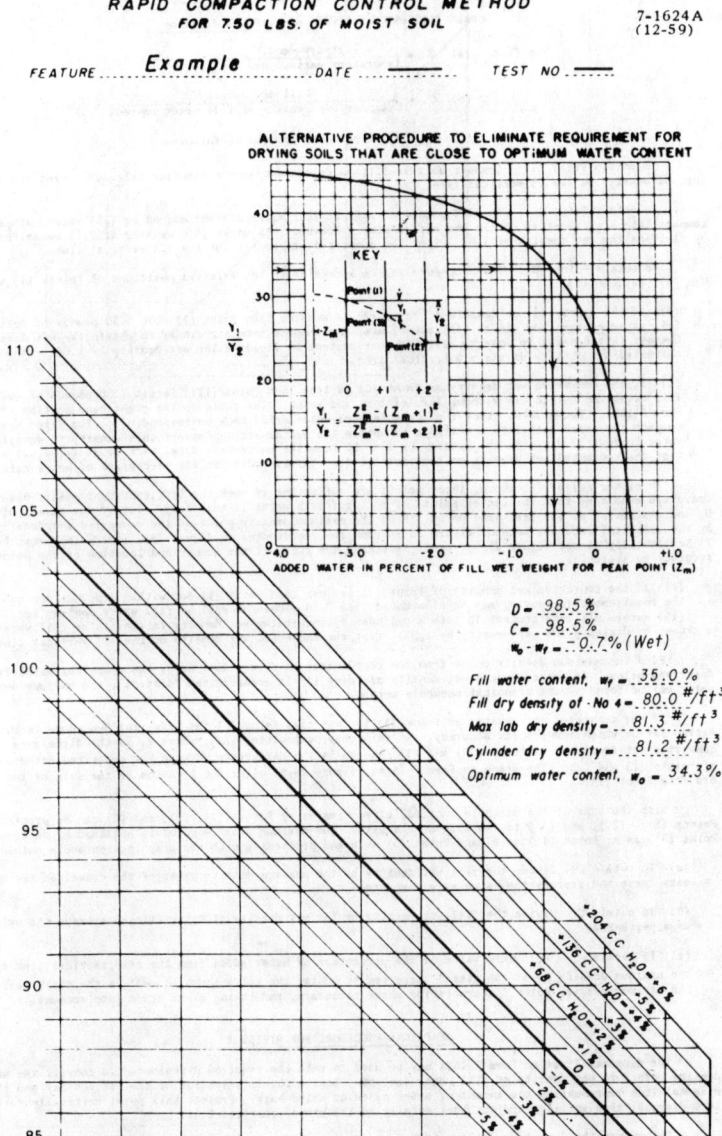

ALTERNATIVE PROCEDURE TO ELIMINATE REQUIREMENT FOR
DRYING SOILS THAT ARE CLOSE TO OPTIMUM WATER CONTENT

$$\frac{Y_1}{Y_2} = \frac{Z_m^2 - (Z_m + 1)^2}{Z_m^2 - (Z_m + 2)^2}$$

ADDED WATER IN PERCENT OF FILL WET WEIGHT FOR PEAK POINT (Z_m)

D = 98.5 %
C = 98.5 %
$w_o - w_f$ = – 0.7 % (Wet)

Fill water content, w_f = 35.0 %
Fill dry density of -No 4 = 80.0 #/ft³
Max lab dry density = 81.3 #/ft³
Cylinder dry density = 81.2 #/ft³
Optimum water content, w_o = 34.3 %

Figure 25–3.—Rapid compaction control method for 7.50 pounds of moist soil (low density range). (Sheet 2 of 2.) 101-D-325.

PROCEDURE FOR RAPID COMPACTION CONTROL METHOD

Obtain fill wet density of minus No. 4 fraction

To find: (a) $D = \dfrac{\text{Fill dry density}}{\text{Laboratory maximum dry density}} \times 100$

           (b) $C = \dfrac{\text{Fill dry density}}{\text{Cylinder dry density at fill water content}} \times 100$

           (c) $w_o - w_f$ in percent; proceed as follows:

To obtain Point (1): Compact soil at fill water content into a standard cylinder. Plot the resulting wet density on the 0% vertical line.

To obtain Point (2): To 7.50 pounds of soil at fill water content add 68 cc (2%) water, mix and compact into a cylinder to obtain the wet density. Convert this density to density at fill water content by dividing the wet density by 1.02. Plot the converted wet density on the +2% vertical line.

To obtain Point (3): The procedure varies according to the relative positions of Points (1) and (2). Use (d), (e), or (f).

(d) If the converted wet density of Point (2) is greater than Point (1): To 7.50 pounds of soil at fill water content add 136 cc (4%) water, mix, and compact into a cylinder to obtain the wet density. Convert this density to density at fill water content by dividing the wet density by 1.04. Plot the converted wet density on the +4% vertical line.

(e) If the converted wet density of Point (2) is less than Point (1): Permit 7.50 pounds of soil at fill water content to dry without loss of soil and weigh. The table on the right-hand portion of Form 7-1624, figure 25-2, gives the percentage of water dried back corresponding to the dried weight. Compact the dried soil into a cylinder to obtain the wet density. Convert this density to density at fill water content by dividing the density by 1 + the negative percentage dried back (e.g. 1+(-0.02) = 0.98). Plot the converted wet density on the vertical line corresponding to the percentage of water dried back.

In either (d) or (e) above, three points are sufficient if both the left and right points have lower converted densities than the center point; if not, a fourth point is necessary. Letter the three points A, B, and C starting at the left-hand point. Find the maximum density point of the converted wet density curve by the graphical parabola method, figure 25-5; from the coordinates on figure 25-6 (the points must be spaced 2% horizontally to use figure 25-6); or by sketching the curve if the number and location of the points permit accuracy by sketching.

(f) If the converted wet density of Point (2) is less than Point (1) but within 3 pounds per cubic foot, the requirement for drying may be eliminated: To 7.50 pounds of soil at fill water content add 34 cc (1%) water, mix, and compact it into a cylinder. Convert the wet density to density at fill water content by dividing the wet density by 1.01. Plot the converted wet density on the +1% vertical line.

Find the maximum density point from the coordinates on figure 25-7 or by the graphical parabola method described below. If the converted wet density of Point (3) is greater than Point (1), the maximum density point may be found by the graphical parabola method.

If the converted wet density of Point (3) is less than Point (1) the graphical procedure requires extrapolation which reduces its accuracy. In this case, calculate Y_1/Y_2, where Y_1 is the difference in converted wet densities of Points (1) and (3), and Y_2 is the difference between the converted wet densities of Points (1) and (2). The graph on Form 7-1624A, figure 25-3, gives the location of the axis of the parabola from Point (1).

With the axis of the parabola located, mirror images of Points (1), (2), and (3) can be plotted as Points (1'), (2'), and (3') to the left of the axis. The maximum density point is obtained by designating Point (3') as A, Point (1') as B, and Point (3) as H and proceeding with the graphical parabola method.

(a) To obtain D: Divide the fill wet density by the maximum density point of the converted wet density curve and express the value as a percentage.

(b) To obtain C: Divide the fill wet density by the wet density of Point (1) and express the value as a percentage.

(c) To obtain $w_o - w_f$: This value is the percentage of water added from the zero vertical line to the maximum density point (abscissa) corrected by adding the value shown in red* on the chart nearest the maximum density point, interpolating where necessary, and taking minus signs into account.

GRAPHICAL PROCEDURE FOR DIVISION

The diagonal lines on Form 7-1624 may be used to make the required divisions. To convert the wet density points to wet density at fill water content: Locate the intersection of the wet density and the diagonal line representing the amount of water added or dried back. Project this point vertically to the zero diagonal line and then project horizontally to the proper vertical line.

To perform the divisions required for D and C: Plot the fill wet density on the zero vertical line and project this value horizontally across the form. Project the maximum density point and the density of Point (1) horizontally to the zero diagonal line. Project vertically to the value of the fill wet density. D = 100 plus the interpolated percentage on the diagonal lines given by the intersection of the fill wet density and the maximum density point, taking minus signs into account. C = 100 plus the interpolated percentage on the diagonal lines given by the intersection of the fill wet density and the density of Point (1) taking minus signs into account.

_____

* Note: These lines are shown as dashed lines on the examples, figures 25-2 and 25-3.

Figure 25-4.—Procedure for rapid compaction control method. 101-D-521.

GRAPHICAL PARABOLA METHOD

Graphical solution for the maximum density point, O, of the converted wet density curve, given three points A, B, and C. If more than three points are available, use the three closest to optimum.

NOTE: Solution based on the assumption that converted wet density curve is a parabola whose axis is vertical.

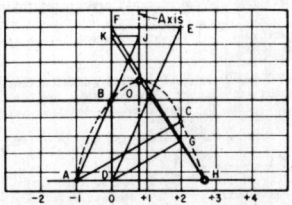

1. Draw horizontal base line through the left point, A, and draw vertical lines through points B and C.
2. Draw line DE parallel to AB, point E lies on the vertical line through point C; project E horizontally to establish point F on the vertical line through B.
3. Draw line DG parallel to AC, point G lies on the vertical line through point C.
4. Line FG intersects the base line at H. Axis of parabola bisects AH; draw the axis.
5. Intersection of line AB with the axis is at J; project J horizontally to K, which lies on the vertical line through point B.
6. Line KH intersects the axis at O, the maximum density point (vertex).

NOTE: If points A, B, and C are equally spaced horizontally (this is true when 2 points are obtained by adding water or when soil is dried exactly 2 percent) steps 2 and 3 above are eliminated. Point F coincides with point B and point G is halfway between the base line and point C. Hence, point H is obtained by drawing BG and point O is obtained by steps 5 and 6 as usual. See graph below.

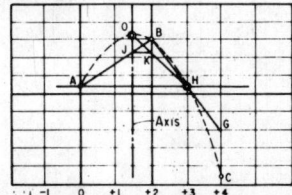

Completion of Test for Record Purposes

Ovendry a sample of minus No. 4 fraction to obtain fill water content, w_f. Then:

Fill dry density of-No. 4 = Fill wet density of - No. 4 / ($1 + w_f$)
Lab. max. dry density = Maximum density point of converted wet density curve / ($1 + w_f$)
Cylinder dry density = Wet density of Point (1) / ($1 + w_f$)
Optimum water content = $w_f + [(1 + w_f)$ (Added water at maximum density point)]

NOTE: In the foregoing calculations, use decimals for water contents or added water (e.g. 0.16, 0.02).

Figure 25-5.—Graphical parabola method. 101-D-522.

designation E–24) by the maximum density point of the converted wet density curve and express the value as a percentage. This division can be done graphically by using the diagonal lines on the forms as described in figure 25–4.

(f) To obtain *C*, the ratio of fill dry density to cylinder dry density at fill water content, divide the fill wet density by the wet density of point (1) and express the result as a percentage. This division can be done graphically by using the diagonal lines on the forms.

Figure 25–6.—Rapid method of compaction control—Tabulation of coordinates of maximum density point. (Sheet 1 of 2.) 101-D-518.

(g) To obtain a close approximation of the value of the difference between the optimum water content and the fill water content, $(w_o\text{-}w_f)$, find the percent of added water at the maximum density point (abscissa) on figure 25–2 or 25–3 and correct it by adding the value shown on the curved dashed line nearest to the point, interpolating where necessary and taking minus signs into account. (*Note:* The curved dashed lines on figures 25–2 and 25–3 are red on the actual forms.) Added water and correction values should be estimated to the nearest 0.1 percent.

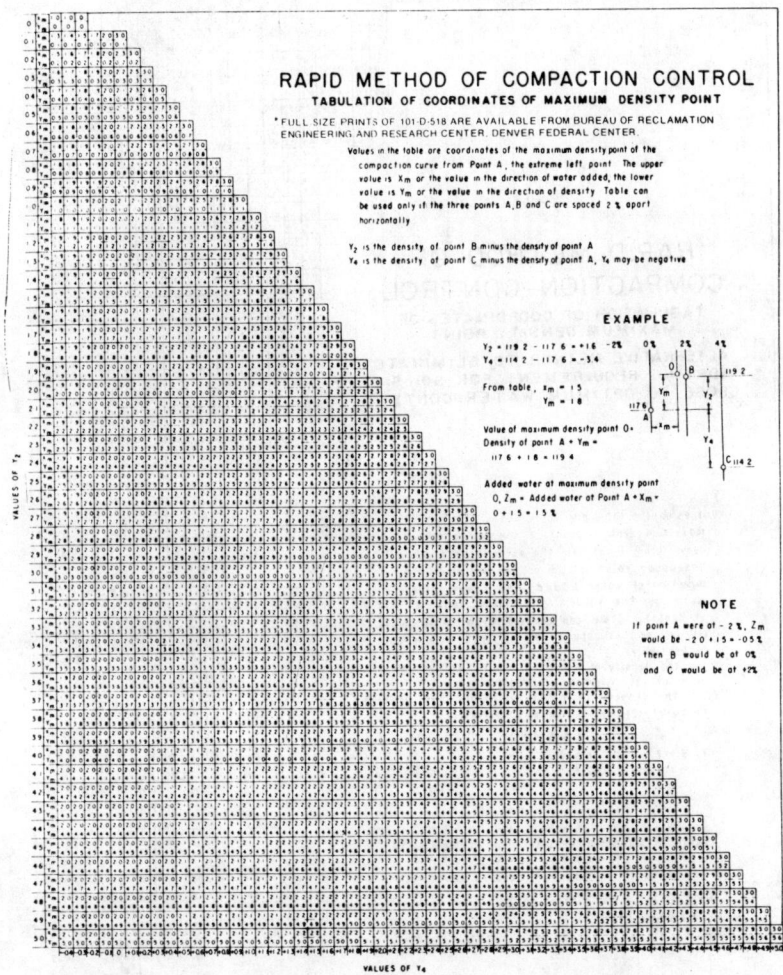

Figure 25–6.—Rapid method of compaction control—Tabulation of coordinates of maximum density point. (Sheet 2 of 2.) 101–D–518.

(h) Rapid control values obtained in (e), (f), and (g) above, yield sufficient information to either accept or reject the work. For record purposes, determine the fill water content of the minus No. 4 fraction of the soil, (w_f), by the procedure given in designation E–9. When the result of the fill water content determination is known, compute the fill dry density of minus No. 4 material, the laboratory maximum dry density, the cylinder dry density at fill water content, and the optimum water content by the following formulas:

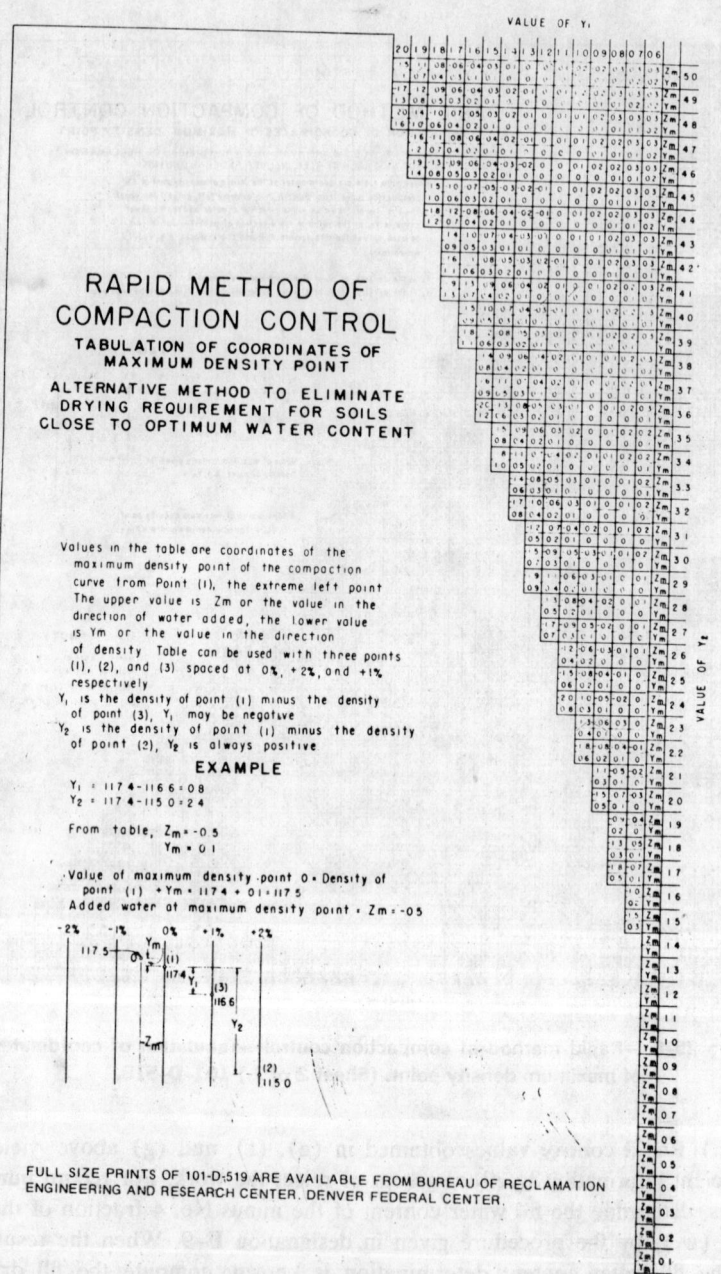

Figure 25–7.—Rapid method of compaction control—Alternative method to eliminate drying requirement for soils close to optimum water content. (Sheet 1 of 2.) 101–D–519.

VALUE OF Y_1

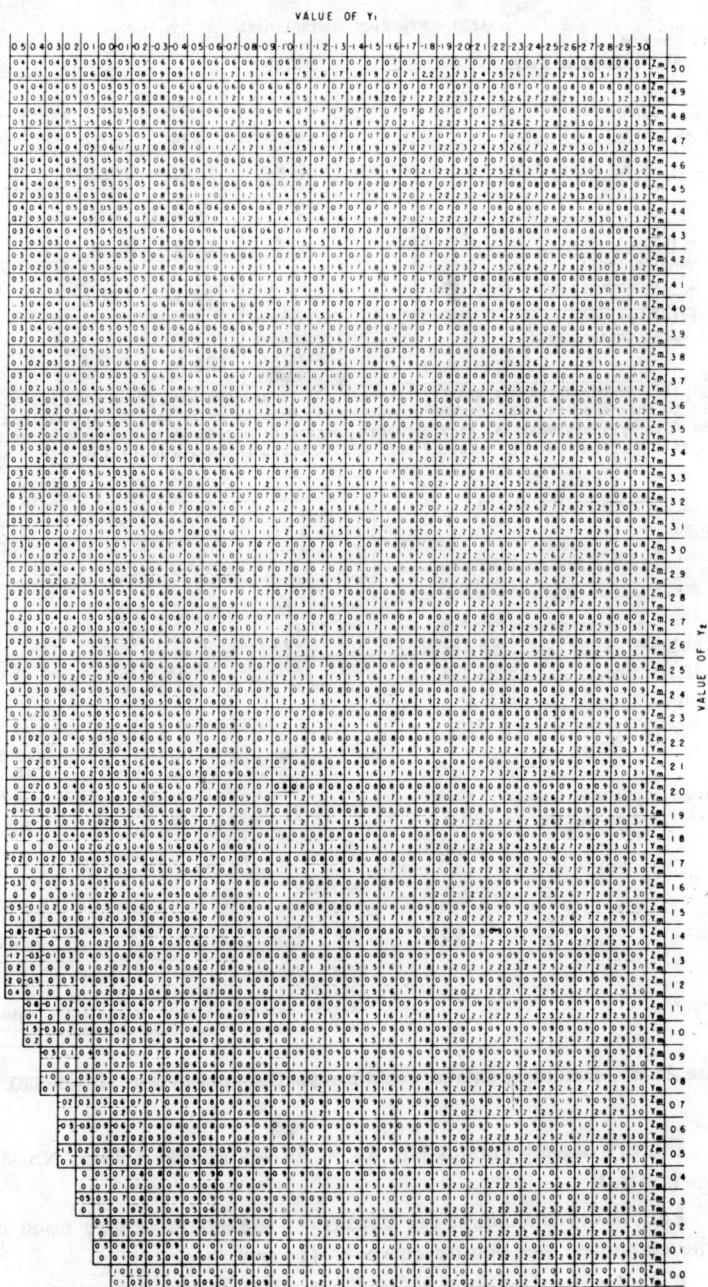

Figure 25–7.—Rapid method of compaction control—Alternative method to eliminate drying requirement for soils close to optimum water content. (Sheet 2 of 2.) 101–D–519.

7-1911 (11-71)
Bureau of Reclamation

RAPID COMPACTION CONTROL DATA

TEST NO. *Example (a)*   FEATURE   TESTED BY   DATE:

Point				FIELD (1)	(2)		
Water Added (%) as decimal (a)				0.00	0.02	0.04	
Wt. Cyl. & Wet Soil (b)				11.58	11.84	11.64	
Wt. Cyl. (c)				5.41	5.41	5.41	
Wt. of Wet Soil (d) = (b)-(c)				6.17	6.43	6.23	
Wet Density (pcf) (e) = $\frac{(d)}{Cyl.\ Vol.}$				123.4	128.6	124.6	
Converted Wet Den. at Fill Water Content (pcf) $\frac{(e)}{1 + (a)}$				XXXXXXXXX XXXXXXXXX XXXXXXXXX XXXXXXXXX	126.1	119.8	

NOTE: If the converted wet density of point (2) is greater than point (1), add 4% water to obtain point (3). If the converted wet density of point (2) is less than point (1), dry back material at fill water content to obtain point (3). If point (2) is less than point (1) but within 3 pcf, add 1% water to obtain point (3).

FIELD MOISTURE DETERMINATION	
Dish No.	
Wt. Wet Soil + Dish	505.00
Wt. Dry Soil + Dish	460.00
Wt. Dish	160.00
Wt. of Water	45.00
Wt. of Dry Soil	300.00
Moisture (%)	15.0
Dry Density (pcf)	110.9

Fill Wet Density of minus
No. 4 fraction (γ_{w_f}) =   127.5

$D = \dfrac{\gamma_{w_f}}{Max.\ Dens.}$ =   101.0%

$C = \dfrac{\gamma_{w_f}}{Pt.\ (1)}$ =   103.3%

$(w_o - w_f)$ =   + 1.8%

Y_1/Y_2 CHART (Figure 25-7)

Point 1	Point 1
Point 3	Point 2
Y_1 (1-3)	Y_2 (1-2)
Z_m	Y_m
Red Line Correction	Point 1
$(w_o - w_f)$ $(Z_m + Red\ Line)$	Max. Den. $(Y_m + Pt.\ 1)$

NOTE: Use when points (1), (2), and (3) are spaced at 0%, 2%, and 1%, respectively.

Y_2/Y_4 CHART (Figure 25-6)

Point B	126.1	Point C	119.8
Point A	123.4	Point A	123.4
Y_2 (B-A)	2.7	Y_4 (C-A)	-3.6
X_m (From Chart)	1.6	Y_m (From Chart)	2.9
Z_m = Added water at Point A + X_m 1.6			
Red Line Correction	+0.2	Point A	123.4
$(w_o - w_f)$ $(Z_m + Red\ Line)$ +1.8		Max. Den. $(Y_m + Point\ A)$	126.3

NOTE: Use when points are spaced at 2% horizontally. Letter the three points nearest optimum Point A on the left, Point B in the center, Point C on the right.

Figure 25–8.—Recording data for rapid compaction control test. 101-D-520.

Fill dry density (minus No. 4) = fill wet density (minus No. 4) $\div (1+w_f)$

Laboratory maximum dry density = maximum density point of the converted wet density curve $\div (1+w_f)$

Cylinder dry density at w_f = wet density of point (1) $\div (1+w_f)$

Optimum water content = w_f + [$(1+w_f) \times$ (percent of added water at the maximum density point (abscissa))], where w_f = fill water content. Both w_f and added water at the maximum density point should be expressed as decimals.

Enter these values on form 7–1425, figure 24–3, to complete the record and to make the values readily available for the periodic earthwork progress report, form 7–1352, figure 98, for earth dam embankments or form 7–1581B, figure 114, for miscellaneous structures.

9. Alternative Method to Eliminate Drying Requirement for Soils Close to Optimum Water Content.—The procedure given in paragraph 8(c) for obtaining point (3) on the converted wet density curve requires drying a sample when the converted wet density of point (2) is less than point (1). When the difference between the converted wet densities of points (1) and (2) is 3 pounds per cubic foot or less, the time-consuming operations of drying and accurate reweighing may be eliminated in the large majority of cases by the procedure in the following paragraph. The use of this method requires that the soil is close enough to the optimum water content so the portion of the compaction curve developed is represented by a parabola. This is generally true if the difference between the wet densities of points (1) and (2) is 3 pounds per cubic foot or less. This method should not be used if the difference is greater unless prior experience with the soil indicates that it is still in the parabolic section.

Find point (3) by adding 34 cubic centimeters (1 percent) of water to 7.50 pounds of soil at fill water content, mixing, and compacting into a cylinder to determine the wet density. Convert the wet density to wet density at fill water content by dividing the wet density by 1.01. Plot the converted wet density on the +1-percent vertical line. This division can be done graphically by use of the diagonal lines on the forms. The maximum density point of the converted wet density curve can be found from the coordinates obtained on figure 25–7. This table gives the coordinates from point (1).

The maximum density point of the converted wet density curve can also be found by the graphical parabola method as described below: If the converted wet density of point (3) thus obtained is greater than point (1), the maximum density point of the converted wet density curve can be obtained graphically by the parabola method shown in figure 25–5. If the converted wet density of point (3) is less than point (1), the graphical procedure for obtaining the maximum density point requires extrapolation which reduces its accuracy. Instead, calculate Y_1/Y_2, where Y_1 is the difference in converted wet densities of points (1) and (3) and Y_2 is the difference in converted wet densities of points (1) and (2). The graph on form 7–1624A, figure 25–3, gives the location of the axis of the parabola (distance Z_m from the origin) for the ratio of Y_1/Y_2.

With the axis of the parabola determined, the mirror images of points (1), (2), and (3) can be plotted as points (1′), (2′), and (3′) to the left of the axis. The maximum density point of the converted wet density curve, point 0, is obtained by designating point (3′) as A, point (1′) as B, and point (3) as H, and proceeding with the parabola method construction. An example of this procedure is given in paragraph 10(c) below. (*Note:* If the value Y_1/Y_2 exceeds 0.38, or if the Z_m obtained from figure 25–7 is less than −1.0, this method is considered inapplicable and should not be used. Instead, a fourth point on the converted wet density curve should be obtained by drying a 7.50-pound sample of soil at fill water content as described in paragraph 8(c); the maximum density point is then determined as described in paragraph 8(d) using the three points closest to optimum which are spaced 2 percent horizontally.)

10. Examples.—(a) Fill Water Content Less than Optimum (see figure 25–2). — Given: Fill wet density (minus No. 4) = 127.5 pounds per cubic foot.

The procedure described in paragraphs 8(a), (b), and (c) results in the following data. The data and calculations are also shown on figure 25–8, form 7–1911.

Point	Wet density, pounds per cubic foot	Added water in percentage of fill wet weight	Converted wet density, pounds per cubic foot
(1)_____	123.4	0	123.4
(2)_____	128.6	+2	126.1
(3)_____	124.6	+4	119.8

The maximum density point of the converted wet density curve can be found with the coordinates from figure 25–6 as follows:

$$Y_2 = B-A = 126.1 - 123.4 = 2.7$$
$$Y_4 = C-A = 119.8 - 123.4 = -3.6.$$

From the table, $X_m = 1.6$ and $Y_m = 2.9$, which give the following values for the maximum density point 0: (+1.6, 126.3).

The maximum density point may also be found by the graphical parabola method shown on figure 25–5.

In accordance with paragraphs 8(e), (f), and (g), the following control values are obtained:

$$D = \frac{127.5}{126.3} = 101.0 \text{ percent}$$

$$C = \frac{127.5}{123.4} = 103.3 \text{ percent}$$

$$(w_0 - w_f) = +1.6 + 0.2 = +1.8 \text{ percent (dry of optimum).}$$

After the fill water content has been determined by drying a sample to constant weight at 110° C., the field density test is completed for record purposes as follows:

Fill water content (from procedure given in designation E-9) = 15.0 percent

Fill dry density (minus No. 4) $= \dfrac{127.5}{1.150} = 110.9$ pounds per cubic foot

Laboratory maximum dry density $= \dfrac{126.3}{1.150} = 109.8$ pounds per cubic foot

Cylinder dry density at fill water content $= \dfrac{123.4}{1.150} = 107.3$ pounds per cubic foot

Optimum water content $= 0.150 + (1.150)(0.016) = 0.168$ or 16.8 percent.

(b) Fill Water Content Greater than Optimum.—Given: Fill wet density (minus No. 4) = 125.8 pounds per cubic foot.

The procedure described in paragraphs 8(a), (b), and (c) results in the following data:

Point	Wet density, pounds per cubic foot	Added water in percentage of fill wet weight	Converted wet density, pounds per cubic foot
(1)	128.4	0	128.4
(2)	124.2	+2	121.8
(3)	123.7	−2.3 (dried)	126.6

By the graphical parabola method shown in figure 25–5, the maximum density point of the converted wet density curve is (−0.7, 128.9).

Note: This problem must be solved by the graphical parabola method since the points are not equally spaced at 2-percent intervals horizontally.

In accordance with paragraphs 8(e), (f), and (g), the following control values are obtained:

$$D = \frac{125.8}{128.9} = 97.6 \text{ percent}$$

$$C = \frac{125.8}{128.4} = 98.0 \text{ percent}$$

$(w_o - w_t) = -0.7 - 0.1 = -0.8$ percent (wet of optimum).

After the fill water content has been determined by drying a sample to constant weight at 110° C. the field density test is completed for record purposes as follows:

Fill water content (from procedure given in designation E-9) = 18.0 percent

Fill dry density (minus No. 4) = $\frac{125.8}{1.180}$ = 106.6 pounds per cubic foot

Laboratory maximum dry density = $\frac{128.9}{1.180}$ = 109.2 pounds per cubic foot

Cylinder dry density at fill water content = $\frac{128.4}{1.180}$ = 108.8 pounds per cubic foot

Optimum water content = 0.180 + (1.180)(-0.007) = 0.172 or 17.2 percent.

(c) **Fill Water Content Close to Optimum, but Converted Wet Density of Point (2) Less than 3 Pounds per Cubic Foot Smaller than Wet Density of Point (1) (See figure 25-3).** — Given: Fill wet density (minus No. 4) = 108.0 pounds per cubic foot.

Point	Wet density, pounds per cubic foot	Added water in percentage of fill wet weight	Converted wet density, pounds per cubic foot
(1)_____	109.6	0	109.6
(2)_____	109.0	+2	106.9
(3)_____	109.8	+1	108.7

The maximum density point of the converted wet density curve can be found with the coordinates from figure 25-7 as follows:

$$Y_1 = (1) - (3) = 109.6 - 108.7 = 0.9$$

$$Y_2 = (1) - (2) = 109.6 - 106.9 = 2.7$$

From the table, $Z_m = -0.5$ and $Y_m = 0.1$, which give the following values for the maximum density point, 0: (-0.5, 109.7).

The maximum density point can also be found by the graphical parabola method as follows: Find the ratio Y_1/Y_2.

$$Y_1/Y_2 = \frac{0.9}{2.7} = 0.333.$$

From the graph on figure 25-3, the axis of the parabola is found to be at $Z_m = -0.5$ percent. The mirror image points (1'), (2'), and (3') are plotted to the left of the axis of the parabola at the same density as points (1), (2), and (3), respectively. Point (3′) is designated A, point (1′) is designated B, point (3) is designated H, and the parabola construction is used to obtain the maximum density point 0. The coordinates of this point are (−0.5, 109.7).

In accordance with paragraphs 8(e), (f), and (g), the following control values are obtained:

$$D = \frac{108.0}{109.7} = 98.5 \text{ percent}$$

$$C = \frac{108.0}{109.6} = 98.5 \text{ percent.}$$

$$(w_o - w_f) = -0.5 - 0.2 = -0.7 \text{ percent (wet of optimum).}$$

After the fill water content has been determined by drying a sample to constant weight at 110° C., the field density test is completed for record purposes as follows:

Fill water content (from procedure given in designation E–9) = 35.0 percent

Fill dry density (minus No. 4) $= \dfrac{108.0}{1.350} = 80.0$ pounds per cubic foot

Laboratory maximum dry density $= \dfrac{109.7}{1.350} = 81.3$ pounds per cubic foot

Cylinder dry density at fill water content $= \dfrac{109.6}{1.350} = 81.2$ pounds per cubic foot

Optimum water content $= 0.350 + (1.350)(-0.005) = 0.343$ or 34.3 percent.

11. Proctor Needle Readings.—Penetration resistance needle readings required for purposes of maintaining moisture control by means of the needle-moisture test (designation E–22) may be obtained during the rapid method test. The needle readings in the compaction cylinder at fill water content (column 9, form 7–1352, figure 98) may be obtained directly after weighing for point (1) of the rapid method. To obtain the Proctor needle value at optimum water content, needle readings for points (2) and (3), and any other necessary points, should be obtained after weighing. All needle readings may then be plotted on figure 25–2 to any convenient ordinate scale. Needle value at optimum is the intersection of a vertical line through the maximum density point of the converted wet density curve and the Proctor needle curve. The penetration resistance curve will require extrapolation when the method illustrated by example of paragraph 10(c) is used and the soil is wet of optimum. This value can be entered in column 15, form 7–1352, figure 98.

VERTICAL LOAD-SETTLEMENT RELATIONSHIP FOR INDIVIDUAL PILES

Designation E–26
(Adapted from ASTM Designation D 1143–69)

1. Scope.—This method covers a procedure for testing individual foundation piles to determine the relationship between the static load applied to the pile and the settlement of the pile. (*Note:* This method describes only a procedure for testing a single pile. It does not cover the application of the test results to the carrying capacity of a group of piles or to foundation design in general.)

2. Apparatus.—This test is not required for construction control; therefore, the apparatus is not listed in the field laboratory equipment list, designation E–4. The test requires apparatus for applying known vertical loads to the top of the pile (par. 4) and apparatus for measuring the settlement of the pile (par. 5).

3. Pretest Information.—At least one test boring shall be advanced extending a minimum of 10 feet or 5 tip diameters, whichever is greater, below the bottom of the test pile, within 20 feet of the test pile but no closer than 10 feet to determine the soil conditions at the test site. Borings from previous investigations satisfying the foregoing requirements are acceptable. If a good soil profile has been developed from previous investigations, a continuous standard field penetration test may be substituted instead of a test boring, in which case it should be located within 20 feet of the test pile but no closer than 5 feet. Penetration tests referred to above shall be performed in the anticipated bearing layers at intervals not exceeding 5 feet in accordance with designation E–21. Included as information required shall be a log (penetration log, if used) of the soil profile, position of water table, and results of tests indicating the in situ densities, water contents, and shear strengths of the materials encountered. The type of soil sampler used, the weight and drop of the hammer, and its resistance to penetration shall be recorded in accordance with designation E–21.

4. Loading Device.—The apparatus for applying the vertical loads shall consist of one of the following devices:

(1) Setup with load applied to pile by hydraulic jack acting

against anchored reaction member.—Two or more piles shall be used as anchor piles and driven as far from the test pile as practicable but in no case less than 5 feet or 5 butt diameters, whichever is greater, face to face from the test pile. Girders of sufficient strength shall be attached to act as a reaction beam by means of a connection designed to carry the reaction load to the upper ends of the anchor piles. A steel bearing plate of appropriate thickness not less than 1 inch shall be placed to distribute the load on the head of the test pile. A hydraulic jack or system of hydraulic jacks with a calibrated pressure gage shall be interposed between the bearing plate and the underside of the reaction beam to apply the test load to the pile by the jack. A typical setup for this type of test is shown in figure 26–1. The system of jacks and the pressure gage shall be calibrated and tested immediately prior to the test so that the load applied is controlled to within 5 percent of the applied load on the pile. Where a greater degree of accuracy is required, a calibrated load cell or an equivalent device may be used. Jacks equipped with spherical bearing shall be used so that the jacks will bear firmly and concentrically against the pile bearing plate and the reaction beam. The entire system shall be constructed so that the load is applied along the vertical axis of the test pile without producing eccentricity or lateral load components.

(2) Setup with load from weighted box or platform applied by hydraulic jack.—A test box or test platform resting on cribbing shall

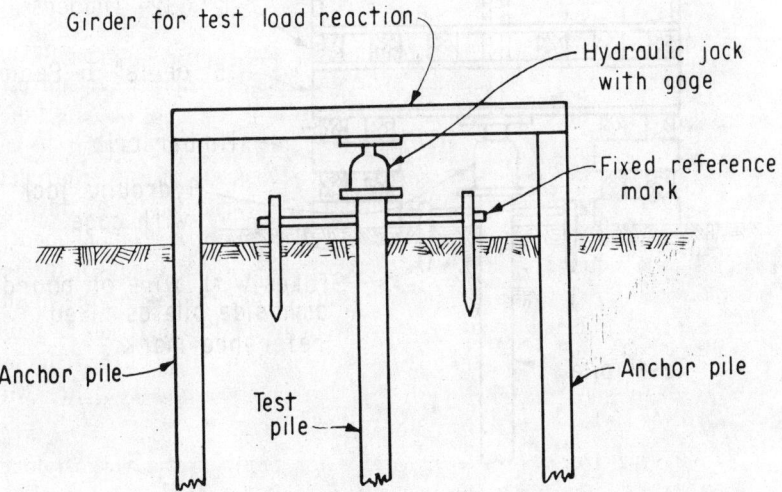

Figure 26–1.—Pile test with load applied by means of a hydraulic jack acting against a reaction member held by anchor piles. 101–D–288.

be constructed over the pile and loaded with earth, sand, concrete, water, pig iron, or other suitable material with a total weight at least as great as the anticipated maximum test load. A calibrated system of hydraulic jacks with a pressure gage shall be interposed between a bearing plate on the pile head and the load box, and load applied to the pile by the jack. The supports for the load box shall be placed as far from the test pile as practicable, preferably so that there is a clear distance of at least 7 feet between the test pile and the supports to provide adequate working space. A typical setup for a test of this type is shown in figure 26–2. Details and accuracy of the jacks, pressure gage, and loading system shall be in accordance with paragraph 4(1) and shall apply the load along the vertical axis of the test pile.

(3) Setup with load supported directly on pile.—A box or platform shall be supported on top of the pile and loaded with earth, sand, concrete, water, pig iron, or other suitable material. The con-

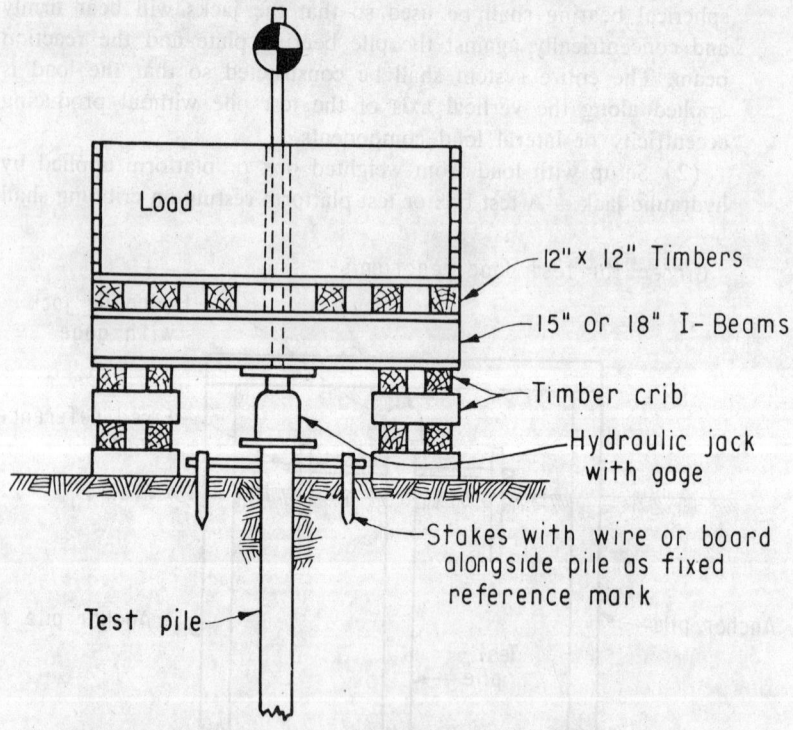

Figure 26–2.—Pile test with load from weighted box or platform applied to pile by means of a hydraulic jack. 101–D–287.

struction of the box and the application of the loads shall be such that no lateral forces will be applied to the top of the pile and no impact will occur as the loads are placed. Each load increment shall be applied gradually to avoid impact. A suitable type of construction is shown in figure 26–3. Loads shall be adjusted so that any wedges will remain loose as settlement occurs. The weight of the box or platform system shall be included in the calculated load on the pile so that the supporting beams and box or platform are in place, but not in contact with the pile, when the "no load" reading is made. In cases where the test pile is in an excavation below the natural ground surface, an extension column of structural steel or steel pipe may be used to extend from the pile head up to the test box or beam. (*Note:* This type of load setup should be used only when the load need not be known within 5 percent. If greater accuracy is required, a calibrated load cell or equivalent load-measuring device may be used. This procedure could be hazardous and proper safety precautions should be taken to avoid tilting or collapse of the box.)

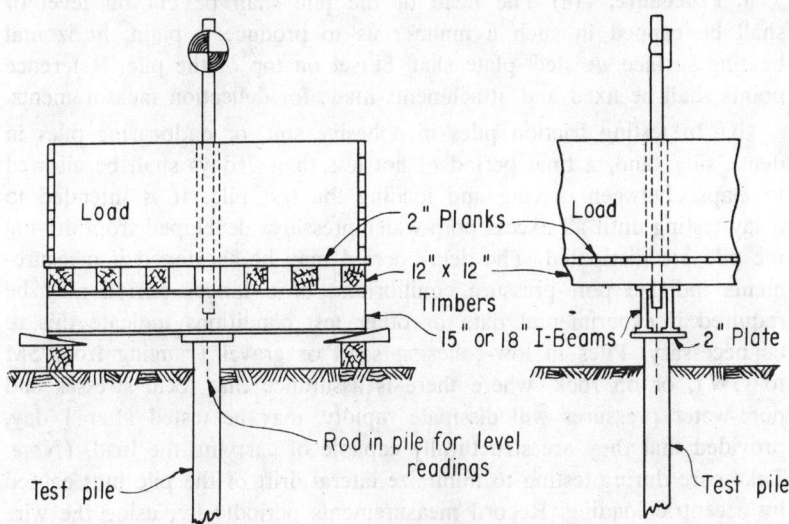

Figure 26–3.—Pile test with load from weighted box or platform supported directly by the pile. 101–D–286.

5. Apparatus for Measuring Settlement.—The apparatus for measuring settlement shall consist of one of the following devices:

(1) Surveyor's level and target rod.—A surveyor's level and target rod reading to 0.001 foot may be used as illustrated in figure 26–3. Two benchmarks shall be established on permanent objects

near the test pile location, and settlement shall be determined by readings made on these benchmarks and on the bolt or rod set in the pile head.

(2) Wire mirror and scale.—A wire shall be stretched between two stakes, each driven into the ground at a distance not less than 10 feet from the center line of the test pile. The wire shall be passed across the face of a scale attached to a mirror which is fixed to the pile, so that settlement readings can be made directly from the scale by lining up the wire and its image in the mirror. Read the scale to the nearest 0.01 inch. Use some suitable method to maintain constant tension in the wire throughout the test.

(3) Dial indicator gage.—A beam shall be attached to two stakes, each driven into the ground at a distance not less than 8 feet from the centerline of the test pile. A dial indicator gage with its stem resting on top of the pile shall be attached to this fixed beam to record the movements of the pile head.

6. Procedure.—(a) The head of the pile shall be cut off level or shall be capped in such a manner as to produce a plain, horizontal bearing surface. A steel plate shall be set on top of the pile. Reference points shall be fixed and attachments made for deflection measurements.

(b) In testing friction piles in cohesive soil, or endbearing piles in dense silty sand, a time period of not less than 7 days shall be allowed to elapse between driving and loading the test pile. It is intended to delay testing until all excess pore-water pressures developed from driving the pile are dissipated. The delay period may be shortened if measurements indicate pore-pressure equilibrium, or a longer period may be required if experimental data or other test conditions indicate this to be necessary. Piles in low-cohesion sand or gravel (ranging from SM to GW), or on rock, where there is assurance that local stresses and pore-water pressures will dissipate rapidly, may be tested after 1 day, provided that they are structurally capable of carrying the load. (*Note:* Take care during testing to minimize lateral drift of the pile butt caused by eccentric loading. Record measurements periodically, using the wire and scale method, dial gages, or an equivalent system to assure minimization of lateral drift.)

(c) The total test load shall be twice the anticipated working load on the pile and shall be applied in increments amounting to 25, 50, 75, 100, 125, 150, and 200 percent of the anticipated working load. (*Note:* Instead of applying a definite amount of load to the test pile, it is recommended that, whenever practicable, loading of the test pile should be continued until rapid progressive settlement occurs.)

Settlement readings made to an accuracy of 0.001 foot shall be taken before and after the application of each new load increment. Additional load shall not be applied until the rate of settlement under the previous increment is less than 0.001 foot in 1 hour or until 2 hours have elapsed, whichever occurs first. As an alternative method of loading, only when specified, the specified load increments may be added in constant time intervals, preferably of not less than 1 hour. Settlement readings shall be made immediately before and after the addition of each load increment and at not less than three specified times between load increments. When loading has been completed, the full test load shall remain on the pile for 24 hours and settlement readings shall be taken during and at the end of that period. (*Note:* When specifically advised, the maximum load shall be maintained for a longer period of time.)

(d) During the unloading of the pile, the rebound shall be measured by means of readings taken when the load remaining on the pile amounts to 50, 25, 10, and 0 percent of the full test load.

7. Report.—The results of the load test shall be reported on data sheets as shown in figures 26–4, 26–5, and 26–6. The following information shall also be reported:

(1) A description of soil conditions at the location of the test pile (include field penetration test or other field data, if available.)

(2) A description of the pile and its driving record, including

PILE DRIVING TEST, DATA SHEET I

PROJECT ___*EXAMPLE*___   FEATURE _____   PILE IDENTIFICATION *BT-19*

DRIVING INSPECTOR _____   DATE _____

RECORDER _____   TIME *3:48 to 4:06 PM*

TYPE OF HAMMER (IF GRAVITY HAMMER, RECORD WEIGHT. THE AVERAGE HEIGHT OF DROP FOR EACH FOOT OR INCH OF PENETRATION SHOULD BE INDICATED ON DATA SHEET 2) *single acting steam*   WEIGHT *5000 lbs.*

BLOWS PER MIN. *55 to 60*   STROKE OF PISTON, FEET *3.0'*   INCHES _____

KIND OF PILE *Timber 30'*   LENGTH DRIVEN *23'*   LENGTH AFTER CUTOFF *25'*

DIAMETER OF PILE TIP *9 1/2"*   DIAMETER AT BUTT *14"*

DIAMETER AT CUTOFF *11 1/2"*   (FOR CONCRETE PILE - DIAMETER INCREASE _____ IN. PER _____ FT)

REMARKS AS APPROPRIATE PILE CONDITION, DIFFICULTY IN DRIVING, AMOUNT DRIVEN AND PULLED, IF PREBORED, DESCRIPTION OF PRE-BORING AND AMOUNT, IF JETTED GIVE DESCRIPTION, DESCRIPTION AND LOG OF SUBSOIL, GROUND WATER ELEVATION, RESULTS OF FIELD PENETRATION TEST OR OTHER FIELD TESTS: GIVE DEPTH AND LENGTH FIGURES FOR ALL REMARKS:

30' pile - straight smooth, few knots and splits, pointed with steel shoe. Exposed length deflected noticeably during first 15' of driving. Ground vibrated a radius of 25'. Reference stake 5' from pile. 2167.509 before driving, 2167.637 after; 0.128' increase in ground surface elevation. (Soil profiles, logs of exploration holes and field penetration given in investigation report.)

Figure 26–4.—Recording information relative to pile driving. 101–D–289.

PILE DRIVING TEST, DATA SHEET 2

PROJECT **EXAMPLE**   FEATURE _____   PILE IDENTIFICATION *BT-19*

DRIVING INSPECTOR _____ DATE _____

RECORDER _____ TIME *3:48 to 4:06 PM*

WEATHER CONDITIONS _____ *Clear, warm breeze* _____

ORIGINAL GROUND ELEVATION _____ *2167.51* _____

NOTE: * WHEN HIGH PENETRATION RESISTANCE IS REACHED, RECORDINGS MAY BE MADE IN TERMS OF INCHES INSTEAD OF FEET. IT IS VERY IMPORTANT TO INDICATE WHICH IS USED.

GRAVITY HAMMER	RECORDED DATA		CALCULATED DATA		GRAVITY HAMMER	RECORDED DATA		CALCULATED DATA	
AVG. HEIGHT OF DROP (FT)	PENETRATION (FT or IN.)*	TOTAL NO. BLOWS AT EACH FOOT or INCH*	BLOWS PER FOOT or INCH*	ELEVATION OF PILE TIP (FT)	AVG. HEIGHT OF DROP (FT)	PENETRATION (FT or IN.)*	TOTAL NO. BLOWS AT EACH FOOT or INCH*	BLOWS PER FOOT or INCH*	ELEVATION OF PILE TIP (FT)
	0			2167.51		21	695	95	2146.51
	1	0 ·				22	811	116	
	2	5	5			23	990	179	2144.51
	3	12	7			23.1	1025		2144.41
	4	20	8		Driving stopped at 1025 blows				
	5	30 ·	10		21 blows/inch				
	6	42	12						
	7	56	14						
	8	72	16						
	9	90	18						
	10	111	21						
	11	136 ·	25						
	12	164	28						
	13	197	33						
	14	235	38						
	15	278	43						
	16	327	49						
	17	384	57		(Observations for 1" penetration)				
	18	446	62		5 blows/inch at 17'				
	19	517	71		6 " " " 18'				
	20	600	83	2147.51	7 " " " 19'-6"				
					9 " " " 21'				
					11 " " " 22'				
					12 " " " 22'-6"				
					14 " " " 22'-9"				
					20 " " " 23'				

Figure 26–5.—Recording pile driving test data. 101-D-290.

PILE LOADING TEST, DATA SHEET 3

PROJECT __EXAMPLE__   FEATURE _____ PILE IDENTIFICATION _BT-19_

PERSONNEL _____ DATE _____

TYPE OF PILE (DESCRIPTION GIVEN ON SHEET I) _____

DATE PILE DRIVEN _____

CONDITION OF PILE _butt at cutoff and exposed 2', not splintered or broomed._

BUTT ELEVATION ___2169.41___   TIP ELEVATION ___2144.41___

GROUND SURFACE ELEVATION (ADJACENT TO PILE AT TIME OF TEST) ___2167.51___

WEATHER CONDITIONS PERTINENT TO TEST (RAIN, WIND, TEMPERATURE) _cold, windy, clear to cloudy_

DESCRIPTION OF ADJACENT STRUCTURES, PILES, IRREGULARITIES IN GROUND SURFACE, AS APPLICABLE. (SKETCH WITH DIMENSIONS, PICTURES, OR OTHER) _Ground surface level over construction area._
Nearest pile 15'

DESCRIPTION OF LOADING EQUIPMENT. (SKETCH WITH DIMENSIONS, AS APPLICABLE, OR OTHER) _____
Reaction load supported on cribbing,
load on pile by hydraulic jack.

METHOD OF MEASURING SETTLEMENT ___Engineers level___

DATE LOADING BEGAN _____   DATE LOADING ENDED _____

DATE AND TIME	ELAPSED TIME (hrs,- min)	LOAD (tons)	SETTLEMENT OBSERVED READINGS	SETTLEMENT COMPUTED PILE MOVEMENT	REMARKS*
Dec. 7, 8:00 A.M.	00:00	0.0		0.000	
	0:02	12.5		"	
	:10	"		"	
	:20	"		"	
	:40	"		"	
	1:00	"		"	
	:01	25.0		0.003	
	:10	"		"	
	:20	"		"	
	:40	"		"	
	2:00	"		"	
	3:00	37.5		0.006	
	4:00	50.0		.008	For illustration:
	5:00	62.5		.011	time intervals, 01
	6:00	75.0		.013	10, 20 40 not shown.
	7:00	100.0		.018	
	30:00	"		.018	No readings
	30:01	50.0		.012	between 7 & 30 hours.
	31:00	"		.012	
	32:00	25.0		.010	Time intervals
	33:00	10.0		.007	not shown.
	34:00	0.0		.005	

*NOTE AND MEASURE LATERAL DEFLECTIONS IF GREATER THAN 1/4 INCH

Figure 26-6.—Recording pile loading test data. 101-D-291.

the date driven, the number of hammer blows per foot throughout the pile length, and the final driving resistance in blows per inch for the last 3 inches of driving.

(3) A description of the driving equipment and driving energy including: (1) for a conventional hammer (steam, air, or diesel), the make, the model, the weight of hammer and ram, the height of fall, the weight and dimensions of drive cap, the type and size of cushion and cap block, the steam, air, or hydraulic pressure supplied to the hammer, and its actual rate of operation during the final driving of the test pile, or (2) for a vibratory hammer, the length and weight of the rotating parts and the relationship of the force-output characteristics to the frequency.

(4) A description, and photographs if possible, of the loading method and measurement procedure and condition of pile.

(5) A tabulation of the loads and settlement readings during the loading and unloading of the pile indicating times involved.

(6) A graphic representation of the test results in the form of a load-settlement curve with a time-settlement curve as an aid in interpretation.

(7) Remarks concerning any unusual occurrences during the driving or loading of the pile.

INSTRUCTIONS FOR INSTALLING AND READING HYDRAULIC-TYPE TWIN-TUBE PIEZOMETERS IN EARTH DAMS

Designation E–27

1. Description of Installation.—Observations on the hydraulic-type twin-tube piezometer installation provide information on pore-water pressures in the embankments and in foundations of earth dams. The apparatus also can be installed in embankments other than earth dams. The installation is comprised of twin-tube piezometer tips, connecting plastic tubing, Bourdon-tube-type pressure gages, accessory valves, air trap, water supply, water filter, and pump. As placing operations progress, piezometer tips are placed in the foundation and in the embankment and are connected by pairs of tubes to gages in the piezometer terminal well generally located near the downstream toe of the dam. Pressures are transmitted from the piezometer tip through the water-filled tubes and are observed on the Bourdon-tube compound-hydrostatic gages mounted in the terminal well. The valves, pump, water supply, and manifold system are used to circulate water in either tube towards the piezometer tip. By circulating liquid through the system, the tubes are completely filled with fluid and air is removed. The manifold in the terminal well allows individual pressures to be recorded without disturbing the equilibrium at the piezometer tip. For an example of the design of a hydraulic-type twin-tube piezometer installation, see figure 27–1.

Pore pressure readings define the areas where pore-fluid pressures will affect the stability of the structure during and after construction. They also provide an indication of embankment control. Since the piezometer installation provides necessary information regarding the stability of a structure under construction and for years thereafter, the apparatus must be installed with great care.

2. History.—Pore-water pressures have been observed and studied by the Bureau of Reclamation since 1935. One of the first practical sys-

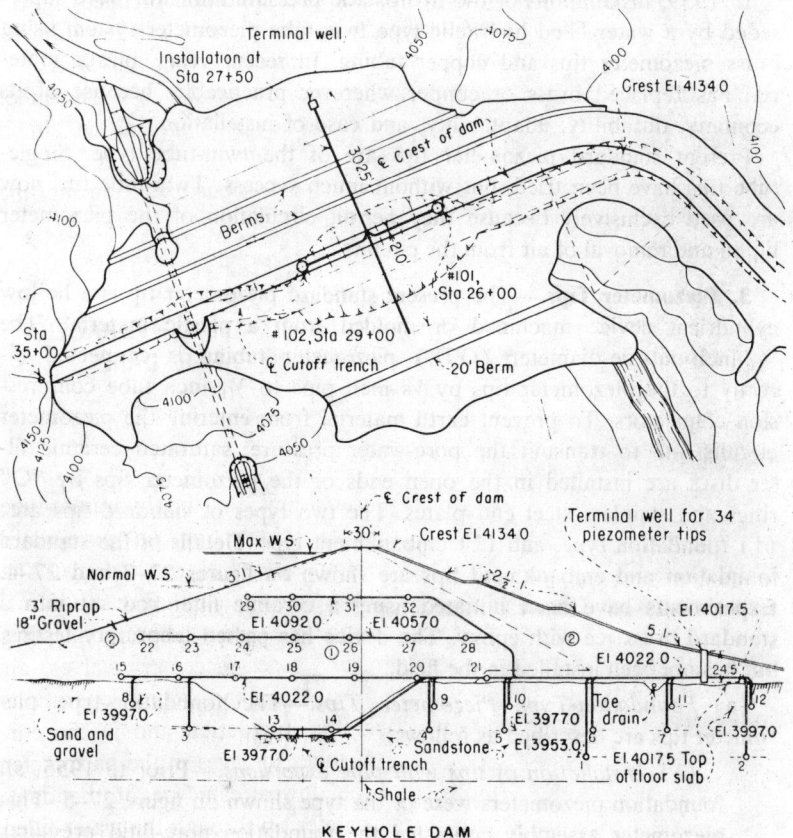

Figure 27–1.—Hydraulic-type twin-tube piezometer installation. 101–D–292.

tems developed to record pressures in earth embankments employed a modified Goldbeck cell which is a type of transducer. This device, known as a hydrostatic pressure indicator, used air pressure to balance the pore-water pressure and break an electrical contact in the cell. Comparative pressure readings are made at both the break and the make points of the contact. Figure 27–2, form 7–1360, shows an example of readings from an indicator installation. Although this system was generally satisfactory, careful machining was required on all the operating parts, and extreme caution was necessary when obtaining readings to prevent rupturing the diaphragm in the cell by application of excessive air pressure. Most failures of this cell are believed to be caused by rupture of the diaphragm or by corrosion resulting from condensation of moisture from the compressed air used in making readings.

In 1939, installations of the hydrostatic pressure indicator were superseded by a water-filled hydraulic-type twin-tube piezometer system using brass piezometer tips and copper tubing. In recent years, plastic material has replaced brass or copper wherever practicable, because of its economy, durability, adaptability, and ease of installation.

Present standard piezometer tips are of the twin-tube type. Single-tube tips have been tried, but without much success. Twin-tube tips now are used exclusively because they permit circulation of the piezometer liquid and removal of air from the circuits.

3. Piezometer Tips.—The present standard piezometer tip is a hollow cylindrical device machined or molded from a plastic material. The $5/16$-inch-outside-diameter (O.D.) piezometer tubing is connected directly to the piezometer tips by $1/8$-inch pipe to $5/16$-inch tube compression connectors. To prevent earth material from entering the piezometer circuits and to transmit the pore-water pressure, saturated ceramic filter discs are installed in the open ends of the piezometer tips by "O" rings and stainless steel end plates. The two types of standard tips are: (1) foundation type, and (2) embankment type. Details of the standard foundation and embankment tips are shown on figures 27–3 and 27–4. Experiments have been initiated using a ceramic filter rod set into a standard brass tee with epoxy. The device has passed laboratory testing, but has not been installed in the field.

(a) *Foundation-Type Piezometer Tips.*—The foundation-type piezometer tips are described as follows:

(1) *Installation of tips with pipe extensions.*—Prior to 1955, all foundation piezometers were of the type shown on figure 27–5. This piezometer assembly contacted the foundation pore-fluid pressures through a pipe extending vertically downward or laterally into the dam foundation. This tip contained a single porous disc. The section

FORM 7-1360 (8-70)
Bureau of Reclamation

TO : DIRECTOR OF DESIGN & CONST
ENGRG & RESEARCH CENTER
DENVER FEDERAL CENTER
DENVER, COLORADO 80225

EXAMPLE

HYDROSTATIC PRESSURE INDICATOR READINGS

Dam .... ISLAND PARK .... Date of Observations .... October 3, 1970

Project .... Minidoka .... Observer .... John E. Williams ....

Ref. Dwg. .... 42-D-338 .... Sheet .... 1 .... of .... 1 ....

Reservoir Water El. .... 6295.36 .... Tailwater El. .... --- ....

INDICATOR NUMBER	PRESSURE		INDICATOR NUMBER	PRESSURE		INDICATOR NUMBER	PRESSURE	
	BREAK	MAKE		BREAK	MAKE		BREAK	MAKE
	Station 4+50			Station 6+10				
(A-Line--215' U/S)			(A-Line--215' U/S)					
175	66.0	64.5	197	60.0	59.5			
176	53.0	52.0	198	*--	*--			
(B-Line--155' U/S)			(B-Line--155' U/S)					
177	63.5	63.0	199	60.5	60.0			
178	50.5	50.0	200	50.0	48.5			
179	40.0	39.0	201	*--	*--			
(C-Line--110' U/S)			(C-Line--110' U/S)					
180	62.0	62.0	202	60.0	58.0			
181	*--	*--	203	47.0	46.5	*-- Indicator plugged		
182	*--	*--	204	*--	*--	or shorted and		
183	*--	*--	205	25.0	24.5	cell abandoned.		
(D-Line--70' U/S)			(D-Line--70' U/S)					
184	56.5	56.0	206	53.5	53.0			
185	35.0	34.5	207	42.5	42.0			
186	28.0	27.5	208	19.0	18.0			
187	19.0	18.0	209	25.0	24.5			
(E-Line--25' U/S)			(E-Line--25' U/S)					
188	49.0	48.5	210	*--	*--			
189	32.0	31.5	211	*--	*--			
190	6.5	6.5	212	*--	*--			
191	*--	*--	213	13.0	12.5			
192	9.0	9.0	214	*--	*--			
(F-Line--20' D/S)			(F-Line--20' D/S)					
193	42.0	41.5	531	27.0	26.5			
194	27.0	26.0	533	14.5	14.5			
195	*--	*--	534	2.5	2.5			
196	*--	*--	535	1.5	1.5			
(G-Line--65' D/S)			(G-Line--65' D/S)					
541	25.0	24.5	536	*--	*--			
542	12.0	11.5	537	11.0	10.0			
543	*--	*--	538	3.0	3.0			
(H-Line--105' D/S)			(H-Line--105' D/S)					
544	*--	*--	539	9.5	8.0			
545	*--	*--	540	*--	*--			

All readings in feet of water.
Record data to nearest 0.5 foot.

Figure 27-2.—Hydrostatic pressure readings. 101-D-510.

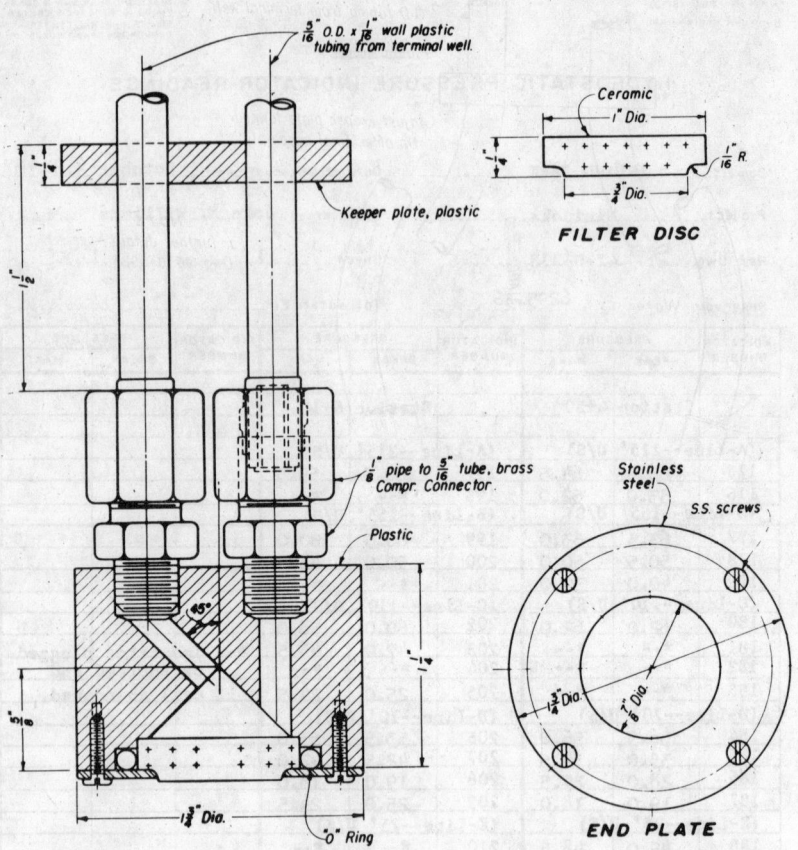

Figure 27–3.—Foundation-type piezometer tip (with ceramic disc). 101–D–497

of the tip below the disc was extended to enclose a 4-inch length of $7/8$-inch O.D. plastic pipe. Plastic pipe in 20-foot sections, having $7/8$-inch O.D. and $1/8$-inch-wall thickness, formed the extensions from these tips to a predetermined elevation or location in the foundation. Sections of the pipe were cut to length with a hacksaw and coupled together by special plastic connections cemented to the pipe sections.

(2) *Present tip.*—The present standard foundation-type piezometer tip uses a single ceramic disc and connects directly to the $5/16$-inch O.D. by $1/16$-inch-wall plastic tubing to extend the lines to a desired elevation in the dam foundation. This tip is shown on figure 27–3.

(b) *Embankment-Type Piezometer Tip.*—The standard embankment tip has two 1-inch-diameter by $1/4$-inch-thick ceramic discs fastened to

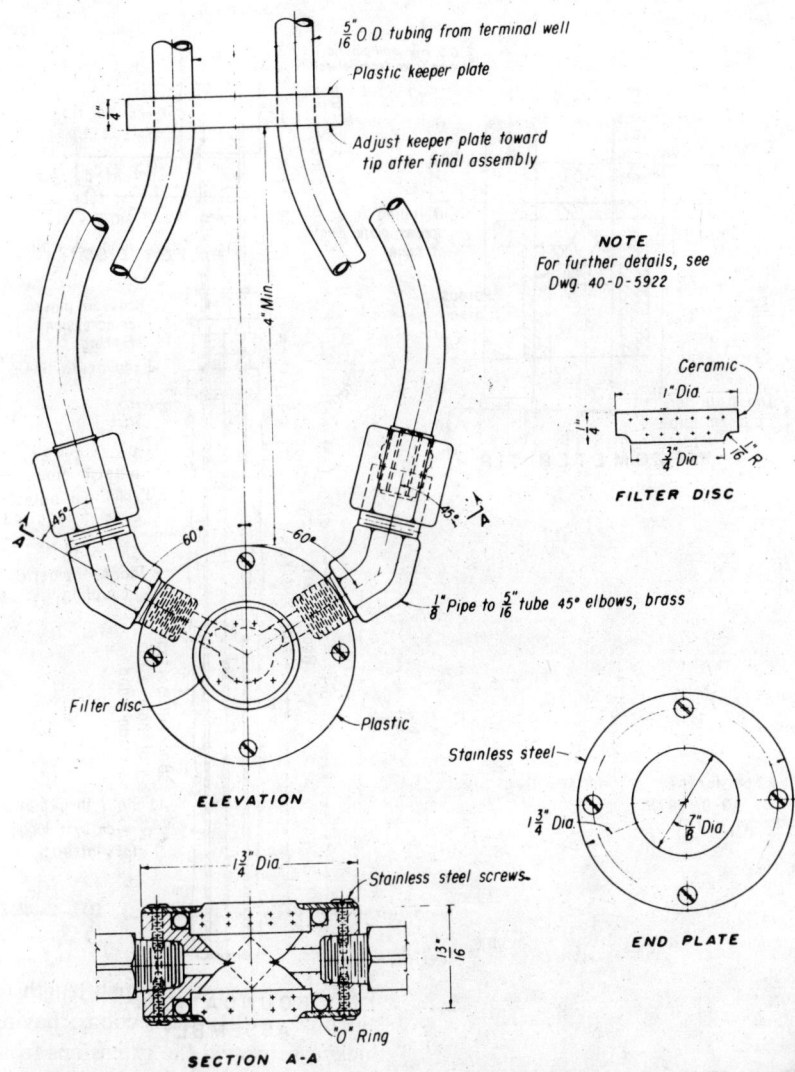

Figure 27–4.—Embankment-type piezometer tip (with ceramic disc). 101–D–499.

the tip as shown in figure 27–4. When these tips are installed in the dam embankment, the flat sides of the discs must be placed horizontally.

4. Protection and Testing of Piezometer Tips.—(a) *Protection of Piezometer Tips.*—Care must be exercised during the storage of the piezometer tips and all materials for the instrument installations, to avoid plugging the tips with dirt and debris and contaminating the filter discs with

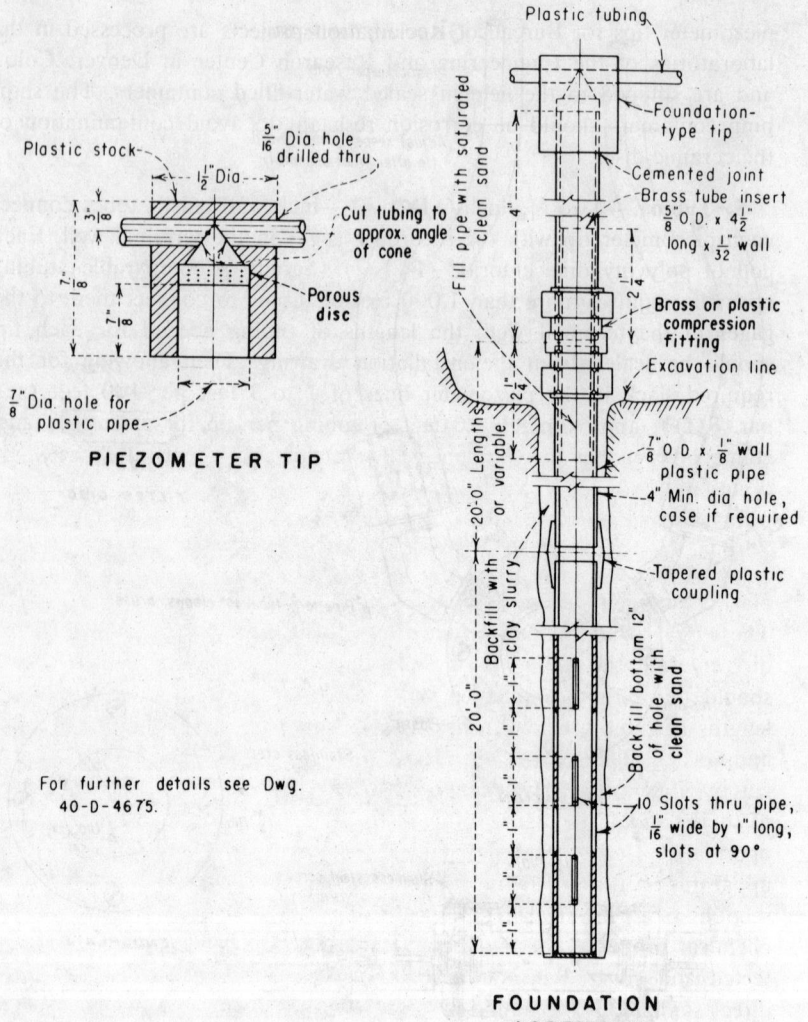

For further details see Dwg. 40-D-4675.

Figure 27–5.—Foundation-type piezometer tip (with pipe extension). 101–D–294.

oils, including oils from handling. The connection between the piezometer tips and the ⁵⁄₁₆-inch-O.D. tubing can be broken by rough handling.

(b) *Testing of Piezometer Tips.*—Before assembly of the pipe-to-tube compression fittings on the piezometer tips, the tips should be visually inspected to assure an open passage through the tips. Prior to installation, all piezometer tips should be boiled in distilled water for at least 15 minutes to saturate and purge the air from the filter discs. All piezometer tips for Bureau of Reclamation projects are processed in the laboratories of the Engineering and Research Center in Denver, Colo., and are shipped to the field in sealed water-filled containers. The shipping containers should be corrosion resistant to avoid contamination of the ceramic discs.

5. Tubing.—Two $5/16$-inch-O.D. by $1/16$-inch-wall plastic tubes connect each piezometer tip with the recording gages in the terminal well. Each coil of polyvinylidine chloride (PVNC) (Saran, or comparable) tubing normally requires more than 1,000 feet of tubing to connect them to the gages in the terminal well; the lengths of tubing needed for each tip should be scaled from the installation drawing. Then allowing for the required slack in the piezometer lines of 1 to 3 feet per 100 feet (see par. 8(f)), approximately 20 feet of tubing per tip for extensions and connections within the terminal well, and for contingencies, each length of tubing (two for each tip) should be marked, coupled, and recoiled prior to installation in the embankment.

(a) *Protection and Care of Tubing.*—Each coil of PVNC or other plastic tubing is crimped or sealed by the extruder prior to shipment. These seals should not be removed until required. To protect against dirt entering the tubing during placement, all exposed ends of tubing should be crimped and taped until finally installed. Before connecting lengths of tubing, a few inches must be removed from the sealed ends and wasted. The material in plastic tubing is relatively soft and can be cut by angular rock fragments or other sharp objects. Therefore, coils of tubing should be stored away from or above the embankment placing operations. Coils of tubing that are being used for installation of the apparatus on the embankment should be stored in timber boxes or within sections of steel barrels. Exposure to air will not impair the physical or chemical properties of the tubing. However, the tubing should be protected and covered insofar as practicable from prolonged exposure to direct sunlight. If the tubing is stored on a rack on the downstream edge of the embankment during construction of the dam, the tubing should be covered with a tarpaulin. Most plastic materials will burn or char when exposed to direct flame or to sparks from a welding torch; thus, it is necessary to protect the tubing from fire. Ordinary care also must be exercised to protect against rodents when the tubing is stored in warehouses.

(b) *Testing.*—A foot-by-foot inspection should be made of each coil or length of $5/16$-inch-O.D. by $1/16$-inch-wall plastic tubing to check for surface irregularities prior to installation. To insure that each length of tubing will withstand pressure, the tubing should be tested under air or gas pressure of approximately 100 pounds per square inch to check for possible leaks. During this pressure check, one end of a length of tubing should be plugged by a compression-type connector, equipped with a pipe cap, and the other end adapted and connected to a source of gas pressure. A bottle of commercial nitrogen or compressed air and the manifold gage assembly from a gas welding outfit can be used for this pressure testing. Likewise, to insure that each tube is open, water should be pumped through its length. However, the water should be blown from the tube if there is a possibility of its freezing before installation.

(c) *Identification of Tubing.*—Each length of tubing should be marked or identified with permanent markers every 50 feet by the use of metal or plastic bands stamped with appropriate piezometer numbers. Suitable identification should include the piezometer number and an "I" or "O" to indicate inlet or outlet tubes. Permanent identification is necessary to prevent errors in making up the proper terminal well connections, and to insure correct splicing if the tubing is cut or broken. The use of adhesive tape for identification is not recommended because moisture can loosen the adhesive. Temporary identification can be obtained by writing the tip number on a strip of paper and covering with transparent plastic tape.

6. Compression-Type Couplings.—Tubing extruded from plastic material such as Saran (polyvinylidine chloride—PVNC) and from various other thermoplastics has no field solvent and cannot be cemented. To provide practicable field connections, lengths of tubing are connected by compression-type brass couplings. A special brass insert must be installed inside the plastic tubing to form a suitable connection. Figure 27–6 illustrates a proper procedure for assembling a compression-type connection between lengths of $\frac{5}{16}$-inch-O.D. tubing. Note that hot water is recommended to soften the plastic tubing before the brass insert is installed. In addition, a special fixed-wheel tube cutter is recommended for cutting the plastic. Figure 27–7 shows the details and procedures to be followed in making up the connections between the plastic tubing and the gages and valves in the piezometer terminal well. A pair of 6-inch open-end or crescent wrenches should be used to tighten the compression couplings. Since the torque applied to wrenches of a given length will vary from person to person, it is recommended that a moderate amount of torque be applied when making these connections. Generally, the coupling nut should be screwed into the union until the threads on the coupling nut are nearly hidden. When assembling and tightening the compression couplings, considerable care must be exercised to install and maintain the compression ferrule approximately one tube diameter back from the end of the tube. Trial connections between lengths of tubing and from the tubing to the valves and gages should be made before actual installation, as described in the procedures given on figures 27–6 and 27–7.

Compression-type coupling
1. Full union, brass
2. Coupling nut and
   compression ferrule, brass
3. Insert, brass, $\frac{3}{16}$" O.D. by
   $\frac{1}{2}$" long

NOTE
Refer to Dwg. 40-D-5007
for further details.

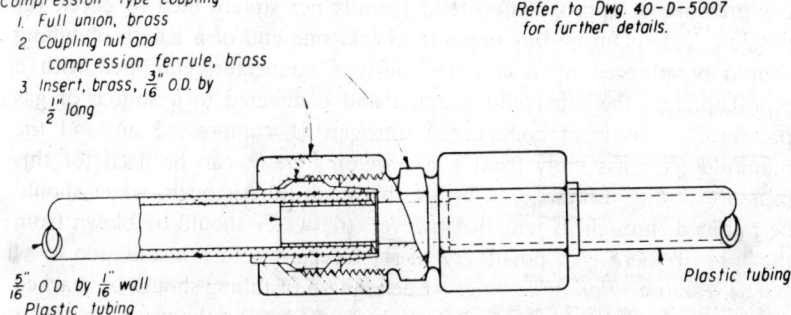

$\frac{5}{16}$" O.D. by $\frac{1}{16}$" wall
Plastic tubing

Plastic tubing

PROCEDURE

1. Square end or ends of plastic tubing to be connected. A special tube cutter for plastic tubing is recommended.
2. Inspect brass ferrule and brass tubing insert for burred edges.
3. Dip end of plastic tubing in hot water. Install brass-tube insert.
4. Slip brass coupling nut and compression ferrule over end of plastic tubing.
5. Push end of plastic tubing into brass union, hold in position, and tighten coupling nut.

Figure 27–6.—Compression-type coupling. 101–D–500.

7. Placing Piezometers in Foundation.—Foundation piezometers are installed in 4-inch minimum-diameter holes bored into the dam foundation. Casing usually is required to maintain the holes during installation of the piezometer tip assembly. These holes may be bored by jetting, or by use of some type of drilling equipment such as a percussion or diamond drill, depending on local conditions and the type of equipment available. Insofar as practicable, no drilling mud should be added to the water during drilling operations, especially in drilling the bottom few feet of hole where the piezometer tip will be installed. If casing is required to maintain the hole, it should be removed after the foundation piezometer tip assembly is placed. Individual holes are required for each foundation piezometer assembly to insure proper functioning of the piezometer at a desired elevation. Each hole also should be logged, when practicable, for its entire depth.

(a) *Location of Foundation Piezometers.*—Unless special or unusual foundation conditions are present, the foundation piezometers should be installed along a single cross section of the embankment as shown on figure 27–1. The elevations and locations for the foundation tips should be established to the nearest one-tenth foot and the data recorded on the

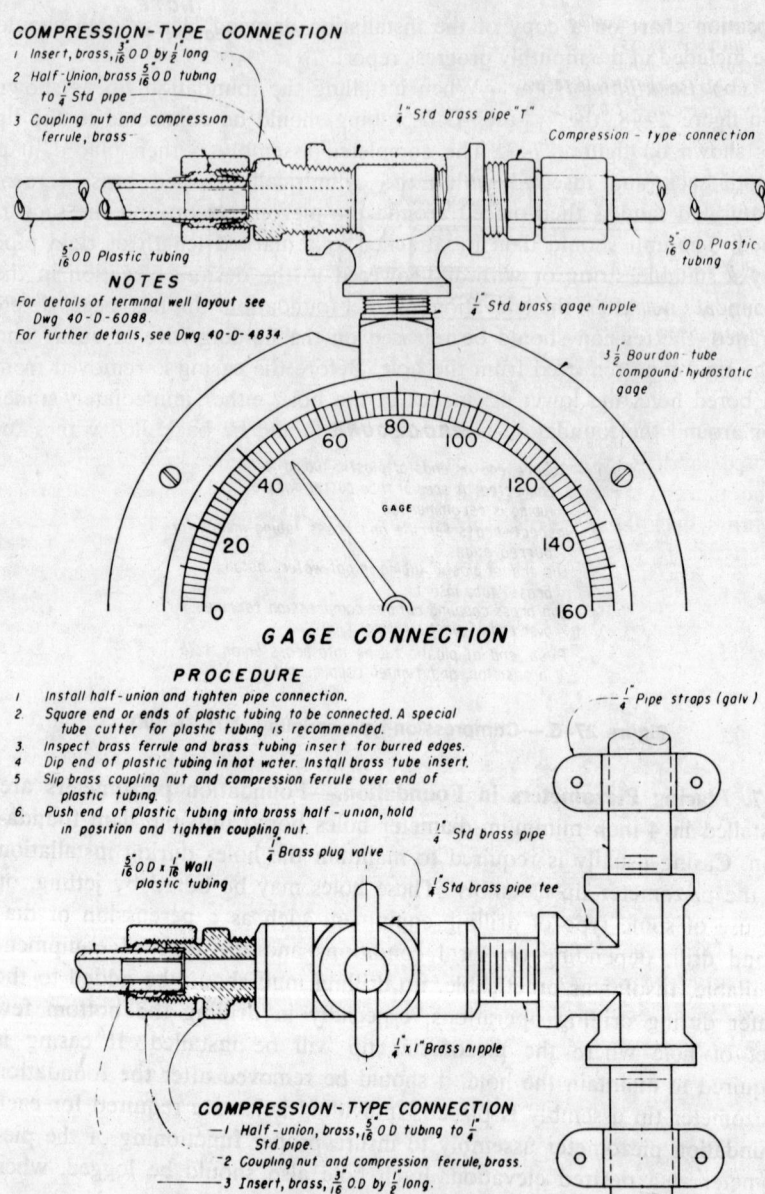

COMPRESSION-TYPE CONNECTION

1 Insert, brass, $\frac{3}{16}$ OD by $\frac{1}{2}$ long
2 Half-Union, brass $\frac{5}{16}$ OD tubing to $\frac{1}{4}$ Std pipe
3 Coupling nut and compression ferrule, brass

$\frac{1}{4}$" Std brass pipe "T"

Compression - type connection

$\frac{5}{16}$ OD Plastic tubing

$\frac{5}{16}$ OD Plastic tubing

NOTES

For details of terminal well layout see Dwg. 40-D-6088

For further details, see Dwg. 40-D-4834

$\frac{1}{4}$" Std brass gage nipple

$3\frac{1}{2}$ Bourdon-tube compound-hydrostatic gage

GAGE CONNECTION

PROCEDURE

1. Install half-union and tighten pipe connection.
2. Square end or ends of plastic tubing to be connected. A special tube cutter for plastic tubing is recommended.
3. Inspect brass ferrule and brass tubing insert for burred edges.
4. Dip end of plastic tubing in hot water. Install brass tube insert.
5. Slip brass coupling nut and compression ferrule over end of plastic tubing.
6. Push end of plastic tubing into brass half-union, hold in position and tighten coupling nut.

$\frac{5}{16}$ OD x $\frac{1}{16}$ Wall plastic tubing

$\frac{1}{4}$ Brass plug valve

$\frac{1}{4}$" Pipe straps (galv.)

$\frac{1}{4}$" Std brass pipe

$\frac{1}{4}$" Std brass pipe tee

$\frac{1}{4}$" x 1" Brass nipple

COMPRESSION-TYPE CONNECTION

1 Half-union, brass, $\frac{5}{16}$ OD tubing to $\frac{1}{4}$" Std pipe.
2 Coupling nut and compression ferrule, brass.
3 Insert, brass, $\frac{3}{16}$ OD by $\frac{1}{2}$ long.

VALVE CONNECTION

Figure 27-7.—Gage and valve connections. 101-D-501.

location chart on a copy of the installation drawing. These data should be included in the monthly progress report.

(b) *Backfilling Holes.*—When installing the foundation tip as shown on figure 27–8, the $\frac{5}{16}$-inch-O.D. tubing should be connected to the tip as shown on figure 27–3. The completed assembly is then placed in a cloth sack and inserted within the cylindrically shaped brass screen. Saturated sand is then placed around the piezometer tip and the sack is tied. The unit should then be attached to a marked length of rigid pipe by a suitable string or wire and lowered to the desired elevation in the foundation. After the elevation of the foundation tip has been determined, the tension should be released on the holding wire or string and the rigid pipe removed from the hole. Before the casing is removed from a bored hole, the lower 12 inches of the hole, either immediately under or around the foundation piezometer tip, should be backfilled with saturated sand. Removal of the casing from the hole and backfilling around the piezometer assembly should be performed in short increments to permit backfilling around the tip with saturated sand and the remainder

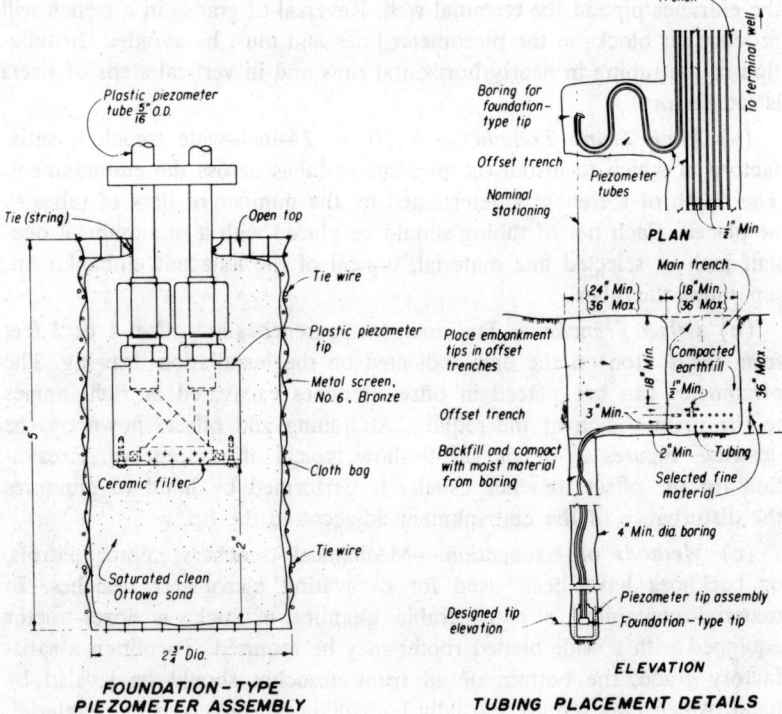

FOUNDATION–TYPE
PIEZOMETER ASSEMBLY

TUBING PLACEMENT DETAILS

Figure 27–8.—Foundation-type piezometer tip and installation details. 101–D–498.

of the hole with a clay slurry. The completed assembly for all foundation piezometers should be covered by a minimum of 18 inches of power-tamped earth materials at the foundation excavation line before heavy equipment is permitted to pass over the immediate area.

8. Placing Piezometers in Embankment.—Pore-water pressures in the foundation and embankment are contacted by the piezometer tips and transmitted through water-filled tubing to the terminal well which is located near the downstream slope of the completed dam embankment. For protection the tubing is placed in shallow trenches excavated into the embankment as it is constructed and the individual tubes are surrounded with compacted selected fine material. The lengths of tubing between the piezometers and the terminal well are placed in main trunk trenches which are offset 1 to 2 feet from the location of the piezometer tips.

To avoid interference with embankment placing operations, trenches for the piezometer tubing should be excavated only as far ahead of trench backfilling operations as construction conditions permit. All trenches must be on a level grade, or on a grade either ascending or descending from the entrance pipe at the terminal well. Reversal of grades in a trench will produce air blocks in the piezometer lines and must be avoided. Installation of the tubing in nearly horizontal runs and in vertical steps or risers is satisfactory.

(a) *Main Trunk Trenches.*—A 20- to 24-inch-wide trench is satisfactory in which to install the piezometer tubes across the embankment. The depth of a trench is determined by the number of tiers of tubes to be placed. Each tier of tubing should be placed with a minimum of one-half inch of selected fine material, typical of the adjacent embankment, separating the tubes.

(b) *Offset Trenches.*—The main trunk trenches are offset 1 or 2 feet from the station on the dam indicated on the installation drawing. The piezometer tips are placed in offset trenches excavated at right angles to the main trench at the required stationing and offset shown on the drawing. Figures 27–8 and 27–9 show typical offset trenches. Excavation for the offset trenches usually is performed by hand to minimize the disturbance of the embankment adjacent to the tip.

(c) *Methods of Excavation.*—Mechanical trenchers, motor patrols, or backhoes have been used for excavating piezometer trenches. In material containing a considerable quantity of rock, a dozer-tractor equipped with a wide-bladed rooter may be required. To obtain a satisfactory grade, the bottom of all trunk trenches should be leveled by hand, which may require partially backfilling with selected fine material.

(d) *Locating Embankment Piezometer Tips.*—When an embankment-

Figure 27–9.—Offset trench. P245–704–5549.

type piezometer tip is installed at an elevation higher than the terminal well, the tip should be placed slightly above the bottom elevation of the main trench. Conversely, an embankment-type tip at a lower elevation than the terminal well should be placed slightly lower than the elevation of the bottom of the main trench. To remove air from the tubes, an attempt should be made to fill each piezometer circuit with warm water before the tip is buried in the embankment or placed in a hole drilled into the dam foundation. The piezometer lines can be filled by circulating fluid from the terminal well or by connection to a temporary pumping setup, using a hand pump.

(e) *Backfilling and Compacting Trenches.*—Impervious or semipervious backfill material immediately adjacent to the embankment-type tips and adjacent to the piezometer tubing should be carefully selected or screened, if required, to eliminate material larger than the No. 4 sieve. No pockets of porous material should be permitted in the backfill of the trenches. This backfill should have a water content similar to, and should be compacted to densities equivalent to those obtained in the adjacent embankment material. The backfill around the embankment piezometer tips should be of selected fine material typical of the adjacent embankment and should be compacted in approximately 4-inch layers. No sand or other porous material should be compacted around the embankment-type piezometer tips. Pneumatic or gasoline-powered portable hand tampers are recommended to compact the backfill.

(f) *Backfill in Trenches.*—After the trench has been cleaned and leveled, a 2-inch-thick protective cushion of selected fine material, typical of the adjacent embankment, should be spread in the bottom of the trench and the first layer of tubes placed. Individual tubes should be separated by approximately one-half inch. The tubes can be evenly spaced and identified by fabricating a rake made of wood or metal similar to a hay or garden rake. A suitable rake is pictured in figure 27–10. The clearance between the tines of the rake should be approximately three-eighth inch. A hinged top board will permit access to the device and keep the tubes in order as the rake is dragged along the trench.

Tubing must not be stretched taut prior to backfilling the trench. A slack of from 1 to 3 feet should be uniformly distributed through a 100-foot length of each tube to compensate for differential settlement across the embankment. This slack can be introduced by laying the tubing in a meander from one side to the other within the trench. Succeeding layers of tubing in the trench should be separated from those previously placed by a minimum of 3 inches of selected fine material. A 3-inch minimum thickness of selected fine material should be placed and backfilled over the uppermost layer or tier of tubing to protect the tubing from angular rock. The remainder of the trench should be backfilled and compacted in 4-inch layers, using typical embankment material. Compaction of the 4-inch lifts of typical material above the layers of tubing can be accomplished by the wheels of rubber-tired equipment, by power tamping, or by other suitable methods. To prevent injury to the tubing by the sheepsfoot rollers, a minimum of 18 inches of compacted typical embankment material must cover the tubing before the rollers are permitted over the trench.

(g) *Bentonite Cutoff.*—During the backfilling of the trenches a plug, approximately 1-foot thick, made of a mixture of 5 percent bentonite (by volume) and 95 percent embankment material should be placed in the trenches at intervals of approximately 50 feet. The bentonite must be obtained from an approved source and exhibit a free-swell-factor of approximately 600 percent. Figure 27–11 depicts the excavation for this plug. These bentonite plugs or cutoffs reduce the possibility of seepage water passing through the embankment along the backfilled trenches.

(h) *Riser Pipes.*—When a group of piezometer tubing is to be extended vertically within the embankment, two methods are possible. When the rise is a matter of a few feet, the tubing can be bundled together and installed within a length of pipe made from standard corrugated metal pipe or from oil barrels with their ends removed and connecting straps welded to the outside of the barrels. However, when

Figure 27–10.—Use of rakes to aline piezometer tubes during installation— Trinity Dam. TD–3105–CV.

the vertical rise is more than 10 feet, it is generally better to fabricate a collar around the upper end of an 8- to 10-foot length of thin-wall steel pipe of from 15 to 18 inches in diameter and then jack the pipe to progressively higher elevations as required. A continuous pipe in the riser section is not required and the enclosure pipe may be removed and reused. The plastic piezometer tubing within the riser should be bundled and taped together, inserted inside the pipe, and the tubing

Figure 27-11.—Bentonite cutoff. P245-704-5550.

installed with approximately a 1-foot radius both below and above the riser section. The tubing must not be pulled taut in these riser sections. In addition, selected fine materials typical of the adjacent embankment should be hand-tamped around the bundle of tubes to completely fill the annular space within the riser section. Mounding of the embankment materials around the riser is required as the metal enclosure pipe is jacked to progressively higher elevations and such mounding should be compacted by pneumatic tampers in approximately 4-inch lifts.

(i) *Embankment Placement over Trenches.*—Specified embankment placement methods shall be used over the completed piezometer trenches.

9. Protection against Freezing.—In areas where piezometer installations are subject to severe freezing temperatures, precautions must be taken to protect the installations.

(a) *Terminal Well.*—When conditions require, an airtight wood frost floor and frame should be constructed as shown in figure 27-12. To provide additional insulation, the backfill around the terminal well should be of a selected sandy or impervious material for a minimum distance of 5 feet from the well. If permanent power is available, two

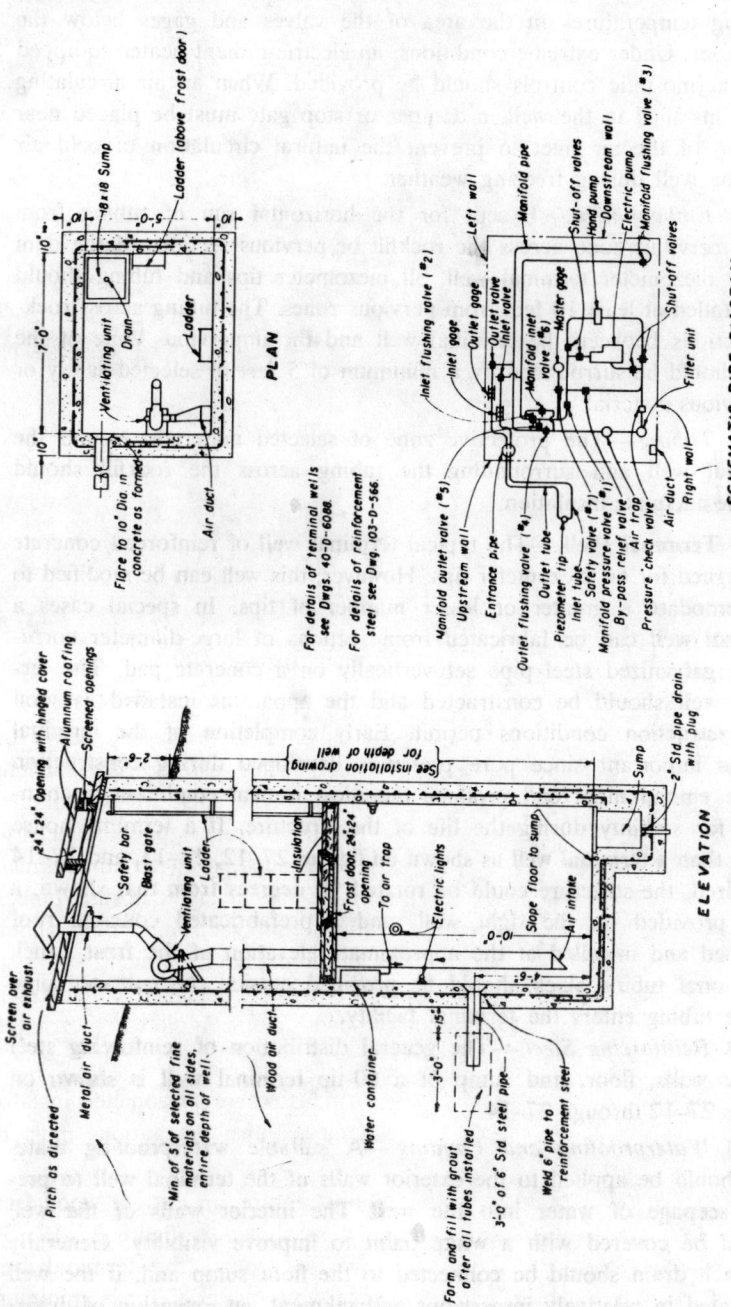

Figure 27-12.—Piezometer terminal well layout. 101-D-502.

or more 100-watt light bulbs should be installed in the well to prevent freezing temperatures in the area of the valves and gages below the frost door. Under extreme conditions, an electric radiant heater equipped with thermostatic controls should be provided. When an air circulating fan is installed in the well, a damper or stop gate must be placed near the top of the air duct to prevent the natural circulation of cold air into the well during freezing weather.

(b) *Embankment.*—Except for the horizontal run of tubing from the impervious zone across the rockfill or pervious sections of the dam to the piezometer terminal well, all piezometer tips and tubing should be installed at least 10 feet from pervious zones. The tubing across rockfill sections between the terminal well and the impervious zone of the dam should be surrounded by a minimum of 5 feet of selected sandy or impervious material.

(c) *Tubing.*—The protective zone of selected materials around the terminal well and surrounding the tubing across the rockfill should provide adequate insulation.

10. Terminal Well.—The typical terminal well of reinforced concrete is designed for 60 piezometer tips. However, this well can be modified to accommodate a greater or lesser number of tips. In special cases a terminal well can be fabricated from sections of large-diameter corrugated, galvanized steel pipe set vertically on a concrete pad. The terminal well should be constructed and the apparatus installed as soon as construction conditions permit. Early completion of the terminal well is important since pore pressures developed during construction of the embankment can produce the most critical conditions encountered for stability during the life of the structure. If a terminal house rather than a terminal well as shown on figures 27–12, 27–13, and 27–14 is desired, the structure could be rotated 90 degrees from that shown, a door provided on the right wall, and a prefabricated concrete roof designed and installed at the approximate elevation of the frost panel. Additional tubing slack should be provided outside the entrance pipe as the tubing enters the terminal facility.

(a) *Reinforcing Steel.*—The general distribution of reinforcing steel in the walls, floor, and sump of a 60-tip terminal well is shown on figures 27–12 through 27–14.

(b) *Waterproofing and Painting.*—A suitable waterproofing material should be applied to the exterior walls of the terminal well to prevent seepage of water into the well. The interior walls of the well should be covered with a white paint to improve visibility. Generally a 2-inch drain should be connected to the floor sump and, if the well is located in relatively impervious embankment, an extension of drain-

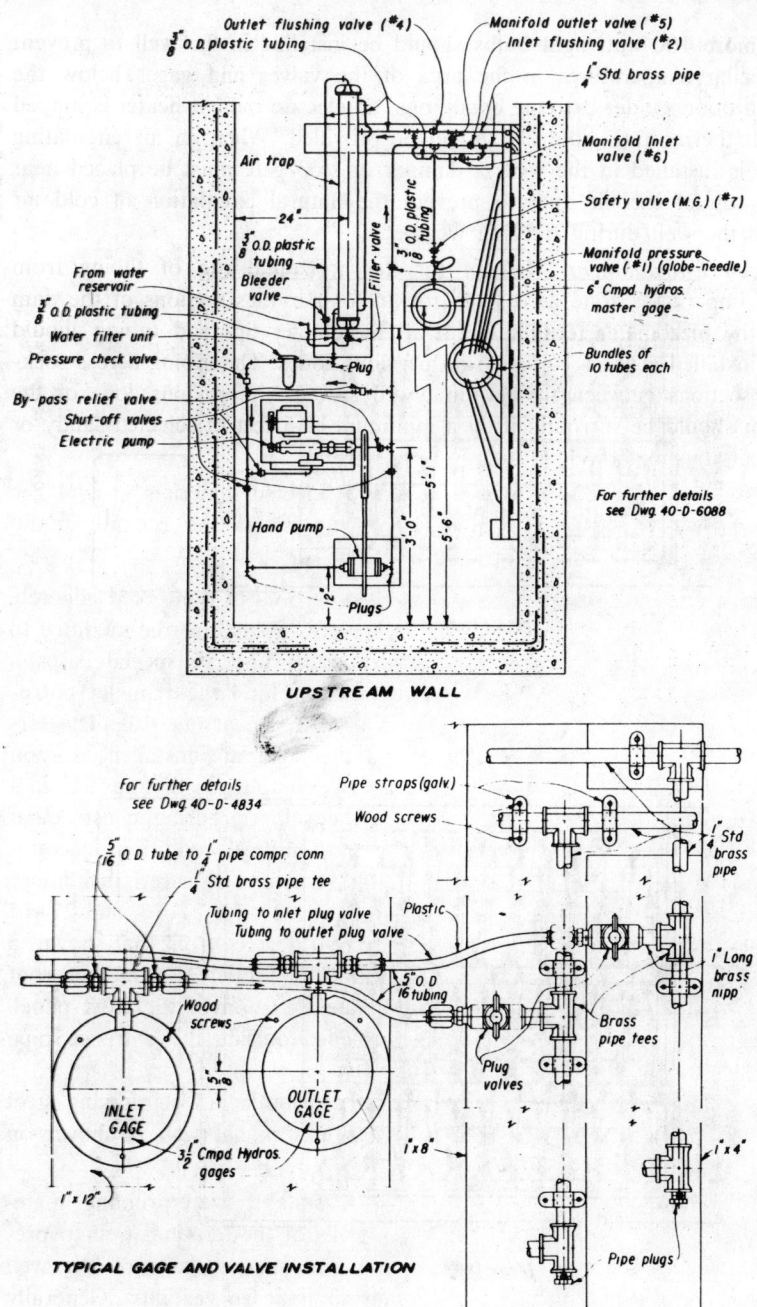

Figure 27-13.—Piezometer terminal well—Elevation of upstream wall and typical gage and valve installation. 101-D-503 and 101-D-504.

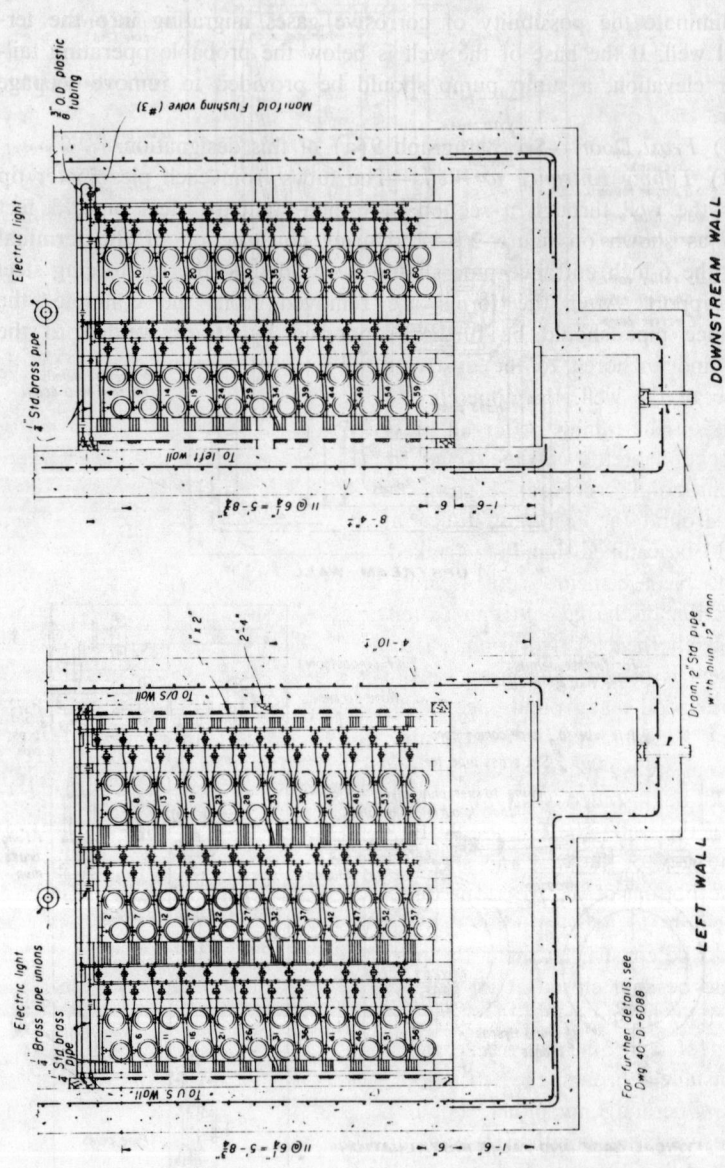

Figure 27-14.—Piezometer terminal well—Elevations of left wall and downstream wall and layout of gages. 101-D-505.

age pipe she.ld be provided from this sump drain to the downstream face of the dam or to a suitable foundation drain. A suitable water trap should be installed on the 2-inch drain immediately below the sump *to eliminate the possibility of corrosive gases migrating into the terminal well. If the base of the well is below the probable operating tailwater elevation, a sump pump should be provided to remove seepage water.

(c) *Frost Door.*—See paragraph 9(a) of this designation.

(d) *Tubing Entrance to Well.*—The tubes from each piezometer tip enter the well through a section of 6-inch standard steel pipe, 3 feet long, as shown on figure 27–12. During construction of the terminal well, the 6-inch entrance pipe should be welded to the reinforcing steel for support. After the forms are removed from the concrete, the entrance pipe should be further supported by straps welded to the pipe and anchored to the upstream wall of the terminal well. On the inside of the well, the concrete should be flared to allow wide-radius bends on the tubing. After all piezometer tubing is installed, the external section of the entrance pipe should be encased in a sand-portland cement grout or bentonite slurry. Prior to this encasement, the openings around the incoming tubes in the entrance pipe inside the well should be caulked and then packed with oakum to form a seal for the grout. Neat cement grout is not recommended because the heat of hydration of the cement can soften and collapse the plastic tubing. To reduce the heat of hydration, all tubes should be filled with fluid before the seal is placed. The grout or slurry should be thoroughly puddled to form a good seal around each of the tubes.

(e) *Ventilation.*—A ventilating system should be provided in all terminal wells. When electrical power is available, a ventilating system should be installed with an exhaust fan situated inside and near the top of the well. The fan should be mounted to exhaust through the duct in the roof of the well and should pull air from the lower working area at the bottom of the wood air duct at about 1 foot above the floor slab. To eliminate convection of air from above, a damper or stop gate should be constructed near the top of the air duct. The damper or gate should be kept closed at all times when the well is not being used, and especially during the cold winter months (see para. 9(a)) when circulation of cold air may freeze the manifold system. An air duct should be installed through the frost floor even though electricity is not available. Operating personnel should be made aware of the possibility of explosive gases (methane) and disabling gases (carbon monoxide and carbon dioxide) in these wells. Ventilation should be required before operating personnel enter the well.

(f) *Roofing.*—Roofing material can be a heavy-gage sheet metal, preferably aluminum, laid over a timber frame, or a prefabricated concrete roof can be installed. If the roof is covered by metal sheeting, skid strips coated with abrasive material should be installed.

11. Apparatus in Terminal Well.—Equipment to be installed in the well includes the manifold system, water pumps, air trap, gages, valves, water filter, auxiliary fittings, and electrical equipment. All connections must be made leakproof and should be tested, where possible, under an air pressure of 50 pounds per square inch before being placed in operation. Figures 27–13 and 27–14 show details of a representative terminal well installation.

(a) *Manifold System.*—Brass pipe (¼-inch standard iron-pipe size) is used to connect the $\frac{5}{16}$-inch-O.D. piezometer tubing to the valves, pumps, gages, and air trap. The system is arranged so that each piezometer circuit is independent and the manifold is common to all units. All pipe and tubes should be blocked away from the walls on either wood framing resistant to dry rot or plastic planking which is rigidly fixed to the concrete. All threaded connections should be made leakproof by applying a nonhardening pipe-thread compound in stick form or by the use of Teflon tape to the threads. The manifold is a double circulating system. By regulation of the proper sequence of valves, water can be pumped towards the tip through any desired piezometer tube. At certain locations in the manifold pipe system, the ¼-inch brass pipe is replaced by ⅜-inch-O.D. plastic tubing for observing the passage of air bubbles in the system.

(b) *Pumps.*—When electrical power is available, both an electrically driven and a hand-operated pump are provided. Figure 27–13 shows both pumps mounted on the upstream wall of the terminal well. The hand-operated test pump is assembled for horizontal mounting. To adapt it to a vertical mounting, remove the bolts holding the flanged cases and turn the cases 90 degrees. Install the pressure and vacuum portals of the pump in the same relative positions shown on figures 27–12 and 27–13. Both hand and electrically operated pumps should be rigidly fixed to the wall.

(c) *Air Trap.*—Air trap assemblies made from steel pipe and from plastic are shown in figure 27–15. When high operating heads are likely in the terminal well, the steel air trap is recommended. The plastic air trap is favored for visibility and when corrosive water exists at the site. During operations of the installation, the plastic air trap should be cleaned with a detergent solution and a nonabrasive scouring pad. Hydrocarbons, such as gasoline or naphtha, should never be used. When fluid is circulated through the system, air bubbles can be observed as they pass through the sections of plastic tubing connected to the inlet and outlet of the air trap. Air from the trap can be released through the ¼-inch bleeder valve. Piezometer fluid should be supplied to the system from a 10- or 20-gallon tank bracketed to the wall above the manifold system and just below the frost floor. This supply tank may be fabricated in the field from a rust-resistant material such as galvanized metal sheeting, but a small glass-lined hot water tank is preferred. A filler pipe or tube should be installed between the supply tank and the top of the terminal well. Very little fluid is needed after the tubes are once filled.

(d) *Master Gage.*—A 6-inch dial Bourdon-tube compound-hydrostatic gage, calibrated in feet of water and mounted for top connection, is used to measure both positive and negative piezometer pressures. This master gage is used to calibrate and furnish datum for the small separate gages. To meet the minimum flow requirements of the installation, any air in the Bourdon-tube elements must be replaced with fluid. The inlets to all gages, both the master gage and the individual separate gages, are therefore placed at the top of the gage case (top connection) where the circulating fluid will displace any entrapped air. All gages for Bureau of Reclamation projects are purchased dry and then filled with piezometer fluid in the Denver laboratories by use of a vacuum pump before being shipped to the projects.

(e) *Separate Gages.*—All existing piezometer gages on Reclamation projects are 3½-inch dial type "B" or type "A" gages. These 3½-inch dial Bourdon-tube compound-hydrostatic gages, calibrated in feet of water and mounted for top connection, are installed on each of the two incoming tubes from a piezometer tip. After stable conditions have been reached (i.e., an air-free system obtained), pressure at each tip can be observed on its pair of gages. This twin-gage installation permits a cross-checking of pressures for each tip. Discrepancies in the observed pressures on the pair of gages will indicate breaks, leaks, air in the system, or faulty gages. As the gages are installed, ½-inch-wide Teflon tape should be cut from a roll dispenser and applied by finger pressure to the threads on the gage nipple. The tape should be cut with shears and the length of tape should provide a minimum overlap on the gage threads. Since the 3½-inch dial gages are being discontinued by the gage manufacturers, and because of the cost and space requirements for mounting 4½-inch dial replacement gages in the existing terminal wells, 2½-inch dial type "A" gages with stainless steel cases and works will be specified for new installations and replacements. After installing the gages, caulking rope or a nonhardening adhesive compound should be applied to the openings around the base of the gage nipple to retard corrosion inside the gage cases.

(f) *Fittings.*—All connections between the plastic tubing and the

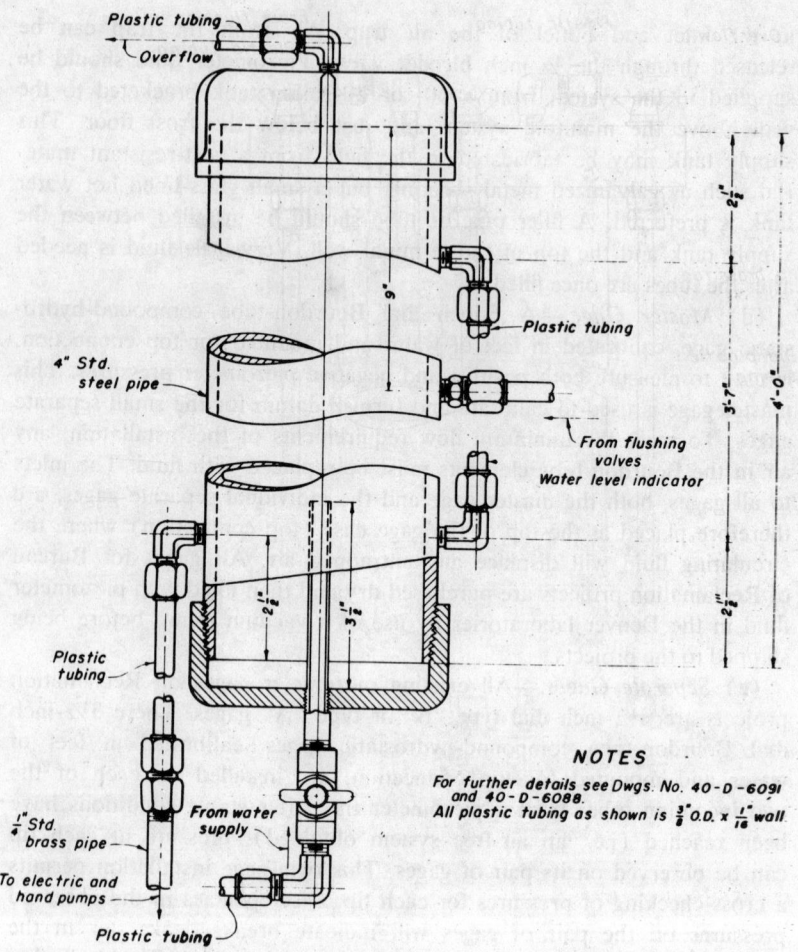

STEEL AIR TRAP
PIEZOMETER TERMINAL WELL

Figure 27–15.—Air traps for piezometer terminal wells. (Sheet 1 of 2.)
101-D-506.

gages and valves within the terminal well are generally made with compression-type brass fittings. For details, see figures 27–6, 27–7, and 27–13, and paragraph 6 of this designation. However, in some remodeling work at older piezometer installations, special polypropylene (PP) plastic fittings have been used.

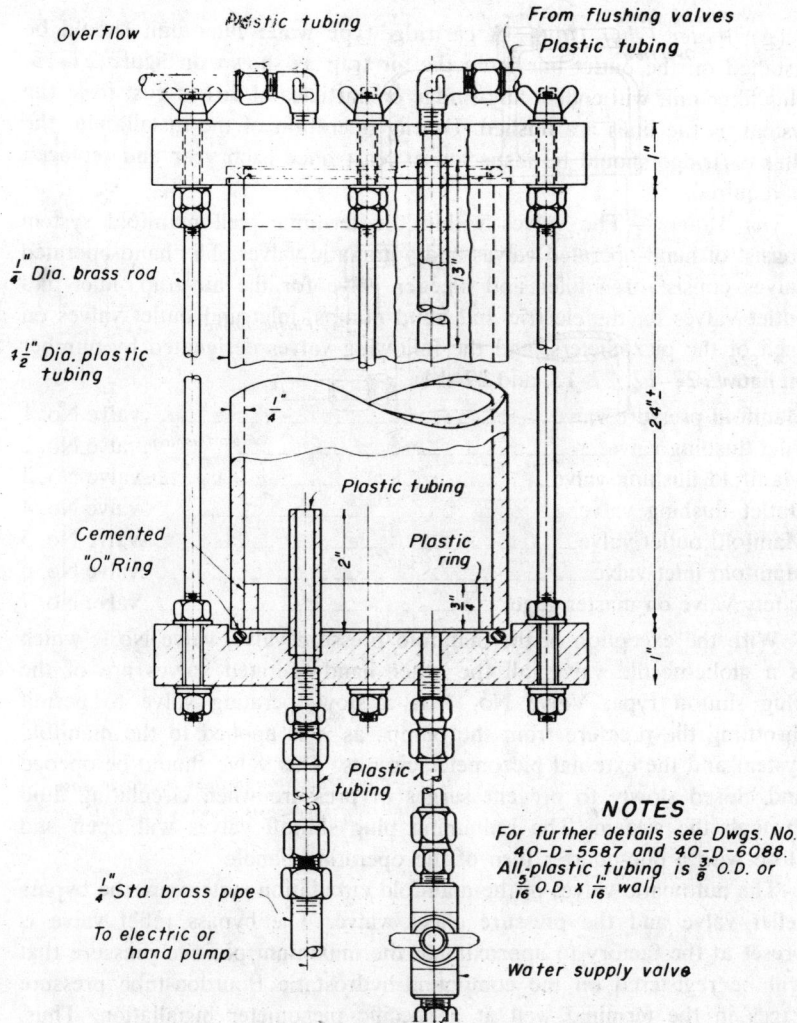

Overflow

Plastic tubing

From flushing valves
Plastic tubing

$\frac{1}{4}"$ Dia. brass rod

$\frac{1}{2}"$ Dia. plastic tubing

Plastic tubing

Plastic ring

Cemented "O" Ring

$\frac{1}{4}"$ Std. brass pipe

To electric or hand pump

NOTES
For further details see Dwgs. No. 40-D-5587 and 40-D-6088. All-plastic tubing is $\frac{3}{8}"$ O.D. or $\frac{5}{16}"$ O.D. x $\frac{1}{16}"$ wall.

Water supply valve

PLASTIC AIR TRAP
PIEZOMETER TERMINAL WELL

Figure 27–15.—Air traps for piezometer terminal wells. (Sheet 2 of 2.)
101-D-507.

(g) *Water Filter Unit.*—A cartridge-type water filter unit should be installed on the outlet line from the air trap as shown on figure 27–13. This filter unit will collect any sludge or particles of dirt or rust from the system as the lines are flushed. During operation of the installation, the filter cartridge should be inspected at least once each year and replaced as required.

(h) *Valves.*—The valves within the terminal well manifold system consist of hand-operated valves and automatic valves. Th hand-operated valves consist of a filler and bleeder valve for the air trap, inlet and outlet valves on the electric and hand pumps, inlet and outlet valves on each of the piezometers, and the following valves designated by number on figures 27–12, 27–13, and 27–14:

Manifold pressure valve_____valve No. 1
Inlet flushing valve_____valve No. 2
Manifold flushing valve_____valve No. 3
Outlet flushing valve_____valve No. 4
Manifold outlet valve_____valve No. 5
Manifold inlet valve_____valve No. 6
Safety valve on master gage_____valve No. 7

With the exception of the manifold pressure valve, valve No. 1, which is a globe-needle valve, all the other hand-operated valves are of the plug shutoff type. Valve No. 1 is a slow-operating valve to permit throttling the pressure from the pumps as it is applied to the manifold system and the external piezometer circuits. This valve should be opened and closed slowly to prevent surges of pressure when circulating fluid through the system. The remaining plug shutoff valves will open and close with a one-quarter turn of the operating handle.

The automatic valves in the manifold circulation system are the bypass relief valve and the pressure check valve. The bypass relief valve is preset at the factory to approximate the maximum positive pressure that will be registered on the compound-hydrostatic Bourdon-tube pressure gages in the terminal well at a specific piezometer installation. Thus, 50-foot-capacity gages will require a relief valve which will release to the bypass line at approximately 25 pounds per square inch; a 200-foot-capacity gage would require a valve setting of about 90 pounds per square inch, and a 500-foot-capacity gage would require a setting of 220 pounds per square inch. The relief valve will prevent excess pumping pressure from being applied directly to the Bourdon-tube gages. When this valve is first installed, sufficient pressure should be applied to open (crack) this valve and assure its proper operation.

The check valve is located between the pumps and the cartridge-type water filter unit. It will prevent back pressure into the filter unit, and into the plastic air trap when the bypass relief valve operates.

12. Calibrations on Gages.—As stated in paragraphs 11(d) and 11(e), both the 6-inch dial master and the 2½-inch or 3½-inch dial separate (small) Bourdon-tube gages are compound-hydrostatic. Both vacuum and pressure sides of the gages are calibrated in feet of water to permit correlation with the hydrostatic head imposed by reservoir operations. Thus, hydrostatic potential from the reservoir can be compared with potential (elevation plus pressure) at a piezometer tip. All compound-hydrostatic pressure gages used by the Bureau of Reclamation are furnished with a zero marker, since the resolution of pressure conditions starts with a condition of atmospheric pressure. Considering the anticipated pressures at a given installation, gages can be supplied with positive gage calibrations for 50, 200, or 500 feet. It is recommended that all gages, both the 6-inch dial master and the 2½-inch or 3½-inch dial gages, be furnished with the same calibration within a terminal well.

13. Fluid for Piezometer System.—Clear water containing a minimum of soluble salts is recommended for filling the piezometer system. When distilled water is available within practicable limits, it should be used. Air-free water would be preferable for the system, but under field conditions it generally cannot be obtained. For convenience in handling, fluid for the piezometer installation should be prepared in quantities of approximately 5 or 10 gallons. To each container holding 10 gallons of clear water, add 2 teaspoonfuls of a 25 percent solution of quaternary ammonium compound (QAC). This resultant solution is recommended for initial introduction into the twin-tube piezometer system as a bacterial inhibitor and as a wetting agent. However, during the recommended annual flushing of operating piezometer systems, it is recommended that 4 teaspoonfuls of the 25 percent QAC solution be added per 10 gallons of replacement water to the water reservoir or tank within the terminal well. Various quaternary ammonium compounds produced by Rohm & Haas Co. under the trade name of Hyamine, and by Onyx Oil & Chemical Co. under the trade name of BTC or Onyxide can be used. For Bureau of Reclamation installations the QAC solution will be furnished on request by the Engineering and Research Center, Division of Design, Denver, Colo. Only the specified amounts of QAC solution should be added to the piezometer fluid. As practicable, the solution of QAC and water should be warm when it is introduced into the piezometer lines. However, the temperature of the solution should not exceed 120° F., since water at too hot a temperature can soften the plastics.

Permanent-type antifreeze solutions (ethylene glycol—Prestone or equal) combined with water are not recommended for the piezometer fluid since greaselike inhibitors found in many of the products tend to plug the filter discs in the piezometer tips. The newer antifreeze products, especially those containing antileak additives, are not recommended. With cellulose plastic material, osmotic conditions also can develop. On existing systems where Tenite ÌI plastic tubing has been installed, it is recommended that the fluid be circulated through the individual piezometer lines twice each year to overcome the loss of fluid. Under extreme conditions where an antifreeze solution may be required, recommendations should be requested from the Engineering and Research Center.

14. Filling Gages with Fluid.—The Bourdon-tube element in the gages must be completely filled with liquid to operate properly. To eliminate troublesome field procedures, gages are processed and calibrated in the laboratories of the Engineering and Research Center, Den·'er, Colo., before shipment to field projects. The gages are obtained with t.;· Bourdon-tube element dry. They are then attached to a manifold arrangement which is connected to a vacuum pump. After pulling a vacuum, the gages are first filled with a detergent solution to remove any oil from inside the Bourdon-tube elements; then flushed with tap water. The tap water is removed by vacuum and the gages are finally filled with distilled water and QAC of the standard solution as described in paragraph 13. After filling, the needles on individual gages are set on the zero mark and the gages are calibrated at –20, –10, 0, and at five equal increments on the positive dial graduation, including the maximum reading for the gage. Calibration is made against large laboratory control gages. A correction curve is prepared for each 6-inch dial master gage and sent to the field with the gage. An average correction curve is prepared for the 2½- or 3½-inch dial gages for a given order of gages.

After adjustment, filling with standard fluid, and calibration, the gages are capped with a standard ¼-inch pipe cap. Teflon tape is used on the gage nipple threads to provide a seal as described in paragraph 11(e) of this designation. The stainless steel Bourdon-tube elements require a steel pipe cap, while the brass tubes require a brass cap. A capping pressure of about 15 percent of the positive dial graduation is set on each gage during capping. The pipe caps must not be removed until immediately before the gages are connected to the pipe tees and piezometer tubing within the terminal well. If necessary, additional fluid can be added at the time of installation by using a hypodermic needle and syringe. Since the gages are filled with a water solution, they must be protected against freezing temperatures (stored in a heated warehouse) until such time as they are installed in the terminal well.

15. Introducing Fluid into Manifold System.—The following procedures are recommended for introducing piezometer fluid into the manifold system:

(1) Prepare approximately 30 gallons of piezometer fluid, consisting of two teaspoonfuls of 25 percent QAC added to each 10

gallons of distilled or best available water as described in paragraph 13 of this designation. Fill water container (reservoir tank) in terminal well.

(2) Check the entire manifold system in the terminal well to be certain that all valves are closed. Then open the filler valve and the bleeder valve to the air trap. Fill the air trap with fluid to within 3 inches from the top of the plastic air trap (fig. 27–15) or to the top of the water level tubing on the steel air trap. Then close these valves.

(3) Open the inlet and outlet valves to the electric or hand pump (whichever pump is used) and, in sequence, valves No. 7, 2, 3, 4, and 1 (see figs. 27–13, 27–14, and par. 11(h) of this designation). Start pumping and continue circulation until no further air bubbles appear at the air trap. In the same sequence, close all valves except the valves on the pump being used.

(4) Open in sequence valves No. 7, 5, 3, 6, 1, and flush. Continue pumping to purge the air from the circuit, and in the same sequence close all valves except the valves on the pump.

(5) Open in sequence valves No. 7, 2, 6, 1, and flush. Continue pumping to purge the air from the circuit, and in the same sequence close all valves.

(6) Add fluid to the air trap when the water level shows less than one-half full.

(7) The above directions will fill the manifold system.

16. Introducing Fluid into External Piezometer Circuits.—To extend the filling and flushing to the external piezometer tips, follow the procedures given in paragraphs 15(1) and 15(2), and proceed as follows:

(1) Open pump valves and valves No. 1, 2, and 4. Also, slowly open the inlet and outlet valves for one piezometer (see left and downstream walls of the terminal well, figures 27–13 and 27–14) and flush. Then close valves No. 2 and 4. This is direct flushing.

(2) Open valves No. 5 and 6 and continue flushing. Close all valves. This is reverse flushing.

17. Flushing Procedures.—The flushing of the external circuits as directed in paragraphs 16(1) and 16(2) should be continued until no further air bubbles appear at the air trap. However, pumping should be continued for not more than 15 minutes on any one piezometer circuit during a single flushing operation. After flushing is completed, close all valves in the terminal well, including the pump valves.

From the explanation given in paragraph 11(h), valve No. 1 should be used to throttle or control the pressure being applied during the flushing operation. Initial pumping pressures should be limited to approximately 30 feet (on the gages) in excess of the tip constant for a piezometer tip.

For example, if a piezometer is located at 100 feet above the terminal well, pumping pressures should be limited to 130 feet of gage pressure. For tips located lower than the terminal well, initial pumping pressure should be limited to 30 feet of gage pressure. For examples of tip constants, see figures 27–16 and 27–17.

7-1346 (8-70)
Bureau of Reclamation

PIEZOMETER READINGS
(SEPARATE GAGES)

TO   DIRECTOR OF DESIGN & CONST
ENGRG & RESEARCH CENTER
DENVER FEDERAL CENTER
DENVER, COLORADO 80225

EXAMPLE

Dam
Project
Ref Dwg 622–D–152, –D–153
Reservoir Water El 2884.96

Date of Observation 11–15–70
Observer
Sheet 1 of 2
Tailwater El 0

PIEZ NO	GAGE READINGS			TIP CONSTANT	AVERAGE PRESS AT TIP	PIEZ NO	GAGE READINGS			TIP CONSTANT	AVERAGE PRESS AT TIP
	INLET	OUTLET	AVERAGE				INLET	OUTLET	AVERAGE		
	Composite Section Stations 36+10 and 37+00						Station 45+06				
101	+24.0	+21.0	+22.5	+96.5	+119.0	201	+16.0	+16.0	+16.0	+96.5	+112.5
102	+18.0	+18.0	+18.0	+69.5	+87.5	202	+1.0	0.0	+1.0	+96.5	+97.5
103	+6.0	+8.0	+7.0	+58.5	+65.5	203	+3.0	+2.0	+2.5	+96.5	+99.0
104	+2.0	0.0	+2.0	+55.0	+57.0	204	-2.0	+1.0	-1.0	+96.5	+95.5
105	-6.0	-6.0	-6.0	+96.5	+90.5	205	+21.0	+20.0	+20.5	+66.5	+87.0
106	+32.0	+31.0	+31.5	+66.0	+97.5	206	+17.0	+17.0	+17.0	+66.0	+83.0
107	+25.0	+26.0	+25.5	+66.0	+91.5	207	+10.0	+10.0	+10.0	+66.0	+76.0
108	+10.0	+12.0	+11.0	+66.0	+77.0	208	+6.0	+4.0	+5.0	+66.0	+71.0
109	+7.0	+6.0	+6.5	+66.0	+72.5	209	+2.0	0.0	+2.0	+66.0	+68.0
110	+5.0	+6.0	+5.5	+66.0	+71.5	210	-6.0	-5.0	-5.5	+66.0	+60.5
111	-4.0	-2.0	-3.0	+65.5	+62.5	211	-4.0	-2.0	-3.0	+65.5	+62.5
112	-2.0	-2.0	-2.0	+65.5	+63.5	212	+7.0	+6.0	+5.5	+65.5	+72.0
113	-12.0	-9.0	-10.5	+65.5	+55.0	213	+22.0	+20.0	+21.0	+45.5	+66.5
114	+30.0	+28.0	+29.0	+45.5	+75.0	214	+16.0	+17.0	+16.5	+45.5	+62.0
115	+27.0	+32.0	+29.5	+45.5	+75.0	215	+12.0	+12.0	+12.0	+45.5	+57.5
116	+16.0	+18.0	+17.0	+45.0	+62.0	216	+12.0	+12.0	+12.0	+45.0	+57.0
117	+14.0	+12.0	+13.0	+45.0	+58.0	217	+6.0	+6.0	+6.0	+45.0	+51.0
118	+10.0	+8.0	+9.0	+45.0	+54.0	218	+4.0	+4.0	+4.0	+45.0	+49.0
119	+4.0	+8.0	+6.0	+45.0	+51.0	219	+4.0	+2.0	+3.0	+45.0	+48.0
120	0.0	0.0	0.0	+45.0	+45.0	220	-6.0	-4.0	-5.0	+45.0	+40.0
121	-4.0	-2.0	-3.0	+44.5	+41.5	221	0.0	-3.0	-3.0	+44.5	+41.5
122	-4.0	-7.0	-5.5	+44.5	+39.0	222	+26.0	+25.0	+25.5	+24.5	+50.0
123	+35.0	+36.0	+35.5	+29.5	+65.0	223	+22.0	+24.0	+23.0	+24.5	+47.5
124	+30.0	+30.0	+30.0	+29.5	+59.5	224	+10.0	+10.0	+10.0	+24.5	+34.5
125	+18.0	+20.0	+19.0	+29.5	+48.5	225	+4.0	+4.0	+4.0	+24.5	+28.5
126	+4.0	+6.0	+5.0	+29.0	+34.0	226	-6.0	-6.0	-6.0	+24.0	+18.0
127	-2.0	-1.0	-1.5	+29.0	+27.5	227	-7.0	-8.0	-7.5	+24.0	+16.5
128	-3.0	-3.0	-3.0	+29.0	+26.0	228	+12.0	+9.0	+10.5	+15.5	+26.0
129	+30.0	+30.0	+30.0	+14.0	+44.0	229A	+5.0	+2.0	+3.5	+14.0	+17.5
130	+26.0	+24.0	+25.0	+14.0	+39.0	230A	+5.0	+2.0	+3.5	+14.0	+17.5
131	+4.0	+5.0	+4.5	+13.5	+18.0	231	+10.0	+10.0	+10.0	+13.5	+23.5
132	+12.0	+10.0	+11.0	+13.5	+24.5	232	+10.0	+10.0	+10.0	+13.5	+23.5
133	+10.0	+12.0	+11.0	+13.5	+24.5	233	+9.0	+6.0	+7.5	+13.5	+21.0
134	+10.0	+12.0	+11.0	+13.5	+24.5	234	0.0	0.0	0.0	+13.5	+13.5
135	+2.0	0.0	+2.0	+13.5	+15.5	235A	+14.0	+17.0	+15.5	0.0	+15.5
						236A	+4.0	+4.0	+4.0	-0.5	+3.5
						237A	+20.0	+19.0	+19.5	-17.0	+2.5
						238A	+22.0	+25.0	+23.5	-17.0	+6.5
						239A	+30.0	+32.0	+31.0	-17.0	+14.0
						240A	+29.0	+26.0	+27.5	-17.0	+10.5

*--Tube plugged or broken and piezometer line abandoned

The tip constant is the difference in elevation between a pair of gages and the corresponding tip, measured to the nearest 0.5 foot. Use a plus (+) for tip constants when the tip is below the elevation of the gages, and minus (–) when the tip is above gages. All readings in feet of water.

Figure 27–16.—Example—Separate gage readings. 101-D-508.

7-1347 (8-70)
Bureau of Reclamation

PIEZOMETER READINGS
(MASTER GAGE)

TO DIRECTOR OF DESIGN & CONST
ENGRG & RESEARCH CENTER
DENVER FEDERAL CENTER
DENVER, COLORADO 80225

EXAMPLE

Dam				Date of Observation 12-15-70			
Project 662-D-152, -D-153				Observer			
Ref Dwg				Sheet 1 of 4 El Master Gage			
Reservoir Water El 2884.96				Tailwater El 0			

El. Master Gage 2824.5 (Sta. 35+60)

PIEZ NO	MASTER GAGE READINGS				AVERAGE GAGE READING	TIP (2) CONSTANT	AVERAGE PRESSURE AT TIP
	INLET		OUTLET				
	(1) SETTING	READING	(1) SETTING	READING			
	Composite Section Stations 36+10 and 37+00						
101	+24.0	+23.0	+21.0	+22.0	+22.5	+94.5	+117.0
102	+18.0	+20.0	+18.0	+21.0	+20.5	+67.5	+ 88.0
103	+ 6.0	+ 8.0	+ 8.0	+ 9.0	+ 8.5	+56.5	+ 65.0
104	+ 2.0	+ 3.0	0.0	0.0	+ 3.0	+53.0	+ 56.0
105	- 6.0	- 5.0	- 6.0	- 5.0	- 5.0	+94.0	+ 89.0
106	+32.0	+32.0	+31.0	+33.0	+32.5	+64.5	+ 97.0
107	+25.0	+27.0	+26.0	+27.0	+27.0	+64.5	+ 91.5
108	+10.0	+10.0	+12.0	+14.0	+12.0	+64.5	+ 76.5
109	+ 7.0	+ 9.0	+ 6.0	+ 7.0	+ 8.0	+64.5	+ 72.5
110	+ 5.0	+ 6.0	+ 6.0	+ 6.0	+ 6.0	+64.5	+ 70.5
111	- 4.0	- 3.0	- 2.0	- 2.0	- 2.5	+64.5	+ 62.0
112	- 2.0	- 2.0	- 2.0	- 2.0	- 2.0	+64.5	+ 62.5
113	-12.0	-10.0	- 9.0	- 9.0	- 9.5	+64.5	+ 55.0
114	+30.0	+30.0	+28.0	+30.0	+30.0	+44.5	+ 74.5
115	+27.0	+28.0	+32.0	+32.0	+30.0	+44.5	+ 74.5
116	+16.0	+16.0	+18.0	+18.0	+17.0	+44.5	+ 61.5
117	+14.0	+13.0	+12.0	+12.0	+12.5	+44.5	+ 57.0
118	+10.0	+ 9.0	+ 8.0	+ 9.0	+ 9.0	+44.5	+ 53.0
119	+ 4.0	+ 4.0	+ 8.0	+ 6.0	+ 5.0	+44.5	+ 49.5
120	0.0	0.0	0.0	0.0	0.0	+44.5	+ 44.5
121	- 4.0	- 5.0	- 2.0	- 3.0	- 4.0	+44.5	+ 40.5
122	- 4.0	- 4.0	- 7.0	- 6.0	- 5.0	+44.5	+ 39.5
123	+35.0	+34.0	+36.0	+36.0	+35.0	+29.5	+ 64.5
124	+30.0	+28.0	+30.0	+28.0	+28.0	+29.5	+ 57.5
125	+18.0	+19.0	+20.0	+21.0	+20.0	+29.5	+ 49.5
126	+ 4.0	+ 4.0	+ 6.0	+ 4.0	+ 4.0	+29.5	+ 33.5
127	- 2.0	- 2.0	- 1.0	- 2.0	- 2.0	+29.5	+ 27.5
128	- 3.0	- 4.0	- 3.0	- 5.0	- 4.5	+29.5	+ 25.0
129	+30.0	+28.0	+30.0	+29.0	+28.5	+14.5	+ 43.0
130	+26.0	+26.0	+24.0	+23.0	+24.5	+14.5	+ 39.0
131	+ 4.0	+ 3.0	+ 5.0	+ 3.0	+ 3.0	+14.5	+ 17.5
132	+12.0	+11.0	+10.0	+ 8.0	+ 9.5	+14.5	+ 24.0
133	+10.0	+10.0	+12.0	+10.0	+10.0	+14.5	+ 24.5
134	+10.0	+ 9.0	+12.0	+10.0	+ 9.5	+14.5	+ 24.0
135	+ 2.0	+ 2.0	0.0	0.0	+ 2.0	+14.5	+ 16.5

✦--Tube plugged or broken and piezometer line abandoned.

(1) Pressure setting on manifold system before making readings on respective piezometer lines.
(2) The tip constant is the difference in elevation between the master gage and the tip, to the nearest 0.5 foot, plus (+) when tip is below elevation of gage, and minus (-) when tip is above gage. All readings in feet of water.

Figure 27-17.—Example—Master gage readings. 101-D-509.

After individual piezometers or the entire piezometer system has been placed in operation, pumping pressures should be limited to 30 feet in excess of the average of the pressures observed on the separate (2½-inch or 3½-inch dial) gages for an individual piezometer.

During construction of the dam, flushing should be performed to place the manifold system in operation as soon as practicable. After individual piezometer circuits are placed in operation, the circuits should be flushed at approximately monthly intervals until the circuits to the respective piezometers are free of air. After the dam is in operation, individual piezometers should be flushed once each year, both to eliminate any air from the circuits and to dislodge and inhibit bacterial growth. Flushing of piezometer circuits should be performed at least two days before regular readings are obtained.

Both direct and reverse flushing should be performed on external piezometer circuits as described in paragraphs 16(1) and 16(2) to complete the flushing on each tip. After flushing has been completed, check the residual pressures on the pair of gages (2½-inch or 3½-inch dial) for each piezometer. If pressures differ by more than 2 feet on 50-foot-capacity gages, by 5 feet on 200-foot-capacity gages, or by 10 feet on 500-foot-capacity gages, repeat the flushing on the line showing the lower pressure to bring the gage pressure within the desired limits. For operating piezometers, it will prove helpful to throttle (partially close) the operating valve No. 4 or No. 6 on the return line to the air trap, depending on the direction of flow, to maintain some pressure on the return line.

An approximation of the fluid being pumped into the piezometer lines on initial filling can be obtained by marking off an interval of 18 inches on the steel air trap (4-inch standard pipe) shown on figure 27–15. When the water level shows a drop of 18 inches, 1 gallon of fluid has been pumped into the lines. A gallon of fluid will fill approximately 680 feet of $\frac{3}{16}$-inch-inside-diameter tubing.

18. Procedures for Obtaining Pressure Readings with Separate Gages.

(1) Read the pressures on the individual (separate) 2½-inch or 3½-inch dial gages for each piezometer and record in the appropriate columns (inlet and outlet gage readings) on form 7–1346, figure 27–16.

(2) If the inlet and outlet gage pressures agree within the desired limits (2 feet for 50-foot gages, 5 feet for 200-foot gages, and 10 feet for 500-foot gages), determine the arithmetical average and record the average gage pressure. Readings should be obtained to the nearest 0.5 foot on the 50-foot capacity gages, to the nearest 1.0 foot on the 200-foot gages, and to the nearest 2.0 feet on the 500-foot gages.

(3) The tip constant for the separate gages is the difference in elevation between the pair of gages for a tip and the corresponding piezometer tip. Where 50-foot gages are installed, the elevations for the piezometer tip and the corresponding pair of gages should be established to 0.5 foot, for 200-foot gages these elevations should be established to 1.0 foot, and for 500-foot gages to 2.0 feet. A plus constant results when the tip is below the elevation of the gages, and a minus constant results when the tip is above the level of the gages.

(4) Arithmetically add the average gage pressure to the tip constant and record in the appropriate column on form 7–1346. Figure 27–16 provides an illustration of separate gage piezometer readings.

19. Procedures for Obtaining Pressure Readings with Master Gage.—

(1) Observe the average gage pressure on the pair of separate gages for a particular piezometer.

(2) Follow the procedures for flushing the manifold system, using the electric or hand pump, as described in paragraph 15, Introducing Fluid into Manifold System. Then close all valves except those to the electric or hand pump.

(3) Continue pumping. Open in sequence valves No. 1, 2, 6, and 7. By throttling valve No. 1 and slowly closing valve No. 6, set the desired pressure (from the small gages) on the master gage. After setting the desired balancing pressure, close all valves, except No. 7 and No. 2, and stop pumping. Record this inlet setting pressure on form 7–1347, figure 27–17, and crack (slightly open) the desired inlet valve (to the tip) for a piezometer. Read and record the responding pressure observed on the master gage. Then close the inlet valve (to the tip).

(4) As an alternative to the procedures given in paragraph 19(3), proceed with the flushing of the manifold system as given in 19(2), except that after the flushing is completed, close all valves, including those on the electric pump, and shut off the pump. Then open in sequence the outlet valve on the hand pump and valves No. 1, 2, and 7. Throttle valve No. 1 and apply a pressure stroke on the hand pump. When the desired pressure (from the small gages) is reached on the master gage, close valve No. 1. Record this inlet setting pressure on form 7–1347, and crack (open) the desired inlet valve (to the tip) for a piezometer. Read and record the responding master gage pressure; then close the inlet valve (to the tip).

(5) To obtain the outlet master gage reading, first resume pumping, and then open in sequence valves No. 1, 4, 5, and 7. Throttle valve No. 1 as before and slowly close valve No. 4 to obtain the outlet setting pressure on the master gage. Then close all valves, except

No. 7 and No. 5, and stop pumping. Record the outlet setting pressure and then crack the desired outlet valve (from the tip) for a piezometer. Read and record the responding pressure observed on the master gage, using the appropriate column on form 7–1347. Then close the outlet valve.

(6) As an alternative to the procedures given in paragraph 19(5) for obtaining the outlet master gage readings, follow the procedures given in paragraph 19(4), except open in sequence the outlet valve on the hand pump and valves No. 1, 5, and 7. Obtain the desired setting pressure on the master gage (from the small gages) and close valve No. 1. Record this pressure, and crack the desired outlet valve (to the tip) for a piezometer. Read and record the responding master gage pressure; then close the outlet valve (to the tip).

(7) Arithmetically add the average inlet and outlet master gage pressures to the tip constant for the master gage and the specific piezometer tip (difference in elevation between the master gage and the tip) and obtain the average pressure at the tip. Record as appropriate on form 7–1347. The tip constant should be determined with the same relative accuracy as for the separate gages as described in paragraphs 18(2) and 18(3).

(8) When a vacuum pressure is indicated on the small gages for a tip, close the inlet valve on the hand pump, open valves No. 1 and 7, and valve No. 2 or No. 5, depending on the desired access line to the tip, and apply a suction stroke on the hand pump; then close valve No. 1. Record this setting as observed on the master gage on form 7–1347.

(9) Crack the desired inlet or outlet valve (to or from the tip) for the desired piezometer on which vacuum readings are required. Record the responding reading observed on the master gage and close all valves.

(10) Repeat as in the paragraphs 19(8) and 19(9) for the opposite valve (inlet or outlet) and record the responding vacuum reading on the master gage. Close all valves.

(11) Figure 27–17 provides an illustration of the master gage readings. Please note the correlation of pressure readings obtained on the same piezometers by separate gages and the master gage (see fig. 27–16).

(12) After the dam is completed, readings from the average of the inlet and outlet separate gages should be compared with those obtained by use of the master gage at least once each year. These comparative readings should be obtained, preferably on the same day, and both sets of readings should be transmitted in duplicate with the regular periodic report on earth dam instrumentation (L–23).

20. Adjustment and Calibration of Gages.—No repairs should be made in the field to the movements or to the Bourdon-tube elements of any of the gages. As the gages are mounted in the terminal well, the brass pipe caps must be removed from the 2½-inch or 3½-inch dial gages to release the capping pressure. After mounting, each gage should be tapped lightly to assure a no pressure condition. Then the gages should be zeroed, if required, using a needle puller and a small plastic hammer. If, after positive checking, a gage is found defective, it should be removed and replaced. This checking should include a companion reading by use of the master gage. Extra small-dial gages are purchased when the equipment is procured. When calibration or repairs are required on the gages, they should be removed from the installation, capped with a brass pipe cap, and shipped to the Bureau of Reclamation, Engineering and Research Center, Division of Design, Warehouse, Building 53, Denver Federal Center, Denver, Colo. 80225.

21. Variations in Gage Readings.—(1) The average pressure for a piezometer as obtained from separate gages and from the master gage should agree within 2-foot limits for the 50-foot positive capacity gages, within 5 feet for the 200-foot-capacity gages, and within 10 feet for the 500-foot-capacity gages.

(2) If reflushing of the circuits will not bring the gage readings within the required limits, the questionable gage or gages should be removed and replaced with gages previously calibrated.

(3) When calibration cannot be obtained between gages, a leak is indicated either within the terminal well or on an external line.

22. Frequency of Observations.—The frequency of readings during construction and for 2 years following completion of the piezometer installation is given in figure 103. When the trend of operating pressures is established for a dam, a form DC-622, Schedule for Periodic Readings, is prepared by the Engineering and Research Center. Any change in scheduling must be done by concurrence with that office.

23. Record Tests.—Record tests shall be made of the embankment material surrounding each piezometer tip at the time the tip is installed. The procedures for record tests are described in section 71(d).

24. Monthly and Final Reports.—Each progress report (L–15) should include the pressure readings taken during the preceding month and a record of elevations and locations for all piezometers installed to date. A revision to the installation drawing corrected in the field should accompany the report to show sections of the embankment to the nearest foot for the date of the latest readings. Within 6 months after completion of the dam embankment, a Final Report (L–16) shall be prepared in the field as described in section 71(d) and forwarded to the Division of Design, Engineering and Research Center.

25. Miscellaneous Implements and Tools.—Although most twin-tube piezometer installations presently being constructed by the Bureau of Reclamation include procurement and installation of equipment by the contractor under Government supervision, the following list of equipment and tools may be useful for twin-tube piezometer installations:

(1) Power equipment, including a trenching machine, motor patrol, and hand-operated power tampers for use in excavating, backfilling, and compacting the earth material.

(2) A spacing rake to provide ½-inch spacing of tubes.

(3) A jacking platform and racks for riser sections.

(4) Sections of oil drums or wood boxes to cover tubing.

(5) Hand tools, including shovels, picks, and tamping bars.

(6) Star drills, masonry drills, and masonry nails.

(7) Pipe die and handle, crescent or open-end wrenches.

(8) Engineer's level and level rod; steel tape.

INSTRUCTIONS FOR INSTALLING AND READING POROUS-TUBE PIEZOMETERS

Designation E–28

1. Description of Apparatus.—The porous-tube piezometer is a device for measuring pore-water pressure in an embankment or a foundation. In its more general application, it is installed in foundations. The porous-tube piezometer is more sensitive to foundation pressures or ground-water fluctuations and is more resistant to plugging due to silting than a conventional observation well.

The installation consists of a length of porous carborundum or alundum tube set into either a drilled or jetted hole in the foundation to a predetermined elevation to intercept ground water or pore pressure in the foundation. The porous tube is plugged at one end, is surrounded by sand, and has a small-diameter plastic riser pipe extended to and slightly above the ground surface. A typical installation is shown on figure 28–1.

The pressure of the pore water surrounding the porous tube causes a flow through the porous tube of the piezometer until the pressures are equalized by the head of water in the standpipe. The elevation of

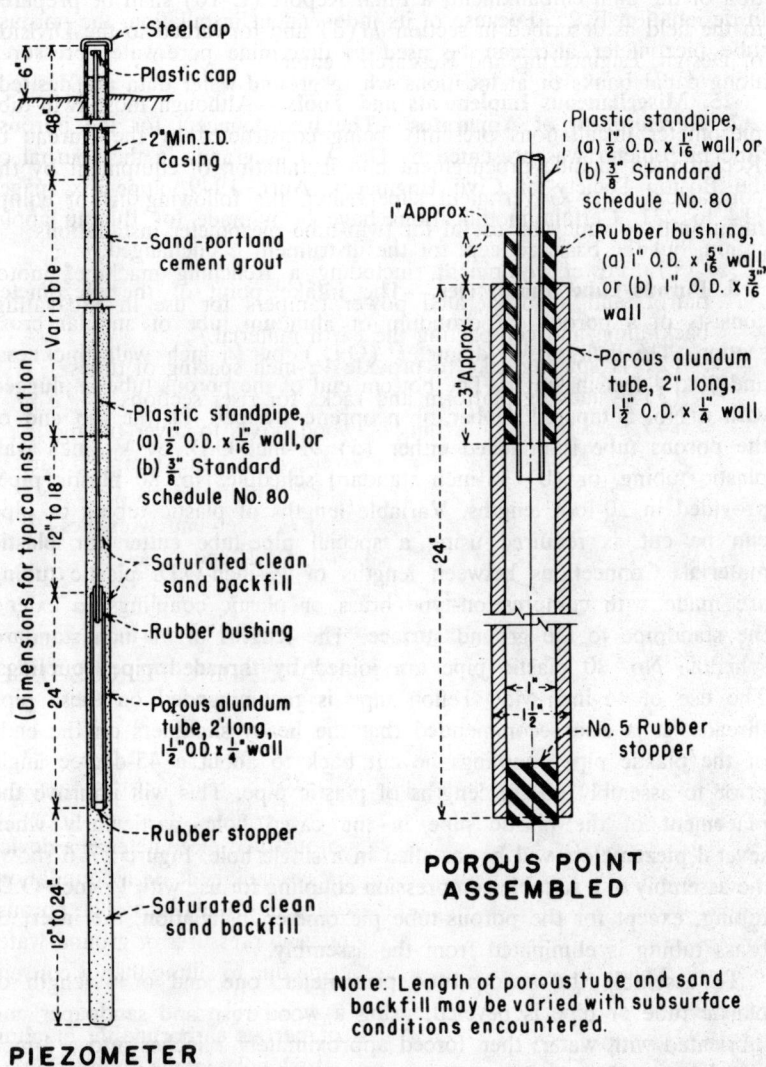

PIEZOMETER INSTALLED

POROUS POINT ASSEMBLED

Note: Length of porous tube and sand backfill may be varied with subsurface conditions encountered.

Figure 28–1.—Porous-tube piezometer. 101–D–302.

water in the plastic tube is determined by an electrical sounding device, lowered from the top of the installation.

Each porous-tube piezometer is an independent installation and, therefore, may be utilized to provide foundation pore pressure data in areas around structures at locations otherwise inaccessible or impracticable to contact with the twin-tube piezometer installation described in designation E–27. Because of its independent installation, the porous-tube piezometer also can be used to determine pore-water pressures along canal banks or at locations where ground-water data are desired.

2. Development of Apparatus.—The basic concept for the porous-tube piezometer was presented by Dr. A. Casagrande in the Journal of the Boston Society of Civil Engineers, April 1949, appendix, pages 214 to 221. Certain modifications have been made for Bureau application, but the basic concept for the instrument is unchanged.

3. Porous-Tube Piezometer.—The intake point of the piezometer consists of a porous carborundum or alundum tube of annular cross section, 1½-inch-outside-diameter (O.D.) by ¼-inch wall thickness, and 2 to 4 feet in length. The bottom end of the porous tube is plugged with a No. 5 tapered rubber or neoprene stopper. To the top end of the porous tube is attached either (a) ½-inch-O.D. by $\frac{1}{16}$-inch wall plastic tubing, or (b) ⅜-inch standard schedule No. 80 plastic pipe, provided in 20-foot lengths. Variable lengths of plastic tubing or pipe can be cut as required using a special pipe-tube cutter for plastic material. Connections between lengths of ½-inch-O.D. plastic tubing are made with compression-type brass or plastic couplings to extend the standpipe to the ground surface. The lengths of ⅜-inch standard schedule No. 80 plastic pipe are joined by threaded pipe couplings. The use of ½-inch-wide Teflon tape is recommended on these pipe threads. It is also recommended that the heavy shoulders on the ends of the plastic pipe couplings be cut back to about a 45-degree angle prior to assembly on the lengths of plastic pipe. This will improve the placement of the plastic pipe in the cased hole, particularly when several piezometers will be installed in a single hole. Figure 27–6 shows the assembly of a suitable compression coupling for use with ½-inch-O.D. tubing; except for the porous-tube piezometer installation, the insert of brass tubing is eliminated from the assembly.

To assemble the porous-tube piezometer, one end of a length of plastic tube or pipe is beveled, using a wood rasp and sandpaper and lubricated with water; then forced approximately 2 inches into a 4-inch length of rubber bushing. The rubber bushing is cut from a length of rubber hose having a suitable outside diameter and inside diameter as required for the plastic tubing or pipe and for the porous tube. Details are shown on figure 28–1. The assembly is then inserted into the porous tube and the bushing is forced about 3 inches into the porous tube with a twisting action, using a strap wrench.

4. Tamping Rod.—A length of reinforcing steel rod or pipe is used to tamp the wet sand below and above the porous tube. Considerable care must be exercised to avoid damaging the fragile porous tube during the tamping operation. The steel rod or pipe should be marked off in 5-foot intervals, starting at the lower end, so that it can be used to check the elevation at which tamping is performed and the final elevation of the top of the porous tube.

5. Water Level Sounder.—Details of the water level sounder are shown on figure 28–2. This device is comprised of a suitable length of ¼-inch-O.D. coaxial cable (S–9U), fitted at the ends to contact the water surface in the plastic riser tube and for connection to an ohmmeter. The ohmmeter is a Simpson Model No. 372–2 or equal. The components of a suitable water level sounder are pictured in figure 28–3. When the water surface in the plastic riser tube or pipe is contacted by the probe, a considerable deflection will result on the ohmmeter needle as shown in figure 28–4. A sonalert (beeper) fitted with a neck strap can be substituted for the ohmmeter. The sounder device shown in figure 28–3 can provide readings to 0.01 foot when measuring water levels. After the probe end and the upper end have been fitted to the coaxial cable, the cable should be marked with paint stripes at 1-foot intervals from the probe end. Readings to 0.01 foot are made with a hand-held rule from the known elevation of the top of the riser tube or pipe. An alternative sounding device can be made from two insulated wires bared at the contact end and weighted with strips of lead. This sounder also requires the use of an ohmmeter or sonalert.

6. Installation.—A cased hole is advanced below the ground surface by jetting or by standard drilling procedures to the elevation planned for the bottom of the porous space. Most of the casing usually is removed from the hole following installation of the apparatus, but if the casing is considered expendable and left in the hole, a sufficient length should be pulled during installation of the porous tube so that the sand-cement or bentonite grout will have direct contact, if possible, with an impermeable stratum.

The successive steps required in the installation of the porous-tube piezometer are illustrated on figure 28–5. The hole is kept full of water in operations ① through ⑤. The lengths of porous tube and sand backfill shown in figure 28–1 are those of a typical installation; but the porosity and length of the porous tube, the length of the sand backfill, and also the diameter of drilled hole may be varied with the subsurface conditions encountered at the site.

(1) After the casing has reached the desired depth, it should be washed clean to the bottom ①. For a drilled hole, clean water is circulated through the bit until the discharge is clear. For a jetted hole, the pump is reversed and the jet pipe is pulled up a few inches from the bottom of the hole, to be used as an intake. The

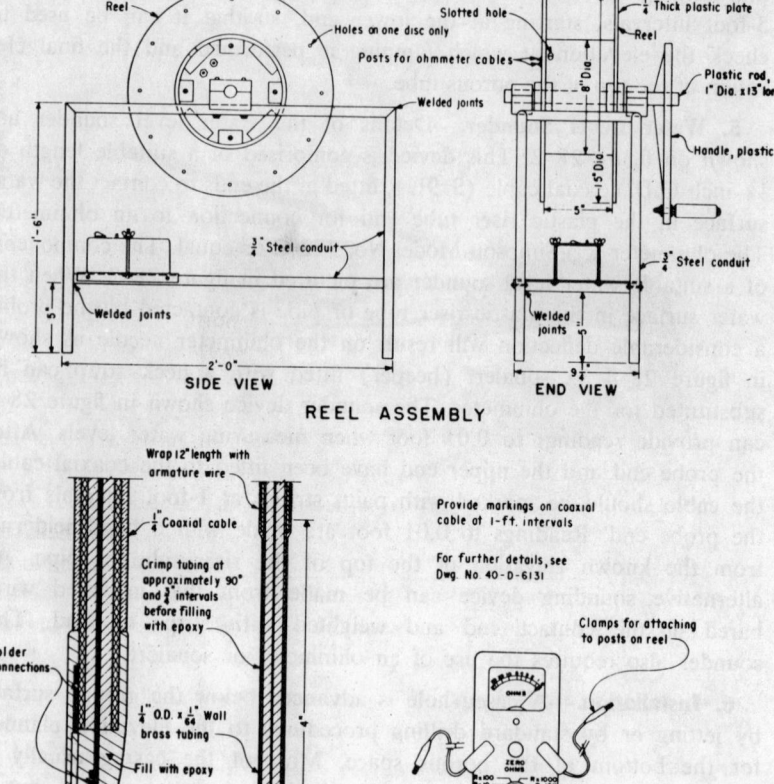

Figure 28–2.—Sounder device and reel for porous-tube piezometer. 101–D–541.

casing should be kept full by pouring in clear water until all cloudiness disappears from the effluent.

(2) After the hole is cleaned, saturated sand is poured into the casing to fill the bottom of the porous space ②. However, the length of hole to be filled with sand will depend on the relative tightness of the natural soil surrounding the hole; that is, the lower the

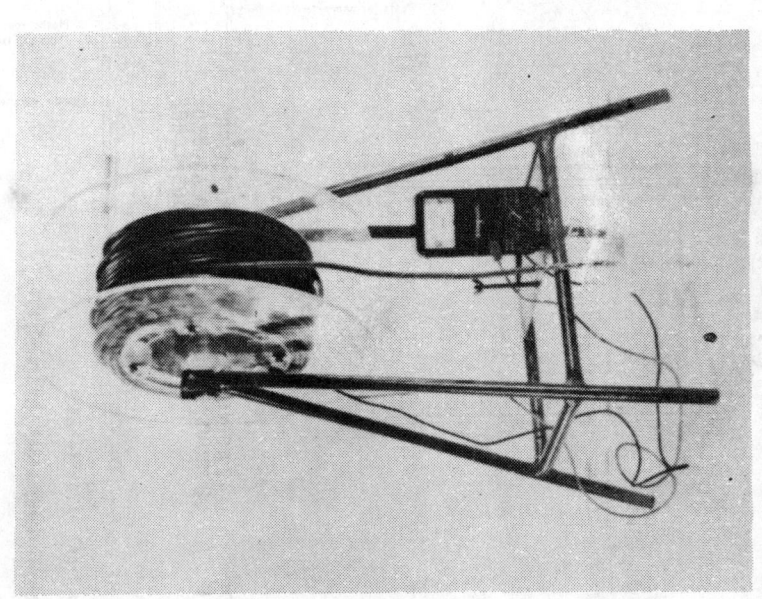

Figure 28–4.—Ohmmeter deflection when probe contacts water surface. E-2263-4NA.

Figure 28-3.—Components of the water level sounder. E-2263-3NA.

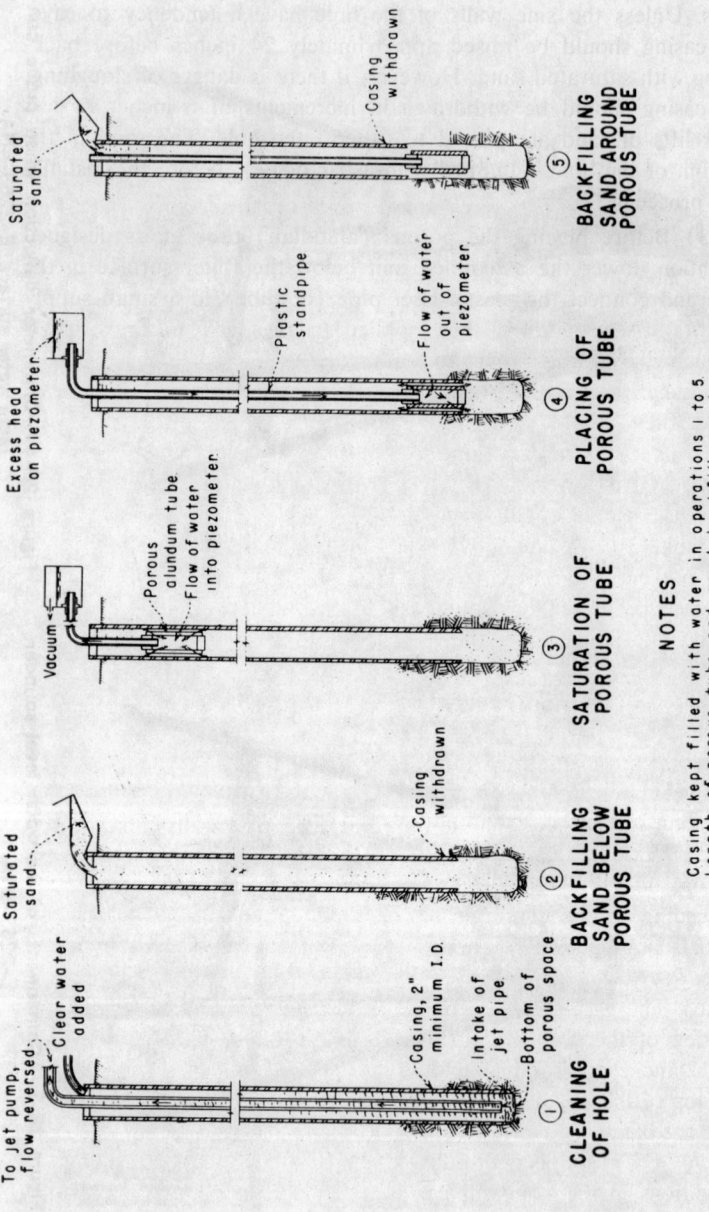

NOTES

Casing kept filled with water in operations 1 to 5
Length of porous tube and sand backfill may be
varied with subsurface conditions encountered

Figure 28–5.—Installing porous-tube piezometer. 101-D-303.

permeability, the greater the length of hole for intake area to the piezometer. Generally the sand backfill should be clean and well-graded of No. 4 to No. 200 mesh, United States standard sieve sizes. Unless the side walls of the hole have a tendency to cave, the casing should be raised approximately 24 inches before back-filling with saturated sand. However, if there is danger of sloughing, the casing should be withdrawn in increments of 6 inches or less after lifts of sand are placed to support the hole. The sand in the bottom of the hole is tamped with a bar or pipe before the installation proceeds.

(3) Before placing the porous (alundum) tube at its designed elevation, lower the assembled unit below the water surface in the hole and connect the plastic riser pipe (or tube) to a small supply tank ③. A vacuum is then applied to the tank to draw water through the porous tube to eliminate air from the system. To expedite saturating and removing air from the porous (alundum) tubes, they should be soaked in warm water for several hours or boiled in water for 15 minutes before installation.

(4) When lowering the assembled porous tube and standpipe into the hole, a small positive pressure should be maintained in the tank to cause an outward flow of water from the tip ④. This will prevent the movement of fines into the porous tube. The designed or original elevation for the porous tube is the elevation of the mid-point along the length of the tube. The length of the porous tube, including the projecting rubber bushing, should be measured before the apparatus is lowered into the hole. Measurements for the original elevation of the porous tube should be taken to the nearest one-tenth foot. However, after the installation is completed, elevations on the tops of each installation and water elevation in each pipe should be made to the nearest 0.01 foot.

(5) With the assembled porous tube resting on the sand in the bottom of the hole, the casing is withdrawn in small increments, depending on the condition of the walls of the hole, and saturated sand is poured into the hole to the level of the top of the porous tube ⑤.

(6) The casing is then pulled approximately 12 inches and that portion of the hole is backfilled with saturated sand. A minimum of 12 inches of sand should be backfilled above the elevation of the top of the porous tube. After all the sand is placed, it is tamped with the bar.

(7) The casing is then pulled approximately 3 feet or as sloughing conditions in the hole will permit and the hole is backfilled with a sand-portland cement grout having a volume ratio of one part

cement to four parts sand, or with a bentonite slurry. Balls of bentonite, equal to Pi-Pellets, have been used successfully to accomplish this required seal. A sealing grout made from sand and an expansive-type cement also has been used to seal above porous-tube piezometers installed in shales. A tamping bar should be lowered into the hole at this stage to puddle and compact the grout or slurry. The plastic standpipe should be maintained in the center of the hole during each increment of backfill.

(8) The casing, thereafter, is pulled in 3- to 5-foot increments or as conditions in the hole will permit and the hole is backfilled with sand-cement grout or a clay-silt slurry. When approximately 3½ feet of casing remains in the hole, the casing should be cut off at about 6 inches above the ground surface. The top of the casing should be provided with a metal pipe cap. The plastic standpipe is cut off flush with the top of the steel casing and capped with a removable pipe cover. The annular space between the steel casing and the plastic standpipe should be filled with grout or slurry to within approximately 2 inches of the top of the pipe.

(9) Upon completion of the installation, a protective tripod or fence made from sections of pipe or reinforcing steel should be constructed and set into the ground over the system to protect the installation from damage.

7. Advantages of the Porous-Tube Piezometer.—The porous-tube piezometer has the following advantages over the ordinary water-level observation well installation which utilizes a perforated pipe in sand backfill:

(1) The large intake area of the installation and the small diameter of the standpipe minimize the flow of water required for equalization of pore-pressure change.

(2) The permeable intake space can be positioned to isolate water pressures occurring in a stratum of limited thickness, even though such pressures are in excess of hydrostatic. (If excess pressure at the porous tube should raise the water level above the top of the plastic standpipe, riser sections or a hydrostatic Bourdon-tube gage can easily be installed.)

(3) The installation is fabricated from durable and inert materials which are unaffected by deterioration or corrosion.

(4) Simple tests can be performed on the piezometer after its installation to appraise its sensitivity and to determine the permeability of the surrounding soil.

8. Limitations of the Porous-Tube Piezometer.—The principal disadvantage of the installation in clay is the tendency of fines from the material surrounding the hole to penetrate into the sand backfill. Although the standpipe should not become clogged, the effective dimensions of the porous space may be decreased, reducing the sensitivity of the piezometer. The sand backfill below, surrounding, and above the porous tube should meet the filter requirements as closely as possible without including any silt sizes. The porosity of the alundum or carborundum tube can be chosen from three grades available, coarse, medium, and fine, in order to inhibit the movement of particles from the surrounding soil through the tube. Sedimentation can be partly controlled by raising the water level in the standpipe when an increase in pore pressure is anticipated, thereby requiring an outward flow to achieve equilibrium.

Experience indicates that the performance of the porous tube is not entirely satisfactory in soils having an appreciable air content. Air bubbles in the soil stratum decrease its permeability from the value for a saturated condition. Changes in the volume of air which accompany pore-pressure fluctuations also tend to retard the flow of water into the standpipe and thus slow the piezometer's response. Careful saturation of the sand backfill and of the porous tube should provide an air-free installation.

To isolate possible hydrostatic pressures in a thin aquifer, it may be necessary to limit the length of the porous space, sacrificing sensitivity. This can be offset in part by using a hole of larger diameter.

9. Sensitivity of the Porous-Tube Piezometer.—The length of time required for the piezometer to equalize after a change of pressure is a measure of its sensitivity. This time lag is directly proportional to the cross-sectional area of the standpipe and varies inversely with the permeability of the surrounding soil. The sensitivity of the piezometer can be increased by lengthening the porous space and to a lesser degree by increasing its diameter.

The time required for the standpipe water level to reach 95-percent equalization after a change in pore pressure is shown on figure 28–6 where the length of the permeable space for installations A, B, and C is 5, 3, and 2 feet, respectively. The diameters of the drilled holes are those of an NX casing, a BX casing, and a 2-inch-diameter standard pipe. The ½-inch O.D. by 1/16-inch-wall plastic tubing, providing an I.D. of 3/8-inch, was assumed for illustration. The table given on figure 28–6 indicates that each of the three designs is sufficiently sensitive to record pore-water pressure fluctuations in soils having permeabilities of from 0.001 to 0.1 foot per year.

Usually the value of permeability assigned to the soil stratum beneath the ground surface is obtained from laboratory tests performed on samples of undisturbed soil oriented in a vertical direction. However, in a sedimentary material the horizontal permeability is often many

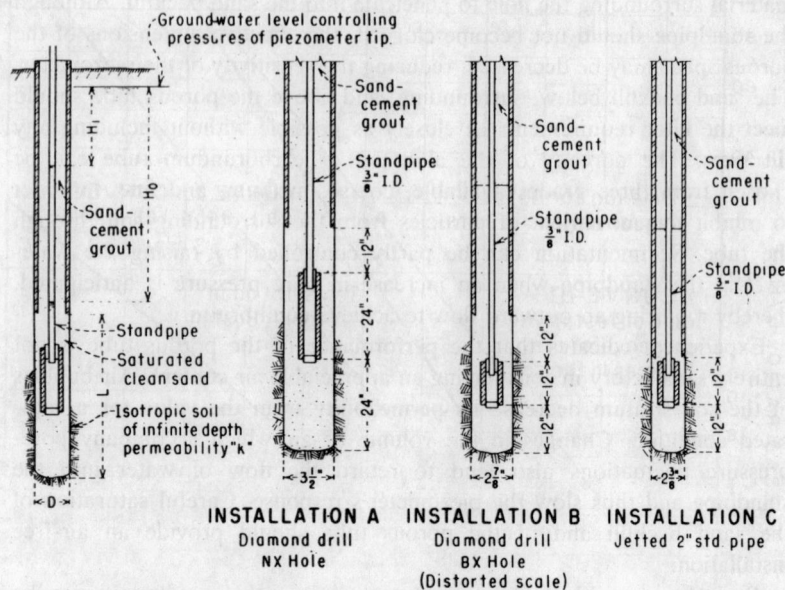

INSTALLATION A
Diamond drill
NX Hole

INSTALLATION B
Diamond drill
BX Hole
(Distorted scale)

INSTALLATION C
Jetted 2"std. pipe

TIME LAG FOR 95% EQUALIZATION

GENERAL SOIL TYPE	SAND	SILT			CLAY		
k in cm./sec.	10^{-3}	10^{-4}	10^{-5}	10^{-6}	10^{-7}	10^{-8}	10^{-9}
k in ft./yr.	1000	100	10	1	0.1	0.01	0.001
Installation A	8 sec.	1.3 min.	13 min.	2.2 hrs.	22 hrs.	9.4 days	94 days
Installation B	12 sec.	2.0 min.	20 min.	3.4 hrs.	34 hrs.	14.2 days	142 days
Installation C	17 sec.	2.9 min.	29 min.	4.8 hrs.	48 hrs.	20.2 days	202 days

Note: Time lag for 99% equalization is approximately
1.5 times the lag for 95% equalization.

Figure 28–6.—Time lag for 95-percent equalization. 101–D–304.

times greater than the vertical. Therefore, the actual sensitivity of a piezometer placed in varved or stratified soil may be greater than that estimated from nominal (laboratory) permeability.

10. Testing the Porous-Tube Piezometer.—The piezometer can be tested in place to evaluate its sensitivity and to determine the permeability and anisotropy of the surrounding soil for the design of further installations. After the water surface in the standpipe has reached a practical level of equilibrium with external pore pressure, the water level in the standpipe can be raised or lowered a substantial distance. The elevation of water in the standpipe is then measured at various times as it returns to its original position. The tests should be performed in a period when there is no important change in the external pore pressure or in the loading conditions.

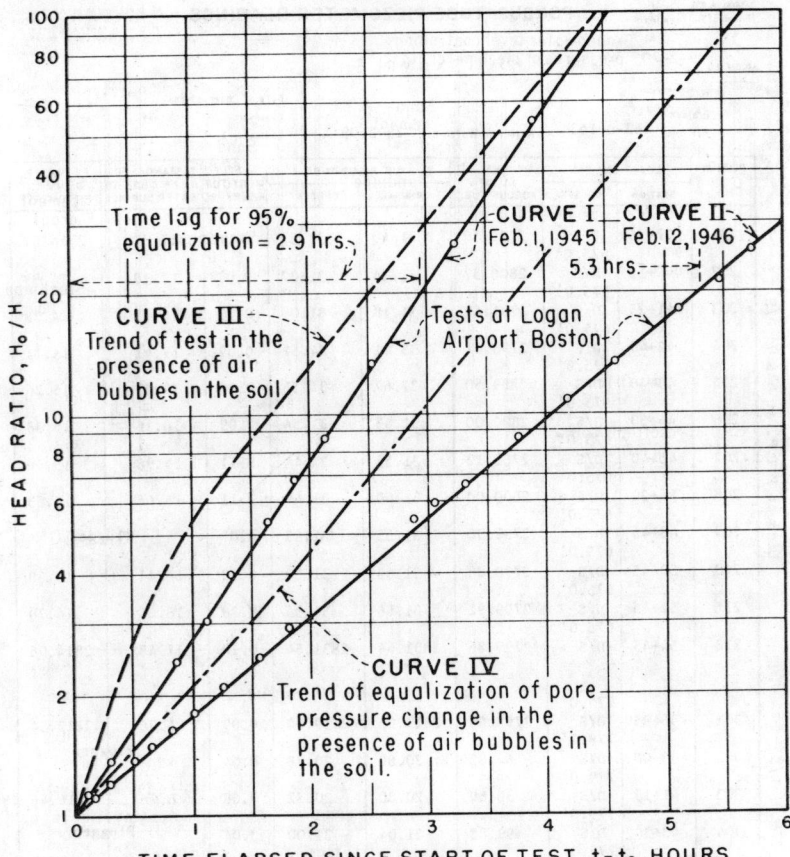

Figure 28-7.—Time lag tests. 101-D-305.

Figure 28-7 presents the results of tests performed on a porous-tube
piezometer placed in the glacial clay beneath Boston's Logan Airport.
According to observations taken 3 months after installation (see curve I),
the time interval for 95-percent equalization was 2.9 hours. In a test
made a year later, 5.3 hours were required, as shown on curve II. This
decrease in sensitivity was probably due to the movement of clay, median

7-1600 (8-70)
Bureau of Reclamation

POROUS-TUBE PIEZOMETER READINGS

TO DIRECTOR OF DESIGN & CONST
ENGRG & RESEARCH CENTER
DENVER FEDERAL CENTER
DENVER, COLORADO 80225

MODIFIED 9-2-70 FOR **EXAMPLE**

Dam ....
Project ....
Ref. Dwg. 662-D-153 Reservoir Water El.[1] 2905.99

Date of Observations 9-17-70
Observer   Sheet 2 of 3
Tailwater El.[1]   - - -

POROUS TUBE NO.	LOCATION [2] STATION	LOCATION [2] OFFSET	ORIGINAL [3] EL. OF POROUS TUBE	EL - TOP RISER TUBE [4] ORIGINAL	EL - TOP RISER TUBE [4] PRESENT	SETTLEMENT-TOP RISER TUBE	DISTANCE - TOP RISER TUBE TO WATER SURFACE	EL. WATER IN PIEZOMETER
201	37+95	675.0' D/S	2799.50	2831.43	2831.37	0.06	17.30	2814.07
202	40+15	675.0' D/S	2800.37	31.34	31.27	0.07	17.18	14.09
203	41+25	675.0' D/S	2731.59	31.18	31.03	0.15	16.15	14.38
204	42+40	675.0' D/S	2798.71	33.61	34.52	0.09	19.41	15.11
205	43+40	675.0' D/S	2799.50	31.67	31.55	0.12	16.29	15.26
206	44+30	675.0' D/S	2800.00	31.63	31.54	0.09	16.16	15.38
207	45+40	675.0' D/S	2765.12	31.55	31.44	0.11	15.98	15.46
208	46+30	675.0' D/S	2800.00	31.60	31.49	0.11	15.66	15.83
209	48+45	675.0' D/S	2799.90	31.55	31.48	0.07	Plugged	
210	50+45	675.0' D/S	2799.80	31.63	31.50	0.13	16.11	15.39
211	52+45	675.0' D/S	2799.92	31.54	31.34	0.20	16.64	14.70
212	54+45	675.0' D/S	2799.86	2831.64	2831.54	0.10	17.48	2814.06
301	36+85	779.0' D/S	2798.66	2819.96	2819.90	0.09	6.36	2813.54
302	39+00	779.0' D/S	64.85	20.80	20.76	0.04	Plugged	
303	41+15	779.0' D/S	30.69	20.80	20.72	0.08	6.26	14.46
304	43+30	779.0' D/S	99.83	21.04	21.00	0.04	Plugged	
305	45+30	779.0' D/S	64.91	21.14	21.09	0.05	5.88	15.21
306	47+30	779.0' D/S	29.47	20.48	20.41	0.07	4.17	16.24
307	49+30	779.0' D/S	64.79	20.64	20.56	0.08	5.22	15.34
308	51+30	779.0' D/S	99.68	20.98	20.88	0.10	6.46	14.42
309	53+30	779.0' D/S	27.39	20.56	20.42	0.14	4.43	15.99
310	55+30	779.0' D/S	2765.05	2820.68	2820.63	0.05	7.15	2813.48

Line of P.T.P. at 675.0' D/S — *Line of P.T.P. at 779.0' D/S*

XXXXXXXXXXXXXXXXXXXXXXXXXXXXXXXXXXXXXXXXXXXXXXXXXXXXXXXXXXXXXXXXXXXXXX (3) Taken as mid-point on length of porous tube. (4) Record all elevations and distances to 0.01 foot. * Use minus (-) to indicate heave.

Figure 28–8.—Example—Porous-tube piezometer readings. 101–D–306.

size 0.002 mm., into the sand backfill which averaged 0.6 mm. Although the nominal permeability of this varved clay was 0.03 foot per year, the value for the horizontal direction appeared to be approximately 0.7 foot per year. Because of this discrepancy the sensitivity of the piezometer in actual operation was much greater than predicted from laboratory information.

If air is present in the soil surrounding the porous tube, the results of such a test will plot in the shape of the dashed curve III (fig. 28–7). The initial change in volume of air resulting from the rise or fall of the standpipe water level accelerates the recovery of equilibrium and decreases the apparent time lag. However, in actual operations where the pore pressure fluctuates, change in the volume of air slows the flow of water through the porous point, increasing the time lag. The trend of the equalization curve for operating conditions in the presence of air in the soil is shown in curve IV (fig. 28–7).

11. Frequency of Observations.—The schedule for readings on porous-tube piezometer installations to be submitted during construction and operation of the apparatus is given in figure 103. Observations should be reported as shown on figure 28–8 (form 7–1600). Elevations for the tops of the pipes and of the water elevations in the pipes should be reported to the nearest 0.01 foot. Elevations of the reservoir and of the tailwater also should be reported to the nearest 0.1 foot if appropriate.

12. Final Report.—Pertinent information regarding the installation of the porous-tube piezometers should be included with the Final Report on Earth Dam Instrumentation (L–16).

INSTRUCTIONS FOR INSTALLING AND READING INTERNAL VERTICAL MOVEMENT DEVICES

Designation E–29

1. Purpose.—The internal vertical movement (crossarm) devices provide a means of determining the volume change within the embankment and settlement of the foundation of an earth embankment. Compression and settlement data, together with the characteristics of the earth materials, can be correlated with pore-fluid pressure observations to check the initial design assumptions. To provide the desired correlation between the pressures and compression, the internal vertical movement apparatus is placed within 10 feet of the piezometer installation. For typical installations, see figure 29–1.

2. Description.—(a) *General.*—All present crossarm installations on Reclamation projects have been fabricated from lengths of standard steel pipe, protected by suitable paint against corrosion. However, considerable corrosion has occurred within the vertical pipe column after years of service, particularly within the limits of the changing free water

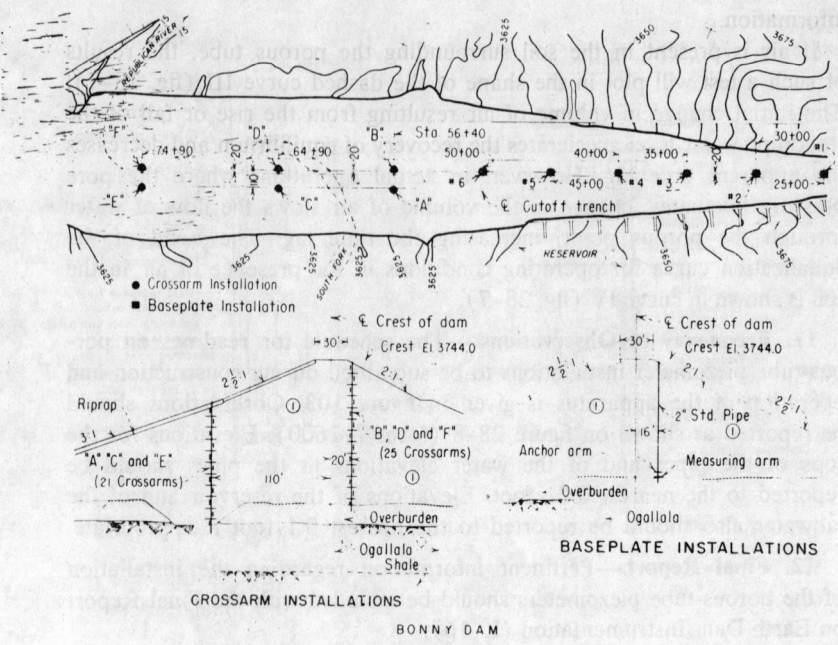

Figure 29–1.—Vertical movement and foundation settlement installation.
101–D–307.

surface. In addition, plant growth has been troublesome when obtaining readings at dams where conditions of continuing high temperature and relative humidity occur. This fibrous plant growth can be retarded by periodic application of the QAC solution used as an inhibitor in the piezometer fluid (see paragraph 13, designation E–27). To overcome the problem of corrosion within the pipe system, consideration is being given to fabricating the pipe sections from ABS (acrylonitrile-butadienestyrene), from PP (polypropylene), or from glass-epoxy plastic pipe extruded to standard pipe sizes (Schedule No. 40 in plastics). However, to provide a sufficiently hard surface for engagement of the pawls on the measuring torpedo, the base or measuring point end of the 1½-inch plastic pipe sections should be provided with a collar of stainless steel.

Although many early installations were placed at 5-foot vertical increments, present practice provides for placement of the crossarm units at 10-foot increments in earth dams. Only the details of the 10-foot spacing of crossarm units in rock-free soils are shown in figure 29–2. The system is comprised of a series of sections of 1½- and 2-inch standard pipe. The installation consists of a base pipe extension, crossarm units, intermediate sections, and a top extension. Metal parts installed permanently in the

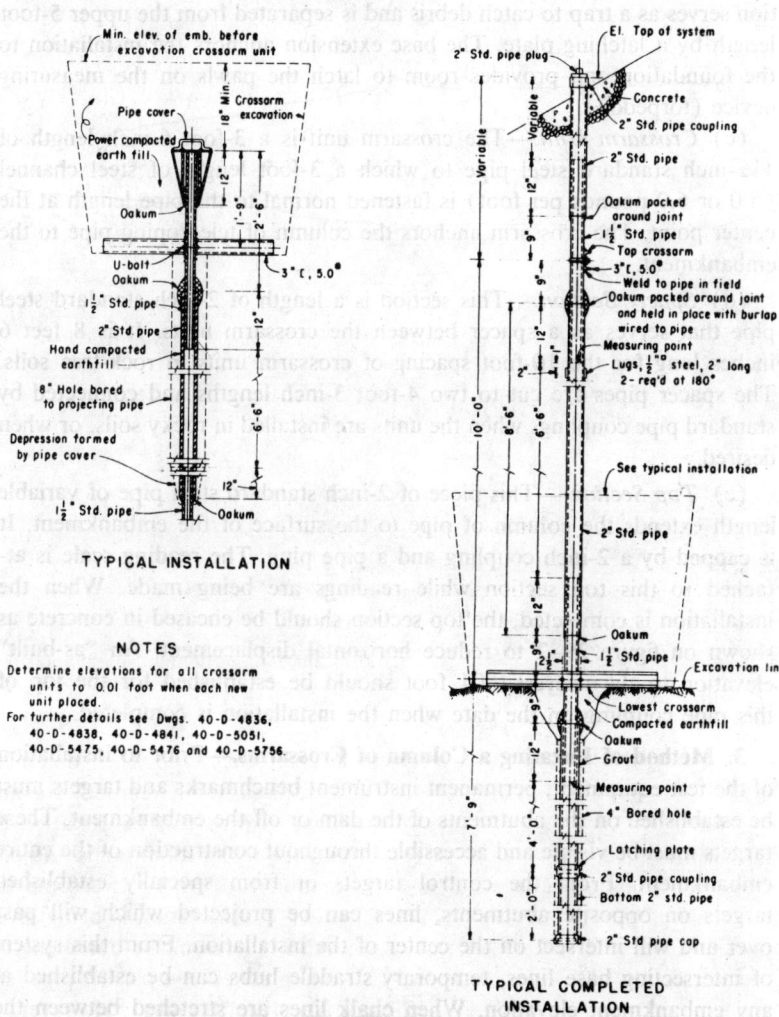

TYPICAL INSTALLATION

Min. elev. of emb. before
excavation crossarm unit

Pipe cover

Power compacted
earth fill

Oakum

U-bolt
Oakum

$1\frac{1}{2}$" Std. pipe
2" Std. pipe
Hand compacted
earthfill

8" Hole bored
to projecting pipe

Depression formed
by pipe cover

$1\frac{1}{2}$" Std. pipe   Oakum

Crossarm
excavation

3" [, 5.0°

12"

El. Top of system

2" Std. pipe plug

Concrete
2" Std. pipe coupling
2" Std. pipe

Oakum packed
around joint

$1\frac{1}{2}$" Std. pipe
Top crossarm
3" [, 5.0°
Weld to pipe in field
Oakum packed around joint
and held in place with burlap
wired to pipe
Measuring point
Lugs, $1\frac{1}{2}$"ø steel, 2" long
2 - req'd at 180°

See typical installation

2" Std. pipe

Oakum
$1\frac{1}{2}$" Std. pipe   Excavation lin.

Lowest crossarm
Compacted earthfill
Oakum
Grout
Measuring point
4" Bored hole
Latching plate
2" Std. pipe coupling
Bottom 2" std. pipe

2" Std pipe cap

**TYPICAL COMPLETED
INSTALLATION**

NOTES

Determine elevations for all crossarm
units to 0.01 foot when each new
unit placed.
For further details see Dwgs. 40-D-4836,
40-D-4838, 40-D-4841, 40-D-5051,
40-D-5475, 40-D-5476 and 40-D-5756.

**Figure 29–2.—Vertical movement device in rock-free soils—10-foot spacing.
101-D-542.**

dam should be coated with protective paint to reduce corrosion. The details of the crossarm units at 10-foot spacing in rock-free soils are as follows:

(b) *Base Extension.*—The base pipe extension consists of two lengths of standard 2-inch steel pipe which extends the installation 7 feet 9 inches below the middle of the bottom crossarm which is placed on the foundation excavation line. The lower 2-foot length of 2-inch pipe in the foundation serves as a trap to catch debris and is separated from the upper 5-foot length by a latching plate. The base extension anchors the installation to the foundation and provides room to latch the pawls on the measuring device (torpedo).

(c) *Crossarm Unit.*—The crossarm unit is a 3-foot 6-inch length of 1½-inch standard steel pipe to which a 3-foot length of steel channel (5.0 or 6.0 pounds per foot) is fastened normal to the pipe length at the center point. The crossarm anchors the column of telescoping pipe to the embankment.

(d) *Spacer Section.*—This section is a length of 2-inch standard steel pipe that serves as a spacer between the crossarm units. It is 8 feet 6 inches long for the 10-foot spacing of crossarm units in rock-free soils. The spacer pipes are cut to two 4-foot 3-inch lengths and connected by standard pipe couplings when the units are installed in rocky soils, or when desired.

(e) *Top Section.*—This piece of 2-inch standard steel pipe of variable length extends the column of pipe to the surface of the embankment. It is capped by a 2-inch coupling and a pipe plug. The reading scale is attached to this top section while readings are being made. When the installation is completed, the top section should be encased in concrete as shown on figure 29–2 to reduce horizontal displacement. An "as-built" elevation to the nearest 0.01 foot should be established for the top of this pipe coupling on the date when the installation is completed.

3. Method of Locating a Column of Crossarms.—Prior to installation of the test equipment, permanent instrument benchmarks and targets must be established on the abutments of the dam or off the embankment. These targets must be visible and accessible throughout construction of the entire embankment. From the control targets or from specially established targets on opposite abutments, lines can be projected which will pass over and will intersect on the center of the installation. From this system of intersecting base lines, temporary straddle hubs can be established at any embankment elevation. When chalk lines are stretched between the straddle hubs, the point of intersection will be the center of the installation.

4. Measuring Equipment.—The equipment required for obtaining crossarm readings includes the torpedo (fig. 29–3), a steel measuring tape, approximately 5/16-inch wide by 0.015-inch thick, tape tension handles, tension spring, and reading scale (see fig. 29–4). A suitable case should be fabricated to hold the torpedo (fig. 29–3), the tape reel, the reading scale and the pipe adaptor. This carrying case will protect the measuring apparatus when not in use. On Reclamation projects, this case will be provided when the apparatus is sent to the field.

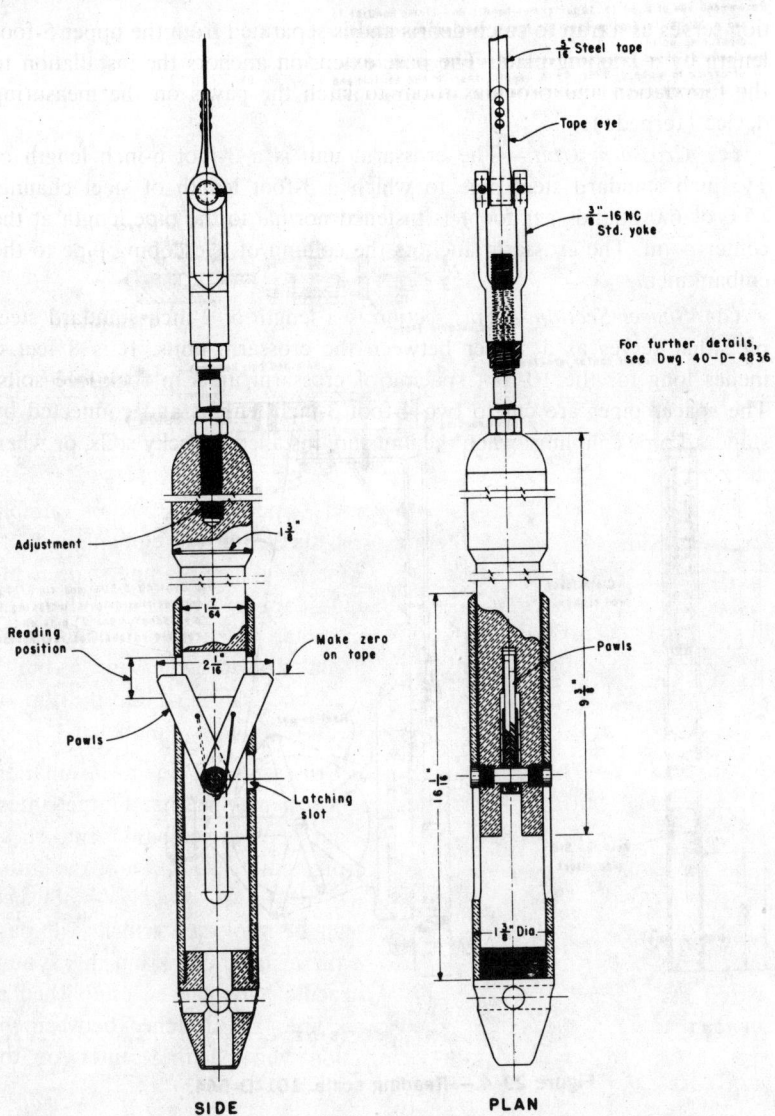

Figure 29-3.—Measuring torpedo. 101-D-543.

Water depth in the pipe system is determined by the indicator device shown on figure 29–5. When the device is lowered down the pipe column by the attached electrical cable, contact with the water causes a float to lift within the device, closing the electrical circuit at the contact points and activating the ohmmeter needle. Distance to the water surface is measured from the top of the system along the cable. When the water level indicator

Recommend use of a 30-lb. tape tension handle and clamp handles to
set a 20-lb. tension on tape for taking readings on crossarm
measuring points.

Use device as shown during construction. Thereafter cut off bottom
of device to make a 3-ft. 8-in. length; then thread bottom end
of 2-inch std. pipe to fit top of installation shown on figure 29-2.

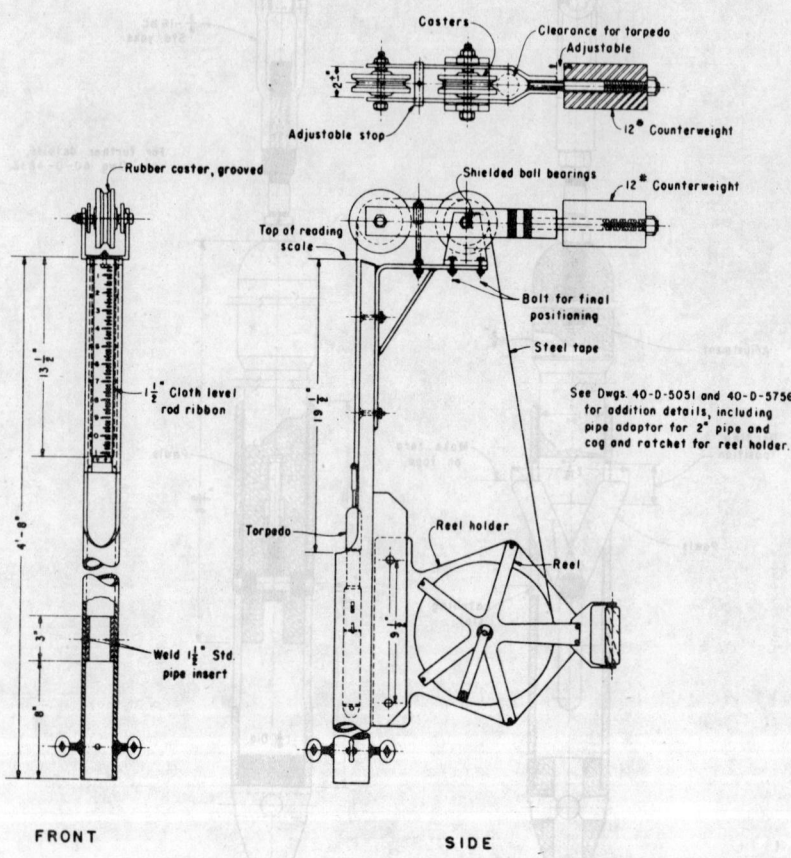

Figure 29–4.—Reading scale. 101–D–544.

is received at a project, the cable should be marked at 1-foot intervals to
facilitate readings. Measurements for water level in the pipe system to 0.1
foot from the top of the system to the nearest foot mark on the cable are
made by use of a hand rule. A suitable reel for holding the indicator and
the electrical cable should be fabricated as illustrated in figure 29–6. This
reel will be fabricated for Reclamation projects before the device is
shipped to the field.

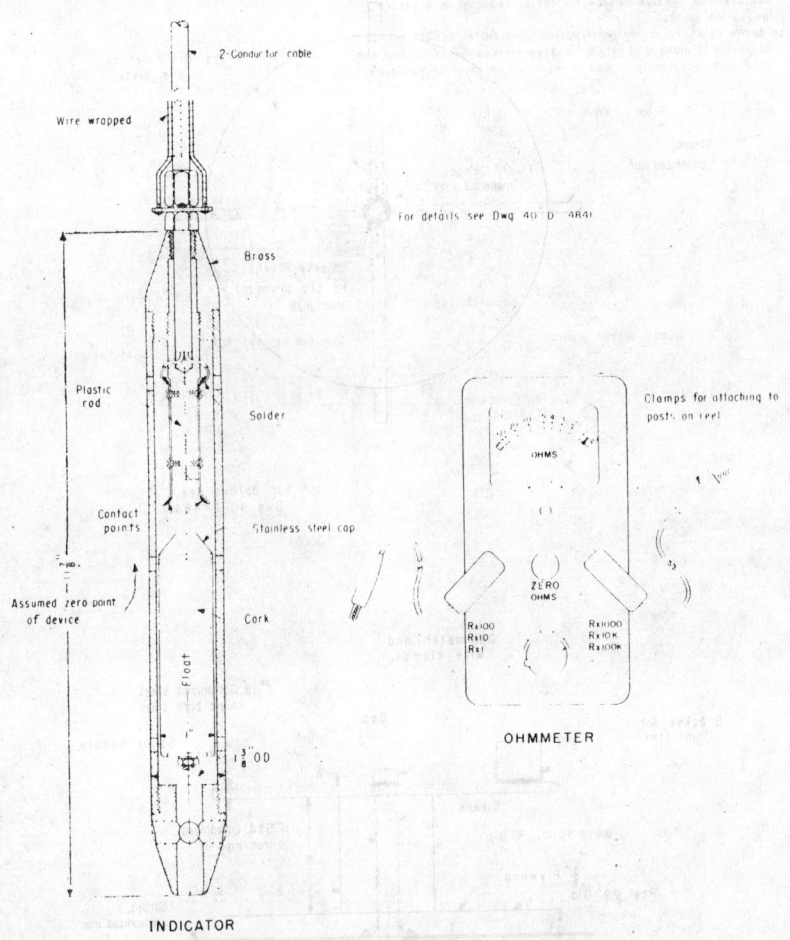

Figure 29–5.—Water level indicator. 101–D–545.

5. Determining Original Elevations.—The original elevation for a crossarm unit is obtained when backfill around the unit is placed to within approximately 12 inches of the top of the 1½-inch pipe. At that time, remove the pipe cover, attach the adaptor and reading scale to the riser pipe, and determine the elevation of the reference plate on top of the reading scale to the nearest 0.01 foot. The elevation for this reference plate must be determined by leveling from a benchmark located off the embankment. Then lower the torpedo which is attached to the measuring tape through the reading scale into the pipe system and read and record the present distances to each of the measuring points (base of 1½-inch

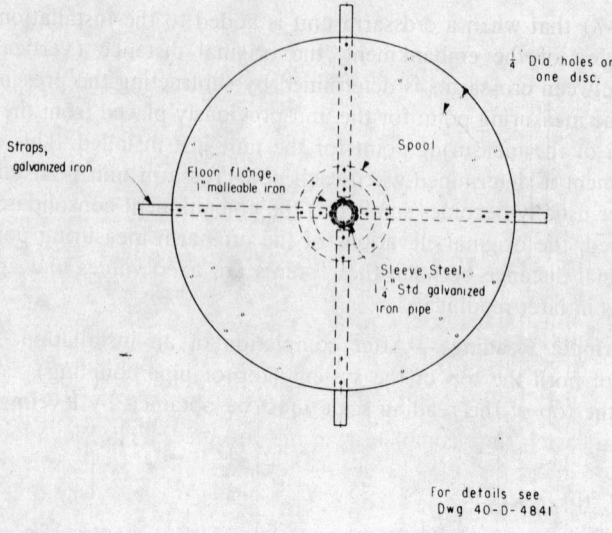

For details see
Dwg 40-D-4841

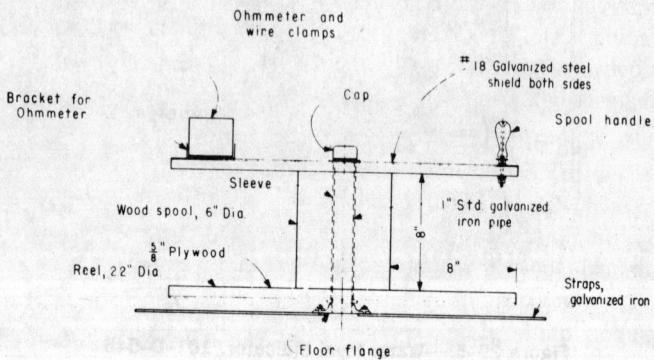

REEL ASSEMBLY

Figure 29–6.—Reel for water level sounder—For use on internal movement device. 101–D–546.

pipe sections) in the system. After determining the distance to the bottom crossarm unit in the system, drop the torpedo to the latching plate to latch the pawls on the torpedo, and withdraw the torpedo and tape from the system.

It should be noted in the subscripts on the bottom of form 7–1348 (fig. 29–7) that when a crossarm unit is added to the installation during construction of the embankment, the original distance (vertical increment) between crossarms is determined by subtracting the present elevation of the measuring point for the unit previously placed from the present elevation of the measuring point for the unit just installed. This distance or increment is determined when each new crossarm unit is installed and thereafter usually becomes smaller as the embankment consolidates. Once established, the original elevations of the crossarm measuring points and the original distance between these points are fixed values and appear as constants in later readings.

6. Periodic Readings.—After completion of an installation, the elevations of both the top of the system (top of pipe coupling) as well as that of the top of the reading scale must be obtained by leveling from a benchmark each time complete readings are made. The elevation for the foundation crossarm (No. 1) in a system should never be assumed to be constant and should not be used for a control elevation.

An example of a set of periodic (L–23) readings made after completion of a dam is shown in figure 29–7. Note that after an installation is completed, the current elevation for the top of system (top of pipe coupling) is required on the reporting form and not the elevation for the top of reading scale. The reading scale is a portable device and the increment between the level for the top of pipe coupling and the top of scale can change slightly each time the reading scale is attached.

7. Reading Intervals.—A complete set of readings must be made on the entire existing installation each time an additional crossarm unit is installed. When construction on the embankment is suspended, as during winter shutdown, the internal vertical movement system should be adapted so that readings can be made at monthly intervals. The first set of periodic readings should be made within 6 months after completion of the dam. Thereafter, periodic readings generally should be made every 2 years. The required frequency of readings is shown on figure 103. All readings should be transmitted in duplicate. Within 2 years after completion of the dam or when the behavior trend for the structure can be determined, a new schedule for periodic readings should be arranged. On Reclamation projects the new schedule for periodic readings will be prepared by the office of the Director of Design and Construction, Denver, Colo., and presented on form DC–622. At dams where normal behavior conditions occur, the interval between periodic readings may be extended to a maximum of every six years. When practicable, such periodic readings should be obtained and analyzed prior to the regular 6-year inspections of major structures.

8. Installation of Crossarm Units.—The vertical movement appa-

7-1348 (8-70)
Bureau of Reclamation

INTERNAL VERTICAL
MOVEMENT READINGS

TO: DIRECTOR OF DESIGN & CONST
ENGRG. & RESEARCH CENTER
DENVER FEDERAL CENTER
DENVER, COLORADO 80225

MODIFIED...9-2-70...FOR   **EXAMPLE**

416-D-226,-D-227;

Dam................................Project Central Valley, Calif....Ref. Dwg 40-D-5475,-D-5477

Installation "D" ....Location....70.0 feet U/S from Axis Station 7+00

El. top of (1) ams. (1) pipe #2375.50....Reservoir Water El. 2354.34................Date Sept. 27, 1970

Observer....Kaiser................El. of water surface in system 2330.82................Sheet...1...of...2

MEAS. POINT NO.	ORIG. EL. OF MEAS. (2) POINT OR ORIG. DIST. BETWEEN M.P. AND GNTRWT IN H.M.B.	PRESENT EL. OF (2) MEAS. POINT OR ORIG. DIST. BETWEEN M.P. AND GNTRWT IN H.M.B.	SETTLEMENT OF MEAS. POINT OR MOVEMENT OF COUNTERWEIGHT	ORIG. DIST. (3) BETWEEN MEAS. POINTS OR HOR. MOV. PLATES	PRESENT DIST. BETWEEN MEAS. POINTS OR HOR. MOV. PLATES	CHANGE IN (4) DISTANCE
X-28	2366.44	2364.38	2.06			
				10.35	10.34	0.01
X-27	2356.57	2354.04	2.53			
				10.13	9.96	0.17
X-26	2346.76	2344.08	2.68			
				10.50	10.28	0.22
X-25	2336.76	2333.80	2.96			
				10.25	10.16	0.09
X-24	2326.75	2323.64	3.11			
				10.38	10.30	0.08
X-23	2316.79	2313.34	3.45			
				10.46	10.40	0.06
X-22	2306.77	2302.94	3.83			
				10.31	10.07	0.24
X-21	2296.79	2292.87	3.92			
				10.35	10.15	0.20
X-20	2286.75	2282.72	4.03			
				10.32	10.12	0.20
X-19	2276.80	2272.60	4.20			
				10.24	9.98	0.26
X-18	2266.78	2262.62	4.16			
				10.26	10.03	0.23
X-17	2256.83	2252.59	4.24			
				10.30	10.02	0.28
X-16	2246.77	2242.57	4.20			
				10.21	9.89	0.32
X-15	2236.80	2232.68	4.12			
				10.31	10.06	0.25
X-14	2226.77	2222.62	4.15			
				10.22	9.97	0.25
X-13	2216.77	2212.65	4.12			
				10.20	9.87	0.33
X-12	2206.76	2202.78	3.98			
				10.14	9.85	0.29
X-11	2196.73	2192.93	3.80			
				10.20	9.79	0.41
X-10	2186.74	2183.14	3.60			
				10.21	9.86	0.35
X-9	2176.71	2173.28	3.43			
				10.05	9.67	0.38
X-8	2166.76	2163.61	3.15			
				10.09	9.69	0.40
X-7	2156.75	2153.92	2.83			

(1) During construction use El. of amb., On completion use El. top of pipe. (2) Elevations are for meas. points of crossarms, horiz. movement units or counterweights. (3) Dist. between meas. points of successive units when top unit was placed. (4) Indicate increase in dist. between crossarm units by minus (-). Indicate decrease in dist. between horiz. movement plates by minus (-). Report data to 0.01 foot.

Figure 29-7.—Example—Internal vertical movement readings. (Sheet 1 of 2.)
101-D-598.

7-1348 (8-70)
Bureau of Reclamation

INTERNAL VERTICAL
MOVEMENT READINGS

TO: DIRECTOR OF DESIGN & CONST
ENGRG & RESEARCH CENTER
DENVER FEDERAL CENTER
DENVER, COLORADO 80225

MODIFIED....9-2-70.._FOR   **EXAMPLE**

                                                        416-D-226,-D-227;
Dam..Project Central Valley, Calif...Ref. Dwg. 40-D-5475,-D-5476...

Installation..."D"..............Location....70.0 feet U/S from ℄ Crest Station 7+00.....

El. top of ⁽¹⁾ ᵃᵐᵇ, ⁽¹⁾ pipe #2375.50......Reservoir Water El. 2354.34...............Date Sept. 27,1970...

Observer... Kaiser............El. of water surface in system. 2330.82...............Sheet....2....of....2.........

MEAS. POINT NO.	ORIG. EL. OF MEAS. (2) POINT OR ORIG. DIST. BETWEEN M.P. AND CNTRWT. IN H.M.D.	PRESENT EL. OF (2) MEAS. POINT OR ORIG. DIST. BETWEEN M.P. AND CNTRWT. IN H.M.D.	SETTLEMENT OF MEAS. POINT OR MOVEMENT OF COUNTERWEIGHT	ORIG. DIST. (3) BETWEEN MEAS. POINTS OR HOR. MOV. PLATES	PRESENT DIST. BETWEEN MEAS. POINTS OR HOR. MOV. PLATES	CHANGE IN (4) DISTANCE
X-7	2156.75	2153.92	2.83			
				10.09	9.72	0.37
X-6	2146.76	2144.20	2.56			
				10.00	9.53	0.47
X-5	2136.84	2134.67	2.17			
				10.11	9.68	0.43
X-4	2126.79	2124.99	1.80			
				10.03	9.48	0.55
X-3	2116.78	2115.51	1.27			
				10.02	9.61	0.41
X-2	2106.77	2105.90	0.87			
				10.01	9.36	0.65
X-1	2096.77	2096.54	0.23			
	2364.38					
	2096.54					
	267.84					
	#El. 2376.84, top of pipe at completion on 10-13-60.					
	64.645.81	64.555.53	90.28	275.74	267.84	7.90

(1) During construction use El. of emb; On completion use El. top of pipe. (2) Elevations are for meas. points of crossarms, horizontal movement units or counterweights. (3) Dist. between meas. points of successive units when top unit was placed. (4) Indicate increase in dist. between crossarm units by minus (-). Indicate decrease in dist between horiz. movement plates by minus (-). Report data to 0.01 foot.

Figure 29-7.—Example—Internal vertical movement readings. (Sheet 2 of 2.)
101-D-599.

ratus must be installed as embankment placement operations progress. Each pipe section must be placed in a vertical position. The operation of all heavy equipment should be prohibited in the immediate vicinity of the installations when sections of the apparatus are being placed. If, for some reason, the upper surface of the embankment in the vicinity of the installation must be reworked before an additional unit is completed, flag or mark the installation so that it will not be damaged or displaced by earthmoving equipment.

(a) *Base Extension.*—The base extension for the device consists of two pieces of pipe which form the extension into the foundation. This extension is assembled and set into a 4-inch-minimum-diameter hole which has been drilled to the required depth into the foundation. The hole may be drilled by wash boring, churn drill, hand or power auger, or diamond drill, depending upon the type of foundation material and the equipment available. After setting the pipe extension, the backfilling around the 2-inch pipe should be done with sand-cement grout to within 12 inches of its top. If the foundation crossarm (No. 1) is not installed immediately, place a special temporary pipe cover over the extending 2-inch pipe and cover with an 18-inch-minimum-thickness of compacted embankment material.

(b) *Placing Crossarm Units.*—The vertical movement apparatus can be installed either in "rock-free" soil or in "rocky" soil, or in a combination of soils as found in a zoned type of earth and rockfill embankment. A rock-free soil is defined as a soil that can be penetrated by a 10-inch-diameter power auger. Installation of the apparatus in rock-free soils is illustrated in figure 29–8. Figure 29–9 shows the installation in rocky soils. These two figures provide recommended step-by-step procedures for installation of the crossarm units at 10-foot spacing; however, the procedures are equally applicable for a 5-foot spacing of crossarm units.

(c) *Methods and Procedures for Installation in Rock-Free Soils.*— A practical method for boring the required holes below the excavated trench for a crossarm unit by the use of a power auger is shown in figure 29–10. Figure 29–11 shows the cleaning of the hole for the spacer pipe between crossarm units prior to removal of the protective pipe cover. Figure 29–12 shows the hand tools that are required to accomplish the excavation for the installation of a crossarm unit. After the crossarm unit is installed, the bored hole between crossarm units should be backfilled with selected fine material and the soil compacted by hand rodding around the telescoping pipe section as illustrated in figure 29–13. Figure 29–14 shows a completed crossarm unit prior to backfilling around the 1½-inch pipe.

(d) *Special Instructions for Installation in Rock-Free Soils.*—In

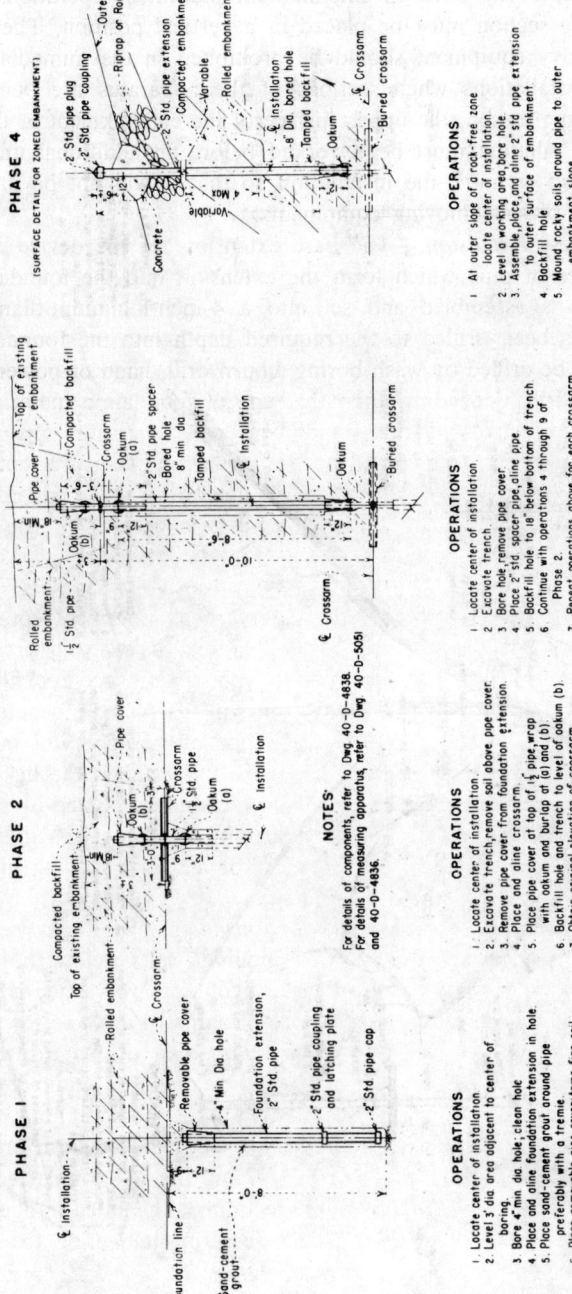

Figure 29-8.—Vertical movement device—Installation in rock-free soils. 101-D-313.

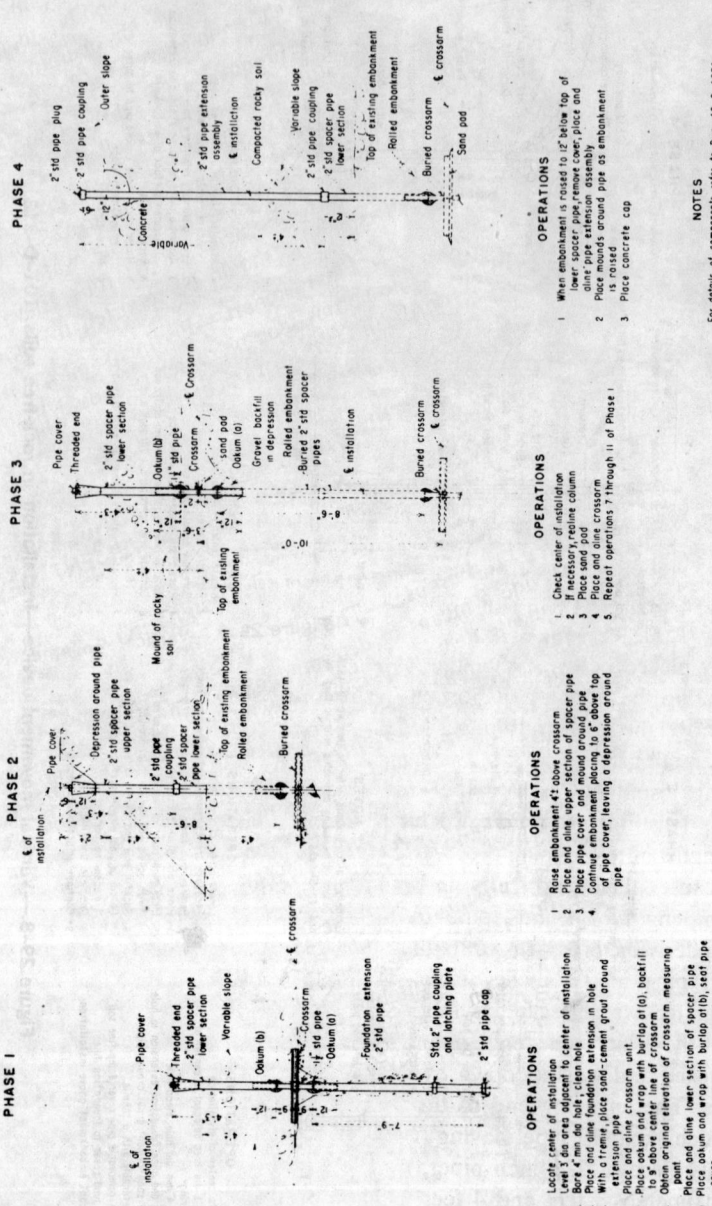

Figure 29-9.—Vertical movement device—Installation in rocky soils. 101-D-314.

Figure 29–10.—Using power auger (air) for drilling 8-inch-diameter holes to top of 2-inch riser pipe of vertical movement device. 245–704–3421.

rock-free soils the embankment must reach an elevation of approximately 15.0 feet above the measuring point of a crossarm unit previously placed before excavating the trench for the next crossarm. After excavating the trench and boring the hole for a crossarm unit, a section of lightweight pipe or tubing should be placed in the hole to keep dirt from entering the pipe system when the pipe cover is removed from the existing installation. The pipe cover should be removed and replaced with a twisting motion in order to retain the oakum packing between the lengths of telescoping pipe.

Because of the difficulty in working the bottom end of the 8-foot 6-inch length of 2-inch pipe over the projecting 1½-inch pipe from the underlying crossarm unit, lugs are welded to opposite sides of the 2-inch pipe at 13 inches from its top as shown on figure 29–2. A special tee wrench can be fabricated on the job to slip over the top end of the 2-inch pipe, engaging the lugs and permitting the 2-inch pipe to be twisted and worked down over the 1½-inch pipe for the required 12 inches. The engaging end of the wrench consists of a 16-inch length of 2½-inch standard pipe, having two slots ¾-inch wide and 3-inches deep. To this piece of 2½-inch pipe, two pieces of 1-inch standard steel pipe, approximately 3 feet and 4 feet long, respectively, are welded to form a T-shaped handle. A suitable wrench is shown in figure 29–15.

The 12 inches of overlap of the 2-inch spacer pipe over the projecting 1½-inch pipe, deep in the augered hole, can be determined by

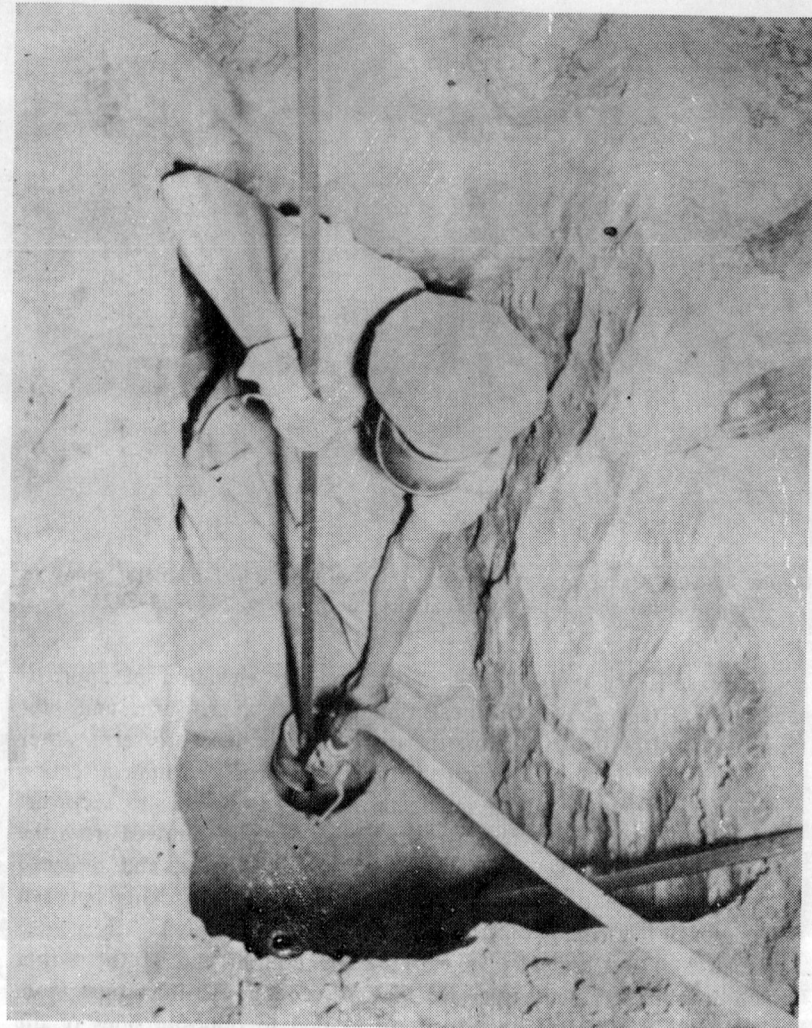

Figure 29–11.—Using air to remove sand from around previously placed pipe cover of vertical movement device. The small bucket contains sand. 245–704–3423.

lowering a T-shaped stick which has been notched at a length of 7 feet 6 inches (see fig. 29–2) inside the 2-inch pipe. When the zone 1 (impermeable zone) embankment contains a considerable amount of rock, the augering of holes for installation of crossarm units can be expedited if the backfill over the pipe cover is selected fine material. The fine material can be placed and compacted inside a cylinder of metal pipe or tubing having an inside diameter of approximately 10 inches. The pipe

Figure 29–12.—Hand tools used in setting units of vertical movement device. From left to right are heavy tamping bar, blow pipe, hook bar, spud bar, spoon bar, tile spade, pipe cover, and bucket. 245–704–3427.

can be progressively jacked to a higher elevation. When the hole is augered from succeeding crossarm units, most of the fine material is removed.

Placement of the 2-inch spacer pipe between crossarms can be expedited if the 8-foot 6-inch lengths of pipe are cut in half and installed in two 4-foot 3-inch lengths. The cut ends will require threading and attachment by a 2-inch pipe coupling. Side lugs will be required on both 4-foot 3-inch lengths of spacer pipe. An 18-inch cover of power-tamped earth material will be required over the partial installation prior to regular embankment placement.

(e) *Backfilling Trenches for Crossarm Units in Rock-Free Soils.*— When the trench has been backfilled with typical embankment materials to within approximately 12 inches of the top of the 1½-inch pipe (bottom of pipe cover), the initial elevation of the crossarm measuring point should be recorded as described in paragraph 5. Thereafter, backfill should continue until a minimum of 18 inches of typical embankment material has been compacted over the 1½-inch riser pipe from the crossarm before embankment placing operations are resumed.

(f) *Special Instructions for Installation in Rocky Materials.*—When

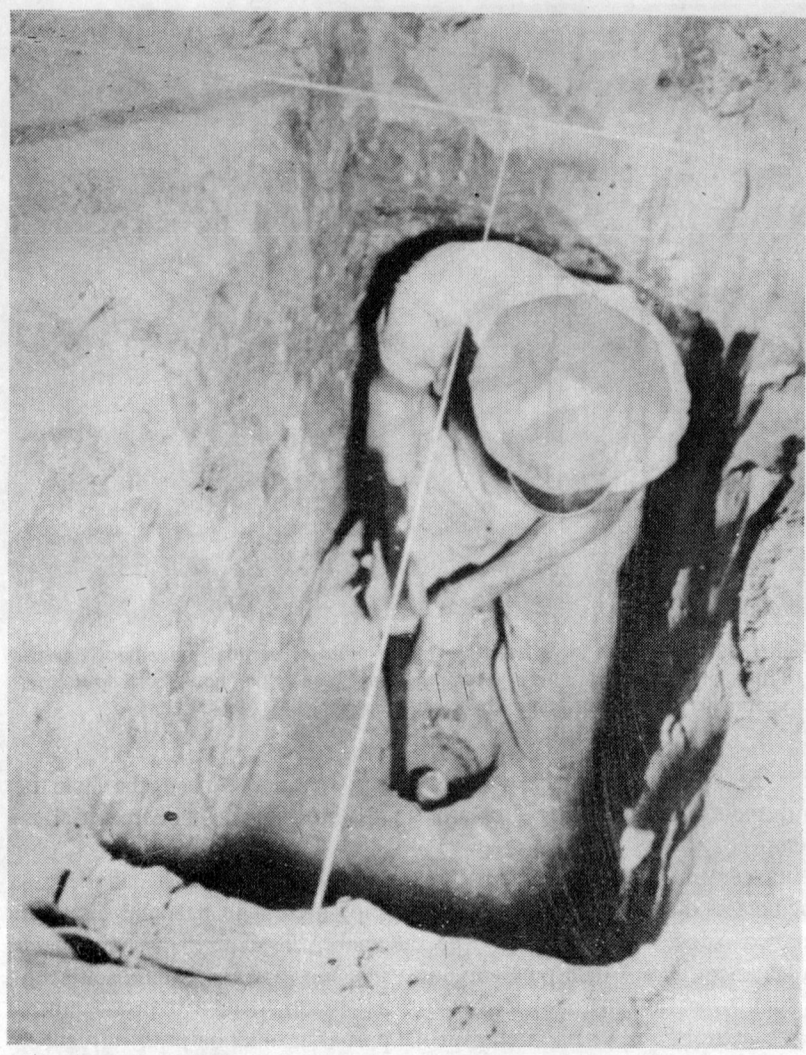

Figure 29–13.—Tamping of backfill around 2-inch riser pipe on vertical movement device. Strings from cross-reference survey hubs give center of installation. 245–704–3425.

crossarm units are to be installed in zoned earth embankment containing considerable rock, special paragraphs will be included in the construction specifications to permit mounding over the crossarm units and special construction drawings will be provided (see fig. 29–9). A typical installation of a crossarm unit in rocky materials is shown in figure 29–16.

(g) *Correction for Alinement.*—After installing the 8-foot 6-inch

Figure 29-14.—1½-inch crossarm pipe being placed inside 2-inch riser pipe of vertical movement device. Crossarm is resting on undisturbed embankment material. 245-704-3426.

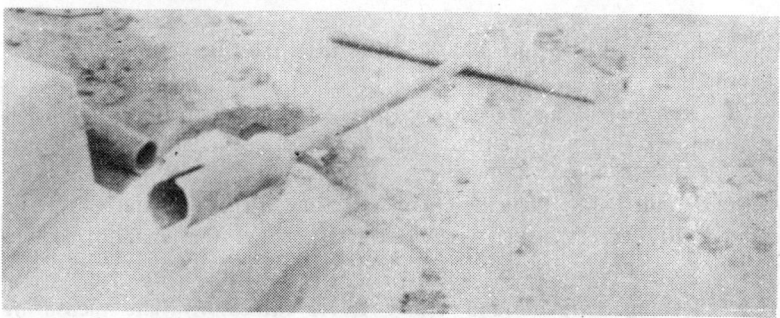

Figure 29-15.—Special wrench fabricated by contractor for installing 2-inch-diameter pipe extensions in vertical movement device and in foundation settlement installation. 245-704-6362.

length of spacer pipe in the bored hole (fig. 29-2), and again after the crossarm assembly has been installed, remove the protective pipe cover and check the vertical alinement (plumb) of the system by lowering a plumb bob from the straddle hub strings. If a deviation from vertical alinement is found, an attempt should be made to correct the error. If it is impractical to correct the error, succeeding pipe sections

Figure 29-16.—Installation of vertical movement device in rocky soils—
Trinity Dam. TD-2979-CV.

should be placed vertically and the offset from the true vertical alinement recorded.

(h) *Rotation of Crossarms.*—Each crossarm section must be placed in a manner that will distribute the weight of the channel section and prevent misalinement of the vertical column of pipe. When each new crossarm unit is installed, the channel section of the new unit should be rotated 90 degrees (clockwise) with respect to the channel section previously placed.

9. Cleaning Torpedo and Water Level Indicator.—The measuring devices must be kept clean and free of grit. It is suggested that each measuring instrument be disassembled, insofar as practicable, and either cleaned or flushed with clear water following completion of readings. The tape should be carefully dried and inspected for kinks and breaks. The application of silicone grease by a cloth on the measuring tape will tend to retard corrosion. If the pawls on the torpedo refuse to latch when the device reaches the bottom of the pipe column, a section of ½-inch pipe, 2-feet long or longer, can be slipped over the tape and lowered by means of a wire into the installation until it rests on top of the torpedo. This added weight should latch the torpedo.

10. Record Tests.—Record tests should be made in the rolled embankment as the trench is excavated for each crossarm unit, and again in the tamped material as the trench is backfilled. Instructions for these tests are given in section 70 (e).

11. Progress Reports.—Copies of form 7–1348, figure 29–7, when properly filled out will show the current elevation of the embankment at the installation, the number of crossarms placed, the elevations of these crossarms, the compression of the embankment, and the settlement of the foundation. The narrative portion of the progress report should discuss variations from the instructions and installation procedures. Photographs, drawings, and charts pertinent to the installation should also be included.

12. Miscellaneous Implements and Tools.—The tools needed to install the settlement apparatus include:

(1) Hand or power augers.
(2) Long-handled spoons.
(3) Shovels.
(4) Picks.
(5) Tamping bars.
(6) Hand-operated mechanical tamper with spade attachment.
(7) Steel engineer's chain.
(8) Reading scale.
(9) Engineer's level and level rod.
(10) Water level indicator.
(11) Measuring device (torpedo).
(12) Straddle hubs (fabricate from suitable material).
(13) Chalk line.
(14) Tape clamp handles.
(15) Tape tension handles.
(16) Tape reel.
(17) A ¼-inch-diameter by 10-foot-long steel rod, hooked on one end.

INSTRUCTIONS FOR INSTALLING AND READING INTERNAL HORIZONTAL MOVEMENT DEVICES

Designation E–30

1. Purpose.—The horizontal movement device was developed to provide data on the horizontal movement within an earth embankment. Horizontal movement is transformed into vertical movement by a linkage system and is obtained by using the measuring torpedo and reading scale as shown on figures 29–3 and 29–4. In addition, vertical movement of the device is measured. Units of this device can be installed as a part of the telescoping vertical movement (crossarm) installation described in designation E–29. Thus, the vertical and horizontal strain may be determined from field measurements of changes in elevations of pipes within an embankment.

 2. **Components.**—The parts and assembly of the horizontal movement device are shown on figure 30–1. The device contains two assemblies: (1) the fixed components and, (2) the moveable components of the counterweight well assembly.

 (a) *Fixed-Component Assembly.*—This assembly consists of the counterweight well (a 6-foot 2-inch length of 4-inch standard steel pipe), the counterweight supports, the top extension, the bottom extension, the extension arms, and necessary reducing couplings.

 (b) *Moveable Components.*—The moveable components (relative to the fixed components) consist of the 24- by 36-inch plates, tie rods, cable linkages, counterweights, and felt spacers.

 3. **Assembly.**—Except for the counterweights, the cable linkages, and the counterweight support units, assembly of the apparatus should be made in a field warehouse prior to installation. To facilitate assembly, similar parts such as counterweights, plates, and fittings are identified, packaged and shipped in the same container. When the parts are received at the project, they should be inspected for the proper quantity and for any damage that might have been incurred during shipment. The following is a suggested assembly sequence:

 (1) Coat the surfaces of the counterweights, the grooved blocks, and the cables with the silicone grease that will be provided with the shipment.

 (2) Attach the cable linkages to the counterweights.

 (3) Place the 6-foot-2-inch length of 4-inch standard pipe (counterweight well) on a working area and attach the 4- by 2-inch standard reducing coupling to the bottom.

 (4) Slide counterweight No. 2 into position, extending the cable to the top of the pipe.

 (5) Slide counterweight No. 1 into position so that its fins straddle the cable attached to counterweight No. 2.

 (6) Tie the cables together and place within the counterweight well.

 (7) Screw the counterweight support unit (the 4-inch standard pipe cross and attached counterweight supports) onto the counterweight well.

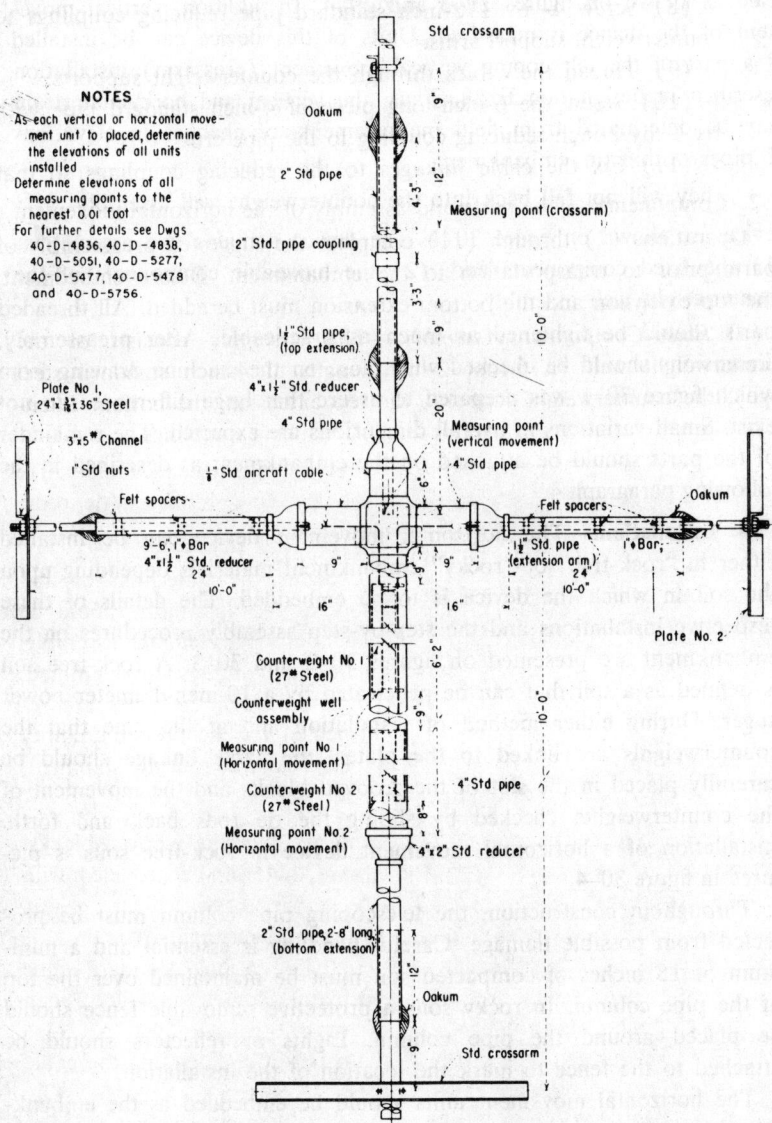

NOTES

As each vertical or horizontal movement unit to placed, determine the elevations of all units installed

Determine elevations of all measuring points to the nearest 0.01 foot

For further details see Dwgs 40-D-4836, 40-D-4838, 40-D-5051, 40-D-5277, 40-D-5477, 40-D-5478 and 40-D-5756.

Std crossarm

Oakum

2" Std pipe

Measuring point (crossarm)

2" Std. pipe coupling

1½" Std pipe, (top extension)

4"x 1½" Std reducer

4" Std pipe

Measuring point (vertical movement)

4" Std pipe

Plate No 1, 24" x ⅜ x 36" Steel

3" x 5" Channel

1" Std nuts

⅝" Std aircraft cable

Felt spacers

Oakum

9'-6", 1"∅ Bar

4" x 1½" Std reducer

1½" Std pipe (extension arm)

1"∅ Bar

Plate No 2

Counterweight No 1 (27" Steel)

Counterweight well assembly

Measuring point No 1 (Horizontal movement)

Counterweight No 2 (27" Steel)

Measuring point No 2 (Horizontal movement)

4" Std pipe

4" x 2" Std. reducer

2" Std pipe, 2'-8" long (bottom extension)

Oakum

Std crossarm

Figure 30–1.—Horizontal movement device. 288–D–2616 (from 101–D–5277).

(8) Screw 4- by 1½-inch standard pipe reducing couplings to counterweight support arms.

(9) Thread the cables through the counterweight supports.

(10) Screw the 6-inch-long piece of 4-inch standard pipe with its 4- by 2-inch reducing coupling to the pipe cross.

(11) Fix the cable linkages to the reducing couplings so that they will not fall back into the counterweight well assembly.

Operations (1) through (11) complete the suggested preassembly of parts prior to transportation to the embankment. Before embedment, the top extension and the bottom extension must be added. All threaded parts should be tightened as much as practicable. After preassembly, dimensions should be checked with those on the machine drawing from which figure 30–1 was prepared to assure that large differences do not exist. Small variations in overall dimensions are expected. The remainder of the parts should be attached on the embankment as described in the following paragraph 4.

4. Installation.—The horizontal movement device can be installed either in "rock-free" or "rocky" embankment materials depending upon the soil in which the device is to be embedded. The details of these respective installations and the step-by-step assembly procedures on the embankment are presented on figures 30–2 and 30–3. A rock-free soil is defined as a soil that can be penetrated by a 10-inch-diameter power auger. During either method of installation and at the time that the counterweights are linked to the plates, the cable linkage should be carefully placed in the slot of the grooved blocks and the movement of the counterweights checked by sliding the tie rods back and forth. Installation of a horizontal movement device in rock-free soils is pictured in figure 30–4.

Throughout construction, the telescoping pipe column must be protected from possible damage. Careful handling is essential and a minimum of 18 inches of compacted soil must be maintained over the top of the pipe column. In rocky soils a protective removable fence should be placed around the pipe column. Lights or reflectors should be attached to the fence to mark the location of the installation.

The horizontal movement units should be embedded as the embankment is raised.

5. Locating.—Initially, targets, hubs, and benchmarks should be established off the structure so that lines between them will intersect at the centerline of the vertical column of telescoping pipe comprising the combined horizontal and vertical movement installations (see par. 3 of designation E–29). Targets, hubs, and benchmarks must be established so that the locations of the installations will be visible throughout construction of the dam. As units of the installation are placed, the top of the pipe column can be found by means of straddle hubs set on the embankment. Straddle hubs are set using the previously established targets, hubs, and benchmarks.

6. Backfilling.—Earth material placed around the horizontal movement unit should be compacted to approximate the same placement conditions for moisture and density as the adjacent embankment material. The compaction should be performed in a manner that will least disturb the installation and still achieve the desired density. Backfill behind the plates and around the tie rods should be placed carefully to avoid disturbing these components. For installation in rocky materials, the sand bed and sand backfill (fig. 30–3) should be reasonably well-graded sand. The sand bed is placed to eliminate possible stress concentration at points where large cobbles and boulders are present in the soil. For installations in rock-free soils, earth removed from the trench can be used as backfill, with the finer material being placed around the components.

7. Alinement.—The telescoping column of pipe must be maintained plumb (in a vertical position) throughout construction. Hence, each unit must be placed vertically, and the alinement of the movement unit previously placed should be checked. This may be accomplished by properly locating the column and making frequent checks of position with a carpenter's level or with the vertical crosshair of a surveying instrument (transit) during embedment. If the previously placed units have been displaced, they should be removed from the embankment, if practicable, and reinstalled to the correct alinement.

The horizontal movement units should be placed so that the centerlines of the units are normal to the centerline of the crest and the vertical plates parallel to the centerline of the crest of the dam. Plate No. 1 should be placed nearest to the outer slope of the embankment so that the direction of movement can be determined.

8. Identification.—The horizontal movement device is designed to be installed as a part of the vertical movement (crossarm) system. Figure 30–5 (form 7–1348) shows a suitable method for identifying the horizontal movement units within a telescoping pipe column. The horizontal movement device is numbered in the order, from bottom to top, of its location in the column of units in the combined installation.

9. Reading Intervals.—Readings should be made as directed on figure 103 and in sections 70(d) and (e). During construction, the elevations of each measurement point in the horizontal movement units are obtained as the unit is installed and each time a vertical movement unit

PHASE I

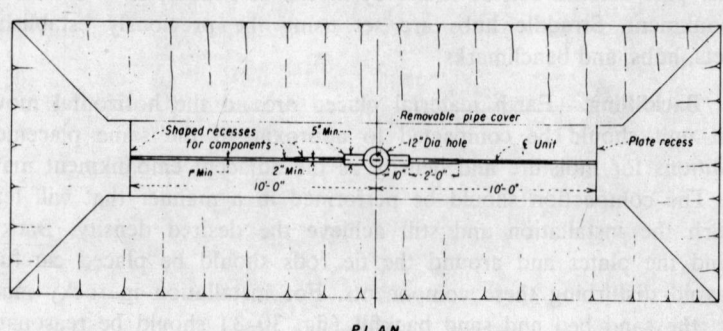

PLAN

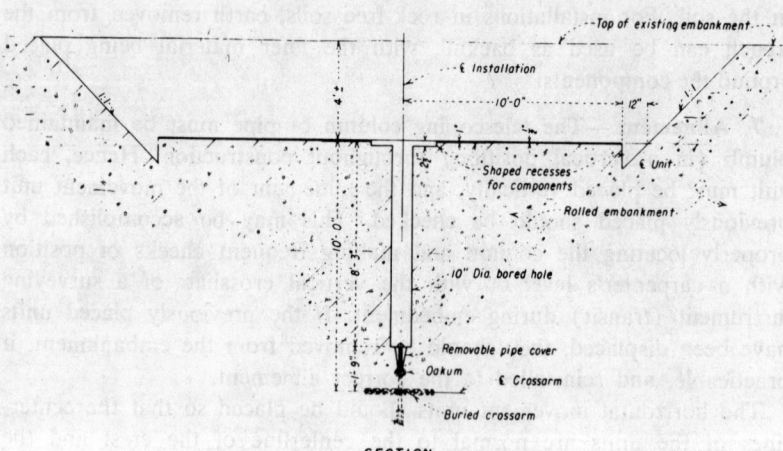

SECTION

OPERATIONS

1. Locate center of installation
2. Excavate trench.
3. Bore 10" min. dia. hole to crossarm beneath
4. Excavate recesses for components.
5. Remove pipe cover.

NOTES

For details of the horizontal movement device, refer
   to Dwg. 40-D-5277.
For details of the vertical movement device, refer to
   Dwg. 40-D-4838.
For details of measuring apparatus, refer to Dwgs.
   40-D-4836, 40-D-4841, 40-D-5051, and 40-D-5756.

Figure 30–2.—Horizontal movement device installation in rock-free soils.
(Sheet 1 of 2.) 101–D–315 (from 40–D–5477).

PHASE 2

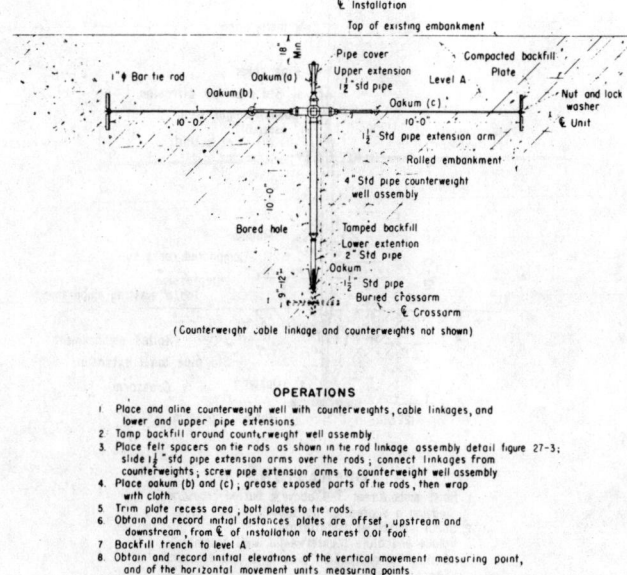

(Counterweight cable linkage and counterweights not shown)

OPERATIONS

1. Place and aline counterweight well with counterweights, cable linkages, and lower and upper pipe extensions
2. Tamp backfill around counterweight well assembly
3. Place felt spacers on tie rods as shown in tie rod linkage assembly detail figure 27-3; slide 1½" std pipe extension arms over the rods; connect linkages from counterweights; screw pipe extension arms to counterweight well assembly
4. Place oakum (b) and (c), grease exposed parts of tie rods, then wrap with cloth
5. Trim plate recess area, bolt plates to tie rods.
6. Obtain and record initial distances plates are offset, upstream and downstream, from ₵ of installation to nearest 0.01 foot
7. Backfill trench to level A
8. Obtain and record initial elevations of the vertical movement measuring point, and of the horizontal movement units measuring points
9. Place pipe cover; place oakum and wrap with burlap at (a) backfill to existing level of embankment

PHASE 3

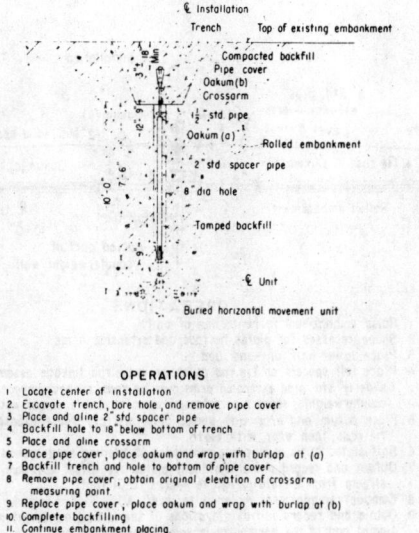

Buried horizontal movement unit

OPERATIONS

1. Locate center of installation
2. Excavate trench, bore hole, and remove pipe cover
3. Place and aline 2" std spacer pipe
4. Backfill hole to 18" below bottom of trench
5. Place and aline crossarm
6. Place pipe cover, place oakum and wrap with burlap at (a)
7. Backfill trench and hole to bottom of pipe cover
8. Remove pipe cover, obtain original elevation of crossarm measuring point
9. Replace pipe cover, place oakum and wrap with burlap at (b)
10. Complete backfilling
11. Continue embankment placing.

Figure 30–2.—Horizontal movement device installation in rock-free soils. (Sheet 2 of 2.) 101–D–315 (from 40–D–5477).

PHASE 1

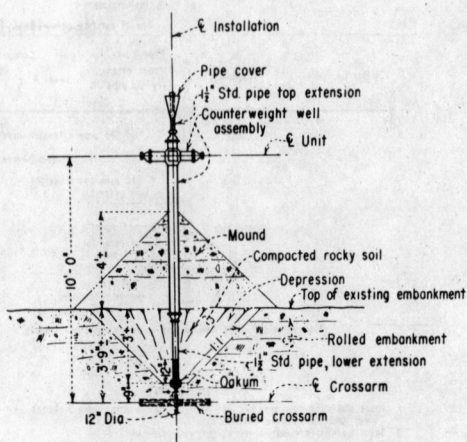

OPERATIONS

1. Raise embankment 3'-9" above ℄ buried crossarm, leaving a depression as shown.
2. Check center of installation; place and aline counterweight well assembly, filling the depression, placing a mound around the pipe as shown.
3. Raise embankment to ℄ horizontal movement unit. (A tripod may be required to maintain the counterweight assembly well in a vertical position.)

PHASE 2

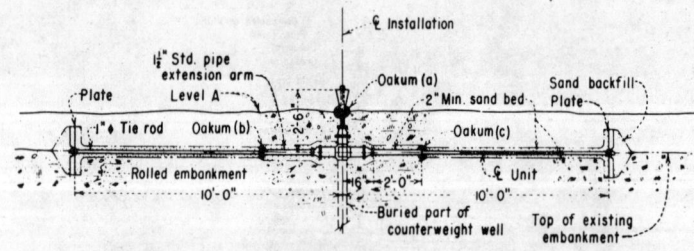

OPERATIONS

1. Raise embankment to centerline of unit.
2. Shape recesses for plates, tie rods, and extension arms.
3. Place lower half of sand bed.
4. Place felt spacers on tie rod as shown in tie rod linkage assembly detail; slide 1¼" std. pipe extension arms over tie rods; connect linkages from counterweights; screw pipe extension arms to counterweight well assembly.
5. Place oakum and wrap with burlap at (b) and (c); grease exposed parts of tie rods; then wrap with cloth.
6. Bolt plates to tie rods; place remainder of bedding sand around components.
7. Obtain and record initial distances plates are offset, upstream and downstream from ℄ installation, to nearest 0.01 foot.
8. Compact embankment material to level A.
9. Obtain and record initial elevations of the vertical movement measuring point and of the horizontal movement measuring points.
10. Place pipe cover; place oakum and wrap with burlap at (a).

Figure 30–3.—Horizontal movement device installation in rocky soils. (Sheet 1 of 2.) 101–D–316 (from 40–D–5478).

PHASE 3

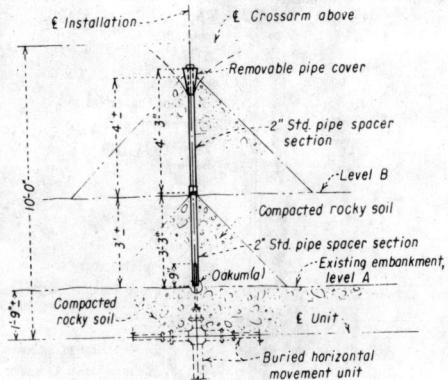

OPERATIONS

1. When embankment is at level A, place and aline 3'-3" std. pipe spacer section, place mound around pipe
2. Raise embankment to level B.
3. Place and aline 4'-3" std. pipe spacer section, placing mound around pipe.
4. Raise embankment to the level of the next crossarm above.
5. Place and aline crossarm as shown on drawing 40-D-5476.

NOTES

For details of the horizontal movement device, refer to Dwg. 40-D-5277.
For details of the vertical movement device, refer to Dwg. 40-D-4838.
For details of measuring apparatus, refer to Dwgs. 40-D-4836, 40-D-4841, 40-D-5051, and 40-D-5756.

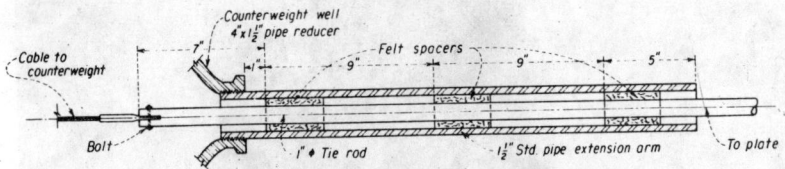

DETAIL
TIE ROD LINKAGE ASSEMBLY

Figure 30–3.—Horizontal movement device installation in rocky soils. (Sheet 2 of 2.) 101–D–316 (from 40–D–5478).

Figure 30–4.—Installation of horizontal movement device in rock-free soils—Trinity Dam. TD–3348–CV.

7-1348 (8-70)
Bureau of Reclamation
MODIFIED 10-2-70 FOR

INTERNAL VERTICAL AND HORIZONTAL
MOVEMENT READINGS
EXAMPLE

TO DIRECTOR OF DESIGN & CONST
ENGRG. & RESEARCH CENTER
DENVER FEDERAL CENTER
DENVER, COLORADO 80225

Dam .... Project San Luis Unit, C.V. ..... Ref. Dwg. 805-D-283 .....
Installation "B" .... Location 20.0 feet D/S from C crest, Sta. 86 + 10
El. top of pipe (1) pipe 554.34   Reservoir Water El. 533.84 .... Date 1-18-71
Observer B. Jones .... El. of water surface in system 344.46 .... Sheet 2 of 2

MEAS. POINT NO.	ORIG. EL. OF MEAS (2) POINT OR ORIG. DIST. BETWEEN M.P. AND CNTRWT. IN H.M.D.	PRESENT EL. OF (2) MEAS. POINT OR ORIG DIST. BETWEEN M.P. AND CNTRWT. IN H.M.D.	SETTLEMENT OF MEAS. POINT OR MOVEMENT OF COUNTERWEIGHT	ORIG. DIST. (3) BETWEEN MEAS. POINTS OR HOR. MOV. PLATES	PRESENT DIST. BETWEEN MEAS. POINTS OR HOR. MOV. PLATES	CHANGE IN (4) DISTANCE
H-10	*370.35	361.69	8.66			
(H-10)-(H-10+1)	3.14	3.14	0.00			
H-10-1	**367.21	358.55	8.66	***20.05	20.03	-0.02
(H-10)-(H-10+2)	5.82	5.84	-0.02			
H-10+2	**364.53	355.85	8.68			
				12.43	12.15	0.28
X-9	358.32	349.54	8.78			
				10.32	10.06	0.26
X-8	348.44	339.48	8.96			
				10.62	10.34	0.28
X-7	338.21	329.14	9.07			
				10.21	10.01	0.20
X-6	328.25	319.13	9.12			
				10.44	10.00	0.44
X-5	318.11	309.13	8.98			
				10.06	9.55	0.51
X-4	308.30	299.58	8.72			
				10.11	9.67	0.44
X-3	298.38	289.91	8.47			
				8.43	7.97	0.46
H-2	*290.40	281.94	8.46			
(H-2)-(H-2+1)	3.13	3.16	-0.03			
H-2-1	**287.27	278.78	8.49	***19.94	19.91	-0.03
(H-2)-(H-2+2)	5.89	5.89	0.00			
H-2-2	**284.51	276.05	8.76			
				11.97	11.23	0.74
X-1	278.48	270.71	7.77			

* El. Vert. Meas. Point for H.M.D.
** El. Counterweights for H.M.D.
**** Sum of data from Vert. Meas. Pts only

	****10856.70	10,663.76	192.94	260.34	253.14	7.20

*** Distance between plates of H.M.D. at time of installation

(1) During construction use El. of amb. On completion use El. top of pipe. (2) Elevations are for meas. points of crossarms, horiz. movement units or counterweights. (3) Dist. between meas. points of successive units when top unit was placed. (4) Indicate increase in dist. between crossarm units by minus (-). Indicate decrease in dist. between horiz. movement plates by minus (-). Report data to 0.01 foot.

Figure 30-5.—Example—Vertical movement readings. 101-D-600.

is placed. During shutdowns and following construction, readings should be made as shown on figure 103, or as directed in paragraphs 6 and 7, designation E–29.

10. Record Tests.—Record tests, both of rolled and tamped material, should be obtained in the soil near the plates at the time the devices are installed. For further details see section 70(e).

11. Reports.—The data from the horizontal movement devices should be submitted in duplicate along with data from the other test apparatus installations as described in section 71(d).

12. Tools and Equipment.—The use of special tools described in paragraph 12 of designation E–29 for the vertical movement devices and shown in figure 29–12 are recommended for placing units of the horizontal movement device. It is suggested that a wooden A-frame be constructed and used to hold units of the horizontal movement device during their installation.

INSTRUCTIONS FOR INSTALLING AND READING FOUNDATION SETTLEMENT APPARATUS

Designation E–31

1. Purpose.—Foundation settlement (baseplate) apparatus is installed on the foundation under an earth dam or dike to permit measurements of vertical settlement of the foundation during construction of the embankment and later during operation of the reservoir. This type of apparatus generally is used where the embankment is relatively shallow (low) and where the foundation is a sandy, silty, or clayey material subject to consolidation under the embankment loading. For a typical installation, see figure 31–1.

2. Description.—The foundation settlement apparatus consists of a baseplate placed on the foundation excavation line and a vertical column of steel pipe. The apparatus is comprised of a base section of 2-inch standard steel pipe, a crossarm measuring unit (baseplate), an anchor crossarm, and an extension of 2-inch pipe to the surface of the embankment. This type of installation is similar to the vertical movement apparatus. For possible substitution of plastic pipe for metal pipe, refer to paragraph 2(a) of designation E–29.

(a) *Base Section.*—The base section is made of two pieces of 2-inch pipe, one 2-feet long and one 5-feet long, joined by a pipe coupling.

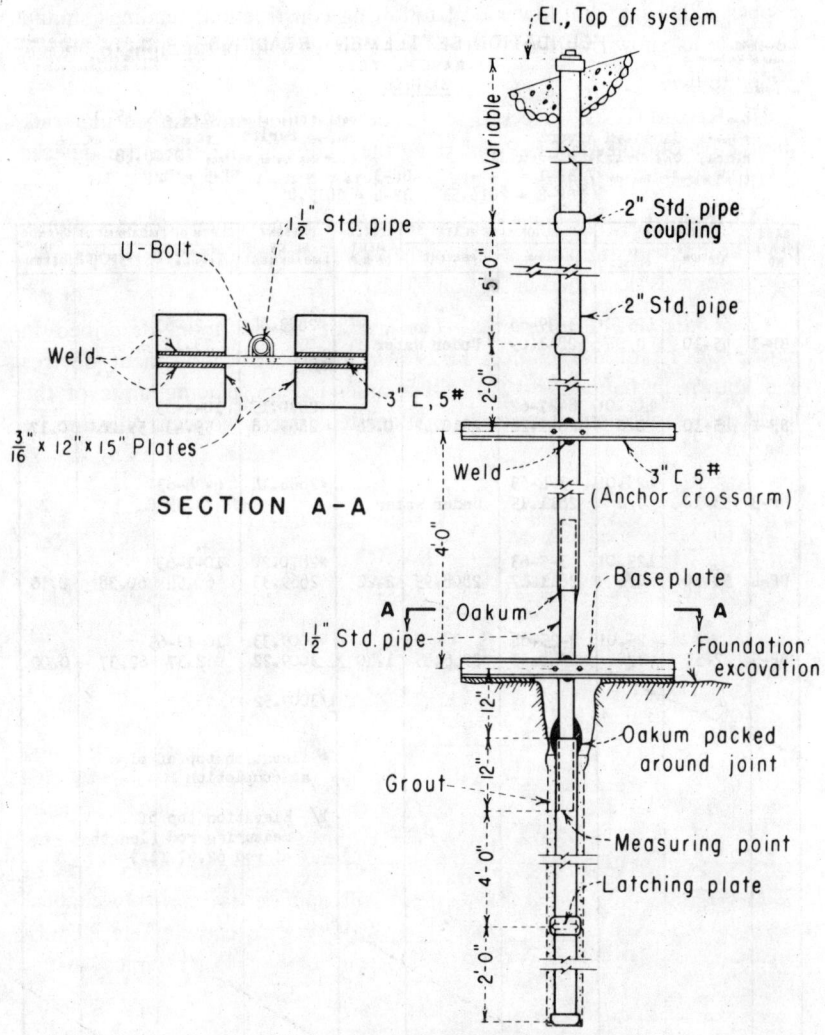

Figure 31–1.—Baseplate installation. 101–D–318.

A latching plate is inserted into the coupling between the two sections of pipe.

(b) *Baseplate*.—The baseplate is a modified crossarm with two 12- by 15-inch steel plates welded to the base of the channel section. The baseplate unit is bolted and welded to 1½-inch pipe which telescopes between lengths of 2-inch pipe in the system. Settlement readings are made only on the baseplate measuring point which is the bottom of the 1½-inch pipe.

7-1359 (8-70)
Bureau of Reclamation

FOUNDATION SETTLEMENT READINGS
(BASE PLATES)
EXAMPLE

TO DIRECTOR OF DESIGN & CONST
ENGRG & RESEARCH CENTER
DENVER FEDERAL CENTER
DENVER, COLORADO 80225

MODIFIED 10-4-70 FOR

Dam .... Date of Observation Nov. 20, 1970
Project Canadian River Observer Surles Sheet 1 of 1
Ref Dwg. 622-D-153, 40-D-4631 El. of Res. Water Surface 2908.48
El. of Water in System BP-1 = - - - - BP-3 = - - - - BP-5 = Dry
　　　　　　　　　　BP-2 = 2816.51 BP-4 = 2815.06

BASE PLATE NO.	LOCATION		EL. OF BASE PLATE (2)		SETTLE-MENT OF B.P.	PRESENT EL. OF (3) EMBANKMENT	DEPTH OF EMBANKMENT		CHANGE IN (6) DEPTH
	STATION	(1) OFFSET	ORIGINAL	PRESENT			ORIGINAL (4) (AT COMP)	PRESENT (5)	
BP-1	45+10	425.0' U/S	1-29-63 2811.09	Under Water		*2883.84	6-24-63 73.13		
BP-2	45+10	402.0' D/S	8-27-62 2811.26	2810.42	0.84	*2870.15 2869.68	10-1-63 59.43	59.26	0.17
BP-3	53+10	425.0' U/S	3-4-63 2811.15	Under Water		*2884.14	6-24-63 73.79		
BP-4	53+10	425.0' D/S	3-5-63 2811.17	2808.95	2.22	*2870.20 2869.33	10-1-63 60.54	60.38	0.16
BP-5	22+50	25.0' U/S	3-25-64 2948.25	2946.85	1.40	*3009.33 3009.22 1/3009.52	10-13-65 62.37	62.37	0.00

* Elevation top of pipe at completion

1/ Elevation top of measuring rod (length of rod 62.67 ft.)

(1) Record U/S or D/S from ℄ Crest of dam. (2) Report data to 0.01 foot. (3) On completion of each installation, use El. of top of pipe in lieu of present El. of embankment. (4) Determine when the installation completed; subtract existing El. of base plate measuring point from El. of top of pipe. (5) Subtract present El. of base plate from present El. of top of pipe. (6) Indicate increase in depth by minus (–).

Figure 31-2.—Example—Foundation settlement readings. 101-D-601.

(c) *Anchor Crossarm.*—At 4 feet above the elevation of the baseplate, an anchor crossarm is placed to prevent the vertical pipe column from sliding down on the baseplate. The anchor crossarm is a 3-foot section of 5-pound steel channel, bolted and welded to a 5-foot section of 2-inch pipe.

(d) *Extension Pipe.*—To complete the installation, 5-foot sections of 2-inch pipe are joined by pipe couplings to extend the column of pipe to the surface of the embankment.

3. Location, Installation, and Reading the Installation.—The foundation settlement apparatus can be installed as described in paragraph 8 of designation E–29. When installing the 2-inch riser or pipe extensions, the crossarm trench is not required. The surface of the embankment is leveled, the center point located, and an 8-inch-diameter hole augered down to the pipe section previously placed. For information describing the measuring equipment, methods of reading the base crossarms, and miscellaneous data concerning alinement, record tests, progress and periodic reports, refer to designation E–29 and sections 70 and 71 of chapter III.

4. Reporting Data.—Readings should be made as shown on figure 103 or as described in paragraphs 6 and 7, designation E–29. During construction, readings for the elevations of the baseplate measuring points should be made each time a 5-foot vertical extension is added or at monthly intervals. When the installation is completed, the elevation of the top of the final pipe coupling should be established to the nearest 0.01 foot and the date of completion noted on form 7–1359. An example of a set of periodic (L–23) readings made on a completed installation is shown on figure 31–2.

INSTRUCTIONS FOR INSTALLING AND READING MEASUREMENT POINTS

—EMBANKMENT—

Designation E–32

1. Purpose.—Embankment measurement points are installed near the edges of the crest and on the upstream and downstream slopes of an earth dam or dike to determine cumulative settlement and horizontal deflections or movement normal to the centerline of the crest of the structure. Figure 32–1 shows an installation of these measurement points.

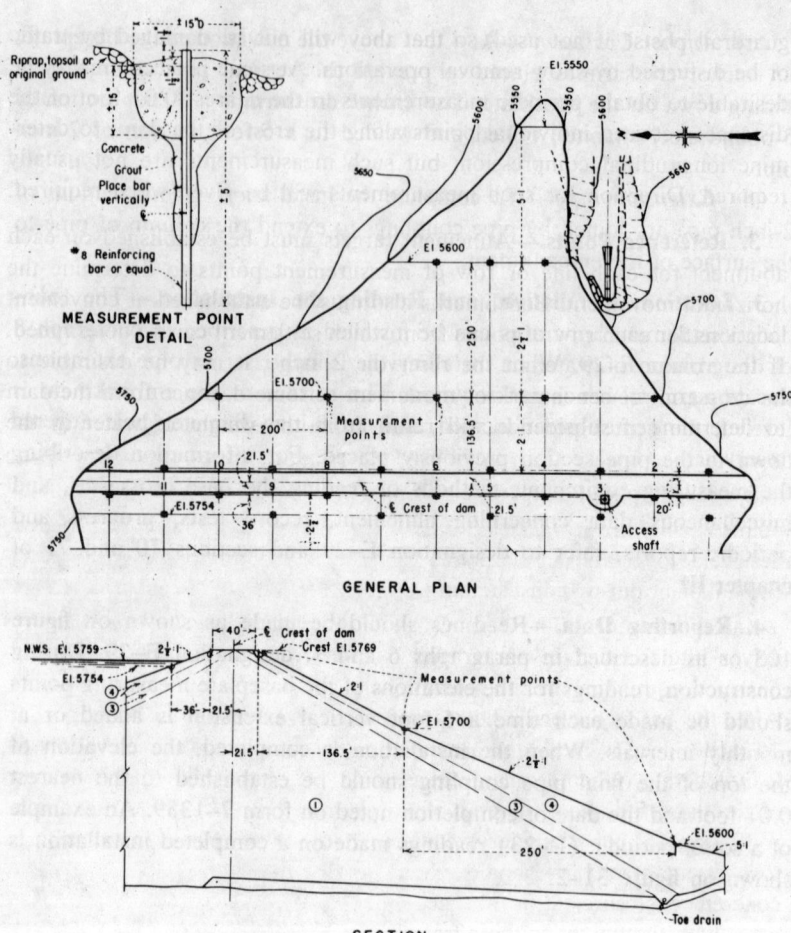

Figure 32-1.—Embankment measurement point installation. 101-D-320.

2. Description.—Embankment measurement points are established at a predetermined spacing, usually on 200- to 400-foot centers, on lines parallel to the centerline of the crest of the dam. The points consist of 5- or 6-foot lengths of #8 reinforcing bars (approximately 1-inch diameter) which are driven vertically into the completed surface of the embankment. Each bar should be crossmarked on its top so that horizontal movement may be measured. Embankment measurement points are installed on the surface of the completed embankment either following construction or as soon as possible after a portion of the embankment is completed. The points along the crest of the dam are established on lines outside the guardrail posts if used, or outside "the usual location for guardrail posts" if not used, so that they will not be damaged by traffic or be disturbed by snow-removal operations. At some projects, it may be desirable to obtain periodic measurements to the nearest 0.01 foot of the distances between individual points along the crest of the dam, to determine longitudinal compression, but such measurements are not usually required. Direction for such measurements will be given when required.

3. Reference Points.—Alinement targets must be established on each abutment for each line or row of measurement points to determine the horizontal movement. Benchmarks also must be established at convenient locations for each row of points from which settlement can be determined. If the volume of water in the reservoir is large, it may be desirable to locate a control benchmark or monument at some distance from the dam to determine if subsidence will result from the weight of water in the reservoir.

4. Installation.—Although permanent embankment measurement points usually are installed after the embankment is completed, it may be desirable to establish additional rows of points along the toes of the embankment during construction to observe possible movement.

(a) *Placement.*—The 5- or 6-foot lengths of reinforcing steel bars are driven into the surface of the embankment either by hand or by power-operated equipment at the locations shown on the installation drawing. After the bar is driven, a hole approximately 15 inches in diameter and 15 inches deep is excavated around the top of the bar, and backfilled with lean concrete. When a measurement point is installed in a rockfill section of the dam, sand-cement grout should be poured around the base of the rod in the 15-inch-diameter excavation prior to placing the concrete. The tops of the steel bars should extend approximately 3 inches above the concrete enclosure. It is not necessary that each point be established exactly on the line or to the offset from the stationing along the centerline of the crest of the dam as shown on the drawing. However, each point should be installed within 3 feet of the designated crest of dam stationing and within 1 foot of the desired elevation. This procedure will permit the placement of the measurement point in a crevice between rocks in the surface of the rockfill.

(b) *Location.*—After a row of points or a series of points along a given row are installed, the "as-built" elevations and offsets (original) from established line and from centerline of crest of dam should be determined to the nearest 0.01 foot. The dates should be recorded when these as-built elevations and offsets are obtained. Control targets for a line of measurement points should be referenced to the centerline of the crest of the dam. Offsets for individual measurement points from an established line between control targets should be recorded. However, when the centerline of the crest is curved, as-built elevations should be established to the nearest 0.01 foot for all points, but original alinement control should be established only for selected points which can be referenced conveniently to targets which may or may not be parallel to the centerline of the crest. All as-built or original data should be recorded on the installation drawing.

5. Reporting Data.—Both as-built and current elevations and deflections should be recorded on form 7–1355 as illustrated on figure 32–2. Readings should be made on the installation as directed on figure 103, or as described in paragraphs 6 and 7, designation E–29. All cumulative readings should be reported to the nearest 0.01 foot, as practicable. Form 7–1355 should be used for reporting both cumulative settlement and cumulative deflection readings.

(a) *Sign Conventions.*—

(1) *Settlement.*—The cumulative settlement of a point is the vertical movement (downward). Cumulative settlement is indicated on form 7–1355 without a prefixing sign. Heave (vertical movement upward) should be indicated by a minus sign.

(2) *Deflection.*—The cumulative deflection of a point is the horizontal movement or deflection, either upstream or downstream, normal to the centerline of the crest of the dam. Use the abbreviation U/S for upstream and D/S for downstream as appropriate.

6. Final Report.—The Final Report on Earth Dam Instrumentation (L–16), should include a revised copy of the installation drawing, corrected to include the as-built elevations and original offsets, upstream or downstream from the centerline of crest of dam, for each point and the dates when each row of points was established. It should also show the locations and elevations of all benchmarks, targets, and monuments used for the installation. The report should include a description of the methods used for installation of the measurement points.

7. Controls.—Permanent benchmarks and targets should be established for use in obtaining future periodic (L–23) behavior readings for all measurement points on the embankments and concrete structures. These points should be located in accessible locations which can be found in later years and which will not be disturbed by the natural elements.

MEASUREMENT POINTS
CUMULATIVE SETTLEMENT
AND
DEFLECTION READINGS
EMBANKMENT

TO: DIRECTOR OF DESIGN & CONST
ENGRG & RESEARCH CENTER
DENVER FEDERAL CENTER
DENVER, COLORADO 80225

MODIFIED. 10-4-70 FOR

EXAMPLE

Dam

Project. 203-621-41

Ref Dwg

Date of Observation. Nov. 20, 1970

Observer.

Sheet 1 of 1

CENTERLINE STATION	SETTLEMENT			OFFSET FROM CENTERLINE DATE POINTS SET OR RESET	DEFLECTION		
	ORIG. EL	PRES. EL	CUM. DIFF		(1) ORIG. OFFSET	PRES. OFFSET	CUM. DIFF.
				96.5' U/S			
4+98.36	2098.39	2098.27	0.12	April 1950	96.50	96.55	0.05 u/s
7+03.76	91.84	91.46	0.38		96.50	96.41	0.09 d/s
9+00.56	91.86	91.41	0.45		96.50	96.36	0.16 d/s
11+00.56	91.61	91.41	0.20		96.50	96.50	0.00
13+01.23	2098.56	2098.52	0.04	April 1950	96.50	96.53	0.03 u/s
				18.5' U/S			
3+98.23	2124.62	2124.57	0.05	June 1950	18.52	18.47	0.05 d/s
8+09.13	24.53	23.87	0.66		18.52	18.44	0.08 d/s
10+04.26	24.32	23.85	0.47		18.52	18.36	0.16 d/s
11+98.98	24.58	24.37	0.21		18.52	18.53	0.01 u/s
14+09.20	2124.45	2124.38	0.07	June 1950	18.52	18.48	0.04 d/s
				18.5' D/S			
5+04.16	2124.62	2124.51	0.11	June 1950	18.52	18.65	0.13 d/s
7+04.20	24.56	24.11	0.45		18.52	18.68	0.16 d/s
8+99.05	24.32	23.73	0.59		18.52	18.79	0.27 d/s
11+09.19	24.88	24.61	0.27		18.52	18.67	0.15 d/s
12+88.79	2124.79	2124.71	0.08	June 1950	18.52	18.66	0.14 d/s
				118.0' D/S			
4+99.24	2081.31	2081.24	0.07	April 1950	118.00	118.03	0.03 d/s
6+99.34	75.70	75.32	0.38		118.00	118.03	0.03 d/s
8+99.10	75.52	74.86	0.66		118.00	118.16	0.16 d/s
10+99.23	75.21	74.98	0.23		118.00	118.09	0.09 d/s
12+99.68	2076.01	2075.93	0.08	April 1950	118.00	118.00	0.00
				230.5' D/S			
5+98.14	2030.02	2029.97	0.05	April 1950	230.50	230.50	0.00
7+97.52	29.68	29.45	0.23		230.50	230.53	0.03 d/s
10+00.14	32.89	32.93	0.06		230.50	230.55	0.05 d/s
11+99.55	2046.16	2046.10	0.06	April 1950	230.50	230.44	0.06 u/s

(1) Record u/s or d/s from ℄ Crest or Axis of Dam
Report original elevations and offsets to 0.01 foot.
Report cumulative settlement and deflections to 0.01 foot.

Use Minus (-) sign to report heave.
Use u/s or d/s to report horizontal movement.

Figure 32-2.—Embankment measurement point readings. 101-D-602.

INSTRUCTIONS FOR INSTALLING AND READING MEASUREMENT POINTS—CONCRETE STRUCTURES

—OUTLET WORKS CONDUITS—

—CONDUIT-TYPE SPILLWAYS—

Designation E–33

1. Purpose.—Measurement points are installed along the centerline of the invert or at other locations such as walkways inside the outlet works or conduit-type spillways to furnish data regarding the differential settlement and possible elongation of the various sections under the loading of the embankment and during progressive saturation of the foundation.

2. Description.—These measurement points consist of stainless steel carriage bolts or comparable rust-resistant metal rods embedded in the concrete surface or set in holes drilled into the concrete. The example on figure 33–1 shows ¼-inch-diameter by 2½-inch-long stainless steel carriage bolts, although bolts to ⅜-inch-diameter and as short as 2 inches may be used, depending on availability. Spacing of the points usually is controlled by the distance between construction joints; however, additional points may be required in the upper surfaces of the outlet works trashrack or spillway intake structure. Locations of the measurement points for conduits will be shown on drawings prepared by the Engineering and Research Center.

3. Recording.—Each conduit measurement point should be numbered, starting with the point located furthest upstream and progressing in a downstream direction, its as-built elevation and station on the structure determined to 0.01 foot, and the date when established should be recorded on form 7–1355A. Figure 33–2 presents an example of measurement points installed along outlet works and spillway conduits and includes as-built elevations and locations recorded to the nearest 0.01 foot.

4. Readings.—Periodic readings should be made on all established conduit measurement points at monthly intervals during the construction of the dam and submitted in duplicate on form 7–1355A. An outline of the embankment should be drawn on the reporting form, and existing embankment elevations should be shown for the date on which the readings were made. After construction of the dam, readings should be made and submitted as directed on figure 103, or as described in paragraphs 6 and 7 of designation E–29. Periodic readings (L–23) are required to be

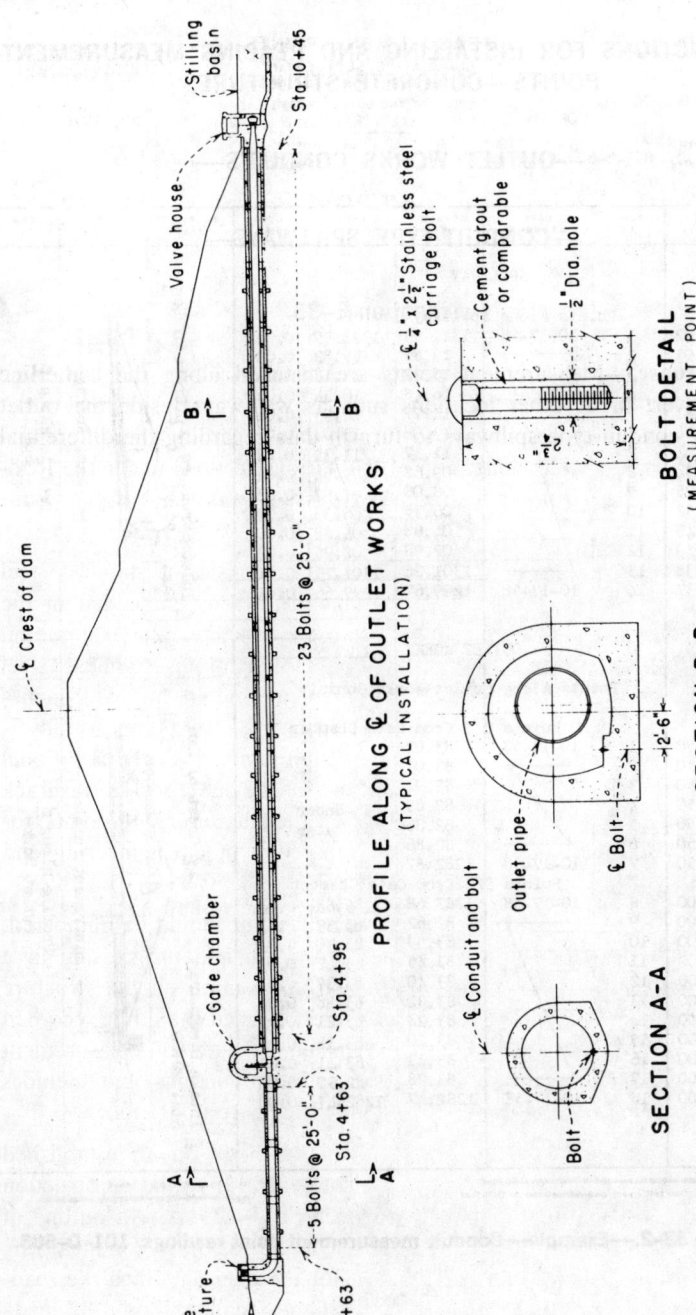

PROFILE ALONG ℄ OF OUTLET WORKS
(TYPICAL INSTALLATION)

BOLT DETAIL
(MEASUREMENT POINT)

SECTION A-A

SECTION B-B

Figure 33-1.—Conduit measurement point installation. 101-D-322.

7-1355A (8-70)
Bureau of Reclamation

MEASUREMENT POINTS
CUMULATIVE SETTLEMENT
AND DEFLECTION READINGS
SPILLWAY AND OUTLET WORKS

TO DIRECTOR OF DESIGN & CONST
ENGRG & RESEARCH CENTER
DENVER FEDERAL CENTER
DENVER, COLORADO 80225

MODIFIED 10-4-70 FOR

Dam   OF

Project

Ref Dwg 853-D-6, -D-10

EXAMPLE   Date of Observation Nov. 20, 1970

Observer

Sheet 1 of 3

STATION	POINT NO	DATE AS-BUILT	SETTLEMENT			DEFLECTION		
			AS-BUILT EL.	PRES EL.	CUM DIFF	AS-BUILT OFFSET	PRES OFFSET	CUM DIFF
		SPILLWAY						
		Points Along Spillway Conduit						
1+36.92	1	10-17-58	1321.53	1321.49	0.04			
1+61.91	2		19.85	19.82	0.03			
1+86.86	3		18.22	18.18	0.04			
2+11.83	4		16.48	16.44	0.04			
2+36.81	5		14.77	14.73	0.04			
2+61.76	6		13.08	13.03	0.05			
2+87.34	7		11.37	11.34	0.03			
3+12.34	8		09.68	09.64	0.04			
3+37.25	9		08.00	07.97	0.03			
3+61.52	10		06.35	06.33	0.02			
3+86.43	11		04.68	04.68	0.00			
4+11.43	12		02.98	02.96	0.02			
4+36.38	13		1301.26	1301.25	0.01			
4+61.37	14	10-17-58	1299.67	1299.66	0.01			
		OUTLET WORKS						
		Points Along Outlet Works Conduit						
		Points U/S from Gate Chamber						
20+25.50	1	10-17-58	1283.02					
20+50.50	2		83.01					
20+75.50	3		82.94					
21+00.50	4		82.89	Under				
21+25.50	5		82.86	Water				
21+50.50	6		82.86					
21+75.50	7	10-17-58	1282.82					
		Points D/S from Gate Chamber						
22+40.00	8	10-17-58	1283.68	1283.64	0.04			
22+65.00	9		83.62	83.59	0.03			
22+90.00	10		83.53	83.50	0.03			
23+15.00	11		83.45	83.42	0.03			
23+40.00	12		83.40	83.37	0.03			
23+65.00	13		83.32	83.28	0.04			
23+90.00	14		83.24	83.21	0.03			
24+15.00	15		83.20	83.18	0.02			
24+40.00	16		83.13	83.11	0.02			
24+65.00	17		83.98	82.97	0.01			
24+90.00	18	10-17-58	1282.47	1282.47	0.00			

Report original elevations and offsets to 0.01 foot
Report cumulation settlement and deflections to 0.01 foot

Use (-) sign to report heave
Use L or R to report horizontal movement

Figure 33-2.—Example—Conduit measurement point readings. 101-D-603.

made during the same survey as when the embankment measurement point readings are obtained.

5. Reporting Data.—Cumulative settlement of the conduit measurement points should be indicated by the same sign conventions as for the embankment measurement points described in paragraph 5 of designation E–32, in that the minus sign should be used to indicate heave. The appropriate abbreviation, U/S or D/S, should be used to indicate movement of the conduit structure. Control targets to be used for obtaining deflection readings on measurement points located on top of the chute and stilling basin walls of the outlets works and spillway should be located to the left and right of the respective walls (high side) so that a reference line can be established and periodic readings made and referenced to the original centerline of the chute and stilling basin (see paragraph 7 of designation E–32 for location of permanent benchmarks and targets, as required).

6. Final Report.—The Final Report on Earth Dam Instrumentation should include a copy of form 7–1355A which has been revised to include the section of the dam embankment over the conduits, the as-built elevations and locations, and the dates when each point was established.

INSTRUCTIONS FOR INSTALLING AND READING MEASUREMENT POINTS—CONCRETE STRUCTURES

—CHUTE AND STILLING BASIN OF OUTLET WORKS—

—CHUTE-TYPE SPILLWAYS—

Designation E–34

1. Purpose.—Measurement points are installed on top of the walls of outlet works and spillway chutes and stilling basins and near the corners of the outlet works and spillway floor slabs to determine differential settlement. Deflection readings also are obtained for the top of walls before and after backfill and periodically after construction.

2. Description.—These measurement points consist of stainless steel carriage bolts as described in paragraph 2 of designation E–33 or of comparable rust-resistant metal rods which are embedded during construction or drilled into the concrete surfaces.

3. Installation.—The measurement points are installed in a manner

similar to that described for the points for outlet works conduit and
conduit-type spillways. The points on top of the walls usually are installed about 6 inches from the construction joints while points in the
floor slabs are installed about 9 inches from corners of each slab.

4. Location.—Design locations for all spillway and outlet works
measurement points will be shown on drawings prepared by the Engineering and Research Center prior to construction. However, before the
points are installed in the field, a sketch drawing should be prepared to
show the orientation for the points in each installation. Points on top of
the walls should be designated by number, starting with the lowest number at the point furthest upstream and progressing downstream. These
points also should be accompanied by a left (L) or right (R) to indicate the appropriate wall when facing downstream. The spillway floor
slabs should be designated by number starting left (L) or right (R) at
the upstream slab with respect to the spillway centerline and progressing
downstream. Then letters should be assigned to each corner of each
spillway floor slab, such as a, b, c, and d, starting with letter *a* in the
upstream left corner (facing downstream) and rotating clockwise.

5. Recording.—When the measurement points are installed in each
floor slab, a tabulation should be maintained throughout construction of
the spillway to show its point designation and slab number, its as-built
elevation to the nearest 0.01 foot, and the date when this elevation was
established. Points installed on top of the spillway or outlet works walls
should be tabulated by station and elevation to the nearest 0.01 foot, by
offset from centerline of structure to the nearest 0.01 foot, and by date
when established. When deflection readings for points on top of the walls
are to be made, the present offsets should be measured to the nearest
0.01 foot left (L) or right (R) from centerline.

6. Periodic Readings.—Periodic readings should be submitted in
duplicate for all measurement points as directed on figure 103, or as
described in paragraphs 6 and 7 of designation E–29. Included with
these readings should be a copy of the sketch drawing as described in
paragraph 4 of this designation. Current readings should show the present elevation and deflection (offset) as appropriate for each point, and
the cumulative change. All data should be presented on form 7–1355A
or 7–1355B as required. Copies of periodic readings on completed structures are shown on figures 34–1 and 34–2.

7. Final Report.—The Final Report on Earth Dam Instrumentation
should include a copy of the design drawings for the structure's measurement points, a sketch map on which all floor slabs are numbered and
corners designated, a complete tabulation of as-built elevations and
locations including offsets, and the dates when each point was installed.

7-1355A (8-70)
Bureau of Reclamation

MEASUREMENT POINTS
CUMULATIVE SETTLEMENT
AND DEFLECTION READINGS
SPILLWAY ~~AND OUTLET WORKS~~

TO DIRECTOR OF DESIGN & CONST
ENGRG & RESEARCH CENTER
DENVER FEDERAL CENTER
DENVER, COLORADO 80225

MODIFIED 10-4-70 FOR
    Dam
    Project
    Ref Dwg   853-D-8

EXAMPLE   Date of Observation   Nov. 20, 1970
                Observer
                Sheet   2   of   3

STATION	POINT NO	DATE AS-BUILT	SETTLEMENT AS-BUILT EL	SETTLEMENT PRES EL	SETTLEMENT CUM DIFF	DEFLECTION AS-BUILT OFFSET	DEFLECTION PRES OFFSET	DEFLECTION CUM DIFF
				SPILLWAY				
		Points on Top of Stilling Basin Walls Left Wall (Looking D/S)						
4+65	A-19-L	3-14-59	1314.98	1314.97	0.01	7.75 L	7.69 L	0.06 R
	B-19-L		05.03	05.02	0.01	10.26 L	10.25 L	0.01 R
5+00	A-20-L		1304.62	1304.62	0.00	10.37 L	10.36 L	0.01 R
	B-20-L	3-14-59	1294.68	1294.67	0.01	12.89 L	12.87 L	0.02 R
5+35	A-21-L	3-20-59	94.51	94.49	0.02	12.97 L	12.94 L	0.03 R
	B-21-L		94.54	94.52	0.02	15.46 L	15.38 L	0.08 R
5+70	A-22-L		94.54	94.52	0.02	15.51 L	15.42 L	0.09 R
	B-22-L		94.54	94.52	0.02	15.47 L	15.33 L	0.14 R
6+10	A-23-L		94.55	94.52	0.03	15.48 L	15.33 L	0.15 R
	B-23-L	3-20-59	94.54	94.52	0.02	15.58 L	15.44 L	0.14 R
6+50	A-24-L	3-14-59	94.53	94.52	0.01	15.67 L	15.52 L	0.15 R
	B-24-L		94.52	94.52	-0.01	20.39 L	20.31 L	0.08 R
6+78.75	A-25-L		94.52	94.52	0.00	20.66 L	20.60 L	0.06 R
	B-25-L	3-14-59	1291.13	1291.13	0.00	25.41 L	25.40 L	0.01 R
7+07.50								
				Right Wall (Looking D/S)				
4+65	A-19-R	3-14-59	1314.95	1314.95	0.00	7.73 R	7.66 R	0.07 L
	B-19-R		04.98	14.98	0.00	10.26 R	10.24 R	0.02 L
5+00	A-20-R		1304.60	1304.60	0.00	10.40 R	10.38 R	0.02 L
	B-20-R	3-14-59	1294.71	1294.71	0.00	12.85 R	12.81 R	0.04 L
5+35	A-21-R	3-20-59	94.52	94.51	0.01	12.97 R	12.94 R	0.03 L
	B-21-R		94.52	94.51	0.01	15.47 R	15.37 R	0.10 L
5+70	A-22-R		94.51	94.50	0.01	15.50 R	15.40 R	0.10 L
	B-22-R		94.52	94.50	0.02	15.52 R	15.35 R	0.17 L
6+10	A-23-R		94.52	94.50	0.02	15.53 R	15.36 R	0.17 L
	B-23-R	3-20-59	94.53	94.51	0.02	15.59 R	15.45 R	0.14 L
6+50	A-24-R	3-14-59	94.52	94.52	0.00	15.68 R	15.55 R	0.13 L
	B-24-R		94.51	94.51	0.00	20.40 R	20.37 R	0.03 L
6+78.75	A-25-R		94.52	94.52	0.00	20.69 R	20.65 R	0.04 L
	B-25-R	3-14-59	1291.12	1291.12	0.00	25.37 R	25.35 R	0.02 L
7+07.50								

Report ~~original elevations and offsets to 0.01 foot~~
Report cumulation settlement and deflections to 0.01 foot

Use (-) sign to report heave
Use L or R to report horizontal movement

GPO 852-873

Figure 34-1.—Example—Measurement point readings—Chute-type spillways.
101-D-604.

MEASUREMENT POINTS
CUMULATIVE SETTLEMENT READINGS
SPILLWAY FLOOR SLABS

7-1355H (8-70)
Bureau of Reclamation

TO DIRECTOR OF DESIGN & CONSTRUCTION
ENGRG & RESEARCH CENTER
DENVER FEDERAL CENTER
DENVER, COLORADO 80225

MODIFIED 10-4-70   FOR   EXAMPLE

Dam

Project

Ref Dwg   492-D-73
(See Sketch)

Date of Obs*   Nov. 20, 1970

Observer

Sheet   1   of   1

SPILLWAY STATION	SETTLEMENT ORIG EL	SETTLEMENT PRES EL	CUM DIFF	Σ--Looking D/S Points LEFT	Points RIGHT	SETTLEMENT ORIG EL	SETTLEMENT PRES EL	CUM DIFF
17+10	2283.51	2283.48	0.03	2	1	2283.49	2283.46	0.03
17+54	83.08	83.06	0.02	4	2	83.08	83.06	0.02
+54	83.08	83.06	0.02	3	1	83.06	83.05	0.01
17+98	82.65	82.64	0.01	4	2	82.61	82.60	0.01
+98	82.61	82.59	0.02	3	1	82.62	82.60	0.02
18+42	82.17	82.15	0.02	6	3	82.19	82.17	0.02
+42	82.23	82.21	0.02	5	2	82.20	82.18	0.02
+42	82.18	82.16	0.02	4	1	82.20	82.18	0.02
18+86	81.77	81.75	0.02	6	3	81.74	81.73	0.01
+86	81.74	81.73	0.01	5	2	81.74	81.73	0.01
+86	81.73	81.72	0.01	4	1	81.74	81.73	0.01
19+30	80.63	80.61	0.02	6	3	80.61	80.59	0.02
+30	80.59	80.57	0.02	5	2	80.60	80.58	0.02
+30	80.60	80.58	0.02	4	1	80.61	80.59	0.02
19+74	70.12	70.10	0.02	6	3	70.12	70.10	0.02
+74	70.10	70.09	0.01	5	2	70.10	70.09	0.01
+74	2270.10	2270.09	0.01	4	1	2270.07	2270.05	0.02

Spillway completed September 1964; measurement points on spillway floor
slabs set February 14, 1967. These measurement points are No. 12
brass screws set in lead anchors at 1 foot U/S and 1 foot to the right
(looking D/S) of contraction joint intersections of spillway slabs.

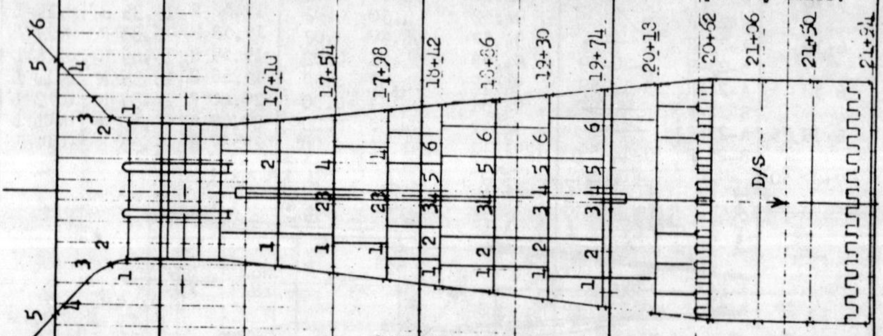

SPILLWAY
Norton Dam

Report original elevations to 0.01 foot
Report cumulative settlement to 0.01 foot

Use (−) sign to report heave

**Figure 34-2.—Example—Measurement point readings—Chute-type spillways.
101-D-605.**

RECORDING EARTHQUAKE VIBRATIONS

Designation E–35

1. Purpose.—Where local seismic conditions warrant, strong-motion accelerographs may be installed in special housing on or adjacent to an earth dam (see figure 35–1). These instruments will provide information on the behavior of embankments subjected to earthquake loadings.

Figure 35–1.—Special housing located on an embankment in which strong-motion accelerograph is installed. PX–D–69901.

2. Description.—The accelerograph operates only during periods of strong earthquake shocks. The instrument measures two horizontal components and one vertical component of acceleration. Horizontal acceleration components are measured in orthogonal directions. Acceleration records may be recorded either on film or paper depending on the type of strong-motion accelerograph used. An accelerograph is shown in figure 35–2.

Figure 35–2.—An example of an accelerograph used on structures located in seismic areas. PX–D–69902.

3. Installation and Operation.—The National Ocean Survey (Survey), formerly Coast and Geodetic Survey, will purchase the accelerographs with funds provided by the Bureau of Reclamation and will perform routine maintenance on the instruments. In addition to supplying funds for purchasing the instruments, the Bureau will also furnish instrument power and housing. Accelerograph records will be removed and retained by the Survey, which in turn will furnish copies to the Bureau as required. Any questions regarding accelerograph maintenance and housing should be referred to the Engineering and Research Center.

FIELD PERMEABILITY TEST (SHALLOW-WELL PERMEAMETER METHOD)

Designation E–36

1. Scope.—This test method is intended for determining the coefficient of permeability of soil linings for canals or other comparatively thin layers of soil which are of appreciably lower permeability than the underlying soil.

2. Definition.—The shallow-well permeability test consists essentially of maintaining a constant head of water in a shallow uncased well excavated in the soil layer to be tested, and measuring the rate of water outflow after a steady state flow condition has been reached. The result of the test is reported as the coefficient of permeability of the soil (in units of feet per year), which is a measure of the quantity of water that will flow through a unit area of soil under a hydraulic gradient of unity. With this coefficient, seepage through a canal lining or other soil layer of given dimensions can be computed.

(*Note:* This test should be distinguished from the field permeability test (well permeameter method), designation E–19, which has a test well with a minimum depth five times the diameter and is used primarily for estimating seepage from unlined canals.)

3. Apparatus.—The apparatus (fig. 36–1) shall consist of the following:

(1) A small metal drum reservoir of about 2-cubic-foot capacity with a plastic gage tube on which the volume of the drum is calibrated in cubic feet and fractions thereof.

(2) Bob-float apparatus for maintaining a constant head of water in the test well by regulating the flow of water from the drum reservoir.

(3) A float guide and casing with a ½-inch-diameter water-supply tube.

(4) A centigrade thermometer.

(5) Hand or power auger or other hand tools for excavating the test well.

(6) Tank truck or other means of conveying water to the permeability test site.

4. Calibration of Reservoir.—Calibrate the drum reservoir by filling it with water and drawing it out in 0.10-cubic-foot increments which are measured by weight. Mark the gage tube after each increment is drawn

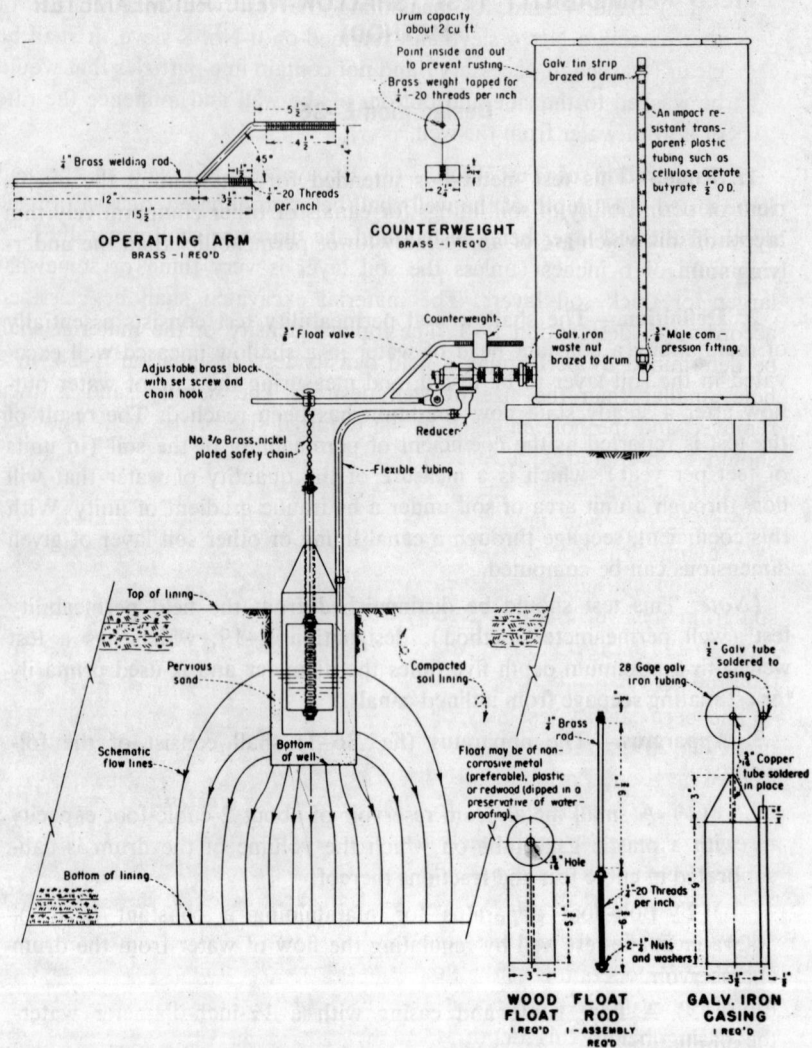

Figure 36–1.—Shallow-well permeameter test equipment. 101–D–352.

off and subdivide the marks by measurement so the water volume can be read to the nearest 0.01 cubic foot.

5. Materials.—The following materials will be required:

(1) Water.—The water used for the permeability test shall be clean and free from sediment. Where possible, the water should be from the same source as that expected to be used in the canal or which will be permeating through the soil layer.

(2) Sand.—Sand for backfilling the test well shall consist of that passing a No. 4 sieve and retained on a No. 8 sieve. It shall be clean (washed, if necessary) and not contain fine particles that would be washed to the side and bottom of the well and influence the rate of flow of water from the well.

6. Preparation of Test Well.—(a) Excavate a test well in the soil to be tested. The depth of the well shall be approximately one-half of the depth of the soil layer being tested, and the diameter shall generally be a minimum of 6 inches (unless the soil layer is very thin) or somewhat larger for thick soil layers. The material excavated shall be classified according to designation E–3. The inplace density of the material shall be determined by performing a field density test, designation E–24, in a hole smaller than the size of well desired. The density sand is then removed and the hole enlarged to the required size of the well. The well should be prepared with care in order to cause as little disturbance to the surrounding soil as possible. It may be excavated with a hand or power auger or with hand tools. After the well is excavated, where necessary, lightly brush or shave the side to remove any soil accumulation or compaction caused by the excavation procedure, and remove the loose soil from the well bottom. Record the radius and depth of the well on form 7–1710, figure 36–2, which is provided for this and other data required for the test.

The permeability test may be conducted in canal side-slope lining by first excavating a horizontal shelf in the lining before starting the vertical well excavation. For this case, the top of the well should be located a minimum of 6 inches back from the original side-slope surface. The reservoir can be set on the shelf in the lining, or it can be set on a wooden support resting on the canal side slope opposite the well (fig. 36–3).

(b) By pouring, refill the well with the sand described in paragraph 5(2), except for the space at the top for the casing which forms a water chamber for operation of the float. Position the casing in the sand so the top of the casing extends about 3 inches above the top of the well and so the casing below this depth is surrounded by sand. The float is also positioned at this time.

7. Test Procedure.—(a) Set up the water reservoir in a level position adjacent to the test well. Assemble the bob-float apparatus extending from the reservoir to the test well. To prevent water evaporation from the well and surrounding area, cover the area to a radius of 3 feet from the well with an impervious material such as a treated canvas or plastic sheet. Fill the reservoir, keeping the bob-float valve closed during the filling.

(b) After recording the volume of water in the full reservoir and the time, open the valve, fill the well with water; record the time, the volume

7-1710 (8-71)
Bureau of Reclamation

FIELD PERMEABILITY TEST (SHALLOW–WELL PERMEAMETER METHOD)

Test No. **Example**_____ Made by _____ Date _____

Test location **South Platte supply canal, Sta. 1477+48 ₵**

_____ Reservoir No. **4**

Canal dimensions: Bottom width, (b) **12'** Side slopes **2:1** Water depth $\frac{1}{}$ **3.4'**
(From figure 36-6)

Lining thickness: Bottom, (t_1) **1.5'** Slopes, (t_2) **1.5'**
(from figure 36-6)
Description of materials tested **Clayey sand (SC). Blended clay and sand**

_____

Radius of well, r = (αH) **0.71'** Depth of water in well, d_w = (βH) **1.11'**

Depth of test well **1.52'** Depth from water surface to bottom of lining, H, in feet. **2.03'**

$a = \dfrac{r}{H} =$ **0.35** . $\beta = \dfrac{d_w}{H} =$ **0.55** . C (From figure 36-5) = **0.283**

V (From figure 36-4) = **0.91**

$Q = \dfrac{col.\ (6)}{col.\ (3)}\ 8760\ \frac{2/}{} =$ **0.036** cu. ft. /yr. $k = CV\dfrac{Q}{H^2} =$ **20.3** ft./yr.

$\frac{1/}{2/}$ d = water depth to be used for canal seepage computations. (See figure 36-6.)
$\frac{2/}{}$ The value, Q, is to be determined on the basis of the final 24 hours of test. The conversion factor, 8760, represents the number of hours in an ordinary year of 365 days.

DATE (1)	CLOCK (2)	ACCUM (HRS) (3)	RES VOL (CU FT) (4)	VOL DIFF (5)	ACCUM DIFF (6)	T °C (7)	DATE (1)	CLOCK (2)	ACCUM (HRS) (3)	RES VOL (CU FT) (4)	VOL DIFF (5)	ACCUM DIFF (6)	T °C (7)
3-2	15.25	–	1.87	–	–	24							
3-3	7.75	16.50	1.13	0.74	0.74	"							
3-3	15.50	24.25	0.78	0.35	1.09	"							
3-3	15.75	–	1.95	–	–	–							
3-4	8.00	40.50	1.25	0.70	1.79	24							
3-4	15.00	47.50	0.96	0.29	2.08	"							
3-5	7.75	64.25	0.30	0.69	2.77	"							
3-5	8.00	–	1.98	–	–	–							
3-5	15.75	72.00	1.67	0.31	3.08	24							
3-6	8.25	88.50	1.03	0.64	3.72	"							
3-6	15.50	95.75	0.75	0.28	4.00	"							
3-6	15.75	–	1.98	–	–	–							
3-7	7.75	111.75	1.39	0.59	4.59	24							
3-7	15.50	119.50	1.11	0.28	4.87	"							
3-8	8.00	136.00	0.52	0.59	5.46	"							
3-8	15.50	143.50	0.25	0.27	5.73	"							
3-8	15.75	–	1.95	–	–	–							
3-9	7.75	159.50	1.37	0.58	6.31	24							
3-9	16.00	167.75	1.08	0.29	6.60	"							

Figure 36–2.—Field permeability test (shallow-well method). 101-D-537.

Figure 36–3.—Shallow-well permeameter installed in the compacted earth lining of a canal side slope. P328–D–18396.

of water in the reservoir, the water temperature in the well, and the depth of water in the well. (The depth and temperature of water can be found by making a small hole in the sand.)

(c) Continue the permeability test, with the well full at all times until a steady state condition is reached (usually about 5 days). For the usual earth linings which are relatively impervious, it should be sufficient to record the reservoir volume, time, and well water temperature twice daily, refilling the reservoir as necessary to keep the well continuously full.

(d) After the test has been discontinued, auger a hole from the well bottom through the lining and determine the distance from the top of the well water to bottom of lining. Record this as the value of H used in computing coefficient of permeability.

8. Calculation[1] and Report.—(a) Determine the average rate of flow to the test well for the final 24 hours of test; this is Q and should be expressed in cubic feet per year.

(b) The coefficient of permeability of the soil can be computed by the following formula:

$$k = CV \frac{Q}{H^2}$$

---

[1] "Coefficients for Field Permeability Tests of Soil Canal Linings," Photoelastic Analysis Unit Report No. 33, Bureau of Reclamation, November 1, 1955 (unpublished).

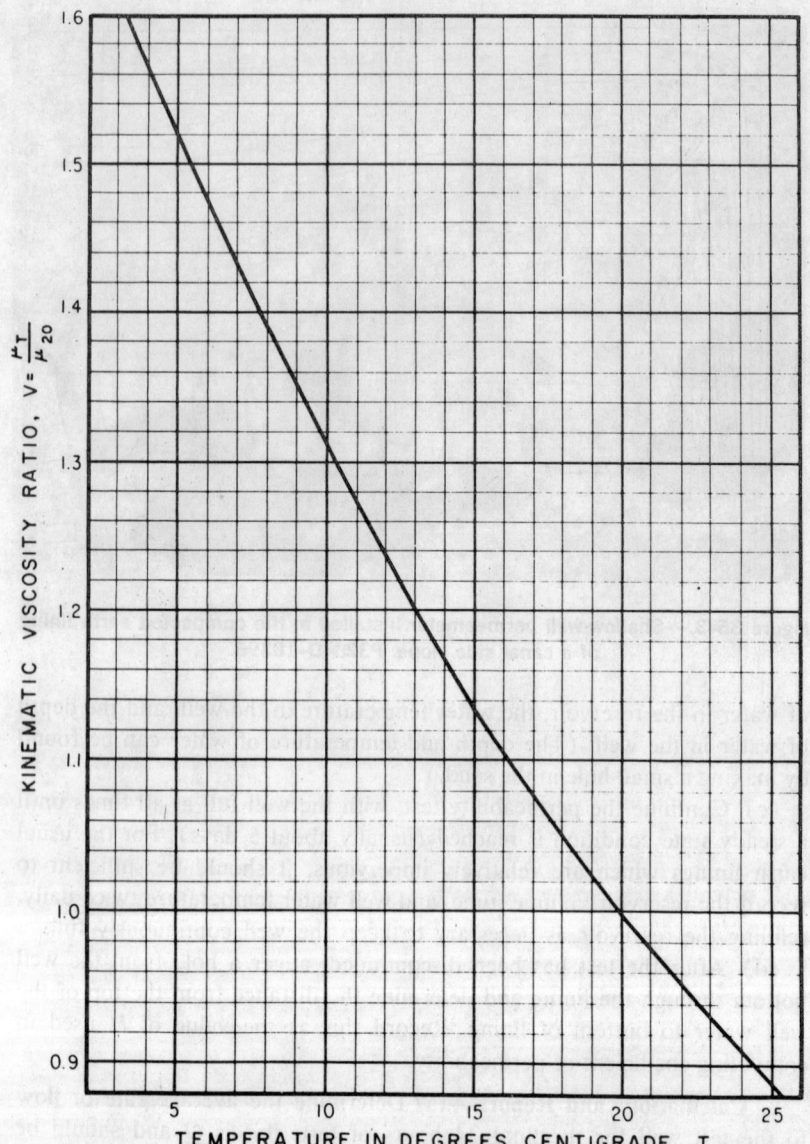

Figure 36-4.—Curve showing relationship between kinematic viscosity ratio of water and temperature in degrees centigrade. 101-D-388.

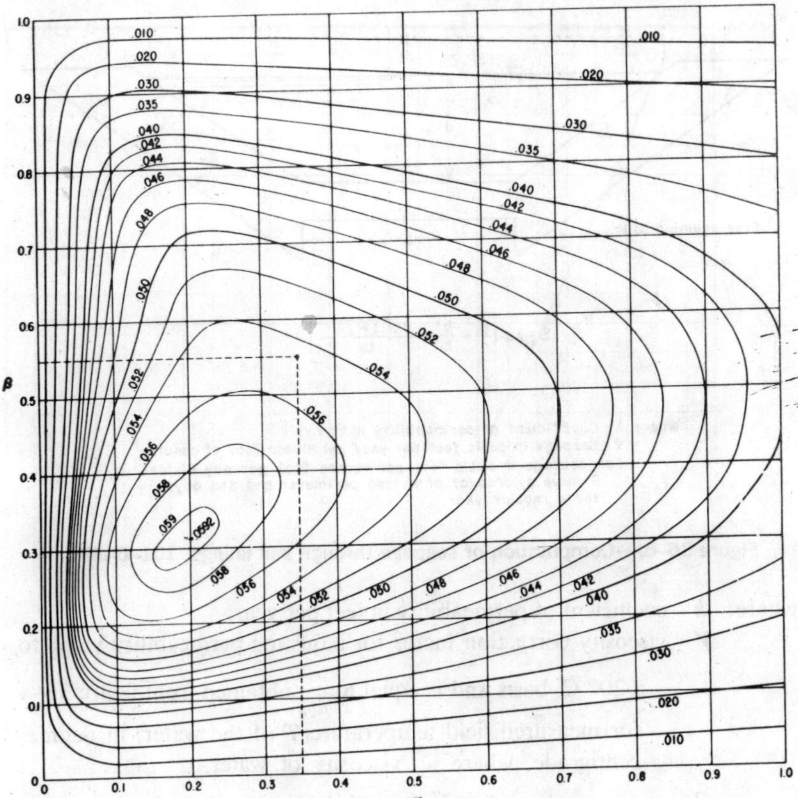

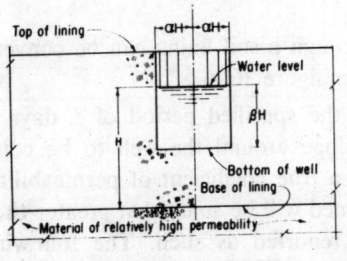

CHART OF $(\alpha\beta C)$ AS A FUNCTION OF α AND β

NOTES

As α approaches 0 the value of C increases without limit.
At α = 0 the test becomes meaningless since no well exists.
As β approaches 0, C approaches $1/\pi\alpha^2$.
For β = 1, C = 0 and the test is again meaningless since
the well would bottom in the pervious material of
relatively high permeability.

EXAMPLE CALCULATION

Known values: α = 0.35, β = 0.55, T = 24° Centigrade
H = 2 feet, *Q = 315 cubic feet per year.

From chart above (dotted lines), for α = 0.35 and β = 0.55,
$(\alpha\beta C)$ = 0.0545, and solving for C,

$$C = \frac{0.0545}{(0.35)(0.55)} = 0.283.$$

From figure 36-4, for T = 24°C., V = 0.91.

Then, $k = VC \dfrac{Q}{H^2} = (0.91)(0.283) \dfrac{315}{(2)^2}$

　　　　= 20.3 feet per year.

*Other units for Q can be used; however, proper
conversion factors must be used so that coefficient of
permeability, k, is in feet per year.

DEFINITION SKETCH

Figure 36-5.—Permeability determination of canal lining by permeameter well. 101-D-538.

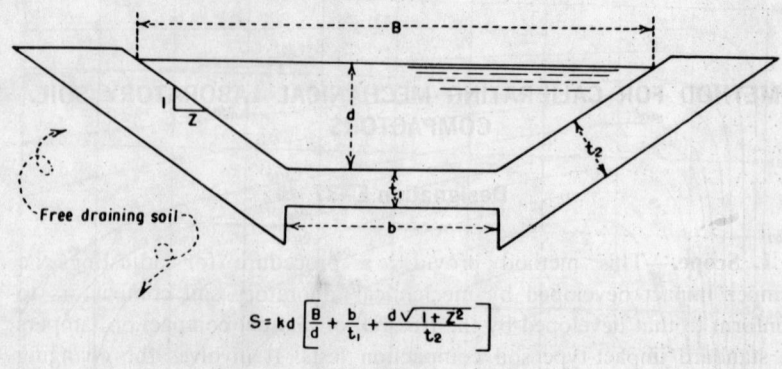

$$S = kd \left[\frac{B}{d} + \frac{b}{t_1} + \frac{d\sqrt{1+Z^2}}{t_2} \right]$$

Where k = Coefficient of permeability in ft. / yr.
S = Seepage in cubic feet per year per linear foot of canal.
For seepage in cubic feet per square foot per day divide
    S above by product of wetted perimeter and 365 days
    for a regular year.

Figure 36–6.—Computation of seepage through soil linings. 101-D-389.

where: k = coefficient of permeability in feet per year,

   V = viscosity correction factor for adjusting permeability value to

a 20° C. basis and is equal to $\frac{\mu_T}{\mu_{20}}$ obtained from figure 36–4
for measured field temperature T of the water, in degrees
centigrade, where μ = viscosity of water,

   Q = average rate of flow from well during last day of test in cubic
feet per year,

   H = depth in feet from water surface in well to bottom of lining,
and

   C = well coefficient determined from the value of H above and
dimensions of the well, as shown in the example in figure
36–5.

(c) The computation of seepage through a soil lining can be conveniently calculated by the formula shown on figure 36–6.

(d) For soils of low permeability, the specified period of 5 days is not sufficiently long for the flow envelope around the well to be completely developed for determination of a true coefficient of permeability. Therefore, the permeability value obtained will be somewhat greater than the true permeability and should be reported as such. The following formula shows the approximate time in years, t, required for determination of a true permeability coefficient:

$$t = \frac{H}{k} (1 - \beta + 2.303 \log_{10}\beta)$$

METHOD FOR CALIBRATING MECHANICAL LABORATORY SOIL COMPACTORS

Designation E–37

1. Scope.—This method provides a procedure for adjusting the tamper impact developed by mechanical laboratory soil compactors to conform to that developed by the free fall of manual compaction tampers in standard impact-type soil compaction tests. It involves the changing of tamper weight to compensate for mechanical variations, such as friction, in accordance with the deformation of small lead cylinders which are struck by the manual and mechanical tampers.

2. Apparatus.—The apparatus shall consist of the following (see fig. 37–1):

(1) Lead deformation device consisting of anvil, guide collar, and striking pin.

(2) A dial comparator for measuring deformation of lead cylinders.

(3) A pedestal sleeve to be used with hand tamper and height-of-drop guide as shown in figure 37–1(d).

(*Note:* The Proctor compaction test, designation E–11, specifies an 18-inch height of drop and a 5.5-pound tamper.)

(4) Equipment required for Proctor compaction test, as listed in designation E–11.

3. Materials.—The following materials are required:

(1) Lead cylinders.—A supply of small lead cylinders described as 38-caliber, 158-grain lead alloy cores which are manufactured for swaging into bullets. These are available at the larger stores handling gun supplies and should be purchased in minimum lots of 1,000. Burrs or irregularities on the ends of these lead cylinders should be removed by sanding with fine emery cloth.

(2) Soil.—A plastic soil, preferably a tough clay with a liquid limit greater than 40 and plasticity index greater than 25, may be used to compare the densities obtained by the manual compaction and those obtained by the calibrated mechanical compactor. Soils of lesser plasticity may approach a maximum density condition with considerably less than the compactive effort provided by the Proctor

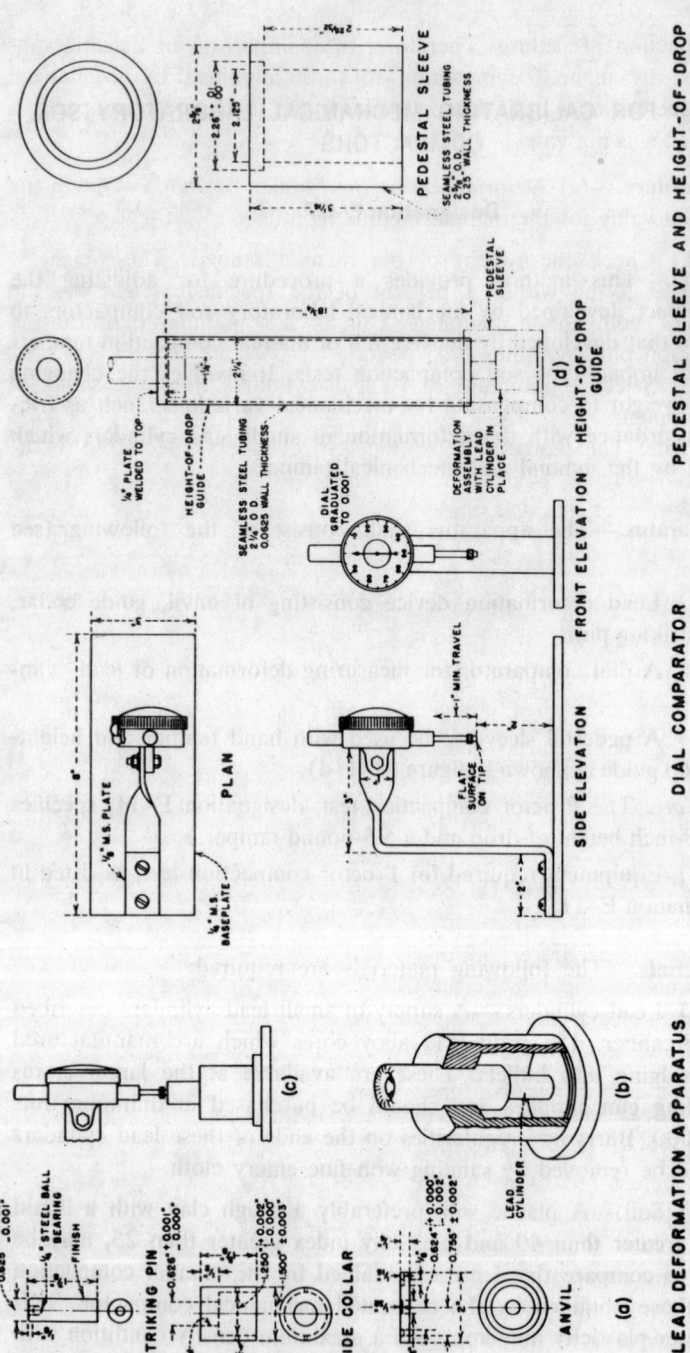

Figure 37-1.—Apparatus for calibrating mechanical laboratory soil compactors. 101-D-366.

compaction procedure. Therefore, the comparison of densities obtained by manual compaction to those obtained by mechanical compactors as a basis for determining the mechanical compactor's efficiency is not valid.

4. Procedure.—(a) *Deformation by the Manual Method.*—Obtain the deformation value for the manual method as follows:

(1) Check the weight of the manual tamper. The weight is required to be within 1 percent of the specified weight for the Proctor compaction hammer. This is 5.5 pounds, permitting a variation plus or minus 0.05 pound.

(2) Select a minimum of 10 lead cylinders from the same lot or shipment.

(*Note:* The temperature of the lead cylinders affects the deformation value, so a fairly constant temperature should be maintained for all the cylinders used in each set of comparative values obtained during calibration operations.)

(3) Assemble the lead deformation apparatus with the lead cylinder in place as shown in figure 37–1(b). Care should be taken so that the lead cylinder is placed precisely in the center of the small recess in the anvil base. Exercise every precaution to assure that it will remain in this position while moving the deformation apparatus. A slight variation from center, in most instances, has resulted in an erroneously high deformation value.

(4) Place the assembled deformation apparatus on the base of the dial comparator with the tip of the dial stem on the top center of the ball bearing to give a maximum dial reading, as shown in figure 37–1(c). Read the dial and record as initial reading.

(5) Place the baseplate of the compaction mold on the 200-pound compaction cylinder foundation and place the assembled deformation apparatus, figure 37–1(b), on the baseplate. Place the pedestal sleeve and height-of-drop guide with tamper over the assembly, as shown in figure 37–1(d), not allowing the tamper to touch the striking pin. Raise the tamper to the specified height of drop and release to apply one blow on the top of the striking pin.

(*Note:* It is advisable to perform several practice deformation trials before recording the information for record. The intent is to develop a technique for the individual operator which will assure that a solid square blow is delivered to the striking pin. The drop guide and tamper should be held in a plumb position centered over the striking pin, noting each blow as being either a solid blow or a glancing blow, if the tamper fell freely, or if it slid against the guide sleeve.)

(6) Return the deformation apparatus to the dial comparator, read the dial, and record as final dial reading. The difference between the initial dial reading of paragraph 4(a)(4) and the final reading of this paragraph is the deformation value. Observe the lead cylinder. Does the determination indicate a square solid blow? Is the lead cylinder barrel shaped uniformly? Was the cylinder in the center of the anvil or offcentered as indicated by a slight ring at the bottom of the cylinder? These observations will assist in determining whether a good test was performed.

(7) Repeat steps (3), (4), (5), and (6) above, using an unused lead cylinder for each determination until at least 10 lead cylinder deformation values are judged to be good; that is, obtained with a square blow free from drag. With good technique the total variation of the deformation values will be less than 0.015 inch. Use the numerical average of at least 10 deformation values judged to be good for comparison with the mechanical compactor.

(b) *Deformation by the Mechanical Compactor.*—Assemble and adjust the mechanical compactor in accordance with the manufacturer's instructions. Operate the compactor for a period of time, not less than 10 minutes, as necessary to cause friction in the parts to become constant and insure consistent deformation values, allowing the tamper to fall on soil or other soft material. Obtain the deformation for the mechanical compactor method as follows:

(1) Check the height of drop of the tamper. The height of drop is required to be within 1 percent of the specified height.

(2) Obtain the initial dial reading, as described in paragraphs 4(a) (3) and (4).

(3) Place the deformation assembly on the compaction mold baseplate which is on the base of the mechanical compactor. Position the assembly and place the tamper lightly on the striking pin in such a manner that for the next blow the tamper will be centered on the striking pin.

(*Note:* Mechanical tampers are designed to distribute blows over the cross-sectional area of the soil specimen. One way of doing this is to have the tamper offset from its shaft which rotates during the "pickup" part of the cycle. For this type of compactor, the magnitude of the revolution of the tamper between blows must be determined and the deformation apparatus so positioned that the tamper blow will be centered on the striking pin. For another type of compactor the tamper moves in a vertical direction only and the baseplate is rotated. In this case, the rotating mechanism of the baseplate should be disconnected during the calibration operation.)

(4) Start the machine and allow the tamper to apply one blow on top of the striking pin.

(*Note:* Observe carefully the height of drop of the tamper.)

(5) Obtain the final dial reading as described in paragraph 4(a)(6).

(6) Repeat steps in paragraphs 4(b) (2), (3), and (4) using an unused lead cylinder for each determination until at least 10 lead cylinder deformation values are judged to be good; that is, a well-centered solid blow. Use the numerical average of these values for comparison with the manual method.

(c) *Adjustment of the Mechanical Compactor.*—If the deformation produced by the manual tamper is greater than that produced by the mechanical compactor, add weight to the tamper on the mechanical compactor in an amount necessary to produce the same average deformation value as produced by the manual method. This added weight shall be securely fastened to the tamper. The deformation values shall be considered to be the same if the average value for the mechanical compactor does not vary more than 2 percent from the average value by the manual method (see paragraph 4(a)(7)). For the Proctor compaction test using the lead cylinders as prescribed, this will be a permissible difference of about 0.005 inch. If the deformation produced by the manual tamper is smaller than that produced by the mechanical compactor, remove weight from the tamper on the mechanical compactor in an amount necessary to produce the same deformation value as produced by the manual method.

(d) *Comparison of Manual and Mechanical Methods Using Plastic Soil.*—Using the plastic soil indicated, conduct one compaction test each using the manual method, designation E–11, and the mechanical compactor and compare.

(*Note:* Maximum density and optimum moisture by each method should be within 2 percent and 0.5 percent, respectively.)

5. Frequency of Check.—The calibration of the mechanical compactor shall be checked using the lead cylinders as described in paragraphs 4 (b) and (c) once each week during operation or once for every 500 soil specimens compacted, whichever occurs first. It is not necessary to compare results of the compaction tests on plastic soil as described in paragraph 4(d) unless the mechanical compactor requires repair or adjustment as specified in paragraph 4(c).

COMPACTION TEST FOR SOIL CONTAINING GRAVEL
(MOISTURE-DENSITY RELATIONS)

Designation E–38

1. Scope.—This designation describes a method for determining the relationship between the water content of gravelly soils of 3-inch-maximum size and the resulting dry densities, when the soil is compacted by a large-scale compaction apparatus in the laboratory. This apparatus is designed to impart the same energy per unit volume (12,375 foot-pounds per cubic foot) to the soil as the Proctor compaction apparatus described in designation E–11. The water content corresponding to the maximum dry density obtained with the large-scale apparatus is referred to as the optimum water content for the gravelly soil.

2. Apparatus.—The apparatus shall consist of the following (see designation E–4 and figs. 38–1, 38–2, and 38–3). Drawings for the large-scale mechanical compactor may be obtained from the Engineering and Research Center.

> Large-scale mechanical compactor.
> Compaction mold, item 25.
> Porous discs, item 32.
> Sponge-rubber liner, ¼-inch thick.
> Chain hoist, ½-ton capacity, item 38A.
> Yoke.
> Four-point sling.
> Scale, 500-pound-capacity, item 74A
> Loading plate.
> Nine-inch ring gage.
> Dial indicator gage, item 35A.
> Modified spanner wrench.
> Spatula or scraper, item 87.
> Yardstick.
> Straightedge, 2-foot, item 93.
> Large tamping rammer.
> Small tamping rammer.
> Air pressure line.
> Rubber mallet.
> T-bar with chisel point.
> Rod, ½-inch diameter, 2 feet long.
> Digging bar, approximately 4 feet long by 1½ inches in diameter.
> Large mixing pan.
> Shovel.

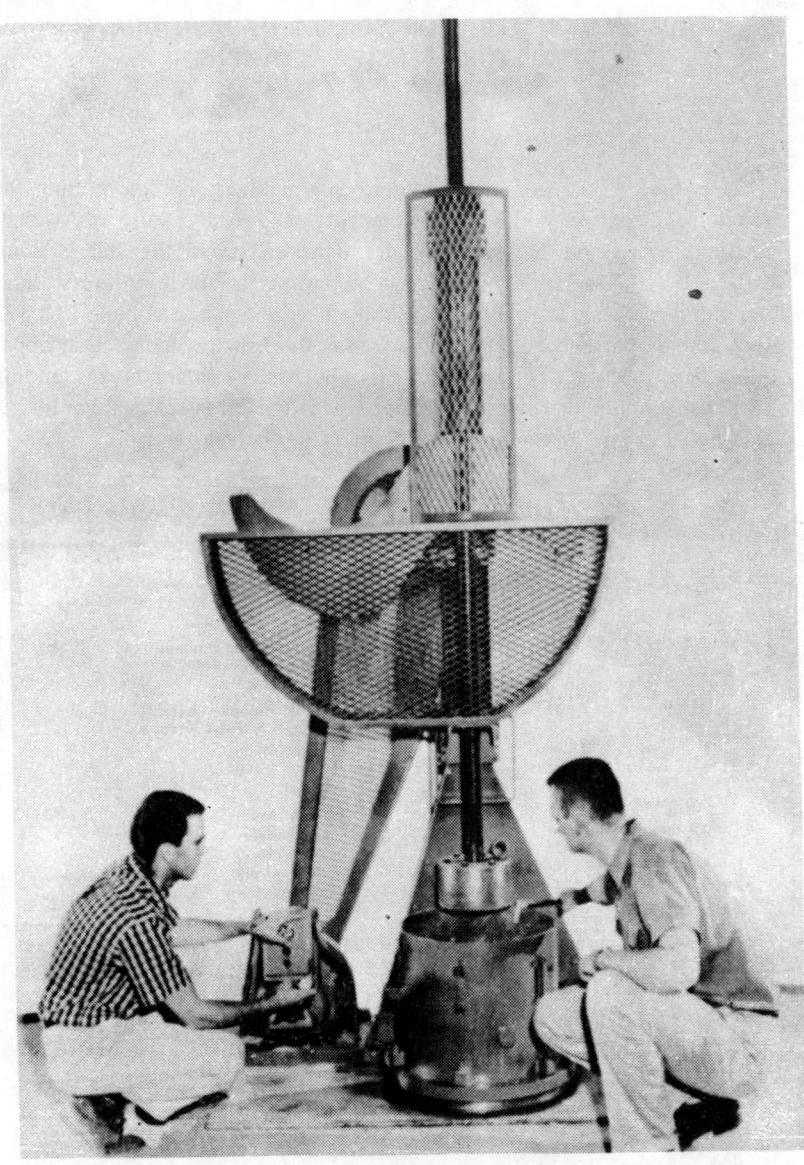

Weight of hammer_____ 185.7 pounds	Area of compaction cylinder without
Height of hammer drop_____ 18 inches	sponge-rubber liner____291 square inches
Area of hammer_____ 70.9 square inches	

Figure 38–1.—Compaction machine for soil containing gravel. PX–D–20361.

IDENTIFICATION LIST

1. Compaction mold.	9. Large tamping rammer.
2. Porous disc.	10. Digging bar.
3. Loading plate with porous disc attached.	11. Yardstick.
	12. Modified spanner wrench.
4. Nine-inch ring gage.	13. Spatula.
5. Dial indicator with case.	14. Rubber mallet.
6. Yoke—used to lift compaction mold.	15. Scraper.
7. Four-point sling—used to lift mixing pan.	16. T-bar.
	17. Straight rod.
8. Small tamping rammer.	18. Straightedge.

(*Note:* The porous disc may be attached to the loading plate with weather-stripping adhesive.)

Figure 38–2.—Compaction apparatus for soil containing gravel. PX–D–31822.

Sprinkler can.

Spray gun.

Towels.

Moistureproof bags or other moistureproof material.

Large moisture pans, same as mixing pan of item 60.

3. Preparation of Sample.—If new material is used for each specimen, approximately 600 to 800 pounds dry weight will be required to complete the compaction test. Although it is recommended that new material be used for each specimen, the material may be reused if a shortage exists, providing that the coarse-grained fraction is not friable and providing that the material is ovendried and screened. The directions

IDENTIFICATION LIST

1. Sprinkler can.   5. Sponge-rubber liner.
2. Moistureproof bag.   6. Large moisture pan.
3. Spray gun.   7. Mixing pan.
4. Shovel.   8. Towel.

(*Note:* The sponge-rubber liner can be placed inside the compaction mold by cutting a strip 8¼ inches wide and long enough to exactly fit inside the mold and attaching it to the mold with rubber cement.)

Figure 38–3.—Sample preparation equipment for compaction test for soil containing gravel. PX–D–31823.

below describe a method for estimating the quantities needed for each specimen. The following properties of the components of the gravelly soil must be known prior to the preparation of the first large-scale compaction specimen:

(1) Gradation of total material, designation E–6.

(2) Optimum water content and Proctor density for the minus No. 4 fraction, designation E–11.

(3) Bulk specific gravity (ovendry) of plus No. 4 fraction,[1] designation E–10.

(4) Percent absorption of the plus No. 4 fraction,[1] designation E–10.

---

[1] Data shall be recorded on figure 38–5 (form 7–1825).

(5) An assumed value of maximum dry density for the total material.[2]

---

[2] The maximum dry density of the total material may be estimated by use of the following formula:

$$\gamma_t = \frac{100}{\dfrac{P_s}{(D)(\gamma_4)} + \dfrac{P_g}{62.4G}} \tag{1}$$

where: γ_t = maximum dry density of total material,

       γ_4 = Proctor maximum dry density of minus No. 4 fraction,

       G = bulk specific gravity (ovendry) of gravel computed on line 13 of fig. 38–5,

       P_s = percent passing No. 4 screen,

       P_g = percent retained on No. 4 screen, and

       D = percent of Proctor density, expressed as a decimal, $= \dfrac{1}{\gamma_4} \times$ (density of minus No. 4 fraction in the compacted large-scale specimen).

If more precise data are not available, the value of D may be estimated by using the curves on figure 38–4. The actual value of D shown on line 29 of figure 38–5 applies to a particular specimen and cannot be calculated until after the specimen has been compacted and the dry density of the total material has been determined.

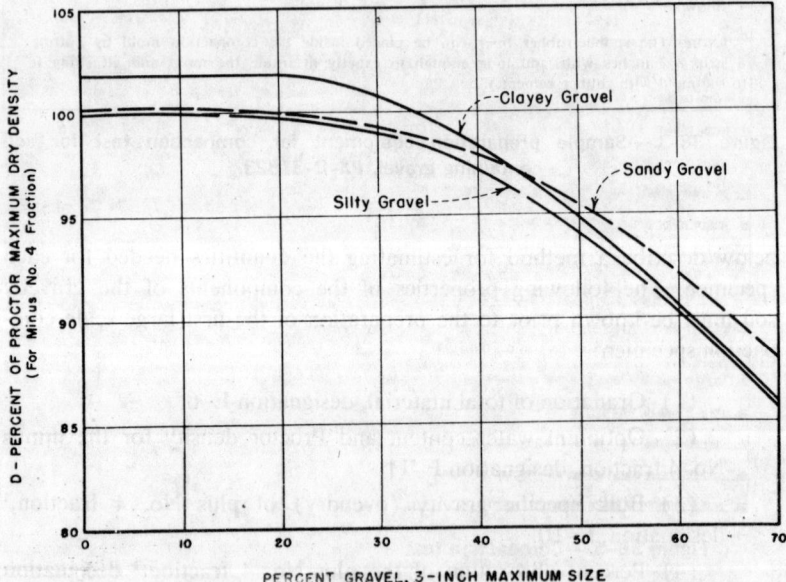

Figure 38–4.—Percent of Proctor maximum dry density obtainable with standard compactive effort in relation to percent of gravel. 101–D–390.

7-1825
(3-66)
Bureau of Reclamation　　　　**LARGE-SCALE COMPACTION TEST - DATA SHEET 1**

PROJECT _Example_ _____ FEATURE _____ SAMPLE NO. _____
DATE _____ TESTED BY _____ COMPUTED BY _____ CHECKED BY _____

SPECIFIC GRAVITY AND PERCENT ABSORPTION OF THE PLUS NO. 4 FRACTION				
TRIAL NO.		1	2	
1. WT OF SATURATED SURFACE-DRY SAMPLE + PAN IN AIR	G	3276.8	3278.8	
2. WT OF PAN	G	485.0	485.0	
3. WT OF SATURATED SURFACE-DRY SAMPLE IN AIR, (1)-(2)	G	2791.8	2793.8	
4. WT OF SAMPLE IN WATER	G	1647.0	1660.8	
5. WT LOSS OF SATURATED SURFACE-DRY SAMPLE IN WATER, (3)-(4)	G	1144.8	1133.0	
6. WT OF OVENDRIED SAMPLE + PAN IN AIR	G	3530.0	3552.8	
7. WT OF PAN	G	885.0	894.8	
8. WT OF OVENDRIED SAMPLE IN AIR, (6)-(7)	G	2645.0	2658.0	
9. BULK SPECIFIC GRAVITY (OVENDRY), (8) ÷ (5)		2.31	2.35	
10. APPARENT SPECIFIC GRAVITY, (8) ÷ [(8)-(4)]		2.65	2.67	
11. WT OF WATER IN SATURATED SURFACE-DRY SAMPLE, (3)-(8)	G	146.8	135.8	
12. ABSORPTION, 100(11) ÷ (8)	%	5.55	5.11	
13. AVG BULK SPECIFIC GRAVITY (OVENDRY)			2.33	
14. AVG APPARENT SPECIFIC GRAVITY			2.66	
15. AVG ABSORPTION	%		5.3	

DRY DENSITY AND MOISTURE CONTENT OF THE MINUS NO. 4 FRACTION							
SPECIMEN NO.		1					
16. PROCTOR DRY DENSITY OF MINUS NO. 4	PCF	115.2					
17. PERCENT OF GRAVEL	%	59.7					
18. PERCENT OF MINUS NO. 4	%	40.3					
19. DRY DENSITY OF TOTAL MATERIAL (FROM LINE 16 OF FORM 7-1826)	PCF	125.3					
20. MOISTURE CONTENT OF TOTAL MATERIAL (FROM LINE 17 OF FORM 7-1826)	%	8.4					
21. (20)-(15)	%	3.1					
22. 100(21) ÷ (18)	%	7.7					
23. DRY WT OF GRAVEL PER CU FT OF TOTAL MATERIAL, [(17) + 100]x(19)	LBS	74.8					
24. VOLUME OF GRAVEL PER CU FT OF TOTAL MATERIAL, (23) ÷ [(13)x 62.4]	CU FT	0.514					
25. VOLUME OF MINUS NO. 4 PER CU FT OF TOTAL MATERIAL, 1-(24)	CU FT	0.486					
26. DRY WT OF MINUS NO. 4 PER CU FT OF TOTAL MATERIAL, [(18) ÷ 100]x(19)	LBS	50.5					
27. COMPUTED DRY DENSITY OF MINUS NO. 4 FRACTION, (26) ÷ (25)	PCF	103.9					
28. COMPUTED MOISTURE CONTENT OF MINUS NO. 4 FRACTION, (22) + (18)	%	13.0					
29. VALUE OF D (PERCENT OF PROCTOR DRY DENSITY), 100(27) ÷ (16)	%	90.2					

**Figure 38-5.—Compaction test for soil containing gravel—Data sheet 1.
101-D-606.**

The optimum water content of the total material is usually near the calculated water content obtained by combining the plus No. 4 fraction saturated surface dry with the minus No. 4 fraction at its optimum water content. Therefore, it is recommended that the material for the first compaction point be prepared so that these conditions are satisfied. The assumed dry density multiplied by the volume of a 9-inch-high specimen gives the total dry weight of material to be prepared for one compaction point. The formula for the total moisture content of the prepared sample is:

$$w_t = \frac{P_s w_0}{100} + \frac{P_g w_g}{100} \qquad (2)$$

where: w_t = moisture content of total material, in percent, when minus No. 4 fraction is at its optimum water content and plus No. 4 fraction is saturated surface dry,

P_s = percent passing No. 4 screen,

P_g = percent retained on No. 4 screen,

w_0 = optimum water content, in percent, for the minus No. 4 fraction, and

w_g = absorption, in percent, (moisture content of plus No. 4 fraction when it is saturated surface dry).

The individual gravel sizes are normally stored separately in an air-dried condition. The moisture contents of the air-dried gravel [3] and the minus No. 4 material must be known prior to weighing quantities for each lift, so specified dry weights can be obtained in accordance with the gradation of the total material. The correct dry weight of each gravel size shall be weighed separately for each lift and saturated in accordance with part C of designation E–10. These weights are calculated on form 7–1826B (fig. 38–8), and are based on the estimated compacted dry density. Therefore, about 5 percent additional material of each gravel size should be saturated so the lifts can be adjusted if a higher density is obtained. This additional material should be soaked in separate cans and will not be needed unless the lift weights W_2 and W_3 have to be increased. The total minus No. 4 material for all the lifts plus about 10 percent additional material shall be mixed to the desired water content. The material shall be stored in moistureproof cans or sacks and the water content shall be adjusted until it is within 0.5 percent of that desired. The additional material shall be used for adjusting lift weights and moisture content checks. In establishing the proper moisture for the minus No. 4 fraction, it is usual practice to provide overnight curing to insure adequate moisture distribution. This is particularly important for clayey fines.

---

[3] In many cases the moisture content of the air-dried gravel may be small enough to be neglected without serious error.

4. Procedure.—(a) The porous disc shall be placed inside the compaction mold and wiped with a very wet towel to prevent significant loss of moisture from the specimen to the disc. Then the combined weight of the compaction mold and the porous disc shall be determined and recorded on line 8 of figure 38–6, form 7–1826.

(b) The 9-inch ring gage shall be placed inside the compaction mold and the loading plate shall be placed on top of it. The dial indicator shall then be placed between the handles of the loading plate and the shoulders on the compaction mold. The readings shall be recorded as initial dial readings on line 1 of figure 38–6. The depth of the compaction mold shall be determined with the straightedge and the yardstick. Subsequent measurements of layer thicknesses will be made with the straightedge and yardstick. With the use of the hoist apparatus, the compaction mold shall be positioned under the compaction hammer in the place provided.

(c) Based upon an assumed value of maximum dry density, the first layer shall be composited by mixing the gravel fractions in a saturated surface-dry condition with the minus No. 4 material. The material shall be thoroughly mixed, shoveled into the compaction mold, and spread evenly over the bottom. The large pieces of gravel should be evenly distributed throughout the layer. The compaction hammer shall then be lowered onto the soil. The collar on the stem of the compaction hammer determines the height of drop of the hammer and should be set to obtain an 18-inch drop. The collar should be lowered one groove each for the following two layers. The mechanical compactor shall then be started with the electric switch. The modified spanner wrench shall be used to rotate the mold, as shown in figure 38–1, approximately one-sixth of a turn after each blow.

If uncompacted soil appears around the perimeter of the mold, a spatula may be used to scrape it back toward the center of the specimen where it will be compacted by the hammer. If large amounts of material stick to the compaction hammer, the mechanical compactor should be stopped and the hammer should be cleaned with the spatula or scraper. The compaction process shall then be resumed until a total of 22 blows has been delivered to the layer. After 22 blows on the first layer, the mechanical compactor shall be stopped with the electric switch. After the hammer has been hoisted out of the way, the straightedge and yardstick shall be used to measure the thickness of the layer. An average of approximately 10 measurements, made at equally distributed points on the surface of the layer, shall be obtained. The desired thicknesses of the specimen after the first, second, and third layer have been compacted are 3 inches, 6 inches, and 9 inches, respectively. If the thickness of the specimen after any layer does not fall within the limits given below, the specimen shall be recompacted with new material.

7-1820
(3-66)
Bureau of Reclamation

LARGE-SCALE COMPACTION TEST - DATA SHEET 2

Project **Example**   Feature _____   Sample No. _____

Date _____   Tested by _____   Computed by _____

Checked by _____   Compaction Mold No. **5**   Volume Factor { with sponge-rubber liner **1.982** / without sponge-rubber liner _____

VOLUME DETERMINATIONS

Specimen No.		A	B	A	B	A	B
1 Initial dial	in	0.471	0.041				
2 Final dial	in	0.250	0.258				
3 Difference	in	+0.221	-0.217				
4 Avg difference	in	+0.002					
5 Spec height (9" ± avg diff)	in	9.002					
6 Specimen volume	cu ft	1.487					

WET DENSITY DETERMINATIONS

7 Wt compaction mold + wet soil	lbs	504.5				
8 Wt compaction mold	lbs	302.5				
9 Wt wet soil	lbs	202.0				
10 Wet density	pcf	135.8				

MOISTURE AND DRY DENSITY DETERMINATIONS

Material	Total	-4	Total	-4	Total	-4
Pan No.	20	21				
11 Wt pan + wet soil	lbs	127.0	125.2			
12 Wt pan + dry soil	lbs	119.1	117.4			
13 Wt pan	lbs	25.2	25.0			
14 Wt water	lbs	15.7				
15 Wt dry soil	lbs	186.3				
16 Dry density	pcf	125.3				
17 Moisture content	%	8.4				

Figure 38–6.—Compaction test for soil containing gravel—Data sheet 2. 101-D-607.

Thickness after first layer—3 inches ± 0.25 inch.
Thickness after second layer—6 inches ± 0.25 inch.
Thickness after third layer—9 inches ± 0.35 inch.

(*Note:* For permeability specimens, refer to paragraph 4(k).) The weight of the material in the first layer, divided by the thickness of the layer, gives a factor which can be used to calculate the weight of the second lift. (This calculation is outlined on form 7–1826A, figure 38–7.) The weight of the second lift shall be calculated to bring the total thickness to 6 inches. After the second lift has been placed and spread evenly over the mold, the hammer shall again be lowered onto the soil. Then the collar shall be lowered one groove. The second layer shall then be compacted in the same manner as the first layer was compacted. The combined thicknesses of the first two layers shall be measured with the straightedge and yardstick, and the new factor thus obtained shall be used to calculate the weight of the third and final lift. (See figure 38–7.) The last lift shall be calculated to bring the total thickness of the specimen to 9 inches. After the hammer has been lowered onto the soil, the collar shall be lowered one groove. The last layer shall then be compacted in the same manner as the first two layers were compacted.

(d) After the final layer has been compacted, the compaction mold shall be placed on the scales with the hoist apparatus. The weight of the compaction mold plus wet soil shall be determined and recorded on line 7 of figure 38–6.

(e) A moisture determination shall be made on the minus No. 4 fraction. The data shall be recorded on figure 38–6.

(f) The surface of the specimen must be leveled before the total height can be measured. The small tamping rammer and the spatula can be used to level the surface, but light taps of the rammer must be used because significant compactive energy must not be added to the specimen during the leveling process. After the specimen has been leveled, the loading plate shall be placed on top of the specimen and in the same position it occupied on the 9-inch ring gage. The loading plate shall be struck three light blows with the large tamping rammer to seat it, and the rammer shall be left on the loading plate while the dial readings are taken. The dial indicator shall then be placed in position and the final dial readings shall be obtained and recorded on line 2, figure 38–6.

(g) With the use of the hoist apparatus, T-bar, and straight rod, the compaction mold shall be inverted over a large moisture pan and lowered into the pan. If the specimen cannot be removed by striking the mold with the rubber mallet, the air pressure line shall be engaged in the bottom of the mold and air pressure shall be used to extrude the specimen.

(h) The T-bar and the digging bar shall then be used to break the specimen into small pieces. A moisture content determination shall be

7-1826A (6-70)
Bureau of Reclamation

Large Scale Compaction Test – Data Sheet 3

Project _Example_   Feature _____   Sample No. _____

Computed by _____   Checked by _____   Date _____

P_s 40.3%	Est D 92%	G 2.33	W_g 5.3
P_g 59.7%	γ_d 115.2	W_o 13.1	

Estimate of Compacted Dry Density

$$\gamma_t = \frac{100}{\dfrac{P_s}{D\gamma_d} + \dfrac{P_g}{62.4\,G}} = \frac{100}{\dfrac{40.3}{(.92)(115.2)} + \dfrac{59.7}{(62.4)(2.33)}} = \frac{100}{.38 + .41} = 126.6$$

Container Area .1982 sq. ft.   Gage Ring Length 9 in.   Container Vol. 1.486 cu. ft.

Estimated Dry Weight per Specimen = Vol × γ_t = 1.486 × 126.6 = 188.1 lbs.

Measurements Top of Container to:

(a) Bottom Stone   14

(b) Top of 1st Lift   11

h_1 (height of 1st Lift) = (a)-(b) = 3

(c) Top of 2nd Lift   7 15/16

h_2 (height of 2nd Lift) = (b)-(c) = 3 1/16

(d) Top of 3rd Lift   5

h_3 (height of 3rd Lift) = (c)-(d) = 2 15/16

Weight per Lift

1st Lift = W_1 = $\dfrac{\text{Total Dry Weight}}{\text{No. of Layers}}$ = $\dfrac{188.1}{3}$ = 62.7

2nd Lift = W_2 = $\dfrac{W_1}{h_1}(6-h_1)$ = $\dfrac{62.7}{3}(6-3)$ = 62.7

3rd Lift = W_3 = $\dfrac{W_1 + W_2}{h_1 + h_2}(9-h_1-h_2)$ = $\dfrac{125.4}{6.06}(9-6.06)$ = 60.8

Figure 38-7.—Compaction test for soil containing gravel—Data sheet 3. 101-D-608.

7-1425B (5-70)
Bureau of Reclamation

Large Scale Compaction Test - Data Sheet 4

Project __Example__

Computed by _____   Checked by _____

Feature _____   Sample No. _____   Date _____

Lift - Number _____
Dry Weight

Screen Size	Percent Retained (e)	Water Content (f) 1/	1 $W_1 = 62.7$			2 $W_2 = 62.7$			3 $W_3 = 60.8$		
			Dry Weight (g) =(e)(w$_1$)	Wet Weight (1+f)(g)	Cum. Wet Weight	Dry Weight (j) =(e)(w$_2$)	Wet Weight (1+f)(j)	Cum. Wet Weight	Dry Weight (k) =(e)(w$_3$)	Wet Weight (1+f)(k)	Cum. Wet Weight
3"	0.0	—	0.0	0	0	0	0	0	0	0	0
1-1/2"	17.9	5.3	11.22	11.8	11.8	11.22	11.8	11.8	10.88	11.5	11.5
3/4"	15.4	5.3	9.66	10.2	22.0	9.66	10.2	22.0	9.36	9.9	21.4
3/8"	14.4	5.3	9.03	9.5	31.5	9.03	9.5	31.5	8.75	9.2	30.6
No. 4	12.0	5.3	7.52	7.9	39.4	7.52	7.9	39.4	7.30	7.7	38.3
-No. 4	40.3	13.1	25.27	28.6	68.0	25.27	28.6	68.0	24.50	27.7	66.0
Totals	100.0		62.7	68.0		62.7	68.0		60.79	66.0	

1/ Percent absorption for gravel sizes, actual water content of -No. 4

NOTES:

Figure 38–8.—Compaction test for soil containing gravel—Data sheet 4. 101-D-609.

made using the entire specimen as a moisture sample. Two large moisture pans will be required to contain the sample. The moisture content determination data shall be recorded on figure 38–6.

(i) After the moisture samples have been weighed, the final weight of the compaction mold and the porous disc shall be determined and recorded on line 8 of figure 38–6.

(j) Additional specimens shall be compacted in the same manner as the first specimen, except that the total moisture content shall be varied in increments of 2 percent. It is recommended that the second specimen be compacted 2 percent wetter than the first specimen. If the dry density of the second specimen is higher than that of the first, the third specimen can be compacted 4 percent wetter than the first specimen. If the dry density of the second specimen is lower than that of the first specimen, then the third specimen can be compacted 2 percent drier than the first specimen. If the first and second specimens have approximately the same dry density, then four specimens, one 2 percent drier than the first specimen and one 4 percent wetter than the first specimen, must be compacted to obtain a satisfactory compaction curve.

(k) If a permeability test is to be performed on the compacted specimen, a compaction mold lined with rubber sheeting must be used (refer to figs. 38–2 and 38–3), and the first and second layers must be scarified before placing the next layer. If the specimen is expected to have extremely high permeability, it may be desirable to use a layer of sand or gravel on the top and bottom of the specimen in lieu of the porous discs. The layer of sand and gravel must meet the filter criteria given in section 79 of chapter III. A layer of No. 4 gravel with a No. 8 hardware cloth between the specimen and gravel layer is also suitable.

5. Calculations.—The specimen height shall be determined from the initial and final dial readings on figure 38–6. If the final reading is smaller than the initial dial reading, the difference in dial readings is positive because the specimen is more than 9 inches high. The specimen height shall be calculated by algebraically adding 9 inches to the average difference in dial readings. The specimen volume shall be calculated by multiplying the appropriate volume factor by the specimen height. The volume factor depends upon the cross-sectional area of the compaction mold, and its units are cubic feet per inch of height. The wet weight of the specimen shall be calculated by subtracting the initial weight of the compaction mold from the weight of the compaction mold plus wet material. If the final weight of the compaction mold is significantly larger than the initial weight, the final weight should be used in the calculations, since the increase in weight was probably due to moisture which passed from the specimen to the disc. The wet density shall then be calculated by dividing the wet weight of the specimen by the specimen volume.

After the dry density and water content of the total specimen have been calculated on figure 38–6, the water content and the dry density of the minus No. 4 fraction may be calculated by performing the indicated operations on figure 38–5. The water content of the minus No. 4 fraction obtained from line 17 of figure 38–6 will serve as a check on the water content calculated on line 28 of figure 38–5.

6. Moisture Density Relationship.—From the data obtained in paragraph 5, dry density values shall be plotted as ordinates with corresponding water contents as abscissas on figure 38–9 (form 7–1414). A smooth curve shall then be drawn through the points. The following values shall be determined from the curve.

   (1) *Optimum water content.*—The water content corresponding to the peak of the curve shall be termed the optimum water content for the large-scale compaction.

   (2) *Large-scale maximum dry density.*—The dry density in pounds per cubic foot corresponding to the optimum water content shall be termed maximum dry density for the large-scale compaction.

It is often desirable to plot the following three moisture-density curves on the same sheet for the sake of comparison:

   (1) Total material.

   (2) Minus No. 4 fraction in the large-scale specimens (values computed on lines 27 and 28 of figure 38–5).

   (3) Standard Proctor.

7. Theoretical Curve at Complete Saturation (Zero Air-Voids Curve). —A curve termed the "curve of complete saturation" or "zero air-voids curve" shall also be drawn on figure 38–9. This curve represents the relationship between dry densities and corresponding moisture contents, assuming the voids are completely filled with water. Values of dry densities and corresponding water contents for plotting the zero air-voids curve shall be computed as follows:

$$\gamma_d = 62.4G\left(1 - \frac{n}{100}\right) \tag{3}$$

$$w = \frac{62.4n}{\gamma_d} \tag{4}$$

where: γ_d = dry density of total material in pounds per cubic foot,
   62.4 = weight of 1 cubic foot of water, in pounds,
   G = apparent specific gravity of total material,[4]
   n = porosity, in percent, and
   w = water content, in percent.

---

[4] Apparent specific gravity of total material is calculated by the following formula:

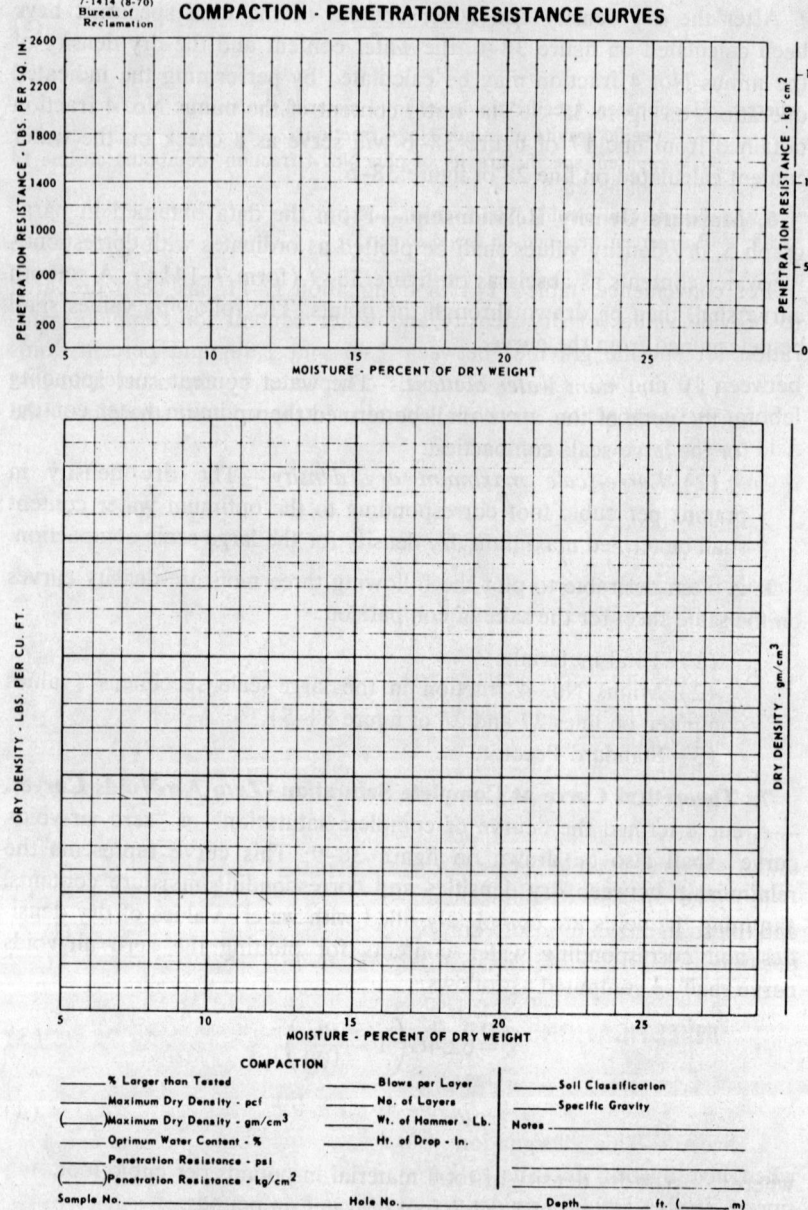

Figure 38–9.—Form for plotting compaction test curves. 101-D-510.

$$G = \frac{100}{\dfrac{P_s}{G_s} + \dfrac{P_g}{G_g}} \qquad (5)$$

where: G = apparent specific gravity of total material,

G_s = specific gravity of minus No. 4 fraction,

G_g = apparent specific gravity of plus No. 4 fraction (computed on line 14 of fig. 38–5),

P_s = percent passing No. 4 screen, and

P_g = percent retained on No. 4 screen.

For convenience, table 11–1 in designation E–11 has been prepared to provide values of dry density and water content for complete saturation for specific gravities between 2.45 and 2.90, and percent voids between 10 and 60. (*Note:* Materials compacted by the usual field and laboratory compaction methods contain entrapped air; therefore, the field roller-density curves and laboratory density test curves may approach but never cross the zero air-voids curves.)

8. Safety Precautions.—The following safety precautions should be given special attention when performing the large-scale compaction test.

(a) The compaction mold must be seated firmly in place before the machine is started, to insure that the hammer will not strike the edge of the mold.

(b) Under no circumstances shall one's hand be placed under the compaction hammer while the machine is in operation, even if the hammer is moving upward.

(c) The compaction mold and the full mixing pan shall always be lifted with the hoist apparatus.

(d) Care must be taken to insure that the yoke is fastened to both sides of the compaction mold before lifting is begun.

(e) The suspended compaction mold shall be rotated with the T-bar and the straight rod to insure control of the mold when it is in an inverted position.

INVESTIGATIONS FOR ROCK SOURCES FOR RIPRAP

Designation E–39

1. Scope.—This designation describes procedures to be followed when investigating deposits to be used as potential sources of rock for riprap. It also includes guides to be followed in the sampling and testing of rock materials, and a guide to be used when reporting the results of the investigation.

2. General.—The complexity and extent of investigations conducted to determine suitable sources of riprap material will be governed by the size and design requirements of the project features and the quantity and quality of material required. Normally, project development occurs in three stages: (a) reconnaissance, (b) feasibility, and (c) specifications. The extent and intensity of materials investigations are usually divided into the corresponding stages. Additional investigations, (d) construction, are sometimes required immediately prior to or during the construction stage. These investigations may be distinct, separate operations or may overlap one another.

Data obtained during the reconnaissance stage are primarily descriptive, based on surface geological examinations, and should be secured when the geological reconnaissance is made for the project feature.

In the feasibility stage, data are primarily qualitative and are obtained to confirm and expand data concerning potential material sources, disclosed during the reconnaissance stage, for preparation of project cost estimates. A complete survey of possible material sources located within economical haul range of the jobsite is made at this time. Quality evaluation tests on representative samples of materials may be required to determine the most promising sources. Field work should be done jointly, if possible, by a geologist and a materials engineer. Sufficient data should be obtained at this time to determine whether the Government should acquire a source or to establish those sources which may be cited in the specifications but which require contractor acquisition and development.

In the specifications stage, data are mainly quantitative and specific for final designs of the various project features such as riprap for upstream slopes of earth dams or structure protection. Sources selected are thoroughly investigated to determine quantities and uniformity of material available and, if necessary, representative samples taken for quality evaluation testing.

Materials investigations during the construction period are generally of a confirmative nature to clarify conditions which could not be verified or resolved during the specifications stage.

Rock sources must satisfy two main requirements: the rock source should produce rock fragments in suitable sizes for the required usage and the rock fragments should be hard, dense, and durable enough to withstand both the processes involved in procurement and placement, and weathering. If material of required quality is available in sufficient quantity in the immediate vicinity of the jobsite, it will be unnecessary to investigate more distant sources. If, however, there is a deficiency of suitable rock in the immediate area, it will be necessary to explore further. In this case, prospecting for rock should extend radially outward from the damsite until a deposit of rock is located which is suitable in quality and sufficient in quantity to fulfill the anticipated requirements.

3. Investigation.—(a) *Reconnaissance.*—This initial or preliminary exploration involves field surface reconnaissance using topographic, geologic, and agricultural soil maps and aerial photographs with supplemental information provided by records of known developed sources of material. A study of maps and aerial photographs may reveal possible sources of material. Contours are often an indication of the type of material: sharp breaks usually indicate hard rock and slopes below cliffs often have talus deposits. During field reconnaissance the countryside should be examined for exposed rock outcrops or cliffs. Road cuts and ditches may also reveal useful deposits. Data obtained should define the major advantages or disadvantages of potential materials sources within reasonable haul distance to the jobsite. Reporting of accumulated data and information at this stage of investigation is described in paragraph 6.

(b) *Feasibility.*—Information accumulated during this stage is needed to prepare preliminary designs and cost estimates. Sufficient information concerning potential sources should be gathered to determine whether the Government should acquire the source or if the rock should be furnished by the contractor. Selection of sources should be limited to those which may eventually be cited in specifications, and core drilling or blast tests may be required to confirm fragment size and quantity of material available in the sources. The potential material sources are examined to determine size and character, and particularly to observe joint and fracture spacing, resistance to weathering, and variability of the rock. The spacing of joints, fractures, and bedding planes will control the size of rock fragments obtainable from the deposit. Observation of weathering resistance of rock in situ will provide a good indication of its durability. Particular attention should be given to location and distribution of unsound seams or strata which must be avoided or wasted during quarrying operations. A general location map and a report describing the potential sources and containing estimates of available quantities, overburden, haul roads, and accessibility are prepared.

Representative samples of riprap material from the most promising potential sources are sometimes required to be submitted to the Engineering and Research laboratories in Denver for quality evaluation tests. The extent and detail of information necessary at this stage is described in paragraph 6.

(c) *Specifications.*—The purpose of investigations at this stage is to furnish design data and information required for specifications preparation. Sources determined by feasibility investigation data to be of suitable quality for project feature work are thoroughly surveyed and investigated to establish the quantity of material available and to determine its uniformity.

Core drilling may be required, if dictated by geologic conditions. Such core drilling should be done on a grid system, if appropriate, and should include both vertical and angled holes as directed by the geologist or materials engineer. Blast testing should also be done at this time if not performed previously. After the blast testing, the data secured therefrom should be submitted to the Engineering and Research Center in a form suitable for inclusion in the specifications. Sampling and testing should also be completed during this stage.

If any additional deposits are considered at this stage, it is imperative that they be investigated as thoroughly as the originally considered source or sources. The required information necessary is described in paragraph 6.

(d) *Construction.*—This investigation stage is sometimes required to provide field and design personnel with additional detailed information for proper source development. This information should be obtained sufficiently ahead of quarrying or excavation operations to provide for proper processing and placing of material. If unforeseen changes occur in quality of material being removed from the source, sampling and quality evaluation testing of the rock may be required to confirm material suitability or delineate unsuitable rock areas.

4. Sampling.—Sampling can often be a weak link in the chain of investigative procedures. Thus, it must be carefully performed by qualified, experienced personnel.

The sample size should be at least 600 pounds and represent proportionally the quality range from poor to medium to best as found at the source. If the material quality is quite variable, it may be preferable to obtain three samples which represent respectively the poorest, medium, and best quality material available. The minimum size of individual fragments selected should be at least ½ cubic foot in volume, if possible. An estimate of the relative percentages of each material quality should be made and included as information relating to the source. Samples from undeveloped sources must be very carefully chosen so that the material selected will, as far as possible, be typical of the deposit and include any significant rock-type variations.

Representative samples may be difficult to obtain. Overburden may limit the area from which material can be taken and obscure the true character of a large part of the deposit. Surface outcrops will often be more weathered than the interior of the deposit. Samples obtained from loose rock fragments on the ground or collected from weathered outer surfaces of rock outcrops are seldom representative. Fresh material may be obtained by breaking away the outer surfaces, or by trenching, blasting, or core drilling. In stratified deposits such as limestones or sandstones, vertical and horizontal uniformity must be evaluated as strata often differ in character and quality.

The dip of stratified formations must also be considered. Strata inclination with respect to surface slope will expose different strata at the surface in different parts of the area. Attention should be directed to the possibility of zones or layers of undesirable material. Clay or shale seams may be so large or prevalent as to require selective quarrying or excessive wasting of undesirable material.

5. Testing.—Quality evaluation investigations performed in the Engineering and Research Center laboratory on representative samples submitted from the field usually include detailed petrographic examination and physical properties tests.

(a) *Petrographic Examination.*—The pieces of rock comprising the sample are examined individually and different rock facies and rock types, if present, segregated. Size range is described and characteristic fragment shape studied, particularly to determine if the fragment shape is determined by joints, fractures, or shear. Surface weathering, and secondary deposits of alkali salts or clay, are noted. Fracture or vein systems are described as well as the ease with which fractures or veins can be opened. Hardness, toughness, or brittleness, and visible void or pore characteristics, with their variations, are noted. Rock pieces representative of the various facies and rock types may be selected for detailed petrographic examination. The texture, internal structure, and mineralogy of the various rock facies and rock types are determined. Special attention is given to internal voids and fractures, and to the nature and character of cementing material in sedimentary rocks. Thin-section studies are made as required. For freeze-thaw durability testing, 2⅞-inch rock cubes are sawed from rock fragments selected to represent the poorest, medium, and best quality rock (for each facies or rock type) on the basis of visual inspection. The actual number of rock cubes tested may vary from sample to sample. The petrographic data are included in their entirety in the final materials report.

(b) *Physical Properties.*—After the rock cubes have been obtained, they are weighed in ovendry condition and photographed. The cubes are immersed in water for 24 hours, saturated surface dry weights and weights in water obtained, and wet bulk, dry bulk, and apparent specific gravities determined. The cubes are then inserted in 3-inch-square rubber sheaths, sufficient water is added to cover the specimens, and the rubber sheaths containing the specimens are placed in automatically controlled freezing and thawing cabinets where the cubes are alternately frozen and thawed at the rate of 50 cycles per week by circulating calcium chloride brine around the sheaths. Each cycle consists of 1½ hours freezing at 10° F. and 1½ hours thawing at 70° F. Throughout the test, the appearance and manner of deterioration of the cubes are noted. Termination of the test is 250 cycles or when the rock fails (failure criterion is 25 percent weight loss), whichever is sooner. Type of failure—splitting or crumbling—is noted, photographs taken, and weight loss determined. Weight loss (in percent) is computed as difference in ovendry weight between the largest piece of the cube remaining after testing and original ovendry weight of the cube. The weight of material lost by splitting of the rock cube along fractures, seams, and bedding planes is considered weight loss and appropriate notation is made to aid in obtaining minimum-size riprap required by specifications.

Material remaining after petrographic examination of the rock sample (excluding any pieces selected for more detailed petrographic analysis and freeze-thaw durability tests) is crushed, separated into 1½- to 3-inch, ¾- to 1½-inch, ⅜- to ¾-inch, and No. 4 to ⅜-inch-size fractions, and representative samples obtained for further physical properties tests. Representative samples are also obtained for petrographic examination of the material if used as crushed coarse concrete aggregate.

Samples consisting of different rock types or radical facies changes should be tested and examined separately. Physical properties tests performed in accordance with the pertinent test designation described in the Seventh Edition of the Concrete Manual are: (1) Specific gravity and absorption—designation 10; (2) Sodium sulfate soundness—designation 19; and (3) Los Angeles abrasion—designation 21.

Upon completion of quality evaluation tests shown in subparagraphs 5a and 5b, suitability of the rock is determined.

6. Reporting.—Reporting of information and data accumulated during any investigation stage is most important. Although detailed information requirements increase with each successive stage, adequate information must be available by the feasibility stage to develop realistic cost estimates and properly select sources for possible use. In some instances, required data may be obtained earlier than needed and should not be withheld but should be submitted when available. For feasibility studies, the designers should have sufficient information to supplement laboratory test data to determine whether the Government should acquire the source, whether the rock should be furnished by the contractor, or whether other types of embankment protection should be considered. A suggested outline for riprap reports for rock obtained from quarry is as follows:

    a.   Ownership
    b.   Location, indicated by map, with reference to section, township, and range
    c.   General description
    d.   Geologic type and classification
    e.   Joint spacing and fracture systems

   f.   Bedding and planes of stratification

   g. Manner and sizes in which rock may break on blasting as affected by jointing, bedding, or internal stresses

   h. Shape and angularity of rock fragments

   i.   Hardness and density of rock

   j.   Degree of weathering

   k. Any abnormal properties or conditions not covered above

   l.   Thickness, extent, estimated volume, and average depth of deposit—type, extent, and thickness of overburden

   m. Accessibility (roads affording access to highways or railroad, giving distance, load limitations, required maintenance, whether privately owned, and other pertinent information)

   n. Photographs and any other information which may be useful or necessary.

If commercial quarry deposits are considered, the following information should be obtained and included in the report:

   a. Name and address of plant operator—if quarry is not in operation, a statement relative to ownership or control

   b. Location of plant and quarry

   c. Age of plant (if inactive, approximate date when operations ceased)

   d. Transportation facilities and difficulties

   e. Deposit extent, plant, and stockpile capacity

   f.   Plant description (type and condition of equipment for excavation, transporting, crushing, classifying, loading, and restrictions, if any)

   g. Approximate percentages of various sizes of material produced by the plant

   h. Location of scales for weighing shipments

   i.   Approximate prices of materials at the plant

   j.   Principal users of plant output

   k. Service history of material produced

   l.   Any other pertinent information.

When rock deposits other than quarries are considered for riprap use, the rock properties and deposit should be described in the same manner as for quarry rock where applicable and, in addition, the deposit description should indicate shape, average size, and variation in sizes of the rock.

CONVERSION FACTORS

Some conversion factors commonly used in earth construction

To reduce units in column 1 to units in column 4, multiply column 1 by column 2
To reduce units in column 4 to units in column 1, multiply column 4 by column 3

Column 1	Column 2	Column 3	Column 4
LENGTH			
Inches_____	2.54	0.3937	Centimeters.
	0.0254	39.37	Meters.
Feet_____	0.3046	3.2808	Meters.
Miles_____	1.609	0.621	Kilometers.
AREA			
Square inches_____	6.4516	0.1550	Square centimeters.
Square meters_____	10.764	.0929	Square feet.
Square miles_____	27.8784×10^6	0.3587×10^{-7}	Square feet.
	640.0	$.15625 \times 10^{-2}$	Acres (1 section).
	30.976×10^5	$.3228 \times 10^{-6}$	Square yards.
	2.59	.386	Square kilometers.
	259.0	$.386 \times 10^{-2}$	Hectares.
Acres_____	43,560.0	0.22957×10^{-4}	Square feet.
	4,046.9	$.2471 \times 10^{-3}$	Square meters.
	4,840.0	$.2066 \times 10^{-3}$	Square yards.
	0.4047	2.471	Hectares.

CONVERSION FACTORS—Continued

Column 1	Column 2	Column 3	Column 4
VOLUME			
Cubic inches_____	16.387	6.102×10^{-2}	Cubic centimeters.
Cubic feet_____	1,728.0 7.4805 6.2321	0.5787×10^{-3} .13368 .16046	Cubic inches. Gallons. Imperial gallons.
Cubic meters_____	35.3145 1.3079	0.028317 .76456	Cubic feet. Cubic yards.
Gallons_____	231.0 3.7854	0.4329×10^{-2} .26417	Cubic inches. Liters.
Million gallons_____	133,681.0 3.0689	0.74805×10^{-5} .32585	Cubic feet. Acre-feet.
Imperial gallons____	1.2003	0.83311	Gallons.
Acre-inches_____	3,630.0	$.27548 \times 10^{-3}$	Cubic feet.
Acre-feet_____	1,233.5 43,560.0	0.81071×10^{-3} $.22957 \times 10^{-4}$	Cubic meters. Cubic feet.
Inches on 1 square mile.	232.32×10^4 53.33	0.43044×10^{-6} .01875	Cubic feet. Acre-feet.
Feet on 1 square mile.	278.784×10^5 640.0	0.3587×10^{-7} $.15625 \times 10^{-2}$	Cubic feet. Acre-feet.
VELOCITY AND GRADE			
Miles per hour_____	1.4667	0.68182	Feet/second.
Meters per second__	3.2808 2.2369	.3048 .44704	Feet/second. Miles/hour.
Fall in feet per mile_	189.39×10^{-6}	5.28×10^3	Fall/foot.

CONVERSION FACTORS—Continued

Column 1	Column 2	Column 3	Column 4
FLOW			
	60.0	0.016667	Cubic feet/minute.
	86,400.0	$.11574 \times 10^{-4}$	Cubic feet/day.
	31.536×10^6	$.31709 \times 10^{-7}$	Cubic feet/year.
	448.83	$.2228 \times 10^{-2}$	Gallons/minute.
	646,317.0	$.15472 \times 10^{-5}$	Gallons/day.
	1.98347	.50417	Acre-feet/day.
Cubic feet/second	723.98	$.13813 \times 10^{-2}$	Acre-feet/365 days.
(second-feet)	725.78	$.13778 \times 10^{-2}$	Acre-feet/366 days.
(sec.-ft.).	55.54	.018005	Acre-feet/28 days.
	57.52	.017385	Acre-feet/29 days.
	59.50	.016806	Acre-feet/30 days.
	61.49	.016262	Acre-feet/31days.
	0.028317	35.31	Cubic meters/second.
	1.699	0.5886	Cubic meters/minute.
	0.99173	1.0083	Acre-inches/hour.
Cubic feet/minute__	7.4805	0.13368	Gallons/minute.
	10,772.0	$.92834 \times 10^{-4}$	Gallons/day.
	1.5472	0.64632	Cubic feet/second.
10^6 gallons/day____	694.44	$.1440 \times 10^{-2}$	Gallons/minute.
	3.0689	.32585	Acre-feet/day.
Acre-feet/day_____	226.24	0.442×10^{-2}	Gallons/minute.
PERMEABILITY			
k, coefficient of per-	2.74×10^{-3}	3.65×10^2	k, feet/day.
meability (feet/	9.67×10^{-7}	1.035×10^6	k, centimeters/second.
year).	1.37×10^{-3}	7.30×10^2	k, inches/hour.

CONVERSION FACTORS—Continued

Column 1	Column 2	Column 3	Column 4
PRESSURE			
Feet water at maximum density.	62.425	0.01602	Pounds/square foot.
	0.4335	2.3087	Pounds/square inch.
	.0295	33.93	Atmospheres.
	.8826	1.133	Inches mercury at 30° F.
	773.3	0.1293×10^{-2}	Feet air at 32° F. and atmospheric pressure.
Feet average sea water.	1.026	0.9746	Feet pure water.
Atmospheres, sea level, 32° F.	14.697	.06804	Pounds/square inch.
Millibars_____	295.299×10^{-4}	33.863	Inches mercury.
	75.008×10^{-2}	1.3331	Millimeters mercury.
Atmospheres_____	29.92	33.48×10^{-3}	Inches mercury.
WEIGHT			
Parts per million___	0.00136	735.29	Tons/acre-foot.
	.0584	17.123	Grains/gallon.
	8.345	0.1198	Pounds/10^6 gallons.
Pounds_____	453.59	2.2046×10^{-3}	Grams.
Kilograms_____	2.2046	.45359	Pounds.
Pounds water at 39.1° F.	27.6812	0.03612	Cubic inches.
	0.11983	8.345	Gallons.
	.09983	10.016	Imperial gallons.
	.453617	2.204	Liters.
	.01602	62.425	Cubic feet pure water.
	.01560	64.048	Cubic feet sea water.
Pounds water at 62° F.	0.01604	62.355	Cubic feet pure water.
	.01563	63.976	Cubic feet sea water.
TEMPERATURE			

$$^\circ C. = \frac{5}{9}(^\circ F. - 32^\circ) \qquad ^\circ F. = \frac{9}{5}\,^\circ C. + 32^\circ$$

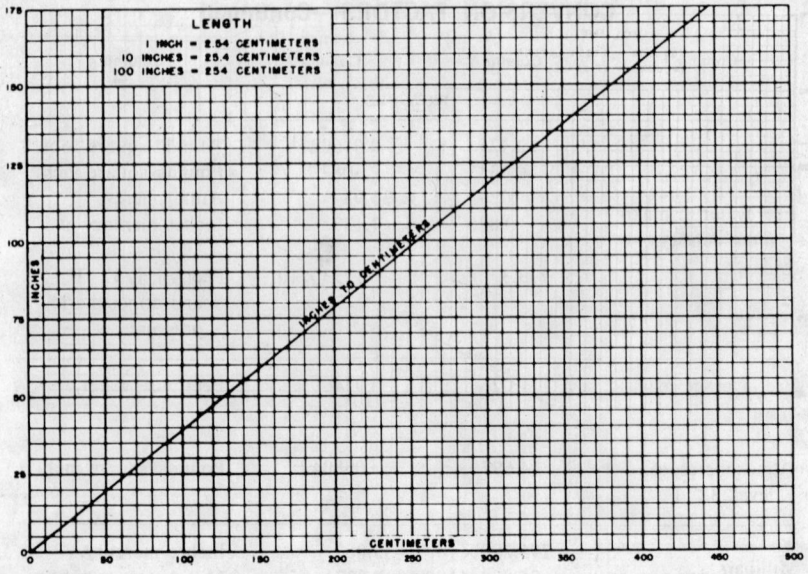

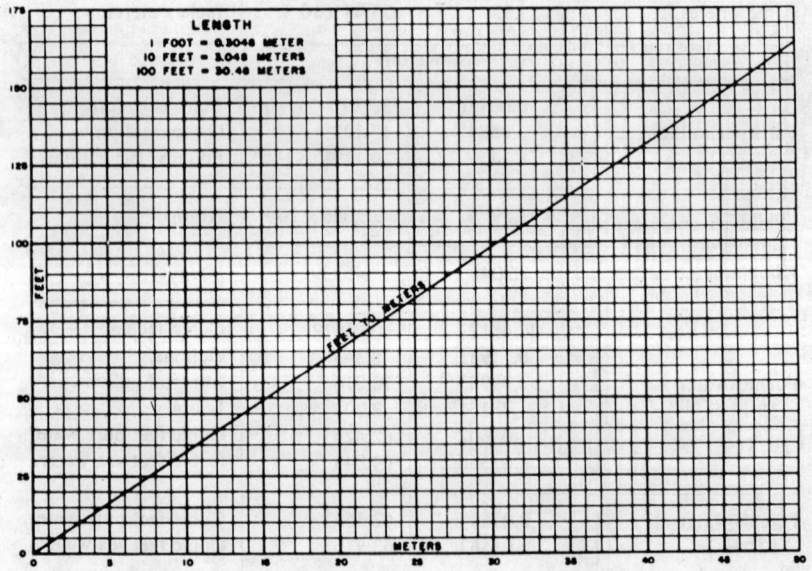

Conversion curves to convert inches to centimeters and feet to meters.

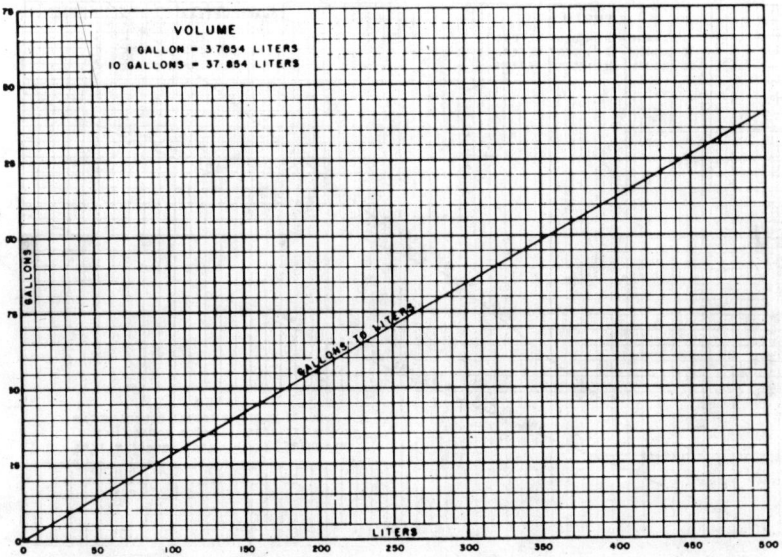

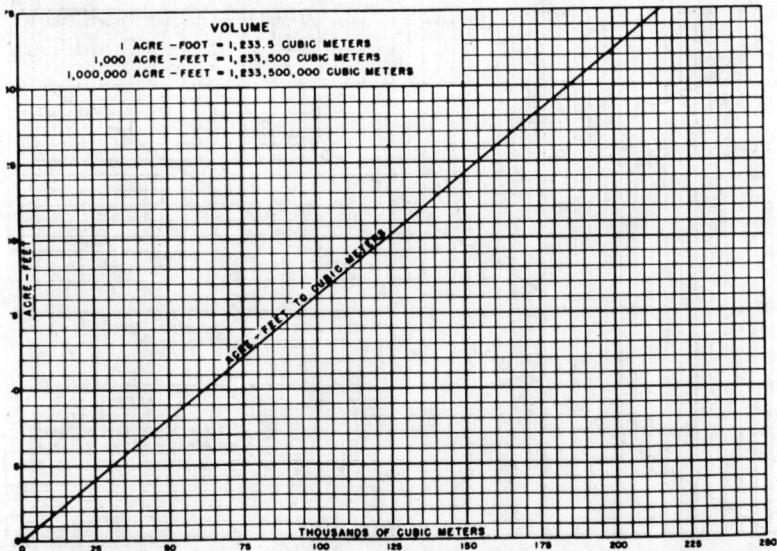

Conversion curves to convert gallons to liters and acre-feet to cubic meters.

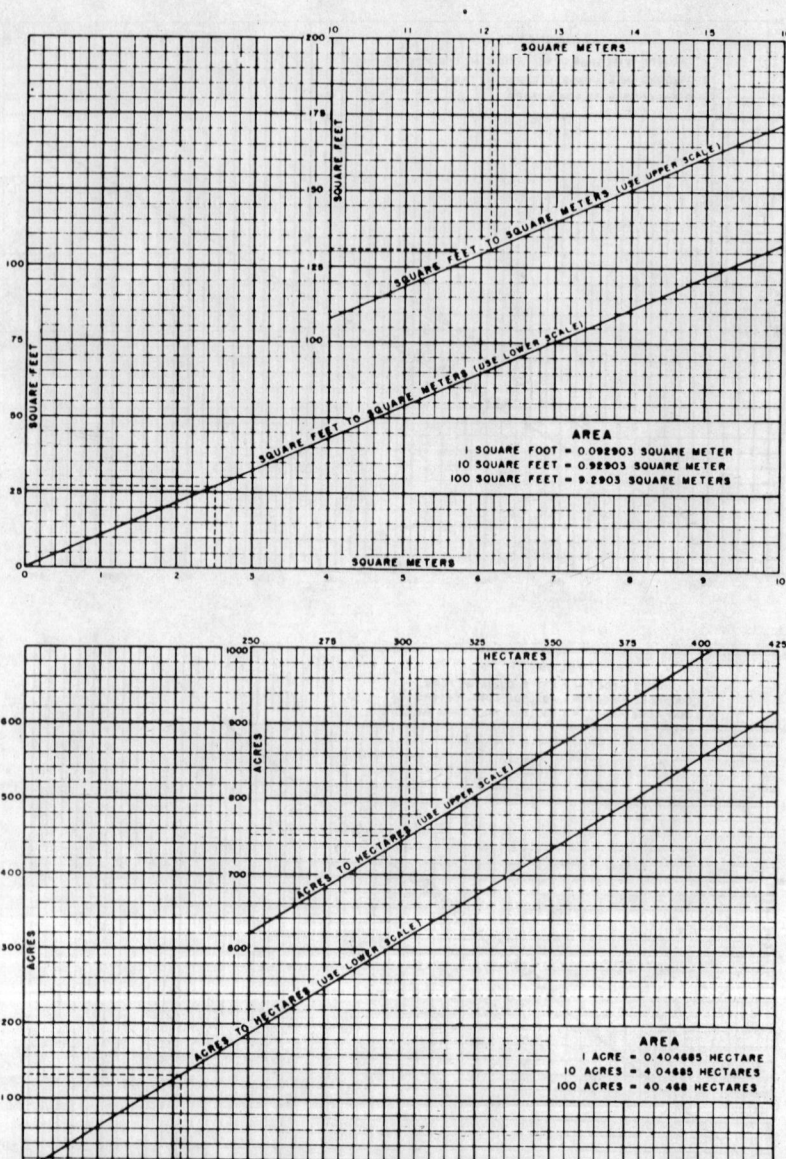

Conversion curves to convert square feet to square meters and acres to hectares.

(B) SOIL SPECIMEN IN RUBBER MEMBRANE

(A) SOIL SPECIMEN WITH END PLATES

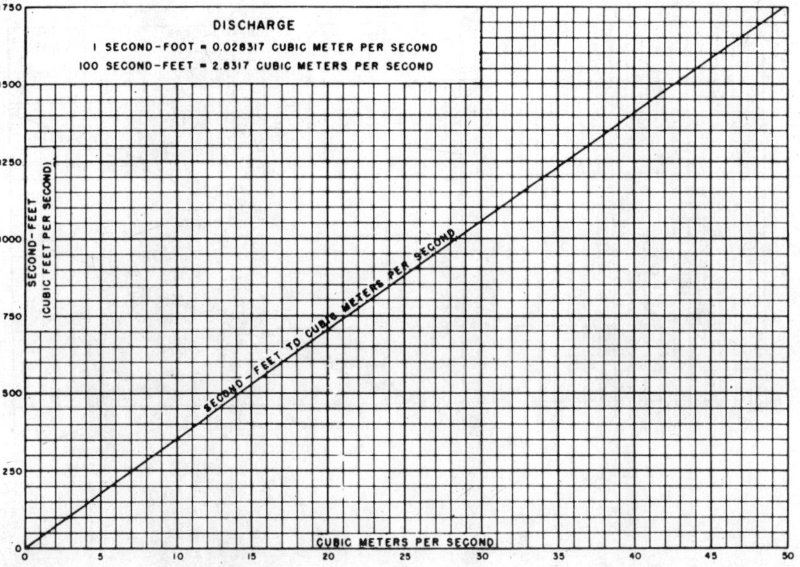

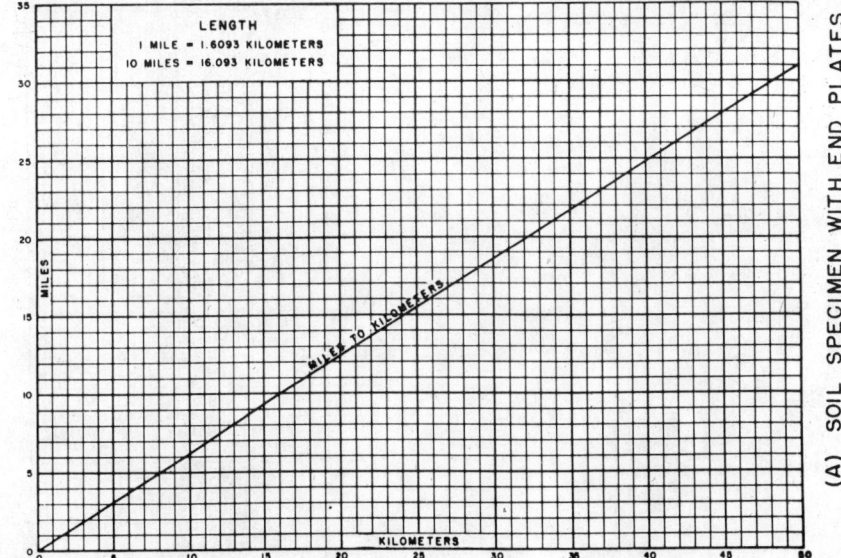

Conversion curves to convert second-feet to cubic meters per second and miles to kilometers.

INDEX*

"A" line, 16, 17, 27, 28, 156
Absorbed moisture, ω_a, 33
Access roads (see also Highways and railroads)
  backfilling operations over culvert, 295
  construction of, 289
Additives (Admixtures), 59, 112, 119, 120, 176, 183, 262, 266, 309, 310, 326
Admixture-stabilized fill, 183, 266
Aerial photographs, 75, 80, 84–87
  coverage of, United States, 85
  interpretation of, 85–87
  mosaics, 85
  orthophotographs, 85
  photo index maps, 85
  types of, 84
  use of, 84–87
Air content, 50
Air void ratio, e_a, 23
Airport construction, 180
Algae growth, 113
Alkali soils, 87
Alluvial cones, 88
Alluvial fans, 88
Alternate bids, 111
American Association of State Highway Officials' Manual (ref.) 288
American Railway Engineering Association Manual, (ref.) 289
Appurtenant structures, 106
  subsurface exploration for (see Subsurface exploration, damsites)
Artesian pressure, 213
ASCE, ix
Asphalt, 183, 201, 202, 262, 299
Asphaltic concrete, 322–326 (see also Stabilized soils)
  construction control, 323–326
  materials for, 324
  testing, 323
  uses for, 322
ASTM,
  Designation D–653–58T, (ref.) viii
  Designation D–1074, (ref.) 324

Designation D–1075, (ref.) 325
Designation D–2172, (ref.) 325
Glossary of Terms and Definitions in Soil Mechanics, (ref.) ix
Atterberg, A., 25
Atterberg limits, 8, 10
Atterberg limits tests, 16, 17, 25, 26, 211 (see also Soil consistency tests)
Auger borings (holes), 17, 129–133, 231 (see also Test holes)
Automatic data processing, 77

Backfill, 176, 190–191, 225, 283, 286, 304–305
  description of, 176, 190
  types of, 191
Backfill about structures, 304–305
  required density of, 304–305
Backfill in pipe trenches, 283–288
  materials for, 283–286
Backhoe, 124, 126
Backhoe trench, 126
Ballasting, 186
Base courses (see Highways and railroads)
Bearing capacity, 191, 195–196
  foundations, evaluation of, 195–196
Bearing strength (see Foundations)
Bedrock, 71, 74, 75, 99, 153, 157, 194, 202, 203, 213, 215
  contours, 71
  structures, 71
Benchmarks, 76
Benefit-cost ratios, 69
Bentonite, 123, 262, 266
Berm, 261, 268, 293
Blading, 182
Blanket, 60, 181, 184–190, 198, 199, 203, 212, 215, 269
  compaction requirements, 186
  description, 176
  impervious (see Impervious blanket)
  materials, earth, for dams, 113 (see also Impervious soil (materials))
  materials selection, importance of, 186
  protective, 269–274

---

*Asterisks denote page references to appropriate test designations in the appendix.

791

Blankets of rock fragments (see Rockfill and Rockfill blankets)
Blankets under riprap, 113, 114–115, 184, 187, 188
Blasted rock, 186
Blasting, 137, 220, 240, 267, 291
Blended earthfill, 176, 181–183, 292
control of, 181–183, 194
Blended materials production, 181–183, 268, 278–279
Block sample, 110 (see also Samples of soil, undisturbed)
Boils, 217
Bonny Dam, 218, 236, 242
Borderline classifications, 15–17, 155 (see also Soil classification)
symbols for, 15–17
Bore holes (see Test holes)
Borings (see Test holes)
Borrow areas, 66, 67, 106–108, 120, 142, 143, 178, 190, 223, 237, 278, 279, 310
downstream from dam, 91
investigation of, 66, 67, 100
grid system, 101, 108
location map and section, example of, 109
moisture control costs, 63
restoration of, 190
recommended cut slopes, 200
subsurface exploration for, 106–108
kinds of materials available, 108
specific kind of material, 108, 120
Borrow materials, 17–18
Borrow pit, 62, 63, 191, 237, 286, 293
cut, depth of, 194
evaluation of, 62, 63
excavation, 181–182, 191, 192, 194, 229–230
operations, to obtain proper water content, 223, 231
Boulders, 2, 89, 92, 124, 129, 137, 154, 186, 267, 291
size of, 2
Bridge piers, 101
subsurface exploration for (see Subsurface exploration, point structures)
Bridges (see Miscellaneous structures)
Brooming (brushing), 312, 316
Bucket augers, 129
Buildings, 74, 100, 101
subsurface exploration for (see Subsurface exploration, point structures)
Bulldozer pits, 124, 126 (see also Test pits)

Bulldozer trench, 126
Bulls liver, 13 (see also Quick silts and Silt)
Bureau of Reclamation, vii, 1, 71, 74, 80, 110, 115, 166, 211, 260, 325
Buried membranes, 112

Caissons, 201, 205, 300–303
Canal banks, 207
Canal construction, 112–113, 176, 177, 178
control techniques, 274–282
dumped fill, 176
embankments, 178, 179, 207
equipment-compacted embankment, 178–179 (see also Equipment-compacted embankment)
selected impervious fill, 177
Canal embankments, 267–269, 294
compacted, 261, 270–271, 275
moisture requirements for, 278, 279, 280
earth, construction control, 274–282
foundation inspection for, 274–278
foundation treatment for, 266–267
uncompacted, 260–261, 270–271
Canals, 67, 68, 75, 100, 101, 201, 202, 206, 260–282
blanket materials, subsurface exploration for, 106, 108
blankets, impervious (see Impervious blankets)
blankets to prevent erosion, 188, 266 (see also Sand and gravel materials)
borrow areas, 269, 278 (see also Borrow areas)
subsurface explorations for, 108
construction control techniques, 274–282
design features, 260–266
embankments (see Canal embankments)
excavation, classification for payment, 267
filter (layer), 266, 305–309 (see also Filters)
gravel fill, 269–272
gravel subbase, 269–274
investigation, use of permeability test, 60, 578*
linings (see also Earth-lined canals; Canals, lined; and Linings)
use for, 112
materials and placing requirements for various fills, 270–271

---

*Asterisks denote page references to appropriate test designations in the appendix.

Canals—Continued
  monthly construction progress report
      (L-29), 275, 282
  operation and maintenance roads for,
      261
  pervious foundation, lining for, 205
  riprap for, 119, 269, 272
  rock foundation, inspection of, 274–
      275
  seepage losses, 143, 262
  specifications, provisions for, 266–274
  structure backfill, 270–271
  subsurface exploration for (*see* Sub-
      surface exploration, line struc-
      tures)
Canals, lined, 261, 262, 263, 322, 323
  asphaltic concrete, 322, 323
  buried asphalt membrane, 262, 266
  buried bentonite membrane, 262
  concrete, 202, 276
  description of, 262
  earth (*see* Earth-lined canals)
  flexible, 201
  hard surfaced, 112, 262
  materials used, 262–263
  membrane, 112, 113, 262, 266, 275
  protection from uplift, 201 (*see also*
      Uplift)
Canals, unlined, 261
  excavated in rock, 267, 274
Capillarity, 63
Capillary stresses, 48
Carbide bomb moisture meter, 315, 319
Casagrande, Arthur, vii, 1, 387
Casing steel (pipe), 122, 130
Cement, 120, 183, 288, 310, 315, 318,
    319, 322
  grout, 123, 204
  hydration of, 318
Chalk, 63
Changed condition, 166
Chemical titration, 318
Chutes, 201
Clamshell, 124
Classification chart, facing 12, 18,
    facing 390 (*see also* Soil classifi-
    cation)
Classification group, 2 (*see also* Soil
    groups)
Classification system (*see* Soil classifi-
    cation *and* Soil classification sys-
    tems)
Clay (clay soil), 3, 13, 18, 25, 50, 51,
    53, 55, 202, 206, 282, 288, 298
  characteristics of, 11–13
  consistency, field penetration relation-
      ship, 298
  contraction of, 13

expansive, 13, 50, 55, 56, 61, 75,
    211, 212, 275, 282, 294, 298,
    299
  criteria for identification of, 211,
      212
  failures, 58
  load curves, 56
  methods for reducing movement of
      structures due to, 201
  moisture and density effect on, 57
fat, H, 13, 15, 298
foundations of, 211
laterite, 61
lean, L, 13, 15
sensitive, 61, 211
Clay balls, 120, 315, 318
  definition of, 120, 315
Clay shales, 53, 187
Coarse-grained soils, 3, 10, 11, 14, 15,
    16, 25, 28, 37, 40, 42, 288
Coarse grains, 2, 3
Cobble blanket, 186
Cobbles, 2, 129, 137, 154, 177, 224,
    238, 267, 291
Coefficient of curvature, C_c, 4, 16, 24
Coefficient of permeability, 43, 55–61,
    115, 236
Coefficient of uniformity, C_u, 3, 16, 24,
    306
Cofferdams, 74, 212
Cohesion, c, 44, 48
Cohesive, clay, 13
Cohesive strength, 25
Colloids, 211, 212
Compacted backfill for bedding in pipe
    trenches, 286–288, 319
Compacted backfill in pipe trenches,
    286–288
Compacted backfill of clayey and silty
    soils, 190, 191, 294, 299, 304
Compacted backfill of cohesionless free-
    draining soils, 190, 191, 286, 304,
    305
  construction methods, 191, 286
  materials, selection of, 191, 286
Compacted earthfill for rolled earth
    dams, 221–234
  control techniques for, 225–234
  design considerations, 221–223
  impervious earthfill zones, design cri-
      teria for, 221–223
  placement water content, specifica-
      tions provisions for control of,
      225
  specifications provisions for earth
      dams, 223–225
  suitable excavated materials from
      foundation, use of, 223

Compacted embankments (*see* Canal
embankments *and* Compacted
earthfill)
Compacted layer, 178, 231–232
Compacted pervious fill for rolled earth
dams, 234–238
control criteria for construction of,
234
control techniques, 237–238
design considerations, 234, 236
specifications provisions for, 236–237
Compacted soil-cement, 310–318 (*see
also* Soil-cement *and* Stabilized
soils)
clay balls, 315, 318
compaction requirements, 312
control techniques, 315–316
control tests (testing), 315, 316–318
construction provisions, 310–312
curing, 312
description of, 310
design considerations, 310–312
materials for, 310
record cores, 318
sampling, 315, 316–318
uses for, 310
Compaction curves, 51
Compaction equipment, 179, 184, 279,
316
Compaction of soil, viii, 36, 37, 51, 53,
176–181
curves, 35, 228, 229
field and laboratory, 228, 229
definition of, 50
degree, 81
degree required, 178, 232, 281, 292
effort, 32, 33
deviation from, 37
test (*see* Proctor compaction test)
Compaction test for soil containing
gravel, 760*
Compactive effort, 180, 238, 279
Bureau of Reclamation standard, 36–
37
Compressibility of soil, 16, 44, 49–55,
62, 75, 92 (*see also* Soil, compressibility)
fine sand, 55
Compression, 48, 50
compacted embankments, characteristics, 52
control of, 51–53
definition of, 50
index, C_c, 54

*Asterisks denote page references to appro-
priate test designations in the appendix.

Compression specimens for plastic soil-
cement, 321
Compression specimens for soil-cement,
318
Concrete, 112, 203, 262, 295, 299
Concrete aggregate, 89, 106, 114, 124,
162
sources of, 89, 91, 94
procedures for investigation of, (ref.)
114
Concrete dams, 74
Concrete grout cap, 215, 220
Concrete Manual, (ref.) 43, 157, 162,
171
Conduits (*see* Pipelines)
Consistency of fine-grained foundation
soils, 298
Consolidated fill, 184 (*see also* Compacted backfill of cohesionless free-
draining soils)
Consolidation, 50, 51, 89, 184, 191
definition of, 50
Consolidation tests, 195
Construction control, 32, 38, 43, 48,
165–176, 181–183, 315, 316, 323,
324–325, 326
changed condition, 166
importance of, 165–166
letter instructions for, 180
order for changes, 166
organization, functions of, 166–167
principles of, 165–176
relation to design requirements, 165
when specified, 178
Construction engineer, 66, 167, 196,
197, 202, 225, 229
Construction materials, locating sources
of, 87, 92
Construction stage, 66, 67
investigations, purpose of, 66, 67
small structures, investigative work
for, 67
Contract administration, 166–167
Contracting officer, 166, 237, 241, 266,
267, 292, 293, 295, 304, 305
Control testing (*see* Testing, construc-
tion control)
Core barrel, 136, 138, 139, 140, 141
Core barrel samplers, 123, 138
Core box, 139, 158
Core catcher, 157
restricted use of, 157
Core recovery, 136, 138
Corps of Engineers, Department of the
Army, vii, 1
Coulomb, 44
Coulomb's equation, 44, 46, 48
Court of law, 147

Cross sections, 71, 72, 100
Crushed rock blanket, 186–189 (*see also* Sand and gravel materials)
Crushed zones, 123
Crystal Dam, 289, 290
Cumulative grain-size curve, 3 (*see also* Gradation)
Curtis, H. G., ix
Cut slopes, 196–200
  customary slopes in earth, 197
  erosion protection, 199
  evaluation of during construction, 196
  ground-water effect on, 197, 198
  loessial soils, 197, 200
  sloughing, prevention of, 198
  when open to public recreation, 200
Cutoff trenches, earth, 176, 203, 213–214, 219, 220
  sloping-side trenches, 213
  stoped cutoffs, 213–214
  vertical-side trenches, 213–214
Cutoff walls, 213–214, 220, 299
Cuts, 75

D-ratio, 191, 260
Daehn, W. W., v, ix
Damsites
  subsurface exploration for, 106
  surface exploration for, 89, 91
Darcy's law, 56
Davis Dam, 199
Davis, F. J., iv, ix
Degree of compaction (*see* Compaction, degree required)
Degree of saturation, S, 23, 30
De Groot, G., v
Deltas, 207
Denison sampler (sampling), 140 (*see also* Sampling of soil)
  procedure, 341*
Density, 22, 23, 33–41
  compacted fill, 38, 40, 255
  compacted, gravel, effects on, 41
  dry, γ_d, 36–37
    theoretical, γ_t, 40
  fill, 255, 278
  inplace, 143–144, 207, 245, 298
  maximum dry (*see* Density, Proctor or laboratory)
  minimum acceptable for canals, 280–281
  natural or inplace, 36, 74, 100, 110, 111
  Proctor maximum dry (laboratory maximum dry), 36, 37, 38, 40,
51, 288 (*see also* Proctor maximum compacted density)
  reduction from frost action, 202
  relative, D_d (*see* Relative density)
  remolded natural, 36
  test (field) (natural) (inplace), 143–144, 195 (*see also* Field density test)
    procedure for, 613*
    procedure for dry, gravel-free soils, 610*
  wet, γ_{wet}, 36
Denver sampler, 140, 341*
Department of Agriculture, 80, 81
Design Considerations, Unit, 167, 225
Desiccation, 61
Diamond drilling, 136
  core sizes, 136
Diatomaceous silts, 13 (*see also* Silt)
Dilatancy test, 10, 11, 12, 13, 15
  procedure for, 12, 387*
Direct exploration methods, 73
Direct shear, 45–46
Double-tube sampler, 110, 123, 130,
  consolidated-quick test (Q_c), 45–46
  quick test (Q), 45
  slow test (S), 45
Disc augers, 129
Ditching machines, 126
  139, 140 (*see also* Denison, Denver, and Pitcher sampler)
Dragline, 124, 126
Dragline-placed fill, 177
Drainage wells, 220
Drains, 60, 101, 113, 114, 211, 212, 219, 275
  filter materials, subsurface exploration for, 106
Drifts, 127 (*see also* Tunnels)
Drill casing, standard dimensions, 139
Drill cores, standard dimensions, 139
Drill holes (*see* Test holes)
Drill rig (machine), 137
Drill rod, 136, 139, 140, 380*, 382*
  standard dimensions, 139, 380*, 382*
Driller's report, 70
Drilling machine, 137
Drilling mud, 122, 123, 130, 136, 138, 139, 341*
  approximate proportions of mixtures, 354*
  stabilizer compounds, 136
Drive samplers, 123, 141
Drive-tube boring, 140–142

---

*Asterisks denote page references to appropriate test designations in the appendix.

Dry density (dry unit weight) γ_d, 23, 30, 36, 255
Dry strength test, 10, 12, 13, 15
  procedure for, 12, 387*
Dry unit weight (*see* Density)
Dumped fill, 176-177
Dune sands, 96

Earth, 1, 262 (*see also* Soil)
Earthquake, 207
Earth dams (*see* Rolled earth dams)
Earth-lined canals, 189, 201, 261-262, 267-271, 274, 278-282 (*see also* Linings)
  blended linings, mixing two soils, 113, 278-279
  compacted, thick, 112-113, 263, 265
    construction of, 263, 267-269
    materials for, 191-192, 270-271
    typical sections, 264
  compacted, thin, 262
  construction control techniques, 278-282
  foundation treatment for, 265, 266-267, 275
  gravel cover for, 262. 265-266, 267
  materials, subsurface exploration for, 106, 108
  recommended thickness for frost prevention, 265
  refill above, material for, 270-271
  riprap or coarse gravel cover, 263, 265
  silting, description of, 262
  stabilized soils for, 266
  thin, loose earth blankets, 262, 266, 270-271
  types of, 262
Earth Manual, scope, etc., iii-ix
Ecological and environmental requirements, iv, 190
"Effective" size of soil particles, 24, 25
Elastic rebound, 4
Elastic silts, 13 (*see also* Silt)
Ellis, W., v
Embankment, 176-184, 191, 227
  construction, factors affecting cost of, 62-63
  construction, factors affecting placement of,
  description of, 176
  materials for, 191
  test section, 227
  types of, 176
Embedded instruments, 226, 233
Enclosed augers, 130
End dumping, 177, 292
Engineering and Research Center

approval required, 68, 110, 147, 168, 179, 196, 198, 211, 221, 231
assistance, 106, 111, 143, 147, 175, 183, 299
density and moisture limits specified, 225, 227
determination of cut slopes, 196
field review and inspection, 206, 221
information and advice provided, 143, 147, 167, 175, 179, 198, 205, 211, 225, 227, 275
laboratory studies or testing, 122, 159, 171, 246, 299, 318
publications, nomographs, and forms, 30, 176, 324
reports (reporting) required, 168, 171, 175, 176, 197, 234, 249, 251, 253, 282
request for field testing, 106, 111
review of specifications by construction engineer, 167
sample (sampling) requirements, 106, 110, 122, 158, 159, 196, 246, 299, 318
special instructions, 111, 122, 143, 184, 211, 326
Engineering and Research Center laboratory, viii, 171, 174, 299
  function of, 171
Engineering properties of soil, 10, 11, 18, 27, 33, 43-64, 66, 74, 108, 158, 159, 176, 178-179
  development of, 178
  improvement of, 176, 178, 179
Engineering use chart for soils, 20-21, 28, 62, 111, 189
Environmental Science Services Administration, 76
Eolian deposits, 94-96 (*see also* Loess)
Equipment-compacted embankment, 176, 178-179, 278, 292, 294, 304
Erosion protection, 120, 189, 199, 240, 266
  from flowing water, 199, 263, 310
  from wave action, 199, 310
  from wind and rain, 189
Erosional features, 76
Esmiol, E. E., ix
Excavation, 62, 63, 181, 191-195, 283, 288
  factors affecting unit cost of, 62-63
  requirements for, 194
Expansion, definition of, 50
Expansive clays, 55, 56, 57, 58, 211, 212, 282 (*see also* Clay (clay soil))
Expansive soils, 55, 56, 57, 58, 201, 202, 208 (*see also* Clay (clay soil))
Exploratory holes (*see* Test holes)

Exploratory methods, 122–144
  accessible, 123
  purpose of, 123
Exploratory procedures, 70
Explosives, 124, 129, 217
Extra work orders, 174

Failures, 51, 55, 61, 197, 207
Feasibility stage of investigation, 66, 68–70
  data for, 66
  importance of water loss investigation, 68–69
  land forms, knowledge, importance of, 87, 89
  objectives, 68–69
  organization of, 69–70
  purpose of, 68–69
  report, preparation of, (ref.) 70
  request for waiving, 68
  test hole requirements for, 101, 108
  unfavorable geologic conditions, 68
  variations by feature, 68
Feldspar, 4
Field classification of soils, 14–17, 387*
Field control laboratories, 169, 171–174
  concrete laboratory facilities, (ref.) 171
  equipment for, 174, 408*
  mobile type, 171, 173
  purpose of, 171
  reports, 171, 174–176
  satellite, 171
  size and type of, 171
  test data, use of, 171
Field density test, 30, 33, 110, 111, 143–144, 227, 232–234, 245, 247, 255, 275, 278, 281, 286, 298, 318
  frequency of, for canal construction, 281
  frequency of, for compacted earthfill, 233–234
  frequency of, for highway and railroad fill construction, 294
  location for compacted earthfill, 233, 234
  location for compacted soil-cement, 318
  minimum number required, 281
  procedure for, 613*
Field identification of soil (see Identification of soils)
Field investigations for feasibility stage, 66
  by contract, 70
  cost of, 68, 70
  initial planning, 65
  performance of, 70

  program of work, 70
  responsibility for authorizing and reviewing, (ref.) 70
Field investigations, recording, and reporting data, 144–164
  maps, 144–146
  logging of exploration holes, 146–157
    (see also Logging (logs) of exploration holes)
Field penetration test, 111, 136, 195, 274, 296, 297, 302–303
  consistency relationship for clays, 298
  driving weights required, 136
  precautions in evaluating, 298
  relative density relationship for sands, 297
Field penetration test with split-tube or split-liner sampler, 22, 42, 53, 74, 129, 141, 296
  procedure for, 42, 603*
  use, 42 (see also Field penetration test)
Field permeability test, 274, 279 (see also Permeability)
Field tests, 174–175, 318, 319, 321–322
  compacted earthfill for rolled earth dams, summary of, 233–234
  compacted pervious earthfill for rolled earth dams, summary of, 239
  plastic soil-cement, 319, 321–322
  reports of, 174–175
  soil-cement, 318
Field tests for exploring foundations, 74, 111, 143–144
  inplace density, 111, 143–144 (see also Density test)
  penetration, 143 (see also Field penetration test with split-tube sampler)
  permeability, 143 (see also Permeability field test)
  vane, 143 (see also Inplace vane shear test)
Fill construction, types of, 176
Fills, 75
Filters, 24, 60, 113, 114, 215, 234, 305–309
  blankets, 186, 203, 265, 267, 275
  criteria, 188
  two-layer, 188
Filters for canals and miscellaneous structures, 305–309
  construction of, 309

---

*Asterisks denote page references to appropriate test designations in the appendix.

Filters for canals and miscellaneous structures—Continued
criteria for selection and grading of material, 305–309
material requirements for, 306–307
requirements for graded filters, 307–309
Fine-grained soils, 3, 14, 15, 16, 17, 27, 28, 42, 51, 92, 280, 288
definition of, 14
source of, 91, 92, 94
Fine grains, 2, 3
Fines, 180, 191 (see also Fine grains)
clayey, GC or SC, 15, 16
influence of, 25
limitation in amount of, 180
silty, GM or SM, 15, 16
Fixed-piston type samplers, 123
Flake off, 61 (see also Shale, slaking)
Flaming Gorge Dam, 289, 295
Flap valves, 201
Flood-plain or valley fill deposits, 91–92
Fluvial soils, 88–92
principal types of, 88–92
Foundations and subgrades (evaluation and control of during construction), 191, 195–212, 216, 296–301
adequacy of, 195, 201
bearing capacity, 191, 195–196 (see also Bearing capacity)
excavation below ground water, methods of, 197–198, 212
grouting, 203, 215, 216
inadequate (poor), 205–212
judgment in evaluation of, 195
permeability, 195, 202–205 (see also Permeability)
piles and caissons, used for, 300–303
prewetting, 201
relative density criteria, 296–297
rock, 274
settlement and uplift, 201–202, 207 (see also Settlement, uplift)
solution channels, open holes, or cracks, treatment of, 204
stability, 196–200 (see also Stability)
surface deterioration on exposure to air, 202
tests for evaluation of, 195, 296–297
unwatering of, 212, 299
water movement, damage to, 202
Foundations for miscellaneous structures, 300–303
Foundations for rolled earth dams, treatment of, 212–221
abutment contacts, 214
construction control techniques, 220–221
design features, 212–218

seepage control, 212–215
stability control, 213, 215–218
diversion and care of stream, 218
earth surface, preparation of, 221
embankment, 219, 220, 221
excavation and stockpiling of suitable construction materials, 223
firm formation, preparation of, 220
shale surfaces, preparation of, 219
specifications provisions for, 218–220
stability methods for improving, 215–218
stabilizing fills, 218
toe drains, 215, 217–218
type of pipe for, 217
unwatering, 215, 217–218, 219, 224
Foundations, inadequate, treatment of, 205–212
clays, 211 (see also Clay (clay soils))
dry fine-grained, treatment, density criterion for, 210
loess, density criterion for, 209 (see also Loess)
loose sands or silts, treatment of, 207, 211
soft or saturated materials, 211–212
Foundations (investigations), 17, 18, 61, 62, 69, 72, 73, 100, 122, 133
appraisal of, 87
bearing strength, 111
compressibility, 51–53
cutoff, 91
description of, 100
evaluation of, 100
inplace strength of, 141
investigation, 100, 101, 133
overexcavation and refill, 295
poor, 92
questionable, 91
satisfactory, 94, 100
settlement, 74, 89
soils, description of, 18
stability of, 123
treatment, factors affecting cost of, 62, 63
uplift, 74
Fowler, H., v
Freezing index, 201
Frequency distribution, 260
Friant-Kern Canal, 58
Friction factor, tan ø, 44, 45, 46, 48
relative density relationship for coarse-grained soils, 47
Frost action, 11, 201, 202, 207, 265, 275
conditions for, 201, 202
definition of, 63
factors affecting, 63, 113

Frost heave, 11, 13, 50, 63, 64, 177
  (*see also* Uplift)
  structures affected, 63
Frozen embankment, 190, 223, 237
Frozen soil, 223, 237, 267, 291
Funnel test, 319

Geologic data, 71
Geologic formation, 77
Geologic (soil) profile, 102
Geologic section, 107, 154
Geologic structures, 71
Geological information, comprehensive
  list of sources, (ref.) 80
Geological investigation, 69
Geologist, 66, 72
Geophysical methods of subsurface ex-
  ploration, 142–143
Gibbs, H. J., iv, ix
Glacial deposits (glacial drift), 92–94
Glacial outwash plains, 94
Glacial till, 92–94
Glasco, R. E., v
Gradation, (Grain-size distribution),
  viii, 3–4, 22, 23–25, 113, 177, 178
  application of, 25
  description of, 23–24
  gap, 24
  in the field, 3
  laboratory test procedure, 424*
  processing to improve, 177
  skip, 24, 89
  10-percent-finer-than size, D_{10}, 4, 24
  30-percent-finer-than size, D_{30}, 4, 24
  60-percent-finer-than size, D_{60}, 4, 24
  upper size limits, 24
Gradation analyses tests, 16, 238
Granby Dam, 216
Gravel, 2, 10, 11, 113, 129, 177, 178,
  189, 265, 267, 292
  coarse, 2
  fine, 2
  surfacing, 273
  well-graded, 15
Gravel blanket, 184–189, 198–199, 262,
  265, 266, 267, 269
  placement, 186
  rock sizes for various thickness of,
    for canals, 272
Gravel soil, 14–17, 79
  clean, definition of, 16
  definition of, 14–15
  dirty, definition of, 16
  poorly graded, GP, definition of, 16
  uniform and skipgraded, 15
  well graded, GW, definition of, 16
Gravel surfacing for roads and parking
  areas, 273

Ground water, 73, 79, 91, 124, 129,
  197, 199, 212, 217
  artesian zones, 73
  control of, 219–220
  level, 70, 130, 140
  removal of, 197
  table, 73, 91, 124, 129, 198, 199
    normal, 73
    perched, 73
Grout (grouting), 213, 215, 310
  bituminous, 215
  chemical, 215
Grout curtain, 215, 216
Grout-stabilized soils, 310
Grouting test holes, 124, 140

Hand augers, 129
Hand tests for field identification of
  soils, 10, 12, 14, 15, 387*
Halloysite, 4, 10
Hardpan, 81
Hard-surfaced linings, 112, 266
Hatcher, R. C., v
Hazen, 24
Heart Butte Dam, 171
Heave, definition of, 50 (*see also* Clay,
  expansive *and* Frost heave)
Heaving, 74, 201, 217
Helical augers, 129, 130
High compressibility, soils of, H, 13
Highly organic (peaty) soils, Pt, 14
Highways and railroads, 68, 75, 101,
  106, 177, 178, 180, 206, 207, 286,
  288–294
  base course material, 186, 188
  borrow pits, specifications provisions
    for, 293
  control techniques, 293–294
  design features of, 288–289
  earthwork, specifications provisions
    for, 289–293
  embankment, construction control of,
    293–294
  embankments, construction specifica-
    tions provisions for, 291–293
  preparation of surfaces under road-
    way embankments, 290
  State highway specifications, 292, 325
  subgrade, preparation of, 273
  subsurface exploration for (*see* Sub-
    surface exploration, line struc-
    tures)
  surfacing, base course, ballast, 106,
    113, 115, 186, 188, 273
  subsurface explorations for, 106

*Asterisks denote page references to appro-
priate test designations in the appendix.

Hilf, J. W., v, vi
Holtz, W. G., v
Hotplate method, 315, 319
Hvorslev-type sampler, 369*
Hydraulic fill, 176, 183–185
  engineering assistance, 184
Hydraulic gradient, 56, 59, 60, 203
Hydrometer analysis, 3

Ice lenses, 202
Identification of soils, 1–22
  in the field, 14–17
  tests, 141
Impervious blankets, 112, 186, 188, 189, 203, 213, 241
Impervious linings, 189
Impervious, range of permeability, 59
Impervious soil (materials), 112–113, 184
  sources of, 92, 94
  use for, 92, 111, 112–113
Inadequate foundation conditions, 205–212 (see also Foundations)
  common methods of treatment, 205–212
Index properties of soils, 23–43, 66, 73
Index tests, 22, 66, 70, 108, 111, 159, 195
  use for, 22–23, 42, 111
Indirect methods of exploration, 73
  geological interpretation, 73
  geophysical, 73
  soundings, 73
Initial planning of a project, 65
Inplace density test (see Density and Field density test)
Inplace moisture (see Natural water content)
Inplace permeability tests (see Permeability, field test)
Inplace vane shear test, 45, 53, 74, 111, 195
  procedure for, 593*
Inspection, 166, 169–170, 179
  final, 221
  types of, 169
  report, 232
Inspection trenches, 269, 275, 279
Inspector, 169, 275, 279–280, 315, 318
  borrow pit, 229, 230
  chief, 229
  duties of, 169, 230, 274, 275, 278–279, 296, 299, 301, 304, 315, 318
  embankment, 229, 230
    compacted earthfill, duties of, 230–234, 274
    compacted pervious fill, duties of, 237–238

foundation, 220
  miscellaneous fills, 243
  riprap and rockfill, 240–241
  soil-cement, 315–318
Instrument installation for rolled earth dams, 243–246
  accelerographs, 243, 745*
  embankment piezometers, 650*
  foundation piezometers, 650*
  foundation settlement baseplates, 730*
  frequency of readings, 248
  horizontal movement devices, 719*
  instructions, 243, 245
    final report on earth dam instrumentation (L–16), 244, 246
    inspection, 244
    observations, 244–245
    record tests, 245–246
  porous-tube piezometers, 686*
  surface points for settlement and deflection, 733*, 738*, 741*
  transducers, 243
  vertical movement devices, 699*
Interior Board of Contract Appeals, 167
Internal friction, 25, 42, 44
Internal vibrators, 191, 280, 286, 305
Inverted filter, 217
Irwin, W. H., ix

Johnson, A. F., ix
Jones, C. W., v, ix
Justin, J. P., 166

Kirwin Dam, 64
Kisselman, H. E., v
Knodel, P. C., v

Laboratory (see Field control laboratories and Engineering and Research Center laboratory)
Laboratory classification of soils, 16, 17, 387*
Laboratory equipment
  field control, 408*
  other (see designation for particular test or procedure)
Laboratory technician, 229
Laboratory testing for construction control, 177, 315, 318, 319, 321–322
  when required, 177, 178
Lagging, 127
Lake sediments or lacustrine deposits, 92
Land forms or topographic features, 76, 87

---

*Asterisks denote page references to appropriate test designations in the appendix.

Landslides, 86, 101, 106, 197
Larson, F. H., ix
Laterals, 75, 260, 261, 282, 325
Layers, thickness of, 180, 181, 182, 278, 280, 286, 292
  adjusted for oversize, 180
Lean clay, CL, (see Clay)
Ledzian, R. R., v
Lifts, thickness of, 179, 286
  half-lift, 181
  loose, 278
Lime, 183
Limestone, 61, 89
Limiting moisture tests, 225
Limits of consistency, 8
Linings, 112, 113, 176, 181, 184–190, 191, 205, 322, 323 (see also Canals, lined and Earth-lined canals)
  description, 176
  hard surfaced, 112, 113
  materials selection, importance of, 186, 191
Liquefaction, 61, 207 (see also Quicksand)
Liquid limit, LL, 4, 13, 14, 26, 41, 208, 273
  definition of, 26
  determination of, 435*
Liquid state, 4
Load-consolidation testing, 110 (see also One-dimensional consolidation test)
Load-expansion characteristics, 53–55 (see also Clay (clay soil), expansive)
Loess, 59, 61, 96, 97, 98, 155, 200, 206, 209
  aerial view of, 97
  characteristics of, 96
  density and moisture versus settlement on saturation, criteria of, 209
  gradation and plasticity, data for, 156
Logging (logs) of exploration holes, 1, 2, 17, 18, 70, 72, 124, 126, 146–157, 245, 278
  computerized log, 147
  definition of, 147
  identification of, 146–147
    symbols, 147
  information required, 146–147, 154
  log forms, 147–155
    geologic log of drill hole, 72, 148, 152
    log of test pit or auger hole, 149–151, 153
    penetration resistance log, 152, 153

requirements, 18, 32
rock cores
  arrangement of, in core box, 139
  description of, 155–157
  soils, description of, 155
  when submitted, 72
Low compressibility, soils of, L, 13
Low-cost canal linings, 262

Machine-driven augers, 129
Major revisions to Earth Manual, discussion of, iii–iv
Manual tests, 15, 155, 179 (see also Hand tests)
Maps, 75
  administrative instructions, (ref.) 75
  agricultural soil maps, 80–83
    extent of, in United States, 80 (facing)
  geologic maps, 77–80
    areal, 77
    bedrock, 78
    example of, 107
    photogrammetric methods, use of, 77
    status of, in United States, 80 (facing)
    structural, 78
    surficial, 78, 106
  river surveys, 76
  sources of information, 75–87
  strip, 76
  techniques and standards, (ref.) 75
  topographic maps, 75–77
    coverage, 75–76
    index to, for States, (ref.) 76
    scales, 75–76
    status of, in United States, 76 (facing)
Mat footings, 102
Materials (soils) investigation, 72, 73–74
Materials of specific properties, 111–119, 288, 310, 319, 324
  exploration for, 111–119, 310
  sources, economics of, 111, 112
Mats, 51
Maximum dry density, 33
McClaren, C. E., ix
Mechanical compactor
  calibration procedure, 755*
  laboratory, soil, 468*, 469*
Mechanical tamping, 232
Mica, 4, 10
Miscellaneous fills in dam embankments, 241–243
  record tests for, 243
Miscellaneous structures, 206, 294–309

*Asterisks denote page references to appropriate test designations in the appendix.

Miscellaneous structures—Continued
    backfill for, 304–305
    filters for, 305–309
    foundation excavation, specifications
        provisions for, 294–300·
    foundations, checking properties of,
        296–300
    pile and caisson foundations for,
        300–303
Mixing, 176
Mixing machines, 182, 268
Modified soil, 108, 119–122, 176, 183,
    326 (see also Stabilized soils)
    construction control, 326
    inspection of, 183
    special sampling instructions, 122
Mohawk Canal, 58
Moisture content, 32–35, 230, 231, 255,
    315 (see also Water content)
Moisture control, 40, 41, 42, 51, 92,
    176, 178, 179, 225, 230, 231, 255,
    315, 319
    carbide bomb moisture meter, use of
        for soil-cement, 315, 319
    costs of, 63
    hotplate method, use of for soil-
        cement, 315, 319
    nuclear devices, use of, 230, 231
Monthly reports, 168
Montmorillonite, 51
Montmorillonoid, 53
Morainal deposits, 92–94
Mud, 61, 206
Mud flows, 207

Natural water content, 18
    dry, definition of, 18
    moist, definition of, 18
    moisture tests, 111
    saturated, definition of, 18
    wet, definition of, 18
Needle-density test, 42
Needle-moisture test, 610*
Niobrara chalk, Ft. Hays member, 64
Nomographs, availability of, 30
Nonhydraulic structures, 59
Nonplastic, definition of, 8
Nonplastic fines, 11 (see also Silt)
Nonsampling borings, 142
    jetting, 142
    percussion or churn drilling, 142
    wash boring, 142
Normal stress, σ, 44
Normally consolidated, definition of, 53
Nuclear devices for determining mois-
    ture content, 230, 231
Number of passes, 179, 225, 232, 279

One-dimensional consolidation test, 53
    procedure for, 509*

Open-end drive sampling, 110, 141
    procedure, 341*
Operation and maintenance stage, 66
    investigations during, 67
Optimum water content, w₀, 32, 33,
    62, 179, 223, 225, 278, 280, 305
    (see also Proctor optimum mois-
    ture)
    procedure for determination of, for
        compaction, 466*
Orders for changes, 166, 174, 180
Organic clay, OH, 14, 15
Organic material (soils), 3, 17, 61, 192,
    275
Organic matter, 13, 14, 32
Organic silt, OL, 14, 15
    identification of, 17
    ovendrying, effects of, 17
Osterberg-type sampler, 369*
Outcrops, 76
Outwash fans, 101
Ovendrying, 32, 318
    effects of, 32
    temperature, 32
Overburden, 71, 106, 116, 123, 153,
    215
Overburden fines, 180
Overburden load, 53
Overconsolidation, 53
Overexcavation, 288, 295
Overhaul, 268
Oversize, 25, 62, 179, 224
Oversize materials, 180, 223, 232, 233
    238, 310
    removal of, 280, 310
    special embedding of, 238, 240

Palisades Dam, 171
Parking areas, 273
Particle shape, 4, 8, 114
    bulky, 4
    diameter, D, 24
    elongated grains and fibers, 4, 10
    flaky, 4, 9
    rounded, subrounded, subangular,
        and angular, 8, 89, 92
Pavers, slip-form, 322, 324, 325
Pay items, 176, 178
Paylines, 283
Peat, 4, 206
Peat or peaty soils, 4, 14
    identification of, 14
Pedological system of soil classifica-
    tion, 80–83
    description of, 80–81
    divisions of, 81

Pedological system of soil classification
—Continued
example of, 83
limitations and value for engineering purposes, 83
soil series of, 81
soil type, 81
textural classification of "A" horizon, 81–82
triangle of basic classes, 82
Penetration needle, 179, 466*
Penetration resistance, 22, 40–42, 142, 298
field (see Field penetration test with split-tube or split-liner sampler)
needle curves, examples of, 34, 35
procedure for, 466*
Percolation, 56
control of, 59
Permeability, 16, 22, 25, 44, 53, 55–61, 60, 62, 70, 71, 75, 91, 180, 184, 195, 200–205, 211, 224, 236, 240, 241, 326
control of, 59–60
definition of, 55
effect of gravel, 60
field test for, 74, 111, 140, 143
in bore holes, 573*
shallow-well method, 747*
well permeameter method, 578*
horizontal, 59
in embankments, 59, 60
laboratory test for, 491*, 505*, 509*, 521*
ranges of, 59
ratio between soils for parts of a dam, 59
sand-gravel mixtures, 59, 60
tests, when required, 181
through foundations, 59–60
vertical, 59
Permeability and settlement test
large-scale, 182, 505*
standard test, 491*
Permeable soils, 179, 180, 236, 280, 281
compacted layer, thickness of, 237
compaction of, 179, 237–238
Personal diaries, 175
Pervious blankets under concrete structures, 188 (see also Sand and gravel, blankets)
Pervious materials, 89, 113–115, 184, 186
construction control of, 280–281
gradation, 269
sources of, 89, 91–92, 94
Pervious soils, range of permeability of, 59

Piers, 51, 201, 205
Piles, 51, 201, 205, 206, 300–303
bearing power, formula for, 300–301
driving tests, 111
forms for recording, 642*
estimating length of timber piles prior to driving, 302
forms for recording the penetration of, 301
jetting systems for, 301
loading tests, 74, 111, 301
procedure for, 642*
number of blows required for driving, 301
soil structure destroyed, due to driving, 11
specifications provision for, 301
Pipe bedding, 121, 286–288, 318–319
plastic soil-cement, 318–319
soil-cement, 286–288
Pipe trenches (see Pipelines)
Pipelines, 100, 101, 102, 121, 191, 206, 282–288, 294, 318–319
alternate types of pipe, 282
backfill bedding requirements, 283–288
backfill materials and placement requirements for, 270–271, 283–288
design features, 282–283
excavation for, 282, 283
plastic soil-cement bedding for, 318–319
safety requirements, 283
soil-cement bedding for, 286–288
construction control, 288
materials for, 288
unconfined compressive strength limits, 288
specifications provisions for, 283–288
trenches, 282–288
Piping, 59, 113, 177, 201, 202, 215, 236, 274, 280, 299
control of, 203–204
Piston-type (drive) sampling, 110, 123, 141
procedure, 241*
Pitcher sampler, 140, 341*
Pit-run material, 188, 265, 269
Plastic fines, 13 (see also Clay)
Plastic limit, PL, 8, 26, 27
definition of, 27
determination of, 441*
Plastic soil-cement, 121, 318–322
control testing, 319, 321
uses for, 318
Plastic state, 8

---

*Asterisks denote page references to appropriate test designations in the appendix.

Plasticity, 8, 10, 25
  chart, 12, 17, 27, 28
  index, *PI*, 8, 13, 14, 27
  needle (Proctor), 40, 466*
Plasticity index, 27, 202, 211, 212, 273, 286, 299
Plowing, 182
Pneumatic paving breakers, 124
Pneumatic rollers, 191, 280, 312, 316
Pneumatically-applied mortar, 112, 202, 262, 299
Poorly graded, definition of, 3
Pore fluids, 2, 48
  freezing of, 63
Pore pressure, 49, 51, 223, 253, 255
  buildup in cohesive soils, 253
  construction, 253, 255
Pore-air pressure, 46, 47, 48
Pore-water pressure, *u*, 46, 47, 48, 217, 225, 236, 255
Porosity, *n*, 23, 28–30
  definition of, 31
Post hole augers, 129
Powerplants (*see* Miscellaneous structures)
Preconsolidated soil, 53
Preconsolidation load, 54
Preirrigation, 207
Preliminary investigations, 178
  objectives of, 178
Pressure pipe, 282
Pressure relief wells, 60
Prewetting, 200
Proctor compaction test, 33, 179, 234, 286
  procedure for, 466*
Proctor maximum compacted density, 191, 207, 223, 255, 260, 280–281, 282, 286, 293, 304, 305
Proctor needle-moisture test, 42, 230
  procedure for, 610*
Proctor optimum moisture, 191, 230
  (*see* Optimum water content)
Proctor, R. R., 32, 40
Project development stages, 66
Project laboratories, viii
Project plan, 65
Puddled fill, 184
  inspection of, 184
Pugmill, 310, 315
Pump sumps, 212
Pumping plants, 101, 206 (*see also* Miscellaneous structures)
  subsurface exploration for (*see* Subsurface exploration, point structures)

Quarrying, 240
Quartz, 4
Quick silts, 11 (*see also* Silt)
Quicksand, 61, 207 (*see also* Liquefaction)

Railroads (*see* Highways, roads and railroads)
Rapid compaction control method, 42, 179, 230, 231, 232, 280, 282, 294, 318
  procedure for, 621*
Rebound, 50, 53
Reclamation Instructions, (ref.) 70, 72, 75, 160, 175
Recompression curve, 54
Reconnaissance design, 69
Reconnaissance mapping, 85
Reconnaissance stage of investigation, 66, 67, 84, 87, 99, 100
  conveyance systems, 67
  data secured, 66
  foundations, descriptions of, 67
  investigated areas, sizes and depths of, 67
  land forms, knowledge, importance of, 87–89
  objectives, 67
  purpose, 67, 69
  report, 70
  surface exploration, 67, 87–100
Record cores, for soil-cement, 318
Record sample, 245, 246
Record tests, 233, 234, 243, 316–318, 321
Refill, foundation, 295
Relative density, 40, 42, 46, 53, 55, 60, 142, 184, 189, 237, 238, 239, 286, 305 (*see also* Density, relative)
  chart, 38
  definition, 479*
  effects on friction factor, 47
  examples of, 39
  field penetration relationships for sand, 297
  maximum, 37, 38
  minimum, 37, 237
  test for, 479*
Relative density tests, 181, 184, 280, 281, 282, 286, 296, 298
  frequency for compacted pervious fill, 238
  materials applicable for, 280–281
  procedure for, 479*
Relative firmness, 42

---

*Asterisks denote page references to appropriate test designations in the appendix.

Relief wells, 204, 215, 217
Remolded soil, 54
Reports of field laboratories, annual survey of, 175
Reports of investigations, 159–162
  construction materials, 161–162
    forms for, 163, 164
  contents of, 160
  feasibility stage, 70
  final, 159–160
  foundations, 160–161
  general requirements for, (ref.) 160
  grouting test holes, 123
  progress, 159
  reconnaissance stage, 70, 159
  specifications stage, 72
Reports for construction control, 168, 174–176, 247–253, 322
  canals, 282
  earth dam instrumentation, 253
  final embankment construction, 250–253
  inspectors' daily, 175, 247
  monthly administrative, 175
  monthly construction progress, 322
  rolled earth dams, 247–253
  summary construction progress (L-29), 175, 249–250
  summary of earthwork construction data, 249
  transmittal of, 253–254
Required excavations, 74, 191, 192
  materials, sources of, 74
Reservoir floor, 91
Reservoir sites, 106
  subsurface exploration for, 106 (see also Subsurface exploration, damsites)
Reservoirs, 112, 113, 202
  protective covers, materials for, 112, 113, 120
  water loss from, 106
Residences (see Miscellaneous structures)
Residual soils, 61, 87, 96–100
  field test sections, 100
  special laboratory tests, 99, 100
Retaining walls, 102
Rigid structures, 207
Riprap, 30, 32, 106, 111, 115–119, 178, 184, 186, 187, 238–251, 265, 266–274, 322
  asphaltic-concrete, substitution for, 322
  blast test for, 116, 117
  competitive sources for, 118
  fragments, size of, 115

investigative procedures for, 119, 775*
laboratory tests for, 116, 119
other sources of, 116, 118, 119
payment for, 240
pervious blankets, materials for, 187–188
placing of, 187, 240–241, 296
quarrying and blasting for, 241
requirements for, 187
rock sizes for various thickness of, for canals, 272
satisfactory materials for, 116
specified sizes, 240
subsurface exploration for, 106
River basin report, 65
Roads, 177, 178 (see also Highways, roads and railroads)
Rock, 2, 111, 177, 186
Rock blanket, 186–189
Rock classification, 157
Rock fines, 114, 186
Rock foundation, 74
Rock (firm) foundation surfaces, preparation of, 220–221, 274
Rock fragments, 2, 123, 186, 224, 238, 291, 292
  grading for, 186
  requirements for, 186
Rock sources
  investigative procedures for, 119
  requirements for, 111, 115–116
Rockfill, 30, 32, 115–119, 178, 186, 238–241
  placing of, 240–241
Rockfill (rock) blankets, 119, 186–189, 198, 199
  construction of, 186–187
  materials for (see Rock fragments)
Rocks, 177, 186
Rocky Mountains, 92
Rolled earth dams, 66, 67, 84, 178, 179, 180, 212–260
  abutments for, 89
  compacted earthfill for (see Compacted earthfill for rolled earth dams)
  compacted pervious fill for (see Compacted pervious fill for rolled earth dams)
  concrete cutoff walls, 213, 214
  control criteria for embankment, 253, 260
  cutoff trench, 213
  downstream drainage blankets, 113, 115, 188

*Asterisks denote page references to appropriate test designations in the appendix.

Rolled earth dams—Continued
  downstream slopes, protection of, 187, 189, 190
  foundation, pervious, treatment for, 203, 204
  foundation treatment (see Foundations for rolled earth dams, treatment of)
  instrument installations (see Instrument installation for rolled earth dams)
  materials investigation for, 66, 67, 75, 193
    depth of, 66, 67
    grid system for, 101
  miscellaneous fills for, 241–243
  riprap for (see Riprap)
  rockfill for (see Rockfill)
  slurry trench, 213
  subsurface investigation for, 101 (see also Subsurface exploration, damsites)
  upstream slopes, protection of, 111, 115, 116, 119, 120, 121, 186, 187, 240, 310, 322
Rolled earthfill, 176, 178–179, 188, 189, 224
  compactive effort for, 179
  construction of, 178–180, 224–225
    inspection (see Inspection and Inspector)
  variables affecting, 228–229
  water content for, 179
Rolled embankment, 176
Roller, 179, 189, 191, 286, 312, 316 (see also Tamping rollers)
Roller data, reporting of, 250
Rotary drilling, 136, 140, 341*
  sampling procedure, 136–140
  speed of, 140

Safety (safety requirements), 282, 283
Safety and Health Regulations for Construction, (ref.) 169
Salts (soluble salts), 32, 43, 59, 61, 73, 183, 202
  determination of soluble, 448*
  effect on concrete and earth structures, 43
Sampling, during subsurface exploration, 158–159
Sampling for construction control, 170, 315, 316–318
Sampling (samples) of soil, 25, 70, 73–74, 122–144, 158–159, 315, 316–318
  cohesionless sands, 131
  contamination of, 122
  continuous, 141

disturbed, 73, 109, 110, 122, 129, 131, 141
  procedure for, 327*
  size of, 158, 159, 327*, 328, 330
  for laboratory testing, 120, 123, 141
  purpose of, 110, 158
  representative, 110, 159
  riprap, 159
  selection of, 108–110
  soil-cement, 315, 316–318
    compression test specimens, 318
  undisturbed, 36, 54, 74, 110, 129, 140, 144, 158, 195, 196, 245, 299
    importance of, 45
    procedure for, 341*
    requirements for, 158, 341*
    size of, 344*
    when used, 144
  visual examination of, 108, 122, 140, 141, 158
Sand, 2, 10, 11, 16, 60, 113 (see also Silt or sand)
  clean, 15
  coarse, 2
  dirty, 15
  fine, 2
  medium, 2
  poorly-graded clean, SP, 15, 24
    limits for, 24
  well-graded clean, SW, 15, 24
    limits for, 24
  with clayey fines, SC, 15
  with silty fines, SM, 15
Sand and gravel (mixtures) (soils), 180, 181, 186–189, 191, 202, 234, 265, 281, 282, 305
  blankets, 186–189, 201
  compaction of, 180, 189
  excavation from below water table, 181
  filter, 265, 267
  natural deposits of, 188, 193
  permeability, effect of gravel and density, 60
  under riprap, 186, 269
  uses for, 111
  water content for compaction of, 181
  well-graded, 269, 305
Sand dunes, 206
Sandbars, 206
Sandstone, 61
Saturation collapse, 51
Saturation curve, 35, 57
Scarifying or loosening, 50, 221, 232, 267

---

*Asterisks denote page references to appropriate test designations in the appendix.

Scraper-placed fill, 177
Sealing wax, 346
Sedimentary rocks, folding of, 88
  containing clay, 116
Sedimentation, 3 (*see also* Gradation)
Seeding, 189, 190
Seepage, 91, 197, 236, 260, 261, 266
  control of, 203-204, 266
  losses in canals, 260, 261
    control of, 261
  water, 177, 202
Selected fill, 176, 177-178
Selected impervious fill, 177
Selected material, 186, 205, 206
Selected sand and gravel fill, 177-178
Selective excavation, 177
Semipervious materials, sources of, 89
Semipervious soils, permeability range
  of, 59
Separation plant, 224
Settlement, 74, 195, 196, 201-202, 207-
  209, 210, 240, 255, 295, 299, 305
  differential, 255
Shale, 61, 63, 119, 202, 219, 299
  slaking, removal or prevention of,
    219, 299
Shear planes, localized, 255
Shear resistance, inplace, 143
Shear strength, 16, 22, 25, 37, 42, 44-
  49, 51, 74, 91, 92, 94, 195, 222,
  255
  apparent, 45, 48, 49, 51
  definition of, 44
  inplace, 45
    direct measurement of, 45 (*see
      also* Inplace vane shear test)
    indirect measurement of, 45 (*see
      also* Field penetration test)
Shear testing, 110 (*see also* Triaxial
  shear)
Shear tests, 195 (*see also* Triaxial
  shear)
Sheepsfoot roller (*see* Tamping rollers)
Sheet piling, 214
Shrinkage limit, *SL*, 26, 27, 212, 299
  definition of, 27
  determination of, 446*
Shrinkage of soil, 50, 51
  definition of, 50
Shrinking, 189, 211
Sierra Nevada Mountains, 92
Sieving, 2
Silt 3, 11, 15, 17 (*see also* Silt or
  sand)
  characteristics of, 11-13
  compressibility of, 13
    High, MH, 15
    Low, ML, 15
  elastic, 13
  micaceous, 13
  organic, 14
Silt or sand, 205, 206-211, 288
  consolidation of, 207
  dry, low-density, 207
  foundations of, 206-211, 275
Single-tube sampler (sampling), 138
Sinkholes, 89, 203
Sliding resistance, 48-49
Slope protection, 111, 115, 116, 119,
  120, 121, 186, 187, 240, 310, 322
  alternate methods, 111, 120
  downstream of dam embankments,
    115, 116, 119, 187, 189, 190
  downstream of stilling basins, 186
Slopes (*see* Cut slopes)
Slopewash deposits, 115
Sluiced fill, 183-185 (*see also* Hydraulic fill)
  construction of, 184
  inspection of, 184
Sluicing fines, 240, 241
Slurry trench, 213
Sod blankets, 119
Soil, 1
  blending (mixing) two soils, 62
  chemical composition of, 81
  classification, 1-22, 51, 73, 155
    advantages of, 1
    use of, 1
    visual and laboratory methods for,
      387*
  components, 2-4, 10
    properties of, 14-20
  compressibility, 1, 22, 25
  definition of, 1
  description of, 1, 17-19
  moisture, 4-10, 23, 32
  mortar, 40, 41
  physical properties of, 1, 14-20, 31
  upper size limits for various appli-
    cations, 25
Soil-cement, 111, 112, 116, 266, 286-
  288 (*see also* Stabilized soils *and*
  Modified soil)
  compacted (*see* Compacted soil-
    cement)
  materials for, 120
  plastic, 121 (*see also* Plastic soil-
    cement)
  testing required, 120
Soil-cement bedding for pipelines, 286-
  288
  construction control, 288
  materials for, 288

*Asterisks denote page references to appro-
priate test designations in the appendix.

Soil classification systems, 100
  pedological (*see* Pedological system of soil classification)
  textural (*see* Pedological system of soil classification)
  Unified (*see* Unified Soil Classification System)
Soil consistency, 25–28, 42, 298
  degree of, 28
  relative, C_r, 28
  states (stages) of, 4, 26
  tests, 22, 26
    procedure for, 435*
Soil group, 15, 155
  properties of, 18–19
  symbols, 12, 16–17
Soil mechanics, 45
Soil profile, 80, 81, 101, 103, 126
Soil properties, 31, 61, 66, 100
  changes in, 61–62
Soil stratum, description of, 73
Soil Survey Manual, U.S. Department of Agriculture, (ref.) 83
Soils engineer, 66, 72
Soils (materials) investigation, 72–73
  objectives, 73–74
  purpose of, 73
Solid state, 4, 8
Special compaction, 221, 225 (*see also* Backfill *and* Rolled earthfill)
Special processing, 178
Specific gravity, G_s, 23, 28–32
  absolute, 29, 30
  apparent, 29, 30
  bulk specific gravity, 30–31, 40
  determination of, 453*
  use of, 30
Specifications, 167–168, 190, 310, 319, 324, 325
  "as directed by the contracting officer" type of, 168
  definite requirement type of, 167–168
  performance type of, 167–168
  precedence, viii
  procedures type of, 167–168
  requirements, 169, 178–179
Specifications stage of investigation, 66, 71–72
  data for, 66
  exploration, modification of, 71
  exploration requirements, 72, 91
  field tests for, 111
  land forms, knowledge, importance of, 87
  materials data, when submitted, 72

programing the exploration, 72
  purpose, 71
  report, 72
  selection of samples, 108–110
  test hole requirements for, 101, 108
Spillway apron slabs, 202
Split-tube penetration sampler (*see* Field penetration test with split-tube sampler)
Spoil piles (*see* Talus and spoil piles)
Spread footings, 51, 102
Spreader, 312
Stability, 120, 177, 231, 240
  of compacted earthfill, 231
  of cut slopes, (*see* Cut slopes)
  of foundations, 196–200
Stabilization, 176, 183
Stabilized fill, 183
  assistance of an engineer, 183
Stabilized soils, 108, 112, 119–122, 262, 266, 309–326
  asphaltic-concrete, 322–326
  compacted soil-cement, 310–318
  description of, 119
  materials for, 119–122
  modified soil, 326
  plastic soil-cement, 318–322
  testing, 120
  uses of, 120, 309, 310
Stages of investigation, 65–72
Standard compaction curve (Proctor compaction), 35, 57
Standard test procedures, viii
  variations from, viii
State highway specifications, 292
Statistical analysis, 260
Stilling basins, riprap for, 118–119
Stockpiling, 190, 223, 236, 268, 315
Stratification of soils in place, 100
Stream deposits, 91
Strength of soil, 61 (*see also* Shear strength)
Stripping, 221, 267, 274, 275
Structure foundations, 74–75
  classes of, 74–75
  on rock, 74
Subgrades (*see* Foundations)
Subgrades, heaving of, 63
Sublaterals, 260
Subsidence, 75
Substations (*see* Miscellaneous structures)
Subsurface exploration, 67, 84, 100–111, 142, 143
  damsites, 106
  geologic features, 106
  geophysical techniques of, 143
  line structures, 101–102

*Asterisks denote page references to appropriate test designations in the appendix.

Subsurface exploration—Continued
    point structures, 101, 102
    purpose for, 100
Subsurface sections, 157–158
    reports, use in, 158
Superimposed load, 53
Surface exploration, 67, 87–100
Surface sodding, 199
Surface vibrators, 191, 280, 305
Surface water, 282
Swamp muck, 205, 206
    foundations of, 206
Swell, 50, 51
    definition of, 50
Swelling, 189, 211
Swelling soils, 51, 61
    embankments of, 51
Switchyards, 183, 188
Symbols, ix
Synthetic plastics, 262

Talus and spoil piles, 101, 115, 116,
    118, 205
    foundations of, 211
Tamping rollers, 191, 221, 225, 232,
    241, 250, 265, 279, 286, 304
    curves for, 228, 229
    form for recording data on, 247, 250
Terminal moraine, 94, 95
Terrace deposits, 91–92
Test holes, 70, 71, 72, 101, 104, 106,
    108, 110, 122, 123–125, 274, 296
    depths of, 70, 101, 104, 126
    designation of, 147
    exploratory, 104
    for foundation, 122
    for outlet works and spillway struc-
        tures, 106
    for tunnels, 106
    intervals of, 101
    location of, 70, 76, 101, 108
    log of (see Logging (logs) of ex-
        ploratory holes)
    number required, 101
    order of, 70
    protection of, 123
    sizes of, 122, 136
    support of, 122–123, 130–131
    use of, 122
Test pits, 17, 18, 123–125, 221, 234
    cribbing of, 124, 125
    depth of, 124
    hand-dug, 124
    log of (see Logging)
    photographs of, 234
    pumping system for, 124–125
    safety, 124
    unusual conditions, reporting of, 234

Test sections of earthfill material, 226–
    227
Test sections of residual soils, 100
Testing, construction control, 183, 225–
    227, 315, 316, 319, 323, 326
    for asphaltic concrete, 323
    for compliance with requirements,
        167
    for modified soil, 183, 326
    for plastic soil-cement, 319
    for soil-cement, 315, 316
Tests on soils, viii, 108
    purpose of, 108
Textural classification of soils (see
    Pedological system of soil classi-
    fication)
Toe support fill, 241
Topsoil, 205, 206, 219
    foundations of, 206
Topsoil blanket or zones, 189–190
    construction and inspection re-
        quirements, 190
    material sources, 190
Torrential outwash deposits, 88–89, 94
Total material, determination of con-
    sistency, 26
Toughness test, 10, 13, 15
    procedure for, 12, 387*
Tractor-compacted embankment, 176,
    179–181, 184, 305
    construction, inspection of, 180
Tractors, 191, 238, 280, 286, 312
    one pass of, 237
Transmission lines, 206
Transmission tower footings, 303–304
    criteria for the selection of tower
        footing types, 303
    specifications provisions for excava-
        tion and unwatering, 303–304
Transmission towers, 101
    subsurface exploration for (see Sub-
        surface exploration, point struc-
        tures)
Trench, 275, 291, 299
    pipe, 282–288 (see also Pipelines)
Trenches, 123, 125–127, 146, 154, 158
Trenton Dam, 209, 289, 290, 296
Triaxial shear, 42, 49
    test, example of, 49
    test procedure for, 545*
Trinity Dam, 171, 227, 228
Truck-dumped fill, 177
Tunnels, 67, 106, 123, 127–129, 158
    test holes for, 106
Turbidity, 113, 266

*Asterisks denote page references to appro-
priate test designations in the appendix.

Unconfined compressive strength, 22, 42, 288
determination of, 545*
Underdrains, 201
Underseepage, 201
Unified Soil Classification System, vii, 2, 3, 12–17, 18, 22, 81, 96, 155
(see also Soil classification)
development, vii
procedure for, 387*
Unit Design Considerations, 167, 225
Unit weight (see Density)
Unlined waterway, 205
Uplift, 60, 75, 201–202
expansive soils, cause of, 201–202
frost action, cause of, 201
hydrostatic, 201, 212, 275
pressure, 55
U.S. Agricultural Stabilization and Conservation Service, 89, 95, 97
U.S. Coast and Geodetic Survey, 76, 145
U.S. Department of the Navy, 132
U.S. Geological Survey (USGS), 75, 76, 79, 85, 145
bulletins of, (ref.) 79
professional papers of, (ref.) 79
quadrangle maps, (ref.) 79
U.S. Soil Conservation Service, 80 (facing), 81, 82

Vane feeder, 315
Vane test (see Inplace vane shear test)
Vibratory rollers, 191
Virgin compression curve, 53, 54
Visual classification of soil, 66, 73
(see also Soil classification)
Visual classification tests, 170
Visual examination for identification of soils, 14, 155 (see also Soil classification)
Visual inspection of construction, 178
Void ratio, e, 23, 28–31, 53, 54, 55
definition of, 29
densest state, e_{min}, 37
load curve, 54

loosest state, e_{max}, 37
volume, V_v, 30
Volcanic ash, 4

Wagner, A. A., ix
Walker, F. C., ix
Wall stabilizer, 123
Warehouses (see Miscellaneous structures)
Waste banks, 268
Water-cement ratio, 288, 319, 322
Water content (moisture content), w, 23, 25, 26, 27, 30–34, 50, 74, 142, 179, 183, 225, 230, 231, 255, 316, 319
descriptive terms, 32
determination of, 450*
gravel, w_g, 33
inplace, 74, 100
Water conveyance system, 205, 282
Water loss from reservoirs, 106
Water table, 59, 91, 122, 124, 141, 154, 157, 194, 213, 217, 282, 299
Water testing, in test holes, 71, 122, 123
Water void ratio, e_w, 23
Water volume, V_w, 30
Wave erosion, 177, 310
Wearing surface, 177
Weathering, 61, 96
Well graded, definition of, 3
Well-graded sands and gravels, 92
gradation, 269
sources of, 92
Well points, 197, 217, 219, 220, 299
Wellton-Mohawk Canal, 264, 265
Wire-line drill rod, 136
Workability of materials, 22, 44, 62–63, 75, 180, 324

Zoned embankments, 215
Zoning, 60

*Asterisks denote page references to appropriate test designations in the appendix.